Optimization of Structural Systems and Industrial Applications

SECOND INTERNATIONAL CONFERENCE
ON
COMPUTER AIDED OPTIMUM DESIGN OF STRUCTURES
OPTI/91

SCIENTIFIC COMMITTEE

Acknowledgement is made to Y. Yamazaki for the use of Figure 2 on page 146, which appears on the front cover of this book.

Computer Aided Optimum Design of Structures 91

Optimization of Structural Systems and Industrial Applications

Editors: S. Hernandez, University of Zaragoza, Spain
C.A. Brebbia, Wessex Institute of Technology, U.K.

Computational Mechanics Publications
Southampton Boston

CMP

Co-published with

Elsevier Applied Science
London New York

ELSEVIER

S. Hernandez
Department of Mechanical Engineering
University of Zaragoza
Maria de Luna
50015 Zaragoza
Spain

C.A. Brebbia
Computational Mechanics Institute
Wessex Institute of Technology
Ashurst Lodge
Ashurst
Southampton SO4 2AA
UK

Co-published by

Computational Mechanics Publications
Ashurst Lodge, Ashurst, Southampton, UK

Computational Mechancics Publications Ltd
Sole Distributor in the USA and Canada:

Computational Mechanics Inc.
25 Bridge Street, Billerica, MA 01821, USA

and

Elsevier Science Publishers Ltd
Crown House, Linton Road, Barking, Essex IG11 8JU, UK

Elsevier's Sole Distributor in the USA and Canada:

Elsevier Science Publishing Company Inc.
655 Avenue of the Americas, New York, NY 10010, USA

British Library Cataloguing in Publication Data
Optimum structural design.
1. Structures. Design. Optimisation
I. Brebbia, C.A. (Carlos Alberto) II. Hernandez,
S. (Santiago) *1951-*
624.17771

ISBN 1-85312-140-1

ISBN 1-85166-669-9 Elsevier Applied Science, London, New York
ISBN 1-85312-140-1 Computational Mechanics Publications, Southampton
ISBN 1-56252-068-7 Computational Mechanics Publications, Boston, USA

Library of Congress Catalog Card Number 91-71738

Printed and Bound by Billings and Sons Ltd, Worcester

PREFACE

This book contains the edited version of the papers presented at the Second International Conference on Computer Aided Optimum Design of Structures sponsored by the Wessex Institute of Technology and held at Cambridge, Massachusetts in June 1991.

The range of the papers in this volume covers a wide field of applications of structural optimization. An important group are the papers dealing with new optimization techniques, improvements to existing methods; as well as techniques based on approximations which have important computational advantages.

The range of practical applications is varied and encompasses aerospace, mechanical, civil and structural engineering and others. The behaviour of structures has been studied when subjected to static and dynamic loading, including seismic effects and fatigue. The case of different materials such as steel, concrete and composite materials is also analysed in several of the papers. This wide range of applications is testimony to the growing importance of structural optimization in all fields of engineering design. Of special importance in this regard are several papers dealing with expert systems and the integration of software packages.

The editors are grateful to the members of the International Scientific Advisory Committee for their support of this meeting, i.e. R.A. Adey, Computational Mechanics Institute; J-F.M. Barthelemy, Langley Research Center; L. Chibani, New Jersey; R.B. Corotis, Johns Hopkins University; D.E. Grierson, University of Waterloo; P. Hajela, University of Florida; N. Kikuchi, University of Michigan; H.G. Natke, University of Hannover; G. Rozvany, Universität Essen; E. Schnack, Karlsruhe University; A.B. Templeman, University of Liverpool; G. Thierauf, Universität-GHS-Essen; P. Thoft-Christensen, Aalborg University Centre; G.N. Vanderplaats, University of California.

The book describes a field of research in rapid development and one which is of increasing importance in engineering.

The Editors
June 1991

CONTENTS

SECTION 1: IMPROVED TECHNIQUES FOR MATHEMATICAL OPTIMIZATION

SECTION 4: SENSITIVITY ANALYSIS AND SHAPE OPTIMIZATION

SECTION 5: STRUCTURAL OPTIMIZATION IN MECHANICAL AND AIRCRAFT INDUSTRIES

SECTION 8: EXPERT SYSTEMS AND INTEGRATED PACKAGES IN OPTIMUM DESIGN

SECTION 1: IMPROVED TECHNIQUES FOR MATHEMATICAL OPTIMIZATION

Augmented Lagrangean Techniques for the Integrated Design and Analysis Problem in Structural Optimization

T. Larsson, M. Rönnqvist

Department of Mathematics, Linköping Institute of Technology, S-581 83 Linköping, Sweden

ABSTRACT

Solution procedures in structural optimization are commonly based on a nested approach where approximations of the analysis and design problems are solved alternately in an iterative scheme. In this paper we study an integrated design and analysis formulation which enables both these problems to be solved simultaneously. A major advantage of the integrated approach compared to the nested one, is that the dependence between the design and analysis variables is stated explicitly. Earlier solution techniques for the integrated approach mostly utilize penalty function reformulations; we make use of two augmented Lagrangean schemes. These schemes give rise to Lagrangean subproblems with somewhat different properties, and two suitable techniques are adopted for their solution; the first is a projected Newton method applied directly to the subproblem, and the second is a simplicial decomposition method combined with the projected Newton method. Computational results show that the schemes we present are viable approaches to structural optimization.

INTRODUCTION

Structural optimization (or optimal design) deals with the problem of designing a mechanical structure in an efficient way with respect to some criterion, such as weight minimization, subject to design restrictions. A mathematical model of a structural optimization problem consists of two connected optimization problems: the design problem that determines sizes, materials or shapes, and the analysis problem that describes the response of the structure when affected by loads. This problem can be formulated as

$$[\mathbf{P}] \qquad \min_{x} \ f(\mathbf{x},\mathbf{u})$$

$$\text{s.t.} \begin{cases} \mathbf{g}(\mathbf{x},\mathbf{u}) & \leq & \mathbf{0} \\ \mathbf{x} & \in & X \end{cases}$$

$$\text{where } \mathbf{u} \text{ solves } \min_{\mathbf{u}} \ \Pi(\mathbf{x},\mathbf{u}).$$

Here, $\mathbf{x}$ are the design variables, $\mathbf{u}$ the state variables, $f(\mathbf{x}, \mathbf{u})$ is a function measuring some efficiency of the structure, $\mathbf{g}(\mathbf{x}, \mathbf{u})$ a set of constraint functions, X a set of admissible values for the design variables, and $\Pi(\mathbf{x}, \mathbf{u})$ the function of potential energy.

This two level mathematical problem is, in general, prohibitively difficult to solve directly. Therefore, it is common to use an iterative nested scheme where both the analysis and the design problems are approximated and solved alternately. Given a design $\mathbf{x}^{(k)}$, an approximate analysis problem is formulated by assuming linearly elastic behaviour of the material and discretizing the structure by finite elements; this is the well known Finite Element Method (FEM). The FEM generates an analysis problem which is equivalent to finding a solution to a system of linear equations,

$$\mathbf{K}\left(\mathbf{x}^{(k)}\right)\mathbf{u} = \mathbf{p}\left(\mathbf{x}^{(k)}\right), \qquad (1)$$

where $\mathbf{K}(\mathbf{x}^{(k)})$ is the structural stiffness matrix, $\mathbf{u}$ is the vector of nodal displacements, and $\mathbf{p}(\mathbf{x}^{(k)})$ is a vector of external nodal loads. (Most commonly $\mathbf{p}(\mathbf{x})$ is assumed constant, $\mathbf{p}$.) This system of equations is known as the state, or equilibrium, equations. (In the case of nonlinear analysis, the resulting equations become nonlinear.) The next step is to formulate an approximate design problem in order to find an improved design. This is accomplished by approximating the constraint functions $\mathbf{g}(\mathbf{x}, \mathbf{u})$ with respect to the design variables by means of Taylor series expansions; however, the calculation of derivatives is generally computationally expensive due to their implicit dependence of state (or analysis) variables via the state equations (1). The approximate problem is then solved by a suitable optimization method, see e.g. Belegundu and Arora [2, 3]. The general availability of analysis systems is one of the main reasons for the popularity of the nested approach.

An alternative to the nested approach is to state and solve the design and analysis problems in an integrated formulation; this approach has so far not been extensively studied. One advantage of using the integrated formulation is that the implicit dependence between the design and analysis variables is removed. A disadvantage is that a larger problem, involving both groups of variables simultaneously, must be solved. The idea of integrating the structural design and analysis was first introduced by Fox and Schmit [7, 8]. They studied linear analysis problems and used a conjugate gradient (CG) minimization technique for solving the integrated problem reformulated by an exterior penalty function technique. Later Fuchs [9, 10] studied truss-type structures and formulated an explicit integrated formulation with the use of an interior penalty function; the sequence of unconstrained problems was solved by CG techniques. Haftka [11, 12] used two solution approaches: a penalty function technique where a preconditioned CG method was used to solve the resulting subproblems, and a sequential quadratic programming algorithm. They considered both linear and nonlinear analysis. Smaoui and Schmit [21] used the integrated approach for finding a minimum weight design of geometrically non-linear three-dimensional truss structures. The algorithm used was the generalized reduced gradient method.

In this paper we develop algorithms for the integrated design and analysis problem based on the augmented Lagrangean concept. Similar approaches have successfully been applied to many classes of nonlinear problems; the main principle is to create a sequence of Lagrangean subproblems, each fairly easily solved. The augmented Lagrangean approach is closely related to penalty function techniques as well as ordinary Lagrangean schemes, but avoids the deficiencies of both; in particular, the ill-conditioning inherent in penalty function methods is not present. For solving the Lagrangean subproblems, we have chosen to adopt a projected Newton method and a simplicial decomposition scheme. These fully exploit the underlying structures of the subproblems. The contents of the paper is as follows. First we state an integrated formulation of the problem under consideration, namely linear elastic truss-type structures. In the following section we give the theoretical background to the solution procedures, which thereafter are stated together with numerical results. Finally, conclusions are drawn and future research areas are suggested.

PROBLEM FORMULATION

In the problem we consider, the objective is to minimize the structural weight and the design variables are transverse sizes of elements. Under this assumption the structural weight, $f(\mathbf{x})$, is a linear function of the design variables,

$$f(\mathbf{x}) = \sum_{j=1}^{n_1} c_j x_j = \mathbf{c}^T \mathbf{x}, \tag{2}$$

where n_1 is the number of elements in the structure, and c_j, $j = 1, \ldots, n_1$, are given constants. The design variables are restricted to be within given bounds,

$$X = \left\{ \mathbf{x} \mid \underline{x}_j \leq x_j \leq \overline{x}_j,\ j = 1, \ldots, n_1 \right\}. \tag{3}$$

Moreover, the structural stiffness matrix used in the equilibrium equations (1) can be stated as a linear function of the design variables,

$$\mathbf{K}(\mathbf{x}) = \mathbf{K}_0 + \sum_{j=1}^{n_1} x_j \mathbf{K}_j, \tag{4}$$

where $\mathbf{K}_j$, $j = 0, \ldots, n_1$, are constant positive semi-definite symmetric matrices. For each $\mathbf{x} \in X$, the structure is assumed to be nondegenerate so that the resulting matrix $\mathbf{K}(\mathbf{x})$ is positive definite and symmetric. Additional constraints are bounds on stresses and nodal displacements; the latter, which are the analysis variables, are restricted by given bounds,

$$U = \left\{ \mathbf{u} \mid \underline{u}_j \leq u_j \leq \overline{u}_j,\ j = 1, \ldots, n_2 \right\}, \tag{5}$$

where n_2 is the number of nodal displacement variables. Given the nodal displacements, $\mathbf{u}$, the stresses in the elements can be calculated as $\boldsymbol{\sigma} = \mathbf{E}\boldsymbol{\epsilon} = \mathbf{E}(\nabla\mathbf{N})\mathbf{u}$, where $\mathbf{N}$ are the so called form-functions; furthermore, for trusses, $\mathbf{E}$ and $\nabla\mathbf{N}$ are constant matrices. The stress constraints may therefore be expressed as a set of linear inequalities, $\mathbf{Q}\mathbf{u} \leq \mathbf{b}$, where $\mathbf{Q} = (\mathbf{q}_1, \ldots, \mathbf{q}_{m_1})^T$, $\mathbf{b}$ are given limits on stresses, and m_1 is the number of stress constraints. The integrated design and analysis problem can now be stated as follows.

$$\text{[IP]} \qquad \min_{x,u} \; f(\mathbf{x}) = \mathbf{c}^T\mathbf{x}$$
$$\text{s.t.} \begin{cases} \mathbf{Qu} & \leq & \mathbf{b} \\ \mathbf{K(x)u} & = & \mathbf{p} \\ \mathbf{u} & \in & U \\ \mathbf{x} & \in & X \end{cases}$$

Due to the state equations, this formulation is, in general, a nonconvex and nonlinear optimization problem, which may have multiple local optima. The fact that all derivatives needed in solution schemes can be explicitly and analytically calculated is an important property of problem **[IP]**. Worth noting is that displacement constraints handled implicitly in the nested approach to problem **[P]** are now stated explicitly as lower and upper bounds on the analysis variables.

PREREQUISITES FOR THE SOLUTION PROCEDURES

The augmented Lagrangean method

Consider the general nonlinear programming problem,

$$\text{[NLP]} \qquad \begin{array}{ll} \min_{z} & f(\mathbf{z}) \\ \text{s.t.} & \mathbf{h}(\mathbf{z}) = \mathbf{0}, \end{array}$$

where the functions $f(\mathbf{z})$ and $\mathbf{h}(\mathbf{z})$ may be nonlinear. Many different methods for solving this problem have been developed, often divided into primal and transformation (or projection) methods, see e.g. Belegundu and Arora [2]; to the latter class belong penalty methods and methods based on Lagrangean duality.

In penalty methods, the original problem is replaced by a sequence of unconstrained minimization problems, where the constraints are included in an extended objective function by means of a penalty term, e.g. the Euclidean norm.

$$\text{[PNLP]} \qquad \min_{z} \; f(\mathbf{z}) + \frac{1}{2} c_k \|\mathbf{h}(\mathbf{z})\|^2$$

The penalty parameter c_k is a positive scalar chosen such that $c_k < c_{k+1} \; \forall \; k$ and $c_k \rightarrow \infty$. The merits of this approach lie in its simplicity and the possibility of using efficient unconstrained minimization methods; a disadvantage is the need to let $c_k \rightarrow \infty$ which leads to numerical difficulties. One possibility for overcoming this disadvantage is to use preconditioning techniques.

The penalty function approach can be combined with Lagrangean duality; this was originally suggested by Hestenes [15] and Powell [19] who independently developed the method of multipliers. (See Bertsekas [4] for an extensive treatise on Lagrangean multiplier methods.) A penalty term is included in an extended objective function without removing the constraints, after which the constraints are Lagrangean dualized with multipliers $\boldsymbol{\lambda}$. The resulting augmented Lagrangean function becomes

$$L_c(\mathbf{z}, \boldsymbol{\lambda}) = f(\mathbf{z}) + \boldsymbol{\lambda}^T \mathbf{h}(\mathbf{z}) + \frac{1}{2} c \|\mathbf{h}(\mathbf{z})\|^2. \quad (6)$$

The augmented Lagrangean dual problem is

$$\textbf{[LD]} \qquad \max_{\lambda} \ \Phi(\boldsymbol{\lambda}),$$

where the dual function $\Phi(\boldsymbol{\lambda})$ is defined through the Lagrangean subproblem as

$$\Phi(\boldsymbol{\lambda}) = \min_{z} \ L_c(\mathbf{z}, \boldsymbol{\lambda}). \quad (7)$$

Despite the fact that the primal problem **[NLP]** might be nonconvex, the dual problem **[LD]** is convex, i.e. the unconstrained maximization of a concave function $\Phi(\boldsymbol{\lambda})$. Another important property is that the dual function $\Phi(\boldsymbol{\lambda})$ is differentiable. One possible strategy for solving the dual problem is to solve a sequence of Lagrangean subproblems and update the Lagrangean multipliers according to

$$\boldsymbol{\lambda}^{(k+1)} = \boldsymbol{\lambda}^{(k)} + c_k \mathbf{h}(\mathbf{z}^{(k)}), \quad (8)$$

where $\mathbf{z}^{(k)}$ is the minimizer of $L_{c_k}(\mathbf{z}, \boldsymbol{\lambda}^{(k)})$. In addition, the penalty parameter c_k is increased in each iteration. This strategy was proposed by Hestenes [15] and Powell [19]. A possibility for enforcing faster convergence in the dual space is to use a second order update formula for the Lagrangean multipliers, i.e. perform a Newton step in the dual space. This will, of course, require more calculations, including matrix inversions.

By combining the penalty and dual approaches, the disadvantages of both can be removed. In particular, convergence can be attained without increasing the penalty parameter c_k to infinity, and the multiplier iteration tends to converge to a Lagrangean multiplier vector much faster than in the ordinary dual approach. As far as the size of the penalty parameter is concerned, one can show that if a point $(\mathbf{z}^\star, \boldsymbol{\lambda}^\star)$ satisfies the Karush-Kuhn-Tucker conditions as well as certain second order conditions, a finite threshold level $c^\star$ exists, such that $\mathbf{z}^\star$ is an isolated minimizer of $L_{c^\star}(\mathbf{z}, \boldsymbol{\lambda})$, that is $\mathbf{z}^\star = \mathbf{z}(\boldsymbol{\lambda}^\star)$, see e.g. Fletcher [6], Theorem 12.2.1. The rate of convergence is heavily dependent on the penalty parameter, which should eventually become larger than the threshold level, in order to exploit the positive features of the multiplier iteration. However, it should not be increased too fast enforcing ill-conditioning upon the subproblem minimization; in particular, the initial value c_0 is not allowed to be too large.

If inequality constraints, $\mathbf{g}(\mathbf{z}) \leq \mathbf{0}$, are considered, the augmented Lagrangean function may be stated, see e.g. Bertsekas [5], as

$$L_c(\mathbf{z}, \boldsymbol{\xi}) = f(\mathbf{z}) + \sum_{i=1}^{m} \left(\xi_i \max\left\{ g_i(\mathbf{z}), -\frac{\xi_i}{c} \right\} + \frac{1}{2} c \left[\max\left\{ g_i(\mathbf{z}), -\frac{\xi_i}{c} \right\} \right]^2 \right), \quad (9)$$

where m is the number of inequality constraints. The updating formula for the multipliers $\boldsymbol{\xi}$ is

$$\xi_i^{(k+1)} = \max\left\{ 0, \xi_i^{(k)} + c_k g_i(\mathbf{z}^{(k)}) \right\}, \ i = 1, \ldots, m, \quad (10)$$

where $\mathbf{z}^{(k)}$ is the optimal solution to the corresponding Lagrangean subproblem.

In our applications of the augmented Lagrangean concept, we have chosen to dualize only subsets of the constraints; the remaining constraints in the Lagrangean subproblem are selected so that they are easily handled explicitly.

The simplicial decomposition scheme

Simplicial decomposition is a simple and direct solution method for dealing with large-scale pseudo-convex problems, especially when the constraints are linear and structured. The name simplicial decomposition was coined by von Hohenbalken [16, 17]; the method is also known as the extended Frank-Wolfe method, Holloway [18]. We consider the following linearly constrained problem.

[LCP]

$$\min_{z} f(\mathbf{z}) \quad \text{s.t.} \begin{cases} \mathbf{Az} = \mathbf{b} \\ \mathbf{z} \geq \mathbf{0} \end{cases}$$

Here $\mathbf{A}$ is an $m \times n$ matrix, $\mathbf{b}$ is a vector in $\mathbf{R}^m$, and $f(\mathbf{z})$ is a continuously differentiable pseudo-convex function. For simplicity, the feasible set is assumed non-empty and bounded.

The simplicial decomposition method is based on Charathéodory's theorem, see e.g. Bazaraa and Shetty [1], Theorem 2.1.6. Letting S be a set of points in $\mathbf{R}^n$, the theorem states that if a point $\mathbf{z} \in conv(S)$, then $\mathbf{z}$ may be expressed as a convex combination of, at the most, $n+1$ points in S (not necessarily distinct). A consequence of Charathéodory's theorem is the so called representation theorem for polyhedral sets, see e.g. Bazaraa and Shetty [1] Theorem 2.5.7. For bounded polyhedral sets, such as the feasible set in [**LCP**], the representation theorem says that any point in the set may be written as a convex combination of the extreme points of the set. We may therefore equivalently reformulate the problem [**LCP**] in terms of the feasible extreme points as

[MP]

$$\min_{\alpha} f(\mathbf{Z}\boldsymbol{\alpha}) \quad \text{s.t.} \begin{cases} \sum_{i \in P} \alpha_i = 1 \\ \alpha_i \geq 0, \quad \forall\, i \in P, \end{cases}$$

where $\mathbf{Z}$ is a matrix whose columns consist of the extreme points, $\boldsymbol{\alpha}$ are the corresponding convexity weights, and P is the index set of the extreme points. The extreme points should not be enumerated *a priori*, but generated algorithmically as needed. The solution scheme iterates between a master problem and a subproblem; the master problem utilizes the extreme points generated so far and determines a minimizing point with respect to their convex hull, and the subproblem generates a new extreme point to be added to the master problem. The subproblem is obtained by approximating the objective function with a first order Taylor series expansion at the current point $\mathbf{z}^{(k)}$ corresponding to the latest master problem solution, $\boldsymbol{\alpha}^{(k)}$. This gives a linear programming problem, which is easily solved.

The master problem generates an upper bound on the optimal value of problem **[LCP]**, since the obtained solution $\mathbf{z}^{(k)}$ is feasible; additionally, when the objective function is convex, the subproblem generates a lower bound. These bounds can be used as a convergence criterion. Global finite convergence has been shown by von Hohenbalken [17]. Due to the simple representation of the feasible set in the master problem, efficient methods for unconstrained optimization can be adapted for its solution, e.g. Bertsekas [5]. A modified procedure, restricted simplicial decomposition, is developed by Hearn *et al.* [13]; this version controls the maximum size of the set of extreme points maintained in the master problem by a user-defined parameter. When this parameter is at its minimum, which is one, the method reduces to the Frank-Wolfe method; at its maximum, $n+1$, the original simplicial decomposition scheme is obtained.

The projected Newton method

In the solution procedures based on augmented Lagrangeans and simplicial decomposition, a sequence of nonlinear problems with simple constraints must be solved as subproblems. These subproblems may be stated as

$$[\mathbf{SP}] \qquad \begin{array}{ll} \min\limits_{z} & f(\mathbf{z}) \\ \text{s.t.} & \mathbf{z} \in Z, \end{array}$$

where $f(\mathbf{z})$ is twice differentiable, and the set Z is expressed as

$$Z_1 = \left\{ \mathbf{z} \mid \underline{z}_j \le z_j \le \overline{z}_j;\ \ j = 1, \ldots, n \right\} \text{ or} \tag{11}$$

$$Z_2 = \left\{ \boldsymbol{\alpha} \mid \sum_{j=1}^{n} \alpha_j = 1,\ \ \alpha_j \ge 0;\ \ j = 1, \ldots, n \right\}. \tag{12}$$

Since these problems are solved repeatedly, it is crucial to utilize efficient solution methods that are suitable for these kinds of constraints in order to obtain efficiency in the overall procedure. One possibility is to use some standard algorithm of feasible direction type. However, methods especially designed for problems with simple constraints have been developed; we focus on the method by Bertsekas [5], which is a projected Newton method shown to be globally convergent at a superlinear rate. This method has been successfully used by Hearn *et al.* [13] for solving master problems in a simplicial decomposition scheme.

To briefly describe the projected Newton method, we consider the case of a feasible set Z_1 as in (11). Given an iteration point $\mathbf{z}^{(k)}$, introduce the index set

$$\begin{aligned} J_k = \Bigg\{ j \mid\ & \underline{z}_j \le z_j^{(k)} \le \underline{z}_j + \epsilon_k \text{ and } \frac{\partial f(\mathbf{z}^{(k)})}{\partial z_j} > 0 \text{ or} \\ & \overline{z}_j - \epsilon_k \le z_j^{(k)} \le \overline{z}_j \text{ and } \frac{\partial f(\mathbf{z}^{(k)})}{\partial z_j} < 0 \Bigg\}, \end{aligned} \tag{13}$$

where the positive sequence $\{\epsilon_k\}$ is nonincreasing, and renumber the variables so that the ones in J_k become the highest numbered, say from $r+1$ to n. Then, choose a positive definite matrix $\mathbf{D}^{(k)}$ with the structure

$$\mathbf{D}^{(k)} = \begin{pmatrix} \overline{\mathbf{D}} & \mathbf{0} \\ \mathbf{0} & \hat{\mathbf{D}} \end{pmatrix}. \tag{14}$$

The matrix $\hat{\mathbf{D}} = diag(d_{r+1}, \ldots, d_n)$ corresponds to the index set J_k, and $\overline{\mathbf{D}}$ is an arbitrary positive definite matrix; one possibility is to choose this second matrix as the inverse of the Hessian with respect to the variables not in J_k. A generalization of Newton's method to the case of bounded variables is then given by the iteration formula

$$\mathbf{z}^{(k+1)} = \left(\mathbf{z}^{(k)} - t^{(k)}\mathbf{D}^{(k)}\nabla f(\mathbf{z}^{(k)})\right)^{\sharp}, \tag{15}$$

where $t^{(k)}$ is a positive scalar stepsize chosen according to an Armijo-type rule, and where for all $\mathbf{z} \in \mathbf{R}^n$ we denote by $(\mathbf{z})^{\sharp}$ the orthogonal projection of the point $\mathbf{z}$ onto the set Z_1.

Suppose that $\mathbf{z}^{\star}$ is a local optimum. The algorithm can be shown to correctly identify the set of constraints which are active at $\mathbf{z}^{\star}$ after a finite number of iterations. This implies that the algorithm eventually reduces to the Newton method restricted to a subspace which is described by the constraint active at $\mathbf{z}^{\star}$; a superlinear convergence rate follows.

For problems with a constraint set given by Z_2 in (12), a transformation proposed by Bertsekas [5], where one variable is implicitly eliminated, may be used. This results in a problem with, essentially, nonnegativity restrictions on the variables; this transformed problem is solved as described above. An alternative approach is to solve the original problem by considering the nonnegative restrictions on the variables only, giving $\hat{\mathbf{z}}$, after which an Euclidean projection of $\hat{\mathbf{z}}$ onto the set Z_2 is made. The projection problem is analytically solvable, Held *et al.* [14]. We have used this latter alternative.

SOLUTION PROCEDURES AND NUMERICAL RESULTS

The two solution procedures for problem [**IP**] to be presented are based on reformulations via augmented Lagrangeans. In the first procedure, the Lagrangean subproblems are solved directly by the projected Newton method; in the second, the Lagrangean subproblems are transformed into a sequence of lower-dimensional problems by simplicial decomposition. Numerical results for the well known ten-bar truss problem, see e.g. Ringertz [20], are presented in order to show the feasibility of the suggested techniques. The implementation was done in FORTRAN 77 and executed on a SUN 4/390. A crucial part of the computations is the calculation of $\mathbf{K}(\mathbf{x})\mathbf{u}$, which can be done efficiently according to

$$\mathbf{K}(\mathbf{x})\mathbf{u} = \sum_{j=1}^{n_1} x_j(\mathbf{K}_j\mathbf{u}) \tag{16}$$

utilizing the sparsity of each $\mathbf{K}_j$.

An augmented Lagrangean reformulation solved by the projected Newton method

Including all the constraints of problem [**IP**], except for the simple bounds, in the augmented Lagrangean function, it becomes

$$
\begin{aligned}
&L_c^1(\mathbf{x},\mathbf{u},\boldsymbol{\lambda},\boldsymbol{\xi}) = \\
&\quad \mathbf{c}^T\mathbf{x} + \boldsymbol{\lambda}^T\left(\mathbf{K}(\mathbf{x})\mathbf{u}-\mathbf{p}\right) + \frac{1}{2}c_1\|\mathbf{K}(\mathbf{x})\mathbf{u}-\mathbf{p}\|^2 + \\
&\quad \sum_{i=1}^{m_1}\left(\xi_i \max\left\{\mathbf{q}_i^T\mathbf{u}-b_i,-\frac{\xi_i}{c_2}\right\} + \frac{1}{2}c_2\left[\max\left\{\mathbf{q}_i^T\mathbf{u}-b_i,-\frac{\xi_i}{c_2}\right\}\right]^2\right).
\end{aligned}
\tag{17}
$$

Here c_1 and c_2 are the penalty parameters. The Lagrangean subproblem, with $\boldsymbol{\lambda}$ and $\boldsymbol{\xi}$ kept constant, becomes a nonseparable quadratic problem with simple bounds on the variables $\mathbf{x}$ and $\mathbf{u}$. The solution of this subproblem with the projected Newton method requires a positive definite matrix $\overline{\mathbf{D}}$ to be selected in order to determine a search direction. By selecting the inverse of the Hessian matrix the method will give a superlinear convergence rate, assuming positive definiteness. In our application, however, this matrix inversion will, for larger problems, require substantial computational work; moreover, the Hessian matrix, denoted $\mathbf{H}(\mathbf{x},\mathbf{u})$, cannot be guaranteed to be positive definite.

These obstacles can be overcome in several ways. The first possibility is to choose the matrix $\overline{\mathbf{D}} = \mathbf{I}$, where $\mathbf{I}$ is the unity matrix; this gives the steepest descent direction. The second is to choose $\overline{\mathbf{D}}$ as an approximation of the inverse of the Hessian, e.g. updated according to a quasi-Newton scheme. Thirdly, one may choose $\overline{\mathbf{D}} = \{diag(\mathbf{H}(\mathbf{x},\mathbf{u})) + \nu\mathbf{I}\}^{-1}$, where ν is a positive scalar, large enough to ensure positive definiteness. In our numerical experiments the last alternative has been adopted; this choice gives an automatic rescaling of the variables, which is important since there are two different types of variables in the subproblem.

As updating formulas for the Lagrangean multipliers, we have chosen the ones given in (8) and (10); c_k is updated according to $c_{k+1} = c_k * \beta$. We have tested different choices of β and c_0, and Figure 1 shows the typical behaviour of the weight and norm $\|\mathbf{K}(\mathbf{x})\mathbf{u}-\mathbf{p}\|^2$ histories, the latter being a measure of the error in the state equations. It should be noted that the norm history is given in a logarithmic scale. (The norm becomes almost constant after some iterations because of the termination criterion in the projected Newton method.)

An augmented Lagrangean reformulation solved by the simplicial decomposition scheme

If only the state equations are included in the augmented Lagrangean function, the subproblem will become more difficult since the constraints $\mathbf{Qu} \le \mathbf{b}$ are kept explicit; however, stronger Lagrangean subproblems, in general, provide more information. The augmented Lagrangean function will be

$$
L_c^2(\mathbf{x},\mathbf{u},\boldsymbol{\lambda},\boldsymbol{\xi}) = \mathbf{c}^T\mathbf{x} + \boldsymbol{\lambda}^T\left(\mathbf{K}(\mathbf{x})\mathbf{u}-\mathbf{p}\right) + \frac{1}{2}c_1\|\mathbf{K}(\mathbf{x})\mathbf{u}-\mathbf{p}\|^2 \tag{18}
$$

and the subproblem becomes

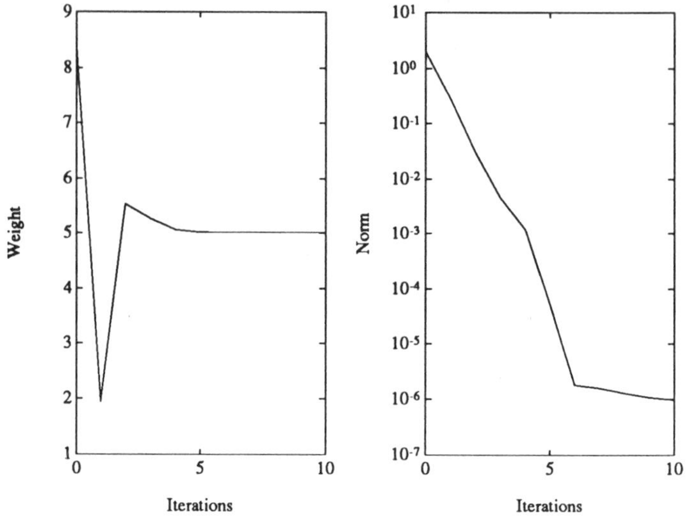

Figure 1: Weight and norm history, ten-bar truss

$$\textbf{[LDS]} \qquad \min_{x,u} \; L_c^2(\mathbf{x}, \mathbf{u}, \boldsymbol{\lambda}, \boldsymbol{\xi})$$
$$\text{s.t.} \begin{cases} \mathbf{Qu} & \leq & \mathbf{b} \\ \mathbf{x} & \in & X \\ \mathbf{u} & \in & U. \end{cases}$$

In the simplicial decomposition scheme for solving [**LDS**], the feasible set is represented as

$$(\mathbf{x}, \mathbf{u})^T = \sum_{l=1}^{N} \boldsymbol{\alpha}_l (\mathbf{x}^{(l)}, \mathbf{u}^{(l)})^T, \;\; \sum_{l=1}^{N} \alpha_l = 1, \;\; \alpha_l \geq 0, \;\; l = 1, \ldots, N, \tag{19}$$

where N is the total number of extreme points. In the master problem the original variables $\mathbf{x}$ and $\mathbf{u}$ are replaced by the weights $\boldsymbol{\alpha}$; furthermore, following the restricted simplicial decomposition scheme, the maximum number of variables in the master problem is preselected. To solve the master problem we use the projected Newton method; in this case we obtain the superlinear convergence rate by choosing the matrix $\overline{\mathbf{D}}$, given in equation (14), as the inverse of the Hessian matrix with respect to the convexity weights. (If the Hessian becomes indefinite, we resort to the technique used in the previous section.)

The performance of this solution technique is essentially the same as that of the direct approach. Figure 2 shows a typical weight and norm history; in this example, we have chosen an initially large penalty parameter c_0 in comparison with the example presented in Figure 1. This makes the algorithm more conservative with respect to the norm $||\mathbf{K}(\mathbf{x})\mathbf{u} - \mathbf{p}||^2$.

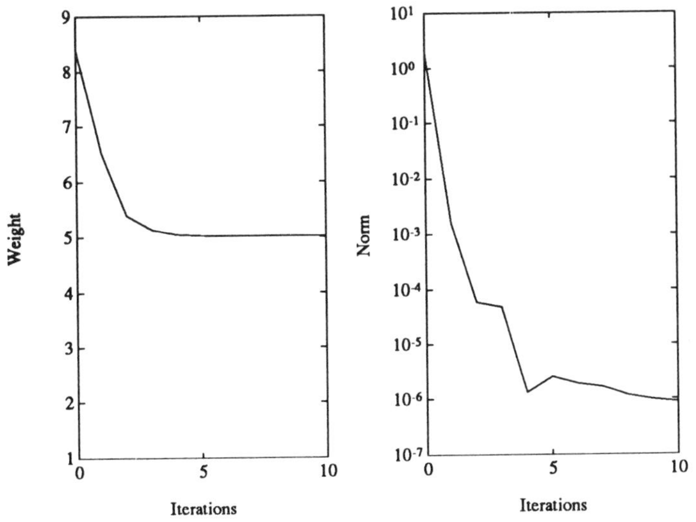

Figure 2: Weight and norm history, ten-bar truss

CONCLUSIONS AND FUTURE RESEARCH

One major advantage of the integrated formulation is that the dependence between the design and analysis variables is explicit; other advantages are that no exact analysis has to be performed during the solution process and the possibility of using the error in the state equations as a convergence criterion. The augmented Lagrangean reformulation technique is known to be an efficient tool for general nonlinear problems; it generates a sequence of unconstrained or simply constrained subproblems and ill-conditioning is avoided when compared to other penalty methods. Our application of the augmented Lagrangean technique is promising. The number of iterations needed to achieve convergence is comparable with the nested approach; moreover, the solution procedure is very insensitive to the quality of the starting solution. Both the projected Newton method and the simplicial decomposition scheme used for solving the subproblems have been found to be stable and efficient.

So far we have chosen to consider relatively simple structures to indicate the potential of the solution procedure; a subject for future research is to take more complicated structures into account. Another extension is to incorporate nonlinear analysis, e.g. geometrical nonlinearities; in a nested approach this requires the solution of a system of nonlinear equations in each iteration, but in an integrated approach the analysis is embedded in the overall procedure. The presented solution techniques involve a substantial part of vector and matrix calculations, all very suitable for parallel computations, also of interest for further study.

REFERENCES

[1] Bazaraa M.S. and Shetty, C.M. Nonlinear Programming, John Wiley & Sons, New York, 1979.

[2] Belegundu, A.D. and Arora, J.S. A Study of Mathematical Programming Methods for Structural Optimization, Part 1: Theory, International Journal for Numerical Methods in Engineering, Vol. 21, pp. 1583-1599, 1985.

[3] Belegundu A.D. and Arora, J.S. A Study of Mathematical Programming Methods for Structural Optimization, Part 2: Numerical Results, International Journal for Numerical Methods in Engineering, Vol. 21, pp. 1601-1625, 1985.

[4] Bertsekas, D.P. Constrained Optimization and Lagrange Multiplier Methods, Academic Press, New York, 1982.

[5] Bertsekas, D.P. Projected Newton Methods for Optimization Problems with Simple Constraints, SIAM Journal on Control and Optimization, Vol. 20, pp. 221-246, 1982.

[6] Fletcher, R. Practical Methods of Optimization, Constrained Optimization 2, John Wiley & Sons, Chichester, 1981.

[7] Fox, F.L and Schmit, L.A. An Integrated Approach to Structural Synthesis and Analysis, AIAA Journal, Vol. 3, pp. 1104-1112, 1965.

[8] Fox, R.L. and Schmit, L.A. Advances in the Integrated Approach to Structural Synthesis, Journal of Spacecraft and Rockets, Vol. 3, pp. 858-866, 1966.

[9] Fuchs, M.B. Explicit Optimum Design, International Journal of Solids and Structures, Vol. 19, pp. 13-22, 1982.

[10] Fuchs, M.B. Explicit Optimum Design Technique for Linear Elastic Trusses, Engineering Optimization, Vol. 6, pp. 218-228, 1983.

[11] Haftka, R.T. Simultaneous Analysis and Design, AIAA Journal, Vol. 23, pp. 1099-1103, 1985.

[12] Haftka, R.T. and Kamat, M.P. Simultaneous Nonlinear Analysis and Design, Computational Mechanics, Vol. 4, pp. 409-416, 1989.

[13] Hearn, D.W., Lawphongpanich, S. and Ventura, J.A. Restricted Simplicial Decomposition: Computation and Extensions, Mathematical Programming Study, Vol. 31, pp. 99-118, 1987.

[14] Held, M., Wolfe, P. and Crowder, H. Validation of Subgradient Optimization, Mathematical Programming, Vol. 6, pp. 62-88, 1984.

[15] Hestenes, M.R. Multiplier and Gradient Methods, Journal of Optimization Theory and Applications, Vol. 5, pp. 303-320, 1969.

[16] von Hohenbalken, B. A Finite Algorithm to Maximize Certain Pseudoconcave Functions on Polytopes, Mathematical Programming, Vol. 9, pp. 189-206, 1975.

[17] von Hohenbalken, B. Simplicial Decomposition in Nonlinear Programming Algorithms, Mathematical Programming, Vol. 13, pp. 49-68, 1977.

[18] Holloway, C.A. An Extension of the Frank and Wolfe Method of Feasible Directions, Mathematical Programming, Vol. 6, pp. 14-27, 1974.

[19] Powell, M.J.D. A Method for Nonlinear Constraints in Minimization Problems, in Optimization (Ed. Fletcher, R.), pp. 283-298, Academic Press, London, 1969.

[20] Ringertz, U.T. On Topology Optimization of Trusses, Engineering Optimization, Vol. 9, pp. 209-218, 1985.

[21] Smaoui, H. and Schmit, L.A. An Integrated Approach to the Synthesis of Geometrically Nonlinear Structures, International Journal for Numerical Methods in Engineering, Vol. 26, pp. 555-570, 1988.

Information Entropy Applications in Structural Optimization

L.M.C. Simões

Departamento de Engenharia Civil, Faculdade de Ciências e Tecnologia, Universidade de Coimbra, 3049 Coimbra, Portugal

ABSTRACT

Two recently developed entropy-based algorithms for the optimization of elastic structures are presented in this work. They are simple to operate and efficient, suggesting their potential use. The first algorithm, which is applied to a structural sizing problem, consists of maximizing a function of a single variable until convergence. Surrogate multipliers are obtained by maximizing Shannon's entropy function. In the second case, the maximum entropy formalism solves a multicriteria optimization via the unconstrained minimization of a nonlinear convex scalar function. Numerical results are given to illustrate the methods.

INTRODUCTION

Methods of optimization currently available seem to have entered a diminishing returns phase in respect of further research potential. Some radically different directions and new approaches are needed for the further development of engineering optimization techniques. Annealing is the physical process of heating up a solid until it melts followed by cooling it down until it crystallizes in a state with a perfect lattice. During the process the free energy of the solid is minimized. Practice shows that the cooling must be done carefully in order not to get trapped in locally optimal lattice structures with crystal imperfections. In nonlinear optimization one can define a similar process by establishing a correspondence between the cost function and the free energy and between the solution and the physical states. Entropy is a natural measure of the amount of disorder (or information) in a system. Entropy is viewed in information theory as a quantitative measure of the information content of a system. In the case of optimization, the entropy can be interpreted as a measure for the degree of optimality.

The main purpose of this paper is to present two recently developed entropy-based techniques. The first method, based on the the maximum entropy principle, is applied to structural sizing problems. Unlike optimality criteria and other more recent algorithms, it does not require an active/passive set strategy. This methodology considers simultaneously all the constraints, assigning to them different weights according to the probabilities given by Shanon's entropy function. The optimization phase reduces to the finding of the parameter which maximizes the (concave) dual volume. The Lagrange multipliers and member sizes are evaluated in terms of this parameter by using a simple algebraic expression.

Entropy is used implicitly in another algorithm in which a group of objectives is minimized. A Pareto solution of this vector problem is obtained by the scalar minimization of a nonlinear convex function involving one control parameter.

Examples are given to demonstrate the potentialities of these methods. The first is a grillage, demonstrating efficient size optimization for this type of structure. The second example is a reliability based configuration optimization of a truss showing the state of the art in discrete element shape optimization.

ENTROPY IN OPTIMIZATION PROCESSES

Constrained Nonlinear Programming

Entropy can be used to deduce desired results when only limited information is available. The general inequality constrained nonlinear programming problem,

$$\text{Min} \quad f(x) \qquad i = 1,\ldots,n \qquad (1a)$$

$$\text{st} \quad g_j(x) \le 0 \quad \text{or} \quad g_j(x) + s_j = 0 \qquad j = 1,\ldots,m \qquad (1b)$$

was examined in Ref.[1]. An initial point was chosen and information is calculated about the objective and constraint functions, typically their numerical values and gradients at the design point. This numerical information was then used in a mathematical programming algorithm to infer where the next trial point should be placed so as to get closer to the constrained optimum of the problem. The new trial generates more information from which another point is inferred and eventually the solution is reached by this process of gathering better information and using it in an inference-based algorithm. The essence of the method consisted in transforming problem (1) into an equivalent surrogate form,

$$\text{Min} \quad f(x) \qquad (2a)$$

$$\text{st} \quad \Sigma_{j=1,M}\ \lambda_j\ g_j(x) = 0 \qquad (2b)$$

$$\Sigma_{j=1,M}\ \lambda_j = 1 \qquad (2c)$$

$$\lambda_j \ge 0 \qquad (2d)$$

and using maximum entropy to obtain least biased estimates of the optimum values of the surrogate multipliers λ_j. In this two phase method the absence of an explicit surrogate dual objective function is overcome by introducing the Shanon entropy as a means of forcing iterations towards a saddle point. Each estimate lead to a new problem in the space of the x variables and generated information upon which to base an improved estimate of the optimum surrogate multipliers.

Truss sizing problem

An initial set of bar cross-sectional areas is chosen to form an initial design which is analyzed to give bar forces and virtual forces for joint displacements. These forces are assumed to remain constant and an optimization problem is set up and solved to give new bar sizes. The structure is reanalyzed with the new bar sizes which are then scaled to ensure feasibility of the new design. Convergence checks are carried out on bar sizes and forces and iterations terminate if the changes are acceptably small. If convergence is not achieved a new optimization is set up with the new bar forces and solution proceeds iteratively until convergence is achieved. The optimization problem which must be solved in each cycle of iteration can be stated as:

$$\min \quad V = \Sigma_{i=1,N}\ l_i\ x_i \qquad (3a)$$

$$\text{st} \quad \Sigma_{i=1,N}\ l_i\ F_{ij}\ \underline{F}_{ik} / (E_i\ x_i) \leqslant u_k \quad ; \quad j = 1,\ldots,J \qquad (3b)$$

$$\sigma_i^L \leqslant \sigma_{ij} = F_{ij} / x_i \leqslant \sigma_i^U \quad ; \quad k = 1,\ldots,K \qquad (3c)$$

$$x_i \geqslant x_i^L \quad ; \quad i = 1,\ldots,N \qquad (3d)$$

The N unknown bar sizes x_i, i=1,...,N comprise the design variable vector x. l_i, E_i are the length and elastic modulus, respectively, of the i-th bar. In the displacement constraints (3b) F_{ij} and $\underline{F}_{ik}$ are the force caused by the j-th load case and the virtual force caused by the k-th joint displacement in the i-th bar and u_k is the maximum permissible displacement of the k-th joint. At each optimization all bar forces are known and are assumed to remain constant, so problem (3) can be stated in a simplified form as:

$$\min \ V = \Sigma_{i=1,N} \ l_i x_i \qquad (4a)$$

$$\text{st} \quad \Sigma_{i=1,N} \, c_{ik}/x_i \leqslant 1 \qquad ; \quad k = 1,...,M \qquad (4b)$$

$$\underline{x}_i/x_i \leqslant 1 \qquad ; \quad i = 1,...,N \qquad (4c)$$

The displacement constraints (3b) correspond with Eq.(4b) with c_{ij} , j=1,...M representing general displacement constants evaluated after each analysis. The stress and size constraints in problem (3) have been merged into Eq.(4c); $\underline{x}_i$ is the largest of either x_i^L or the minimum size necessary to satisfy the stress constraints (3c). Problem (4) has the following Lagrangean function,

$$\mathscr{L}(x,\beta) = \Sigma_{i=1,N} \, l_i x_i + \Sigma_{k=1,M} \, \mu_k [\Sigma_{i=1,N} \, c_{ik}/x_i - 1] + \Sigma_{j=1,N} \, \mu_{M+i} [\underline{x}_i/x_i - 1] \qquad (5)$$

Examining the stationarity of $\mathscr{L}(x,\mu)$ with respect to all x_i, i=1,...,N yields equations in x which may be solved algebraically to give:

$$x_i^{[k]} = \{ [\Sigma_{j=1,M} \, c_{ik} \, \mu_j + \underline{x}_i \, \mu_{M+i}] / l_i \}^{1/2} \qquad (6)$$

If an optimum set of multipliers μ^* exists, then the resulting bar sizes x* will also solve the problem. Such a set of optimal surrogate multipliers μ^* is, of course, not known "a priori" but found iteratively. The problem then becomes one of developing a method whereby the μ may be iteratively updated towards μ^*, thus solving problem (4). Very many engineering optimization problems essentially consist of iteratively sorting out which ones of many constraints are active at the optimum and which are inactive and then of iteratively estimating values for the active constraint multipliers. Though such a strategy is theoretically valid, changes in the active set between iterations change the optimization problem being solved in a discontinuous way and lead to erratic convergence behavior. The maximum entropy-based algorithms avoid these difficulties by retaining and updating all constraints at all times. Problem discontinuities are not introduced and consequently convergence is smooth. Assuming that the Lagrange multipliers μ_j are given by,

$$\mu_j = \lambda_j v_j \qquad (7)$$

where λ_j is a entropy multiplier and v_j is a correction factor, these multipliers may be interpreted probabilistically with each λ_j representing the probability that its corresponding constraint is active at the optimum. With this probabilistic view of the multipliers it is then entirely logical and sensible to calculate most likely or least biased values for them from the Jaynes maximum entropy formalism. An initial set of values for v and λ is chosen such that $v_j^{[o]}=1$ and $\lambda_j^{[o]}=1/(M+N)$, j=1,...,M+N ie: all constraints are equally likely to be active at the optimum.The set of bar cross-sectional areas x obtained from (6) forms an initial design which is analyzed to give bar forces and virtual forces for joint displacements. All bar areas are scaled to ensure that no constraint is violated. The correction factors vector $v^{[1]}$ is assumed a unit vector in this iteration. New estimates of the multipliers $\lambda^{[1]}$ are then obtained by solving the maximum entropy mathematical problem:

$$\text{Max} \quad S = -K\, \Sigma_{j=1,M}\, \lambda_j^{[1]} \ln \lambda_j^{[1]} \tag{8a}$$

$$\text{st,} \quad \Sigma_{j=1,M}\, \lambda_j^{[1]} = 1 \tag{8b}$$

$$\Sigma_{j=1,M}\, \lambda_j^{[1]}\, g_j(x^{[0]}) = \varepsilon \tag{8c}$$

$$\lambda_j^{[1]} \geq 0 \tag{8d}$$

S is the Shannon entropy, K is a positive constant. Equation (8c), that represents the constraints:

$$g_j(x) = c_{ik}/x_i - 1 \qquad \text{for } j = 1,\ldots,M \tag{9a}$$

$$g_j(x) = \underline{x}_i/x_i - 1 \qquad \text{for } j = M+1,\ldots,M+N \tag{9b}$$

has an expected value zero. If the left-hand side had contained $g_j(x^{[1]})$, then the right-hand side would be zero, but since $g_j(x^{[1]})$ values are not yet known $g_j(x^{[0]})$ values are used as the best currently available estimates and this introduces the error term ε into Eq.(8c). The entropy maximization problem has an algebraic solution for $\lambda^{[1]}$:

$$\lambda_j^{[1]} = \frac{\exp[\beta\, g_j(x^{[0]})/K]}{\Sigma_{j=1,M} \exp[\beta\, g_j(x^{[0]})/K]} \tag{10}$$

in which β, the Lagrange multiplier for Eq.(12c), can be found by substituting result (10) into Eq. (8c). However, since ε is not uniquely known and K is an arbitrary constant, $p=\beta/K$ may be viewed as a penalty parameter used to close the duality gap. Eq.(10) with a selected p yields new constraint activity probabilities $\lambda^{[1]}$. At each iteration, it is necessary to search for the value of p that maximizes the truss volume given by (6) and using the new correction factor and multiplier values. The new design is analyzed by the matrix stiffness method and all bar areas are scaled to ensure that no constraint is violated. The correction factors vector are given by:

$$v^{[2]} = F^{[1]}_{\#}\, l = F^{[1]t}\, (F^{[1]}\, F^{[1]t})^{-1}\, l \tag{11}$$

where l represents the member lengths vector and the elements of the matrix F are given by,

$$f_{ij}^{[1]} = \lambda_j\, c_{ij} / x_i^{\,2} \qquad \text{for } j = 1,\ldots,M \qquad i = 1,\ldots,N \tag{12a}$$

$$f_{jj}^{[1]} = \lambda_{M+i}\, \underline{x}_i / x_i^2 \qquad \text{for } j = M+1,\ldots,M+N \tag{12b}$$

Using $g(x^{[1]})$ in place of $g(x^{[0]})$ in Eq.(10) with an appropriate p yields new multipliers $\lambda^{[2]}$. Using $v^{[2]}$ and $\lambda^{[2]}$, values of $x^{[2]}$ follow from Eq.(6) and the dual volume $V^{[2]}$ from Eq.(4a). Substituting $x^{[2]}$ into Eq.(4b) and (4c) yields values for the constraint functions and all bar areas are scaled to ensure that no constraint is violated. In subsequent iterations, this scaled design and the previous scaled design would be compared and checked against convergence criteria and iterations would be either stopped here or continued.

Alternative entropy-based formulation
In ref.1 it is proposed an alternative solution scheme which combines the two phases into a single phase consisting of solving an unconstrained problem. The Lagrangean of problem (2) is augmented with an entropy term and the stationarity conditions reduce to:

$$\text{Min}_x\; f(x) + 1/\rho\, \ln \Sigma_{j=1,m} \exp[\rho\, \alpha\, g_j(x)] \tag{13}$$

that must be solved for an increasing positive parameter $\rho\, \alpha$. A different entropy-based procedure more appropriate for shape optimization will be proposed next.

MINIMAX OPTIMIZATION

Minimax problems are discontinuous and non-differentiable, both of which attributes make its numerical solution by direct means difficult. Ref.2 explores the relationships between the minimax optimization problem and the scalar optimization function and extends the equivalences to general multicriteria optimization. Specifically it is shown that a minimax problem can be solved indirectly by minimizing a continuous differentiable scalar optimization problem. The Shannon/Jaynes maximum entropy principle plays a keyrole in these classes of problems, hence the characterization of these methods as entropy-based. In this section some of the theory behind this approach to minimax optimization is briefly described.

For any set of real, positive numbers Uj, j=1, ..., J, and real $\rho \geq q \geq 1$, Jensen's inequality states that,

$$(\Sigma_{j=1,m} U_j^{\rho})^{1/\rho} \leq (\Sigma_{j=1,m} U_j^{q})^{1/q} \tag{14}$$

Inequality (14) means that the p-th norm of the set U decreases monotonically as its order, p, increases. Another important property of the p-th norm is its limit as ρ tends towards infinity:

$$\lim_{\rho \rightarrow \infty} (\Sigma_{j=1,m} U_j^{\rho})^{1/\rho} = \mathrm{Max}_{j=1,m} < U_j > \tag{15}$$

Consider the minimax optimization problem,

$$\mathrm{Min}_x \ \mathrm{Max}_{j=1,m} < g_j(x) > \tag{16}$$

and Jensen's inequality. Let $U_j = \exp [g_j(x)]$, j=1,...,m thus ensuring that $U_j > 0$, for all positive $g_j(x)$. Then,

$$(\Sigma_{j=1,m} U_j^{\rho})^{1/\rho} = \{ \Sigma_{j=1,m} \exp[\rho \ g_j(x)] \}^{1/\rho} \tag{17}$$

And from (14),

$$\lim_{\rho \rightarrow \infty} \{ \Sigma_{j=1,m} \exp[\rho \ g(x)]\}^{1/\rho} = \mathrm{Max}_{j=1,m} < g_j(x) > \tag{18}$$

Taking logarithms of both sides and noting that,

$$\log \lim(f) = \lim \log(f) \quad \underline{\text{and}} \quad \log \mathrm{Max}(f) = \mathrm{Max} \log(f) \tag{19}$$

Eq.(18) becomes,

$$\lim_{\rho \rightarrow \infty} (1/\rho) \log\{ \Sigma_{j=1,m} \exp[\rho \ g(x)]\} = \mathrm{Max}_{j=1,m} < g_j(x) > \tag{20}$$

Result (20) holds for any set of vectors g(x), including that set which results from minimizing both sides of (16) over x. Thus (20) can be extended to:

$$\mathrm{Min}_x \ \mathrm{Max}_{j=1,m} < g_j(x) > = \mathrm{Min}_x \ (1/\rho) \log\{ \Sigma_{j=1,m} \exp[\rho \ g_j(x)]\} \tag{21}$$

with increasing ρ in the range $1 \leq \rho \leq \infty$. Result (21) shows that a Pareto solution of the minimax optimization problem can be obtained by the scalar minimization,

$$\mathrm{Min}_x \ (1/\rho) \ \log\{ \Sigma_{j=1,m} \exp[\rho \ g_j(x)]\} \tag{22}$$

with a sequence of values of increasingly large positive $\rho \geq 1$.

Truss configuration optimization

The Pareto optimal design of truss geometry and cross sections consists of minimizing a whole set of goals by finding an optimal set of cross sectional areas x, joint coordinates y and corresponding displacements d. All these goals (volume, nodal displacement, etc.) need to be cast in a normalized form. If $\underline{V}$ represents a reference volume, the volume is reduced if,

$$l(y)^t x \le \underline{V} \quad \Rightarrow \quad g_1(x,y) = \frac{l(y)^t x}{\underline{V}} - 1 \le 0 \tag{23a}$$

where the member lengths are functions of joint coordinates. Lower bounds on cross-sectional areas are imposed to avoid topology changes,

$$g_2(x) = -\frac{x}{x^L} + 1 \le 0 \tag{23b}$$

Similarly, one has for the upper and lower bounds on the joint coordinates:

$$g_3(y) = \frac{y}{y^U} - 1 \le 0 \quad ; \quad g_4(y) = -\frac{y}{y^L} + 1 \le 0 \tag{23c}$$

The displacements d are computed for any given design by solving the displacement analysis equilibrium equations. The elements of the load vector R are constants and the elements of the stiffness matrix K are functions of both the variables x and y. The criterium concerning upper bounds on the nodal displacements is,

$$g_5(x,y) = \frac{d(x,y)}{d^U} - 1 \le 0 \tag{23d}$$

For the upper and lower bounds on the stresses:

$$g_6(x,y) = \frac{\sigma = S(y)\, d(x,y)}{\sigma^U} - 1 \le 0 \; ; \; g_7(x,y) = -\frac{\sigma = S(y)\, d(x,y)}{\sigma^L} + 1 \le 0 \tag{23e}$$

where the elements of the stress-transformation matrix S are functions of only the variables y.

Design variable linking is used to meet symmetry requirements and to reduce the number of design variables. In general, upper and lower bounds on design variables and stresses are assumed to be constant. If stability of members is considered, the lower bound σ^L can be defined as,

$$\sigma^L = \max\{\sigma_c, \sigma_b\} \tag{24}$$

in which σ_c is the lower stress limit and σ_b is the allowable stress for Euler buckling. For tubular sections with a nominal diameter to thickness ratio of D/t=10, the buckling stress can be given as,

$$\sigma_{bi} = \frac{10.1\, \pi\, E\, x_i}{8\, l_i(y)^2} \tag{25}$$

which depends on both the sizing and geometric variables.

The problem of finding values for the the cross sectional areas x and joint coordinates y which minimize the maximum of the goals has the form,

$$\text{Min}_{x,y}\ \text{Max}_j\,(g_1, ..., g_j ... g_7) = \text{Min}_{x,\,y}\ \text{Max}_{j \in J} < g_j(x,y) > \tag{26}$$

and belongs to the class of minimax optimization.

Reliability-based elastic design

Consistent with a first-order second-moment reliability approach, the minimum statistical information required for the evaluation of the optimum solution is: (a) The mean values of the loads, the coefficients of variation of the loads Ω_L and the coefficients of correlation between pairs of loads; (b) The coefficients of variation of the admissible stresses Ω_σ and the coefficients of correlation between pairs of admissible stresses. Assuming that the vector σ represents the elastic envelope stress-resultant coefficients obtained by the deterministic analysis and $\mu_{\sigma}U$, $\mu_{\sigma}L$ are the mean elastic capacities, the probability of unserviceability of individual sections is given by,

$$P_{sj} = P[\,\sigma^U_j - \sigma_j \le 0\,] \quad \text{or} \quad P[\,\sigma_j - \sigma^L_j \le 0\,] \quad ; \qquad j=1,..,m \qquad (27)$$

It is assumed that safety with regard to unserviceability of the section j depends only on the reliability index $\beta_j = \Phi^{-1}(P_{sj})$, that is defined as the shortest distance from the origin to the admissible stress surface in the reduced random variables coordinate system:

$$\beta_j = \frac{\mu_{\sigma U_j} - \mu_{\sigma_j}}{\sqrt{(\mu_{\sigma U_j}\,\Omega_\sigma)^2 + (\mu_{\sigma_j}\,\Omega_L)^2}} \qquad (28a)$$

or

$$\beta_j = \frac{\mu_{\sigma_j} - \mu_{\sigma L_j}}{\sqrt{(\mu_{\sigma L_j}\,\Omega_\sigma)^2 + (\mu_{\sigma_j}\,\Omega_L)^2}} \qquad (28b)$$

For completely correlated elastic capacities, the probability of failure is,

$$P_s = \max_{j=1,m} P_{sj} \qquad (29a)$$

and for uncorrelated elastic capacities,

$$P_s = \max_{j=1,m} (\Sigma_{k=1,v_j} P_{sk}) \qquad (29b)$$

where v_j represents the number of critical sections corresponding to all considered loading schemes.

Assuming that the nodal coordinates are deterministic, the reliability-based optimization problem consists of member size selection for given probabilities of failure against unserviceability P_{s*}. If $\underline{V}$ represents an (average) reference volume, (23a) becomes,

$$g_1(x,y) = \frac{l(y)^t\,\mu_x}{\underline{V}} - 1 \le 0 \qquad (30a)$$

Similarly, one has for the bounds on design variables:

$$g_2(x) = -\frac{\mu_x}{x^L} + 1 \le 0 \qquad (30b)$$

$$g_3(y) = \frac{y}{y^U} - 1 \le 0 \quad ; \qquad g_4(y) = -\frac{y}{y^L} + 1 \le 0 \qquad (30c)$$

By defining $\beta_* = \Phi^{-1}(P_{s*})$, individual stress bounds are given by,

$$g_5(x,y) = -\frac{\beta(x,y)}{\beta_*} + 1 \le 0 \tag{30d}$$

Similar expressions can be employed to satisfy the overal probability of failure against unserviceability and the upper bounds on the nodal displacements specified in probabilistic terms. This formulation give a Pareto solution to the multi-objective reliability-based optimization.

Scalar Function Optimization

Problem (22) is unconstrained and differentiable which, in theory, gives a wide choice of possible numerical solution methods. However, since the goal functions $g_j(x)$ do not have explicit algebraic form in most cases, the strategy adopted was to solve (22) by means of an iterative sequence of explicit approximation models. An explicit approximation can be formulated by taking Taylor series expansions of all the goal functions $g_j(x,y)$, truncated after the linear term. The quality of the approximation improves by considering the quadratic term for the geometric variables. This gives Eq.(31):

$$\text{Min } (1/\rho)\log\{\Sigma_{j=1,J} \exp \rho[g_j(x_o,y_o) + \Sigma_{i=1,N}\frac{\partial g_j}{\partial x_i}\bigg|_o (x_i-x_o) + \Sigma_{k=1,\beta}\frac{\partial g_j}{\partial y_k}\bigg|_o (y_k-y_o) +$$

$$+\frac{1}{2}\Sigma_{1=1,N}\Sigma_{k=1,\beta}\frac{\partial^2 g_j}{\partial y_i \partial y_k}\bigg|_o (y_i-y_o)(y_k-y_o) + \Sigma_{i=1,N}\Sigma_{k=1,\beta}\frac{\partial^2 g_j}{\partial x_i \partial y_k}\bigg|_o (x_i-x_o)(y_k-y_o)]\}$$

Problem (36) is an approximation to problem (27) if values of all the $g_j(x,y)$, $(\partial g_j/\partial x_i)$ and $(\partial g_j/\partial y_k)$ are known numerically. Given such values, problem (36) can be solved directly by any standard unconstrained optimization method. This problem must be solved iteratively, x_o and y_o being redefined each time as the optimum solution to the preceding problem. Iterations continue until changes in the design variables x,y become small. During these iterations the parameter ρ must be increased in value to ensure that a minimax optimum solution is found. In the present work, a constant value of $\rho = 100$ was used.

The choice of a large control parameter ρ for a very unfeasible design point may cause overflow problems. To overcome this situation (22) can be replaced by,

$$\text{Min}_x \{g_M(x) + (1/\rho)\log(\Sigma_{j=1,m} \exp\{\rho[g_j(x)-g_M(x)]\}) \tag{32}$$

where $g_M(x)$ is the largest of the goals $g_j(x)$, j=1,..m and ρ is a positive constant.

Sensitivity Analysis

To formulate and solve the scalar function minimization (31) used for the direct design, numerical values are required for all the functions $g_j(x,y)$ and their derivatives with respect to the design variables. The truss volume is known explicitly and its first derivatives are:

$$\frac{\partial V}{\partial x_i} = l_i \quad ; \quad \frac{\partial V}{\partial y_k} = \Sigma_{i=1,N} x_i \frac{\partial l_i}{\partial y_k} \tag{33}$$

in which $\partial l_i/\partial y_k$ is the direction cosine of the bar corresponding to the displacement y_k. The second derivative of V with respect to y_k is given by ratio of the square of the direction sine of the bar corresponding to the displacement y_k divided by the member length.

The derivatives of the reliability indices β with respect to the design variables require an approximation of the member stresses and nodal displacements which are implicit functions of x and y. One way of evaluating the derivatives of σ and d is to calculate them from analytical expressions, as follows. The displacement derivatives $\partial d^o/\partial x_i$ are computed by implicit differentiation of the equilibrium equations:

$$K^o \frac{\partial d^o}{\partial x_i} = - \frac{\partial K^o}{\partial x_i} d^o \tag{34}$$

Since d^o and K^o are known from analysis of the initial design, solution for $\partial d^o/\partial x_i$ involves only calculation of the r.h.s. vector of Eq.(34) and forward and back substitutions. The stress derivatives $\partial \sigma^o/\partial x_i$ are then determined directly by explicit differentiation,

$$\frac{\partial \sigma^o}{\partial x_i} = S \frac{\partial d^o}{\partial x_i} \tag{35}$$

The derivatives $\partial d^o/\partial y_k$ and $\partial \sigma^o/\partial y_k$ are computed in a similar manner; however, it should be remembered that the elements of S are functions of the joint coordinates y. The expressions for $\partial d^o/\partial y_k$ and $\partial \sigma^o/\partial y_k$ are:

$$K^o \frac{\partial d^o}{\partial y_k} = - \frac{\partial K^o}{\partial y_k} d^o \tag{36}$$

$$\frac{\partial \sigma^o}{\partial y_k} = S \frac{\partial d^o}{\partial y_k} + \frac{\partial S^o}{\partial y_k} d^o \tag{37}$$

To compute $\partial K^o/\partial x_i$, only elements of K associated with member i must be considered. Furthermore, the elements of $\partial K/\partial X_i$ are constant, therefore the computation must not be repeated. To find $\partial K^o/\partial y_k$ and $\partial S^o/\partial y_k$, only elements of K and S associated with the kth joint coordinate must be considered.

The second order derivatives with respect to y_k can be calculated in a similar manner,

$$K^o \frac{\partial^2 d^o}{\partial {y_k}^2} = - \frac{\partial^2 K^o}{\partial {y_k}^2} d^o - 2 \frac{\partial K^o}{\partial y_k} \frac{\partial d^o}{\partial y_k} \tag{38}$$

It can be observed that the solution for each of the derivative vectors involves only the calculation of the corresponding right hand side vector and forward and back substitutions.

NUMERICAL EXAMPLES

Least volume design of a grillage

The existence of relative minima in grillage structures can easily be found in two-dimensional design problems. Consider the grillage shown in Fig.1 subjected to a uniformly distributed load of 1.0 on longitudinal beams.The grillage has eight design variables. Moments of inertia and section moduli are derived from the areas using the relationships of eq.(39). Lower bounds on design variables are 5.0. Upper and lower bounds on normal stresses are ± 20

$$I_i = 0.3563\, x_i^{2.65} \quad ; \quad W_i = 0.4899\, x_i^{1.82} \tag{39}$$

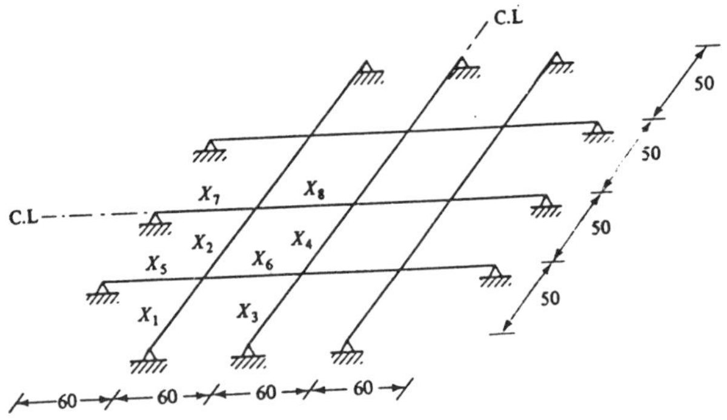

Figure 1

The iteration history is shown in Fig.2, where it can be seen that convergence is achieved with only six analysis.

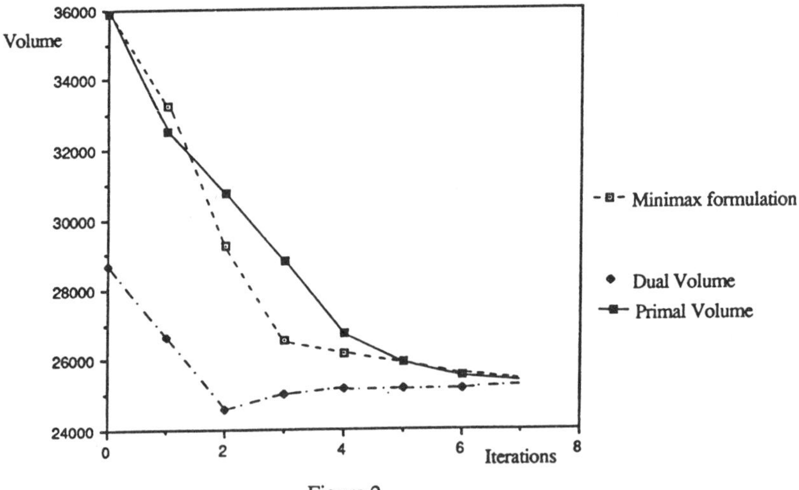

Figure 2

The sensitivity analysis is required in the minimax formulation as opposed to the first algorithm, that only needs the information concerning the current design point. On the other hand, convergence is faster in the second solution scheme. The grillage has several local optima and both methods lead to different solutions according to the starting point and the schedule for the control parameter selection.

Reliability-based truss configuration

Fig.3 shows a 47-bar planar tower subject to three independent loading conditions. This structure is designed for optimum geometry and this reliability-based problem is formulated based on probabilistic requirements for the sizing variables. The average loads are described in Table 1, where $\Omega_L = 0.20$.

	Load condition 1		Load condition 2		Load condition 3	
Joint	17	22	17	22	17	22
Load, x direction	4.5	0.	0.	4.5	4.5	4.5
Load, y direction	-10.5	0.	0.	-10.5	-10.5	-10.5

Average lower and upper bounds on stresses are $\mu_{\sigma}L = -20$, $\mu_{\sigma}U = 27$ and $\Omega_{\sigma} = 0.10$. The members were assumed to be tubular with a constant ratio of diameter-to-wall thickness of 10. Euler buckling was prohibited for all members. The modulus of elasticity was taken as 3×10^4 and the material density, $\rho = 3 \times 10^{-4}$. A minimum allowable area of 10^{-6} was specified. Joints 15, 16, 17 and 22 were held stationary in space and joints 1 and 2 were required to lie on the x axis. Symmetry is imposed and there are a total of 27 independent area variables and 17 independent coordinate variables. The specified probability of failure against unserviceability is $1.5\ 10^{-3}$ and the members are assumed uncorrelated.

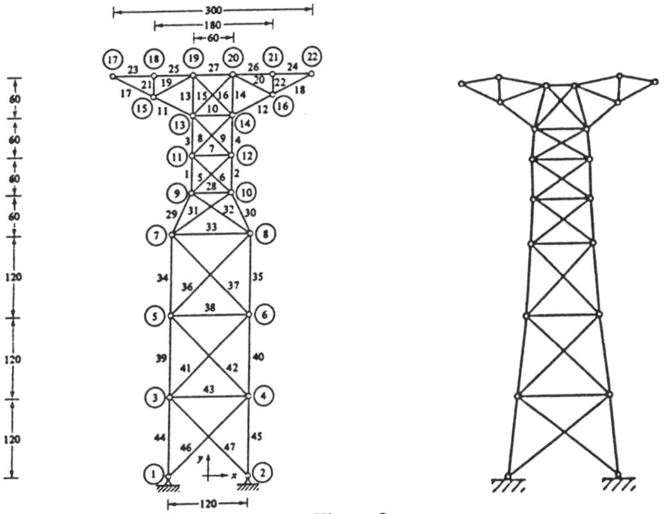

Figure 3

The iteration history is represented in Fig.4, showing a smooth convergence. The solution drawn in Fig.3 required a total of 9 iterations to converge, although results within a 6% margin of error were reached after five analyses. The difference in the geometries obtained after the fifth iteration show that the design space is rather flat in the vicinity of the Pareto solution. As opposed to the algorithms more conventionally used, the minimax formulation is not so heavily dependent on the specified move limits. Buckling constrained problems in which the forces in the members with the minimum allowable areas change as a result of geometry changes may lead to erratic convergence behavior. The minimax formulation prevents this occurrence because it does not look for active constraints (such as the lower bounds imposed on the sizing variables) but rather considers all objectives simultaneously.

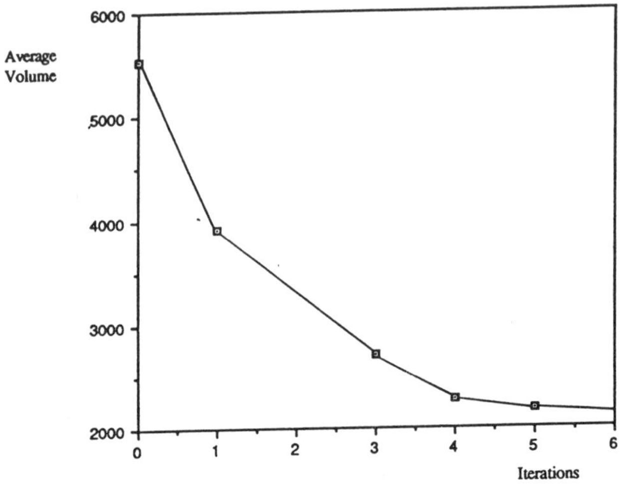

Figure 4

CONCLUSIONS

This paper presents a new class of optimization methods based on informational entropy concepts that are under current development. They are a radically different alternative to existing methods possessing distinctive features and advantages. One of these entropy-based methods, which is applied to structural sizing problems, is computationally extremely simple to implement. The optimization phase is reduced to calculating values for multipliers from an algebraic expression similar in complexity to those used in stress ratio or optimality criteria methods. Unlike optimality criteria and other more recent methods, it does not require an active/passive set strategy. The assignment is done on the basis of the application of Shannon's maximum entropy principle, used to measure the uncertainty in a random process.

Alternatively, the maximum entropy principle is applied to vector and Pareto optimization. Specifically it was shown that the minimax problem can be solved by minimizing a continuous differentiable unconstrained function. The number of iterations required to obtain optimum solutions is small, what makes this algorithm competitive with respect to other more sophisticated methods.

ACKNOWLEDGEMENTS
The author wish to thank the financial support given by JNICT (Junta Nacional de Investigação Cientifica e Tecnológica, Proj. 87 230) and Calouste Gulbenkian Foundation.

REFERENCES

1. A.B. Templeman and Li Xingsi, "A Maximum Entropy Approach to Constrained Non-linear Programming", Eng. Opt., 12, 191-205, 1987.
2. A. B. Templeman "Entropy-based Minimax Applications in Shape Optimal Designs" In Lecture Notes in Engineering, Nº42: Discretization Methods and Structural Optimization (H.A. Eschenauer and G. Thierauf, Ed.), 335-342, Springer Verlag, 1989.

Optimality Criterion Algorithm for Mixed Steel and Reinforced Concrete Seismic Structures

K.Z. Truman (*), F.Y. Cheng (**)
(*) Department of Civil Engineering, Washington University, St. Louis, MO., U.S.A.
(**) Department of Civil Engineering, University of Missouri-Rolla, Rolla, MO., U.S.A.

ABSTRACT

A structural optimization algorithm based upon an optimality criteria approach is presented for three-dimensional statically and dynamically loaded steel and/or reinforced concrete structures. The theoretical work is presented in terms of scaling, sensitivity analyses, optimality criteria, and Lagrange multiplier determination. The structures can be subjected to a combination of static and/or pseudo-dynamic displacement and stress, and natural frequency constraints. The dynamic analyses are based upon the ATC-3-06 provisions or multi-component response spectra modal analyses. Using the algorithm presented, a computer program called ODRESB-3D was developed for both analysis and design of building systems. A design example using ATC-3-06 loads is provided to illustrate the rapid convergence and the practicality of the presented method.

INTRODUCTION

The demand for economical, reliable structures in virtually all fields of endeavor has provided the impetus for the development of rapid, convergent and effective structural optimization algorithms. Various optimization techniques of linear, nonlinear and dynamic programming have been developed for different types of statically and dynamically loaded structures [1,2]. Recently, Cheng et.al. [3,4] have extended structural optimization into the field of seismic design of structural systems. Currently, Truman and Cheng [5,6] have extended structural optimization to seismic design of three dimensional systems. The three dimensional systems can be seismically loaded using equivalent static forces from ATC-3-06 [7] provisions or by using multicomponent modal response spectra procedures. The optimization algorithm to be presented is an optimality criteria based technique which has been implemented in a computer program, ODRESB-3D, for automated, optimal design of 3-D steel and/or reinforced concrete building systems subject to static or pseudo-dynamic loads (equivalent static loads or modal response spectra analyses).

STRUCTURAL MODEL

The structural model is typically three dimensional in ODRESB-3D, but through symmetry it can be used to model two dimensional systems. Each floor is considered to be a rigid slab in the horizontal plane while being considered flexible with regards to out of plane bending. The rigid slab assumption allows each floor to be represented by two translational and one rotational degree of freedom in the horizontal plane plus a vertical and two rotational (about two orthogonal horizontal axes) degrees of freedom at each column node. The two nodal rotational degrees of freedom are eliminated through condensation prior to solving the system response. Each structure can consist of steel beams, beam-columns and braces coupled with concrete beam-columns or shear walls. All analyses are assumed to be linearly elastic.

STRUCTURAL OPTIMIZATION

Structural optimization refers to the development and application of numerical techniques for improving designs with respect to a distinct objective while staying within well-defined constraints. The objective can be weight, cost, reliability, or any combination. The constraints generally represent the structural response and design limitations. Mathematically, the optimization problem for the presented examples can be written as:

$$\textit{Minimize } O(\delta) = \sum_{i=1}^{n} \gamma_i V_i \tag{1}$$

$$\textit{subject } g_j(\delta) \leq 0 \qquad j=1,\ldots,l \tag{2}$$

$$\underline{\delta}_i \leq \delta_i \leq \overline{\delta_i} \tag{3}$$

where $O(\delta)$, is the objective function γ_i are the appropriate constants for the objective function per unit volume for element i, V_i, is the volume of element i, δ_i, is the ith primary design variable, $g_j(\delta)$, are the structural response constraints, $\underline{\delta}_i$ and $\overline{\delta_i}$ are the minimum and maximum sizes for element i, l, is the number of inequality constraints and n, is the number of structural elements.

Design variables can be broken into two categories, primary and secondary design variables. Primary design variables are those structural properties which are directly resized in the optimization algorithm. The secondary design variables are those structural properties which are dependent upon the primary design variables. For example, a beam is represented by its major axis moment of inertia in the optimization algorithm, while its crossectional area is a function of its major axis moment of inertia. This procedure is termed pseudo discrete optimization. The secondary design variables take the form:

$$S_{ij} = C_{1j} \delta_i^{C_{2j}} + C_{3j} \tag{4}$$

where C_{1j} and C_{2j} and C_{3j} are algebraically derived constants for regular crossections such as circular or rectangular, etc. or statistically derived for irregular crossections such as wide flange beams, channels, etc. This technique is used for two reasons: 1) it eliminates a large number of the design variables and 2) it gives reasonable designs since the variables are related to a realistic design shape. Otherwise, the optimal member such as the beam previously mentioned might have an incompatible crossectional area or minor axis moment of inertia compared to the major axis moment of inertia. A complete set of constants for wide flange, rectangular concrete, and regular crossections have been derived by Cheng and Truman [5]. Typical values for the constants for wide flange beams and columns with major axis moments of inertia between 1550 in^4 and 12,100 in^4 for the minor axis moment of inertia, I_y, polar moment of inertia, J, and the crossectional area, A, are:

$$I_y = 0.0265\, I_x + 20.47 \tag{5}$$

$$J = 0.0124\, I_x^{0.905} \tag{6}$$

$$A = 0.5008\, I_x^{0.487} \tag{7}$$

Typical values for rectangular concrete sections based on a working stress model are:

$$I_y = \frac{1}{h^8 D^2} I_y^3 \tag{8}$$

$$A = \frac{P(k + 2n\rho - \rho)}{h^2 D} I_x \tag{9}$$

$$A_N = \frac{1}{h^2 D} I_x \tag{10}$$

where I_x is the major axis moment of inertia, A_N,is the gross concrete area, h, is the depth of the crossection, P, is the percentage of depth to the lumped tensile reinforcement, k, is the percentage of depth for the compression area, n, is the modular ratio, ρ, is the ratio of the area of steel to the gross concrete area, and D, is a constant based on these given properties. The equation for D is:

$$D = \frac{(Pk)^3}{3} + \rho P(n - 1)(P(k + 1) - 1)^2 + nP^3 \rho (1 - k)^2 \tag{11}$$

Constraints, $g_j(\delta)$, represent the restrictions that the structural designer would like to impose while trying to find the optimal design. The inequality constraints are used to place limits on structural response such as displacements, drifts, frequencies, stresses, and buckling loads. Side constraints, Equation (3), are also inequality constraints but are generally not handled in the same manner as the structural responses. The side constraints are used to limit the size of the structural members so they will remain within a practical design range. Linkage constraints are used to force certain structural elements to maintain identical sizes such as all columns on one floor having the same moment of inertia.

The development of optimality criteria methods in the early 70's may be considered as a great contribution to the field of engineering optimization. It provided major improvements over other classical mathematical methods. The major advantage is the significant reduction of iterations required for convergence to an optimum design as shown by Truman and Jan [8]. The optimality criteria used in the optimization algorithm is derived from the Kuhn-Tucker conditions of optimality. The derivative of the Lagrangian function gives:

$$\frac{\partial L}{\partial \delta_i} = \frac{\partial O}{\partial \delta_i} + \sum_{j=1}^{l} \lambda_j \frac{\partial g_j}{\partial \delta_i} = 0 \qquad i=1,\ldots,n \tag{12}$$

with $\lambda_j > 0$ and $\lambda_j\, g_j = 0$ for $j = 1, \ldots, l$ where l is the number of active constraints (constraints which have reached either their upper or lower bounds). Rearranging Equation (12) gives:

$$T_i = -\sum_{j=1}^{l} \lambda_j \frac{\left(\frac{\partial g_j}{\partial \delta_i}\right)}{\left(\frac{\partial O}{\partial \delta_i}\right)} = 1 \qquad i=1,\ldots,n \tag{13}$$

which must be true along with $\lambda_j > 0$ and $\lambda_j\, g_j = 0$ when a local or global optimal solution is obtained. The major difficulty with using the T_i values is their heavy dependence on the unknown values and sets of Lagrange multipliers which represent the active set of constraints.

As the structure reaches its optimal design, T_i approaches unity, therefore T_i is a measure of the error in the design versus the optimal solution. Using this error measurement in a recurrence relationship allows the updating of the member sizes in order to force T_i to unity and an optimal design. A linear recurrence relationship which was derived by keeping the first two terms of the binomial expansion of a power law recurrence relationship gives:

$$\delta_i^{k+1} = \delta_i^k \left(1 + \frac{1}{r}\left(T_i^k - 1\right)\right) \qquad i=1,\ldots,n \tag{14}$$

where the term $(T_i - 1)$ measures the error associated with element i at iteration k and r is a convergence control parameter (common value is 2). Prior to the use of Equation (14) for resizing of the structural elements, an estimate of the optimal Lagrange multipliers must be determined. A set of linear equations can be derived based on the use of a linear recurrence relationship and the assumption that there will be a small change in the active constraint values between iterations. The equations take the form:

$$r g_j - \sum_{i=1}^{n} \frac{\partial g_j}{\partial \delta_i}\delta_i^k = \sum_{p=1}^{l} \lambda_p \Big(\sum_{i=1}^{n} \frac{\dfrac{\partial g_j}{\partial \delta_i}\dfrac{\partial g_p}{\partial \delta_i}}{\dfrac{\partial O}{\partial \delta_i}}\delta_i^k\Big) \qquad j=1,\ldots,l \tag{15}$$

These equations are desirable since an initial estimate of the Lagrange multipliers is not needed. Unlike many Lagrange multiplier seeking techniques, these equations also take into account the dependence of one constraint upon another as evidenced by the double summation in Equation (15). This set of equations is then solved at which time any coefficients which are linked to a negative Lagrange multiplier are removed and the equations resolved. This is done until a set of positive Lagrange multipliers are found. The initial set of potentially active constraints, prior to the iterative solution of Equation (15), is chosen by using:

$$(1 - P_1) \le \frac{u_j}{\bar{u}_j} \le (1 + P_2) \quad , \quad (1 - P_1) \le \frac{\underline{u}_j}{u_j} \le (1 + P_2) \tag{16}$$

for upper bound and lower bound constraints, respectively, where u_j is the structural response being considered, $\bar{u}_j$ is an upper bound and $\underline{u}_j$ is a lower bound to the response. As long as the ratios shown in Equation (16) are below one, the response is in the feasible region. The value $(1 - P_1)$ provides the margin of error on the feasible side of the constraint surface and the value $(1 + P_2)$ provides the acceptable margin of error on the infeasible side of the constraint surface for choosing those constraints which are potentially active (a ratio of one).

Due to the iterative nature of the algorithm, termination criteria must be developed. The criteria must be able to handle several distinct conditions. The primary function is to check for convergence or divergence of the objective function. Secondary conditions are to limit the amount of computing time or iterations. Convergence is considered by comparing successive values of the objective function and claiming convergence if several successive iterations produce less change in the objective compared to a specified change.

As seen by Equations (12-15) an important part of the optimization algorithm is the determination of the constraint gradients with respect to the primary design variables. The stiffness is directly differentiable with respect to the primary and secondary design variables. Therefore, the gradient of the stiffness including secondary design variables becomes:

$$\frac{\partial [K]_T}{\partial \delta_i} = \sum_{j=1}^{t} \frac{\partial [K]_{ij}}{\partial \delta_i} = \sum_{j=1}^{t} \frac{\partial [K]_{ij}}{\partial S_{ij}} \frac{\partial S_{ij}}{\partial \delta_i} \tag{17}$$

where

$$\frac{\partial S_{ij}}{\partial \delta_i} = C_{1j} C_{2j} \delta_i^{(C_{2j}-1)} \tag{18}$$

and $[K]_T$ is the total stiffness matrix, $[K]_{ij}$ is the ith elemental stiffness with the jth secondary design variable terms and t is the number of secondary design variables for the ith elemental type. Note that $[K]_{ij}$ is linear in terms of the secondary design variable S_{ij} which gives the final form for the derivative of the total stiffness as:

$$\frac{\partial [K]_T}{\partial \delta_i} = \sum_{j=1}^{t} \frac{[K]_{ij}}{S_{ij}} C_{1j} C_{2j} \delta_i^{(C_{2j}-1)} \tag{19}$$

The gradients for the displacements, stresses, and drifts are derived by direct differentiation, application of the chain rule and applying techniques similar to a pseudo-load or virtual load approaches. The virtual and pseudo-load approaches are based upon the premise that the displacements, drifts and stresses can be written as a linear combination of the displacements. In equation form this becomes:

$$u_j = \{b\}_j^T \{U\} \tag{20}$$

where $\{b\}_j$, is a vector of terms which enforce this relationship (i.e. a vector of all zeros and one component of one would be appropriate for displacements, whereas a less sparse vector with terms relating to the section modulus are required for stresses as derived by Cheng and Truman [5]), u_j, is the jth global displacement, stresses or drift and $\{U\}$, is the vector of global displacements. Using Equation (20), the gradient for the jth response can be written as:

$$\frac{\partial u_j}{\partial \delta_i} = \frac{\partial \{b\}_j^T}{\partial \delta_i} \{U\} + \{b\}_j^T \frac{\partial \{U\}}{\partial \delta_i} \tag{21}$$

where the gradients of $\{b\}_j$ are zero for displacements and drifts but are nonzero for stress constraints. The gradient for the displacements is found as:

$$\frac{\partial \{U\}}{\partial \delta_i} = -[K]_T^{-1} \frac{\partial [K]_T}{\partial \delta_i} \{U\} \tag{22}$$

In addition to these gradients, Cheng and Truman have derived the gradients of the eigenvectors, generalized mass, and response spectra used for the optimal design of systems subject to multicomponent modal analysis loadings. The gradients of these responses are found directly by using a generalized coordinate solution for the modal analyses.

OPTIMIZATION ALGORITHM

There are eight major steps in the optimization algorithm. After the analyses have been performed (static or pseudo-dynamic), STEP(1) the constraints are then separated into potentially active and passive constraints by using Equation (16). If an incorrect set of active constraints has been chosen, the optimization will have to adjust either during scaling or the Lagrange multiplier determination. STEP(2) scaling is used in several instances within the optimization procedures. First, it can be used to force the most violated constraint to be within the active region by changing all primary design variables by a prescribed factor, basically the ratio of the most violated constraint to the allowable value. Since the three dimensional problem is nonlinear with respect to the primary design variables, the scaling is an iterative process. Secondly, scaling can be used if no constraints are in the active region. STEP(3) termination criteria will then be checked (not applicable during first cycle). It will only terminate if successive cycles of optimization have a smaller percentage of change than the prescribed value or if the allowable number of iterations is violated. STEP(4) the optimization phase begins with the calculation of the gradients of the active constraints with respect to the primary design variables. STEP(5) these gradients are then used to create a set of linear equations, Equation (15) used for the determination of the Lagrange multipliers. Once these equations have been solved, the Lagrange multipliers are checked with regards to their sign. If all Lagrange multipliers are positive a valid set of active constraints were chosen in step (1). If one or more Lagrange multipliers are negative, the constraints associated with those negative Lagrange multipliers are removed from the active set and the reduced set of equations are resolved until all Lagrange multipliers are positive. STEP(6) once the Lagrange multipliers are all positive, the primary design variables can be resized. This is accomplished by using the linear recurrence relationship in Equation (14). STEP(7) if Equation (14) forces any of the elements to violate the side constraints, the algorithm returns to step (5) and regenerates the equations for the Lagrange multipliers using a form of the Equation (15), Cheng and Truman [5], which includes passive element terms. Therefore, steps (5) and (6) are iterative in nature. The algorithm then begins with a new analysis and follows through the same steps until step (3) the termination criteria is satisfied. Then step (8) postprocessing of the data is performed.

RESULTS

The structure shown in Figure 1 will be designed to resist the lateral forces calculated from the ATC-3-06 equivalent lateral force provisions for seismic analysis and design. These forces change during every iteration as they are dependent on the structural mass and period of the system. This example will be used to examine the fast convergence and use of this algorithm for the design of mixed concrete and steel structures subject to pseudo-dynamic loads.

Several element properties have been held constant during each optimization. The eight perimeter columns are rectangular, concrete columns which are linked two stories at a time, i.e. all columns on levels 9 and 10, levels 7 and 8, etc., are forced to have the same crossectional dimensions. The inner core shear walls, can be grouped into two sets, the shear walls parallel to the x-direction and the shear walls parallel to the y-direction. Each of these sets are linked in the same manner as the columns. The steel beams are linked on a per floor basis. The concrete elements have one fixed dimension with respect to the rectangular crossection while the other dimension is allowed to vary and represent the primary design variable, as shown in Tables 1-3. All concrete elements were assumed to have a Young's modulus of 3,000 k/in^2, a modular ratio of 10, a steel percentage of 0.015, and a shear modulus of 1150 k/in^2. The steel beams have a Young's modulus of 30,000 k/in^2. Due to the

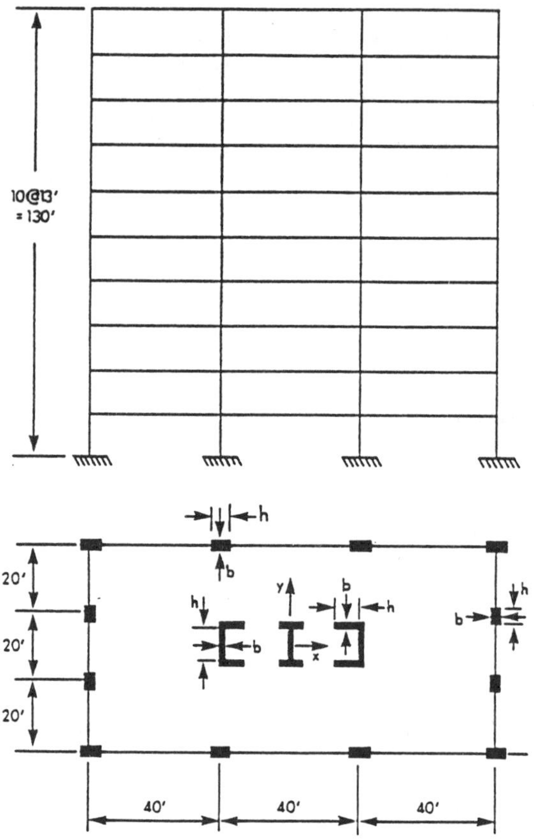

Figure 1. Ten Story Concrete Structure with Steel Beams

fixed dimension of the columns, three designs were completed in order to create more realistic final designs from the previous optimal solutions.

The optimization was based on a set of fixed parameters. Termination of the optimization was to occur if ten cycles of optimization, fifteen analyses, or less than a five percent weight change is achieved. Since the previous optimal designs were to be used only as guidelines for producing better initial concrete dimensions for the next optimization a lower percentage was not justified. The convergence control parameter was two. The displacement constraints were 0.45 in. per floor. Design 1 had a constraint range of 10 percent below and five percent above the allowable displacement, while Designs 2 and 3 had a range of 20 percent below and five percent above the constraint limits.

The ATC-3-06 equivalent lateral force analyses were based upon a constant set of ATC-3-06 and structural parameters. A non-structural mass of 3.26 k-s^2/in, and a non-structural rotational mass of 704,354 k-s^2-in were used for each level. Each level was assumed to be 13.0 feet in height. Map area seven (Los Angeles) was used for both the effective peak acceleration and effective peak velocity-related acceleration. The structure was assumed to be in seismic hazard exposure group 2 with soil condition three (soft soil). The response modification factor (ductility) was considered as 5.5, and the deflection amplification factor (elastic to inelastic deformation) was assumed to be 5.0. The two load cases consisted of ATC-3-06 x-direction loads with a plus/minus five percent eccentricity plus 30 percent of the y-direction loads.

Table 1. Initial and Final Design Sizes for the Vertical Members of Design 1

Levels	Columns (hxb) (in) $b_{min} = 5$ $b_{max} = 240$		Shear Walls Parallel to y-direction (hxb) (in) $b_{min} = 5$ $b_{max} = 240$	Shear Walls Parallel to x-direction (hxb) (in) $b_{min} = 5$ $b_{max} = 240$
	Initial	Final	Final	Final
9-10	12 x 65	12 x 42.1	72 x 18.0	60 x 8.8
7-8	14 x 75	14 x 26.9	72 x 58.9	60 x 10.4
5-6	16 x 90	16 x 16.6	72 x 93.1	60 x 12.7
3-4	18 x 100	18 x 12.1	72 x 137.3	60 x 13.7
1-2	20 x 110	20 x 12.7	72 x 149.8	60 x 12.9

initial design values for shear walls parallel to y-direction for all levels is 72 x 130
initial design values for shear walls parallel to x-direction for all levels is 60 x 130

The results are presented in Tables 1-3 and Figure 2. Obviously, the initial sizes for Design 1 are not within reason, but were determined by scaling in order to satisfy all of the displacement constraints, as shown in Table 1. The initial sizes for the second and third designs were progressively changed based upon the previous optimal results. From the final sizes from Design 1, it was apparent that the shear wall results were undesirable. Therefore, the primary design variable for the shear walls parallel to the x-direction was changed from b to h in Designs 2 and 3. This change was made since the optimization was trying to force the y-direction shear walls to become excessively large (taking advantage of the fixed dimension h). The upper level columns are large while the lower columns are small which is also undesirable. The optimization algorithm is trying to eliminate the large structural mass of the shear walls at the upper levels in order to reduce the pseudo-dynamic loads which are mass dependent. While at the lower levels it is more efficient to use the shear walls rather than the columns to eliminate excessive upper story displacements.

Initial sizes were chosen for Design 2 based on the optimal solution of Design 1, as shown in Table 2. The column's h dimensions were increased in order to reduce the aspect ratio of the crossectional dimensions for the upper level columns. The columns were assumed to have a width, b, of 28.5 in. which is basically the average size of the upper three column widths from Design 1. The initial beam size was chosen as 8510 in^4 which is slightly larger than the average size of the top six beams from the optimal Design 1 beams. The new values for the shear walls were determined to provide nearly equivalent stiffness as Design 1, using 36 in. as the fixed width, b, for the x-direction shear walls, and 72 in. as the fixed width, h, for the y-direction shear walls. The results for Design 2 are considerably better than those for Design 1. The upper columns are reasonable, but the lower columns are not. Now the x-direction shear walls are providing most of the lateral support instead of the y-direction shear walls.

Table 2. Initial and Final Design Sizes for the Vertical Members of Design 2

Levels	Columns (hxb) (in) $b_{min} = 5$ $b_{max} = 36$		Shear Walls Parallel to y-direction (hxb) (in) $b_{min} = 5$ $b_{max} = 240$	Shear Walls Parallel to x-direction (hxb) (in) $h_{min}^* = 5$ $h_{max}^* = 240$
	Initial	Final	Final	Final
9-10	16 x 28.5	16 x 19.9	72 x 7.9	28.8 x 36
7-8	18 x 28.5	18 x 20.1	72 x 8.1	45.1 x 36
5-6	20 x 28.5	20 x 12.4	72 x 8.5	91.4 x 36
3-4	22 x 28.5	22 x 8.5	72 x 8.5	134.3 x 36
1-2	24 x 28.5	24 x 7.5	72 x 8.8	191.1 x 36

*h is the primary design variable for these elements
initial design values for shear walls parallel to y-direction for all levels is 72 x 34.1
initial design values for shear walls parallel to x-direction for all levels is 123 x 36

The initial sizes for Design 3 were determined in a subjective manner which placed more emphasis on having a realistic distribution using side constraints, as shown in Table 3. In order to have a more realistic distribution for the columns the maximum and minimum sizes for the shear walls and columns were adjusted. For example, the shear walls parallel to the y-direction were given the same initial values as Design 2, but had the upper limit on the width reduced to a more realistic value than the previous value of 240 in. The x-direction shear walls were reduced in width in order to force the lower level columns to become more involved. The columns all have reasonable aspect ratios and a reasonable distribution with the exception of levels 7 and 8 which require a slightly larger width. Note that several members have violated their side constraints. This is due to the large number of potentially passive elements. Scaling was used to satisfy all of the constraints which caused some violation of the upper size limits. Relaxing the side constraints and increasing the convergence control parameter would most likely prevent this violation.

The final weights are all within 4,000 to 4,700 kips, as shown in Figure 2, which is a reduction in weight for all designs. The lightest design is Design 2, whereas, the heaviest is Design 3 which in theory is the best design. The increased weight is due to the increase in the number of side constraints which become active. The final set of active constraints are very similar for all three designs. Each design has an active x-displacement constraint at level 10 (4.5 in.) with one additional active constraint

Table 3. Initial and Final Design Sizes for the Vertical Members of Design 3

Levels	Columns (hxb) (in)		Shear Walls Parallel to y-direction (hxb) (in) $b_{min} = 8$ $b_{max} = 24$	Shear Walls Parallel to x-direction (hxb) (in) $h_{min}^* = 48$ $h_{max}^* = 192$	
	Initial	Final	Final	Initial	Final
9-10	16 x 16	16 x 20	72 x 10	123 x 24	60 x 24
7-8	16 x 16	16 x 14	72 x 10	123 x 24	60 x 24
5-6	20 x 20	20 x 25	72 x 10	123 x 30	60 x 30
3-4	20 x 20	20 x 25	72 x 10	123 x 30	194 x 30
1-2	20 x 20	20 x 25	72 x 10	123 x 30	240 x 30

*h is the primary design variable for these elements
initial design values for shear walls parallel to y-direction for all levels is 72 x 34
b_{min} for columns on levels 1-6 is 20 in.
b_{min} for columns on levels 7-10 is 5 in.
b_{max} for columns on levels 1-6 is 36 in.
b_{max} for columns on levels 7-10 is 16 in.

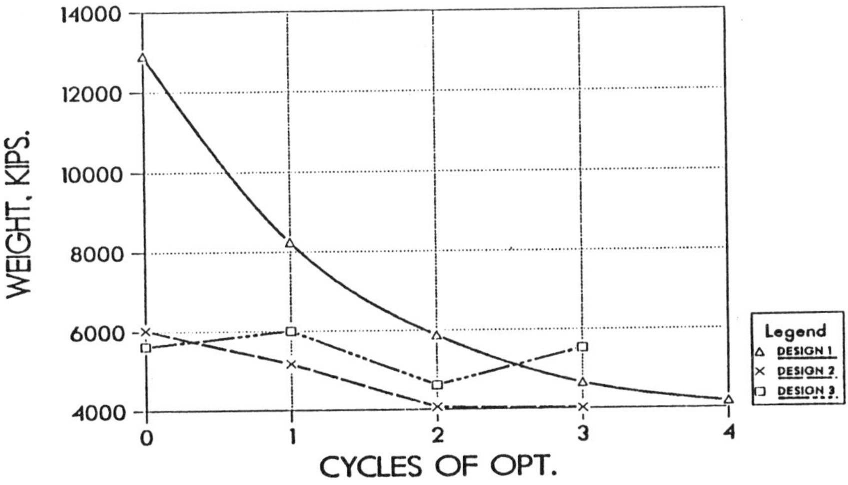

Figure 2. Comparison of Optimal Weights and Convergence

at either level 8 or 9 (3.6 or 4.5 in.). All three designs have x-displacements at the top four levels which are close (seven percent) of the active values. The final designs of the beams for Designs 1 and 2 were to have larger beams for levels 6-9 in order to provide framing action at the upper levels where the shear walls are small. Levels 6-9 have moments of inertia of approximately 13,500 in^4. Levels 1-5 average 5,000 in^4. Design 3 had constant beams of 25,000 in^4 for levels 3-9.

CONCLUSIONS

This algorithm provides an efficient means for producing preliminary designs of three dimensional building systems subject to static and pseudo-dynamic loads. The optimality criteria approach provides rapid convergence and is quite capable of handling steel, concrete or mixed steel/concrete structural systems. The technique presented provides a series of feasible designs which generally cannot be produced by using the conventional technique of applying engineering intuition, reanalysis and design in a continual loop until a final design is chosen. Most often this set of conventional procedures produces several feasible and several infeasible designs. The optimality criteria algorithm presented is fast, efficient and useful for the design of structural systems.

REFERENCES

1. Venkayya, V.B. Structural Optimization: A Review and Some Recommendations, Intl. Journal of Numerical Methods in Engineering, Vol. 13, 1978.

2. Feng, T.T., Arora, J.S., and Haug, J.E. Optimum Structural Design Under Dynamic Loads, Intl. Journal of Numerical Methods in Engineering, Vol. 11, pp. 39-52, 1977.

3. Cheng, F.Y., Srifuengfung,D., and Sheng, L. ODEWS- Optimum Design of Static, Earthquake and Wind Steel Structures, NSF Report, the U.S. Department of Commerce, National Technical Information Service, Virginia, NTIS Access No. PB 81-232738, 1981.

4. Cheng, F.Y.and Botkin, M.E. Nonlinear Optimum Design of Dynamic Damped Frames, Journal of the Structural division, American Society of Civil Engineers, Vol 102, pp. 609-628, March 1976.

5. Cheng, F.Y. and Truman, K.Z. Optimum Design of Reinforced Concrete and Steel 3-D Static and Seismic Building Systems with Assessment of ATC-03, NSF Report, the U.S. Department of Commerce, National Technical Information Service, Springfield, Virginia, NTIS Access No. PB 87-168564/AS (414 pages), 1985.

6. Truman, K.Z., Juang, D.S., and Cheng, F.Y. ODRESB-3D, User's Manual, A Computer Program for Optimum Design of Reinforced Concrete and Steel Building Systems Subjected to 3-D Ground Motions and ATC-03 Provisions, NSF Report, the U.S. Department of Commerce, National Technical Information Service, Springfield, Virginia, NTIS Access No. PB 87-162970/AS (212 pages), 1985.

7. ATC-3-06 Tentative Provisions for the Development of Seismic Regulations for Buildings, Applied Technology Council, National Science Foundation, and National Bureau of Standards, NBS Special Publication 510, June 1978.

8. Truman, K.Z. and Jan, C.T. Optimum Design of Earthquake Resistant Structures Subject to the ATC-3-06, BOCA, and UBC Seismic Provisions, Report 91-1, Department of Civil Engineering, Washington University, St. Louis, MO, 1991.

A New Direction in Cross-Section and Layout Optimization: The COC Algorithm

G.I.N. Rozvany, M. Zhou

FB 10, Essen University, Postfach 10 37 64, D-4300 Essen 1, Germany

ABSTRACT

Basic features of iterative continuum-based optimality criteria (COC) methods are briefly reviewed and some new applications in cross-section, generalized shape and layout optimization are presented. Finally, advantages and disadvantages of this technique are discussed.

INTRODUCTION

The development of optimality criteria methods dates back to the late sixties and has followed, until recently, two separate paths. The so-called "numerical school", consisting mostly of aerospace scientists, developed ***discretized optimality criteria*** (DOC), which are necessary conditions of optimality based on the Kuhn-Tucker condition in a finite dimensional design space, expressed in terms of a discretized (usually FE) system. Extensive reviews of this approach are available (e.g. Khot [1]; Berke and Khot [2], Venkayya [3] and Haftka and Kamat [4, pp. 195-222]).

About the same time, the so-called "analytical school" around W. Prager derived optimality criteria for plastic and later for elastic design using variational principles and energy theorems. These necessary (and sometimes sufficient) conditions of optimality are expressed in an infinite dimensional design space and are generalizations of earlier special-purpose optimality criteria by Michell [5], Foulkes [6] and Heyman [7]. They are particularly relevant to one- and two-dimensional structures (e.g. beams, frames or trusses and plates, disks or shells, respectively), which are idealized as one- and two-dimensional continua. Since the static and kinematic relations for these "continua" are expressed in terms of differential equations (and ***not*** algebraic equations, as in FE methods), the optimality criteria developed by the analytical school are termed ***continuum-based optimality criteria*** (COC). It is important to note that these differential equations are ***not*** referring to the usual three- (or two-) dimensional elastic continua in terms of local stresses (e.g. σ_x, τ_{xy}), but to a formulation in terms of so-called ***generalized*** stresses **Q** and strains **q** which, respectively, represent stress resultants (e.g. bending moment, shear force) on a cross-section and derivatives of the displacements of centroidal axes (e.g. curvature or twist). The static and kinematic conditions for a beam (example of a one-dimensional "continuum") for example, are

$$M'' + p = 0, \qquad u'' + \kappa = 0, \tag{1}$$

where $M(x)$ is the bending moment, $p(x)$ is the load, $u(x)$ is the beam deflection, $\kappa(x)$ is the curvature and primes denote differentiation with respect to the longitudinal coordinate x.

Until the late eighties, continuum-based optimality criteria were used mostly for *analytical solutions* in relatively simple, idealized examples, which, however, are very useful in checking the validity and convergence of numerical methods as well as in providing some insight into fundamental mechanical features of optimal structures.

During the last three years, COC were used in iterative procedures for large FE systems in order to check

- if this approach results in similar criteria and procedures to the earlier discretized optimality criteria (DOC); and
- whether a further improvement of the OC approach can be achieved by using certain features of the continuum formulation.

It should be pointed out that DOC and COC are by no means competing developments, but alternative formulations of the same approach which complement each other.

Several reviews of continuum-based optimality criteria methods are available, e.g. by Prager [8] and Rozvany [9, 10].

BASIC FORMULATION: STRESS AND DEFLECTION CONSTRAINTS

We consider a structure with the following characteristics:

(a) The behaviour is linearly elastic (no material or geometrical nonlinearities).
(b) The cross-sectional dimensions can vary freely along the members (no variable linking).
(c) The structure can be idealized as a one- or two-dimensional continuum.
(d) A single displacement constraint sets an upper limit on the deflection at a single point or on the weighted combination of several deflections.
(e) Stress (strength) constraints are set for all cross-sections of the structure.

Following Prager's notation, $\mathbf{p}(\mathbf{x})$ is the load vector, $\mathbf{u}(\mathbf{x})$ is the displacement vector, $\mathbf{Q}(\mathbf{x})$ is the generalized stress vector and $\mathbf{q}(\mathbf{x})$ is the generalized strain vector, where $\mathbf{x}$ denotes the spatial coordinates.

Then the optimal structure must fulfill the following conditions:

(i) Equilibrium ($\mathbf{Q}$, $\mathbf{p}$), compatibility ($\mathbf{q}$, $\mathbf{u}$) and generalized strain stress ($\mathbf{Q}$, $\mathbf{q}$) relations for the *real structure*. Examples of the first two are given in (1) with $\mathbf{Q} \rightarrow M$, $\mathbf{q} \rightarrow \kappa$, $\mathbf{p} \rightarrow p$ and $\mathbf{u} \rightarrow u$. For Bernoulli beams, the third one becomes

$$\kappa EI = M , \tag{2}$$

where E is Young's modulus and I is the moment of inertia.

(ii) Equilibrium, compatibility and strain-stress relations for a fictitious structure termed *adjoint structure*. The first two of these are the same as for the real structure, whereas the third one is in general different.

(iii) *Static and kinematic boundary conditions or end conditions.* These are the same for the real and adjoint structures if the supports are rigid and the cost of reactions is zero, otherwise the kinematic boundary conditions differ [10, p. 73].

(iv) Optimality criteria giving the relationship between the real ($\mathbf{Q}$) and adjoint ($\overline{\mathbf{Q}}$) generalized stresses and the cross-sectional dimensions ($\mathbf{z}$).

These conditions are summarized graphically in earlier publications [10, p. 370; 12, p. 49; 13, p. 224].

Denoting the stress condition by $S(\mathbf{Q},\ \mathbf{z}) \leq 0$ and the generalized local flexibility matrix by $[\mathbf{F}]$ (giving $\mathbf{q} = [\mathbf{F}]\mathbf{Q}$), we have the following optimality criterion for elastic systems with stress and deflection constraints [10, p. 62]:

$$\frac{\partial \psi}{\partial z_i} + \lambda(\mathbf{x})\frac{\partial S}{\partial z_i} + \nu\overline{\mathbf{Q}} \cdot \left[\frac{\partial \mathbf{F}}{\partial z_i}\right] \mathbf{Q} = 0\,, \tag{3}$$

where $\lambda(\mathbf{x})$ is a variable and ν is a constant Lagrangian multiplier and $\psi(\mathbf{x})$ is the specific cost (cost per unit length or unit area).

Moreover, the generalized stress-strain relation for the adjoint system becomes:

$$q_j = \lambda(\mathbf{x})\frac{\partial S(\mathbf{z},\ \mathbf{Q})}{\partial Q_j} + \nu[\mathbf{F}_j]\overline{\mathbf{Q}}\,, \tag{4}$$

where $[\mathbf{F}]_j$ is the j-th row of the flexibility matrix. In (3) and (4), we have

$$\lambda(\mathbf{x}) \neq 0 \qquad \text{only if} \qquad S(\mathbf{z},\ \mathbf{Q}) = 0\,. \tag{5}$$

CROSS-SECTION OPTIMIZATION, EXAMPLE: FRAME OF VARIABLE WIDTH

This application is due to M. Zhou, who used a variational formulation [11]. Moreover, the same results can be derived using the general formulae (3)-(5) above, in which we have:

$$\mathbf{Q} = (M, N)\,, \quad \overline{\mathbf{Q}} = (\overline{M},\ \overline{N})\,, \quad \mathbf{q} = (\kappa,\ \epsilon)\,, \quad \overline{\mathbf{q}} = (\overline{\kappa},\ \overline{\epsilon})\,,$$

$$[\mathbf{F}] = \frac{1}{zE}\begin{bmatrix} \frac{12}{d^3} & 0 \\ 0 & d \end{bmatrix}, \qquad S = \left(\frac{6|M|}{zd^2} + \frac{|N|}{zd} - \sigma_p\right), \qquad \psi = \gamma z d\,, \tag{6}$$

where κ and $\overline{\kappa}$ are the real and adjoint curvatures, ϵ and $\overline{\epsilon}$ the real and adjoint axial strains, $z(x)$ is the variable beam width, d is the beam depth, σ_p is the permissible stress, ψ is the weight per unit length and γ is the specific weight of the material. Then (3), (4) and (6) imply

$$\gamma d - \frac{\lambda}{z^2}\left(\frac{6|M|}{d^2} + \frac{|N|}{d}\right) - \frac{\nu}{z^2 E}\left(\frac{12M\overline{M}}{d^3} + \frac{N\overline{N}}{d}\right) = 0\,, \tag{7}$$

$$\overline{\kappa} = \frac{12\nu\overline{M}}{E\ zd^3} + \frac{6\lambda \text{ sgn } M}{zd^2}\,, \tag{8}$$

$$\overline{\epsilon} = \frac{\nu\overline{N}}{E\ zd} + \frac{\lambda \text{ sgn } N}{zd}\,. \tag{9}$$

Then for cross-sections with an inactive stress constraint we have by (5) $\lambda = 0$ and then (7) implies

$$z = \nu^{\frac{1}{2}}(E\gamma)^{-\frac{1}{2}}d^{-1}\left(\frac{12M\overline{M}}{d^2} + N\overline{N}\right)^{\frac{1}{2}}\,. \tag{10}$$

ITERATIVE COC METHODS

In these methods, the equations described in the last two sections are solved iteratively in a discretized form. Each iteration uses the following three steps:

(1) Analysis of the real and adjoint structures using an FE program.

(2) Calculation of the value of the Lagrangian multiplier ν from the deflection condition expressed in the form of a work equation:

$$\Delta = \sum_k \overline{\mathbf{Q}}_k \cdot [\mathbf{F}]_k \mathbf{Q}_k L_k \,, \tag{11}$$

where Δ is the prescribed deflection, k refers to the k-th element and L_k is the length or area of that element.

(3) Re-sizing using the optimality condition (3).

Illustrations of this procedure were given previously [12]. In the case of the adjoint analysis in the frame example, for fully stressed elements the value of λ is first calculated from (7),

$$\lambda = \frac{\gamma z^2 d^2 - \nu(12M\overline{M}/d^2 + N\overline{N})/E}{6|\overline{M}|/d + \overline{N}} \,, \tag{12}$$

and then substituted into (8) and (9). In the FE analysis of the adjoint system, the second terms in (8) and (9) are generated as thermal strains.

The iterative procedure is stopped when the absolute value of the change of the total cost (total wieght) Φ and of the design variables z_i is smaller than a specified value.

Considering the above application to frames of variable width, the structural dimensions, deflection constraint and loading for a numerical example are shown in Fig. 1a and the optimal distribution of the width z in Fig. 1b. In the above problem, we have $d = 0.3$ m, $\gamma = 1.0$ (volume instead of weight), $E = 1.1 * 10^7$ kN/m^2, $\sigma_P = 1.1 * 10^4$ kN/m^2 and $\Delta = L/300 = 0.03333$ m, which are values taken from the relevant German design code for laminated timber beams (DIN 1052). The number of elements used was 2000 and the convergence criterion,

$$\frac{|\Phi_{\text{new}} - \Phi_{\text{old}}|}{\Phi_{\text{new}}} \le 10^{-5} \,, \tag{13}$$

was reached after 13 iterations. The iteration history in Fig. 1c shows an apparent instability at the third cycle, but this was caused mostly by larger constraint violations ($> 10^{-3}$; these were for the later iterations smaller than 10^{-5}).

ADVANTAGES AND DISADVANTAGES OF COC METHODS

It was demonstrated in previous publications (e.g. [12]) that the main advantage of COC methods is that (at least in the special test examples considered) they increase the optimization capability by several orders of magnitude, thereby not only eliminating but also reversing the discrepancy that presently exists between analysis capability (10^4 to 10^5) and optimization capability (10^2) if conventional (e.g. primal mathematical programming) methods are used.

In optimizing a clamped beam of variable width and depth for a uniformly distributed load in the vertical direction and a central point load in the horizontal direction, for 50 elements (100 variables) the COC method required only 57 sec (94 iterations) whilst a sequential quadratic programming (SQP) method took

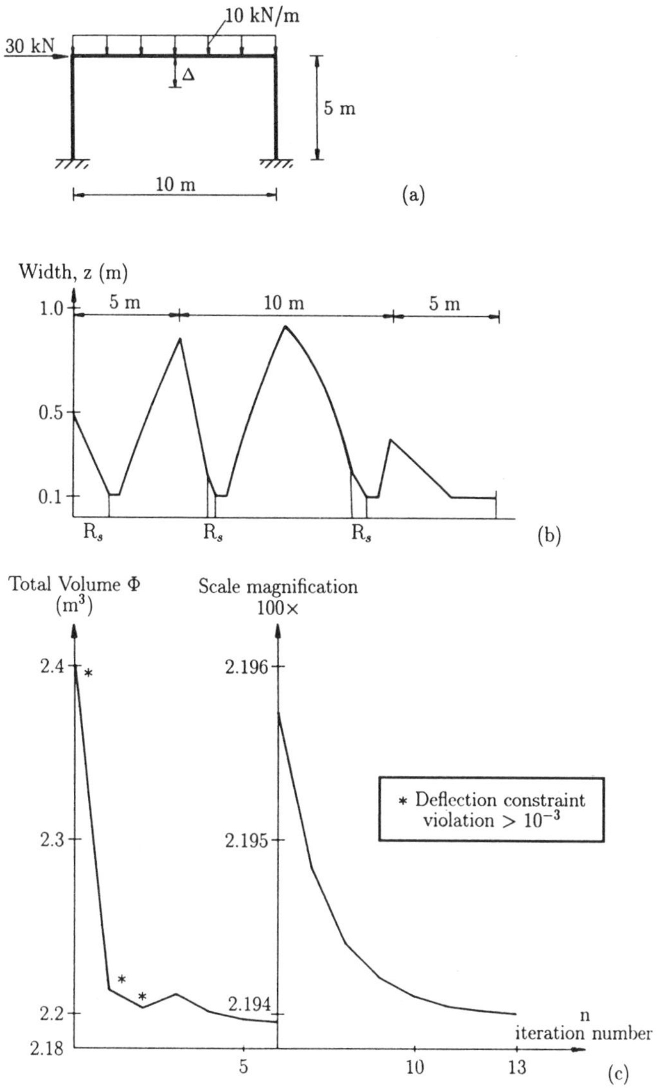

Fig. 1 Frame example.

1507451 sec (17.45 days, 77564 iterations) for a convergence criterion $|\Phi_{\text{new}} - \Phi_{\text{old}}|/\Phi_{\text{new}} \leq 10^{-8}$. It is shown in Fig. 2 that the SQP method indeed converged to the COC result. The same problem was also solved by the COC method with 10000 elements.

The disadvantage of the COC algorithm is that the optimality criteria and formulae for calculating the Lagrangian multipliers must be derived analytically and programmed for each class of design problem. Moreover, a computer algorithm is not available as yet for some design conditions (e.g. multiple displacement conditions) or complex cross-sections. This draw-back could only be eliminated by as-

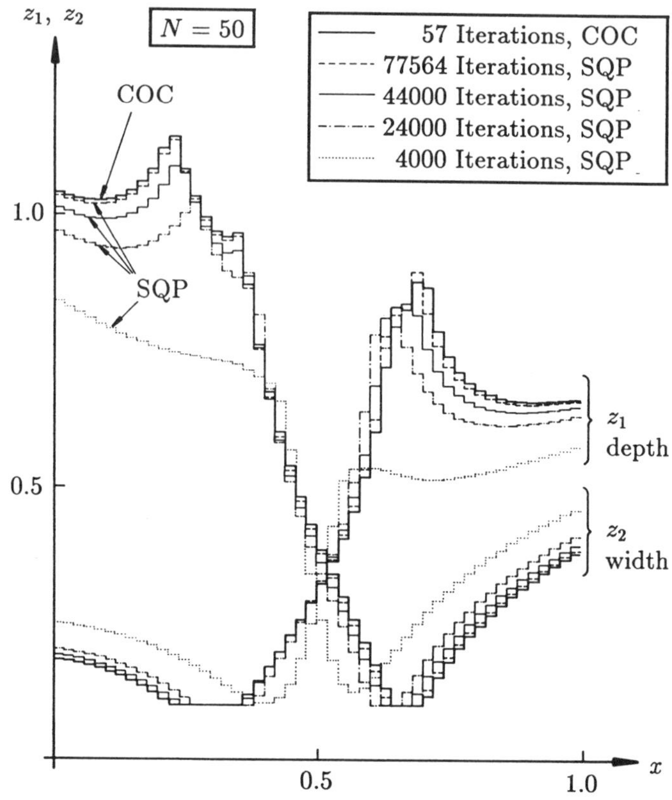

Fig. 2 Beam of variable width and depth: solutions by the COC and SQP methods.

sembling a comprehensive library of formulae and programs for most cross-section topologies and design conditions which are likely to come up in the practice.

ADVANTAGES OF THE PROPOSED APPROACH FOR STRESS CONSTRAINTS

This aspect of the COC approach was pointed out in a recent note [14]. The main advantage of the proposed method over the traditional DOC formulation arises from the fact that in the latter the stress constraints are expressed in terms of relative displacements involving a virtual load system consisting of two unit forces or moments at the ends of the element involved. This means that in the "equivalent displacement" formulation the number of analyses for virtual load systems and the number of Lagrangian multipliers to be calculated equals the total number of active stress and displacement constraints. On the other hand, the proposed approach requires such calculations only for active displacement constraints, the number of which is usually much smaller than that of the active stress constraints (many thousand in some problems solved [12]). The calculation of the Lagrangian multipliers for the stress constraints in the COC method is either unnecessary (e.g. trusses [14]), or can be done using an *uncoupled* (usually explicit) equation, see e.g. (12) herein. In the traditional DOC formulation, all Lagrangian multipliers corresponding to both deflection and stress constraints must be evaluated *simultaneously*, which is a very labourious process. Naturally, the function of the Lagrangian multipliers for stress constraints is entirely different in the two methods since in the COC method they only influence the strains in the adjoint structure.

It is to be noted that at present we can only handle one active displacement constraint and we have also problems with elements for which both a stress constraint and a minimum size constraint are active. However, these difficulties will be investigated in the near future.

GENERALIZED SHAPE OPTIMIZATION BY THE COC METHOD

Most research on shape optimization is concerned with the optimal shape of the boundaries, but the topology is usually excluded from the optimization process. Simultaneous optimization of topology and boundary shape is termed *generalized shape optimization.* Some theoretical studies by mathematicians (e.g. Murat and Tartar [15], Lurie and Cherkaev [16] as well as Kohn and Strang [17]) have shown that the optimal shape of elastic two-dimensional continua for a compliance (given external work) constraint is characterized by regions with an infinite number of internal boundaries, separated by ribs of first and second order infinitesimal width, respectively, in the two principal strain directions. Using the above optimal microstructure, Rozvany, Olhoff, Bendsøe *et al.* [18] derived closed form analytical solutions for least-weight perforated plates. These solutions have shown that the optimal solutions for two-dimensional continua usually consist of three types of regions, namely: (a) *solid regions* filled with material, (b) *empty regions* and (c) *perforated (porous) regions* containing a fine system of cavities. Perforated regions usually occupy a large part of the structural domain.

The topology of internal boundaries in near-optimal designs was investigated by the so-called *homogenization method* (e.g. Bendsøe [19], Kikuchi and Suzuki [20]), in which some assumed class of microstructures (e.g. a system of rectangular cavities) is used for optimizing the generalized shape of an elastic continuum. At a macroscopic level, the perforated region is replaced by an anisotropic, non-homogeneous material whose stiffness characteristics are the same as that of the assumed microstructure and then a finite element method is employed for the analysis stage. Since the limiting cases of the assumed microstructure include empty and solid regions as well, the optimal solutions obtained by this procedure include all three regions mentioned above. However, due to the non-optimality of the assumed microstructure, this procedure tends to penalize and thus suppress perforated regions which tend to occupy only a very small proportion of the structural domain in solutions obtained by this method.

To illustrate the above procedure, we consider a problem which is often referred to in the literature as the "MBB-beam". In this problem, a simply supported rectangular domain in plane stress is subject to a central point load and the total weight is to be minimized within stress and deflection constraints. Using the above homogenization procedure, Olhoff, Bendsøe and Rasmussen [21] obtained the solution shown in Fig. 3a for a compliance constraint which also represents an approximate solution for stress constraints. In the second stage of their method, they extracted a topology from the solution in Fig. 3a and then carried out a conventional shape optimization for the given stress and deflection constraints, obtaining the design in Fig. 3b.

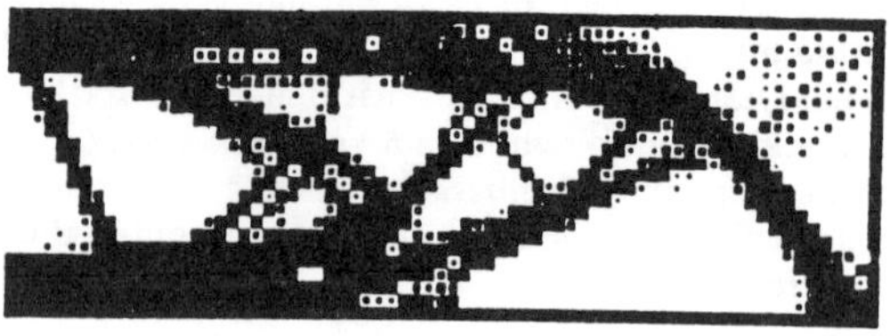

homogenization (a)

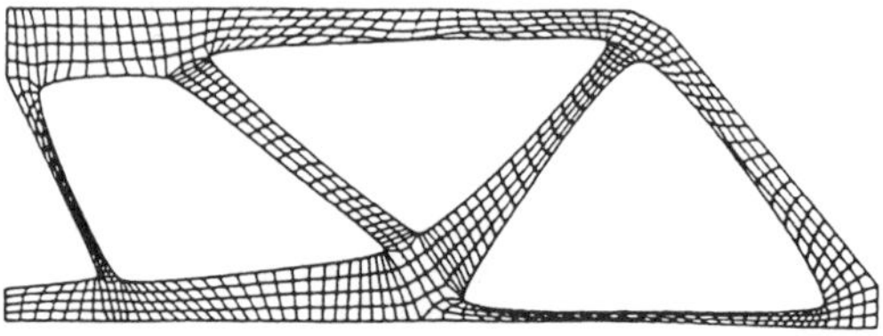

two–stage process (b)

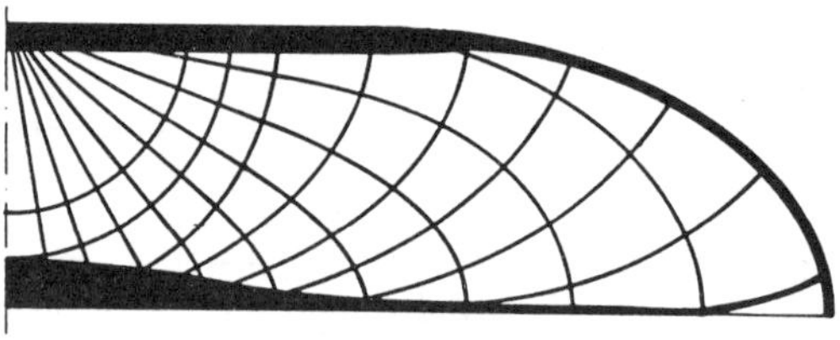

COC (c)

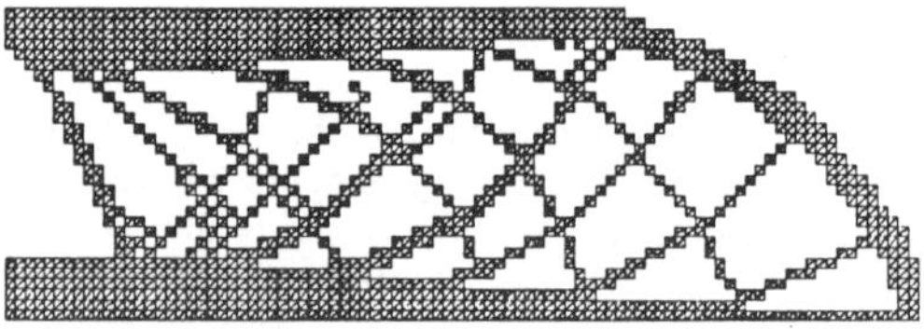

modified homogenization (d)

Fig. 3 A comparison of various methods for generalized shape optimization; (a) and (b) after Olhoff, Bensøe and Rasmussen [21].

Using the COC methods for layout optimization to be described in the next section, the authors have shown that the exact solution for the above compliance problem is likely to tend to the solution in Fig. 3c, in which the top right corner is an empty region, black thick lines indicate solid regions and the net of thin lines in the middle indicates the principal directions in a perforated region which fills most of the structural domain. It can be seen that there is a good agreement between the solutions in Figs. 3a and 3c, but in the former perforated regions are somewhat suppressed.

More recently, the authors developed a modified homogenization method which has two advantages: first, the microstructure is assumed to be isotropic, thereby making the analysis and optimization simpler; second, the cost of fabrication of the cavities is included in the cost function, thereby suppressing perforated regions. The degree of penalty for such regions can be adjusted by increasing or decreasing the fabrication costs. The solution in Fig. 3d, which was obtained by the above procedure, clearly confirms the "exact" topology shown in Fig. 3c.

LAYOUT OPTIMIZATION BY THE COC METHOD

One of the most complex tasks in structural optimization concern layout problems which involve the simultaneous optimization of the (a) topology (spatial sequence of members and joints), (b) geometry (coordinates of joints and shape of member axes) and (c) cross-sectional dimensions.

The theory of optimal layouts, a generalization of Michell's theory of least-weight trusses [5], was developed in the late seventies by W. Prager and the first author and extended considerably by the latter in the eighties. Comprehensive reviews of this field are available, e.g. by Prager [8], Rozvany [9], [10, Chapter 8], [22], [24] and Prager und Rozvany [23]. The above layout theory is based on two underlying concepts, namely

- *continuum-based optimality criteria* (COC), which were reviewed here earlier, and
- the *structural universe* (or "ground structure" or "basic structure") which is the union of all potential members.

Earlier applications of the layout theory involved exact *analytical solutions* for problems in which the structural universe contained an infinite number of members (in all possible directions within the structural domain). This investigation covered least weight grillages (beam systems, [8–10], [22], [23], cable-nets and shell-grids [22], [24] and, quite recently, trusses [25]).

The study of exact layout solutions included the introduction of computer programs which derive and plot the optimal layout using purely analytical operations [26]. Further work in this field is being carried out by the first author and Gerdes.

The most important recent development in the layout field is a computer algorithm for *numerical layout optimization* using the COC method. Earlier work in this field (e.g. Kirsch [27]) was based on a two-stage procedure in which first the topology was optimized and then the geometry and cross-sections for a given topology. In spite of the approximations used, this earlier method was restricted to a relatively small number of members (up to a dozen). The COC layout algorithm uses

- a fine grid of potential members (up to about 20000 at present) in the structural universe, and
- a very small value (e.g. 10^{-12}) for the prescribed minimum cross-sectional area.

As an introductory example, we consider a layout problem, in which a load must be transmitted to a straight support (AB) with all members within the triangle ABC (Fig. 4a). A relatively small structural universe (consisting of only 707 members) is shown in Fig. 4b.

For the latter, the computer output consisted of two bars along the edges AC and BC, having a cross-sectional area of 5.59017 and all other members took on the minimum cross-sectional area of 10^{-12}. The exact analytical solution for

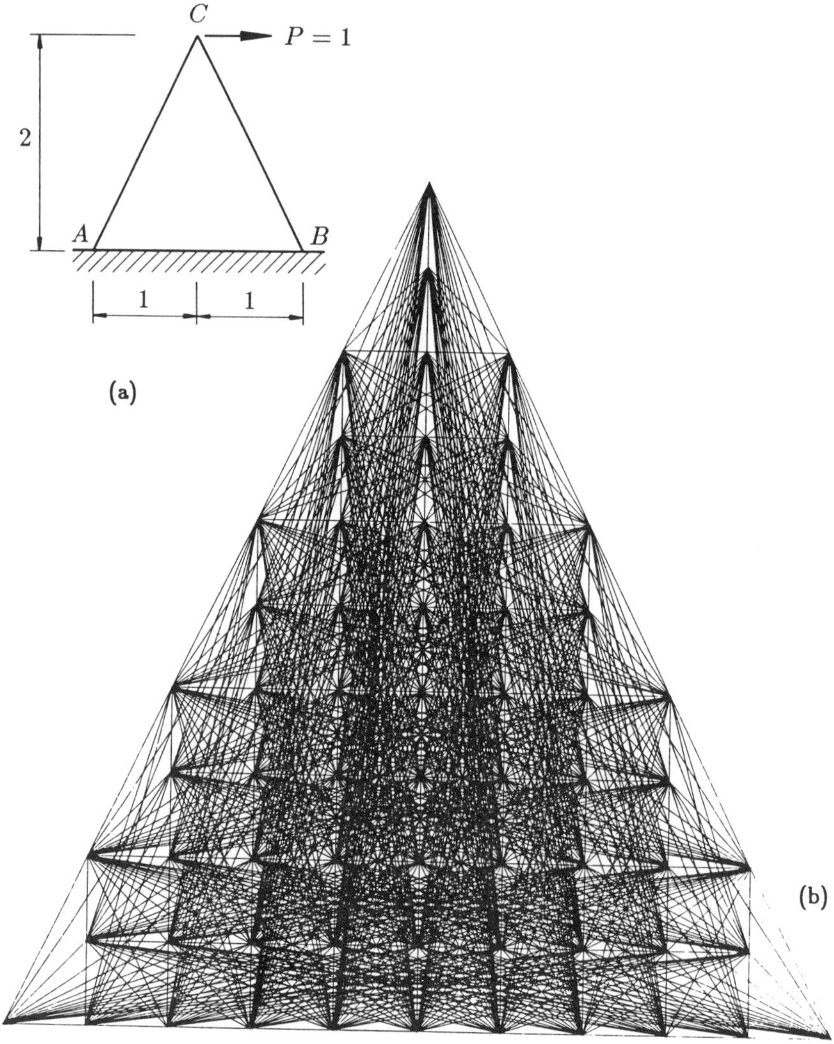

Fig. 4 Analytical solution for a layout problem and the structural universe for deriving an iterative COC method.

this problem also contains two bars along the above edges and has a total non-dimensional volume of $\Phi = 25$ (see [25], Part II). The numerical COC solution gave a total volume of $\Phi = 25.000000001168$.

For a similar loading with a rectangular structural domain the theoretical solution (Fig. 5a) consists of an infinite number of members [28, pp. 97-99], giving a total non-dimensionalized volume of $\Phi = 4.498115$.

Using the iterative COC algorithm with 5055 and 12992 (Fig. 5b) members, respectively, truss volumes of $\Phi = 4.532197$ and $\Phi = 4.518816$ were obtained, which represent 0.76% and 0.46% errors compared to the analytical solution.

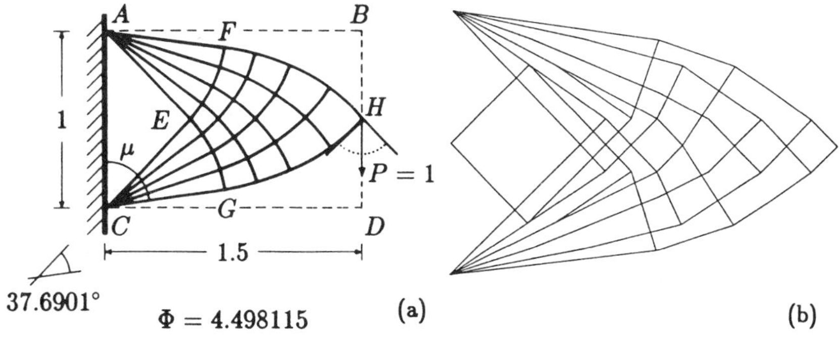

Fig. 5 Exact analytical solution *vs.* iterative COC solution with 12992 members in the structural universe.

The above computer algorithm was used for many other layout problems and it achieves, in effect, a simultaneous optimization of the topology and geometry. The solution in Fig. 3c was also based on one such layout output.

CONCLUSIONS

It was demonstrated in a brief overview of the COC algorithm that this technique can potentially be highly efficient in all three branches of structural optimization, namely cross-section, generalized shape and layout optimization.

REFERENCES

1. Khot, N.S. Optimality Criterion Methods in Structural Optimization. In: Foundations of Structural Optimization: A Unified Approach (Ed. Morris, A.J.) pp. 99-236, Wiley, Chichester, 1982.
2. Berke, L. and Khot, N.S. Structural Optimization Using Optimality Criteria. In: Computer Aided Optimal Design: Structural and Mechanical Systems (Ed. Mota Soares, C.A.) pp. 271-312, Springer, Berlin.
3. Venkayya, V.B. Optimality Criteria: A Basis for Multidisciplinary Design Optimization, Comput. Mech., Vol 5, pp. 1-21, 1989.
4. Haftka, R.T. and Kamat, M.P. Elements of Structural Optimization, Martinus Nijhoff, The Hague, 1985.
5. Michell, A.G.M. The Limits of Economy of Material in Frame-Structures, Phil. Mag., Vol. 8, pp. 589-597, 1904.
6. Foulkes, J. The Minimum-Weight Design of Structural Frames, Proc. Royal Soc., Vol. 223, pp. 482-494, 1954.
7. Heyman, J. On the Absolute Minimum Weight Design of Framed Structures, Quart. J. Mech. Appl. Math., Vol. 12, pp. 314-324, 1959.
8. Prager, W. Introduction to Structural Optimization, Springer, Vienna, 1974.
9. Rozvany, G.I.N. Optimal Design of Flexural Systems, Pergamon, Oxford, 1976; In Russian: Stroiizdat, Moscow, 1980.
10. Rozvany, G.I.N. Structural Design via Optimality Criteria, Kluwer, Dordrecht, 1989.
11. Zhou, M. Iterative Continuum-Type Optimality Criteria Methods in Structural Optimization. Doctoral Thesis, Essen University, FB 10, 1991.

12. Rozvany, G.I.N., Zhou, M., Rotthaus, M., Gollub, W. and Spengemann, F. Continuum-Type Optimality Criteria Methods for Large Finite Element Systems with a Displacement Constraint: Parts I and II, Struct. Optim., Vol. 1, pp. 47-72, 1989, Vol. 2, pp. 77-104, 1990.
13. Rozvany, G.I.N. and Zhou, M. Multi-Purpose Optimal Design of Large Discretized Systems by New Iterative Optimality Criteria Methods. In: Computer Aided Optimum Design of Structures: Recent Advances (Eds. Brebbia, C.A. and Hernandez, S.) pp. 221-230, Springer, Berlin, 1989.
14. Rozvany, G.I.N. and Zhou, M. A Note on Truss Design for Stress and Displacement Constraints by Optimality Criteria Methods, Struct. Optim., Vol. 3, pp. 45-50, 1991.
15. Murat, J. and Tartar, L. Calcul des Variations et Homogénéisation: Les Méthodes de l'Homogénéisation. In: Théorie et Application en Physique. Coll. de la Dir. des Études et Recherches de Elec. de France, Eyrolles, Paris, pp. 319-370, 1985.
16. Lurie, K.A. and Cherkaev, A.V. Optimal Structural Design and Relaxed Controls, Opt. Control Appl. Mech., Vol. 4, pp. 387-392, 1983.
17. Kohn, R.V. and Strang, G. Optimal Design and Relaxation of Variational Problems, I, II and III, Comm. Pure Appl. Math., Vol. 39, pp. 113-137, 139-182, 353-377, 1986.
18. Rozvany, G.I.N., Ong, T.G., Sandler, R., Szeto, W.T., Olhoff, N. and Bendsøe, M.P. Least-Weight Design of Perforated Elastic Plates, I and II, Int. J. Solids Struct., Vol. 23, pp. 521-536, 537-550, 1987.
19. Bendsøe, M.P. Optimal Shape Design as a Material Distribution Problem, Struct. Optim., Vol. 1, pp. 193-202, 1989.
20. Kikuchi, N. and Suzuki, K. Shape and Topology Optimization by the Homogenization Method. In: Proc. CISM Advanced School, Shape and Layout Optimization of Structural Systems, held in Udine, 1990 (Ed. Rozvany, G.I.N.) Springer, Vienna, 1991.
21. Olhoff, N., Bendsøe, M.P. and Rasmussen, J. On CAD-Integrated Structural Topology and Design Optimization, Proc. 2nd World Congress on Comp. Mech., pp. 95-99, Int. Assoc. Comp. Mech., Stuttgart, 1990.
22. Rozvany, G.I.N. Optimality Criteria for Grids, Shells and Arches. In: Optimization of Distributed Parameter Structures (Eds. Haug, E.J. and Cea, J.) pp. 112-151, Sijthoff and Noordhoff, Alphen aan der Rijn, 1981.
23. Prager, W. and Rozvany, G.I.N. Optimization of Structural Geometry. In: Dynamical Systems (Eds. Bednarek, A.R. and Cesari, L.) pp. 265-293, Academic Press, New York, 1977.
24. Rozvany, G.I.N. Structural Layout Theory - The Present State of Knowledge. In: New Directions in Optimum Structural Design (Eds. Atrek, E., Gallagher, R.H., Ragsdell, K.M. and Zienkiewicz, O.C.) pp. 167-195, Wiley, Chichester, 1984.
25. Rozvany, G.I.N. and Gollub, W. Michell Layouts for Various Combinations of Line Supports – I, Int. J. Mech. Sci., Vol. 32, pp. 1021-1043, 1990. Part II is to be submitted.
26. Hill, R.H. and Rozvany, G.I.N. Prager's Layout Theory: A Non-Numeric Computer Method for Generating Optimal Structural Configurations and Weight-Influence Surfaces, Comp. Meth. Appl. Meth. Engrg., Vol. 49, pp. 131-148, 1985.
27. Kirsch, U. On the Relationship between Structural Topologies and Geometries, Struct. Optim., Vol. 2, pp. 39-45, 1989.
28. Hemp, W.S. Optimum Structures. Clarendon, Oxford, 1973.

Structural Optimization by Virtual Strain Energy

C.T. Jan (*), A.B. Laurence (**)

(*) DRC Consultants, Inc., Flushing, New York, U.S.A.

(**) O'Brien-Kreitzberg & Associates, Inc., New York, U.S.A.

ABSTRACT

With the evolutionary advancement in high-rise technologies, structural optimization becomes an efficient access in design phase. In particular, the use of structural optimization offers a great advantage in drift control. So far, there are a number of optimization techniques developed to deal with displacement constrained problems. However, due to the complexity involved in their formulations, most structural optimization techniques are too difficult to be recognized and implemented into the designs of building structures. For mathematical programming methods, the number of numerical iterations and structural analyses required to achieve an optimal solution is often too high. As for the optimality criterion methods, it requires large computational efforts in sensitivity analyses. Therefore, a simple optimization method, based primarily on the use of virtual strain energy, provides a rapid approximation to design structures subject to displacement constraints. In essence, for truss structure with one displacement constraint, this method gives a true optimal solution. For framed structures, this method provides an approximate solution due to axial, shear, and flexural actions. But, the ease of implementation of this method and the fast rate of convergence make it a very convenient tool in the designs of building structures, especially high-rise buildings. The structure under consideration is first treated by means of finite element analysis. Then, the application of virtual strain energy performs its optimization process for a few iterations and reaches to an approximate solution. The optimization is based on having constant strain energy density distributed equally in each member throughout the whole structure. As for multiple displacement constrained problems, it is treated indirectly by a series of approximations. By supplementing the principle of virtual work, the accuracy of such an approach can be readily verified. Clearly, this approach offers a simple means to utilize structural optimization to deal with building structures.

INTRODUCTION

In high-rise building technologies, one major element in design stage is the use of structural optimization to choose lateral load resistant system. The use of structural optimization offers great advantage in easing off the complexity involved in lateral displacement control. In aerospace engineering, structural optimization has been widely accepted together with the advent of powerful computer facilities. Many optimization techniques have been evolved to solve structures with a variety of constraints, such as displacements, stresses, natural frequencies, stability, and shape (geometry). In recent years, a few mathematical programming techniques and optimality criterion methods are proven to be very favorable. However, in building structures, there is a different aspect of viewpoint in accepting structural optimization.

Most of all, in general, there is no urgent need to purchase powerful computer facilities in building industries just for the sake of structural optimization. Hence, it is very difficult for design professions to get started with structural optimization which practically requires ten, twenty, or even more iterations in order to achieve one optimal result. Therefore, it makes the feasibility of most mathematical programming techniques and optimality criterion methods questionable in building industries. This approximate approach is simple and well-known. The basis of its formulations conforms to the Kuhn-Tucker conditions [1] used in mathematical non-linear programming techniques and optimality criterion methods. Its derivations are outlined in detail in this paper.

It is based primarily on the use of virtual strain energy and provides a rapid approximation to design structures subject to a single displacement constraint. For truss structure with one displacement constraint, this method gives a true optimal solution. For framed structures, member forces in a frame element vary with different member sizes, this method provides an approximate solution due to axial, shear, and flexural actions. But, the ease of implementation of this method and the fast rate of convergence make it become a very convenient tool in the designs of building structures. Currently, it is applied to steel structures. The application of virtual strain energy performs its optimization process for having constant strain energy density distributed equally in each member throughout the whole structure.

THEORETICAL FORMULATIONS

The theoretical basis of this paper was first presented by Venkayya, Khot and Burke [2] in different forms. It then started a lot of attentions to optimality criterion methods. Most of the researches and developments were concentrated on different iterative or recursive schemes in order to improve the rate of convergence. In general, there is a lot of number crunching involved in those approaches due to both optimization and analysis.

In the meantime, there are some structural engineers looking at displacement constrained problems without knowing much about structural optimization. The use of principle of virtual work was adopted. For structures having identical material density, the approximate optimization was simplified to keep constant strain energy density, strain energy divided by volume, over the whole structure. This basis sometimes gives misconception for optimizing composite structures. In fact, for composite structures of one displacement constraint, it is to achieve constant ratio of strain energy versus structural weight for each member.

For truss structures of member i = 1, 2, 3, ..., n, the total weight of the structure, W, is

$$W = \sum_{i=1}^{n} W_i = \sum_{i=1}^{n} \rho_i A_i L_i + W_{NON\text{-}STRUCTURAL} \qquad (1)$$

where n is the total number of members, ρ_i is the density of member i, A_i is the cross sectional area of member i, and L_i is the longitudinal length of member i. In addition, $W_{NON\text{-}STRUCTURAL}$ is the total weight of non-structural elements.

In static analysis of truss members, the strain energy of member i subjected to axial force P_i, U_i, can be expressed as

$$U_i = \frac{P_i^2 L_i}{2 E_i A_i} \qquad (2)$$

where E_i is the modulus of elasticity for member i.

As for the total strain energy of the structure, U_E can be represented as

$$U_E = \sum_{i=1}^{n} U_i = \sum_{i=1}^{n} \frac{P_i^2 L_i}{2 E_i A_i} \tag{3}$$

By means of the principle of virtual work, displacement at joint *j* can be expressed as

$$\Delta_j = \frac{\partial U_E}{\partial P_j} = \sum_{i=1}^{n} \frac{P_i L_i \partial P_i / \partial P_j}{E_i A_i} \tag{4}$$

The gradient of displacement Δ_j with respect to area A_i is

$$\frac{\partial \Delta_j}{\partial A_i} = - \frac{P_i L_i \partial P_i / \partial P_j}{E_i A_i^2} \tag{5}$$

For one displacement constraint, based on Equation (1), the Lagrangian can be expressed as

$$L(A_i, \lambda) = W + \lambda (\Delta_j - \bar{\Delta}) \tag{6}$$

where Lagrange multipler $\lambda \geq 0$ and constrained target displacement $\bar{\Delta}$ yields $(\Delta_j - \bar{\Delta}) \leq 0$. They are the familiar formulations in Kuhn Tucker conditions. By substituting Equations (1) and (4) into Equation (6), the Lagrangian can be expressed as

$$L = \sum_{i=1}^{n} \rho_i A_i L_i + W_{NON\text{-}STRUCTURAL} + \lambda \left(\sum_{i=1}^{n} \frac{P_i L_i \partial P_i / \partial P_j}{E_i A_i} - \bar{\Delta} \right) \tag{7}$$

At the optimum, the following relations are valid: $\partial L / \partial A_i = 0$ and $\Delta_j^* = \bar{\Delta}$. Therefore, by taking the first derivative of the Lagrangian in Equation (7) with respect to area A_i, it becomes

$$\rho_i L_i + \lambda^* \left(\frac{P_i L_i \partial P_i / \partial P_j}{E_i A_i^{*2}} \right) = 0 \tag{8}$$

By multiplying Equation (8) with A_i^* on both sides,

$$\rho_i A_i^* L_i + \lambda^* \left(\frac{P_i L_i \partial P_i/\partial P_j}{E_i A_i^*} \right) = 0 \tag{9}$$

and based on Equation (2), the strain energy of member *i* at the optimum can be expressed as

$$U_i^* = \frac{P_i P_j \partial P_i/\partial P_j L_i}{E_i A_i^*} \tag{10}$$

Substitute Equation (10) into Equation (9), it becomes

$$W_i^* - \lambda^* \frac{U_i^*}{P_j} = 0 \tag{11}$$

where W_i^* is the structural weight of member *i* at the optimum.

Therefore, Equation (11) gives the following relationship at the optimum,

$$\frac{U_i^*}{W_i^*} = \frac{P_i}{\lambda^*} \tag{12}$$

Since P_i/λ^* is constant, Equation (12) implies that the ratio of member's strain energy versus member's structural weight is of a constant value for all truss members. For truss structures having the same material density ρ_i, Equation (12) can be further simplified and interpreted as having constant strain energy density, member's strain energy divided by its volume, for each member.

By means of Equation (11) for member *i*, taking summation of all members from 1 to n, it gives,

$$W^* - \lambda^* \frac{U_E^*}{P_j} = 0 \tag{13}$$

Thus, from Equations (12) and (13), at the optimum, it yields

$$\frac{U_E^*}{W^*} = \frac{U_i^*}{W_i^*} = \frac{P_i}{\lambda^*} \tag{14}$$

Note that the relationships in Equation (14) gives the exact solution at the optimum.

In the light of framed structures, in addition to axial force P_i (or F_{Xi}), there are more actions attributed to flexural moments M_{Xi}, M_{Yi}, M_{Zi}, and shear forces F_{Yi}, F_{Zi}. Accordingly, the total strain energy U_E for framed structures is expressed as

$$U_E = \sum_{i=1}^{n} \left(\frac{F_{Xi}^2 L_i}{2 E_i A_{Xi}} + \frac{F_{Yi}^2 L_i}{2 G_i A_{Yi}} + \frac{F_{Zi}^2 L_i}{2 G_i A_{Zi}} + \frac{M_{Xi}^2 L_i}{2 G_i I_{Xi}} + \frac{M_{Yi}^2 L_i}{2 E_i I_{Yi}} + \frac{M_{Zi}^2 L_i}{2 E_i I_{Zi}} \right) \qquad (15)$$

$\Delta_j = \partial U_E / \partial P_j$ still holds true for framed structures, but involves with much complicated forms than Equation (4).

The derivation of $\partial \Delta_j / \partial A_{Xi}$ leads us to the degree of approximation. For framed structures with prominent flexural behaviors,the distribution of member forces varies with different member sizes, the rate of convergence for the proposed method is slower. However, the basis of formulations for framed structures is similar to the ones shown in Equations (1) to (14). Note that the relations shown in Equation (14) hold true, including framed structures, for one displacement constrained problem[2].

NUMERICAL EXAMPLES

A computer program OPTIMA was developed to accommodate the algorithm stated herein. Basically, program OPTIMA serves as a post-processor of a commercial finite element package. The structures are analyzed accordingly to the inputting format of the commercial finite element package. However, in the load cases, there is an additional point load applied at the same location and along the direction of the constrained displacement. The optimal solutions are divided into two portions. First, the input member sizes are considered as side constraints, and the optimal member sizes are greater or equal to their initial sizes. Secondly, it is treated as no size constraint. Thus, some member sizes may turn out to be very small or zero. It implies that those members can be taken out without affecting the structure's performance.

Based on the optimal solutions provided by OPTIMA, the efficiency of each member in satisfying displacement constraint can be easily identified. If the value of one member's strain energy divided by its own structural weight is relatively smaller than others, it implies that this member does not contribute significant resistance to comply with the desired displacement.

Example 1. Fourteen-Story K-Braced Frame

The fourteen-story and one-bay braced K-braced frame shown in Figure 1 was first studied by Velivasakis and others [3] based on principle of virtual work and strain energy density but a different optimization approach. The comparison of some selected member sizes from optimal results were listed in Table 1. With this rigorous approximate optimization, it demonstrates an efficient scheme to optimize structures subjected to one displacement constraint. The relationship of strain energy versus its structural weight offers an option to explore composite structures having different densities in steel and concrete materials. The constrained displacement was 4.5 inches.

Example 2. An Outrigger Structure

The fourteen-story and one-bay braced K-braced frame shown in Figure 2 was studied by adding one outrigger at the top. All members are of steel material. With the same applied forces as Example 1, the constrained displacement was also 4.5 inches. The exterior column lines are numbered as 11 and 12, and the interior core column lines are numbered as 01 and 02. Member numbering for columns are starting from first floor, 111, 101, 102, 112 to fourteen floor 1411, 1401, 1402, 1412 in the increment of one hundred.

The optimal results for column members were listed in Table 2. Again, it demonstrates the application to optimize outrigger structures subjected to one displacement constraint. The initial volume is 245.77 cubic feet, and the final optimal volume is 368.31 cubic feet. The initial total weight is 120.425 kips, and the final optimal weight is 180.47 kips. The difference in weight is 49.86%. The number of constrained members is 43, and the number of unconstrained members is 60. A more refined model ought to be studied to represent an outrigger system. However, this example demonstrates the effectiveness of such approach.

CONCLUSIONS

The use of a simple approximate optimization method to design building structures subjected to one displacement constraint is described in detail in this paper. Its derivations show that this method complies very closely with mathematical programming techniques and optimality criterion methods. The fast rate of convergence is very appealing. To design professions, a far less number of analyses is required to achieve the optimum. As for multiple displacement constrained problems, it can be treated individually as a series of one displacement constrained approximation. Without too much computational effort, the method described herein can serve as an efficient tool in designing building structures.

ACKNOWLEDGEMENTS

The authors appreciate the suggestions from Mr. Chun-Man Chan, Ph.D. candidate at the Univerity of Waterloo in Ontario, Canada; and wish to acknowledge the supports of the Office of Irwin G. Cantor and Weidlinger Associates during the preparation of this research.

REFERENCES

1. Luenberger, D.G., Linear and Nonlinear Programming, Second Edition, Addison-Wesley, Reading, Mass., 1984.

2. Venkayya, V.B., Khot, N.S. and Burke, L. Application of Optimality Criteria Approaches to Automated Design of Large Practical Structures, pp. 301-319, Second Symposium on Structural Optimization, Milan, Italy, AGARD Conference Proceeding No. 123, 1973.

3. Velivasakis, E.E., Joseph, L., Scarangello, T.Z. and DeScenza, R., Design Optimization of Tall Steel Structures for Lateral Loads, pp. 104-120, Special Structures, ASCE Metropolitan Section - The Structures Group, Spring Seminar, 1989.

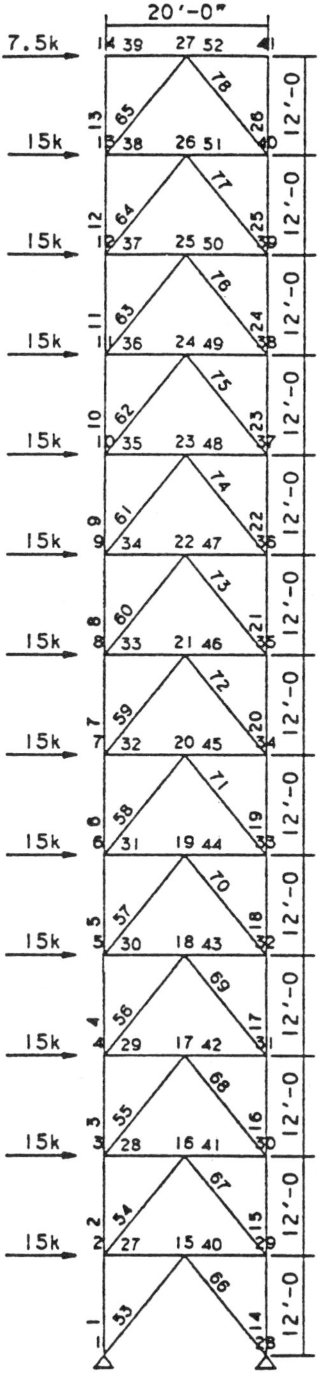

Figure 1. Fourteen-Story K-Braced Frame

Table 1. Fourteen-story One-bay K-Braced Frame

Approach	Velivasakis	OPTIMA
Member Number	Area in^2	Area in^2
1 & 14	72.49	76.42
2 & 15	72.49	67.28
3 & 16	54.50	58.20
4 & 17	54.50	49.80
5 & 18	38.32	41.68
6 & 19	38.32	36.03
7 & 20	24.20	27.15
8 & 21	24.20	20.71
9 & 22	12.50	14.75
10 & 23	12.50	9.71
Initial Weight	19.49 tons	19.53 tons
Optimum Weight	27.69 tons	27.45 tons

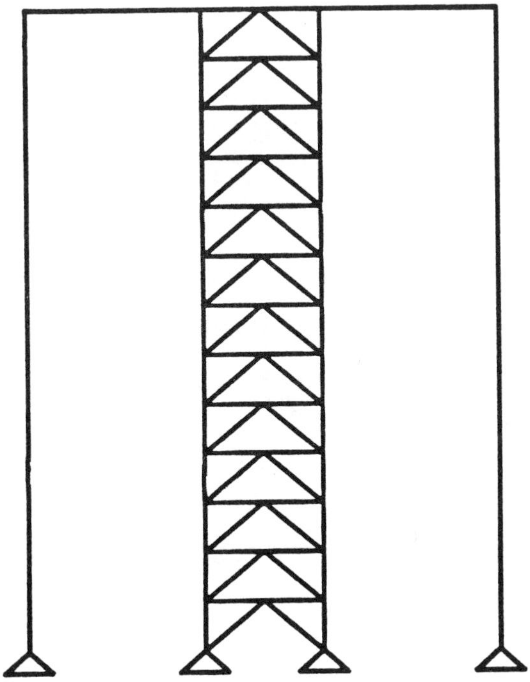

FIGURE 2. OUTRIGGER STRUCTURE

Table 2. An Outrigger Structure

BEAM NO.	OLD SIZE (LB/FT)	NEW SIZE (LB/FT)
101	120	427
102	120	398
201	120	393
202	120	364
301	90	357
302	90	328
311	120	120
401	90	323
402	90	294
411	120	157
501	68	287
502	68	259
511	120	186
601	68	252
602	68	224
611	120	205
701	68	217
702	68	189
711	120	213
801	68	182
802	68	155
811	120	206
901	33	146
902	33	119
911	120	177

Optimal Design of Nonlinear Structures: A Stability Approach

S. Pezeshk

Department of Civil Engineering, Memphis State University, Memphis, TN 38115, U.S.A.

ABSTRACT

This paper suggests an optimization-based methodology for the design of minimum weight structures with nonlinear behavior. Attention is focused on three-dimensional reticulated structures idealized with beam elements under proportional static loadings. An optimization procedure based on an optimality criterion approach is introduced. An example is given to evaluate the validity of the underlying assumptions and to demonstrate some of the characteristics of the proposed procedure.

INTRODUCTION

The minimum weight design subjected to stability constraint is one of the most important problems in structural optimization and has attracted a great deal of interest in the structural mechanics community. Methods for optimum design of structures have progressed rapidly in recent years. In particular, optimality criteria procedures have significantly advanced the state-of-the-art of the minimum weight design of structures involving large finite-element assemblies.

In most optimization procedure for stability problems, the constraints were defined by associated linear buckling eigenvalue problem. The "linear stability" can only give physically significant answers if the linear analysis gives such deformations that are exactly the same when geometric nonlinearity is considered. This only happens in a very limited number of practical situations such as a perfectly straight column under axial load. In real engineering situations where the qualitative nature of the behavior is completely unknown linearized stability represented by critical eigenmodes does not provide an adequate representation of nonlinear behavior of a structure. For such cases, it is more appropriate to optimize a structure on the basis of a "nonlinear stability constraint" when it has an inherent tendency to posses nonlinear behavior.

Virtually all of the methods developed in the area of linear and specially nonlinear stability optimal design procedures have dealt with the optimization of truss elements or truss-like idealized elements (Khot and Kamat, 1983; Levy et. al., 1988), but are seldom engaged with complex stiffness elements such as beam elements. The present procedure develops an optimality criterion approach to determine the optimal minimum weight design of a structure idealized by beam elements with constraint on the nonlinear strain energy density distribution.

In the following sections we will first discuss the formulation of a nonlinear analysis procedure for three-dimensional structures and then we will formally state the problem as a nonlinear stability constraint optimization with single equality constraint. The first order necessary conditions will be derived and employed as the basis of a fixed-point iteration method to search for the optimal design.

NONLINEAR ANALYSIS

The approach employed for the analysis of 3-D structures for this study is based on a second order approximation to the nonlinear equilibrium equations in terms of stress resultants. The kinematic hypothesis employed is based upon the Bernoulli-Kirchhoff hypothesis that plane sections remain planar after deformation. Based on the assumption, the rigid body kinematic transformation can be expressed as (Elias, 1986):

$$u_1(x) = u - y\phi_3 + z\phi_2 \tag{1}$$

$$u_2(x) = v - z\phi_1 \tag{2}$$

$$u_3(x) = w + y\phi_1 \tag{3}$$

where the vector $\mathbf{u}_o = [u, v, w]^t$ is the displacement at the origin of axes (x, y, z) where (y=z=0) and $\phi_o = [\phi_1, \phi_2, \phi_3]^t$ is the vector of rotation of cross section about (x, y, z) axis respectively. Note that superscript t means the transpose of the argument.

We can express Green's strain tensor in terms of Cartesian components of displacement vectors in indicial notation as

$$\varepsilon_{ij} = \frac{1}{2}(u_{i,j} + u_{j,i} + u_{m,i}u_{m,j}) \tag{4}$$

Using the kinematic hypothesis given Eq. (1) through Eq. (3) in conjunction with the definition of strain given in Eq. (4), we can directly compute the components of strain. Defining *strain resultant* $\boldsymbol{\varepsilon}_o = [\varepsilon_1, \varepsilon_2, \varepsilon_3]^t$, $\kappa_o = [\kappa_1, \kappa_2, \kappa_3]^t$ and by noting the vector of strains as $\boldsymbol{\varepsilon} = [\varepsilon_{ij}]$ and letting $\mathbf{e}_1$ be the unit vector pointing in positive x direction, we have:

$$\varepsilon_o = A_o^{-1}\int_A \varepsilon \cdot e_1\, dA \quad \text{and} \quad \kappa_0 = I_0^{-1}\int_A r_0 \times (\varepsilon \cdot e_1)dA \tag{5}$$

where $\boldsymbol{r}_o = (0,y,z)$ points from origin to the point (y,z), $\boldsymbol{A}_o = \text{diag}[1/A, 2/A, 2/A]$, and $\boldsymbol{I}_o = \text{diag}[2/J, 1/I_2, 1/I_3]$ in which A is the cross-sectional area, J is the torsion constant, and I_2, I_3 are the principal moments of inertia about y and z axes, respectively. The strain resultants vector $\boldsymbol{\varepsilon}_o$ represents the actual stretching and shearing. The curvature resultants vector κ_o represents the rate of twist and flexural curvatures.

The next step of the formulation would be to develop the constitutive equations to relate strains and stresses. For this study, the material is assumed to be elastic, such that resultant internal forces $\boldsymbol{Q}(\mathbf{x}) = [N, V_2, V_3]^t$ and resultant internal moments $\boldsymbol{M}(\mathbf{x}) = [T, M_2, M_3]^t$ can be expressed in terms of strain resultants:

$$\begin{bmatrix} T & 0 & 0 \\ 0 & M_2 & 0 \\ 0 & 0 & M_3 \end{bmatrix} = \begin{bmatrix} GJ & 0 & 0 \\ 0 & EI_2 & 0 \\ 0 & 0 & EI_3 \end{bmatrix} \kappa_o \quad and \quad \begin{bmatrix} N & 0 & 0 \\ 0 & V_2 & 0 \\ 0 & 0 & V_3 \end{bmatrix} = \begin{bmatrix} EA & 0 & 0 \\ 0 & GA_2 & 0 \\ 0 & 0 & GA_3 \end{bmatrix} \varepsilon_0 \tag{6}$$

where N is the axial force, V_2 and V_3 are the shear forces acting in y and z directions respectively. T is the torque; M_2 and M_3 are the flexural moment acting about y and z directions respectively.

Using Eqs. (5) through (6), we can easily derive the constitutive equations as:

$$N = EA\lambda_1 \qquad where \qquad \lambda_1 = u' + \frac{1}{2}[(u')^2 + (v')^2 + (w')^2] \tag{7}$$

$$V_2 = GA_2\lambda_2 \qquad where \qquad \lambda_2 = v' - \phi_3(1+u') + w'\phi_1 \tag{8}$$

$$V_3 = GA_3\lambda_3 \qquad where \qquad \lambda_3 = w' + \phi_2(1+u') - v'\phi_1 \tag{9}$$

$$M_2 = EI_2\lambda_4 \qquad where \qquad \lambda_4 = \phi_3'(1+u') - w'\phi_1' \tag{10}$$

$$M_3 = EI_3\lambda_5 \qquad where \qquad \lambda_5 = \phi_2'(1+u') - v'\phi_1' \tag{11}$$

$$T = GJ\lambda_6 \qquad where \qquad \lambda_6 = \phi_1' + \frac{I_3}{J}\phi_2'\phi_3 - \frac{I_2}{J}\phi_2\phi_3' \tag{12}$$

where $\lambda = [\lambda_1, \lambda_2, ..., \lambda_6]^t$ is the vector of strain measures which is conjugate to the stress resultant **R**.

To set the stage for the following developments, we will denote the displacements by $\boldsymbol{u} = [u, v, w, \phi_1, \phi_2, \phi_3]^t$, the stress resultants by $\boldsymbol{R} = [N, V_2, V_3, M_2, M_3, T]^t$, the vector of the applied forces by $\boldsymbol{q} = [p, q_2, q_3, m_2, m_3, t]^t$ with p being the applied axial force, q_2 and q_3 being the applied shear forces in direction of y and z, respectively, m_2, m_3 being the applied moments in direction of y and z, respectively, and t being the applied torque.

Eqs. (7) through (12) furnish the simplest constitutive model in terms of the generalized strains. Clearly, the model (7) through (12) derives the strain energy potential of the ith group, W_i, defined as:

$$W_i = \frac{1}{2}\int_0^L \left[A_i(x_i)E_i\lambda_1^2 + (A_2)_i(x_i)G_i\lambda_2^2 + (A_3)_i(x_i)G_i\lambda_3^2 + E_i\bar{I}_i(x_i)\lambda_4^2 + E_i\underline{I}_i(x_i)\lambda_5^2 + J_i(x_i)G_i\lambda_6^2\right]dx \tag{13}$$

where $A_i(\mathbf{x})$, $(A_2)_i(\mathbf{x})$, $(A_3)_i(\mathbf{x})$, $\bar{I}_i(\mathbf{x})$, $\underline{I}_i(\mathbf{x})$, $J_i(\mathbf{x})$ are the area, major shear area, minor shear area, major moment of inertia, minor moment of inertia, and torsion constant of the cross section for group i, respectively, all which depend on the design variables. The term G_i is the shear modulus of group i.

Finally, the total potential energy of a structure can be simply expressed as

$$\Pi = \sum_{i=1}^{N} W_i - \int_0^L \eta^t q ds \tag{14}$$

Where N is the total number of groups.

EQUILIBRIUM

The equations governing equilibrium of a beam can be expressed in their weak or variational form as a statement of the principle of stationary value of the total potential energy, which are obtained by setting the first variation of total potential energy, Π, to zero. Accordingly, we can define the following functional for all admissible variations, η, of the displacement field as

$$G(\boldsymbol{u},\eta) = \int_0^L \boldsymbol{R}^t \,(D\lambda \cdot \eta\,)\, dx - \int_0^L \eta^t q \, dx = 0 \tag{15}$$

The expression for the variation in strains presented in the first term of Eq. (15) is given by the usual definition of (Hughes and Pister, 1978):

$$D\,\lambda \cdot \eta = \frac{d}{d\alpha}\left[\lambda(\boldsymbol{u} + \alpha\eta)\right]_{\alpha=0} = \Xi(\boldsymbol{u})\boldsymbol{B}(\eta) \tag{16}$$

Defining $\mathbf{B}(\eta) = [u', v', w', \phi_1, \phi_2, \phi_3, \phi'_1, \phi'_2, \phi'_3]^t$ as the strain displacement operator that acts on real displacements $\mathbf{u}$ or their variation η the expression (15) takes the explicit form of

$$G(\boldsymbol{u},\eta) = \int_0^L \boldsymbol{B}^t(\eta)\Xi^t(\boldsymbol{u})\boldsymbol{R}(x)dx - \int_0^L \eta^t q dx = 0 \tag{17}$$

where the $\Xi(\mathbf{u})$ is the matrix of the gradient operator which reflects the effect of geometry on the equilibrium of the internal resisting force $\boldsymbol{R}$ and is approximated to second order.

Equilibrium is satisfied for any configuration, $\boldsymbol{u}$, in which $G(\boldsymbol{u},\eta)=0$ for any admissible virtual displacement η. Thus, $G(\boldsymbol{u},\eta)$ has the physical significance that it measures, in a weak sense, the equilibrium imbalance in the system.

SOLUTION PROCEDURE

The motion of the system involves geometric nonlinearity. Hence, the procedure of linearization about an intermediate configuration must be employed in the solution scheme. Specific details regarding the linearization process are treated extensively by Hughes and Pister (1978) and Marsden and Hughes (1983), Hjelmstad (1986), and Hjelmstad and Pezeshk (1988). Here we simply state the final results and refer the interested reader to the cited works for details.

Linearization of the Weak Form. Employing standard procedures, the weak form of the equilibrium equations can be linearized about an intermediate configuration $\bar{\boldsymbol{u}}$, with incremental motion $\Delta \boldsymbol{u}$. The linear part of $G(\boldsymbol{u},\eta)$ can then be expressed as

$$L[G]_{\bar{u}} = \int_0^L B'(\eta) \left[A_g(R) + \Xi'(\bar{u}) D(x) \Xi(\bar{u}) \right] B(\Delta u) dx + G(\bar{u},\eta) \qquad (18)$$

where $\boldsymbol{D} = \text{diag}[EA, GA_2, GA_3, EI_2, EI_3, EJ]$, for the formulation considered here, the matrix $\boldsymbol{A}_g$ gives rise to the geometric part of the stiffness matrix. The so-called residual $G(\bar{\boldsymbol{u}},\eta)$ is given by Eqs. (16) and (17) as

$$G(\bar{u},\eta) = \int_0^L B'(\eta) \Xi'(\bar{u}) R dx - \int_0^L \eta' q \, dx \qquad (19)$$

NONLINEAR ANALYSIS OF STRUCTURES

The equations for equilibrium can be discretized using the finite element method and can be solved using an incremental procedure with a Newton-Raphson iteration at each step. Ramm (1980) has presented a general summary of algorithms for tracing the response of a structure, including passage through limit load points. A displacement control procedure was used to carry out the analysis of the problems.

FORMULATION AND DEVELOPMENT OF THE OPTIMIZATION PROBLEM

Nonlinear critical load, which may be either a limit or a bifurcation point, can be characterized as the load which results in a loss of positive definiteness of the tangent stiffness matrix. For the optimization problem considered here, the load distribution applied to the structure is specified and is assumed to be proportional. The geometry of the structure is given and the optimization problem seeks to procure the minimum weight design of a structure, such that for the given design load distribution an instability can be achieved. The instability can be a limit or a bifurcation point. The structure is idealized with nonlinear beam elements and in each optimization cycle, the structure is analyzed using the geometrically nonlinear procedure formulated earlier. Using a displacement control procedure such as Ramm (1980) and Batoz and Dhatt (1979), we can reach and pass a limit or a bifurcation point. A limit or a bifurcation point can be traced either by monitoring the positive definiteness of tangent stiffness matrix or using *the current stiffness parameter* developed by Bergan and Soreide (1978). Bergan and

Soreide (1978) showed that the nonlinear behavior of multi-dimensional problems may be characterized by a single scalar quantity called the current stiffness parameter. The current stiffness parameter is implemented and employed in this study. As soon as the load approaches an instability load the current stiffness parameter approaches zero and becomes zero right at the instability load, and changes sign after passing the instability point.

For the present study we will consider the members of the structures to be arranged into M distinct groups. Each group is associated with a set of design variables which describe the geometry of the cross section of that group. For example an I-beam can be described by its depth h, flange width b, web thickness t, flange thickness t_f. Consequently, the I-beam has four design variables. A rectangular cross section has two design variables: the width b and the height h of the cross section. A square cross section can be identified by one design variable which can be either the width or the height of the cross section. The vector of design variables will be designated as $\mathbf{x} = \{x_1, x_2, ..., x_{dv}\}$, where dv is the number of design variables, computed as the sum over all the groups of the number of design variables per group. To simplify notation, we designate the *specific weight* of the mth group as the weight per unit of cross sectional area of the entire group

$$P_m = \sum_{i \in m} \rho_i L_i \tag{20}$$

where the length of member i is L_i, and its density is ρ_i. The sum is taken over all members associated with group m.

With the above definitions, the optimization problem can be posed in the following way

$$\textit{Minimize} \qquad A^t(x)P \tag{21}$$

$$\textit{Such that} \qquad h(x) = \Pi(x) - \bar{\Pi} = 0 \tag{22}$$

$$\textit{and} \qquad \underline{x} \leq x \leq \bar{x} \tag{23}$$

where $\Pi(\mathbf{x})$ is the total potential energy and $\bar{\Pi}$ is the total potential energy associated with the optimum design at the nonlinear critical load. $P = \{P_1, P_2, ..., P_M\}$ is the vector of specific weights, and $A(\mathbf{x}) = \{A_1(\mathbf{x}), A_2(\mathbf{x}), ..., A_m(\mathbf{x})\}$ is the vector of cross sectional areas of M groups. Each function $A_m(\mathbf{x})$ depends only on the design variables from group m. The inequality constraints given in Eq. (23) indicate that each design variable has a minimum permissible size, $\underline{x}_i$, and maximum permissible size, $\bar{x}_i$.

OPTIMALITY CRITERIA

The optimality criteria can be obtained from the first order necessary conditions for a constraint optimum. The Lagrangian functional corresponding to the optimization problem given in Eqs. (21) through (23) can be written as

$$L(x,\xi) = A'P - \xi[\ \Pi - \bar{\Pi}\ ] \tag{24}$$

where ξ is the Lagrange multiplier for the equality constraint. The element size limit constraints are not included in the Lagrangian and hence do not have corresponding Lagrange multipliers. The explicit size constraints can be handled more efficiently with an active-set strategy. Whenever a design variable violates a size constraint, it is assigned its limiting value and is removed from the active set and is no longer considered as a design variable.

The first order necessary conditions for an optimum are obtained by differentiating the Lagrangian functional with respect to design variables **x** and setting the corresponding equation to zero

$$\nabla L(x,\xi) = \nabla AP - \xi\left[\nabla\Pi - D\,\Pi\cdot u\right] = 0 \tag{25}$$

where $[\nabla(.)]_i = \partial(.)/\partial x_i$ is the ordinary gradient operator and $D\Pi.u$ is zero as given Eq. (16). Thus, by simplifying Eq. (25) we arrive at the following expression referred to as the optimality criteria:

$$\frac{\nabla AP}{\xi\ \nabla\Pi} = I \tag{26}$$

where I is the identity matrix.

SOLUTION PROCEDURE

The optimum design must satisfy the optimality criteria and the nonlinear energy density constraint. Since these equations are nonlinear, they must be solved by an iterative procedure. The algorithm suggested here is a fixed point iteration based on the first order necessary conditions (optimality criteria). The fixed point iteration, used in conjunction with a scaling procedure, will move the initial design toward a configuration which satisfies the optimality criteria and the constraints. The algorithm steps are as following:

(1) Choose an initial design;
(2) Perform a nonlinear analysis and determine the nonlinear limit load;
(3) Perform a line search (scaling) to satisfy the constraint and keep the design in feasible region, by assuming the Lagrange multiplier is equal to one;
(4) Select a design vector direction and determine new design variables;
(5) Check the optimality criteria;
(6) If convergence is not achieved go to step 2.

The fixed-point iteration. Various forms of recurrence relations have been developed by various researches and used to update the configuration in an optimization problem. Berke (1970) used a recurrence relation in virtual strain energy formulation for minimizing weight with prescribed displacement constraints. The same recurrence relation was effectively used by Gellaty and Berke (1971) for design problems with stress and displacement constraints. Later, Venkayya, Khot, and Berke (1973) and Khot, Venkayya, and Berke (1973, 1976) derived various forms of the recurrence relations for displacement constraints, stress constraints, and dynamic stiffness requirements.

To make the formulation easier to follow, Eq. (27) can be simplified as

$$Q(x) = I \quad ; \qquad where \quad Q(x) = \frac{\nabla AP}{\xi \nabla \Pi} \tag{27}$$

The optimality criteria are used to modify the design variable vector. Therefore, we generate a new design vector from previous one with the exponential recursion relation

$$x^{\kappa+1} = x^{\kappa} \left[Q(x^{\kappa}) \right]^{\frac{1}{r}} \tag{28}$$

where κ denotes the iteration number and r is the step size parameter. At the optimum the optimality criteria will be satisfied; therefore, the design variables will be unchanged with any additional iterations at optimum. The convergence behavior depends on the parameter r. Depending on the behavior of the constraint, it may be necessary to increase r in order to prevent convergence. A parameter study on the magnitude of the step length and its effects on the convergence of the algorithm can be found in the paper by No and Aguinagalde (1987).

Scaling Procedure. After each iteration, the design variables must be scaled to ensure that the specified load on the structure is equal to the nonlinear critical load of the structure. The problem of scaling is essentially a line search in the direction of the current design vector **x**. We wish to make the limit load or the bifurcation instability load to be the actual load applied for the scaled design variables ς**x**, where the following notation has been proposed to accommodate the active-set strategy

$$[\varsigma x]_i = \begin{bmatrix} \varsigma x_i & \text{for } i \text{ in the active set} \\ x_i & \text{for } i \text{ not in the active set} \end{bmatrix} \tag{29}$$

The scaling equation for n design variables per group is an nth order polynomial. Higher order polynomials can be easily solved by Newton's method. Care must be exercised to adjust the active set to reflect variables which become passive as a result of scaling.

Active-Set Constraint Strategy. After each iteration a new set of design variables is obtained. If a design variable lies within its permissible range, it is placed in the active set, otherwise, it is placed in the passive set so that a proper scaling can be performed before the next iteration. At the start of each iteration, formerly passive variable can either remain in passive set or be reactivated. In general, it is not known *a priori* if a variable will be active at the optimum. Allwood and Chung (1984) have suggested that if a design variable is moved to the passive set in two consecutive iterations, it will probably be passive at optimum. In the investigation reported here, in principle, the method suggest by Allwood and Chung was followed in programming the algorithm.

Sensitivity analysis. Evaluation of the optimality conditions (Eq. 27) requires knowledge of the sensitivity, or rate of change, of the potential energy with respect to the design variables. The sensitivity of the potential energy can be computed by differentiating the potential energy given in Eq. (14) with respect to design variable **x**:

$$\nabla \Pi = \nabla W \tag{30}$$

in which the differentiation of the strain energy involves only differentiation of the scalar cross sectional parameters of Eq. (13)

$$\nabla W = \frac{1}{2}\int_0^L \left[\nabla A(x) E_i \lambda_1^2 + \nabla A_2(x) G_i \lambda_2^2 + \nabla A_3(x) G_i \lambda_3^2 + \nabla \bar{I}(x) E_i \lambda_4^2 + \nabla \underline{I}(x) E_i \lambda_5^2 + \nabla J(x) G_i \lambda_6^2 - \nabla \eta' q\right] dx \tag{31}$$

where $\bar{I} = [\bar{I}_1, \bar{I}_2, \ldots, \bar{I}_m]^t$ is the vector of major moment of inertias of M groups, $\underline{\mathbf{I}} = [\underline{I}_1, \underline{I}_2, \ldots, \underline{I}_m]^t$ is the vector of minor moment of inertias of M group, $\mathbf{J} = [J_1, J_2, \ldots, J_m]^t$ is the vector of torsion constants of M groups. Each function such as $\underline{I}_m(\mathbf{x})$ depends only on the design variables from group m.

EXAMPLE

In the example problem investigated in this report, Young's modulus of elasticity of 10^7 psi and material density, ρ, of the 0.1 lbs/in^3 were used. A square cross section was used for all the members with a minimum allowable size of 0.316 inches for each design variable (or 0.1 in^2 for minimum allowable cross sectional area). Each design variable consists of either height or width of a square cross-section. In addition each structural member is modeled with two three-noded quadratic geometrically nonlinear beam elements.

The analysis and the optimization procedure, used to analyze and design the structures studied here, are implemented in the general finite element program FEAP (Chapt. 24 of Zienkiewicz, 1982).

FOUR-BEAM STRUCTURE

The four-beam structure consists of four beams with four fixed supports and all beams have one common end point. The topology of the structure is given in Figs. 1.a and 1.b. The loading on the structure consists of a single proportional load applied at junction node in negative z-direction. The magnitude of the applied loading is 200 lbs. This structure is to be optimized to determine the optimal design with a minimum weight which exhibits a limit or a bifurcation instability under the applied loading. The initial design was chosen with the cross-sectional areas of each member to be 2.0929 in^2, with a total initial weight of 120.64 lbs. The history of the optimization process is given in Fig. 2. The problem converged in 15 iterations with the cross-sectional properties as given in Table 1. The weight of the structure at the optimum is 77.67 lbs.

The relative strain energy densities of each member are also given in Table 1. One of the members became passive at the 11th iteration, and the other three members stayed active with strain energy densities of almost unity. The member which became passive is the member AD (see Figs. 1.a and 1.b). Member AD is the longest member of the structure, and consequently the most slender member of the system. The optimization problem, converged with member AD becoming passive, physically points out the fact that this member is an unimportant member and in order to have a minimum weight structure to exhibit a limit or a bifurcation instability under the applied loading, we need to take this element out of the active set. This was done by assigning the minimum allowable design variable size to member AD and setting it to be a passive element.

Fig. 3 presents the magnitude of the load factor at limit load versus the optimization cycle. Looking at this figure, it becomes apparent that the structure is reallocating properties to each of the different elements to achieve the minimum weight and unity for the load factor at the limit load. After the second iteration, the optimization problem moves smoothly toward having unity for the load factor at the limit load, which happens at iteration 15.

SUMMARY AND CONCLUSIONS

An optimization-based design procedure that efficiently produces structural designs with minimum weight with geometric nonlinear behavior has been presented. The procedure is based on nonlinear stability for structures idealized with beam elements. The method establishes a rational framework in which to address the nonlinear stability of minimum weight design of structures which exhibit instability under the applied loading conditions.

LIST OF REFERENCES

1. Allwood, R.J. and Y.S. Chung (1984). "Minimum-Weight Design of Trusses by an Optimality Criteria Method." *International Journal for Numerical Methods in Engineering*, **20**, 697-713.
2. Atrek, E., R.H. Gallagher, K.M. Ragsdell, and O.C. Zienkiewics (1984). **New Directions in Optimum Structural Design**. John Wiley & Sons Ltd., New York.
3. Batoz, J.L. and G. Dhatt (1971). "Incremental Displacement Algorithms for Nonlinear Problems." *International Journal for Numerical Methods in Engineering*, **14**, 1262-1267.
4. Berke, L. (July 1970). "An Efficient Approach to the Minimum Weight Design of Deflection Limited Structures." *AFFDL-TM-70-4-FDTR*, Flight Dynamics Laboratory, Wright-Patterson, AFB, Ohio.
5. Bergan, P.G. and T.H. Soreide. (1978). "Solution of Large Displacements and Instability Problems Using the Current Stiffness Parameter." *Finite Elements in Nonlinear Mechanics*, **2**, TAPIR, Norway.
6. Elias, Z.M. (1986). **Theory and Method of Structural Analysis**. John Wiley & Sons. New York.
7. Gellatly, R.A. and L. Berke (April 1971). "Optimal Structural Design." *AFFDL-TR-70-165*, Air Force Flight Dynamics Laboratory, Wright-Patterson AFB, Ohio.

8. Hjelmstad, K. D. and E. P. Popov (July 1983). "Seismic Behavior of Active Beam Links in Eccentrically Braced Frames." Report No. UBC/SESM-83/15, University of California, Berkeley.
9. Hughes, T.J.R. and K.S. Pister (1978). "Consistent Linearization in Mechanics of Solids and Structures." *Computers and Structures*, **8**(2).
10. Kamat, M.P. and P. Ruangsilasingha (1985). "Optimization of Space Trusses Against Instability Using Design Sensitivity Derivatives." *Engineering Optimization*, 177-188, **8**.
11. Khot, N.S. (1981). "Algorithm Based on Optimality Criteria to Design Minimum Weight Structures." *Engineering Optimization*, **5**, 73-90.
12. Khot, N.S., V.B. Venkayya, and L. Berke (Dec. 1973). "Optimization of Structures for Strength and Stability Requirements." *AFFDL-TR-73-89*, Air Force Flight Dynamics Laboratory, Wright-Patterson AFB, Ohio.
13. Khot, N.S., V.B. Venkayya, and L. Berke (1976). "Optimum Structural Design with Stability Constraints." *International Journal for Numerical Methods in Engineering*, **10**, 1097-1114.
14. Khot, N.S., L. Berke, and V.B. Venkayya (Feb. 1979). "Comparison of Optimality Criteria Algorithms for Minimum Weight Design of Structures." *AIAA Journal*, **17**(2), 182-190.
15. Khot, N.S. (Aug. 1981). "Optimality Criteria Methods in Structural Optimization." *Technical Report AFWAL-TR-81-3124*.
16. Khot, N.S. and M.P. Kamat (May 1983). "Minimum Weight Design of Structures with Geometric Nonlinear Behavior." *24th Structural Dynamics and Material Conference*, Lake Tahoe, Nevada.
17. Levy, R. and H. S. Perng (1988). "Optimization for Nonlinear Stability." Computers & Structures, **30**(3), 529-535.
18. Marsden, J.E. and T.J.R. Hughes (1983). **Mathematical Foundation of Elasticity**, Prentice-Hall, Inc. Englewood Cliffs, New Jersey.
19. No, M. and J.M. Aguinagalde (1987). "Finite Element Method and Optimality Criterion Based Structural Optimization." *Computers & Structures*, **27**(2), 287-295.
20. Ramm, E. (1980). "Strategies for Tracing Nonlinear Response Near Limit Points." Europe -U.S.-Workshop on *Nonlinear Finite Element Analysis in Structural Mechanics*, Bochum.
21. Simo, J.C. (1982). "A Consistent Formulation of Nonlinear Theories of Elastic Beams and Plates." *Report No. UBC/SESM-82/06*, University of California, Berkeley.
22. Simo, J.C., K.D. Hjelmstad, and R.L. Taylor (1983). "Finite Element Formulations for Problems of Finite Deformation of Elasto-Viscoplastic Beams." *Report No. UBC/SESM-83/01*, University of California, Berkeley.
23. Venkayya, V.B., N.S. Khot, V.A. Tischler, and R.F. Taylor (1971). "Design of Optimum Structures for Dynamic Loads." *Presented at the Third Air Force Conference on Matrix Methods in Structural Mechanics*, WPAFB, Ohio.
24. Venkayya, V.B., N.S. Khot, and L. Berke (April 1973). "Application of Optimality Criteria Approaches to Automated Design of Large Practical Structures." *Second Symposium on Structural Optimization, AGARD Conference Proceedings*, (123), Milan, Italy, April.
25. Zienkiewicz, O.C. (1982). **The Finite Element Method**. McGraw Hill Book Company (UK) Limited, London, 3rd Edition.

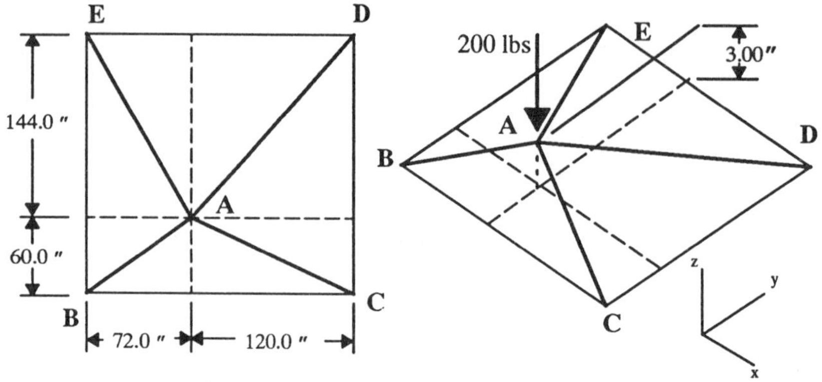

(a) Top View (b) Isometric view

Fig. 1. Topology of Four-Beam Structure

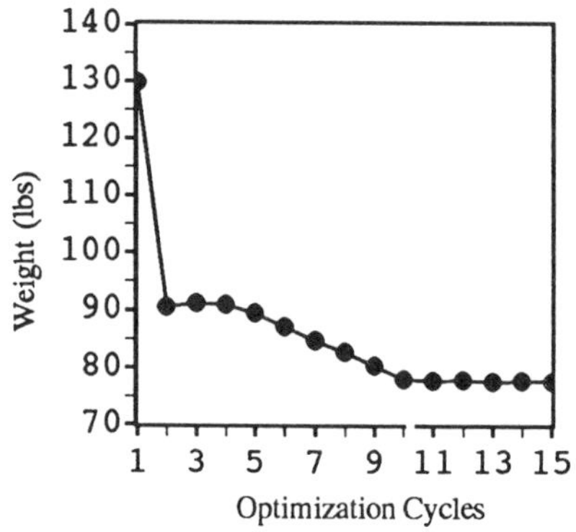

Fig. 2

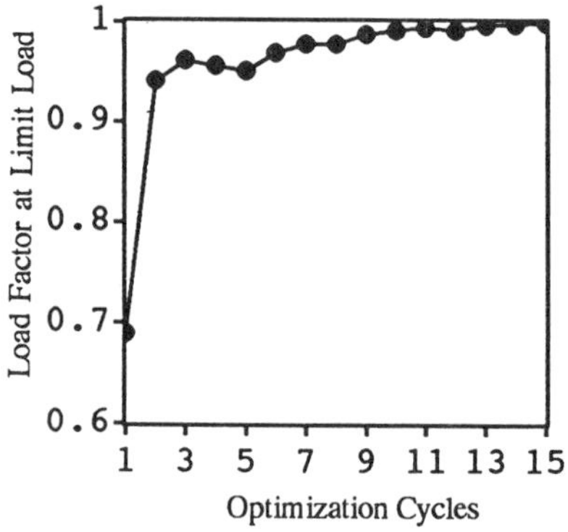

Fig. 3

Table 1. Properties of Four–Beam Structure

Member	Cross Sectional Area (in^2)	Major moment of Inertia (in^4)	Minor Moment of inertia (in^4)	Torsion (in^4)	Relative Energy Densities at Optimum
AB	2.7101	0.61207	0.61207	10.356	1.0000
AC	1.8762	0.29336	0.29336	4.963	0.99786
AD	0.1000	0.00083	0.00083	0.013	0.20986*
AE	1.5665	0.20423	0.20423	3.455	0.99770

Optimization of Spatial Structures for Maximum Dynamic Stiffness Using Optimality Criterion Method

W. Szyszkowski, J.M. King, J. Czyz
Department of Mechanical Engineering, University of Saskatchewan, Saskatoon, Canada, S7N 0W0

ABSTRACT

A method of optimization for maximum fundamental frequency of the vibrations handling multimodal eigensolutions is presented. The design variables (including the modality) at optimum are determined by solving iteratively the set of Kuhn-Tucker conditions. The numerical examples illustrate the convergence of the procedure.

INTRODUCTION

Structural design of any device or apparatus should, in general, secure their dimensional stability when exposed to the normal loading conditions (predicted) as well as to somewhat unpredictable random disturbances. The dimensional stability is especially important for structures operating in outer space for which maintaining various objects (mirrors, antennas, sensors, etc.) in precise positions is crucial. These structures, in the absence of gravity, are exposed to rather small mechanical forces and therefore can be extremely light and flexible. Consequently, any loading variations, planned or unexpected, may bring about vibrations of unacceptable amplitudes lasting for a very long time.

The active control methods, now being extensively developed, try to minimize these unwanted effects by installing sophisticated mechanisms and controllers. Clearly any vibration control method should be easier to implement if the structure itself is as rigid as possible. The dynamic rigidity can be improved by raising the fundamental frequency of the structure to its highest possible level. This is a relatively simple task for a structure for which only one fundamental vibration mode is to be considered. However, most complex structures, designed intuitively for "good" dynamic properties, are characterized by a cluster of vibration modes having similar frequencies. Often, the optimal structure will have several modes vibrating with the same fundamental frequency. Therefore, any mathematical optimization procedure to be used must be able to handle such multimodal problems. Additionally, the unknown modality of the optimal design should be treated as one of the design variables. This multimodality introduces some ambiguity to the eigenvalue problem and makes the optimization difficult to handle numerically. Thus far, only the simplest cases of multimodal structures have been effectively optimized using rather elaborate analytical methods (Olhoff[1], Blachut[2], Plaut[3], Bochenek[4]). Numerous publications (see Grandhi[5], Khot[6-7], for example), report designs of minimum weight structures with different eigenvalues constraints, in which, however, either only one mode is considered or the modality (bimodality) of the problem is assumed a priori.

The method presented here utilizes a multimodal optimality criterion and allows for inclusion of an arbitrary number of vibration modes which potentially might influence the optimization process. All the modes not affecting the optimal design are gradually eliminated during the iterations. This way the modality of the optimal design is determined automatically. Because of a natural use of the FEM technique the method is easy to program and should be helpful in design of large flexible spatial structures.

THE OPTIMALITY CRITERION

Modifications to the structure, when using Direct Optimization Methods, can be determined only after the completion of some sensitivity analysis which, in turn, may necessitate a large number of the structural analyses. Usually, the computational effort required in this phase of optimization increases rapidly with the number of design variables.

Optimization based on Optimality Criteria Techniques attempts to satisfy iteratively the set of Kuhn-Tucker conditions for the problem. However, the modifications in each step of iteration can be found after only one structural analysis, independently of the number of design variables involved. Clearly this makes the optimality criterion approach very attractive numerically, especially in applications to problems with a large number of design variables.

Upon reflection, this superiority of the optimality criterion approach is due to the fact that the sensitivity analysis, which usually must be performed numerically at each iteration step when using the direct approach, is incorporated inherently into the optimality condition and therefore, does not need to be repeated over and over again.

Obviously, the use of the optimality criterion technique is limited to the problems for which numerically tractable optimality conditions are available. In this paper such conditions and the accompanying optimization procedures are discussed for structures optimal with respect to dynamic stiffness.

The vibrations problem is defined by:

$$(K - \lambda_i M)\, x_i = 0 \qquad (1)$$

where K and M are the stiffness and mass matrices, x_i is the i-th vibrations mode and $\lambda_i = (\omega_i)^2$ where ω_i represents the frequency of vibrations.

The purpose of optimization for maximum dynamic stiffness is to maximize the first eigenvalue λ_1 for the structure of an assumed volume. However, when increasing λ_1 it may become equal to λ_2 and then to λ_3 and etc. up to λ_N. Consequently, the optimization procedure finally may need to monitor N modes of vibrations simultaneously.

For the above eigenvalue problem optimization can be formally stated as:

$$\text{maximize: } \lambda_{opt} = \lambda_1 = \lambda_2 \;.....\; = \lambda_N < \lambda_{N+1} \qquad (2a)$$

$$\text{subject to: } V(h) = V_o \qquad (2b)$$

$$\text{and } h_{min} < h_j < h_{max} \qquad (2c)$$

where: V_o is the volume of the structure, h is the vector of size design variables (thickness of the elements, width, area, movement of inertia, etc. ...).

The size constraints (2c) can be easily taken care of numerically. Namely, all the elements violating these constraints are resized to the given limits and excluded from the optimization procedure for the next (one) iteration. This way the size constraints can be satisfied without formally including them into the optimality criterion.

Using the Rayleigh quotient, the eigenvalues of the problems defined by Eqn. (1) can be expressed as:

$$\lambda_i = SE_i/C_i \tag{3a}$$

where:
$$SE_i = \frac{1}{2} x_i^T K x_i; \tag{3b}$$

$$C_i = \frac{1}{2} x_i^T M x_i \tag{3c}$$

The mass matrix M includes structural and nonstructural masses. The nonstructural masses remain constant during the optimization. Clearly, SE_i is the strain energy related to the i-th mode of vibrations, C_i is related to the kinetic energy of vibrations. The eigenmodes x_i can always be normalized so that the magnitudes of C_i become equal to one.

In order to derive an optimality criterion for the problem (2) consider the following Lagrange functional:

$$F(x,h) = \lambda_1 + \sum_{i=2}^{N} \gamma_i (\lambda_i - \lambda_1) + \sum_{i=1}^{N} \eta_i (C_i - 1) + \beta(\Sigma_j V_j - V_o) \tag{4}$$

where V_j is the volume of the j-th element and γ_i, η_i and β are the Lagrange multipliers. The maximum of λ_1 is required which, clearly, should be equal to all λ_i for which $\gamma_i > 0$. The multipliers η_i and β which represent equality constraints can be eliminated analytically [8].

Assume the usual relations holding for the j-th element, namely:

$$K_j = a_j \, [V_j(h_j)]^p \;\; ; \;\; M_j = b_j \, V_j(h_j) \tag{5}$$

where K_j and M_j are the stiffness and mass matrices for this element, a_j and b_j are matrices independent of h_j, and the parameter p depends on the type of optimization, for example, p = 1 for truss elements, p = 2 for cylindrical beams and p = 3 for plate or shell elements.

The Kuhn-Tucker conditions for the functional (4) can be obtained differentiating Eqn. (4) with respect to x and h to obtain:

$$(K - \lambda_i M)x_i = 0 \qquad i = 1 ..N \tag{6a}$$

$$\sum_{i=1}^{N} \gamma_i NE_{ij} - p + 1 = 0 \qquad j = 1 ..EN \tag{6b}$$

$$\gamma_i(\lambda_i - \lambda_1) = 0 \qquad i = 2..N \tag{6c}$$

$$\sum_{i=1}^{N} \gamma_i = 1 \qquad \gamma_i > 0 \tag{6d}$$

Eqn. (6b) represents the stationary condition of the functional (4) with respect to the size design variable h and was derived and discussed in Szyszkowski[8]. For p = 1 a similar condition were used in Grandhi[5]. The functions NE_{ij} (i-th mode, j-th element) denote a combination of normalized strain and kinetic energies where:

$$NE_{ij} = p\, NSB_{ij} - NKE_{ij} - \alpha_i \tag{6e}$$

$$NSB_{ij} = \frac{BE_{ij}}{V_j} \cdot \frac{V_o}{SE_i}$$

$$NKE_{ij} = \frac{KE_{ij}}{V_j} \bullet \frac{V_o}{SE_i} \tag{6f}$$

and where $BE_{ij} = \frac{1}{2} x_{ij}^T K_j x_{ij}$ is the elastic energy and $KE_{ij} = \frac{1}{2} \lambda_i x_{ij}^T M_j x_{ij}$ is the kinetic energy of j-th element due to i-th mode of vibrations.

$SE_i = \sum_{j}^{EN} BE_{ij}$ is the total strain energy stored in the structure for the i-th mode.

The parameter α_i is the ratio of the kinetic energy stored in all non-structural elements to the total kinetic energy due to the i-th mode and is defined as:

$$\alpha_i = \frac{KN_i}{KE_i} \tag{6g}$$

where $KN_i = \frac{1}{2} \lambda_i \, x_i^T \, M_s \, x_i$

and $KE_i = \frac{1}{2} \lambda_i \, x_i^T \, M \, x_i = SE_i$

The mass matrix for all non-structural elements which remain constant during the optimization is denoted here as M_s.

The function NE_{ij} defined by Eqn. (6e) satisfies the relation

$$\sum_j NE_{ij} \, V_j = V_o \, (p - 1) \qquad i = 1 \,..\, N \tag{6h}$$

The optimality conditions as specified by Eqn. (6) have the following interpretations:

- Eqn. (6a) represents the eigenvalue problem which must be solved at the beginning of each optimization step to obtain the set of $\lambda_1 \ldots \lambda_N$ and the vectors $x_i \ldots x_n$.
- Eqn. (6b) must be satisfied for all EN elements.
- The switching conditions (6c) must be satisfied for all N modes.
- Conditions (6d) represent the usual inequalities to secure the maximum of the functional (4).

Note that N in the above equations represents a somewhat arbitrary number of modes which may be included in the analysis. The real modality of the optimal design will be determined by the values of the coefficients γ_i which must be positive for all active modes and zero for the rest of the modes considered.

Any optimization procedure making use of the above conditions will have to determine the magnitudes of the design variables h_j for each element (EN parameters) and the magnitudes of the Lagrange multipliers, γ_i (N parameters), indicating the participation of each mode considered in the optimal design. Formally these EN + N parameters can be determined solving EN equations (6b) together with (N-1) equations (6c) and equation (6d). This is, in general, a highly nonlinear problem and only some iterative procedure can be used when attempting to solve it.

THE ITERATIVE PROCEDURE

The iterative procedure should correct the vector h (size variables) in order to meet the optimality conditions as defined by Eqn. (6a-d).

Clearly by solving the eigenvalue problem for a given structure (and assumed vector h) the condition (6a) is satisfied automatically and the magnitudes of NE_{ij} are found.

First consider the condition (6b). Define the local error, ξ_j, as:

$$\xi_j = \sum_{i=1}^{N} \gamma_i \, NE_{ij} - p + 1 \tag{7}$$

If the size variable vector is changed from h to h + δh, the new magnitude of the local error can be determined as:

$$\xi_j \, (h + \delta h) \simeq \xi_j(h) + \sum_{k=1}^{EN} \frac{\partial \xi_j}{\partial h_k} \Big/ \left(\frac{\partial V_k}{\partial h_k}\right) \delta V_k \tag{8}$$

Choosing the volume corrections, δV_k, in the form:

$$\delta V_k = c\xi_k(h) \; V_k \quad (\text{and } \delta h_k = \delta V_k \,/\, (\frac{\partial V_k}{\partial h_k})) \tag{9}$$

and making use of Eqn. (6f) it can be shown [8], that Eqn. (8) takes the form:

$$\xi_j(h + \delta h) \,/\, \xi_j(h) \simeq 1 - c \tag{10}$$

This relation is valid for any set of γ_i satisfying Eqn. (6d). Clearly for $c > 0$ the iteration process will drive the local error to zero. The magnitude of the parameter c should be small enough (we used $c < 0.2$) to secure the validity of Eqn. (8).

Additionally, when using Eqn. (9) the volume of the structures remains constant since one can prove that:

$$\delta V = \sum_{k=1}^{EN} \delta V_k = c \sum_{k=1}^{EN} \xi_k \; V_k = 0 \tag{11}$$

Thus, no scaling is required between the iteration steps. Clearly, this represents a big advantage over methods for which scaling is necessary, as for example when minimizing weight with assumed eigenvalue constraints (Grandhi[5]).

The switching conditions (6c) can be satisfied via a proper selection of the magnitudes of γ_i.

First introduce the global error defined as:

$$\Omega = \frac{c}{2V_o} \sum_j (\xi_j)^2 V_j + \frac{1}{2} \sum_{i=2}^{N} (\gamma_i)^2 \left(\frac{\lambda_i}{\lambda_1} - 1\right) \tag{12}$$

Minimizing Ω with respect to γ_i one obtains the set of equations:

$$\frac{c}{V_o} \sum_j \xi_j \frac{\partial \xi_j}{\partial \gamma_i} V_j + \gamma_i \left(\frac{\lambda_i}{\lambda_1} - 1\right) = 0 \quad i = 2 \ldots N \tag{13}$$

These equations when using the definition (7) can be transformed into the set of linear equations:

$$\sum_{i=2}^{N} \gamma_i A_{ki} = B_k \qquad k = 2..N \tag{14}$$

where A_{ki} and B_k are easily calculated for known values of NE_{ij} and λ_i. The magnitudes of γ_i are obtained solving Eqn. (14). Note that with the error ξ_j being reduced to zero accordingly to the formula (10), Eqn. (13) becomes identical to the switching conditions (6c).

Summarizing, if the corrections to the volumes or the corrections to the size variables are taken in the form (9) in which the set of γ_i are calculated using Eqn. (14) all the optimality conditions will be finally satisfied. In particular, the number of nonzero γ_i will determine the modality of the optimal design.

It can also be shown [8] that:

$$\frac{\delta\lambda_1}{\lambda_1} = 2\Omega \geq 0 \tag{15}$$

$$\frac{\delta\lambda_i}{\lambda_i} = \frac{c}{V_o} \sum_j NE_{ij} \xi_j V_j \qquad i = 2 \ldots N \tag{16}$$

Eqn. (15) proves that the procedure presented here always maximizes the first eigenvalue λ_1. The subsequent eigenvalues λ_i

will also reach some extreme points as indicated by Eqn. (16), however, the type of this point (max. or min.) is not determined.

SOME RESULTS

The procedure discussed here has been used for optimal design of simple beams, frames and plates, mostly in order to verify the convergence process. The optimization program uses ANSYS for the required FEM analysis and for solving eigenvalue problems. Typically at the beginning, the first nine modes (N=9) were taken into consideration.

Figure 1 presents the example of a beam supporting two large mirrors. Only the beam is optimized, the mirrors and connecting elements are assumed rigid. A tubular cross-sectional area of the beam with constant thickness is considered which constitutes the case p = 3 (see Eqn. (5)). The optimal beam (bimodal) has the fundamental frequency about 45% higher than the frequency of the uniform beam.

Examples of optimization of free flying structures are shown in Figure 2. A flexible boom connecting two small sensors is considered in Figure 1a. The optimization of the boom raises the fundamental frequency by some 26%.

The frame shown in Figure 2b is to represent a part of a large modular space structure. The frequency of this single bay (without any nonstructural masses) has increased by about 17%.

Other examples obtained with the help of the method discussed here were presented in Szyszkowski [9], (including a very similar optimization for maximum stability).

Conclusions

Elastic structures are optimized with respect to the fundamental frequency of free vibrations using the multimodal optimality criterion. Since optimization of almost all the more complex structures would probably involve multimodality, the approach discussed here should be useful for rational design of such structures.

References

1. N. Olhoff an S.H. Ramussen. On single and bimodal optimum buckling loads of clamped columns. Int. J. Solids Struct., **11**, 605-614 (1977).

2. J. Blachut and A. Gajewski. On unimodal and bimodal optimal design of funicular arches. Int. J. Solids Struct., **17**, 653-667 (1981).

3. R.H. Plaut, L.W. Johnson and N. Olhoff. Bimodal optimization of compressed columns on elastic foundations. J. Appl. Mech. **53**, 130-134 (1986).

4. B. Bochenek. Multimodal optimal design of a compressed column with respect to buckling in two planes. Int. J. Solids Struct., **23**, 599-606 (1987).

5. R.V. Grandhi and V.B. Venkayya. Structural optimization with frequency constraints. AJAA Journal, 858-866 (1988).

6. N.S. Khot, V.B. Venkayya, L. Berke. Optimum structural design with stability constraints. Int. J. Numer. Math. in Engrg., **10**, 1097-1114 (1976).

7. N.S. Khot. Optimization of structures with multiple frequency constraints. Computers & Structures, **20**, 868-876 (1985).

8. W. Szyszkowski. Multimodal optimality criterion for maximum fundamental frequency of free vibrations. Computers & Structures (to appear).

9. W. Szyszkowski, J. Czyz and J.M. King. Multimodal optimization for maximum stability or dynamic stiffness. Proceedings of the 3rd AF/NASA Symposium on Recent Advances in Multidisciplinary Analysis and Optimization, San Francisco, CA, 1990. A Collection of Technical Papers, pp. 297-302 (1990).

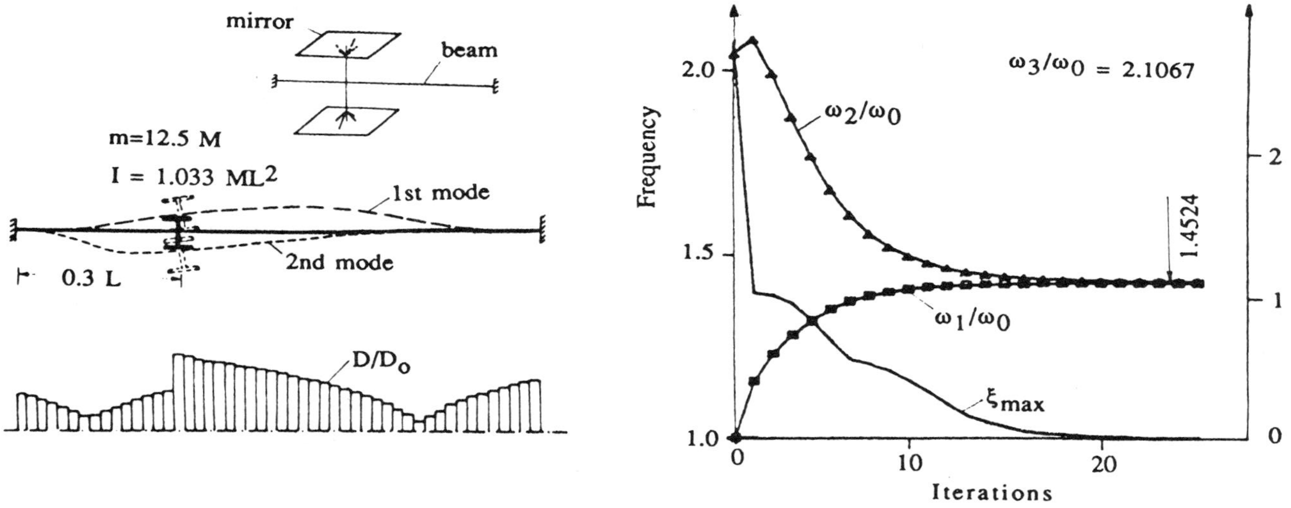

FIG. 1 Optimization of Beam with Non-Structural Mass-Bimodal Case

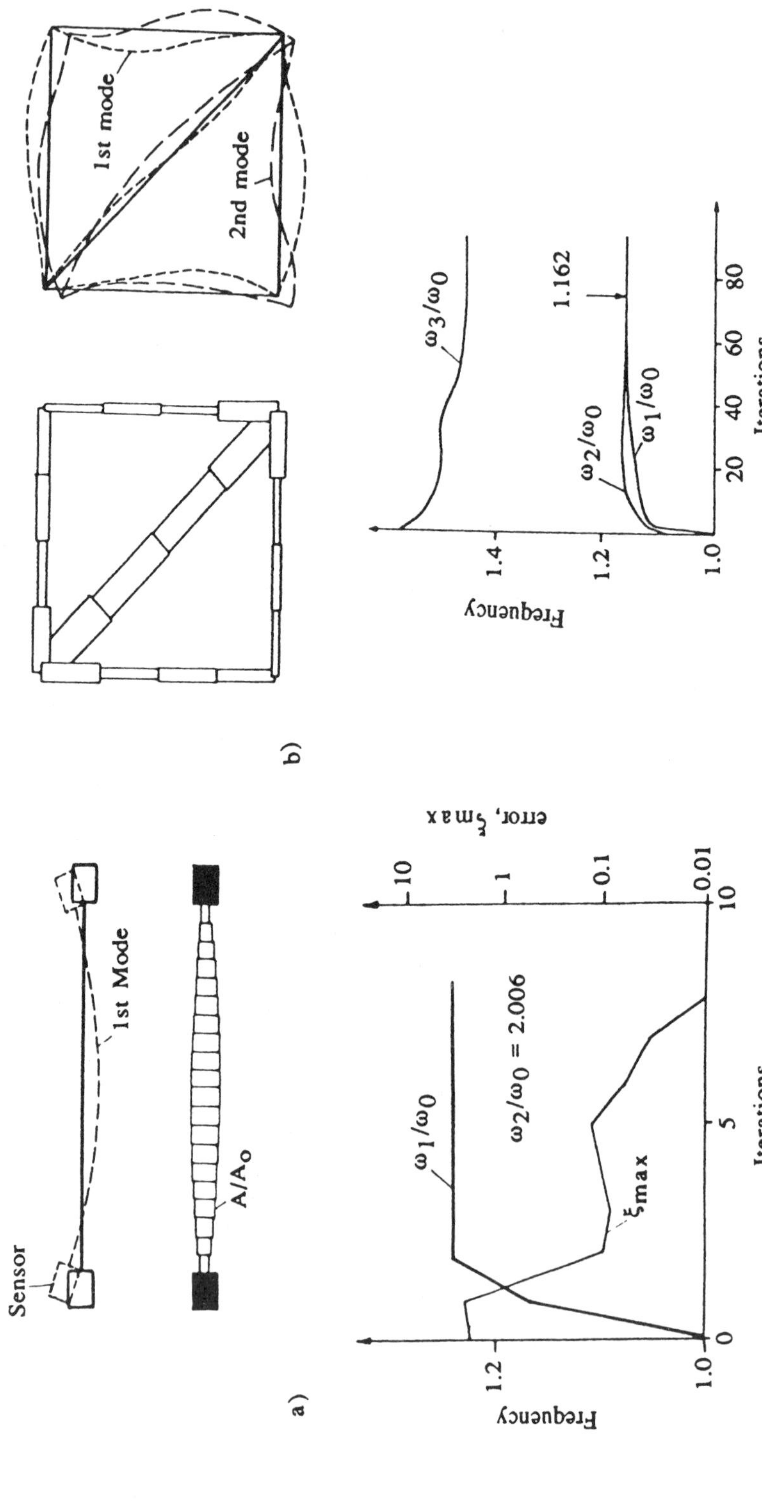

FIG. 2 **Optimization of Free Flying Structures**

Multi-Objective Structural Optimization Via Goal Formulation

M.E.M. El-Sayed, T.S. Jang

Department of Mechanical and Aerospace Engineering, University of Missouri, Columbia, Missouri 65211, U.S.A.

ABSTRACT

Multi-objective formulations are required for most of real-life structural optimisation problems . In this paper we discuss an approach for multi-objective optimisation following the formulation of goal programming techniques. The formulated optimisation problem requires the minimisation of the sum of the deviations of each objective from a target value, subject to nonnegativity requirement on each deviation. This minimisation problem can be solved using any constrained nonlinear optimisation algorithm. To demonstrate the approach and its application to structural optimisation problems some truss examples are presented.

INTRODUCTION

In traditional structural optimisation models, the user selects a single criterion which is to be maximised or minimised, and defines a set of design constraints which delimit a feasible design space. This general formulation does not allow much flexibility and has drawbacks for application in many design situations.

It is often difficult to define the problem exactly in the format required for a traditional structural optimisation algorithm. There are, however, optimisation techniques which address multi-objective optimisation problems and have the capacity to handle rank ordered design objectives or goals. Such techniques are known as goal programming [1-6]. In a goal programming formulation, the design goals are defined and a priority is assigned to each one. The algorithm then attempts to satisfy as many of the goals as possible, starting with the highest priority goal. This removes the difficulty of having to define an objective function and constraints.

One of the main difficulties in using goal programming as a design tool, however, is that the resulting optimisation problem can not be solved using regular nonlinear optimisation routines. Reliable nonlinear goal programming that can be used in design optimisation are hard to find. For this reason linearization methods have been introduced to solve nonlinear goal structural optimisation problems [7].

In this paper we present a method for solving large scale multi- objective structural optimisation problems using goal programming formulation. By avoiding the priority levels, the resulting multi-criteria optimisation problem can be solved using any nonlinear optimisation routine. In the following we discuss the formulation of the nonlinear structural optimisation problem into a multi-objective form. We also demonstrate the application of this approach to structural optimisation problems.

FORMULATION

A detailed discussion of of goal programming terminology, basic elements and the formulation of goal programming problems are given in references [5-7]. In the following we discuss the formulation of the general structural optimisation problem to multi-objective form.

The Goal Programming Form

The goal optimisation problem with positive deviations and without assigned priorities is:

$$\text{Minimise} \quad Z = \sum_{i=1}^{I} w_i \, (d_i)$$

$$\text{Subject to} \quad g_i(\mathbf{x}) - d_i = b_i \qquad (i = 1, 2, \cdots, I) \tag{1}$$

$$\mathbf{x} = [x_1, x_2, \ldots, x_N]^T , \qquad \mathbf{x} \in R^N$$

$$\text{Non-negativity requirement} \qquad d_i \geq 0$$

where Z is the goal programming objective function to be minimised. The differential weights w_i are used to differentiate the deviational variables. d_i is the ith deviational variable that expresses the possibility of deviation from a goal right-hand-side value. g_i is the goal constraint function we desire to minimise its numerical deviation from a stated right-hand-side value b_i in a selected goal constraint. X is a set of decision variables (x_n, n = 1, 2, ..., N) we seek to determine.

The Structural Optimisation problem

The traditional model of the minimum weight structural optimisation problem is:

$$\text{Minimise} \qquad W(\mathbf{x}) , \qquad \mathbf{x} \in R^N$$

$$\text{Subject to} \qquad G_i(\mathbf{x}) \leq 0 \qquad (\text{for } i=1, 2, \cdots, I) \tag{2}$$

$$\text{and} \qquad \mathbf{x}^L \leq \mathbf{x} \leq \mathbf{x}^U$$

where W(x) is the objective function, representing the total weight of the structure. $G_i(\mathbf{x})$ represents the inequality constraint function including the equality constraints and it may be a function of stress, displacement, Euler buckling and frequency, $\mathbf{x}^L$ and $\mathbf{x}^U$ represent the lower and upper bound of design variable respectively.

For simplicity we consider the minimum weight optimisation problem of truss structures with stress constraints only. The structural optimisation problem with the cross-sectional area of the truss members taken as the design variables becomes:

$$\begin{aligned} &\text{Minimise} && W(\mathbf{x}) = \sum_{n=1}^{N} \rho_n l_n x_n, && \mathbf{x} \in R^N \\ &\text{Subject to} && S_i(\mathbf{x})/S_a - 1 \le 0 && (\text{for } i=1, 2, \cdots, I) \\ &\text{and} && \mathbf{x}^L \le \mathbf{x} \le \mathbf{x}^U && \end{aligned} \tag{3}$$

where ρ_n, l_n and x_n are the density, length and cross-sectional area of the nth truss member. The structural optimisation problem, equation (3) can be rewritten, in a goal structural optimisation model, as:

$$\begin{aligned} &\text{Minimise} && Z = \sum_{j=1}^{I} w_i \,(d_i) \\ &\text{Subject to} && W(\mathbf{x})/W_a - 1 - d_1 = 0 \\ & && S_i(\mathbf{x})/S_a - 1 + d_i = 0 \qquad (\text{for } i = 2, \cdots, I) \\ &\text{and} && \mathbf{x}^L \le \mathbf{x} \le \mathbf{x}^U, \mathbf{x} \in R^N, \qquad d_i \ge 0 \end{aligned} \tag{4}$$

where $W(\mathbf{x})$ is the total weight of the structure, $S_i(\mathbf{x})$ is the stress in the ith member. W_a and S_a represent the target weight and allowable stress respectively. From equation (4), the deviational variables can be written as:

$$\begin{aligned} &d_1(\mathbf{x}) = W(\mathbf{x})/W_a - 1 \\ &d_i(\mathbf{x}) = 1 - S_i(\mathbf{x})/S_a \qquad (\text{for } i = 2, \cdots, I) \end{aligned} \tag{5}$$

Using equation (5), the multi-objective structural optimisation problem, equation (4) can be rewritten, as:

$$\begin{aligned} &\text{Minimise} && Z = \sum_{j=1}^{I} w_i \,(d_i(\mathbf{x})) \\ &\text{Subject to} && d_i(\mathbf{x}) \ge 0 \qquad (\text{for } i = 1, \cdots, I) \\ &\text{and} && \mathbf{x}^L \le \mathbf{x} \le \mathbf{x}^U, \qquad \mathbf{x} \in R^N \end{aligned} \tag{6}$$

This structural optimisation problem can be solved using any constrained nonlinear optimisation routine. Stresses in the structure members during the optimisation process can be obtained using the finite element analysis. The gradient information can be obtained using the finite difference approximation.

The optimisation problem of equations (5) and (6) can be applied to other type of constrains such as displacement and frequency constraints. In the following we apply the developed multi-objective formulation to solve the structural optimisation problems of some truss problems and compare the solutions obtained with the published results.

TEST CASES

The following are three structural optimisation examples to test the developed multi-objective formulation. The planar 10 member truss, the 25 member space truss and the 63 member Wing-Carry-Through Structure problems were chosen because their solutions are available in the literature [8,9].

The three problems were solved using the multi-objective formulation for target weights less than the known optimum weight and target stresses equal to the allowable. The resulting optimisation problems were solved using a Generalized Reduced Gradient nonlinear optimisation routine. A finite element analysis routine was used for the stress calculation and the finite difference technique was used for the gradient calculations.

Example (1): Planar 10 Member Truss

Figure (1) shows the geometry and dimensions of the 10 member truss. The truss elements description including initial cross sectional areas, minimum member sizes, material properties and load data are specified in Table (1). The truss was designed to withstand a single loading condition subject to stress constraints. The optimum results for this example using the multi-objective formulation (M. O. F.) in comparison with those of references [8,9] are given in Table (2).

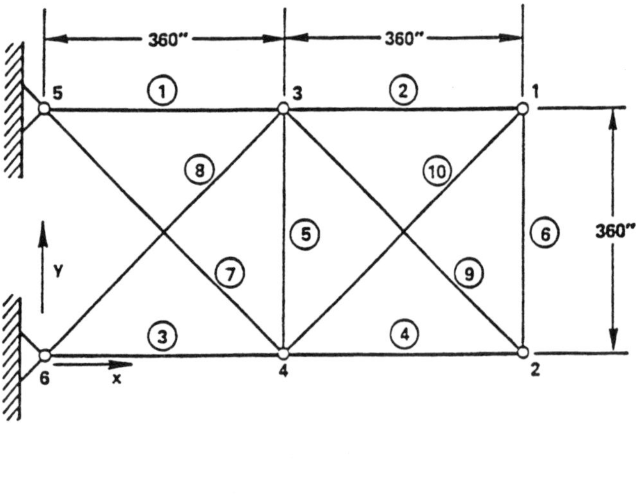

Figure (1) 10 Member Planar Truss

Table (1) Design Data for 10 Member Planar Truss

Modulus of elasticity = 10^4 ksi

Material density = 0.10 lb/in^3

Lower limit on cross-sectional areas = 0.10 in^2

Initial value of design variable = 10 in^2

Target weight = 1500 Ib

Stress limit = $\pm$ 25 ksi

Number of loading conditions = 1

Load Data

Loading Condition	Node	Load component (kips) in Direction x	y	z
1	2	0.0	-100.0	0.0
	4	0.0	-100.0	0.0

Table (2) Results for 10 Member Problem

Optimum Cross-Sectional Area in in.2

Member Number	M. O. F.	Ref. [8]	Ref. [9]
1	7.9379	7.938	7.9379
2	0.1	0.1	0.1
3	8.0621	8.062	8.0621
4	3.9379	3.938	3.9379
5	0.1	0.1	0.1
6	0.1	0.1	0.1
7	5.7447	5.745	5.7447
8	5.5690	5.569	5.5690
9	5.5690	5.569	5.5690
10	0.1	0.1	0.1
Optimum Weight (lb)	1593.18	1593.23	1593.18

Example (2): 25 Member Space Truss

A 25 member space truss is shown in Figure (2). Design data for this truss is given in Table (3). Simple design variable linking was used to impose the double symmetry which exists for this design and resulted in the total of 7 design variables. The truss was designed to withstand double loading condition subject to stress constrains. Table (4) shows the optimum values of design variables and the weight obtained in comparison with those of reference [9].

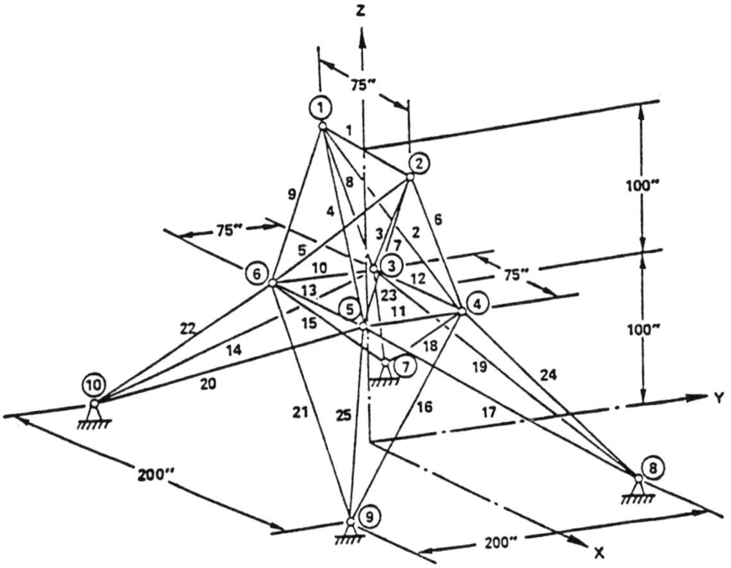

Figure (2) 25 Member Space Truss

Table (3) Design Data for 25 Member Space Truss

Modulus of elasticity = 10^4 ksi

Material density = 0.10 lb/in^3

Lower limit on cross-sectional areas = 0.10 in^2 (stress constraints only)

Initial value of design variable = 0.5 in^2 (stress constraints only)
Target weight = 80 Ib

Stress limit = $\pm$ 40ksi

Number of loading conditions = 2

Load Data

Load Condition	Node	Load Component (kips) in Direction x	y	z
1	1	0.5	0. 0	0. 0
	2	0.5	0. 0	0. 0
	3	1.0	10.0	-5.0
	4	0.0	10.0	-5.0
2	3	0.0	20.0	-5.0
	4	0.0	-20.0	-5.0

Table (4) Results for 25 Member Problem

Member Numbers	Optimum Cross-Sectional Area in in.2 M. O. F.	Ref. [9]
1	0.1	0.1
2,3,4,5	0.3763	0.3755
6,7,8,9	0.4708	0.4734
10,11,12,13	0.1	0.1
14,15,16,17	0.1	0.1
18,19,20,21	0.2772	0.2786
22,23,24,25	0.3821	0.3796
Optimum Weight (Ib)	91.23	91.27

Example (3): 63 member Wing-Carry-Through Structure

Figure (3) shows the truss idealization of the 63 member wing-carry-through structure. Design data for the structure are given in Table (5). The structure is designed to withstand double loading conditions subject to stress constraints. The Optimum results for the wing-carry-through structure, using the multi-objective formulation (M. O. F.) in comparison with those of references [8,9] are given in Table (6).

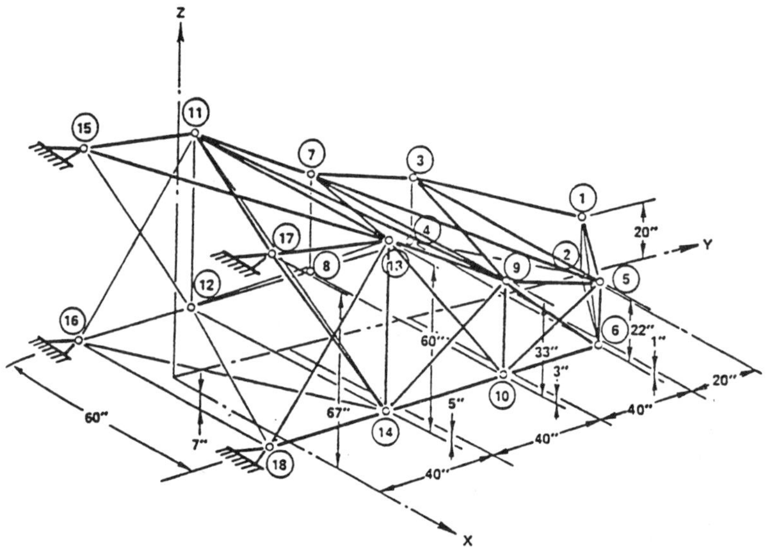

Figure (3) Geometry of 63 Member W.C.T.S

Table (5) Design Data for 63 Member W.C.T.S

Modulus of elasticity = 16000 ksi

Material density = 0.161b/in^3

Lower limit on cross-sectional areas = 0.01 in^2

Initial value of design variable = 20 in^2
Target weight = 3000 Ib

Stress limit = $\pm$ 100 ksi

Number of loading conditions = 2

Load Data

Load Condition	Node	Load Component (kips) in Direction		
		x	y	z
1	1	2500	-5000	250
	2	-2500	5000	250
2	1	5000	-2500	250
	2	-5000	2500	250

Table (6) Results for 63 Member W.C.T.S.

Member Number	Optimum Cross-Sectional Area in in.^{2}		
	M. O. F.	Ref.[8]	Ref.[9]
1	38.5901	38.28	38.0780
2	35.6586	35.93	36.1020
3	51.2843	51.69	51.9230
4	54.8130	54.49	54.2430
5	25.0776	24.98	25.0690
6	28.3508	28.46	28.2220
7	17.9558	17.64	17.4610
8	20.0347	20.52	20.7660
9	25.0328	25.21	24.81
10	27.1615	26.82	26.937
11	7.4038	7.535	7.5854
12	8.8567	8.801	9.0049
13	24.2984	24.21	24.047
14	20.7124	20.63	20.488
15	4.1163	4.123	4.2966
16	2.3234	2.495	2.9538
17	37.1079	37.07	37.06

Table (6) (Continued)

Member Number	M. O. F.	Ref.[8]	Ref.[9]
18	37.2066	37.14	37.171
19	0.01	0.01	0.01
20	0.01	0.01	0.01
21	0.0498	0.151	0.047
22	0.01	0.067	0.01
23	0.0868	0.137	0.1048
24	1.2966	1.085	1.1187
25	0.093	0.065	0.0312
26	2.5769	2.574	2.9
27	0.6841	0.804	1.304
28	4.8123	4.582	4.7738
29	0.6099	0.670	0.8632
30	2.86	2.651	2.6809
31	2.7843	2.580	2.7707
32	5.6472	5.829	5.8034
33	5.6597	5.839	5.5564
34	5.9315	6.122	6.0951
35	5.6603	5.839	5.6651
36	3.0033	2.783	2.815
37	2.7841	2.597	2.766
38	2.8715	2.705	2.692
39	2.7898	2.603	2.7716
40	5.596	5.736	5.7531
41	5.6603	5.821	5.6651
42	17.0739	16.45	16.029
43	19.397	18.8	18.397
44	10.440	11.01	11.49
45	12.8576	13.40	13.864
46	10.9538	11.42	11.735
47	5.4903	5.961	6.1777
48	12.63	12.16	11.793
49	13.7376	14.25	13.988
50	7.343	7.240	7.3508
51	7.7539	7.416	7.7067
52	5.3319	5.501	5.3699
53	0.01	0.566	0.0285
54	3.5541	3.639	3.5992
55	9.8593	9.631	9.7775
56	4.5153	4.375	4.2407
57	0.01	0.310	0.1179
58	0.01	0.051	0.01
59	0.01	0.125	0.01
60	0.01	0.01	0.01
61	0.01	0.01	0.01
62	0.01	0.01	0.01
63	0.01	0.01	0.01
Optimum Weight (lb)	4973.66	4976.00	4975.06

CONCLUSION

A multi-objective formulation following the formulation of goal programming techniques is developed. The formulated optimisation problem requires the minimisation of the sum of the deviations of each objective from a target value, subject to nonnegativity requirement. This optimisation problem can be solved using any constrained nonlinear optimisation algorithm.While the developed formulation was applied to truss structures with stress constraints only, it can be used for the design of other types of structures with different types of constraints.

The numerical examples performed demonstrated the ability of the developed formulation in solving structural optimisation problems. By minimising the difference between the stresses in the members and the allowable stresses the optimum weight obtained from the multi-objective formulation are similar or slightly less than the previously published results.

REFERENCES

1. Ijiri, Y. Management Goals and Accounting for Control, Rand McNally, 1965.

2. Lee, S. Goal Programming for Decision Analysis, Auerback Publishers, 1972.

3. Fisk, J. C. A Goal Programming Model for Output Planning, Decision Sciences, Vol. 10, pp. 593-603, 1979.

4. Wilson, J. M. The Handling of Goals in Marketing Problems, Management Decision, Vol.13, pp. 175-180, 1975.

5. Ignizio, J. P. Linear Programming in Single and Multiple Objective Systems, Prentice-Hall, 1982.

6. Schniederjans, M. J. Linear Goal Programming, Petrocelli Books, 1984.

7. El-Sayed, M. E., Ridgely, B. J., and Sandgren, E. Nonlinear Structural Optimizations Using Goal Programming, Computers and Structures, Vol.32, pp. 69-73, 1989.

8. Schmit, A. and Miura, H. Approximation Concepts for Efficient Structural Synthesis, NASA CR-2552, Univ. of California, Los Angeles, CA, 1976.

9. Haug, E. and Arora, J. Applied Optimal Design, John Wiley and Sons, 1979.

Optimization Applications in Structural Design

H.-R. Shih (*), E. Sandgren (**)

(*) Department of Technology and I.A., Jackson State University, Jackson, Mississippi 39217, U.S.A.

(**) School of Mechanical Engineering, Purdue University, West Lafayette, Indiana 47907, U.S.A.

ABSTRACT

An efficient algorithm based on combining the finite element package NISA and the optimization software package OPT, which implements the generalized reduced gradient method, is developed. This algorithm can be applied to general structures, such as trusses, frames, etc. The pre- and postprocessor are introduced to link NISA with OPT. In order to shorten the computer run time, approximation concepts and approximate analysis methods are used. The approximation concepts are used to replace an original nonlinear and implicit structural optimization problem with a sequence of explicit and high-quality approximate problems. The approximate methods include temporary deletion of noncritical constraints, design variable linking and Taylor series expansions for response variables in terms of design variables. The numerical examples indicate that this algorithm is a powerful and a rather general approach to structural optimization.

1. INTRODUCTION

The history of structural optimization is long and the literature is extensive. The main classification based on the optimization method can be described as either mathematical programming methods or optimality criteria approaches [1]. The fully stressed design is the earliest optimality criteria approach used in structural design [2]. By definition the fully stressed design is the one in which the maximum stress in each of the elements reaches its limiting value at least under one loading condition. This criterion is valid if the structure is statically determinate but is only approximately true for indeterminate structures.

Historically, the application of mathematical programming methods to structural optimization was initiated by Schmit [3]. Moses [4] reviewed the rapid development of mathematical programming approach to structural design optimization and concluded that the application of mathematical programming methods to the optimum design of structural elements and simple components is a well established and profitable technique. He indicates that a major consideration in structural systems optimization will be to reduce the number of analyses.

Haug and Arora [5] used a design sensitivity analysis coupled with a gradient projection method for the optimal design of trusses and frames. The objective function was to minimize the structural weight. In Refs. [6-8], structural optimization with materials selection, large and continuum systems, and optimization of geometry and topology are discussed.

The purpose of this work is to develop an efficient algorithm based on combining a finite element analysis method and a nonlinear programming method for structural optimization problems. This algorithm can be simultaneously applied to general structures. Since a conventional nonlinear programming approach was found to require many finite element analyses and resulted in an extremely long computer run time, approximation concepts and approximate analysis methods are used to improve the efficiency of the optimization process. The approximation concepts are used to replace an original nonlinear and implicit structural optimization problem with a sequence of explicit and high-quality approximate problem.

2. ILLUSTRATIVE EXAMPLES

A simple structural optimization problem that will be used to illustrate basic ideas and formulation is the three-bar truss. The first step in optimizing a structure is to idealize the structure as an assemblage of finite elements, as shown in Figure1. The design objective is to choose the element radii, so that the truss is as light as possible, while satisfying constraints on the stress, buckling, deflection, natural frequency, and member size.

The objective function F(X) is the volume of the structure

$$F(X) = \sum_{i=1}^{3} L_i(X_i * X_i)\,\pi \qquad (1)$$

$$\text{Subject to } G_j(X) \geq o \; j=1,2,....m \qquad (2)$$

$$\text{and } X_i^l \leq X_i \leq X_i^u \; i=1,2,3 \qquad (3)$$

where

G = stress, displacement, buckling and natural frequency constraints

X = design variables (radius of each member)

X_i^l = lower bound for variable i, X_i^u = upper bound for variable i

L_i = specific length of i-th element

Design data for this truss is given in Table1.

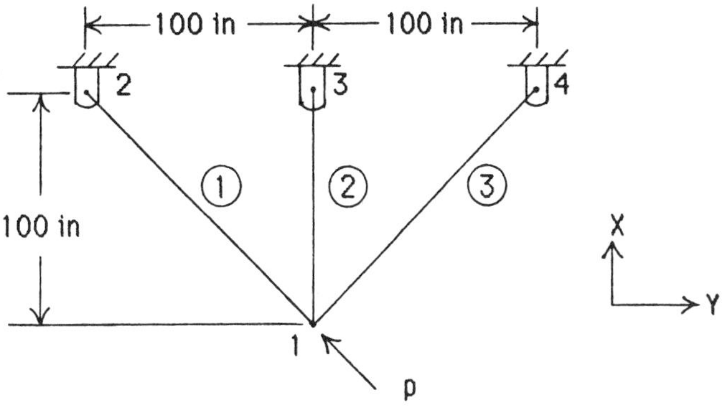

Figure 1 Three-bar truss

Table 1 Design data for three-bar truss

Modulus of elasticity = 2.07×10^4 ksi Lower limit on radius = 0.15 in Displacement limit at node 1 in x direction = ±0.15 in, in y direction = ±0.06 in Stress limit=±25 ksi, Lower bound frequency = 13.316 HZ			
Load Data			
Node Number	Direction of Load, lbf		
	F_x	F_y	F_z
1	-28300	28300	0.0

2.1 Nonlinear Programming Method

Central to optimization scheme are two elements - the optimizer and the analysis processor. The analysis processor accepts values of design variables X, and produce the constraint values. For the structural optimization problem, the analysis processor essentially provides a structural analysis of a given configuration and finds the stresses, deflections, and natural frequencies. The optimizer provides the input to the analysis processor and based on the information returned from the analysis processor, determines what changes should be made in the variables to both reduce the objective function and to remain in the feasible region.

2.1.1 Optimization Algorithm In Equations (1), (2) and (3), either the objective function F(X) or constraints G(X) may be nonlinear functions of the design variables X. These equations are solved by the generalized reduced gradient (GRG) code, OPT[9]. The detailed description of the algorithm is given in [10].

2.1.2 Analysis Processor As was discussed earlier, a finite element code is used as the analysis routine by the optimization program. Given a set of design variables, this routine must return stresses, displacements, and frequencies when requested by the optimizer. Here, the finite element package NISA (Numerically Integrated Elements for System Analysis) is used [11].

2.1.3 Coupling Finite Element And Optimization In order that NISA can become an integral part of the optimization procedure, OPT and NISA must interact with one another. To do so, it is necessary to link NISA with OPT. With the conventional use of finite element programs pre- and postprocessors have been introduced in order to save valuable time in data preparation and evaluation. The optimizer communicates with the analysis processor through a pre-processor and the analysis processor supplies the analysis information to the optimizer through a post-processor. The function of the pre-processor is to convert the variables of the optimization process to a set of input parameters written in the format required by the analysis processor. The function of the post-processor is to compute the constraints and their gradients, if required, and to provide them in the format required by the optimizer. To do so, the post-processor extracts the pertinent behavior variables from the analysis output and compares them with the allowable value of these quantities in the equations specific to the design problem at hand.

The subroutine PREPOR was written to generate the input data for a NISA analysis, and to drive NISA. In the optimization process with OPT, in order to find the minimum, the design variables must be updated before each evaluation. PREPOR must alter the data for NISA corresponding to design variables which are being changed. For the example under consideration, the design variables are the element radii. Thus in the input data, only the data group *D1, which defines the cross sectional properties [11], must be changed. The other data groups remain fixed. In the subroutine PREPOR, we create a new data file which is constructed from the data file which has been generated previously. When reading the *D1 data group, the new variables are used instead of the old ones. This data file is then used to drive the NISA program.

Each time NISA is run, it creates an output file. But this output file has a fixed format, so it can't be accessed by the optimizer. We use a subroutine called CONDS to condense NISA's output file to a uniformly formated file that can be called by subroutine CONST which calculates the constraint functions.

During each iteration of GRG method, finite element solutions are required every time the constraint subroutine 'CONST' is called. After each analysis is complete, subroutine CONST must retrieve the results from CONDS which serve to constrain the design.

2.2 Approximation Techniques

As discussed in the previous section, we consider the design task to be automated structural design generated by combining NISA with OPT. However, this approach can create difficulties which result from too many finite element analyses which result in a long computing run time. In order to reduce the number of detailed analyses required during the optimization, while still maintaining the salient features of the design problem, what we term approximation techniques will be used.

2.2.1 Active Constraint Logic In structural optimization problems, there are a great many constraints that must be considered. However, a large percentage of these constraints may never become active during the design optimization process and so could be excluded from the constraint set. The concept of constraint deletion is to identify constraints which are far from being active and simply omit these from consideration, at least for a portion of the design process. We can't, in general, identify nonactive constraints at the beginning of the optimization and delete them entirely because they may become active as the design progresses. Therefore, we will only temporarily delete constraints from consideration, allowing them to be included later if necessary.

At every iteration, we may delete any constraints whose value is more than some cutoff value, g_0, then, proceed with the optimization including only the retained constraints. Constraint deletion can conserve computer storage, reduce computational effort, and enhance the practical use of optimization techniques.

2.2.2 Linearized Constraints Whenever an analysis was called for, a complete, detailed finite element analysis was performed. These evaluations are generally very expensive. Here we wish to look at ways in which the analysis itself can be simplified. Linearizing the constraints through the use of approximation creates a simplified analysis model of the original complicated analysis program. These approximations are processed using OPT, instead of original program.

Following the approximation concept described above, the constraints given by Equation (2) are approximated by introducing a first-order Taylor series expansion. The expression takes the form of

$$G_j = G_j(X^t) + \nabla G_j(X^t)\,(X - X^t) \qquad (4)$$

at the specified design point X^t.

Using the explicit form of the objective function and the now explicit form of the constraint functions which result from the Taylor series approximation, the function, constraints, and gradient evaluations required during the optimization process are both simple and fast. Because the first-order Taylor series expansion is used to approximate the original behavior constraints, during the optimization process, all the constraints are linear. The subroutine 'CONST' now contains the linear functions.

Now, the approximate method is used to design the three-bar truss, shown in Figure 1. The following two problems were solved:

Case 1: Stress and displacement constrained

Case 2: All constrained (stress, displacement, buckling and natural frequency)

It should be recognized that the linearization is in most instances a very gross approximation, and must be used with great caution. One way to ensure the approximation is adequate is to impose limits on the allowable increments in the variables. That is, for each subproblem constructed around a base point X^t, we impose the bounds

$$-\delta_i \le \bar{X}_i^t - X_i^t \le \delta_i \qquad i=1,\ldots,N \qquad (5)$$

where δ_i is some suitably chosen positive step-size parameter.

The approximation concept consists of the following fundamental procedures:

A) Construct a first-order Taylor series expansion of the response quantities around the current design point.

B) Create an approximate optimization problem by substituting the expansion in (A) into the behavioural constraints.

C) Select pontentially critical constraints in order to temporarily remove all unnecessary constraints.

D) Solve the approximate problem with OPT.

E) Repeat steps (A) to (D) until convergence is attained.

The comparison between the results of conventional nonlinear programming method (CNP) and approximate method (APPRO) are shown in Tables 2 and 3.

Table 2 With stress and displacement constraints

Element Number	Final radius, in	
	CNP	APPRO
1	0.75928	0.75927
2	0.15449	0.15451
3	0.15000	0.15002
Volume, in^3	273.626	273.625
Detailed structural analysis number	232	16

Table 3 With all constraints

Element Number	Final radius, in	
	CNP	APPRO
1	1.44500	1.44499
2	0.81652	0.81662
3	0.91286	0.91280
Volume, in^3	1507.37	1507.36
Detailed structural analysis number	180	20

From the above tables, we can see that the approximate method provides similar results but greatly improves the efficiency of the optimization process.

2.3 Multiple Loading Conditions

In reality, a mechanical system must operate over a range of loading conditions. At every point where stress and displacement constraints are imposed, these constraints must be satisfied for every load condition. Therefore, it is necessary to develop an optimization technique which will allow multiple load cases to be applied to the structure.

Assume that several load conditions are applied to a structure. When each of these loads is applied to the structure, the structure will be subject to different stresses, displacements and any other response quantities. Because the structure must function under each specified load case without failing, it is necessary to choose the worst case to constrain the design. Thus the i-th constraint $G_i \geq 0$ may be expressed as

$$G_i = \min\{G_{ir}(X)\} > 0 \qquad r=1 \text{ to NLC} \tag{6}$$

where NLC is the number of loading conditions. Then 'active constraint logic' can be used in order to eliminate redundant constraints.

For each of these load cases applied to the structure, only the magnitude and direction of loads or points of application are different. The element dimensions and boundary conditions are the same. When a call is made to finite element program (NISA), the decomposed stiffness matrix from the first load case can be reused for other load cases, so that significant recomputation can be avoided. This is accomplished by using the restart capability of NISA.

2.4 Ten-Member Cantilever Truss

A ten-member cantilever truss is shown in Figure 2. This problem has been used in the literature [5,6] to compare various techniques of optimal design. The problem is to find the cross-sectional area of each member of truss in order to minimize its weight, subject to stress, displacement, and member size constraints. Design data for this problem is given in Table 4.

Table 4 Design data for ten-member truss

Modulus of elasticity = 10^4 ksi, Material density = 0.1 lb/in^3
Stress limit = 25 ksi, Displacement limit = 2.0 in
Lower limit on area = 0.1 in^2

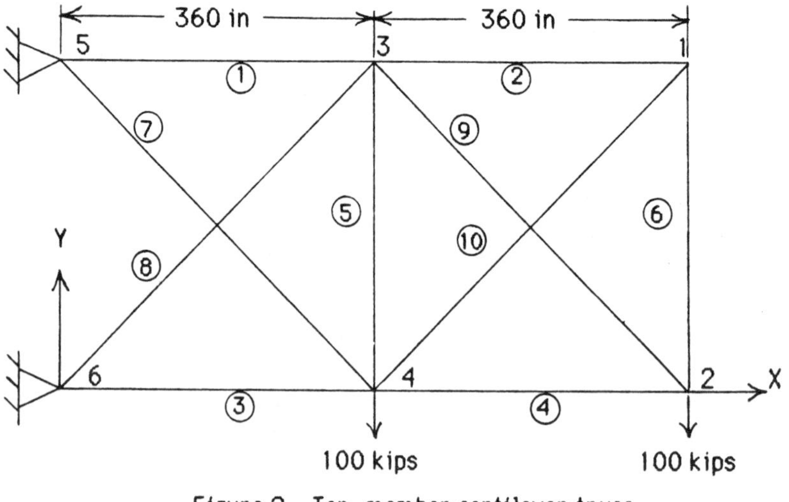

Figure 2 Ten-member cantilever truss

Table 5 Comparison of final designs for ten-bar truss

Member No.	Final cross-sectional area (in^2)			
	Ref. 5	Ref. 6	Ref. 6	OPT-NISA
1	30.031	30.730	30.521	30.352
2	0.100	0.100	0.100	0.100
3	23.274	23.941	23.200	23.031
4	15.286	14.733	15.223	15.203
5	0.100	0.100	0.100	0.100
6	0.557	0.100	0.551	0.570
7	7.468	8.341	7.457	7.282
8	21.198	20.951	21.036	20.636
9	21.618	20.836	21.528	22.259
10	0.100	0.100	0.100	0.100
Weight	5061.6	5076.7	5060.8	5056.5

Table 5 presents a comparison of optimum solution obtained using the algorithm of this paper with results available in literature. The optimum weight found is 5056.5 lb. The downward displacement constraints at nodes 1 and 2, the stress constraint for member 5 and minimum-size constraints on members 2, 5, and 10 are active at the optimum. From the comparison made with the other available programs in this example, it is shown that this proposed method provides the least weight design.

3. PRACTICAL EXAMPLE

We now want to turn our attention to application of the formulation described in this paper to the solution of an actual seat frame structural design problem. This is a space frame problem. This structure, shown in Figures 3,4 and 5, is idealized using 71 beam elements and 60 nodes. Design variable linking is employed, which limits specified groups of finite elements to be the same type with identical design variables. The grouping of elements is shown in Table 6.

In this example both hollow circular and rectangular sections are considered. The element groups of 1, 6, and 8 have rectangular cross sections and the others have cross circular sections. In the optimization all element cross-sectional dimensions including widths, heights, mean radii and thicknesses are considered to be design variables. The frame is optimized under stress and displacement constraints. There are four loading conditions. In the first condition, loads of 326.9 and 138.8 lbf in the positive X-direction and negative Z-direction are applied at node 48. The second loading condition is

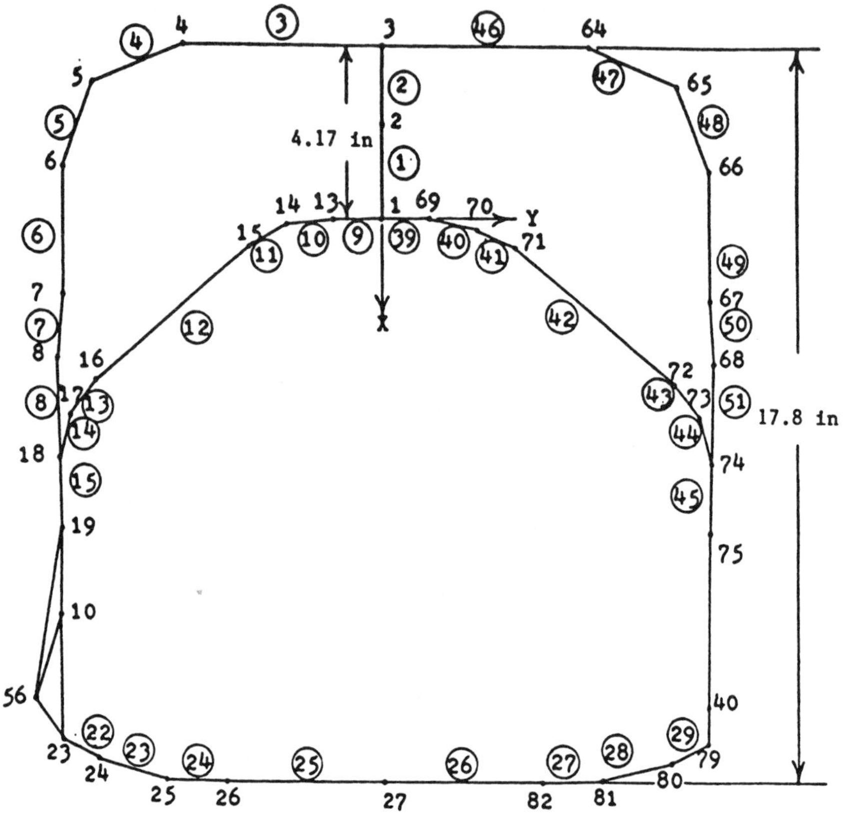

Figure 3 Top view of seat frame

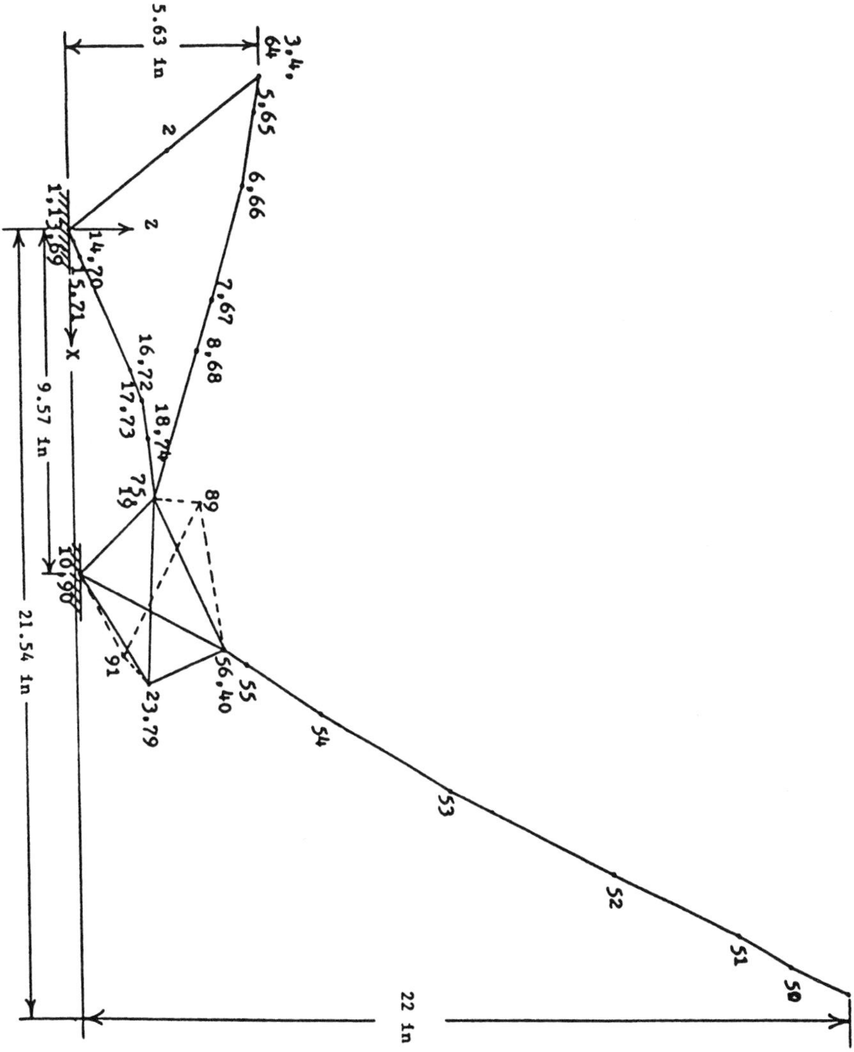

Figure 4 Front view of seat frame

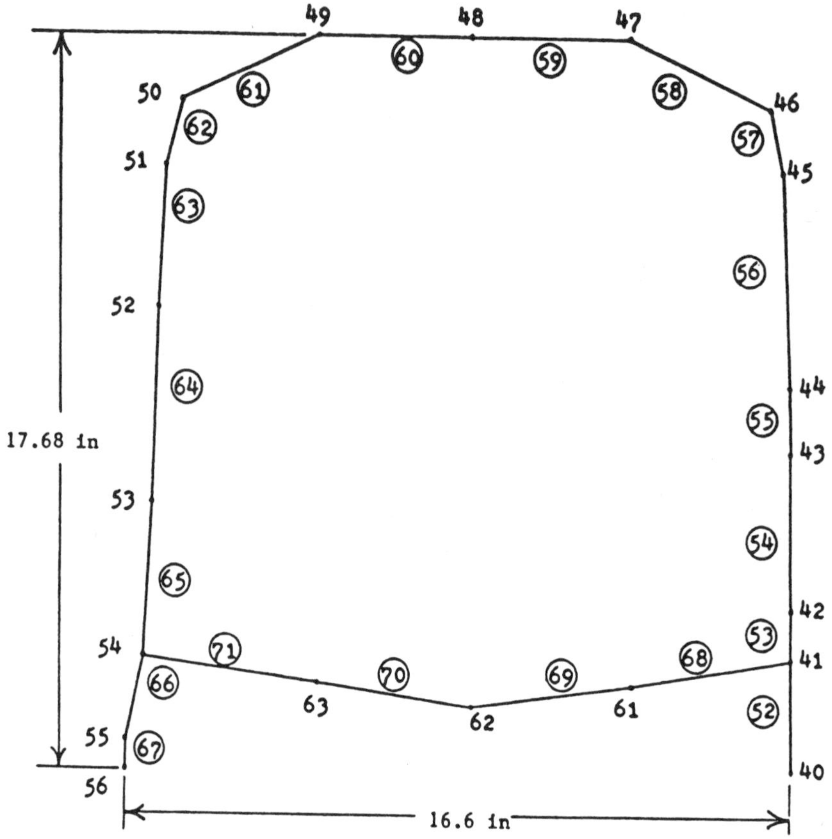

Figure 5 Side view of seat frame

to have 1349 lbf normal force and 3147.3 lbf tangential force distributed over the elements of group 2. The third condition is a load of 285 lbf in the positive X-direction applied at node 48. In the fourth condition, loads of 270 lbf in the positive X-direction and 135 lbf in negative Z-direction are applied at node 46.

Table 6 The element group of seat frame

Group No.	Element Number
1	1,2
2	3-8, 46-51
3	9-15, 39-45
4	22-29
5	52-57, 62-67
6	58-61
7	68-71
8	16-21, 30-38

Table 7 Design data for seat frame

Modulus of elasticity = 2×10^4 ksi, Poisson's ratio = 0.3
Stress limit = 25 ksi, Lower limit on thickness = 0.0472 in
Lower limit on mean radius = 0.0866 in
Displacement limit at node 48 in x-direction = 0.7874 in

Table 8 Results for seat frame structure

Group Number	Optimum design, in			
	Mean Rad.	Height	Width	Thickness
1		0.4102	2.5146	0.0472
2	0.6940			0.0494
3	0.7079			0.0472
4	0.0895			0.0472
5	0.5717			0.0472
6		2.7509	0.7824	0.0472
7	0.2902			0.0472
8		0.7983	0.7983	0.0802
At optimum Volume, in^3		44.0073		

Table 8 gives the results for this example. The optimum volume is 44.0073 in^3. The stress constraints on elements 46 (for load condition 2), 66 (for load condition 1) and 32 (for load condition 4), and the minimum thickness constraint on groups 1 and 3-7 are active at the optimum.

4. CONCLUSION

In this paper, we have presented in detail an efficient automated minimum weight design procedure for the solution of the structural optimization problem. The numerical examples have indicated that the linear sequence of the explicit approximate method which is used to replace the original problem is a powerful and a rather general approach to structural optimization. The approximate methods include temporary deletion of noncritical constraints, design variable linking and Taylor series expansions for response variables in terms of design variables. The subroutines PREPOR and CONDS which are used to link NISA with OPT play an extremely important role in the programming procedure. Although all the examples included in this paper considered only truss and frame structures, extensions can be made to include plate, shell and solid elements which would enhance the potential applications of the method.

REFERENCES

(1) Venkayya, V. B. Structural Optimization: A Review and Some Recommendations, International Journal for Numerical Methods in Engineering, pp. 203-228, 1978.

(2) Berke, L. and Venkayya, V. B. Review of Optimality Criteria Approaches to Structural Optimization, ASME Structural Optimization Symposium, AMD, 7, pp. 23-34, 1974.

(3) Schmit, L. A. Structural Engineering Applications of Mathematical Programming Techniques, Symposium on Structural Optimization, AGARD Conf. Proceed. No. 36, 1969.

(4) Moses, F. Mathematical Programming Methods for Structural Optimization, ASME Structural Optimization Symposium, AMD, 7, pp. 35-48, 1974.

(5) Haug, E. J. and Arora, J. S. Applied Optimal Design, Mechanical and Structural Systems, Wiley-Interscience, 1979.

(6) Morris, A. J. Foundations of Structural Optimization: A Unified Approach, Wiley-Interscience, 1982.

(7) Vanderplaats, G. N. Numerical Optimization Techniques for Engineering Design with Applications, McGraw-Hill, 1984.

(8) Kirsch, U. Optimum Structural Design, Concepts, Methods, and Applications, McGraw-Hill, 1981.

(9) Gabriele, G. A, and Ragsdell, K. M. OPT: A Nonlinear Programming Code in FORTRAN 77 Users Manual, Design Optimization Laboratory, Univ. of

Missouri, Columbia, MO, 1984.
(10) Reklaitis, G. V., Ravindran, A. and Ragsdell, K. M. Engineering Optimization, Methods and Applications, Wiley-Interscience, 1983.
(11) NISA User's Manual, Vol. 1, Engineering Mechanics Research Co., 1983.

An Improved Approximation for Stresses Constraints in Plate Structures

H.L. Thomas, G.N. Vanderplaats

VMA Engineering, 5960 Mandarin Ave., Suite F, Goleta, California, 93117, U.S.A.

ABSTRACT

An improved approximation for stresses (and strains) in plate structures is presented. This approximation is highly accurate because it is based on approximations of the element forces, which are more invariant than the element stresses and strains, and because it captures the cross coupling effect between the design variables. Both sizing and shape design problems are used as examples to show the increased accuracy and robustness of this approximation function.

INTRODUCTION

The usual approach to approximating stresses in plate structures is to form a first order Taylor Series expansion of the stresses with respect to the design variables (plate thicknesses) or their inverse. The weaknesses of this approximation are that (1) it is a separable approximation and does not capture the coupling effects present in statically indeterminate structures and (2) it cannot capture the higher order nonlinearity associated with the stresses.

In the approach presented in this work, the element forces are approximated by a first order Taylor Series and then the approximate stresses are recovered from these approximate element forces using the explicit force-stress relations. This approximation is more accurate because (1) it captures the cross coupling effect between the change in forces in one region of the structure due to changes in the thickness of other regions and the change in stresses caused by the change in thickness in this region and (2) the Taylor Series approximation of the element forces is more accurate than that for the stresses. The approximations of the element forces are more accurate because they are not as nonlinear functions of the design variables as the stresses are. The improved accuracy of this type of approximation allows for the use of larger move limits (which are used to protect the accuracy of the approximation) in each design cycle and therefore will lead to fewer design cycles and a reduction in total design cost. The cost of constructing and evaluating the new approximation per design cycle is comparable to the direct stress approximation since (1) the sensitivity analysis, slightly cheaper and (2) the calculation of the approximate stresses from the approximate element forces is inexpensive, being an explicit relationship.

BACKGROUND

An improved approximation for stress constraints in truss and frame members was presented in reference 1. In that work the element forces are approximated, rather than the element stresses, using a Taylor Series expansion. The approximate stresses (approximate constrained response quantities) are then calculated explicitly from these approximate forces (approximate intermediate response quantities) using the element section properties and cross sectional dimensions. For example in a truss member:

$$\tilde{\sigma} = \frac{\tilde{P}}{A} \tag{1}$$

and in a frame member

$$\tilde{\sigma} = \frac{\tilde{M}c}{I} + \frac{\tilde{P}}{A} \tag{2}$$

where the tilde (~) represents an approximated quantity. Note that since the forces in a statically determinate structure are constant, these approximations are exact for these structures. These approximations are more accurate than direct approximations of stresses because the forces are more constant than the stresses and because the approximations capture the cross coupling effect between the design variables. The capturing of this cross coupling effect can be demonstrated by examining approximations for the stress in one member of a two bar truss. The direct linear approximation of the stress in member 1 is

$$\tilde{\sigma}_1 = \sigma_{01} + \frac{\partial \sigma_1}{\partial A_1}\left(A_1 - A_{01}\right) + \frac{\partial \sigma_1}{\partial A_2}\left(A_2 - A_{02}\right) \tag{3}$$

while the approximation for stress using the approximate element force is

$$\tilde{\sigma}_1 = \frac{F_{01} + \frac{\partial F_1}{\partial A_1}\left(A_1 - A_{01}\right) + \frac{\partial F_1}{\partial A_2}\left(A_2 - A_{02}\right)}{A_1} \tag{4}$$

Note that 2nd cross partial derivative of Eq. 3, $\frac{\partial^2 \tilde{\sigma}_1}{\partial A_1 \partial A_2}$, is zero while for Eq. 4 it is

$$\frac{\partial^2 \tilde{\sigma}}{\partial A_1 \partial A_2} = \frac{\frac{\partial F_1}{\partial A_2} A_{02}}{A_1^2} \tag{5}$$

The use of the intermediate force approximation was applied to the configuration and sizing optimization of truss structures in reference 2 and to arch shape and sizing optimization in reference 3. A large decrease in the number of design cycles due to the increased accuracy of the approximations was noted in both works.

The idea of approximating element forces in order to improve the approximations of element stresses was extended to continuum structures in reference 4, where it was applied to shape optimization of three dimensional structures. In this approach the approximate element displacements were calculated from the approximate nodal forces and then the stresses are recovered from these approximate displacements. Because there is no explicit stress-nodal force relations for three dimensional finite elements the cost of evaluating the approximate stresses is quite large (see reference 4). This increased cost somewhat offset the overall gains, due to a reduced number of design cycles, for three dimensional structures. This method was improved, in terms of cost, in Ref. 5 where the stresses were approximated using a two level approximation.

The approaches developed in references 4 and 5 were applied to stress approximations in plate structures in reference 6. In this work the nodal forces were approximated using a Taylor Series expansion. The nodal displacements were then recovered from these forces using

$$\left\{\tilde{u}^e\right\} = \left[K^e\right]^{-1}\left\{\tilde{f}^e\right\} \tag{6}$$

after the singularities of $\left[K^e\right]$ have been removed. Then the approximate element forces are calculated using

$$\left\{\tilde{M}\right\} = [D][B]\left\{\tilde{u}^e\right\} \tag{7}$$

where [D] is the element material matrix and [B] is the strain-displacement matrix. Finally, the approximate stresses are calculated using these approximate element forces.

In the approach presented in this paper the element forces will be the approximated intermediate response quantities rather that the nodal forces. This formulation has two distinct advantages. Firstly the number of approximated intermediate response quantities is reduced. For example, for a four noded plate element with 6 degrees of freedom per node, 24 nodal forces must be approximated as opposed to 6 element forces. This results in a factor of 4 reduction in sensitivity and approximation costs. The second advantage is that several costly steps needed to evaluate the element forces are not needed. These are the removal of the singularities from $[K^e]$, its inversion, and the calculations involved in Eqs. 6 and 7 which take 756 flops for the aforementioned element.

CALCULATION OF APPROXIMATE STRESSES

The development of the new approximation for stresses and strains in plate elements will now be presented in detail.

The six approximate element forces

$$\{\tilde{M}\} = \begin{Bmatrix} \tilde{N}_x \\ \tilde{N}_y \\ \tilde{N}_{xy} \\ \tilde{M}_x \\ \tilde{M}_y \\ \tilde{M}_{xy} \end{Bmatrix} \tag{8}$$

are first calculated using a Taylor Series expansion in the intermediate design variables (see reference 7 for a discussion of intermediate variables). The intermediate design variables for plate structures are the shape design variables, the plate thicknesses, t, and bending stiffnesses, D, where

$$D = \frac{t^3}{12} \tag{9}$$

The approximate surface stresses are then calculated using

$$\{\tilde{\sigma}\} = \begin{Bmatrix} \tilde{\sigma}_x \\ \tilde{\sigma}_y \\ \tilde{\tau}_{xy} \end{Bmatrix} = \begin{Bmatrix} \dfrac{\tilde{N}_x}{t} - \dfrac{\tilde{M}_x z}{D} \\ \dfrac{\tilde{N}_y}{t} - \dfrac{\tilde{M}_y z}{D} \\ \dfrac{\tilde{N}_{xy}}{t} - \dfrac{\tilde{M}_{xy} z}{D} \end{Bmatrix} \tag{10}$$

Finally the approximate principle, maximum shear, and Von Mises stresses are calculated using

$$\tilde{\sigma}_I = \frac{\tilde{\sigma}_x + \tilde{\sigma}_y}{2} + \sqrt{\left(\frac{\tilde{\sigma}_x - \tilde{\sigma}_y}{2}\right)^2 + \tilde{\tau}_{xy}^2}$$

$$\tilde{\sigma}_{II} = \frac{\tilde{\sigma}_x + \tilde{\sigma}_y}{2} - \sqrt{\left(\frac{\tilde{\sigma}_x - \tilde{\sigma}_y}{2}\right)^2 + \tilde{\tau}_{xy}^2} \tag{11}$$

$$\tilde{\tau}_{max} = \sqrt{\left(\frac{\tilde{\sigma}_x - \tilde{\sigma}_y}{2}\right)^2 + \tilde{\tau}_{xy}^2}$$

$$\tilde{\sigma}_{VM} = \sqrt{\tilde{\sigma}_x^2 + \tilde{\sigma}_y^2 - \frac{\tilde{\sigma}_x \tilde{\sigma}_y}{2} + 3\tilde{\tau}_{xy}^2}$$

Note that calculations in Eqs. 9-11 are inexpensive and that all the nonlinearities associated with these equations are captured explicitly. Also note that the cross coupling between the plate element's design dimension, t, and all the other design variables, of which $\{\tilde{M}\}$ is a function of, is captured.

CALCULATION OF APPROXIMATE STRAINS

As before, the six approximate element forces are first calculated using a Taylor Series expansion. Noting that the forces and midplane strains and curvatures, ε_0, are related by

$$\{M\} = [D][B]\{u\} = [D]\{\varepsilon_0\} \tag{12}$$

The approximate element midplane strains and curvatures are then calculated as

$$\{\tilde{\varepsilon}_0\} = \begin{Bmatrix} \tilde{\varepsilon}_{0x} \\ \tilde{\varepsilon}_{0y} \\ \tilde{\gamma}_{0xy} \\ \tilde{\kappa}_x \\ \tilde{\kappa}_y \\ \tilde{\kappa}_{xy} \end{Bmatrix} = [D]^{-1}\{\tilde{M}\} \tag{13}$$

Since [D] consists of two 3x3 symmetric matrices its inversion is relatively inexpensive. The approximate surface strains are then calculated using

$$\{\tilde{\varepsilon}\} = \begin{Bmatrix} \tilde{\varepsilon}_x \\ \tilde{\varepsilon}_y \\ \tilde{\gamma}_{xy} \end{Bmatrix} = \begin{Bmatrix} \tilde{\varepsilon}_{0x} \\ \tilde{\varepsilon}_{0y} \\ \tilde{\gamma}_{0xy} \end{Bmatrix} - z\begin{Bmatrix} \tilde{\kappa}_x \\ \tilde{\kappa}_y \\ \tilde{\kappa}_{xy} \end{Bmatrix} \tag{14}$$

Finally the approximate principle, maximum shear, and Von Mises strains are calculated using

$$\tilde{\varepsilon}_I = \frac{\tilde{\varepsilon}_x + \tilde{\varepsilon}_y}{2} + \sqrt{\left(\frac{\tilde{\varepsilon}_x - \tilde{\varepsilon}_y}{2}\right)^2 + \tilde{\gamma}_{xy}^2}$$

$$\tilde{\varepsilon}_{II} = \frac{\tilde{\varepsilon}_x + \tilde{\varepsilon}_y}{2} - \sqrt{\left(\frac{\tilde{\varepsilon}_x - \tilde{\varepsilon}_y}{2}\right)^2 + \tilde{\gamma}_{xy}^2} \tag{15}$$

$$\tilde{\gamma}_{max} = \sqrt{\left(\tilde{\varepsilon}_x - \tilde{\varepsilon}_y\right)^2 + \tilde{\gamma}_{xy}^2}$$

$$\tilde{\varepsilon}_{VM} = \sqrt{\frac{4}{9}\left(\tilde{\varepsilon}_x^2 - \tilde{\varepsilon}_y^2 - \tilde{\varepsilon}_x\tilde{\varepsilon}_y\right) + \frac{1}{3}\tilde{\gamma}_{xy}^2}$$

As for the stresses, note that calculations in Eqs. 13-15 are inexpensive, all the nonlinearities associated with these equations are captured explicitly, and that the cross coupling between the plate elements design dimension, t, and all the other design variables is captured.

PROBLEM STATEMENT

The optimization problems presented in this paper are stated as

$$\text{Minimize } W(X) \tag{16a}$$

Subject to;

$$\frac{\sigma_{VM}(X)-\sigma_a}{\sigma_a} \leq 0 \qquad (16b)$$

or

$$\frac{\varepsilon_{VM}(X)-\varepsilon_a}{\varepsilon_a} \leq 0 \qquad (16c)$$

and

$$X_i^L \leq X_i \leq X_i^U \qquad (16d)$$

where W is the structural weight, X is the vector of design variables, X^L and X^U are the bounds on these design variables, σ_{VM} and ε_{VM} are the Von Mises stresses and strains respectively, and σ_a and ε_a are the allowable stress and strain.

OPTIMIZATION PROCEDURE

The overall optimization procedure is:

1. Perform a finite element analysis of the structure.
2. Evaluate the constraints and retain those that are active and potentially active.
3. Perform a sensitivity analysis of the element forces with respect to the intermediate design variables for the elements with retained constraints.
4. Formulate and solve the approximate optimization problem.
5. If the design has not converged then update the analysis model and go to step 1.

The procedure for the solution of the approximate problem in step 4 above is:

a. Calculate the intermediate variables (D in Eq. 9 in this case) from the design variables, t.

b. Evaluate the Taylor Series approximations for the element forces and weight.

c. Calculate the approximate element stresses and strains using Eqs. 10-15.

d. Evaluate the objective function and constraints (Eq. 16).

e. Call the DOT optimizer [8] to determine the new design variable values.

f. If the approximate problem has converged then go to step 5 of the overall optimization procedure, else go to step a.

EXAMPLES

Examples are offered here to demonstrate the new method. All examples were solved by the GENESIS structural design program [9].

Pressure Loaded Plate This example consists of finding the minimum weight design of a 100 x 100 inch clamped plate subject to a uniform pressure load. The plate is made of steel with a Young's Modulus of 29×10^6, Poisson's ratio of 0.208 and weight density of 0.283 lb/in^3. The initial thickness of the plate is 2.0 in. This problem was presented in reference 6 where it was loaded by a 100 psi. pressure load and subject to an allowable Von Mises stress of 10,000 psi. In this paper the pressure load will be reduced to 95.6 psi in order to achieve the same initial maximum stress in the plate. The difference in the initial stresses is due to the use of different finite elements.

The analysis and design model consists of a 25 finite element quarter model of the plate with the appropriate symmetry boundary conditions and is shown in Figure 1. The lower and upper bounds on the plate thicknesses are 0.01 and 10.0 in. respectively. No move limits were used on the design variables.

1 Design Variable Model In this design the plate has uniform thickness. Since there is no force redistribution the approximations are exact and the optimum design is found in 1 design cycle. The optimum design has a quarter model weight of 2954 lb. and a thickness of 4.17 in.

15 Design Variable Model In this design the 25 plate thicknesses are linked so that they are controlled by 15 independent design variables in the configuration shown in Figure 1. This design model was also used in Ref. 6. The final design and design cycle history are presented in Tables 1 and 2 respectively. The final design and design cycle history plot from reference 6 are also shown. If the direct stress approximation is used to solve this problem 17 design cycles are required [6]. The approach presented in this work requires only 12 design cycles. The final thickness distribution is shown in Figure 2. Note that the final design has the thickest elements at the center of the plate and along the middle of the sides, as would be expected.

TABLE 1. Pressure Loaded Plate: Final Designs

Design Variable	THICKNESS		
	15 DV	Ref. 6	6 DV
1	0.01	0.04	0.01
2	0.01	0.86	4.62
3	2.50	2.61	6.35
4	3.64	3.71	4.72
5	4.74	4.92	1.95
6	0.02	0.01	6.06
7	0.01	0.23	6.06
8	0.02	0.34	3.94
9	0.02	0.35	1.52
10	2.06	0.59	4.72
11	0.01	4.46	2.51
12	4.41	2.35	1.65
13	4.33	0.14	1.23
14	4.83	5.94	1.65
15	4.41	0.68	4.32

6 Design Variable Model Although the 15 design variable model shows the accuracy and efficiency that can be gained by using the approximation function presented in this paper, it does not generate a practical design. An approach that will lead to a more realistic design involves the use of combinations of basis designs [10]. The six basis designs used in this example are shown in Figure 3. Note that R is the radial distance out from a corner of the plate. The six design variables are now the participation coefficients of each of the basis designs in the actual design. The initial design variable values are 0.0. This approach has the advantages that it generates a more practical design and has less design variables. It has the disadvantage that the final design will have a larger objective function value due to the decreased size of the design space.

The design cycle history for this example is presented in Table 3. Note the rapid convergence to a practical design in this example. The final design is shown in Table 1. The final design variable values are presented in Table 4.

TABLE 2. 15 Design Variable Pressure Loaded Plate: Design Cycle History

Design Cycle	Weight (LB)	Max. Stress (psi)
0	1,415	43,613
1	2,390	14,661
2	1,885	11,285
3	1,602	10,746
4	1,529	11,982
5	1,517	20,145
6	1,473	10,227
7	1,498	10,093
8	1,445	10,423
9	1,444	10,251
10	1,446	10,190
11	1,448	10,087
12	1,449	9,999

TABLE 3. 6 Design Variable Pressure Loaded Plate: Design Cycle History

Design Cycle	Weight (LB)	Max Stress (psi)
0	1,415	43,613
1	2,620	12,355
2	2,615	11,114
3	2,589	9,435
4	2,417	11,079
5	2,428	10,364
6	2,442	10,007

TABLE 4. 6 Design Variable Pressure Loaded Plate: Final Design Variable Values

Design Variable	Value
1	1.71
2	1.03
3	-1.05
4	0.96
5	-1.17
6	1.58
7	

Plate with a Hole

This example consists of finding the optimum hole size and shape that will produce a minimum weight design of a 20 x 20 in. biaxially loaded plate. The plate has an allowable Von Mises stress of 10,000 psi. The 0.15 in. thick plate is made of steel with a Young's Modulus of 29×10^6, Poisson's ratio of 0.208 and weight density of 0.283 lb/in^3. This example is also taken from reference 6.

The analysis and design model consist of a 30 finite element model of a quarter of the plate with the appropriate symmetry boundary conditions (see Figure 4). The initial hole is circular with a radius of 2.0 in.

The basis vector approach to shape optimization is used in this paper. In this approach the shape and size of the hole are changed and the model remeshed. Each mesh generated in this manner is a basis design and has a design variable controlling is participation in the actual design. The initial design variable values are 0.0. No move limits are used on the design variables in this example.

7 Design Variable Model In this example the plate is loaded with a balanced biaxial tension of 1500 lb. on each side (750 lb. on each side of the quarter model). The design variables are the radial locations of the seven nodes on the hole boundary. Although the load on the structure is symmetric, symmetry is not enforced. The design converges in two cycles (see Table 5) to radii of 3.61-3.64 in. As expected the final design is symmetric and the hole is nearly circular (within less than 1%). In reference 6 it was reported that 11 design cycles were required using a direct approximation of the stresses and 5 design cycles were required using the nodal forces as the approximated intermediate response quantities. It was also reported in reference 6 that the hole boundary took on a sawtooth shape in the optimum design. This is caused by zero stress modes in the finite elements [11]. This usually occurs when the nodal locations are used as design variables. A better approach is to use basis designs which represent different design shapes rather than nodal locations.

4 Design Variable Model - Balanced Load In this example the basis designs represent the four different design shapes pictured in Figure 5. The four shapes are a circular hole, a square hole, a diamond shaped hole, and an oval hole. When the plate is loaded with balanced biaxial tension the design converges in two design cycles to a circular hole with a radius of 3.59 inches (see Table 5). Note that, as expected, the only nonzero design variable is the one that controls the circular hole (see Table 6).

TABLE 5. Balanced Biaxially Loaded Plate with a Hole: Design Cycle History

Design Cycle	7 Design Variables		4 Design Variables	
	Weight (LB)	Max Stress (psi)	Weight (LB)	Max Stress (psi)
0	4.113	8,487	4.113	8,487
1	3.852	9,918	3.854	9,936
2	3.813	10,045	3.818	10,050

4 Design Variable Model - Unbalanced Load If the load in the Y direction is reduced to half of that in the X direction a different optimum design results (see Table 6). The design cycle history for this case is shown in Table 7. The optimum shape is shown in Figure 6. As expected the ratio of major and minor axes is near 2.0 (1.91).

4 Design Variable Model - Strain Constraint In this example the maximum Von Mises strain, due to the unbalanced load, is constrained rather than the maximum stress. The allowable strain value is 0.00016. The final design variable values are shown in Table 6 and the design cycle history is shown in Table 7. As in the previous example rapid convergence is observed (4 design cycles). The final design is shown in Figure 7.

TABLE 6. 4 Design Variable Biaxially Loaded Plate: Final Design Variable Values

Design Variable	Value		
	Balanced Load	Unbalanced Load (Stress Constraint)	Unbalanced Load (Strain Constraint)
1	15.9	30.2	-4.2
2	0.0	-6.2	-2.9
3	0.0	1.1	0.5
4	0.0	-0.1	-0.2

TABLE 7. Unbalanced Biaxially Loaded Plate with a Hole: Design Cycle History

Design Cycle	Stress Constraint		Strain Constraint	
	Weight (LB)	Max Stress (psi)	Weight (LB)	Max Strain x10^4 (psi)
0	4.113	10,027	4.113	2.455
1	3.720	8,718	4.200	1.468
2	3.107	12,242	4.157	1.586
3	3.260	10,446	4.155	1.593
4	3.296	10,026	4.146	1.600

CONCLUSIONS

Approximating the element forces and then explicitly calculating the approximate element stresses (strains) from these forces yields a more accurate approximation function than direct approximation of the element stresses (strains) for plate/membrane problems. The use of these highly accurate approximations leads to rapid convergence in the design process, even with ill-posed problems. Using the element forces as the approximated intermediate response quantity is less expensive and leads to better approximations than using the nodal forces as the approximated intermediate response quantity. This approximation function is highly accurate for both sizing and shape optimization problems.

ACKNOWLEDGEMENTS

The authors would like to thank Dr. Y. K. Shyy of VMA Engineering, Dr. H. Miura of NASA Ames Reseaerch Center and Professor L. A. Schmit of UCLA for their helpful insight and discussions that went into this work.

REFERENCES

1. Vanderplaats, G. N. and Salajegheh, E., "A New Approximation Method for Stress Constraints in Structural Synthesis," AIAA Journal, Vol. 27, No. 3, pp. 352-358, 1989.

2. Hansen, S. R. and Vanderplaats, G. N., "Approximation Method for Configuration Optimization of Trusses," AIAA Journal, Vol. 28, No. 1, pp. 161-168, 1990.

3. Vanderplaats, G. N. and Han, S. H., "Arch Shape Optimization Using Force Approximation Methods. Structural Optimization," Vol. 2, No. 4, pp. 193-201, 1990.

4. Kodiyalam, S. and Vanderplaats, G. N., "Shape Optimization of Three-Dimensional Continuum Structures via Force Approximation Techniques," AIAA Journal, Vol. 27, No. 9, pp. 1256-1263, 1989.

5. Vanderplaats, G. N., Kodiyalam, S. and Long, M. G., "A Two Level Approximation Method for Stress Constraints in Structural Optimization," pp. 541-545, Proceedings of the AIAA/ASME/ASCE/AHS/ASC 30th Structures, Structural Dynamics and Materials Conference, Mobile, Alabama, 1989, AIAA, Washington, D. C., 1989.

6. Moore, G. J. and Vanderplaats, G. N., "Improved Approximations for Static Stress Constraints in Shape Optimal Design of Shell Structures," pp. 161-170, Proceedings of the AIAA/ASME/ASCE/AHS/ASC 31st Structures, Structural Dynamics and Materials Conference, Long Beach, California, 1990, AIAA, Washington, D. C., 1990.

7. Vanderplaats, G. N. and Salajegheh, E., "An Efficient Approximation Technique for Frequency Constraints in Frame Optimization," International Journal for Numerical Methods, Vol. 26, pp. 1057-1069, 1988.

8. DOT User's Manual, Version 2.04, VMA Engineering, Goleta, CA, 1989.

9. GENESIS User's Manual, Version 1, VMA Engineering, Goleta, CA, 1990.

10. Pickett Jr., R. M., Rubinstein, M. F., and Nelson, R. B., "Automated Structural Synthesis Using a Reduced Number of Design Coordinates," AIAA Journal, Vol. 11, No. 4, pp. 489-494, 1973.

11. Braibant, V. and Fleury, C., "Shape Optimal Design using B-Splines," Computational Methods and Applications in Mechanical Engineering, Vol. 44, pp. 246-267, 1984.

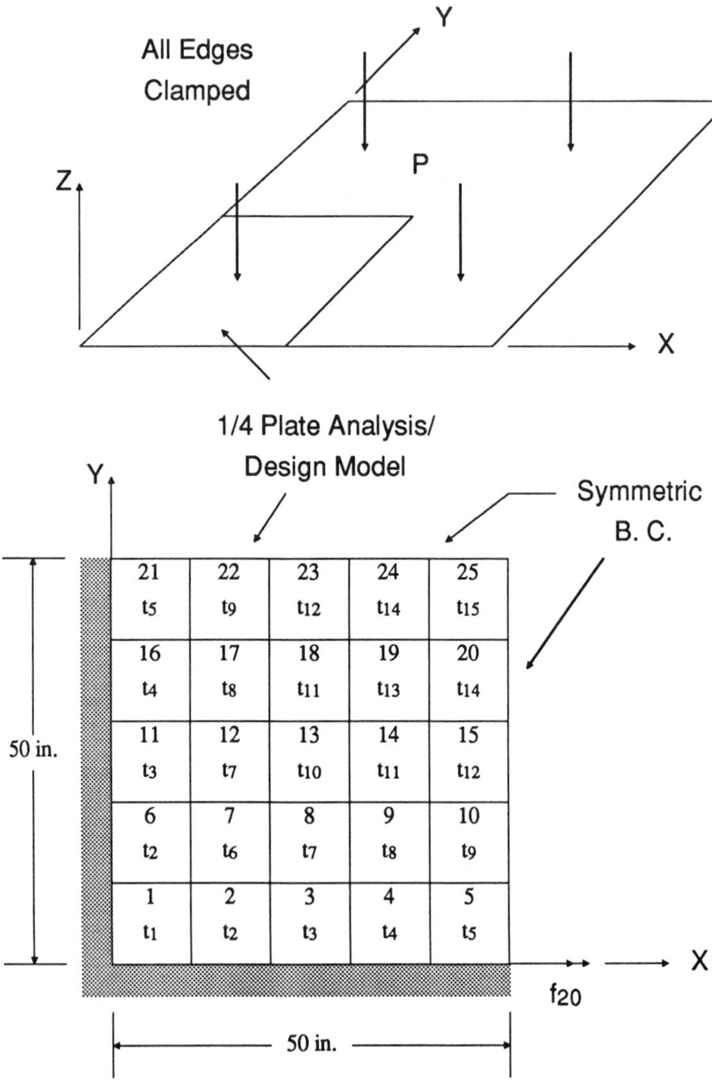

Figure 1. Pressure Loaded Plate

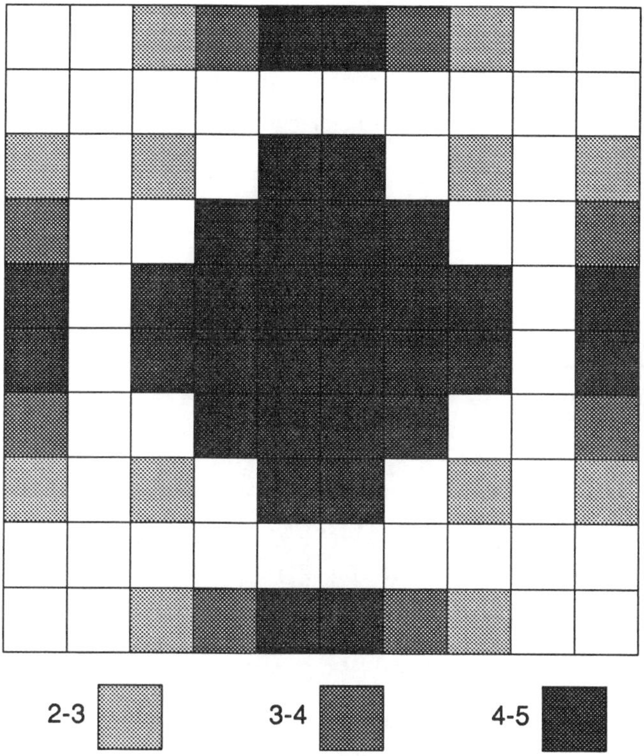

Figure 2. Final Thickness Distribution

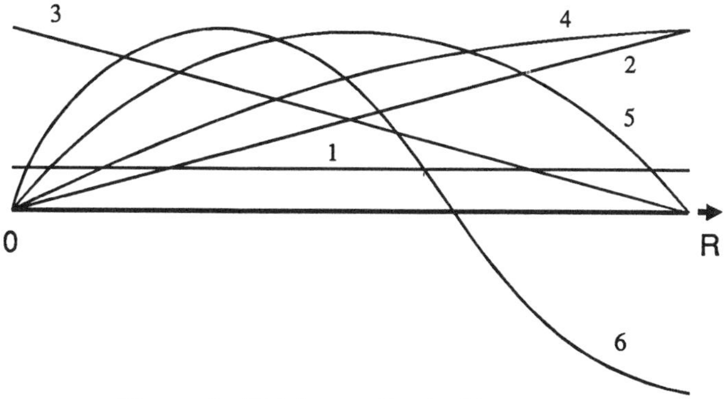

Figure 3. Thickness Basis Shapes

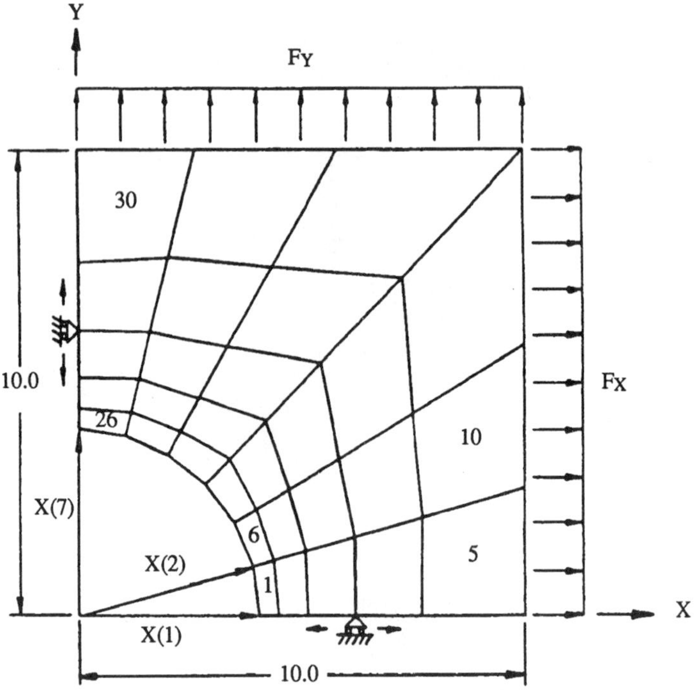

Figure 4. Plate with a Hole

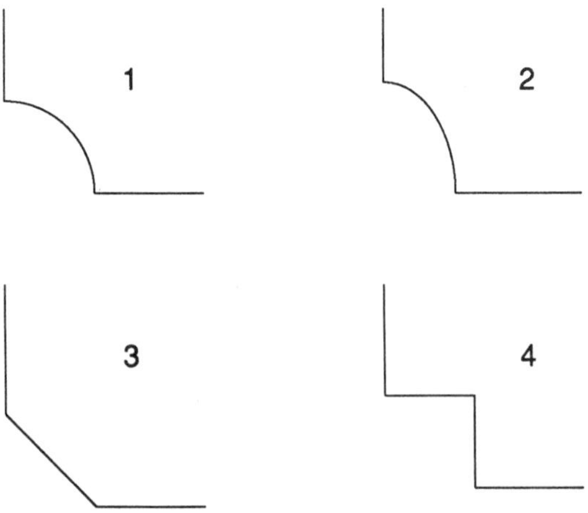

Figure 5. Basis Shapes

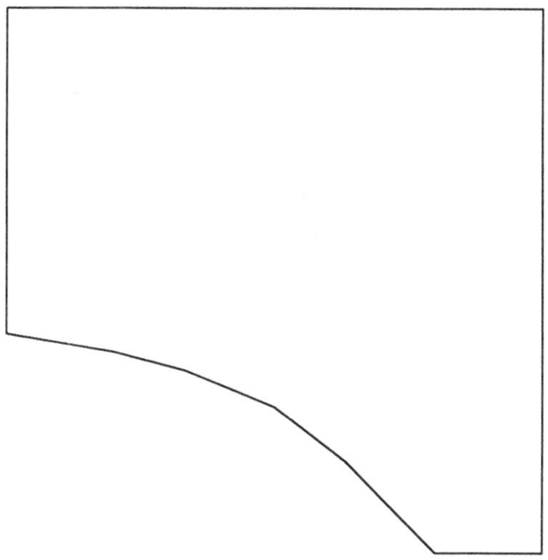

Figure 6. Final Shape. Unbalanced Load

Figure 7. Final Shape. Strain Constraint

Techniques to Minimize the Cost of Shape Optimization of Stress Concentration Regions

J.C. Thompson, O.K. Bedair

Department of Civil Engineering, University of Waterloo, Waterloo, Ontario, N2L 3G1, Canada

ABSTRACT

This paper presents a sequential linear programming-based technique for shape optimization of stress concentration regions. Approximate techniques are derived which dramatically reduce the computation cost involved in the calculation of sensitivity coefficients.

The optimization algorithm is applied to a 90° V-notch with an initially constant root radius. Local loading and geometric similarity criteria are established which permit the generalization of the result to other stress concentration regions.

INTRODUCTION

One of the serious concerns of the designers of engine components, aircraft, off-shore oil rigs, pressure vessels and other structural components subject to cyclic loadings is the potential for initiation and subsequent growth of fatigue cracks in stress concentration regions (SCR). The serious consequences of the development of such cracks often justifies the extra cost of the effort to improve the fatigue life by optimizing the shape of the SCR, i.e., to determine the shape which minimizes the peak stress. Since this optimization problem is inherently non-linear, one must expect that the optimum shape will have to be achieved by an iterative procedure. The purpose of this paper is to show how the cost of this optimization procedure can be minimized through the use of various simplifying, yet very accurate approximations which exploit certain fundamental characteristics of the stress fields in stress concentration regions.

This research forms part of a larger project, the aim of which is to produce design charts, tables and other aids for designers, to allow them simply to " look up " the optimum shape and the associated stress concentration factor once they have carried out a single analysis of the original (non-optimized) SCR and determined the intensities of the symmetric and non-symmetric " modes " of the local stress fields.

PROBLEM FORMULATION

Consider the SCR depicted schematically in Fig.1. This could represent, for example, the welded fillet region between a chord and column members of an off-shore oil rig, the machined fillet of a crankshaft or turbine blade, or the root region of the threaded end closure cap of a high pressure vessel. Manufacturing, design, and/or other considerations usually limit the portion of the boundary which can be modified to a region such as that between points A and B. Because of the analytical intractability of most realistic problems, they are usually solved by approximate techniques, the most common being the boundary and finite element methods. In both cases, the boundary is defined in terms of the locations of the nodal points, N of which lie between A and B. Optimizing the shape of the boundary then becomes the problem of determining the unique set of locations of these $1 \leq j \leq N$ nodes which minimizes a stress σ that is greater than or equal to the boundary stress at each of $1 \leq i \leq M$ preselected locations which may, but need not coincide with the N nodes, i.e.,

minimize $\quad \sigma$

subject to

$$\sigma - \sigma_i \geq 0 \qquad (\text{all } i). \tag{1}$$

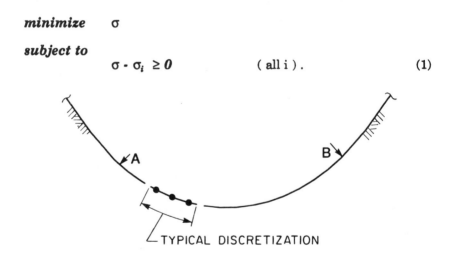

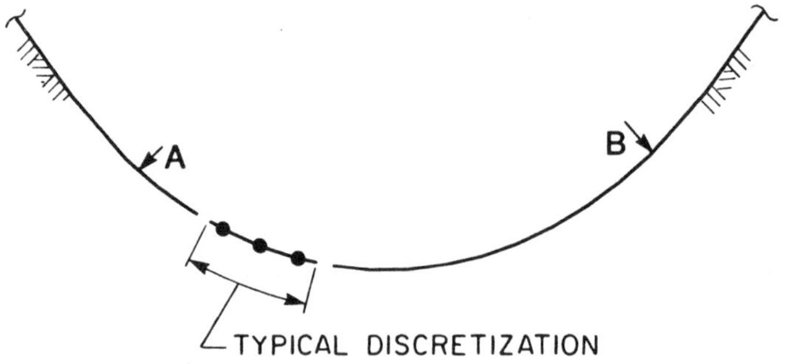

Fig.1 Schematic of a typical stress concentration region, a discretized segment of the boundary, and the segment to be optimized.

Figure 2 depicts, with greatly exaggerated curvature, a three node segment of the discretized boundary. The open and solid circles depict the respective locations of nodes j-1, j and j+1 at the kth and $k+1^{st}$ iterative

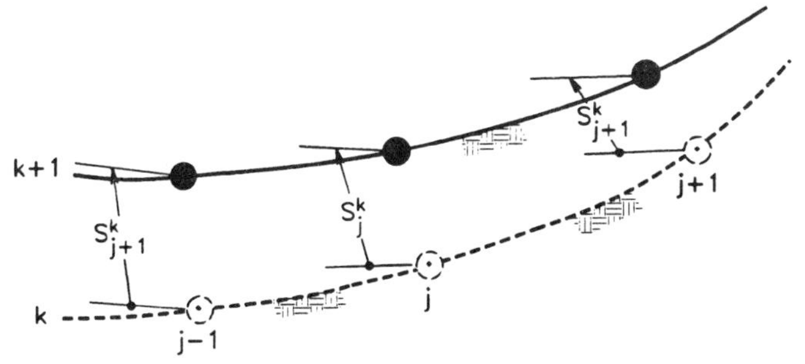

Fig.2 Schematic of a boundary segment at the kth and $k+1^{st}$ stages.

stages. The S^k_{j-1} , S^k_j and S^k_{j+1} denote the movement of these nodes from their position at the kth stage. Usually the movement of each node is restricted to a direction normal or roughly normal to the boundary at that point of the kth stage.

Frequently, there is a manufacturing constraint requiring that the curvature of the boundary remain convex, i.e., no undercuts are permitted. In such cases, this can be met by imposing additional geometrical constraints which ensure that the movement of each of the $1 \leq j \leq N$ nodes along segment AB does not take it beyond the straight line joining nodes j-1 and j+1. These constraints can be denoted symbolically as

$$\tilde{S}_j - S_j > 0 \tag{2}$$

where each $\tilde{S}_j$ is inherently a non-linear function of the movements S_{j-1} and S_{j+1} .

In order to take advantage of any one of the well established linear programming techniques, the inherently non-linear constraints defined by (1) and (2) must be linearized. For sufficiently small displacements S_j of each of the $1 \leq j \leq N$ nodes from their locations in the kth state, the σ_i^{k+1} and $\tilde{S}^k_j$ of (1) and (2) can be approximated as

$$\sigma_i^{k+1} = \sigma_i^k + \sum_{i=1}^{M} \sigma^k_{i,j} \cdot S_j \tag{3}$$

$$\tilde{S}^k_j = \Delta^k_j + S^k_{j,j-1} \cdot S^k_{j-1} + S^k_{j,j+1} \cdot S^k_{j+1} \tag{4}$$

In (3), $\sigma^k_{i,j} = \partial\sigma^k_i / \partial S^k_j$, the sensitivity of the boundary stress at point i to movement of node j. As can be seen from Fig.3, Δ^k_j is the maximum distance node j could move if nodes j-1 and j+1 did not move; the $S^k_{j,j-1}$, $S^k_{j,j+1}$ reflect the sensitivity of $\tilde{S}_j$ to movements of nodes j-1 and j+1.

The S_j of (3) and (4) are now the design variables for the linearized problem.

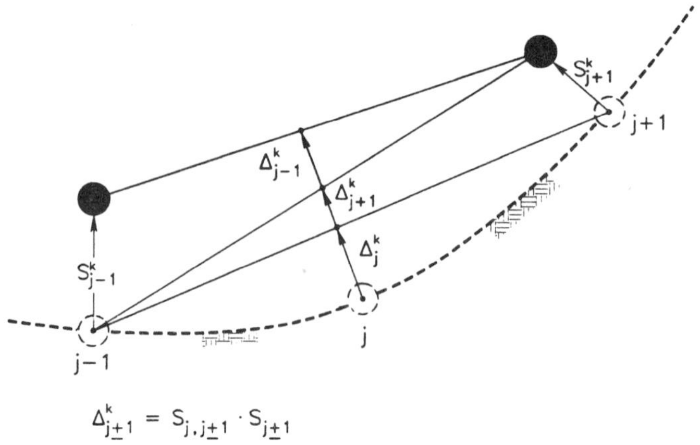

Fig.3 Schematic of the displacements involved in the convexity constraint of node j.

PROPERTIES OF THE SENSITIVITY COEFFICIENTS

From Fig.4 one can readily deduce the properties of the $\sigma^k_{i,j}$ sensitivity coefficients that will be exploited in subsequent sections to achieve an efficient optimization technique. The upper figure depicts, again with greatly exaggerated curvature, a *virtual notch-like* perturbation of the boundary created by a *virtual displacement* of node j by an amount δ_j. The

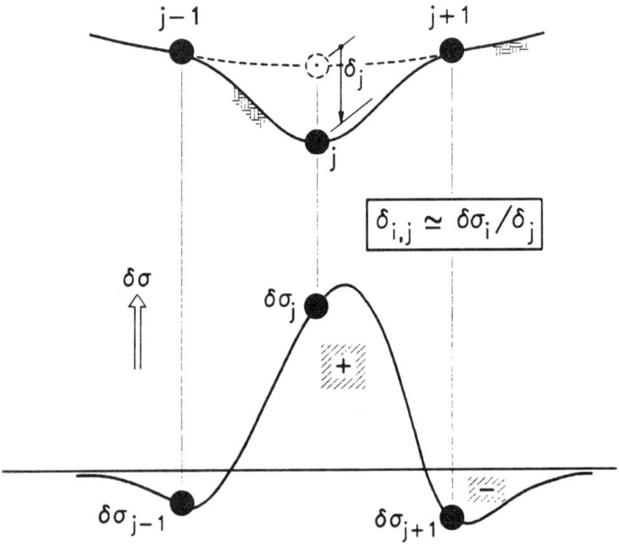

Fig.4 Schematic of the virtual notch-like boundary perturbation due a virtual displacement of node j (upper) and virtual perturbation of the boundary stress (lower).

resulting virtual perturbation of the boundary stress, $\delta\sigma$, has the characteristics shown. Specifically, $\delta\sigma$ has a maximum value at or near the location of the displaced node because of the stress concentration effect. With increasing distance in either direction from this node, $\delta\sigma$ passes through zero to a minimum value in the vicinity of the each end of the *notch* where the " *shielding* " effect of the notch is maximum. Beyond these points, $\delta\sigma$ approaches zero monotonically, reflecting the rapidly diminishing influence of the notch with increasing distance from the perturbed point.

COMPUTATIONAL EFFICIENCY

As mentioned above, the inherent non-linear property of the class of problem being considered requires that a number of iterations (i.e., complete reanalyses) be performed to achieve an optimal solution. Overall efficiency requires that the operations at each iterative stage be minimized, and that the number of iterations also be minimized.

Several efficient techniques [1,2] have been developed to determine the $\sigma^k_{i,j}$ of Eq.(3) for each of the $0 \leq k \leq K$ iterations required to reach a sufficiently optimized shape. The authors have chosen to use a boundary element-based approach, simply because they had access to, and were completely familiar with the source code for a programme that had been thoroughly tested and proven to give very accurate analysis for stress concentration problems.

For linearly elastic problems of the type being considered, the boundary element techniques inherently requires the solution of a set of linear equations of the form

$$\mathbf{A}\,\mathbf{X} = \mathbf{B} \tag{5}$$

where **A** is a non-symmetric fully populated matrix and **B** is the vector of prescribed boundary conditions. Solving Eq.(5) gives a vector **X** from which the σ^k_i are subsequently determined. To obtain the exact values of $\sigma^k_{i,j}$ for all of the N possible nodal perturbations requires the solution of an additional N similar sets of equations, denoted as

$$\mathbf{A}_j\,\mathbf{X}_j = \mathbf{B}_j \tag{6}$$

Any approximation which avoids the necessity of having to solve the N+1 sets of equations defined by (5) and (6) without significant loss in accuracy is highly desirable. To achieve this, it is convenient to note that each of Eq.(6) may be rewritten as

$$(\mathbf{A} + \delta\mathbf{A}_j)\,(\mathbf{X} + \delta\mathbf{X}_j) = (\mathbf{B} + \delta\mathbf{B}_j) \tag{7}$$

where $\delta\mathbf{A}_j$, $\delta\mathbf{X}_j$ and $\delta\mathbf{B}_j$ denote the respective changes in the **A**, **X** and

B of Eq.(5) as a result of the perturbation of node j. If the second order product $\delta A_j \, \delta X_j$ is neglected, a first order approximation of X_j is given by

$$\mathbf{A}\,\mathbf{X}_j \cong \mathbf{B}^*_j \qquad \text{(all j)} \tag{8}$$

where $\mathbf{B}^*_j = (\mathbf{B} + \delta\mathbf{B}_j - \delta\mathbf{A}_j\,\mathbf{X})$. The $\mathbf{B}^*_j$ may be interpreted as fictitious boundary conditions applied to the unperturbed geometry. Thus instead of solving N+1 sets of equations, one has the much less costly task of solving only a single set of equations for N+1 sets of boundary conditions,i.e.,

$$\mathbf{A}\,\{\mathbf{X}, \mathbf{X}_1, ..., \mathbf{X}_N\} \cong \{\mathbf{B}, \mathbf{B}^*_1, ..., \mathbf{B}^*_N\} \tag{9}$$

Note that the entries of $\delta\mathbf{A}_j$ and $\delta\mathbf{B}_j$ are zero except along the rows and columns associated with node j. ***Avoiding the computation and multiplication of these zero entries gives a considerable saving in computational time.*** Furthermore, in the following section we shall show that it is not even necessary to solve for all of the $\mathbf{X}_1, ..., \mathbf{X}_N$. ***From the coefficients computed from a subset of these cases, one can determine the "missing "*** $\sigma^k_{i,j}$ ***by a simple interpolation / extrapolation algorithm.***

Finally, we shall show that the optimized shape can be achieved within three or four iterations, even with the various approximations described above.

CASE STUDIES

GEOMETRY

For reasons that will be explained later, we considered the problem of optimizing the shape of the segment AB at the root of a symmetric, 90° V-notch in the traction free edge of a semi-infinite plate shown in Fig.5.

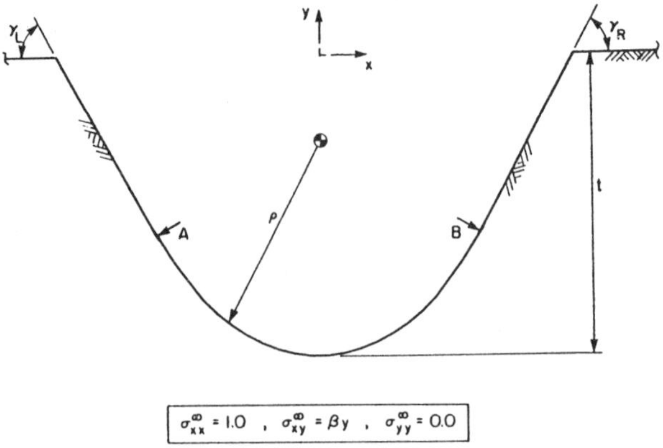

Fig.5 Geometric details of the non-optimized notch; $t/\rho = 6.0$, $\gamma_L = \gamma_R = 45°$.

For an analysis by means of the boundary element method, the initially circular segment AB and each of the flanking sides were approximated by 15 and 19 linear elements, respectively. At all stages in the optimization process, the nodes along AB were required to satisfy the linearized convexity constraint given by Eqs.(2) and (4).

LOADING

The magnitudes of the symmetric and anti-symmetric components of the high gradient stress fields around the root of the notch were varied by means of the prescribed values of the uniform σ^{∞}_{xx} and linearly increasing σ^{∞}_{xy} stresses in the pre-notched plate. Optimized shapes were determined for the values of $\sigma^{\infty}_{xy} / \sigma^{\infty}_{xx}$ ranging between 0.0 and 0.36 at the depth of the notch root.

RESULTS

Sensitivity Analysis The qualitative properties of the sensitivity coefficients anticipated in a previous section are clearly evident in each of the $\sigma^{k}_{i,j}$, j = const. curves of the upper half of Fig.6. Each curve, corresponding to the perturbation of the jth node indicated by the legend, has a maximum value at the $i = j^{th}$ centroid (the centroid adjacent to node j nearer the notch root). With increasing distance on either side of this maximum, the value of $\sigma^{k}_{i,j}$ decreases through zero to a negative minimum somewhere within the range $1 < | i - j | < 3$ and then approaches zero asymptotically, being essentially zero for all $| i - j | > 5$. By assuming that terms beyond this are zero, the cost of calculating them and any subsequent operations can be eliminated.

Probably more important, however, from the economics that can be affected, is the observation that the quantitative variation of the $\sigma^{k}_{i,j}$, j=const. curves is quite smooth. Constructing curves through $\sigma^{k}_{i,j}$ values at the same i - j = const. locations relative to the perturbed node gives the almost linear curves as shown in the lower half of Fig.6. Clearly then, one need not compute all the $\sigma^{k}_{i,j}$ at any stage. From the values computed for a subset of the $1 \leq j \leq N$ nodes, a sufficiently accurate estimation of the " missing " values can be obtained by a simple interpolation / extrapolation algorithm. Similar results were found at all the subsequent k^{th} iterations.

Finally, it should be noted that an insignificant inaccuracy was introduced by using the approximate $\mathbf{X}_j$ of Eq.(8) to compute the $\sigma^{k}_{i,j}$ rather than the exact $\mathbf{X}_j$ of Eq.(6). Tests showed that convergence was invariably reached within three to four iterations.

$\sigma^{\infty}_{xy} / \sigma^{\infty}_{xx}$ ***= 0.0*** The evolution of the boundary stress and shape during the optimization process for the $\sigma^{\infty}_{xy} / \sigma^{\infty}_{xx}$ = ***0.0*** is shown in Fig.7. The magnitudes of the reduction of the peak stress and of the movement of the boundary are greatest in the first iteration; the magnitudes of

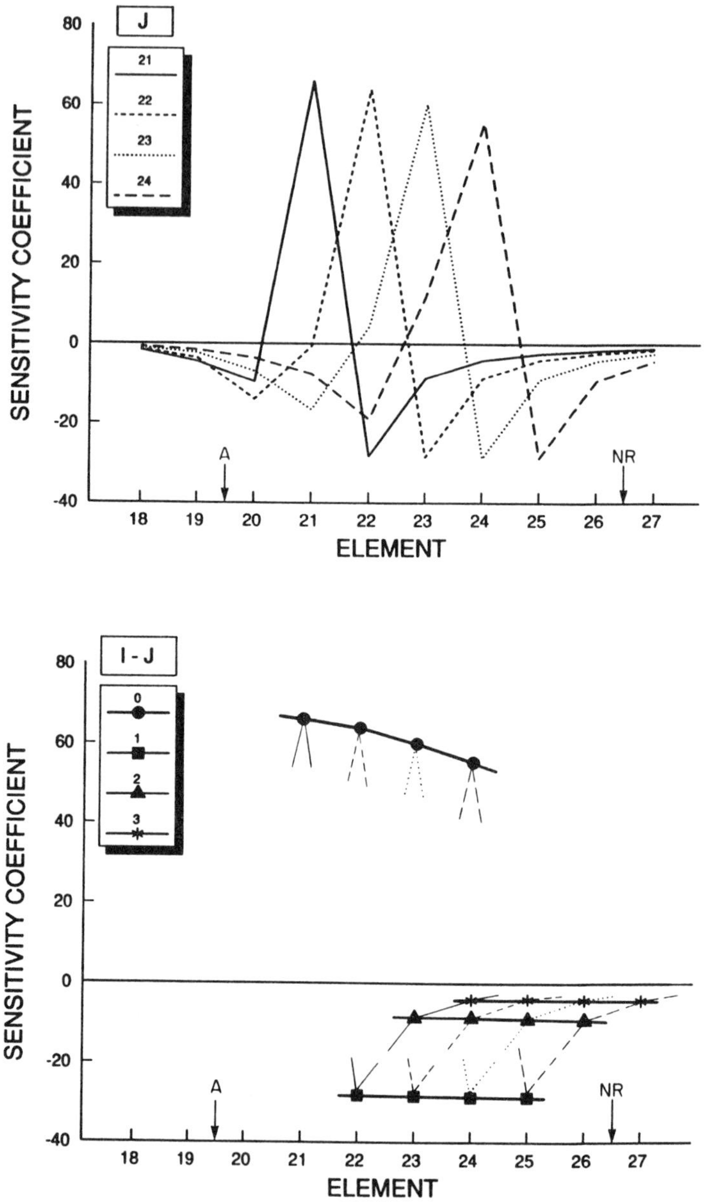

Fig.6 Variation of the stress sensitivity coefficients along the boundary for perturbations, in turn, of four nodes in the stress concentration region between **A** and other notch root (NR); j= const. (upper) , i - j = const. (lower) , for $\sigma^{\infty}_{xy} / \sigma^{\infty}_{xx} = 0.0$, and k = 0 .

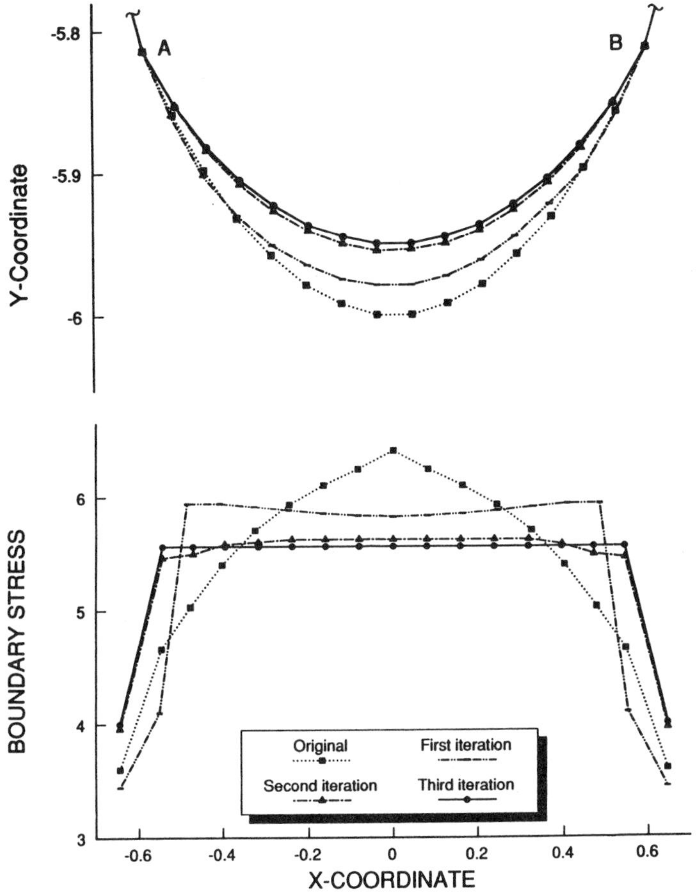

Fig.7 Evolution of the boundary stress and shape during optimization for $\sigma^{\infty}_{xy} / \sigma^{\infty}_{xx} = 0.0$.

subsequent changes decrease rapidly with each successive iteration. For all practical purposes, the optimum shape was achieved by the third iteration.

Note also that the portion of the SCR over which the peak stress has a constant value increases each time the magnitude of the peak stress decreases. These decreases are achieved by reducing the curvature in the initially most highly stressed central segment of the SCR while simultaneously increasing the curvature in the initially lower stressed outer segments. Note also that the movements required to optimize the shape are relatively small, the maximum of these being about 5% of the initial root radius.

$\sigma^{\infty}_{xy} / \sigma^{\infty}_{xx} > 0.0$ The optimized boundary shape and stress distributions for loading cases involving both symmetric and anti-symmetric stress fields are shown in Fig.8. In each case, the boundary shape and stress distribution are skewed, with the regions of minimum curvature and maximum stress occurring on the side of the notch where the boundary stresses from σ^{∞}_{xy} and σ^{∞}_{xx} stress fields are additive.

Note also that with increasing $\sigma^{\infty}_{xy} / \sigma^{\infty}_{xx}$ values, there is a slight decrease in the portion of the notch boundary over which the peak stress is maximum, and an increasing portion of the boundary which has reached the tangent to one of the flanking side of the V-notch.

Finally it should be noted that the convexity constraint of Eqs.(2) and (4) prevented the problem of " wiggling " or " jogging " of the boundary reported recently [3,4].

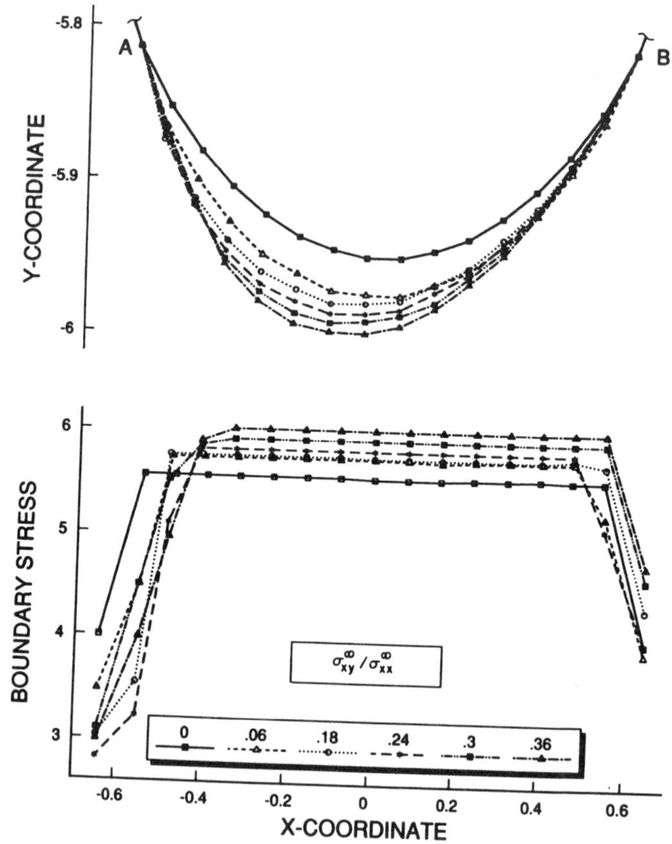

Fig.8 Optimized boundary stress and shape for $0.0 \leq \sigma^{\infty}_{xy} / \sigma^{\infty}_{xx} \leq 0.36$.

In the following section we shall show that the optimum shapes determined for a V-notched plate are identical to the optimum shapes for SCR in other bodies, provided certain local geometric and loading similarity requirements are met.

GENERALIZATION

By means of asymptotic analysis techniques [5], the stress field σ, in the neighbourhood of the root of the V-notch or any other SCR can be expressed in infinite series form as

$$\sigma\,(x,y) = \sum_{i=1}^{\infty} \alpha_i \cdot \hat{\sigma}_i\,(x,y) \tag{10}$$

The α_i and $\hat{\sigma}_i$ may conveniently be interpreted as the local intensity and spatial distribution of the ith mode of σ. The α_i depend principally upon conditions (geometry, load, ... etc.) beyond the immediate neighbourhood of the SCR; as such, they are very insensitive to minor variations of the geometry of the SCR during the optimization process. In contrast, the $\hat{\sigma}_i$ depend almost exclusively on the geometry of the traction-free portion of the boundary in and adjacent to the SCR; as such, they are very sensitive to perturbations within the region.

Since the contribution of the third and higher modes σ is invariably negligible, the only criteria necessary for the optimum shapes of two SCR to be geometrically similar are 1) identical α_1 / α_2 ratios and 2) geometric similarity of the boundaries in and adjacent to the SCR. This means that the results obtained for the 90° V-notch, for example, apply for the fillet region joining any boundary segments that are perpendicular to each other.

Once the optimum shape for a particular notch geometry has been determined for the full range of α_2 / α_1 values, the cost of optimization of all other geometrically similar SCR is dramatically simplified. Data from the notch region of the analysis of the original (non-optimized) geometry is post-processed to obtain α_2 / α_1 intensity. The optimized shape is then simply determined from pre-established design aids or look-up tables, both of which could be stored electronically.

CONCLUSION

Several approximate techniques have been proposed to minimize the cost of the first order sensitivity analysis carried out to optimize the shape of stress concentration regions. Tests have shown the accuracy of these techniques.

A method based on the asymptotic analysis of the stress field in the SCR is applied to generalize the results of a particular case for all problems having geometrically similar notch root regions. This generalization virtually eliminates the computation costs of the optimization process.

REFERENCES

[1] Haftka, R.T. and Adelman, H. M., *Recent Developments in Structural Sensitivity Analysis,* Struct. Opt., Vol.1, pp. 137-151, 1989.

[2] Haug, E. J., Choi,K. K. and Komkov, V., *Design Sensitivity Analysis of Structural Systems,* Academic Press, 1986.

[3] Miyamoto, Y., Iwasaki, S., Deto, H. and Sugimoto, H., *On Shape Opti mization Study With Automatic Mesh Division of Two Dimensional Bodies by BEM,*(Ed. Tanaka, H. and Cruse, T.A.),Proc. of the 1st joint Japan/US symposium on BEM,Tokyo,Japan,1988. Pergamon Press,1988.

[4] Zhao, Z. and Adey, R.A., *Shape Optimization- A Numerical Consider ation,*(Ed. Brebbia, C.A. and Connor, J.J.), pp.195-208, Proc. of the 11th Int. Conf. on BEM, Cambridge, USA, 1989. Springer-Verlag,1989.

[5] Thompson, J.C., *Analysis of Stress Concentration Data by Asymptotic Techniques.,* Proc. of V Int. Cong. on Experimental Mechanics, Montreal, pp. 758-764, 1984.

Approximation Method in Optimization of Stiffened Plate Structure

K. Yamazaki

Department of Mechanical Systems Engineering, Kanazawa University, 2-40-20 Kodatsuno, Kanazawa, 920 Japan

ABSTRACT

A new intermediate variable approximation method for stiffened plate structures will be suggested and minimum weight design under displacement and stress constraints will be discussed. At first, an effective direct differentiation method of sensitivity analysis with isoparametric plate and shell elements for thickness and geometric design variables is suggested and numerical results for a simple typical shell structure is compared with theoretical and finite difference values. Next, the approximation method of intermediate variables is considered for thickness and geometric dimensions of plate and stiffener to construct a wide range approximation model. For the displacement and stress constraints, it is found that the direct series expansion using intermediate variables of thickness and geometric dimensions gives a nice approximation for the typical stiffened plate structures. The approximation method suggested here is applied to the optimization of the practical stiffened plate structures and some numerical results will be shown.

INTRODUCTION

The objective of the structural optimization is to minimize structural weight subject to some maximum(or minimum) allowable stress and displacement as well as gage sizes. The stiffened plate and shell structures may be adopted for constructing large scale structures because of rigidity and light weight. For these type of structures, the optimization procedure requires large storage and computer time even if the layout of the stiffeners will be given in advance and the thicknesses and the dimensions of the plate and stiffeners are taken as the design variables, since the structural analysis and its sensitivity analysis themselves will be large scale ones.

For the efficient implementation of the optimization, an effective and exact sensitivity analysis and a good approximation method are required to reduce the iteration of the mathematical programming procedure and of the structural analysis. The direct differentiation method of the sensitivity analysis is more applicable than the material derivative method for general use. A semi-analytic method in which the derivatives of the stiffness matrix are evaluated by the finite difference is frequently applied because of the convenience of the programming and efficiency of the computation, when the stiffness matrix is evaluated by the numerical integration. However, in this paper we will adopt the analytic direct differentiation procedures which will be formulated for the curved isoparametric shell element by extending

Brockman and Lung's formulas[1] for the flat plate and shell isoparametric elements. The efficiency of the procedures would be competitive with the semi-analytic ones for the isoparametric finite elements.

Taylor series approximation(TSA) has been used to form approximate sub-problems to the actual design problem. Schmit and Miura[2] originally suggested the use of reciprocal variables which can form a wide range approximation for the framework. Starners and Haftka[3], and Fleury and Braibant[4] have shown that a hybrid constraint using mixed variables yields a more conservative approximation. Vanderplaats and Salajegheh[5] have suggested, for stress constraints, to use a Taylor series to approximate the internal loads instead of the stresses themselves.

For the stiffened plate structures we will observe the influence of the design variables such as thickness of the plate, that of the stiffener, stiffener height and flange width into the nonlinearity of the static response constraints. Then a Taylor series approximation by using the reciprocal variables will be suggested for the displacement and stress constraints. Some numerical results using the reciprocal variables will be compared with the results of direct variables to demonstrate the increase of the convergence rate to the optimum solutions.

SENSITIVITY ANALYSIS WITH ISOPARAMETRIC SHELL ELEMENTS

Derivatives of Shape Function and Jacobian Determinant

Let us consider a curved eight-node isoparametric shell element as shown in Fig.1. In the isoparametric finite shell elements, the co-ordinates x_i and the displacement u_i at any point (ξ_1,ξ_2,ξ_3) are interpolated by the same shape function $N_J(\xi_1,\xi_2)$ for node J as

$$x_i(\xi_1,\xi_2,\xi_3) = N_J x_{iJ} + \xi_3 N_J g_{iJ} \qquad (1)$$

$$u_i(\xi_1,\xi_2,\xi_3) = N_J u_{iJ} + \xi_3 N_J \Theta_{iJ} \qquad (2)$$

in which x_{iJ} and u_{iJ} are the co-ordinate x_i and the displacement u_i at nodal point J in an element, and the parametric co-ordinate ξ_3 refers to the perpendicular direction to the mid-surface of the shell. Repeated indices must be summed up as

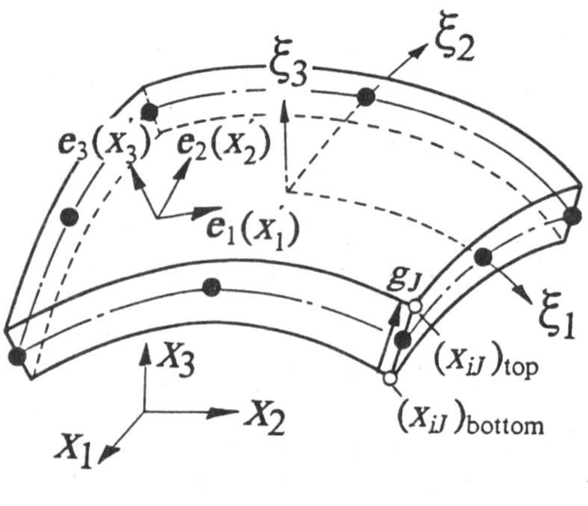

Fig.1 Quadratic isoparametric shell element model.

usual tensor symbolism manner, and lower-case Latin indices refer to the Cartesian co-ordinate directions, upper-case indices to the nodes of an element. The vector g_J is a vector in the ξ_3 direction defined as

$$g_{iJ} = t_J \, v_{3i}{}^J/2 \tag{3}$$

where $v_{3i}{}^J$ is a unit vector component in ξ_3 direction and t_J is the thickness at node J. Θ_{iJ} is a component of rotation vector defined as

$$\Theta_{iJ} = t_J(\theta_{2J} \, v_{1i}{}^J - \theta_{1J} \, v_{2i}{}^J)/2 \tag{4}$$

where θ_{1J} and θ_{2J} denote the rotations of $v_3{}^J$ about the vectors $v_1{}^J$ and $v_2{}^J$ at node J defined as

$$v_1^J = \frac{i \times g_J}{|i \times g_J|}, \qquad v_2^J = \frac{g_J \times v_1^J}{|g_J \times v_1^J|} \tag{5}$$

in which i is a unit vector in x_1 direction (or x_2). Then the Jacobian matrix can be expressed as

$$[J] = [x_{m,\alpha}] = [N_{J,\alpha} x_{mJ} + (\xi_3 N_J)_{,\alpha} g_{mJ}] \tag{6}$$

and the Jacobian determinant is given as

$$|J| = \varepsilon_{ijk} \, x_{i,\xi 1} x_{j,\xi 2} x_{k,\xi 3} = \varepsilon_{ijk} \, N_{I,\xi 1} N_{J,\xi 2} N_K (x_{iI} + \xi_3 g_{iI})(x_{jJ} + \xi_3 g_{jJ}) g_{kK} \tag{7}$$

where ε_{ijk} is a permutation tensor and Greek indices refer to the parametric co-ordinate directions of an element.

For the efficient design sensitivity calculation of the stiffness matrix with respect to design parameter p, it is important to formulate the derivatives of the shape function such as $\partial(N_{I,m})/\partial p$ and $\partial(\xi_3 N_I)_{,m}/\partial p$, and that of the Jacobian determinant $\partial|J|/\partial p$. If the design parameter p is connected to the shell thickness t_J and the nodal co-ordinate x_{iJ} at node J, they are calculated by the chain rule as

$$\frac{\partial(N_{I,m})}{\partial p} = \frac{\partial(N_{I,m})}{\partial t_J}\frac{\partial t_J}{\partial p} + \frac{\partial(N_{I,m})}{\partial x_{iJ}}\frac{\partial x_{iJ}}{\partial p},$$
$$\frac{\partial(\xi_3 N_I)_{,m}}{\partial p} = \frac{\partial(\xi_3 N_I)_{,m}}{\partial t_J}\frac{\partial t_J}{\partial p} + \frac{\partial(\xi_3 N_I)_{,m}}{\partial x_{iJ}}\frac{\partial x_{iJ}}{\partial p},$$
$$\frac{\partial|J|}{\partial p} = \frac{\partial|J|}{\partial t_J}\frac{\partial t_J}{\partial p} + \frac{\partial|J|}{\partial x_{iJ}}\frac{\partial x_{iJ}}{\partial p} \tag{8}$$

Therefore the derivatives of the shape function and the Jacobian determinant with respect to the thickness t_J and the nodal co-ordinate x_{iJ} will be derived from the relationship $J J^{-1} = I$ (identity matrix) and

$$\frac{\partial x_i}{\partial t_J} = \xi_3 N_I \frac{\partial g_{iI}}{\partial t_J} = \xi_3 N_J \frac{v_{3i}^J}{2} \quad (J \text{ is not summed up})$$

$$\frac{\partial x_i}{\partial x_{kJ}} = N_I(\delta_{ik}\delta_{IJ} + \xi_3 \frac{\partial g_{iI}}{\partial x_{kJ}}) = \delta_{ik} N_J + N_I \xi_3 \frac{\partial g_{iI}}{\partial x_{kJ}} \tag{9}$$

where δ_{ik} is Kroneker's delta. Finally we will get the derivatives of the shape function as

$$\frac{\partial(N_{I,n})}{\partial t_J} = N_{I,m}(\xi_3 N_K)_{,n} \frac{\partial g_{mK}}{\partial t_J} = -N_{I,m}(\xi_3 N_J)_{,n} \frac{v_{3m}^J}{2}$$

(J is not summed up)

$$\frac{\partial(\xi_3 N_I)_{,n}}{\partial t_J} = -(\xi_3 N_I)_{,m}(\xi_3 N_J)_{,n} \frac{v_{3m}^J}{2} \quad (J \text{ is not summed up})$$

$$\frac{\partial(N_{I,n})}{\partial x_{kJ}} = -N_{I,k} N_{J,n} - N_{I,m}(\xi_3 N_K)_{,n} \frac{\partial g_{mK}}{\partial x_{kJ}}$$

$$\frac{\partial(\xi_3 N_I)_{,n}}{\partial x_{kJ}} = -(\xi_3 N_I)_{,k} N_{J,n} - (\xi_3 N_I)_{,m}(\xi_3 N_K)_{,n} \frac{\partial g_{mK}}{\partial x_{kJ}} \tag{10}$$

and the derivatives of the Jacobian determinant as

$$\frac{\partial |J|}{\partial t_J} = |J|\,(\xi_3 N_I)_{,i} \frac{\partial g_{iI}}{\partial t_J} = |J|\,(\xi_3 N_J)_{,i} \frac{v_{3i}^J}{2} \quad (J \text{ is not summed up})$$

$$\frac{\partial |J|}{\partial x_{jJ}} = |J|\{N_{J,j} + (\xi_3 N_I)_{,i} \frac{\partial g_{iI}}{\partial x_{jJ}}\} \tag{11}$$

where the derivative $\partial g_{mK} / \partial x_{kJ}$ can be calculated approximately from the derivatives of e_3 vector defined later for the local strain components.

Element-Level Sensitivity

Let the local co-ordinates $(x_{1'}, x_{2'}, x_{3'})$ be referred to the unit vector e_1, e_2 and e_3 defined by

$$e_3 = \frac{x_{,\xi 1} \times x_{,\xi 2}}{|x_{,\xi 1} \times x_{,\xi 2}|}, \quad e_1 = \frac{i \times e_3}{|i \times e_3|}, \quad e_2 = \frac{e_3 \times e_1}{|e_3 \times e_1|} \tag{12}$$

as shown in Fig.1. Then we can define the co-ordinate transformation matrix

$$[Q] = [q_{ij}] = [\ e_1\ e_2\ e_3\ ] \tag{13}$$

and the derivatives of the displacement in the local co-ordinate

$$\frac{\partial u'_i}{\partial x'_j} = q_{im}\, q_{jn} \frac{\partial u_m}{\partial x_n} = q_{im}\, q_{jn} \{N_{I,n}\, u_{mI} + (\xi_3 N_I)_{,n} \Theta_{mI}\} \tag{14}$$

The strain vector $\varepsilon = (\varepsilon_x, \varepsilon_y, \gamma_{xy}, \gamma_{xz}, \gamma_{yz})^T$ will be expressed by the displacement vector $d_I = (u_{1I}\ u_{2I}\ u_{3I}\ \theta_{1I}\ \theta_{2I})^T$ as

$$\varepsilon = B_I d_I \tag{15}$$

where B_I is the strain-displacement matrix defined as

$$B_I = \begin{bmatrix} b_1^u & b_1^\theta \\ b_2^u & b_2^\theta \\ b_3^u & b_3^\theta \\ b_4^u & b_4^\theta \\ b_5^u & b_5^\theta \end{bmatrix} ; \quad \begin{array}{c} b_i^u = (b_{i1}^u\ b_{i2}^u\ b_{i3}^u) \\ b_i^\theta = (b_{i1}^\theta\ b_{i2}^\theta) \\ (i = 1, \cdots, 5) \end{array}$$

$$b_{1m}^u = q_{1m}\, q_{1n} N_{I,n}, \quad b_{2m}^u = q_{2m}\, q_{2n} N_{I,n}, \quad b_{3m}^u = (q_{1m}\, q_{2n} + q_{2m}\, q_{1n}) N_{I,n}$$
$$b_{4m}^u = (q_{2m}\, q_{3n} + q_{3m}\, q_{2n}) N_{I,n}, \quad b_{5m}^u = (q_{3m}\, q_{1n} + q_{1m}\, q_{3n}) N_{I,n} \quad (m,n = 1,2,3)$$
$$b_{1m}^\theta = \omega_{1m}\, q_{1n} (\xi_3 N_I)_{,n}\, t_I/2, \quad b_{2m}^\theta = \omega_{2m}\, q_{2n} (\xi_3 N_I)_{,n}\, t_I/2$$
$$b_{3m}^\theta = (\omega_{1m}\, q_{2n} + \omega_{2m}\, q_{1n})(\xi_3 N_I)_{,n}\, t_I/2, \quad b_{4m}^\theta = \omega_{2m}\, q_{3n} (\xi_3 N_I)_{,n}\, t_I/2$$
$$b_{5m}^\theta = \omega_{1m}\, q_{3n} (\xi_3 N_I)_{,n}\, t_I/2, \quad \omega_{m1} = -q_{mn} v_{2n}, \quad \omega_{m2} = q_{mn} v_{1n}$$
$$(m = 1,2, \quad n = 1,2,3) \tag{16}$$

Then the IJ component of the stiffness matrix K_{IJ}^e in an element may be given by

$$K_{IJ}^e = \int_{-1}^{1}\int_{-1}^{1}\int_{-1}^{1} B_I^T D B_J |J|\, d\xi_1 d\xi_2 d\xi_3 \tag{17}$$

where D is the stress-strain matrix for the plate and shell. In the isoparametric element the above integration is evaluated by the numerical integration except in the ξ_3 direction for the constant thickness.

By differentiating equation (17) with respect to the design parameter p, we get the design sensitivity of the element stiffness matrix components as

$$\frac{\partial K_{IJ}^e}{\partial p} = \int_{-1}^{1}\int_{-1}^{1}\int_{-1}^{1} \{(\frac{\partial B_I^T}{\partial p} D\, B_J + B_I^T D \frac{\partial B_J}{\partial p}) |J| + B_I^T D\, B_J \frac{\partial |J|}{\partial p})\}\, d\xi_1 d\xi_2 d\xi_3 \tag{18}$$

This sensitivity expression contains the derivative $\partial B_I/\partial p$ which can be evaluated from the derivatives of $\partial(N_{I,n})/\partial p$, $\partial(\xi_3 N_I)_{,n}/\partial p$ and $\partial q_{mn}/\partial p$ formulated above. Equation (18) also contains the derivative $\partial |J|/\partial p$ which will be evaluated by the equations (8) and (11) directly.

<u>Static Response Sensitivities</u>

Consider a linear static problem for which a finite element discretization leads to the algebraic system $KU=F$. If the applied forces F are dependent upon the design parameter p, then the sensitivity of the nodal displacement U may be obtained by solving

$$K\frac{\partial U}{\partial p}=\frac{\partial F}{\partial p}-\frac{\partial K}{\partial p}U \tag{19}$$

If we first factor the original stiffness $K=LDL^T$, we can solve the above equation for the displacement sensitivity by forming only the right hand side of equation (19). The second term of the right hand side of (19) should be evaluated element by element for the efficient matrix calculation.

APPROXIMATION METHOD FOR STATIC RESPONSE CONSTRAINTS

Minimum weight Design Problem and its Approximation

Consider the minimum weight design of the stiffened thin plate structures with the webs and the flanges as shown in Fig. 2. When the thickness of the plate t_p, the height h_{wi} and the thickness t_{wi} of the web, and the width h_{fi} and the thickness t_{fi} of the flange are taken as the design variables,

$$x=(t_p,h_{w1},t_{w1},\cdots,h_{wN},t_{wN},h_{f1},t_{f1},\cdots,h_{fN},t_{fN})^T \tag{20}$$

the minimum weight design problem is stated mathematically as minimizing the weight W

$$W(x)=k_pt_p+\sum_{i=1}^{N}(k_{wi}h_{wi}t_{wi}+k_{fi}h_{fi}t_{fi})\rightarrow \min \tag{21}$$

subject to the displacement constraints, the stress constraints,

$$g_j(x)=w_j/w_{aj}-1\leq 0,\quad j=1,\cdots,M_w \tag{22}$$

$$g_k(x)=\sigma_k/\sigma_{ak}-1\leq 0,\quad k=1,\cdots,M_\sigma \tag{23}$$

and the side constraints

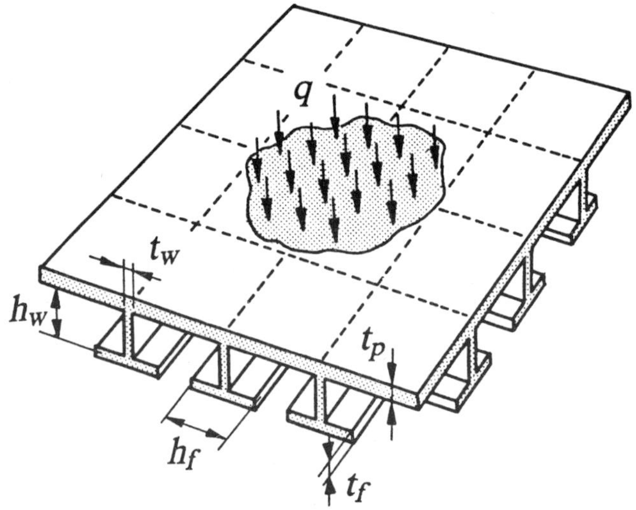

Fig. 2 The stiffened structure model for minimum weight design.

$$x_i^L \leq x_i \leq x_i^U, \quad i=1,\cdots,N_D \tag{24}$$

where the constants k_p, k_{wi} and k_{fi} denote the contribution of the plate, the webs and the flanges to the weight. w_{aj} and σ_{ak} are the allowable upper limits of the deflection w_j and the stress σ_k at the observation point j and k, and M_w, M_σ represent the numbers of the deflection and the stress observation points. x_i^L and x_i^U are the lower and upper bounds of the design variable x_i. The design variables are usually linked to one or more of the physical dimensions of the stiffened plate structures.

For solving the minimum weight design problem defined above, the mathematical programming method will be used which requires the evaluation of the objective function, the displacement and stress constraint function values, and their sensitivities. It is easy and inexpensive to evaluate the objective function value for the design variable change, but the structural analysis and its sensitivity analysis have to be done to evaluate the displacement and stress constraints for each design variable change. Therefore, if we can form an approximated problem for which the optimum solution will be searched directly in the approximated design space, it may be expected that the number of the structural analysis and the sensitivity analysis until getting the final solution will be reduced.

A direct approximation to the actual optimization problem is constructed by estimating the constraints using a first-order Taylor series (DTSA) at the current design variable x_0 as

$$g_j(x) \cong g_j(x_0)+\sum_{i=1}^{N_D}\left(\frac{\partial g_j}{\partial x_i}\right)_{x_0}(x_i-x_{0i}) \quad (j=1,\cdots,M) \tag{25}$$

The approximated subproblem is solved by a nonlinear programming optimization algorithm with appropriate move limits. The move limits are employed to ensure that a new design point remains in the vicinity of the current point x_0 around which the Taylor series was expanded. The move limits are typically specified by a limit factor δ to determine the upper and lower bounds as

$$\delta x_i \leq x_i \leq x_i/\delta, \quad 0<\delta<1 \quad (i=1,\cdots,N_D) \tag{26}$$

The move limits of equation(26) are applied as side constraints instead of equation (24), if they are more restrictive than the minimum and maximum gage constraints.

Since the deflections and the stresses of the stiffened plate structure are usually related nonlinearly with the design variables defined in equation (20), the direct Taylor series expansion of equation (25) gives a poor approximation which frequently causes the optimization subproblem to have no feasible design space at the initial design point. Therefore, suitable intermediate variables $y=(y_1,\ldots,y_{ND})^T$ corresponding to x which will be able to approximate more exactly the static responses of the stiffened plate structures, like a reciprocal variable of cross sectional area for the frameworks, are expected to be found for constructing an approximate subproblem which gives a wide range usability and allows a large value of limit factor to be taken. Then, the intermediate variable approximation to the constraints using first-order Taylor series expansion will be derived as

$$g_j(y) \cong g_j(y_0)+\sum_{i=1}^{N_D}\left(\frac{\partial g_j}{\partial y_i}\right)_{y_0}(y_i-y_{0i}) \quad (j=1,\cdots,M) \tag{27}$$

where y_{0i} denotes the current intermediate variable corresponding to x_{0i}. The constraint sensitivities are usually calculated from the sensitivities with respect to the direct variables as

$$\left(\frac{\partial g_j}{\partial y_i}\right)_{y_0}=\left(\frac{\partial g_j}{\partial x_i}\right)_{x_0}\Big/\left(\frac{\partial y_i}{\partial x_i}\right)_{x_0} \quad (i=1,\cdots,N_D) \tag{28}$$

when the intermediate variables are a function of a single design variable.

Approximation Using Reciprocal Variables

A simply supported plate with stiffeners as shown in Fig.3, which consist of the webs and flanges, under uniform pressure is used to find the intermediate design variable y that gives a wide range approximation. Figure 4 illustrates the difference between the approximation techniques for Von Mises stress at the center of the plate without the stiffeners, when the thickness of the plate t_p is taken as the design variable. DTSA line shows an estimation by the direct variable approximation of equation (25) and RTSA curve by a square of reciprocal variable $1/t_p^2$ approximation using equation (27). RTSA gives a good and conservative estimation of the stress variation for a wide range. The stress variation at the center of a quarter region of the plate for the stiffener height change is illustrated in Fig. 5, where RTSA is formed using Taylor series expansion by a square of reciprocal variable $1/h_w^2$. This figure shows the validity of the reciprocal variable for the stiffened plate design problems. Similar tendency is observed for the web thickness, flange width and the flange thickness. Consequently, it is found that the following intermediate variables give a good Taylor series approximation to Von Mises stress variation.

$$y=(1/t_p^2, 1/h_{w\,1}^2, 1/t_{w\,1}, \cdots, 1/h_f, 1/t_f)^{\mathrm{T}} \tag{29}$$

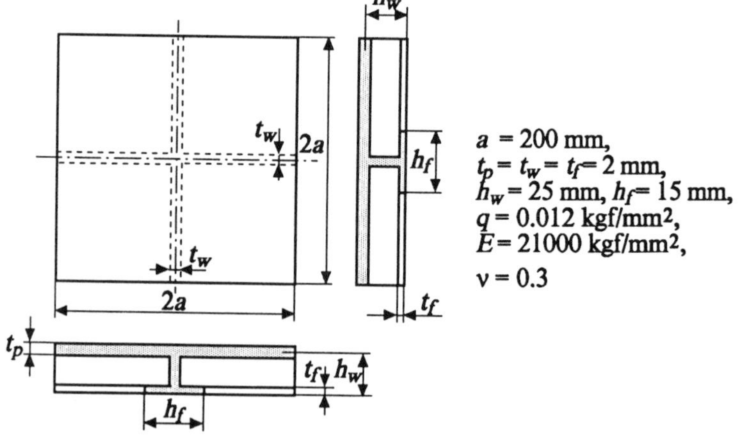

Fig. 3 Simply supported plate with cross stiffeners under uniform pressure.

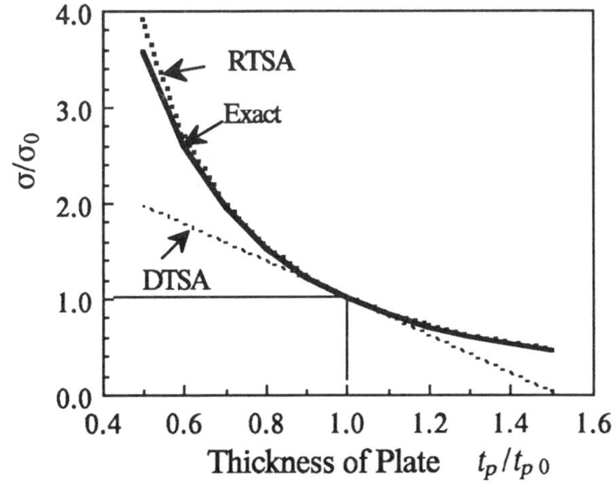

Fig. 4 Taylor series approximation of stress at the center of plate for the plate thickness variation (t_{p0} = 2 mm).

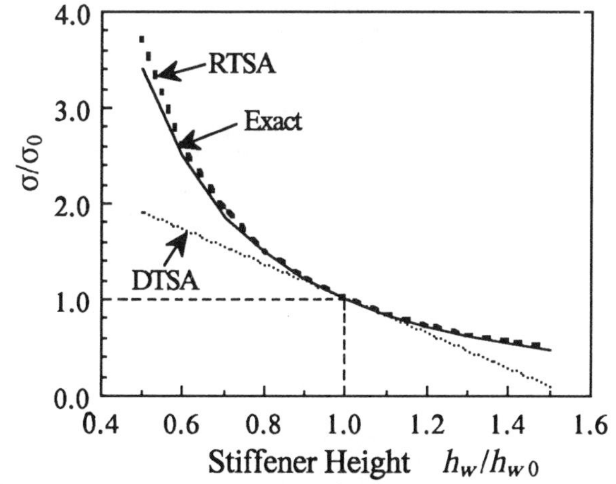

Fig. 5 Taylor series approximation of stress at the center of quarter region for the web height variation (h_{w0} = 25 mm).

On the other hand, the reciprocal variable approximation is also tried for the displacement constraint and it is finally found that the following reciprocal variables give a wide range approximation to the plate deflection.

$$y = (1/t_p^3, 1/h_{w1}^3, 1/t_{w1}, \cdots, 1/h_f, 1/t_f)^T \tag{30}$$

The validity of these intermediate variables will be shown by the numerical examples.

NUMERICAL EXAMPLES

Sensitivity Analysis of Cylindrical Shell

A conical cylinder applied a shear force Q_0 and a bending moment M_0 at one end, shown in Fig.6, is considered for the displacement and the stress sensitivity.

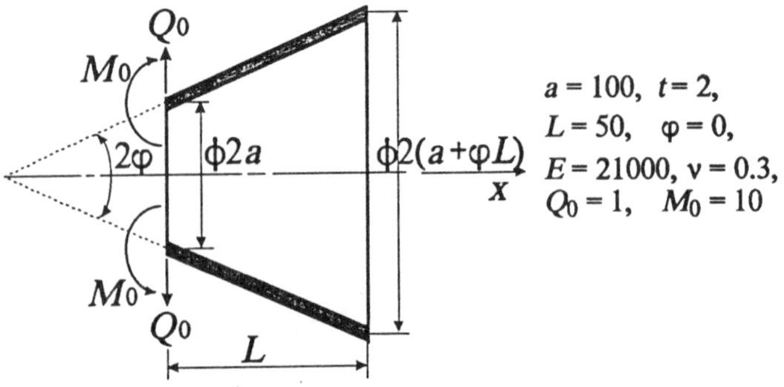

Fig.6 Conical cylinder under shear force and bending moment.

Table 1 Deflection and its sensitivity of cylindrical shell.

x/a		0.0	0.1	0.2	0.3
Deflection	Theory	8.262	0.796	-0.948	-0.599
wx10^2	FEM	8.233	0.844	-0.980	-0.648
Sensitivity	Theory	-71.80	4.713	9.941	1.137
$(\partial w/\partial t)$x10^3	FEM	-70.86	3.510	10.29	1.745
	FDM	-70.87	3.511	10.29	1.745
Sensitivity	Theory	10.43	3.330	-0.854	-1.538
$(\partial w/\partial a)$x10^4	FEM	10.45	3.296	-0.914	-1.632
	FDM	10.45	3.296	-0.914	-1.632
Sensitivity	FEM	17.88	2.448	-11.89	-20.19
$(\partial w/\partial \varphi)$x10^4	FDM	17.40	2.374	-11.58	-19.67

FDM : Finite Difference Method

Table 2 Stress and its sensitivity of cylindrical shell.

x/a		0.0	0.1	0.2	0.3
Stress	Theory	24.14	12.68	3.941	1.272
σ_e	FEM	24.67	11.60	4.219	1.4036
Sensitivity	Theory	-22.23	-8.459	-0.661	-0.642
$\partial\sigma_e/\partial t$	FEM	-22.78	-6.881	-1.552	-0.348
	FDM	-22.79	-6.882	-1.522	-0.349
Sensitivity	Theory	3.831	8.449	6.561	1.259
$(\partial\sigma_e/\partial a)$	FEM	3.952	8.801	5.974	2.289
x10^2	FDM	3.952	8.801	5.975	2.288

analysis suggested in the previous chapter. The radius a, the thickness t and the angle φ of the cone are taken as the design variables. A quarter region of the cylindrical shell is divided into 5 elements in the axial direction and into 4 elements in the circumferential direction. Table 1 and 2 show the displacement and Von Mises stress sensitivities along the cylinder axis when the angle $\varphi = 0$. The result of the direct differentiation method (FEM) and that of the central finite difference (FDM) agree well with the values of the thin shell theory.

Minimum Weight Design of Stiffened Plate Structures under Uniform Pressure

The minimum weight design of the stiffened plate model, as same as the model shown in Fig. 3, except no flanges, is considered with the displacement constraints for the first example. The deflection of the plate at the center of the plate and of the quarter region are restricted less than $w_a = 1.2$ mm. The upper bound of the stiffener heights and the lower bound of the plate and stiffener thicknesses are taken as $h^U = 60$ mm and $t^L = 1$ mm. For the reciprocal variable, the objective function of the weight is also approximated using a second-order Taylor series and a linear complementary pivot method is employed[6] as the nonlinear optimization technique. The optimization using the direct variable approximation is also tried to compare the efficiency of convergence. A quarter region of the model is divided into 8 quadratic plate elements. Table 3 shows the weight, the constrained displacements and the design variables for the initial and final design. Figure 7 illustrates the variations of the objective function and the displacement constraint at the center of the quarter region. The direct variable approximation causes a severe violation of the constraint at the early iteration steps due to the poor approximated constraint. On the other hand, the reciprocal variable approximation without the move limits shows a steady convergence to the optimum solution with less iteration.

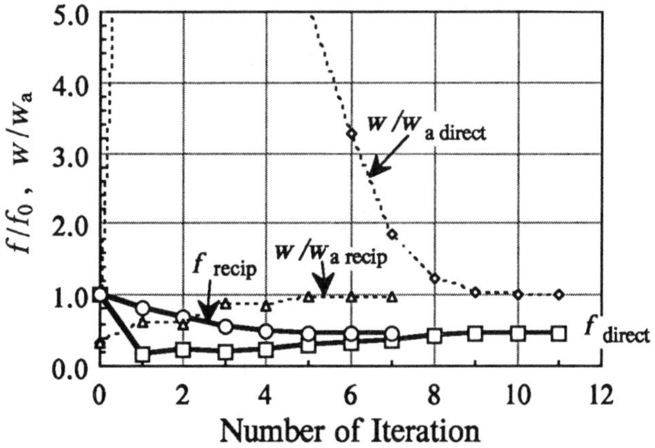

Fig. 7 Variations of objective function and displacement constraint for stiffened plate with cross ribs.

Table 3 The objective function, the displacement constraint values and the design variable of the stiffened plate with cross rib ($w_a = 1.2$ mm, $a = 200$ mm).

	Objective	Constraint w/w_a		Design variable (mm)		
	f/f_0	1/4 ceter	Plate center	t_p	h	t_w
Initial	1.000	0.3245	0.4965	6.000	30.00	4.000
Direct TSA	0.4465	1.0001	0.3714	2.647	60.00	1.000
Reciprocal TSA	0.4497	0.9798	0.3680	2.668	60.22	1.000

A simply supported plate with six stiffeners of different heights and thicknesses with stress constraints is considered as another example, as shown in Fig. 8. The thicknesses t_{w1}, t_{w2} and heights h_1, h_2 of the center stiffeners and the intermediate stiffeners as well as the thickness of the plate t_p are taken as the design variables. Von Mises stresses at the tops of stiffener intersections and at the center of 1/16 region of plate are restricted less than $\sigma_a = 10$ kgf/mm2. The upper bounds of the stiffeners and the lower bounds of the thickness are taken as $h^U = 70$ mm and $t^L = 1$ mm. A quarter region of the design model is divided into 40 elements. Table 4 shows the weight, the constrained stresses and the design variables for the initial and final design. Figure 9 illustrates the variations of the objective function and the displacement constraint at the center of the quarter region. The optimization using the reciprocal variable approximation shows a steady and rapid convergence to the optimum solution. On the other hand, the optimization using the direct variable approximation has not converged to the

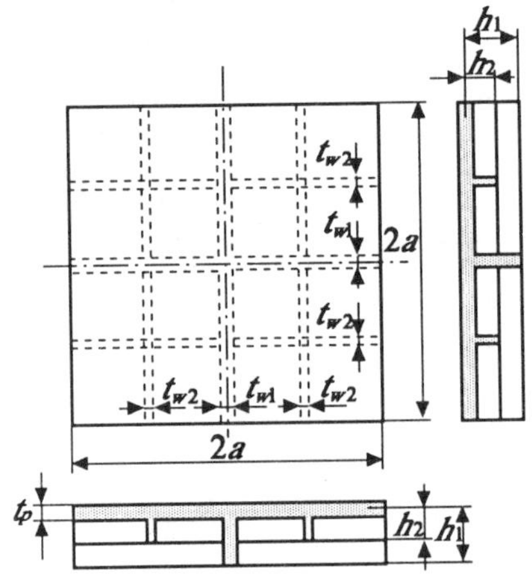

Fig. 8 Design model of stiffened plate with six different stiffeners under stress constraint

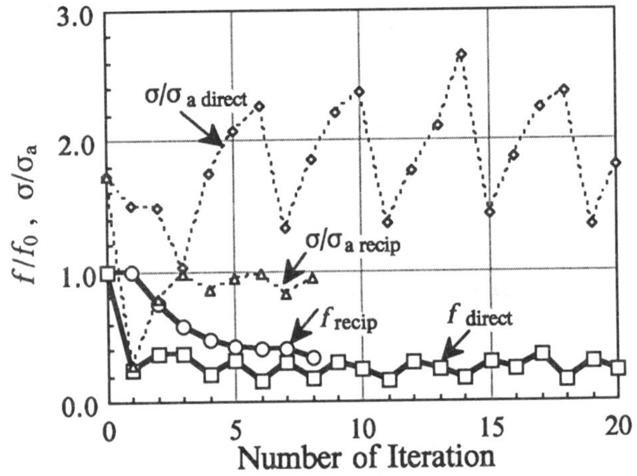

Fig.9 Variations of objective function and stress constraint for stiffened plate with six different stiffeners.

Table 4 The objective function, the stress constraint values and the design variables of the stiffened plate with six stiffeners (a = 200 mm, b = 100 mm, σ_a = 10.0 kgf/mm2).

	Objective function f/f_0	Stress constraint σ/σ_a g_1 g_2 / g_3 g_4	Design variables (mm) t_p t_{w1} t_{w2} / h_1 h_2
Initial	1.000	1.732 0.4158	6.000 4.000 4.000
		0.7510 0.3328	30.00 15.00
Direct TSA*	0.2327	1.803 1.129	1.076 3.222 3.401
		1.354 1.287	16.11 10.00
Reciprocal TSA	0.3993	0.9387 0.9802	2.003 1.537 1.000
		0.9260 1.006	77.43 27.73

* Optimization has not converged.

solution after 20 times iteration, even if the move limits are considered after the initial several steps.

CONCLUDING REMARKS

For the stress and displacement constraints of the stiffened plate structures, a reciprocal variable approximation using Taylor series expansion shows a steady and rapid convergence to the optimum solution. This concept will be applicable to the minimum weight design of the shell structures with the static response constraints.

REFERENCES

1. Brockman, R.A. and Lung, F.Y., Sensitivity Analysis with Plate and Shell Finite Elements, *International Journal for Numerical Methods in Engineering*, Vol.26,pp. 1129-1143,1988.
2. Schmit, L.A. and Miura, H., Approximation Concepts for Efficient Structural Synthesis, *NASA CR*-2552, 1976.
3. Starnes, J.H. and Haftka, R.T., Preliminary Design of Composite Wings for Buckling, Strength and Displacement Constraints, *Journal of Aircraft*, Vol. 16, pp. 564-570, 1979.
4. Fleury, C. and Braibant, V., Structural Optimization-A New Dual Method Using Mixed Variables, *International Journal for Numerical Methods in Engineering*, Vol.23, pp. 409-428, 1986.
5. Vanderplaats, G.N. and Salajegheh, V.A., A New Approximation Method for Stress Constraints in Structural Synthesis, *AIAA Journal*, Vol. 27, pp. 352-358, 1989.
6. Bazaraa, M.S. and Shetty, C.M., *Nonlinear Programming, Theory and Algorithms*, Jhon Wiley, 1973.

Structural Optimization with Adaptive and Two Level Approximation Techniques

R. Xia, H. Huang

Beijing University of Aeronautics & Astronautics, Beijing 100083, China

ABSTRACT

In this paper, a new technique for constructing approximation functions with high quality and adaptive capability to original functions is proposed by using values of the functions and their derivatives at points obtained in the process of optimization. And a new algorithm for solving optimization problems is created by introducing two level approximation concepts. Several typical examples have been optimized to test its power.

INTRODUCTION

The functions involved in structural optimization problems are in general implicit functions with a high degree of nonlinearity with respect to design variables. An efficient way of solving these kind of problems is employing approximation techniques [1-3]. The original functions are expressed by explicit approximation function, and a corresponding approximate problem can be constructed. The sequence of solutions of approximate problems converges to the solution of the original problem. The convergence property of approximation method depends mainly on the quality of the approximate functions. It must be pointed out that the degree of nonlinearity of functions is different for different kinds of optimization problems, and even for a function the degree of nonlinearity at different points of design space is also different. So it does not have general significance for improving convergence if a fixed mathematical approximation model is used to approximate a complicated function. That is the reason why some approximation methods can only be used efficiently for a few kinds of problems but not to others.

The present work considers the approximation of a function based on values of the function and its derivatives at points given in the process of optimization as well as adaptive techniques. Explicit approximation functions with high quality are thus constructed, and a corresponding approximation problem is formed as the first level approximation, which can be solved by many kinds of methods. But it is more efficient if the dual method of nonlinearly mathematical programming is employed, especially for problems with large numbers of design variables and a small number of critical constraints. For this reason the second level approximation problems are created to approximate the first level approximation problem. The sequence of solutions of the second level problems converges to the solution of the first level problem. And the sequence of solutions of the first level problems converges to the solution of the original problem. Only at the beginning of each first level approximation stage is an accurate analysis necessary.

THEORY

The original design problem to be solved can be stated as:
Find the vector of design variables

$$X = \{x_1, x_2, \cdots, x_N\} \tag{1a}$$

such that the objective function

$$f(X) \rightarrow Min \tag{1b}$$

while

$$g_j(X) \leq 0 \qquad j = 1, ..., J \tag{1c}$$

$$x_i^L \leq x_i \leq x_i^U \qquad i = 1, ..., n \tag{1d}$$

where g_j (X) is the behavioral constraint function; x_i^U and x_i^L are upper and lower bounds on x, respectively.

By introducing modified Hermite polynomial as approximation functions, the first level approximation problem of the pth stage can be written as follows: Find vector of variables

$$X = \{x_1, x_2, \cdots, x_n\}^T \tag{2a}$$

such that

$$f^{(p)}(X) \rightarrow Min \tag{2b}$$

while

$$g_j^{(p)}(X) \leq 0 \qquad j = 1, ..., J \tag{2c}$$

$$x_i^L \leq x_i \leq x_i^U \qquad i = 1, ..., n \tag{2d}$$

the constraint function in Eq. (2c) is expressed as:

$$g_j^{(p)}(X) = \sum_{t=1}^{H}\{g_j(X_t)+[\nabla g_j(X_t)-2g_j(X_t)\nabla h_t(X_t)]^T\cdot(X-X_t)\}\cdot h_t^2(X) \quad (3)$$

where

$$h_t(X) = \prod_{i=1 i\neq t}^{H} \frac{(X^r - X_i^r)^T(X^r - X_i^r)}{(X_t^r - X_i^r)^T(X_t^r - X_i^r)} \quad (4)$$

and

$$X^r = \{x_1^r, x_2^r, \cdots, x_n^r\}^T \quad (5)$$

$$X_t^r = \{x_{1t}^r, x_{2t}^r, \cdots, x_{nt}^r\}^T \qquad t = 1,...,H \quad (6)$$

The values of function $g_j(X_t)$ and gradient $\nabla g_j(X_t)$ at point X_t in Eq.(3) can be obtained by structural analysis and sensitivity analysis. And the parameter r in Eq.(4) is controlled by the degree of nonlinearity of the original function g (X), which is determined as follows: By introducing intermediate variables

$$\begin{aligned} Y &= \{y_1, y_2, \cdots, y_n\}^T \\ &= \{x_1^s, x_2^s, \cdots, x_n^s\}^T \end{aligned} \quad (7)$$

If the constraint function $g_j(X)$ is approximated by Taylor series expansion at a given point X_2 in terms of variables Y, one obtains

$$\tilde{g}(X) = g_j(X_2) + \sum_{i=1}^{n} \frac{\partial g_j(X_2)}{\partial y_i}(y_i - y_{i2}) \quad (8)$$

Assuming that the value of function $\tilde{g}_j$ (X) at point X_1 is equal to the value of original function $g_j(X)$ at the same point X_1, that is

$$\tilde{g}_j(X_1) = g_j(X_1) \quad (9)$$

From Eq.(8) and Eq.(9), the following expression is given

$$g_j(X_1) - [g_j(X_2) + \tfrac{1}{s}\sum_{i=1}^{n} x_{i2}^{(1-s)} \cdot \frac{\partial g_j(X_2)}{\partial x_i}(x_{i1}^s - x_{i2}^s)] = 0 \quad (10)$$

The value of function and its derivatives at points X_1 and X_2 in Eq.(10) have been obtained in the optimization process. And the multi-value solutions of s can be given by solving Eq.(10), but the one which is nearest to 1.0 is taken as the solution of s, which indicates the degree of nonlinearity of the original function in the neighborhood of point X_1. Now it is assumed that the approximate function expressed by Eq.(3) and the original function $g_j(X)$ have the same degree of nonlinearity, then the following relation holds

$$r = \frac{s-1}{4(H-1)} \tag{11}$$

It gives the value of parameter r in Eq.(3).

The first level approximation problem expressed in Eq.(2) can be solved in general by mathematical programming method, but it is still not efficient especially for large scale optimization problems. In order to improve the computational efficiency, a second level approximation of the functions involved in Eq.(2) is proposed, and the corresponding approximation problem is established as follows:
Find

$$X = \{x_1, x_2, \cdots, x_n\}^T \tag{12a}$$

such that

$$\bar{f}(X) \rightarrow Min \tag{12b}$$

while

$$\bar{g}_j(X) \leq 0 \qquad \text{j} = 1, ..., \text{m} \tag{12c}$$

$$x_i^L \leq x_i \leq x_I^U \qquad \text{i} = 1, ..., \text{n} \tag{12d}$$

where

$$\bar{f}(X) = f(X_k) - \sum_{i=1}^{n} x_{ik}^2 \cdot \frac{\partial f(X_k)}{\partial x_i}(\frac{1}{x_i} - \frac{1}{x_{ik}}) \tag{13}$$

$$\bar{g}_j(X) = g_j^{(p)}(X_k) + \sum_{i=1}^{n} \frac{\partial g^{(p)}(X_k)}{\partial x_i}(x_i - x_{ik}) \tag{14}$$

The problem expressed in Eq.(12) can be solved by introducing the dual theory [2] and the variable metric method (DFP method) [4] without any difficulty.

EXAMPLES

To illustrate the effectiveness and efficiency of the algorithm presented in this paper, some typical structures have been optimized. The computational results have shown that this method performs satisfactorily when compared to other computing techniques.

Example 1. Ten-bar truss (Fig.1) [2].

One load condition: $P_2 = P_4 = 10^5$ lbs; Young's modulus: $E = 10^7$ lb/in^2; specific weight: $\rho = 0.1$ lb/in^3; allowable stresses: $\sigma_a = \pm 25000$ lb/in; vertical displacement limits at points: $v_a = 2.0$ in.; minimum member limits $x_{imin} = 0.1$ in^2.

The final design and the comparison with other methods are listed in Table 1 and the iteration history is shown in Table 2.

Table 1 Final design for Example 1

Truss member number i	Final cross-sectional areas (in^2)			
	Schmit [2] DUAL 2	NEWSUMT [1]	Xia [3]	Present paper
1	30.52	30.23	30.95	30.62
2	0.10	0.179	0.10	0.10
3	23.20	23.94	23.27	23.28
4	15.22	13.43	15.19	15.13
5	0.10	0.10	0.10	0.10
6	0.551	0.18	0.46	0.529
7	7.457	8.565	7.50	7.503
8	21.04	21.95	21.07	21.10
9	21.53	21.19	21.48	21.40
10	0.10	0.241	0.10	0.10
Final weight (lb)	5061	5096	5062	5062

Table 2 Iteration history data for Example 1

Number of analysis	Weight (lbs)			
	Schmit [2] DUAL 2	NEWSUMT [1]	Xia [3]	Present paper
1	8266	7853	8266	8266
2	5994	6650	6040	6142
3	5793	6161	5720	5739
4	5683	5892	5604	5608
5	5566	5656	5481	5485
6	5438	5427	5343	5384
7	5294	5291	5201	5200
8	5200	5154	5078	5078
9	5112	5107	5067	5072
10	5075	5096	5062	5063
11	5065	5096		5062
12	5061	5096		

Example 2. 25-bar space truss (Fig.2) [2].

Two load conditions (see [2, Table 9b]); Young's modulus: $E = 10^7\,lb/in^2$.; specific weight: $\rho = 0.1\,lb/in^3$.; allowable stress (see[2, Table 9b]): displacement limits on nodes 1-6 in X-,Y-,Z-direction: $v_a = \pm 0.35$ in.; minimum member limitations: $x_{imin} = 0.01in^2$.

After linking (see [2, Table 11b]) this example has eight independent design variables.

The computational results are given in Table 3. Convergence history is shown in Table 4.

Table 3 Final design for Example 2

Member in D.V.group	Final cross-sectional areas (in^2)			
	Schmit[2] DUAL 2	NEWSUMT [1]	Xia [3]	Present paper
1	0.010	0.010	0.010	0.010
2	1.987	1.985	1.971	1.987
3	2.991	2.996	2.998	2.991
4	0.010	0.010	0.010	0.010
5	0.012	0.010	0.010	0.013
6	0.683	0.684	0.685	0.683
7	1.679	1.677	1.689	1.678
8	2.664	2.662	2.662	2.664
Final weight (lb)	545.22	545.17	545.51	545.22

Table 4 Iteration history data for Example 2

Number of analtsis	Weight (lbs)			
	Schmit [2] DUAL 2	NEWSUMT [1]	Xia [3]	Present paper
1	734.4	783.7	734.4	734.4
2	565.3	609.7	563.6	564.9
3	546.8	564.4	548.3	546.9
4	545.8	552.1	546.3	545.7
5	545.4	547.4	545.6	545.4
6	545.2	546.0	545.5	545.2
7		545.4		
8		545.2		

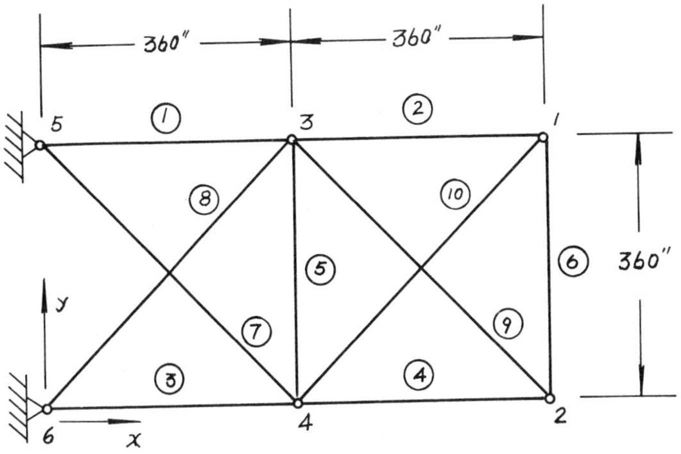

Figure 1. Planar ten-bar cantilever truss

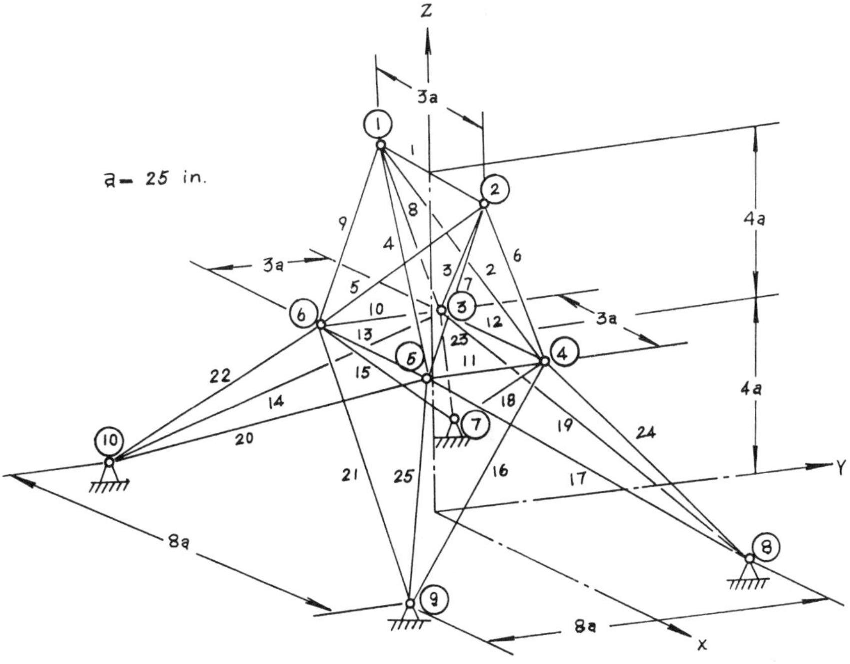

Figure 2. 25-bar truss

Example 3. Bartel Structure (Fig.3) [5]

One load condition: P = 44480 Nt; Young's modulus: E = $2.07 \times 10''\,Nt/m^2$; specific weight: $\rho = 2.77 \times 10^4\,kg/m^3$; allowable stress : $\sigma_a = 2.76 \times 10^8\,Nt/m^2$. This problem considers buckling, stress and side constaints.

The final design and the comparison with other methods are listed in Table 5.

Table 5 Final design for Example 3

Design variables	Final design (cm) Schmit [5] DUAL 2	 Present paper
B_1	6.35	6.35
H_1	6.35	6.35
t_{h1}	0.254	0.254
t_{b1}	0.229	0.229
B_2	6.35	10.67
H_2	25.4	25.4
t_{h2}	0.254	0.254
t_{b2}	1.14	0.707
Final mass (kg)	133.70	131.75
Number of analyses	20	5

Example 4. 2 × 5 Grillage (Fig.4) [5]

One load condition: a 175 Nt/cm distributed load on all members; Young's modulus: E = $2.07 \times 10''\,Nt/m^2$; specific weight: $\rho = 2.77 \times 10^4\,kg/m^3$; allowable stress: $\sigma_a = 1.38 \times 10^8\,Nt/m^2$. In this problem only displacement and side constraints are considered.

The computational results are given in Table 6.

Table 6 Final design for Example 4.

Design variables	Final design (cm) Schmit [5] DUAL 2	Final design (cm) Present paper
B_1	16.0	9.118
H_1	48.0	50.8
t_{h1}	0.127	0.127
t_{b1}	0.114	0.515
B_2	16.8	2.895
H_2	39.9	42.87
t_{h2}	0.127	0.127
t_{b2}	0.114	0.185
B_3	34.3	28.22
H_3	50.8	50.8
t_{h3}	0.127	0.127
t_{b3}	2.09	1.44
B_4	12.4	37.01
H_4	50.8	50.8
t_{h4}	0.127	0.127
t_{b4}	2.52	1.725
Final weight (kg)	3162	2947
Number of analyses	37	6

CONCLUSIONS

In this paper a new method for structural optimization is proposed by using multi-point constraint approximation and adaptive techniques, and two-level approximation concepts. Theory and the computational examples have shown that the main advantages of methods developed in the present paper are the generality in use and the efficency in computation, which are particularly important for large scale problems with complicated functions in engineering applications.

REFERENCES

1. L.A. Schmit and H. Miura, Approximation concepts for efficient structural synthesis, NASA CR-2552.
2. C. Fleury and L.A. Schmit, Dual methods and approximation concepts

in structural synthesis, NASA CR 3226.
3. Xia Renwei and Liu Peng, Structural optimization based on second-order approximations of functions and dual theory, Computer Methods in Applied Mechanics and Engineering 65 (1987) 101-114.
4. David M. Himmelblau, Applied Nonlinear Programming, McGRAW-HILL Book Company, 1972.
5. W.C. Mills-Curran, R.V. Lust and L.A. Schmit, Approximations method for Space Frame Synthesis, AIAA Journal, Vol.21, No.11, pp.1571-1580, 1983.

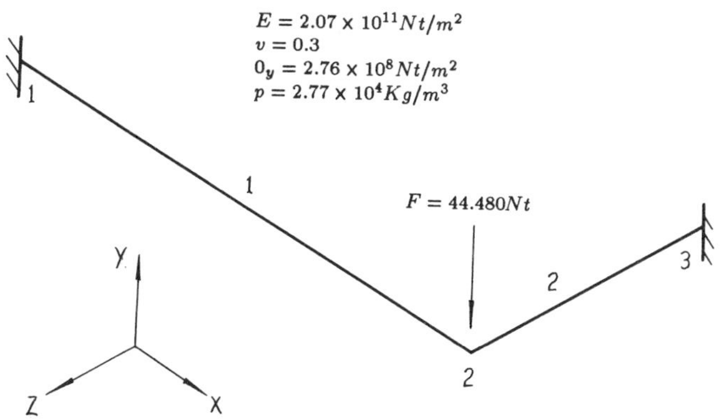

Figure 3. Bartel structure

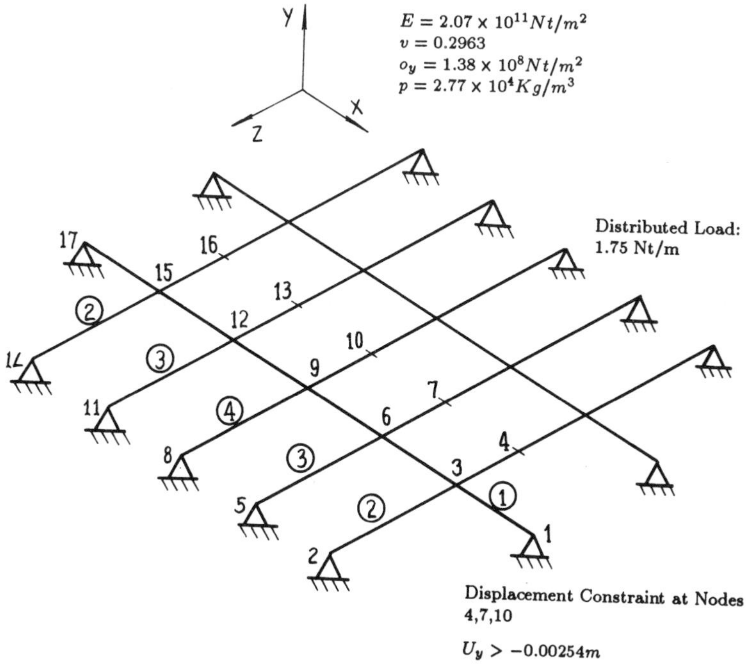

Figure 4. 2 × 5grillage

Appendix A. Solution method of the second level approximation problem expressed in Eq.(12)

According to the dual theory of nonlinear programming, the dual function of the problem expressed in Eq.(12) can be defined as

$$d(\lambda) = \min_{x \epsilon \bar{x}} L(X, \lambda), \tag{A1}$$

where

$$\bar{x} = (x \mid x_i^L \leq x_i \leq x_i^U), \tag{A2}$$

$$L(x, \lambda) = f(X) + \sum_{j=1}^{m} \lambda_j \cdot g_j(X), \tag{A3}$$

and the dual problem has the following form: Find λ such that the dual function

$$\underset{\lambda \in D}{d(\lambda)} \quad \rightarrow \quad \max, \tag{A4}$$

where D denotes the set of all dual points satisfying the nonnegativity conditions

$$\lambda_j \geq 0, \quad j \in J. \tag{A5}$$

It has been proved by the duality theorem that the following relation holds:

$$f(X^*) = d(\lambda^*), \tag{A6}$$

where X^* solves the primal problem and λ^* solves the dual problem.

By introducing Eqs.(13)-(14) into Eqs.(A1)-(A3), the explicit relations between the primal variables and the dual variables can be written as follows:

$$x_i = \begin{cases} x_i^L & \tilde{x}_i < x_i^L \\ \tilde{x}_i & x_i^L \leq \tilde{x}_i \leq x_i^U \\ x_i^U & \tilde{x}_i > x_i^U \end{cases} \tag{A7}$$

where

$$\tilde{x}_i = \left(-\frac{x_{ik}^2 \cdot \dfrac{\partial f(X_k)}{\partial x_i}}{\sum_{j=1}^{m} \lambda_j \dfrac{\partial g_j^{(p)}(X_k)}{\partial x_i}}\right)^{\frac{1}{2}} \qquad \text{i} = 1, ..., \text{n} \tag{A8}$$

The computational algorithm used to solve the dual explicit dual problem expressed in Eq.(A4) involves iterative modifications of the dual variable vectors as follows:

$$\lambda_{t+1} = \lambda_t + \alpha_t \cdot S_t. \tag{A9}$$

In the present paper, the move directions S_t in dual space are determined by using the variable metric method DFP, and the step lengths α_t are obtained by one-dimensional searching along the direction S_t for maximizing the dual function subject to nonnegativity constraints on the dual variables.

Structural Optimization Based on Relaxed Compatibility and Overlapping Decomposition

G. Thierauf

University of Essen, Essen, Germany

Summary: A two stage sequential technique for the solution of a wide class of stuctural optimization problems is proposed. It is based on an overlapping decomposition of the stuctural domain, which is assumed to be discretized by finite elements.

The optimal overlapping decomposition is derived from the nullspace of the equilibrium equations which is computed by a turn–back LU–decomposition minimizing the overlap of the subdomains. The iterative solution of the structural optimization problem is performed by a two level approach where the nodal displacements are the coordination variables.

1 Introduction

The idea to decompose a large and complex mechanical system into a number of smaller subsystems in order to facilitate analysis or design of the original system was born in the early days of the finite element method when computer capacity was limited: In 1955/56 Gabriel Kron [1] published his work on "Solving highly complex elastic structures in easy stages" and on "... the method of tearing" [2], which he applied to 'complex nonlinear plastic structures'.

One reason why this method attained little attention in structural engineering might be the rapid development of the finite element method and also the growing capacity of computers, which at first seemed to make decomposition unneccessary.

Realistic modelling of large structured systems and the demand for automated design together with the development of parallel computers caused a renewed interest in decomposition during the last two decades ([3], [4]).
Many variants of decomposition in structural optimization have been proposed, e.g. multilevel optimization, partitioning, substructuring, dissec-

tion and tearing.

Most of these methods are either based on a decomposition of the governing equations for the state variables or on a dissection of the domain of the mechanical system.

If we understand by decomposition an iterative sequential process which involves the state variables, it becomes obvious that the connectivity of the mechanical system will greatly influence the efficiency of decompositon: For a highly connected system, where changes in one subregion strongly influence all other parts, no benefits can be expected from decomposition.

2 The finite element model and classes of structural optimization

The optimization of stuctural systems considered here is based on a discretization by mixed elements. Mixed finite element approximations are based on the Hellinger–Reissner variational principle [5], [6] and the assembly process results in the following governing equation

$$\begin{bmatrix} -\underline{f} & \underline{a} \\ \underline{a}^T & \underline{0} \end{bmatrix} \begin{bmatrix} \underline{F} \\ \underline{r} \end{bmatrix} + \begin{bmatrix} \underline{G}(\underline{r}) \\ \underline{M}(\underline{F}, \underline{r}) \end{bmatrix} = \begin{bmatrix} \underline{0} \\ \underline{R} \end{bmatrix} , \tag{1}$$

where $\underline{f}$ is the symmetric positive definite system–flexibility with full column rank, $\underline{a}^T$ is the equilibrium matrix and $\underline{G}$ and $\underline{M}$ are approximations for the neighbouring state of equilibrium. $\underline{F}$ is the vector of element forces, $\underline{r}$ the displacements and $\underline{R}$ the loads. $\underline{f}$ is a hyperdiagonal matrix and for most structural systems $\underline{a}^T$ is sparse (70% to 90% zero elements). A wide class of optimization problems can be set up by assuming that the flexibility matrices are functions of the optimization variables $\underline{x}$:

$$\underline{f} = \{ \underline{f}^i(\underline{x}) \} , \tag{2}$$

$$\underline{x} \in X^+ = \{ x_i | x_i^l \leq x_i \leq x_i^u \} \tag{3}$$

In nonlinear analysis (1) is solved by iterative methods, e.g. by first–order Taylor expansion and Newton–Raphson iteration.
In first order structural optimization the second term in (1) for the neighbouring state of equilibrium is neglected. Further on, if only the flexibilities are functions of $\underline{x}$ we obtain optimal design problems. For shape optimization problems, the position of the nodal points is considered to be variable. Further distinctions are not necessary; if zero–values of the optimization variables are accepted, layout optimization and topological optimization with vanishing finite elements can be viewed as special cases.
Second order structural optimization with an approximation of the $\underline{G}$ and $\underline{M}$–functions is not the subject of this paper but the techniques considered seem to be suited for an extension.

3 Decomposition of Structural Domain

According to the finite element model we assume that the optimization variables can be grouped into subvectors

$$\underline{x} = \{\underline{x}_1, \underline{x}_2, \ldots, \underline{x}_k\}$$

and that each of the subvectors relates to one or more parts Ω_i of the structural domain.:

$$\Omega = \bigcup_{i=1}^{k} \Omega_i \ .$$

This decomposition can be either overlapping or non-overlapping (fig. 1). In the second case we have a partitioning of the structural domain; the first case, which will be treated here, results from a decomposition of the governing equations.

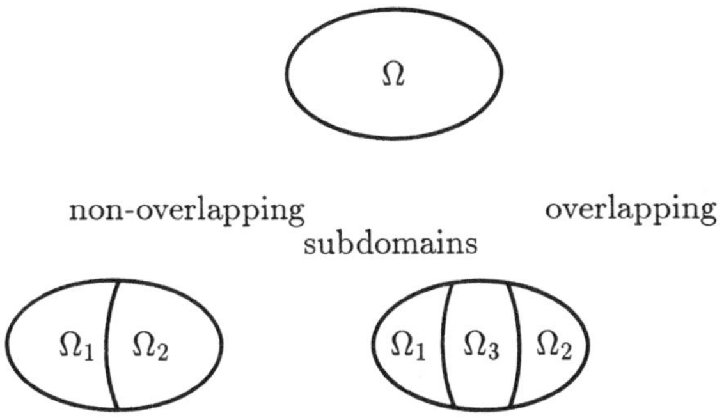

$\Omega_1 \cup \Omega_2 = \Omega$ $\qquad$ $\Omega_1 \cup \Omega_2 = \Omega$

$\Omega_1 \cap \Omega_2 = \emptyset$ $\qquad$ $\Omega_1 \cap \Omega_2 = \Omega_3$ "overlap"

fig. 1: Overlapping and non-overlapping subdomains

The overlapping decomposition follows from a special decomposition (tearing) of the governing equations which will be described in the next section. With regards to the iterative procedure we assume that for all i, j

$$\Omega_i \neq \Omega_i \cap \Omega_j \ , \tag{4}$$

which means, that for a domain as shown in fig. 1 the iteration would be carried out on the subdomains Ω_1 and Ω_2 only, as Ω_3 is completely contained in Ω_1.
In general, we have to find the maximal independent subdomains with the property that none is completely contained in any of the remaining.

4 Structural Optimization for Relaxed Compatibility and Domain Decomposition

4.1 Governing Equations and Relaxed Compatibility

We consider a linear elastic structure with several loading cases $j = 1, 2, \ldots, q$.
The optimization problem can be stated as follows:

$$\min G(\underline{x}), \qquad \underline{x} \in X^+ \tag{5}$$

subject to

compatibility: $\underline{a}(\underline{x})\ \underline{r}^{(j)} - \underline{f}(\underline{x})\ \underline{F}^{(j)} = \underline{0}$,
equilibrium: $\underline{a}^T\ \underline{F}^{(j)} - \underline{R}^{(j)} = \underline{0}$,
structural constraints: $g_i(\underline{x}, \underline{F}^{(j)}) \geq 0$, $\forall i, j$.

Superscript (j) refers to the loading case. It should be noted, that at first constraints are assumed only for the optimization variables and for the internal forces. Displacement constraints will be considered later.

Instead of trying to solve (5) directly we consider an augmented Lagrangian problem.

With

$$\underline{\zeta}^{(j)} = \underline{a}(\underline{x})\ \underline{r}^{(j)} - f(\underline{x})\ \underline{F}^{(j)} \tag{6}$$

this augmented Lagrangian problem can be formulated:

$$\widehat{G}(\underline{x}, \underline{F}, \underline{r}, \underline{\lambda}, \underline{w}) = G(x) - \sum_{j=1}^{q} (\underline{\zeta}^{(j)})^T\ \underline{\lambda}^{(j)} + \sum_{j} w_j \parallel \underline{\zeta}^{(j)} \parallel \tag{7}$$

subject to

$$\underline{a}^T(\underline{x})\ \underline{F}^{(j)} - \underline{R}^{(j)} = \underline{0} ,$$
$$g_i(\underline{x}, \underline{F}^{(j)}) \geq 0 , \quad \underline{x} \in X^+ .$$

For sufficiently large but finite w_j a multiplier algorithm [7] can be used to solve (7):

Starting from the current point $(\underline{x}, \underline{F})_{\nu-1}$ and an updated $\underline{\lambda}_\nu$:
Step 1: Solve (7) to find $(x, F)_\nu$,
Step 2: Evaluate $\| \bullet \|_\nu$ and adjust w_j ,
Step 3: Update λ_ν to $\lambda_{\nu+1}$ and return to step 1.

From a mechanical point of view the augmented Lagrangian approach is *a successive fitting* of the relaxed compatibility equations.
A measure for the 'lack of fit' is an appropriate norm of $\underline{\zeta}^{(j)}$.

4.2 Domain Decomposition and Subproblems

As mentioned earlier, the connectivity of the governing equations plays an essential role for decomposition. The first set of constraints in (7), the equilibrium conditions, are in general unstructured and matrix $\underline{a}^T$ is sparse but can be rearranged to diagonal form (fig. 2).

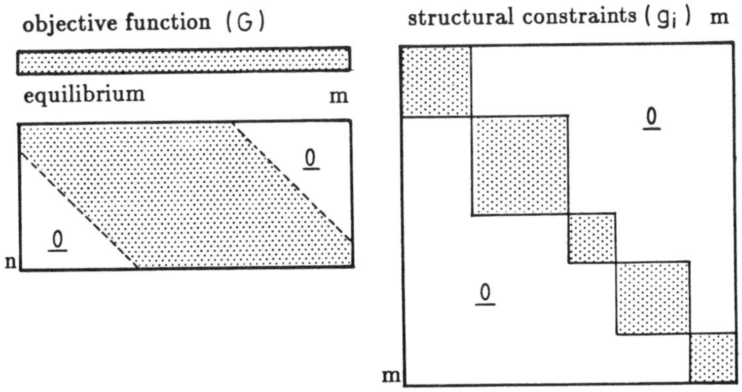

fig. 2: Constraint matrices in symbolic form

The second set of constraints is of diagonal form: Stress constraints are local functions for the structural domain. The decomposition used here is based on a special form of the nullspace of the equilibrium matrix. Originally, this method, known as the turnback LU decomposition ($TBLU$) was developed in the seventies for the purpose of structural optimization [8], [9]. It was further developed and frequently used in the future ([10], [11]).

The original idea emerged from elementary principles of statics. In a statically undetermined structure with redundancy ρ there are infinitely many possibilities to define the redundant forces. It is well known, that not only single actions can be chosen but that all linear independent combinations of redundant forces are possible. Among these 'groups actions' of redundant forces there are certain combinations which lead to compact fields of selfstresses. An example is shown in fig. 3. For simple structures, such as plane frames or trusses, certain non-algebraic methods based on graph theory seem to be suited to find these shortest 'stress-circuits'. However, for the general case these methods are limited: the kinematic conditions can only be verified by algebraic methods. The $TBLU$ decomposition of the equilibrium matrix is described in detail following [10] and [11] in appendix A and will be discussed here only in the context of an optimal decomposition.

Within the limitations of the original numbering (or bandwidth), the $TBLU$ decomposition results in a set of selfstresses $\underline{S}^{(1)}, \underline{S}^{(2)}, \ldots, \underline{S}^{(\rho)}$, the columns of the selfstress matrix $\underline{S}$. The subdomains related to these selfstresses are a maximal decomposition of the structural domain Ω.

Let $\widetilde{\Omega}_j$, $j = 1, \ldots, \rho$, denote the decomposition derived from the $TBLU$ decomposition. For general finite element systems it can not be excluded that for certain subdomains

$$\widetilde{\Omega}_i = \widetilde{\Omega}_i \cap \widetilde{\Omega}_j \; .$$

These subdomains can be found by boolean methods. We first form a boolean matrix

$$\underline{\widetilde{S}} = [\underline{\widetilde{S}}^{(1)}, \ldots, \underline{\widetilde{S}}^{(\rho)}]$$

derived from $\underline{S}$ by

$$\widetilde{S}_{ij} = \begin{cases} 1 & \text{if } S_{ij} \neq 0 \\ 0 & \text{else.} \end{cases}$$

A systematic column elimination by boolean addition results in an optimal (overlapping) decomposition

$$\Omega = \bigcup_{i=1}^{k} \Omega_i \; , \qquad k \leq \rho$$

$$\Omega_j \neq \Omega_j \cap \Omega_l \qquad \forall\, j, l \; .$$

An interesting alternative representation can be derived from the adjacency matrix

$$\underline{\widetilde{A}} = \{\widetilde{A}_{ij}\} \; ,$$

$$\text{where } \widetilde{A}_{ij} = \begin{cases} 1 & \text{if } a_{ji} \neq 0 \\ 0 & \text{else.} \end{cases}$$

An example is shown in fig. 3; for simplicity, here only the edge-adjacent elements are considered to be connected.

The corresponding $(0,1)$ matrix for the selfstresses $\underline{\widetilde{S}}$ is found by $TBLU$ decomposition. The row numbering corresponds to the connections of elements within a selfstress (fig. 3). For this alternative representation we find the optimal decomposition by modulo 2 algebra. All columns $\underline{\widetilde{S}}_j$ where

$$\underline{\widetilde{A}}\; \underline{\widetilde{S}}_j = \underline{0}$$

form a circuit and therefore include other selfstresses. The set of selfstresses not included in these circuits is denoted by $\mathcal{M}$ and the set of circuits by $\mathcal{C}$.

$\mathcal{M}$ and $\mathcal{C}$ form a matroid [12] and the corresponding subdomains are an optimal decomposition of the structural domain.

The equality constraints of the augmented Lagrangian prcblem (7) can now be eliminated; the underdetermined system

$$\underline{a}^T\; \underline{F}^{(j)} - \underline{R}^{(j)} = \underline{0} \tag{8}$$

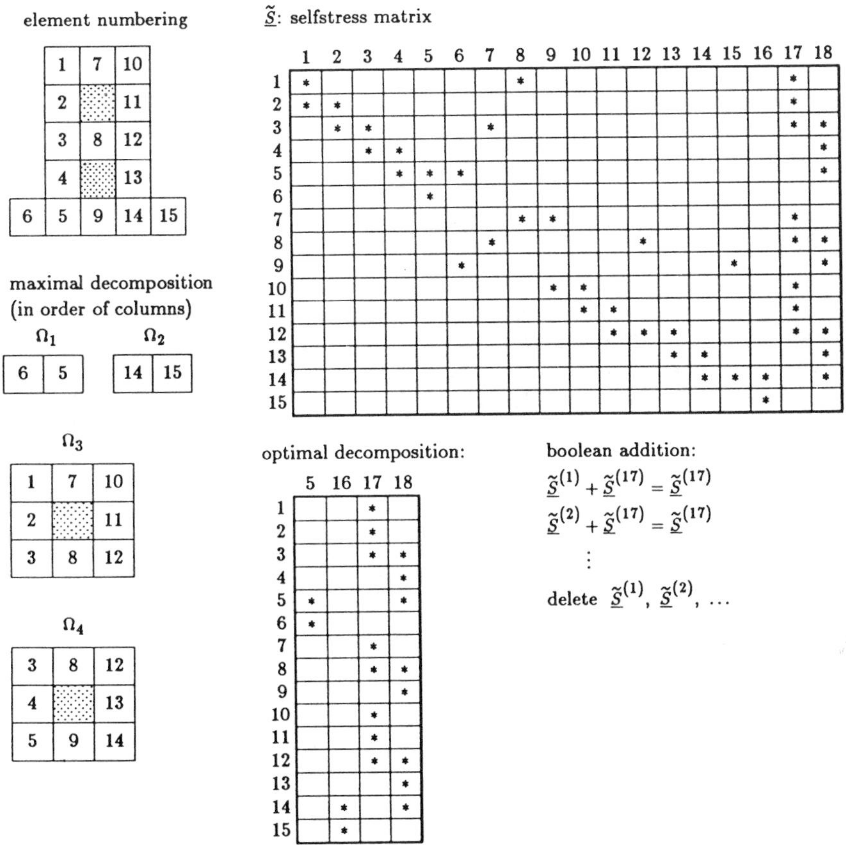

fig. 3: Domain decomposition

can be solved when the $(m \times \rho)$ basis matrix $\underline{S}$ of the nullspace of $\underline{a}^T$ is known. The general solution is given by

$$\underline{F}^{(j)} = \underline{F}_R^{(j)} + \underline{S}\ \underline{y}^{(j)} \tag{9}$$

where $\underline{F}_R^{(j)}$ is any particular solution to (8) and $\underline{y}$ is a $(m \times \rho)$ vector of generalized redundant forces.

The reduced optimization problem can now be stated as follows

$$\text{minimize} \left\{ G(\underline{x}) - \sum_{j=1}^{q} (\underline{\varsigma}^{(j)})^T \, \underline{\lambda}^{(j)} + w_j \parallel \underline{\varsigma}^{(j)} \parallel \right\} \tag{10}$$

$$\begin{aligned} &\text{subject to } g_i(\underline{x}, \underline{y}^{(j)}) \geq 0 \,, \quad \underline{x} \in X^+ \,, \\ &\text{and} \qquad \underline{\varsigma}^{(j)} = \underline{f}(\underline{x})^{-1} \underline{a}(\underline{x}) \, \underline{r}^{(j)} - \underline{S} \, \underline{y}^{(j)} - \underline{F}_R^{(j)} \end{aligned}$$

The optimization variables $\underline{x}$ are local variables (element properties, nodal coordinates, etc.), the generalized redundant forces $\underline{y}_i$ are defined for each of the subdomains Ω_i, $i = 1, \ldots, k$ and the structural constraints are local constraints, which can be formulated for each subdomain independently. Thus we have to solve k subproblems Q_α, $\alpha = 1, \ldots, k$ of the following form

$$Q_\alpha\text{: minimize } \{G(\underline{x}) - H(\underline{x}, \underline{y}^{(j)}, \underline{r}^{(j)})\} \tag{11}$$

$$\begin{aligned} &\text{subject to } g_i(\underline{x}, \underline{y}^{(j)}) \geq 0 \\ &\qquad \underline{x}, \underline{y}^{(j)} \text{ on } \Omega_\alpha \,, \quad x \in X^+ \,, \\ &\qquad \underline{r}^{(j)} \text{ on } \Omega. \end{aligned}$$

4.3 Iterative Solution

For simplicity we delete the superscript (j); the iteration must be performed for every loading case separately.

Step 0: Initialization $(\nu = 0)$
Choose any kinematically admissible $\underline{r}_\nu$
Step 1: Solve $Q_1, \ldots, Q_k$ for $\underline{r} = \underline{r}_\nu$ (fixed) by an iterative method: $\underline{x}_\nu, \underline{y}_\nu$
Step 2: Solve $\underline{K} \; \underline{r} = \underline{R}$ to obtain $\underline{r} = \underline{r}_{\nu+1}$
Step 3: Restart with step 1 until convergence criteria are reached.

The nodal displacements $\underline{r}$ are the *coordination variables* for this stage approach. Side constraints in $\underline{r}$ can be included by modification of step 2.

5 Example

A simple truss with three redundant forces and two subsystems is considered. The total weight is taken as objective function, two different load

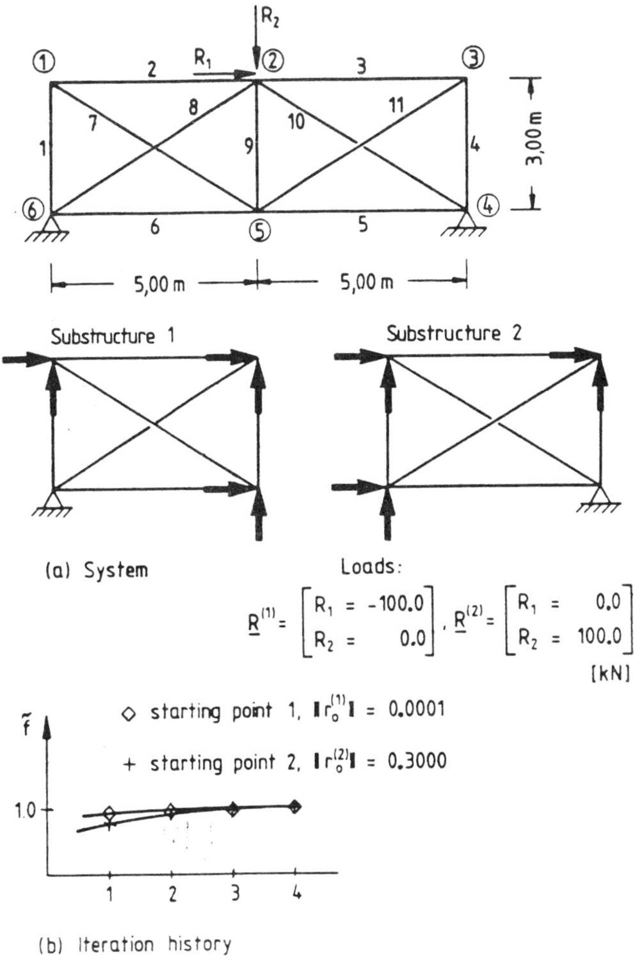

fig. 4: Example for iterative design of overlapping substructures

vectors and stress and displacement constraints are assumed. The optimal decomposition into subdomains is shown in fig. 4 together with the iteration history.

The optimal cross-sections are given in table 1. It is interesting to note, that zero cross-sections for the truss elements can be included. As the variables are statically and kinematically admissible in every step of the iteration the structure degenerates in this case into the two element truss as shown in fig. 5.

table 1: Optimal cross-section (example)

i	$(b) : A_i\,[cm^2]$	$(a) : A_i\,[cm^2]$
1	1.206	1.132
2	2.011	1.821
3	2.010	1.820
4	1.206	1.142
5	0.100	0.100
6	0.100	0.100
7	2.345	2.104
8	3.000*	3.000*
9	2.413	2.107
10	3.000*	3.000*
11	2.345	2.221
	$(9791\,cm^3)$	$(8487\,cm^3)$

$^*)\,\sigma = 24.08\,kN/cm^2\,(0.3\%)$

$(a):\quad |\sigma| \leq 24\,kN/cm^2$

$$.1 \leq A_i \leq 3.\,cm^2$$

$(b):\quad |r_i| \leq 1.2\,cm \qquad$ (in addition to (a))

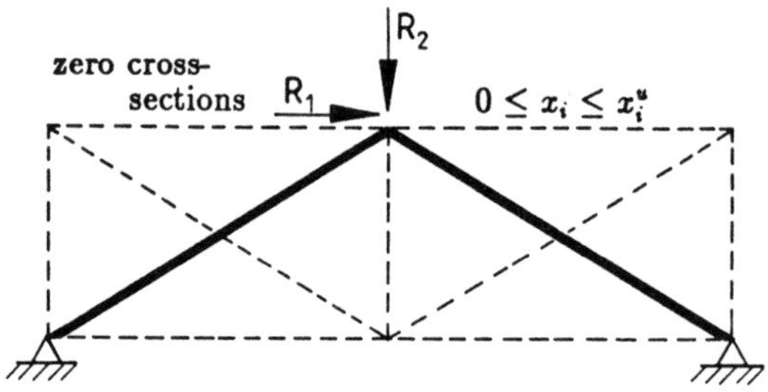

fig. 5: Optimal truss for zero cross-sections

6 Concluding Remarks

For structural optimization problems with a great number of variables, decomposition into a sequence of subproblems, which are easier to solve, seems to be one promising way to overcome the well known difficulties encountered for large scale problems. The method proposed here is based on a special domain decomposition together with an alternating sequential solution of the subproblems generated by this decomposition.

At present the basic decomposition has been tested for a wide class of finite elements and the iterative solution of simple optimization problems has been carried out for a few cases.

It is planned to link an automated "decomposer" to a general purpose finite element programming system which includes an "optimizer" already and to extend the method to second order optimization problems.

Appendix A

$\underline{A}$: a real $n \times m$ matrix, $m > n$
rank $(\underline{A}) = n$
redundancy $\rho = m - n$ (dimension of nullspace)
$\underline{B}$: $m \times \rho$, the basis matrix of the nullspace

$$\underline{A}\,\underline{B}^{(j)} = \underline{0}, \qquad \underline{B} = [\underline{B}^{(1)}, \ldots, \underline{B}^{(\rho)}] \tag{A1}$$

$\underline{A}$ and $\underline{B}$: banded and sparse

LU procedure

$\underline{A}$ is assumed to be of 'minimal' bandwidth, obtained from a column/row permutation algorithm, which minimizes the profile of $\underline{A}\,\underline{A}^T$.

Step 1: triangular decomposition of $\underline{A}$

$$\underline{A}\,\underline{V} = \underline{L}[\underline{U}_1|\underline{U}_2] \tag{A2}$$

by Gauss-Jordan factorization with row pivoting*), where $\underline{V}$ is a permutation matrix.

Step 2: Back substitution, solve

$$\underline{U}_1\,\underline{\hat{B}} = U_2$$

to obtain basis matrix

$$\underline{B}^{LU} = \underline{V}\begin{bmatrix} -\underline{\hat{B}} \\ \underline{I} \end{bmatrix}$$

*) a detailed description is given in [10].

Turnback LU procedure*)

For $j = 1, \ldots, \rho$, we let $\underline{B}^{(j)}$ denote the jth column of $\underline{B}$; i.e.

$$\underline{B} = [\underline{B}^{(1)}, \ldots, \underline{B}^{(\rho)}]$$

For each $\underline{B}^{(j)}$, let $k = k(j)$ be the smallest index such that

$$B_i^{(j)} = 0 \qquad i = k+1, \ldots, m$$

This implies that the first k columns of the matrix $\underline{A}$, i.e.

$$\{\underline{A}^{(1)}, \ldots, \underline{A}^{(k)}\} \tag{A3}$$

are linearly dependent.

We try to find a subset of (A3) which is still linearly dependent. Note that index $k = k(j)$ for each $i \in \{1, \ldots, \rho\}$ can be readily determined by looking at the column permutation $\underline{V}$ in the LU factorization (A2), and thus we never really have to compute $\underline{B}^{(j)}$.

In the turnback scheme, one finds the minimal linearly dependent set of columns of the form

$$\underline{E}_t = [\underline{A}^{(k)}, \underline{A}^{(k-1)}, \ldots, \underline{A}^{(t)}] \qquad t = k-1, k-2, \ldots, 1$$

such that $\underline{A}^{(k)}$ is linearly dependent of the others. That is, $\underline{E}_{t^*}$ is the desired minimal set if $\underline{A}^{(k)}$ is not linearly dependent of $\underline{A}^{(k-1)}$ to $\underline{A}^{(t^*+1)}$ but of $\underline{A}^{(k-1)}$ to $\underline{A}^{(t^*)}$. Note that in $\underline{E}_t$, the columns $\underline{A}^{(i)}$ are arranged in the *opposite* order of indices. This is crucial in reducing the band width of the resulting $\underline{B}$.

Suppose that $\underline{E}_{t^*}$ is the minimal dependent set. It follows that $\underline{A}^{(k)}$ is linearly dependent onthe set of columns

$$\{\underline{A}^{(k-1)}, \ldots, \underline{A}^{(t^*)}\}$$

Thus, there exist scalars $(\beta_k, \ldots, \beta_{t^*})$, such that $\beta_k \neq 0$ and

$$\sum_{i=k}^{t^*} \underline{A}^{(i)} \beta_i = \underline{0}$$

If we let

$$\underline{y}^{(j)} = (\underbrace{0, \ldots, 0}_{t^*-1}, \beta_{t^*}, \beta_{t^*+1}, \ldots, \beta_{k-1}, \beta_k, \underbrace{0, \ldots, 0}_{m-k})^T$$

then $\underline{y}^{(j)}$ is a solution of (A1). This solution $\underline{y}^{(j)}$ compares favourably with $\underline{B}^{(j)}$ as

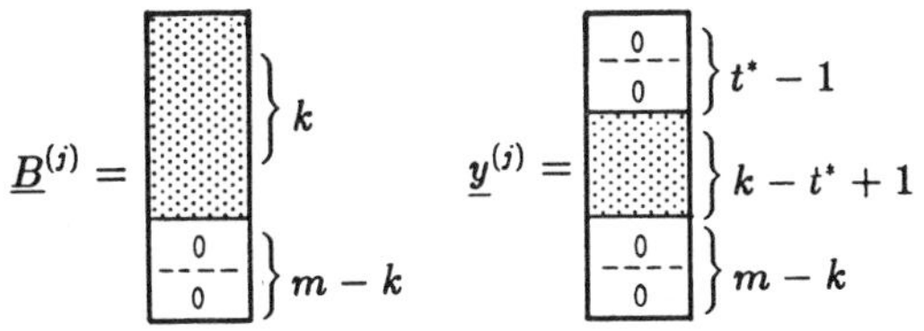

We apply this turn-back scheme to each $j \in \{2, \ldots, \rho\}$ to obtain a solution $\underline{y}^{(j)}$ of (1); for $j = 1$, we use the first column of $\underline{B}^{LU}$. The resulting self-stress matrix $\underline{B} = [\underline{y}^{(1)}, \ldots, \underline{y}^{(\rho)}]$, where $\underline{y}^{(1)} = \underline{B}^{(1)}$, tends to be sparse and banded. Schematically, this $\underline{B}$ compares favourably with $\underline{B}^{LU}$ as

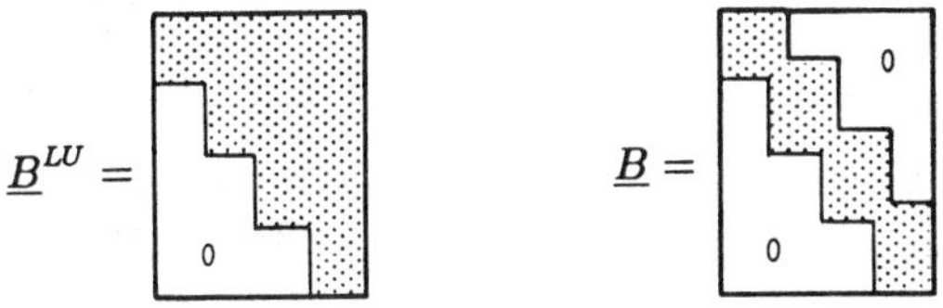

The turnback procedure can be carried out, efficiently, by the LU decomposition (or orthogonal reduction by Givens rotation) applied to the matrix

$$[\underline{A}^{(k)}, \underline{A}^{(k-1)}, \ldots, \underline{A}^{(1)}]$$

The algorithm is terminated, the first time one cannot find a pivot in the next column, where the element in the first row of that column is nonzero. Upon finding the minimal linearly dependent set in this way, the coefficients $(\beta_k, \beta_{k-1}, \ldots, \beta_{t^*})$ can be determinded by solving a triangular system using (a part of) the 'current' values of $\underline{A}$.

References

[1] G. Kron, 'Solving Highly Complex Elastic Structures in Easy Stages', *J. Applied Mechanics*, (June 1955).

[2] G. Kron, 'Solution of Complex Nonlinear Plastic Structures by the Method of Tearing', *J. Aeronautical Sciences*, (June 1956).

'Improved Procedure for Interconnecting Piece-wise Solutions', *J. Franklin Inst.*, 385–392 (Nov. 1956).

[3] D.M. Himmelblau (Ed.), *Decomposition of Large-Scale Problems*, North-Holland, Amsterdam, 1973.

[4] R. Glowinski, G.H. Golub, G.A. Meurant and J. Périaux (Eds.), *Proceedings of the First Int. Symp. on Domain Decomposition Methods for Partial Differential Equations*, SIAM, Philadelphia, 1988.

[5] P.L. Boland, T.H.H. Pian, 'Large Deflection Analysis of Thin Elastic Structures by the Assumed Stress Hybrid Finite Element Method', *Computers & Structures*, Vol. 7, pp. 1–12, 1977.

[6] A.K. Noor, S.J. Hartley, 'Nonlinear Shell Analysis via Mixed Isoparametric Elements', *Computers & Structures*, Vol. 7, pp. 615–626, 1977.

[7] D.M. Greig, *Optimization*, Longman, London, 1980.

[8] A. Topçu, 'Ein Beitrag zur systematischen Berechnung finiter Elementstrukturen mit der Kraftmethode', Dissertation, Universität GH Essen, Essen, 1979.

[9] G. Thierauf and A. Topçu, 'Structural optimization using the force method, presented at World Congress on Finite Element Methods in Structural Mechanics', Bournemouth, England, 1975.

[10] I. Kaneko, M. Lawo and G. Thierauf, 'On Computational Procedures for the Force Method', *Int. J. Num. Meth. Eng.*, **18**, 1469–1495 (1982).

[11] M.W. Berry et.al., 'An Algorithm to Compute a Sparse Basis of the Null Space', *Numer. Math.*, **47**, 483–504 (1985).

[12] F. Harary, *Graph theory*, Addison-Wesely, Reading Massachusetts, 1972.

Approximation Methods in Optimization Using Design and Analysis of Numerical Experiments

A.J.G. Schoofs, D.H. van Campen
Eindhoven University of Technology, The Netherlands

ABSTRACT

Methods for the planning and analysis of physical experiments appear to be also very useful in the field of structural optimization, where they are used to plan numerical experiments for building approximate analysis models describing the behaviour of structures. The classical methods of experimental design, such as latin squares and fractional factorial designs, require much statistical expertise, and they are difficult to apply in the more complicated situations. Research has therefore been shifted towards the optimal experimental design theory. Based on this theory an interactive program, called CADE, has been developed which consists of the modules: model input, optimal design, and model fitting. In this paper, strategies for experimental design and model building are discussed. The application of the proposed method is illustrated with two approximate analysis models.

INTRODUCTION

The constrained optimization problem can be generalized as:

Find the set of design variables $\underset{\sim}{x} = [x_1 \; .... \; x_n]^T$ [1)] (1)

such that the objective function $F(\underset{\sim}{x})$ is minimized (2)

subject to:

Inequality constraints: $g_j(\underset{\sim}{x}) \leq 0 \qquad j = 1,...,m$ (3)

Equality constraints: $h_k(\underset{\sim}{x}) = 0 \qquad k = 1,...,l$ (4)

Side constraints: $x_i^l \leq x_i \leq x_i^u \qquad i = 1,...,n$ (5)

In computer programs for structural optimization the finite element method (FEM) is often used for modeling and analysis of the structure. This offers the possibility to model a great variety of structures in a general, accurate and flexible way. The accuracy can easily be

[1)] We use the following notations: $\underset{\sim}{x}$ is a column matrix
$\underline{x}$ is a stochastic variable
$\hat{x}$ is an estimated variable

controlled by an appropriate modification of the FEM model. In the iterative solution process of optimization problems even slight improvements of a structure are of interest, so a realistic model and accurate analysis results are of great importance. However, these advantages must be paid for with some serious drawbacks. First, the computing time is large, second,the implementation of optimization concepts in FEM packages is non-standard.Hence, to solve structural optimization problems using FEM there is a strong demand for explicit approximations of the problem functions (the objective and constraint functions) in order to reduce the required number of detailed FEM analyses.

Commonly the problem functions are approximated only using data gathered at the current design point based on function values and first derivatives, because such data can be computed rather straightforwardly. Additional use of second derivatives, computed at the current design point, in order to improve the approximations is hardly applied since the CPU-time, required for generating second derivatives using FEM models, is prohibitive. Therefore, if we want to use second order information we are obliged to extract such information from function values and/or first derivatives computed at different design points, for instance, the successive solution points of the optimization process. There is another reason for this approach: using successive design points renders more effect of the invested computational effort, because the information from a previous design point is not discarded as soon as a new design point has been found.

In this paper we describe an approach in which the Experimental Design Theory (EDT), Montgomery [1], is used as a tool in building approximate analysis models. This theory has been developed for the planning and analysis of comprehensive physical experiments in order to reduce the number of required experiments while preserving the amount of information which can be extracted from the experiments. This situation is very similar to that of structural optimization, where the number of expensive FEM analyses has to be minimized, Schoofs [2]. FEM computations can be regarded as numerical experiments, where the design variables are treated as input quantities. All computable properties of the structure, such as weight, displacement, stresses, etc. can be regarded as response quantities of the numerical experiment. The approximating models will be derived for these responses and they can be substituted in the optimization problem given by the relations (1) through (5).

EXPERIMENTAL DESIGN THEORY

Regression model

EDT consists of two main parts. The first part concerns the planning of experiments and ends up with a list of experiments to be carried out. This list is called the experimental design (ED). In the second part the experimental results are analyzed and fitted to some mathematical relationship.

When a structure is determined by n design variables, $\underset{\sim}{x}$, according to (1), we may search for t functions describing the response quantities:

$$y_j = y_j(\underset{\sim}{x}) \qquad j = 1,\ldots,t \tag{6}$$

in a certain limited area according to (5). In the sequel we will consider only one response y_j and for brevity we omit the index j. To find the relation $y = y(\underset{\sim}{x})$ we assume a mathematical model. Mostly a linear model of the form will apply:

$$\underline{y} = \underset{\sim}{f}^T(\underset{\sim}{x})\underset{\sim}{\beta} + \underline{e} = \beta_1 f_1(\underset{\sim}{x}) + \ldots \beta_k f_k(\underset{\sim}{x}) + \underline{e} \tag{7}$$

where the components β_1 ,..., β_k of the column $\underset{\sim}{\beta}$ are unknown parameters;the model is linear in the β's. The functions $f_1(\underset{\sim}{x})$, ...,$f_k(\underset{\sim}{x})$ are the components of the column $\underset{\sim}{f}(\underset{\sim}{x})$. We can choose both linear and non-linear functions for them; in most cases for (7) a polynomial is chosen. The variable $\underline{e}$ accounts for the stochastic and/or deterministic model error that is inherent in every model assumption.

Parameter and response estimation

The formulation of an ED implies:

1. The choice of discrete values (levels) for all design variables x_i.
2. The choice of certain combinations of levels of the different x_i. Each set of design variables determines one specific structure.

For this moment we assume that somehow an experimental design has been determined consisting of N points (N > n) in the design variable space, represented by the sets of design variables $\underset{\sim}{x}_1,\ldots,\underset{\sim}{x}_N$. If we analyze the structure at these points yielding the column of response quantities $\underset{\sim}{y} = [y_1\ y_2\ \ldots.\ y_N]^T$, then by using a least-squares technique the unknown parameters $\underset{\sim}{\beta}$ can be estimated from:

$$\hat{\underset{\sim}{\beta}} = (X^TX)^{-1}X^T\underset{\sim}{y} \tag{8}$$

where X is the (N*k) "design matrix", which is given by:

$$X = [\underset{\sim}{f}(\underset{\sim}{x}_1)\ \underset{\sim}{f}(\underset{\sim}{x}_2)\ \ldots.\ \underset{\sim}{f}(\underset{\sim}{x}_N)]^T \tag{9}$$

Subsequently, for each design point $\underset{\sim}{x}$ the response variable can be estimated from:

$$\hat{y}(\underset{\sim}{x}) = \underset{\sim}{f}^T(\underset{\sim}{x})\hat{\underset{\sim}{\beta}} \tag{10}$$

Accuracy of the estimates

A measure for the accuracy of $\hat{\underset{\sim}{\beta}}$ is the variance-covariance matrix $V(\hat{\underset{\sim}{\beta}})$, which is defined as:

$$V(\hat{\underset{\sim}{\beta}}) = E\left[(\hat{\underset{\sim}{\beta}} - \underset{\sim}{\beta})(\hat{\underset{\sim}{\beta}} - \underset{\sim}{\beta})^T\right] = (X^TX)^{-1}\sigma^2 \tag{11}$$

where σ^2 is the variance of the response variable $\underline{y}$. For the response estimator $\hat{\underline{y}}(\underset{\sim}{x})$ the variance $V(\hat{\underline{y}}(\underset{\sim}{x}))$ is used as a measure for its accuracy. From (10) and (11) follows:

$$V(\hat{\underline{y}}(\underset{\sim}{x})) = \underset{\sim}{f}^T(\underset{\sim}{x})(X^TX)^{-1}\underset{\sim}{f}(\underset{\sim}{x})\sigma^2 \tag{12}$$

Planning of the experiments

The first part of EDT concerns the determination of the list of experiments to be carried out, the experimental design (ED), in such a way that model parameters and responses can accurately be estimated. For this purpose several methods are available. We will treat two of these methods. In the next section a more or less classical method, resulting in so-called 2^n-designs, is discussed. The optimal experimental design theory is the subject of a subsequent section.

2^n-Factorial designs

An ED, where each of the n design variables (here called "factors") is varied on two levels, is called a 2^n-factorial design. Such designs are very popular and are quite suitable to develop regression models. There exists a wealth of literature on 2^n-designs, e.g. Box et al.[3],Montgomery[1], and only the main topics of the method are discussed here.

In 2^n-designs a special notation is adopted. The design variables, i.e. the factors, are indicated by capitals: A, B, C ... etc., and they can be adjusted to two levels: "high" and "low". "High" and "low" are just names; their specific meaning must be properly defined. A specific experiment is indicated by a string of lower case letters. This string only contains those characters that correspond to the factors at the high level. The observation of the response quantity for a given experiment is denoted by a string of the same lower case letters as the string denoting that experiment. For example, observation ab is the result of the experiment ab, where the factors A and B are both on the high level.Fig.1 illustrates the possible experiment in a complete 2^3-factorial design with the factors A, B and C. The vertex (1) in Fig.1 represents the experiment with all factors at the low level.

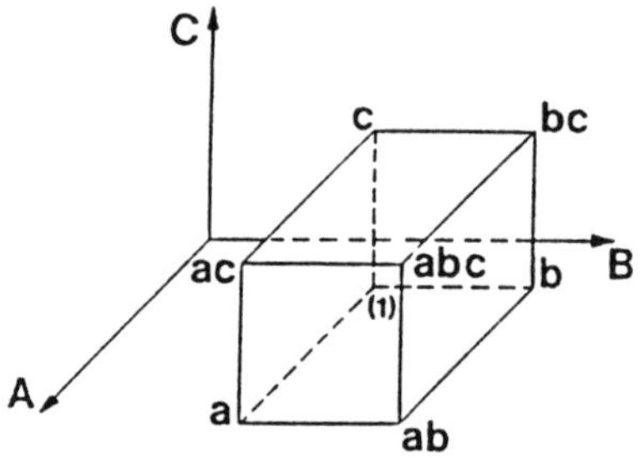

Fig.1. Eight possible experiments in a 2^3-design.

A complete 2^3-design contains eight different experiments, which enables us to estimate up to a maximum of eight parameters in a regression equation, for instance:

$$y = \beta_1 + \beta_A x_A + \beta_B x_B + \beta_C x_C + \beta_{AB} x_A x_B + \beta_{AC} x_A x_C + \beta_{BC} x_B x_C + \beta_{ABC} x_A x_B x_C \quad (13)$$

The parameters are often called effects and are usually indicated by the same uppercase characters as the factors: the effect β_1 is indicated by I, β_A by A etc. Main effects are those effects which are indicated by just one letter, i. e. the effects I, A, B and C in the considered example. Effects indicated by more than one letter, for example AB, AC, etc., are called interactions or interactive effects.

In most cases it is possible to code the levels of the factors to +1 and -1. If for the 2^3-design, per experiment one observation is made, the design matrix, X, is given by:

$$
\begin{array}{c}
\text{Effects} \\
\begin{array}{cccccccc|c}
I & A & B & C & AB & AC & BC & ABC & \text{Observation}
\end{array}
\end{array}
$$

$$
X = \begin{bmatrix}
+1 & -1 & -1 & -1 & +1 & +1 & +1 & -1 \\
+1 & +1 & -1 & -1 & -1 & -1 & +1 & +1 \\
+1 & -1 & +1 & -1 & -1 & +1 & -1 & +1 \\
+1 & -1 & -1 & +1 & +1 & -1 & -1 & +1 \\
+1 & +1 & +1 & -1 & +1 & -1 & -1 & -1 \\
+1 & +1 & -1 & +1 & -1 & +1 & -1 & -1 \\
+1 & -1 & +1 & +1 & -1 & -1 & +1 & -1 \\
+1 & +1 & +1 & +1 & +1 & +1 & +1 & +1
\end{bmatrix}
\begin{bmatrix}
(1) \\ a \\ b \\ c \\ ab \\ ac \\ bc \\ abc
\end{bmatrix}
\qquad (14)
$$

and the parameters β_1 through β_{ABC} can be estimated using (8). Another algorithm to estimate a certain effect can be explained with the help of the above mentioned example. For instance, an estimate AC (= β_{AC}) of the effect AC can be calculated as follows:

1. summing all the observations with $x_A * x_C = +1$ ("high"):

$$h = (1) + a + b + ac + abc$$

2. next summing all the observations with $x_A * x_C = -1$ ("low"):

$$l = a + c + ab + bc$$

3. dividing the difference (h - l) by the total number of observations, results in the estimate

$$\hat{AC} = \{((1) + a + b + ac + abc) - (a + c + ab + bc)\}/8 \qquad (15)$$

For the other effects similarly simple relations can be derived. The right hand side of (15) is called the contrast associated with the effect AC.

In order to estimate all possible effects from a 2^n-design, a complete set of 2^nexperiments should be used and under each experiment at least one observation should be made. However, for increasing n the ED will become impractically large and the number of experiments must be reduced. This is possible since in many problems the great majority of effects are interactions of little importance. To estimate the effects of interest usually much fewer observations are necessary than in a complete 2^n-design. To account for this a (small) number of chosen experiments is grouped in a so-called fractional design and observations are made only for the experiments in this fractional design. The crucial point in this procedure is the choice of the experiments in the fraction. For that purpose comprehensive tabulated fractional designs are available, e.g. Box et al.[3]. Furthermore, computer algorithms have been developed. The program DSIGN, developed by Patterson [4], produces factorial designs for variables at any number of levels. The program CADE, Nagtegaal [5], incorporates besides algorithms for optimal experimental designs, also facilities for the generation of (fractional) 2^n-designs.

Optimal experimental design

In the optimal experimental design theory (ODT), the accuracy of the estimated parameters and/or the variance of the response estimator are used as criteria to search for optimal EDs. ODT is useful in those situations where classic designs are unsuitable or unavailable, that is when:

- the experimental region is irregularly shaped due to constraints on the variables,
- it is needed to augment an existing design,
- the number of levels of the variables varies considerably,
- design have to be constructed for special models, i.e. other than polynomial models,
- designs have to be constructed for simultaneous observation of several responses.

ODT was initiated by Smith [6]. Important contributions to the methods were made by Kiefer [7] and Fedorov [8]. Nagtegaal [5] gave some generalizations for the construction of experimental designs in the case of several regression models being used simultaneously.

In ODT a number of N, not necessarily distinct points, $\underset{\sim}{x}_1$, ..., $\underset{\sim}{x}_N$, are determined from an a priori chosen and fixed set of r discrete candidate points. It is assumed that somehow an appropriate choice for the candidate points can be made. The problem is now to choose N points from the r candidate points, resulting in the best N-point design.

Experimental designs can be evaluated, using the variances of the parameter estimators $V(\hat{\underset{\sim}{\beta}})$ or the variance of the response estimator $V(\hat{y}(\underset{\sim}{x}))$ as a measure, see (11) and (12) respectively. In both cases the quality of the ED is a function of the matrix $(X^TX)^{-1}$ and the objective is to determine that ED among all possible N-point EDs which makes $(X^TX)^{-1}$ minimal. However, the minimum of a matrix is not a well defined concept and a number of operational criteria have been developed. The most important of these criteria are:

D-optimality: minimize $\det(X^TX)^{-1}$ (16)

G-optimality: minimize the maximum response variance (17)

V-optimality: minimize the average response variance (18)

Mitchell [9] developed an efficient algorithm called DETMAX as the most popular in ODT. The algorithm starts with an initial ED. During each iteration step, the candidate point which results in the largest increase of $\det(X^TX)$ is added to the design, and subsequently, the point which results in the smallest decrease of $\det(X^TX)$ is removed from the design. The algorithm generates high quality EDs at relatively low computing costs.

MODEL BUILDING

The building of an accurate regression model for a given system or structure is an iterative process. Initially the following questions have to be resolved to some degree:

- which variables play a role and what is their range of interest,
- which form of functions $f_i(\underset{\sim}{x})$, see (7), may be suitable to describe the searched relationship.

A good strategy is to begin with moderate model demands thus reducing the initial computing costs. The iterative model building process is able to enhance models in a cost efficient way, see Fig.2.

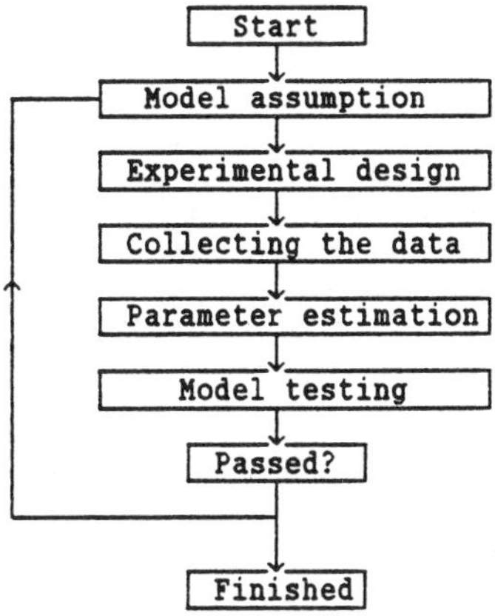

Fig.2. Scheme for model building

At the start of each iteration step a model assumption of the type of (7) must be available. The iteration step then involves generation of an ED, collection of data, followed by estimation of the parameters from the collected data, and the evaluation of the model. Evaluation implies answering questions like:

- Is the model valid?
- Are the estimated parameters accurate enough?
- Are the response predictions accurate enough for all relevant values of $\underset{\sim}{x}$?

If the results of the testing require further model improvement, it is necessary to perform another model building cycle consisting of experimental design, data collection, parameter estimation, and retesting.

COMPUTER PROGRAM FOR EXPERIMENTAL DESIGN

Nagtegaal [5] developed an interactive computer program called CADE, which stands for "Computer Aided Design of Experiments. Besides for experimental design, facilities for the analysis of experiments have also been implemented. For the experimental design part, the core of the program ACED, Welch [10], has been used. In CADE the optimality criteria and algorithms of ACED have been generalized to the case of simultaneous observation of several response quantities. CADE has been coded in Fortran 77, and runs on Apollo D3000 work stations, Vax systems and an Alliant FX40 computer. The program consists of three main modules, being model input, design of experiments and parameter estimation.

In the model input module all kinds of linear models can be entered, stored in a file or read from a previously prepared file without the need for user supplied subroutines.

The experimental design module offers the following facilities:

- Optimal design for a single and for several simultaneous responses.
- Implementation of the D-, G- and V-optimality criteria.
- Implementation of several optimization algorithms including DETMAX.
- Determination of the characteristics of user-supplied designs.

- Augmentation of existing experimental designs.
- Generation of (fractional) 2^n-designs.

Finally, the main characteristics of the parameter estimation module are:
- Parameters can be selected by means of stepwise regression, backward elimination and forward selection; they can also selected "by hand".
- Parameters can be protected against removing from the model.
- Parameters are estimated accurately by means of QR-decomposition, followed by an iterative refinement procedure.

APPLICATIONS

The procedures described in the preceding sections have been applied to several mechanical engineering problems. In this section two applications are presented.

Three-dimensional bearing problem

Dry running journal bearings are important connecting elements in mechanical engineering. Many practical bearing structures can be modeled according to Fig.3a. There is a need for a three-dimensional model due to the inclination of the shaft.

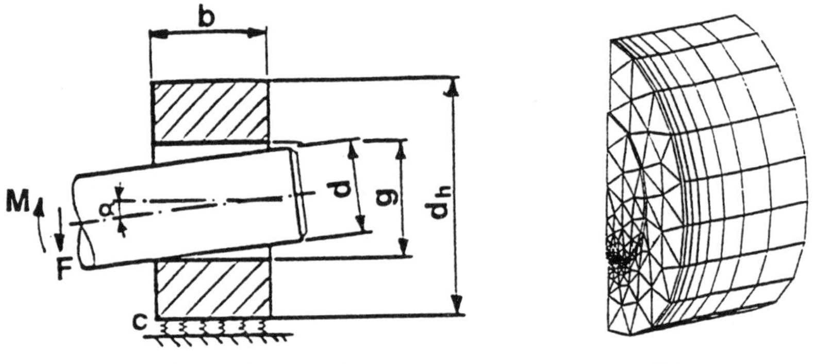

(a) (b)

Fig.3 (a)Modeling and (b) FEM model of journal bearing.

Wouters [11] developed for this situation regression models describing the surface pressure in the contact zone between shaft and journal as a function of a number of design variables. Initially considered design variables are indicated in Fig.3a. After explorative calculations it was decided to develop regression models approximating the following dimensionless relations:

$$\frac{\sigma_v(i)}{E_1} = f\left(\frac{s}{d},\ \alpha,\ \frac{F}{E_1 d^2} \right) \qquad i = 1, 2, \ldots \tag{19}$$

where $\sigma_v(i)$ is the contact pressure in a discrete point, i, of the contact zone; s = g - d is the clearance between shaft and journal.

From the explorative calculations it appeared that the regression equations should contain third-order main effects. All first order interaction terms and a part of the second-order ones were inserted into the models. The design variables were all varied on four levels.

It was decided to use a complete 4^3-design, which was transformed to a complete 2^6-one. The required FEM analyses of 64 different contact problems were carried out using the I-DEAS package [12]. Fig.3b shows a sample of the used three-dimensional element meshes. Each analysis required about 2.5 hours of computing time on a VAX 11/750 computer. Fig.4 represents qualitative pictures of the contact pressure computed from the derived regression models for two different inclination angles of the shaft. From Fig.5 it can be seen that contact pressures computed from the regression models agree very well with results from direct FEM analyses.

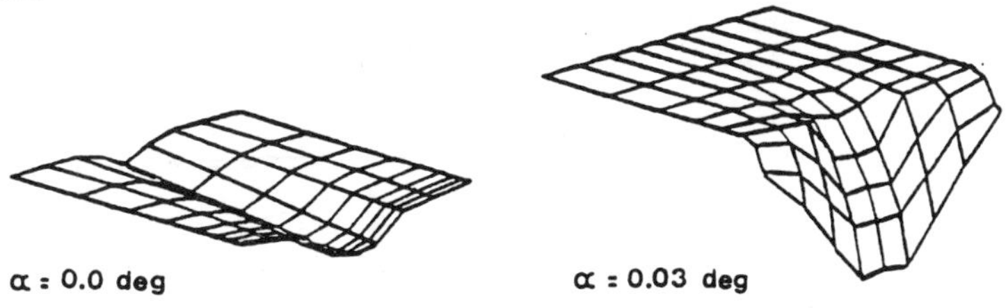

Fig.4 Qualitative pictures of the contact pressure.

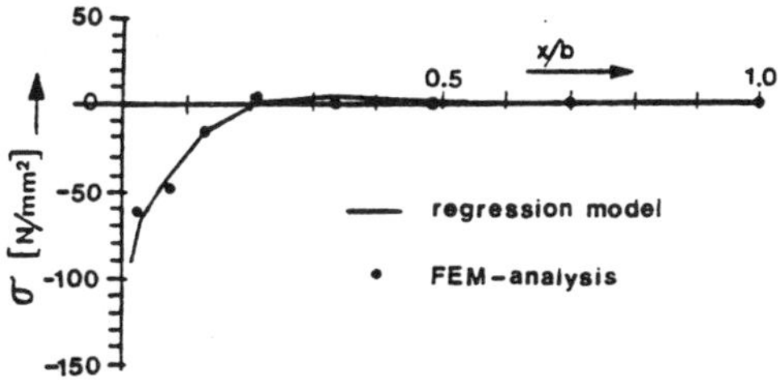

Fig.5 Contact pressure from regression models versus direct FEM results.

Furthermore, Wouters [11] made the following interesting comparison between the use of regression models and a direct FEM analysis.

From one FEM analysis using a mesh of 1000 nodal points and 1290 6-node elements the following results emerge:

number of displacements :	1000 * 3 =	3000
number of nodal point forces:	1000 * 3 =	3000
number of stress components :	1290 * 6 * 6 =	46440
total number of response quantities:		52440

The computing time for one FEM analysis of about 2.5 hours, results in an average computing time of 0.17 sec for one response quantity. The computing time required for one evaluation of a regression model is approximately 0.01 sec on the same computer. So, if regression models were available for all response quantities defined in a particular FEM model, evaluation of all those models would be a factor of 17, faster compared to a direct FEM analysis. This result may be of interest for the derivation of regression models for use in real time computer simulations, especially if a relatively small number of response quantities is relevant.

Stress concentration problem

Van Campen et al.[13], applied the method on a stress concentration problem in a link of a chain of a continuous variable transmission system, see Fig.6a. Each section of the chain contains a number of links of about 0.5 mm in thickness. The pins which transmit the driving force to conical discs are locked up by the links in subsequent sections of the chain. Only a symmetric loading case is considered, allowing to use only one quarter of the link in the FEM model. Fig.6b shows the geometry of the link. The loading force F is 268 N.

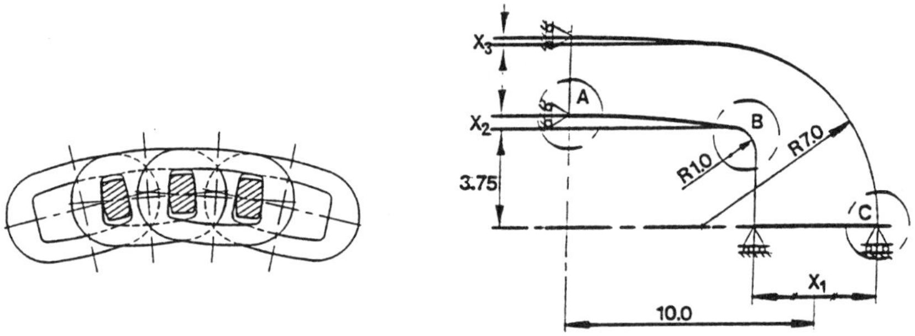

(a) (b)

Fig.6 (a) View of the chain; (b) One quarter of the link.

In Fig.6b three areas, A, B, and C are indicated with potentially high tensile stresses. The maximum tensile stresses are denoted by y_A, y_B, and y_C respectively and the objective is to derive regression models for these quantities. The level of the stresses can be influenced by variation of the geometry parameters x_1, x_2, and x_3. Hence these parameters are used as design variables.

The design variables are subject to the constructive constraints:

$$4.5 \le x_1 \le 6.0 \ , \qquad 0.0 \le x_2 \le 0.6 \ , \qquad 0.0 \le x_3 \le 0.6 \tag{20}$$

Each design variable is varied on four levels. For the set of candidate points from which the experimental design has to be selected, all possible combinations of the levels are used resulting in 4x4x4 = 64 candidate points.One FEM analysis provides 4 observations, namely one value of the stress and three values of its partial derivatives:

$$y_i \ , \quad \frac{\partial y_i}{\partial x_1} \ , \quad \frac{\partial y_i}{\partial x_2} \ , \quad \frac{\partial y_i}{\partial x_3} \ , \qquad i = A, \ ,B \ ,C \tag{21}$$

The model fitting resulted in the following regression models for the maximum tensile stresses in the areas A, B, and C:

$$y_A = 540.3 - 110.1x_1 + 7.7x_1^2 + 201.6x_2 + 10.5x_2^2 - 76.3x_3^2 -$$

$$- 17.1x_1x_2 - 3.0x_1x_2x_3 + 12.6x_1x_3 \ , \tag{22a}$$

$$y_B = 870.9 - 199.3x_1 + 15.2x_1^2 - 263.6x_2 - 46.1x_2^2 + \\ + 41.2x_1x_3 + 3.1x_1^2x_3 - 7.0x_1^2x_3 \,, \tag{22b}$$

$$y_C = 1311.6 - 335.9x_1 + 23.7x_1^2 + 44.2x_2 + 7.7x_2^2 - \\ - 39.2x_3^2 - 0.9x_1^2x_2 + 6.0x_1x_3^2 \,, \tag{22c}$$

In order to test the capability of the procedure one hundred test points were chosen in the design space at random. The FEM observations in these points were compared with the predictions of the models (22). Fig.7 shows the distributions of the residuals.

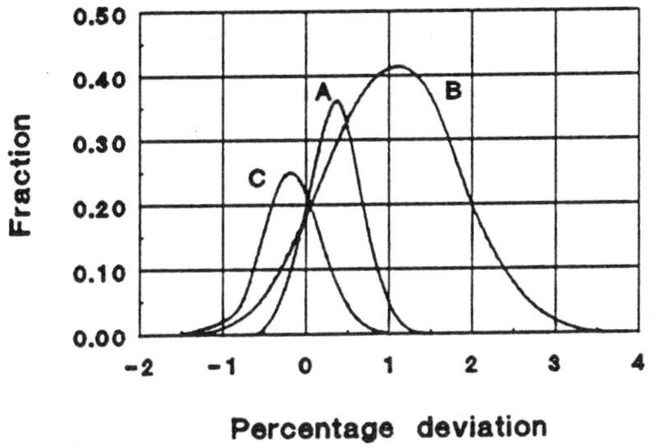

Fig.7 Distribution of residuals of 100 random test points for the approximations in the areas A, B, and C.

We may conclude that, based on as little as five FEM analyses (and using partial derivatives), regression models of good overall fit could be derived.

CONCLUSIONS

We described a method for deriving approximate analysis models as a substitute for time consuming FEM analyses in solving structural optimization problems. FEM analysis are regarded as numerical experiments from which data is extracted as input for the model building process by means of multiple linear regression. The resulting regression models can be used to define the objective function and the constraint functions of the optimization problem.

In a relatively simple situation, the required FEM analyses can be planned using classical experimental design techniques, such as 2^n-designs. The optimal experimental design theory offers a flexible tool to tackle more complicated problems.

Due to the iterative character of the model building process, regression models can be created in a cost effective way.

The proposed method has been tested and illustrated by two practical examples, both described by three design variables. The derived approximate analysis models proved to agree very well with direct FEM analysis results.

REFERENCES

1. Montgomery, D.C. Design and analysis of experiments, John Wiley, New York, 1984.
2. Schoofs, A.J.G. Experimental design and structural optimization, Ph.D. thesis, Eindhoven University of Technology, The Netherlands, 1987.
3. Box, G.E.P., Hunter, W.G., and Hunter, J.S. Statistics for experimenters, John Wiley, New York, 1978.
4. Patterson, H.D. Generation of factorial designs, Journal of the Royal Statistics Society, Ser. B, pp.175-179, 1976.
5. Nagtegaal, R. Computer aided design of experiments. A computer program for experimental design and model building (in Dutch), EUT Report 87-WFW-005, Univ. Press, Eindhoven, 1987.
6. Smith, K. On the standard deviations of adjusted and interpolated values of an observed polynomial function and its constants and the guidance they give towards a proper choice of the distribution of observations, Biometrica,12, pp. 1-85, 1918.
7. Kiefer, J., and Wolfowitz, J. Optimum design in regression problems, Canadian J. Math., 12, pp.363-366, 1959.
8. Fedorov, V.V. Theory of optimal experiments, Academic Press, New York, 1972.
9. Mitchell, T.J. An algorithm for the construction of D-optimal experimental designs, Technometrics, 16, pp.203-210, 1974.
10. Welch, W.J. Algorithm for the construction of experimental designs. User's manual ACED, version 1.6.1, Univ. of British Columbia, Vancouver, Canada, 1985.
11. Wouters, H. The stress distribution in dry running journal bearings (in Dutch), Report WFW 86.012, Univ. Press, Eindhoven, 1986.
12. SDRC Supertab engineering analysis, Model solution and optimization, User's guide, I-DEAS, Level 3, CAE International, Milford Ohio, 1986.
13. Campen, D.H. van, Nagtegaal, R, and Schoofs, A.J.G. Approximation methods in structural optimization using experimental designs for multiple responses, Chapter 5.5, Multicriteria design optimization, Procedures and applications,(Eds. Eschenauer, H., Koski, J., and Osyczka, A.), Springer-Verlag, Berlin, Heidelberg, 1990.

SECTION 2: OPTIMAL DESIGN UNDER DYNAMIC AND EARTHQUAKE CONSTRAINTS

Optimum Design of Dynamically Excited Structural Systems Using Time History Analysis

K. Truman, D. Petruska

Department of Civil Engineering, Washington University, Campus Box 1130, St. Louis, MO 63130, U.S.A.

ABSTRACT

The optimality criteria method for optimizing the design of two dimensional structural systems subjected to static and time variant loads will be presented. The static and dynamic constraints consist of displacement, story drift, stresses, and natural frequencies. In addition to these behavioral constraints, practical design constraints such as side constraints (upper and lower limit on member sizes) and linking constraints (forcing a group of members to be the same size) will be discussed. The objective function is the weight of the structure. Specific areas related to the time domain optimization algorithm which will be presented are the dynamic analysis method, dynamic gradient calculations, convergence criteria, active constraint determination, and meaningful examples.

INTRODUCTION

Numerous optimization algorithms of linear, nonlinear, and dynamic programming have been developed and used to solve structural problems with great success over the past decades. However, thus far much consideration has been given only to static loading. Little consideration has been given to the area of dynamic structural response with most of the work being done only to control frequency constraints (Khan [1], Khot [2]). There has been some work done with dynamic loading, but most often for time variant loads which are sinusoidal or some form of a pseudo earthquake loading (Cassis [3], Fox [4], Cheng [5],). No explicit method has been developed to handle any arbitrary dynamic loading.

The optimization algorithm used to find the optimum solution for plane frame steel structure using a time history analysis will be presented. The optimality criteria method is used with the objective function being the weight of the structure. Possible constraints are displacement, story drift, stress, frequency, and side constraints. The design procedure uses the AISC Manual of Steel Construction (6) wide flange sections with nonlinear expressions relating the major axis moment of inertia to the cross-sectional area and the section modules. The algorithm presented provides rapid convergence. Examples will be presented to show the usefulness of this method to structural engineering applications.

STRUCTURAL MODEL

Structural Elements

The building systems can include beam-columns, along with bracing members. The beam-columns are allowed to have bending, shear, and axial deformation while the bracing members are axially deformed members. All members are assumed to be wide flange steel sections, with the primary design variable for the beam-columns being the major axis moment of inertia and for the bracing, the cross-sectional area. The following relationship is assumed to exist between the primary design variable and the secondary design variables

$$L = C_1 I_x^P + C_2 \tag{1}$$

in which L is the is the cross-sectional area or section modules desired, and C_1, C_2 and P are constants developed statistically from appropriate tables for wide-flange sections (Truman [7]).

Dynamic Structural Analyses

The dynamic analysis is based upon an elastic stiffness and consistent mass system. The equilibrium equations of motion are

$$[M]\{\ddot{U}\} + [C]\{\dot{U}\} + [K]\{U\} = \{R\} \tag{2}$$

where [M], is the mass matrix, [C] is the damping matrix, [K] is the stiffness matrix, {R} is the load vector, {U} is the displacement vector and each dot represents one differentiation with respect to time.

The system of ordinary differential equations can be solved for approximately using any of the available numerical integration methods. The algorithm has been tested using both the central difference method and Newmark's method. Since Newmark method is unconditionally stable thus allowing larger time steps to be used, and greatly reducing the computing time, only this method will be discussed here.

The algorithm for using Newmark Integration Method (Bathe [8]) can be broken into two parts: initial calculations and calculations for each time step. The initial calculations involve forming the mass, damping and stiffness matrix, the initial displacement, velocity and acceleration at time zero, selecting a suitable time step Δt, and parameters α and δ, calculating the integration constants:

$$\beta \geq 0.50; \quad \alpha \geq 0.25(0.5 + \beta)^2$$

$$a_0 = \frac{1}{\alpha \Delta t^2}; \quad a_1 = \frac{\beta}{\alpha \Delta t}; \quad a_2 = \frac{1}{\alpha \Delta t}; \quad a_3 = \frac{1}{2\alpha} - 1;$$

$$a_4 = \frac{\beta}{\alpha} - 1; \quad a_5 = \frac{\Delta t}{2}\left(\frac{\beta}{\alpha} - 2\right); \quad a_6 = \Delta t(1 - \beta);$$

$$a_7 = \beta \Delta t$$

and forming the effective stiffness matrix given by

$$[\hat{K}] = [K] + a_0 [M] + a_1 [C] \tag{3}$$

For each time step, the effective loads at time t + Δt is calculated by

$$\{\hat{R}_{t+\Delta t}\} = \{R_{t+\Delta t}\} + [M] (a_0\{U_t\} + a_2\{\dot{U}_t\} + a_3\{\ddot{U}_t\}) +$$

$$[C] (a_1\{U_t\} + a_4\{\dot{U}_t\} + a_5\{\ddot{U}_t\}) \tag{4}$$

thus the displacements at time t+Δt can be solved for by

$$[\hat{K}] \{U_{t+\Delta t}\} = \{\hat{R}_{t+\Delta t}\} \tag{5}$$

and the accelerations and velocities at time t+Δt are given by

$$\{\ddot{U}_{t+\Delta t}\} = a_0 (\{U_{t+\Delta t}\} - \{U_t\}) - a_2\{\dot{U}_t\} - a_3\{\ddot{U}_t\} \tag{6}$$

$$\{\dot{U}_{t+\Delta t}\} = \{\dot{U}_t\} + a_6\{\ddot{U}_t\} + a_7\{\ddot{U}_{t+\Delta t}\} \tag{7}$$

THE OPTIMALITY CRITERION METHOD

When the structural weight is the objective function to be minimized, the structural optimization problem can be stated as follows

$$minimize\ W_{Total} = \sum_{i=1}^{m} \rho_i A_i l_i$$

subject to

$$\underline{U_j} \leq U_j(\delta, t) \leq \overline{U_j} \quad j = 1, \ldots, k$$

$$\underline{\delta_i} \leq \delta_i \leq \overline{\delta_i} \quad i = 1, \ldots, m$$

where ρ is the density of the material, A is the cross-sectional area, l is the length of the member, U is the constraint which can be displacement, story drift, stress, or frequency, $\overline{U}$ is the upper bound limit on the constraint, $\underline{U}$ is the lower bound limit, δ is the primary design variable, $\underline{\delta}$ is the lower bound value on the design variable, $\overline{\delta}$ is the upper bound value, k is the number of constraints, and m is the number of design variables.

The basis of the optimality criterion is the KHUN-TUCKER conditions. These conditions are needed to describe a local minimum of a function. To form the KUHN-

TUCKER conditions, the Lagrangian is required which may be written as

$$L = W_{Total} + \sum_{j=1}^{k} \lambda_j h_j \tag{8}$$

where λ_j is the Lagrange multiplier for the jth constraint, h_j is the constraint which for displacement, drift, and stress constraints is written as

$$h_j = \left|\frac{U_j}{\overline{U_j}}\right| - 1 \leq 0 \tag{9}$$

and k is the number of active constraints. Taking the derivative of the Lagrangian with respect to the design variable and assuming the derivative of the Lagrange multiplier with respect to the design variable is zero, the KUHN-TUCKER condition is obtained

$$\frac{\partial W_{Total}}{\partial \delta_i} + \sum_{j=1}^{k} \lambda_j \frac{\partial h_j}{\partial \delta_i} = 0 \qquad i = 1,...,m \tag{10}$$

and $\lambda_j(h_j)$ must equal zero, λ_j must be greater than or equal to zero and h_j must be less than or equal to zero for j = 1,...,k. If h_j is zero, the constraint is active, otherwise the constraint is inactive and λ_j must be zero. Equation 10 can be rewritten as

$$T_i = - \frac{\sum_{j=1}^{k} \lambda_j \frac{\partial h_j}{\partial \delta_i}}{\frac{\partial W_{Total}}{\partial \delta_i}} = 1 \tag{11}$$

which is the optimality criteria.

<u>Sensitivity Analysis for Dynamic Loads</u>

The equilibrium equations of motion are solved approximately by using the numerical integration method outlined above. The displacement at time t + Δt is given by Equation 5. Differentiating Equation 5 with respect to the design variable, δ, and noting that the derivatives of the integration constants with respect to δ is zero, gives the gradient of the dynamic displacement for each time step

$$\frac{\partial \{U_{t+\Delta t}\}}{\partial \delta_i} = [\hat{K}]^{-1}(\frac{\partial \{\hat{R}_{t+\Delta t}\}}{\partial \delta_i} -$$

$$[\frac{\partial [K]}{\partial \delta_i} + a_0 \frac{\partial [M]}{\partial \delta_i} + a_3 \frac{\partial [C]}{\partial \delta_i}]\{U_{t+\Delta t}\}) \tag{12}$$

where

$$\frac{\partial\{\hat{R}_{t+\Delta t}\}}{\partial\delta_i} = \frac{\partial\{R_{t+\Delta t}\}}{\partial\delta_i} + \frac{\partial[M]}{\partial\delta_i}(a_0\{U_t\} + a_2\{\dot{U}_t\} + a_3\{\ddot{U}_t\}) +$$

$$[M](a_0\frac{\partial\{U_t\}}{\partial\delta_i} + a_2\frac{\partial\{\dot{U}_t\}}{\partial\delta_i} + a_3\frac{\partial\{\ddot{U}_t\}}{\partial\delta_i}) +$$

$$[C](a_1\frac{\partial\{U_t\}}{\partial\delta_i} + a_4\frac{\partial\{\dot{U}_t\}}{\partial\delta_i} + a_5\frac{\partial\{\ddot{U}_t\}}{\partial\delta_i}) +$$

$$\frac{\partial[C]}{\partial\delta_i}(a_1\{U_t\} + a_4\{\dot{U}_t\} + a_5\{\ddot{U}_t\}) \tag{13}$$

and the derivatives of the acceleration and velocity can be found by directly differentiating equations 6 and 7. With the displacement gradients for each time step calculated, the story drift gradients can be obtained and so can the stress gradients. Cheng [5] gives the equation to find the gradients of the frequency constraints which was used in this paper.

Recurrence Relationship

To find the optimum solution, the optimality criteria given in Equation 10 must be incorporated into a means for redistributing the element sizes. The linear recurrence relationship used is

$$\delta_i^{v+1} = \delta_i^{v}(1 + \frac{1}{r}(T_i - 1)) \tag{14}$$

where r is the convergence control parameter, and v is the iteration number.

Lagrange Multipliers

The value of the optimality criteria for each iteration cannot be determined without estimates of the Lagrange multipliers. The Lagrange multipliers are obtained by developing linear equations by writing the change in an active constraint due to a change in a design variable. This is shown as

$$\Delta h_j = h_j(\delta_i + \Delta\delta_i) - h_j(\delta_i) = \sum_{i=1}^{n}\frac{\partial h_j}{\partial\delta_i}\Delta\delta_i \tag{15}$$

If the value of the active constraint at the point $(\delta + \Delta\delta)$ is assumed to be zero and using the recurrence relationship to find $\Delta\delta$, the linear equations that include linking and side constraints are

$$rh_j - \sum_{i=1}^{n_1}\delta_i^v + r\sum_{i=n_1+1}^{n}\frac{\partial h_j}{\partial\delta_i}(\delta_i^p - \delta_i^v) =$$

$$\sum_{p=1}^{k}\lambda_p\Bigg(\sum_{i=1}^{n_2}\Bigg(\frac{\sum_{q=1}^{s}\frac{\partial h_j}{\partial\delta_q}\frac{\partial h_p}{\partial\delta_q}}{\sum_{q=1}^{s}\frac{\partial W_{Total}}{\partial\delta_q}}\Bigg)\delta_i^v\Bigg) \qquad j = 1,\dots,k \qquad (16)$$

where n_1 is the number of active elements, δ_i is the passive value for the i th design variable, n_2 is the total number of active global design variables, s is the number elemental design variables linked in a given set, and k is the number of active constraints. An equation can be written for each active constraint producing k equations that can be solved for the Lagrange multipliers. If a Lagrange multiplier is negative, the constraint is assumed to be inactive and is removed from the equations. This is repeated until all Lagrange multipliers are positive.

OPTIMIZATION ALGORITHM

The algorithm outlined is designed for dynamic displacement, story drift, stress, frequency, and side constraints. Eliminate appropriate steps if that particular constraint is not considered.

1. Determine initial values for the elements primary design variable. Select suitable time step.
2. Determine dynamic displacements at each time step using numerical integration method.
3. Calculate dynamic stress at each time step.
4. Find largest ratio of structural response to constraint value.
5. If largest ratio is above (1 + P1) or below (1 -P2), where P1 is usually 0.1 and P2 is 0.05, scale the primary design variables by this ratio. If the largest factor is between these values, at least one constraint is active, proceed to step 8.
6. Use scaled primary design variables to find values of the secondary design variables.
7. Go to step 2.
8. Calculate dynamic displacements and derivative of displacements with respect to each design variable. Store values for assumed active constraints (i.e. every constraint whose ratio of actual response to allowed response is above 1+P1).
9. Calculate dynamic stress at each time step. If stress constraint is active, store stress, and associated displacements and displacement gradients required to calculate stress gradient to be performed after dynamic analyses is complete.
10. Calculate story drift and derivatives for active constraints.
11. Calculate frequency and gradients for active constraints.
12. Separate elements into active and passive categories.
13. Form and solve the linear equations to determine the Lagrange multipliers for the assumed active constraints. If negative Lagrange multipliers exist, eliminate these constraints from the active set and resolve until all Lagrange multipliers are positive.
14. Use linear recurrence relationship to evaluate the new primary design variables, and determine secondary design variables.
15. Reanalyze the structure and check for constraint violation.

16. Check the termination criterion. The process stops if a specific number of iteration is exceeded, or if the optimality criteria is within a small tolerance of unity or if the percent weight change between iteration v and v-2 is less than a prescribed tolerance such as 0.5%. If the terminating criteria is not satisfied, go to step 8. If satisfied, stop the process.

EXAMPLE PROBLEMS

The three-story, two bay frame shown in Figure 1 was analyzed to show the effectiveness of the algorithm. The constraints for this design problem are displacement and side. The absolute value of the horizontal displacement at each floor was limited to 1.00, 1.65, and 2.35 inches for the first, second and third floor respectively. The primary design variable is the moment of inertia. The lower limit on the moment of inertia is 290 $in.^4$. Initial values for the member sizes were 1500 $in.^4$ Material constants are E = 29000 ksi and ρ = 490 $lb/ft.^3$ A uniformly distributed non-structural weight totaling 33.33 kips was applied to each beam member of the frame. Damping of the structure was neglected. Three different loadings were examined. At time t = 0, the structure is at rest and the base undergoes a transient acceleration. The first base acceleration is defined by

$$\ddot{a}_g(t) = 135\sin(2\pi t)\ \textit{inches/sec}^2, \quad 0 \le t \le 1.0 \text{ sec}$$

and is zero for t greater than 1.0 seconds. The second base acceleration is defined by the accelogram for the north-south component of the El-Centro, California earthquake on May 18, 1940 and is shown in Figure 2. The third base acceleration is for the north-south component of the San Fernando, California earthquake on February 9, 1971 taken at the Caltech Seismological Lab and is shown in Figure 3. The constant of 135 used in the sine function was chosen to match the peak acceleration of the El-Centro earthquake record. The San Fernando earthquake was also multiplied by a constant of 1.81 so that both earthquakes would have the same peak acceleration. The time step selected was 0.01 seconds. For the examples were linking was used, all columns on a given floor were linked together. For the sine load, the time frame examined was 2.0 seconds and for the El-Centro and San Fernando earthquake, the time frame was 10.0 seconds. Table 1 shows the optimum design, and Figure 4 shows the weight versus iteration.

Comparing Case A to Case B and Case C to Case D shows that the linking example produces a heavier structure since each member is not able to take on its optimum size. The El-Centro Earthquake results in a slightly heavier structure over the San Fernando earthquake. The El-Centro earthquake results in heavier beams and columns on the first two floors but the San-Fernando earthquake produces heavier columns and beams on the top floor. The maximum constraint violation is 1.9%. Figure 4 shows for Case C a zig-zag pattern between iteration 5 through 9 because at iteration 4, the process over shots the constraint, and the optimization process is trying to get near the constraint surface again in iterations 5 through 9. Case D shows a large decrease in weight at iteration 6 due a change in the active constraint thus allowing the optimization process to find a new path to move along that produces a local minimum.

CONCLUSIONS

An optimization algorithm is presented for plane frame building systems subject to time history loading. The associated program can be used directly for practical preliminary design as well as for parametric studies on how different dynamic loads and different set of behavioral and design constraint effect the optimum solution. The example presented demonstrates the efficiency, usefulness and practicality of the algorithm.

ACKNOWLEDGMENT

Support for this study for the second author was provided by the Office of Naval Research through the graduate fellowship program.

REFERENCES

1. Khan, M., Willmert, K., "An Efficient Optimality Criterion Method for natural Frequency Constrained Structures," J. of Comp. Struct., 1981, pp. 501-507.

2. Khot, N., "Optimization of Structures with Multiple Frequency Constraints," J. Comp. Struct., 1985, pp. 869-876.

3. Cassis, J., Schmit, L., "Optimum Structural Design with Dynamic Constraints," J. of Struct. Div., ASCE, Vol. 102, No. ST10, 1976, pp. 2053-2071.

4. Fox, R., Kapoor, M., "Structures Optimization in the Dynamic Response Regime: A Computational Approach," AIAA Journal, 1985, pp. 15-22.

5. Cheng, F., Truman, K., "Optimization Algorithm of 3-D Building Systems for Static and Seismic Loading," Modelling and Simulation in Engineering, edited by W. R. Ames, et al., 1983, pp. 315-326.

6. American Institute of Steel Construction: Manual of Steel Construction, 8th Edition, 1980.

7. Truman, K. "Optimal Design of Concrete and Steel Buildings Subjected to Static and Seismic Loads," Ph. D Dissertation, Dept. of Civil Engr., U. of Missouri-Rolla, 1985.

8. Bathe, K., Wilson, E., Numerical Methods in Finite Element Analysis, Prentice-Hall, Inc., New Jersey, 1976.

TABLE 1. Optimum Solutions for Example Problem

MEMBER	MOMENT OF INERTIA (IN.^{4})				
	CASE A	CASE B	CASE C	CASE D	CASE E
1	1222.1	1631.8	2775.0	1503.0	2450.7
2	2274.1	1631.8	2775.0	5694.3	2450.7
3	1222.1	1631.8	2775.0	1501.2	2450.7
4	381.6	639.8	944.0	621.5	897.6
5	1072.0	639.8	944.0	1827.3	897.6
6	381.6	639.8	944.0	623.7	897.6
7	290.0	351.7	657.5	355.0	834.9
8	543.7	351.7	657.5	1050.7	834.9
9	290.0	351.7	657.5	358.4	834.9
10	1250.8	1287.2	1713.9	1887.0	1548.5
11	1250.8	1287.2	1716.7	1887.1	1548.5
12	740.8	782.4	1076.8	1035.0	1031.5
13	740.8	782.4	1058.0	1042.1	1028.3
14	290.0	290.0	403.9	304.4	501.6
15	290.0	290.0	393.0	307.5	501.6
W^*_{Total} (kips)	9.741	9.946	12.141	11.867	12.038
ITERATION	4	5	11	10	4
CPU (SEC)	483.5	585.5	5660.2	5315.5	2428.2
ACTIVE CONSTRAINTS					
DOF TIME CONSTRAINT VALUE (INCHES)	19,25 t=0.69 2.356	22 t=0.69 2.351	10 t=8.00 1.665 25 t=8.02 2.396	10,16 t=3.58 1.628 22 t=3.59 2.378 19 t=4.22 -2.328	10,16 t=7.23 1.650 19,25 t=7.22 2.368

CASE A Sine Function, no linking
CASE B Sine Function, linking
CASE C El-Centro Earthquake, linking
CASE D El-Centro Earthquake, no linking
CASE E San Fernando Earthquake, linking

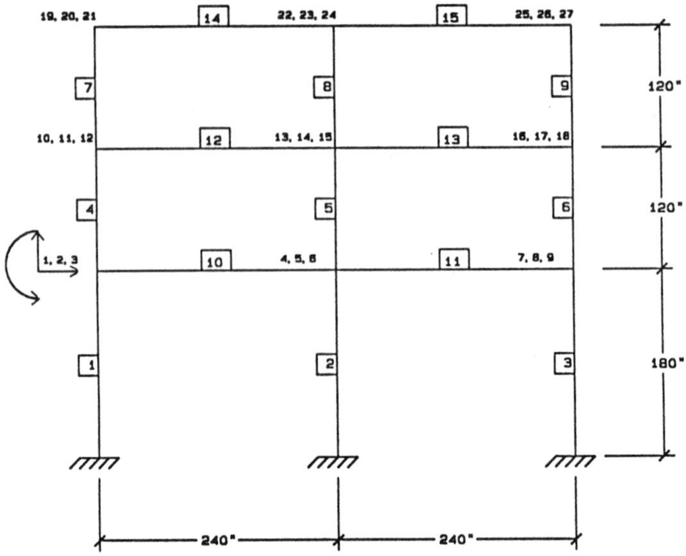

FIGURE 1. THREE-STORY AND TWO BAY FRAME

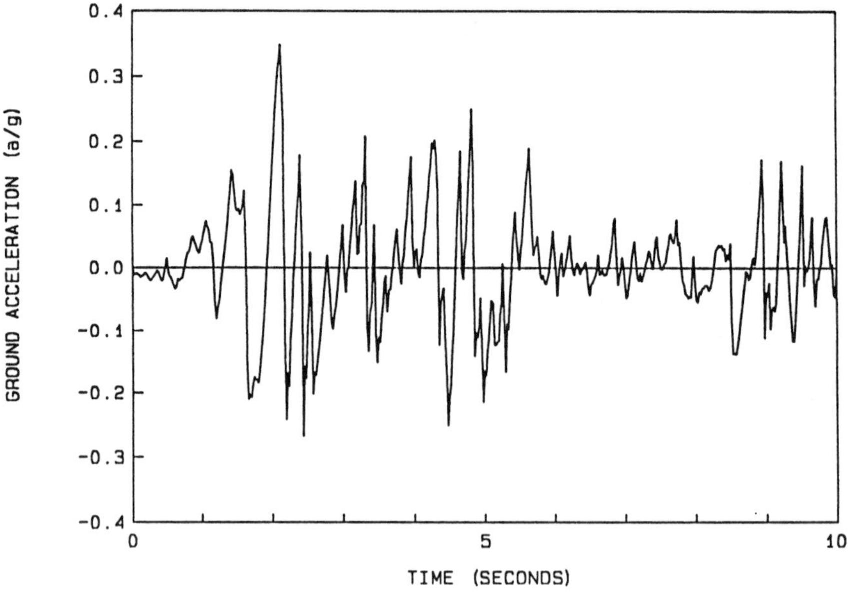

FIGURE 2. EL-CENTRO EARTHQUAKE ACCELERATION RECORD

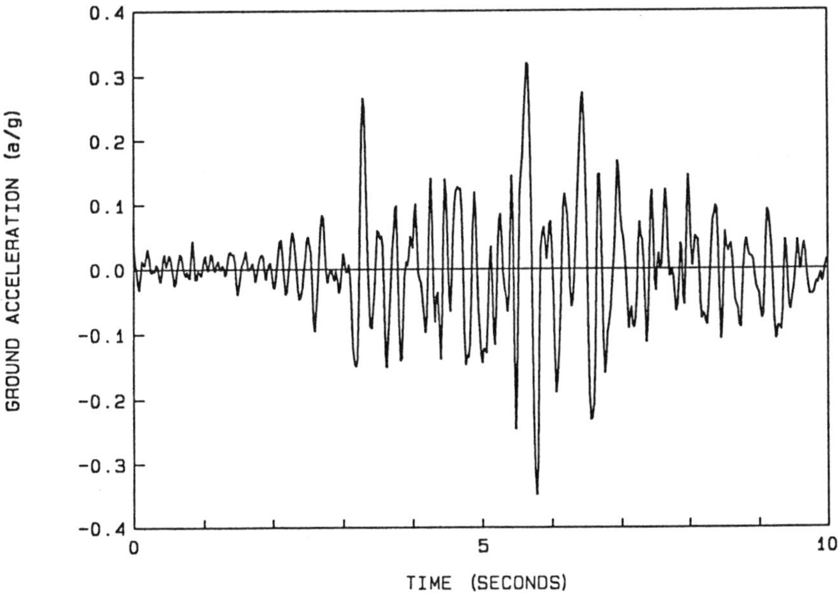

FIGURE 3. SAN FERNANDO EARTHQUAKE ACCELERATION RECORD

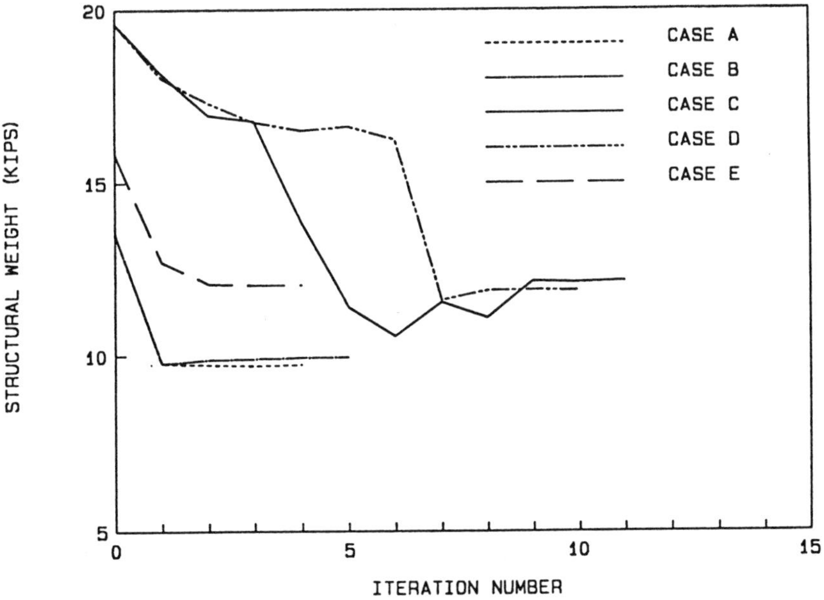

FIGURE 4. WEIGHT VS ITERATION NUMBER

Optimal Design with Multiple Frequency Constraints

C.T. Jan (*), K.Z. Truman (**)

(*) DRC Consultants, Inc., Flushing, New York, U.S.A.

(**) Department of Civil Engineering, Washington University, St. Louis, Missouri, U.S.A.

ABSTRACT

In structural optimization, Venkayya, Khan and a few others, have applied the optimality criterion method to structural design problems of natural frequency constrained problems. Their techniques developed, however, have been unable to handle large nonstructural masses and have been applied to relatively simple structural systems such as trusses and beams. On the other hand, the mathematical programming methods developed for natural frequency constrained problems, in general, have had slower rates of convergence and are dependent on initial estimates of design variables and problem size. This paper presents an optimality criterion method for the design of structural systems under multiple natural frequency constraints. It is based on certain aspects of the structural model under investigation while it maintains the mathematical approach that could be adapted to any system. The design procedures include nonstructural masses and the AISC Manual of Steel Construction wide-flanged shapes having exponentially nonlinear cross-sectional relationships among the area, moments of inertia, and section modulus. In addition, the algorithm is capable of handling any regular shaped cross sections such as tubular, rectangular, or circular members. In essence, the optimization algorithm consists of efficient schemes for iterative scaling, the linking of design variables, the calculation of Lagrange multipliers, the formulation of recurrence relationships, and the choosing of active, passive, and side constraints. Results of benchmark problems demonstrate the efficiency of the optimality criterion method presented in this paper.

INTRODUCTION

Optimality criterion techniques have been applied to many structural problems with reasonable success in the past few years. Khan and Willmert [1,2] have developed several efficient schemes for the designs of trusses, frames, and other complex structures under static loadings. Only minor attention has been given to frequency constrained problems.

Venkayya, Khot, and Burke [3,4,5], and a few others [6,7,8,9,10] have successfully applied optimality criterion techniques to structural designs of this type. However, the techniques developed have had major drawbacks in handling large structures. On the other hand, the mathematical programming methods developed for frequency constrained problems, in general, have had slower rates of convergence. Most of all, those approaches are unable to solve multiple frequency constrained problems.

The optimality criterion method developed here, first proposed by Cheng and Truman [11], is based on certain aspects of the structural model under investigation while it maintains the mathematical approach that could be adapted to any system. This method is utilized to analyze structures under multiple frequency constraints.

OPTIMALITY CRITERION METHOD

Objective functions are the functions to be optimized. The total weight of the structure

$$W = \sum_{i=1}^{N} \rho_i A_i(I_i) L_i + W_{non} \tag{1}$$

in which W is the total weight, ρ_i the density for element i, L_i the length of the ith element, W_{non} the non-structural weight, and N the number of structural elements.

The frequency constraints are given as

$$\underline{\omega}_j \le \omega_j \le \overline{\omega}_j \tag{2}$$

in which ω_j is the jth natural frequency, $\underline{\omega}_j$ the lower limit for the natural frequency, and $\overline{\omega}_j$ the upper limit for the natural frequency.

The side constraints are generally given as only a lower limit when used for structural frames but can be also have an upper limit; such as

$$\underline{I}_i \le I_i \le \overline{I}_i \tag{3}$$

in which I_i is the primary design variable, the moment of inertia about the major axis, for the ith element, $\underline{I}_i$ the lower limit, and $\overline{I}_i$ the upper limit for the primary design variable.

The basis for optimality criterion method, is the Kuhn-Tucher conditions. These conditions are needed to describe a local minimum of a function. If the problem is convex, they become sufficient conditions which ensure that the solution is the global minimum. The Lagrangian is required prior to forming the Kuhn-Tucker conditions and is written as

$$L(I_i, \mu_j) = \sum_{i=1}^{N} \rho_i A_i L_i + \sum_{j=1}^{M} \mu_j h_j + W_{non} \tag{4}$$

in which μ_j equals the Lagrange multiplier for the jth constraints, h_j the displacement, drift, and frequency constraints, and M the total number of constraints. As mentioned previously, the side constraints are handled separately from the other constraints. The Lagrange multipliers must be positive or zero for inequality constraints. Depending on the nature of the constraint equations, along with other optimality conditions are the Kuhn-Tucker conditions. The Kuhn-Tucker conditions become

$$\partial W/\partial I_i + \sum_{j=1}^{M} \mu_j \, \partial h_j/\partial I_i = 0 \qquad i = 1,\ldots,N \tag{5}$$

$$\mu_j h_j = 0 \qquad j = 1,\ldots,M \tag{6}$$

$$\mu_j \geq 0 \qquad j = 1,\ldots,M \tag{7}$$

$$h_j \leq 0 \qquad j = 1,\ldots,M \tag{8}$$

The Kuhn-Tucker conditions can be used to test a point for optimality. These conditions are, in general, not sufficient to ensure a relative minimum. We can observe that the first of the Kuhn-Tucker conditions is a necessary condition for I_i to minimize $L(I_i, \mu_j)$. The optimality criteria can be expressed as

$$\frac{- \sum_{j=1}^{M} \mu_j \, \partial h_j / \partial I_i}{\partial W / \partial I_i} = 1 \tag{9}$$

Calculation of derivatives of the constraint functions with respect to design variables, often called design sensitivity analysis, is required in most of the efficient optimization methods. In dynamic analysis,

$$([K] - \omega_j^2 \, [M]) \, \{X\}_j = 0 \tag{10}$$

in which [M] is the total mass matrix including both structural and non-structural masses, $\{X\}_j$ and the jth mode shape. By arranging and premultiplying by the transpose of the jth mode shape, Equation (10) becomes

$$\{X\}_j^T \, ([K] - \omega_j^2 \, [M]) \, \{X\}_j = 0 \tag{11}$$

Differentiating Equation (11) with respect to design variable I_i gives the gradient of the jth natural frequency in the form of

$$\frac{\partial \omega_j}{\partial I_i} = \frac{\{X\}_j^T \left(\dfrac{\partial [K]}{\partial I_i} - \omega_j^2 \dfrac{\partial [M]}{\partial I_i} \right) \{X\}_j}{2 \, \omega_j \, (\{X\}_j^T \, [M] \, \{X\}_j)} \tag{12}$$

As in the virtual explicit expression for the derivative of the natural frequency is formulated, Equation (12) also requires the explicit expressions for the derivatives of the stiffness matrix and the mass matrix. Note that the only nonzero terms for the partial derivative of the stiffness matrix are those which are dependent on the primary design variable, I_i.

In order to obtain the optimum solution, the optimality criterion stated in Equation (9) must be incorporated into a means for redistributing the element sizes. This is achieved by multiplying each side of Equation (9) by I_i^k and taking the rth root. The exponential recurrence relation becomes

$$I_i^{k+1} = \left[\frac{- \sum_{j=1}^{M} \mu_j \dfrac{\partial h_j}{\partial I_i}}{\partial W/\partial I_i} \right]^{1/r} I_i^k \qquad i = 1,\ldots,N \qquad (13)$$

in which k is the number of iterations previously performed, and r is a convergence control parameter. As r becomes larger, the value of I_i^k is modified to a lesser extent, conversely, as r becomes smaller, I_i^k is modified to a greater extent. Expanding Equation (13) with the use of the binomial theorem provides a linear recurrence relationship such that

$$I_i^{k+1} = \left[1 + \frac{1}{r} \left[\frac{- \sum_{j=1}^{M} \mu_j \dfrac{\partial h_j}{\partial I_i}}{\partial W/\partial I_i} - 1 \right] \right] I_i^k \qquad i = 1,\ldots,N \qquad (14)$$

Only the first two terms of the binomial theorem are retained, because a linear recurrence relationship is desired. Note that the term

$$\left[\frac{- \sum_{j=1}^{M} \mu_j \dfrac{\partial h_j}{\partial I_i}}{\partial W/\partial I_i} - 1 \right] \qquad (15)$$

should be zero at the optimal solution and can be viewed as a measure of the error for k^{th} iteration. Equation (14) is the recursive relationship to be used in the algorithm. The optimality criteria of Equation (9) were derived from linear approximation of the constraints in the neighborhood of the current design point. Thus, the approximate recurrence relations for redesign derived from these criteria are valid only in a small region around this point.

The value for the optimality criteria of each iteration cannot be determined without an estimation of the values for the Lagrange multipliers. There are many different algorithms available that provide these estimates. The most common seem to be the exponential and linear recurrence relationships. They are derived from the optimality criterion in the same manner as for the design variables. These recurrence relations have two major disadvantages. The first is slow convergence, and the second is that initial values for the Lagrange multipliers must be assumed. The recurrence relations do have two advantages over most other methods. The first advantage is minimal computational effort, and the second is that active constraints need not be selected. The recurrence relations numerically remove the passive constraints during optimization. In general, more exact methods of determining the Lagrange multipliers require large computational efforts and necessitate some knowledge as to which constraints are active.

The method of developing linear equations to determine the Lagrange multipliers is one such algorithm. These equations are developed by writing an expression for the change in an active constraint occasioned by a change in one design variable. This is shown as

$$\Delta h_j = h_j(I_i+\Delta I_i) - h_j(I_i) = \frac{\partial h_j}{\partial I_i}\Delta I_i \qquad (16)$$

If the value of the active constraint at the point $(I_i+\Delta I_i)$ is assumed to be zero, the equation becomes

$$- h_j(I_i) = \frac{\partial h_j}{\partial I_i}\Delta I_i \qquad (17)$$

By using the linear recurrence relationship for design variables and taking account of Equation (17), a set of simultaneous equations for solving Lagrange multipliers are expressed as

$$\sum_{p=1}^{M} \mu_p \sum_{i=1}^{N} \frac{\dfrac{\partial h_j}{\partial I_i}\dfrac{\partial h_p}{\partial I_i}\Delta I_i}{\partial W/\partial I_i} = r\, h_j^k - \sum_{i=1}^{N}\left[\frac{\partial h_j}{\partial I_i}\right]_k \Delta I_i \qquad j = 1,\ldots M \qquad (18)$$

If there are N1 active elements and (N-N1) passive elements, Equation (18) becomes

$$\sum_{p=1}^{M} \mu_p \sum_{i=1}^{N} \frac{\dfrac{\partial h_j}{\partial I_i}\dfrac{\partial h_p}{\partial I_i}\Delta I_i}{\partial W/\partial I_i} = r\, h_j^{\,k} - \sum_{i=1}^{N1}\left[\frac{\partial h_j}{\partial I_i}\right]_k \Delta I_i$$

$$+\; r \sum_{i=N1+1}^{N} \frac{\partial h_j}{\partial I_i}(I_i^{\,p} - I_i) \qquad j = 1,\ldots,M \qquad (19)$$

where $I_i^{\,p}$ represents a member size which becomes passive, and N1 is the total number of active elements. Note that as long as a passive element remains at a constant value, the last term is zero. If any Lagrange multiplier is found to be zero or negative after solving Equation (18) or (19), the associated terms with respect to this constraint should be removed from the simultaneous equations and a new set of Lagrange multipliers must be found. This is an iterative procedure until all the Lagrange multipliers are found to be positive or zero.

NUMERICAL EXAMPLES

A computer program, OPFF [12], has been implemented to accommodate the algorithm presented in this paper. Each member in Examples 1 and 2 is considered to be a wide flanged steel section with Young's modulus of 30,000 ksi and a density of 7.25×10^{-4} lb-s^2/in^4. The moment of inertia I_i is related to the cross sectional area A_i by the following expressions used by Khan and Willmert [6].

$$I_i = \begin{cases} 4.6248\, A_i^2, & 0 \le A_i \le 44 \text{ in}^2 \\ 256\, A_i - 2300, & 44 \le A_i \le 88.3 \text{ in}^2 \end{cases} \qquad (20)$$

Example 1. Two-Story and One-Bay Frame

The structural configuration is shown in Figure 1. The first natural frequency was constrained to 78.5 rad/sec. A uniformly distributed weight of 10 lb/in. has been treated as non-structural weight on each beam member of the structure. The initial design cross sectional areas of 40, 80, and 80 in^2 respectively, based on Khan and Willmert [6], and Yu [10], were used for all members in OPFF.

Table 1 provides the final results of OPFF and compares

them with the designs obtained by Khan and Willmert [6], and Yu [10]. It demonstrates the efficiency of the proposed optimality criterion method and provides an economical design compared with other approaches.

Example 2. Seven-Story and One-Bay Frame

The structure is shown in Figure 2. The first natural frequency was constrained to 10.2 rad/sec. A uniformly distributed weight of 10 lb/in has been treated as non-structural weight on each beam member of the structure. The initial cross sectional areas were 80 in^2 for all members. Table 2 shows the final results of OPFF and compares them with the designs obtained by Khan and Willmert [6], Cassis and Schmit [9].

The final design of OPFF gives reasonably lighter structure than other approaches. The results demonstrate the optimum design for a first natural frequency constraint which prefers a system of weak beams and strong columns.

CONCLUSIONS

In analyzing structures with multiple frequency constraints, the algorithm stated herein is shown to converge while controlling the first few natural frequencies. The comparison for the first natural frequency constraint with other approaches demonstrates the efficiency of the proposed optimality criterion method. Being able to constrain the natural frequencies of a structure provides a useful means for keeping the structure away from certain critical frequency ranges which may give peak acceleration or an unwanted resonance.

ACKNOWLEDGEMENT

This study was performed with the financial support of the Department of Civil Engineering at Washington University, their assistance is gratefully acknowledged.

REFERENCES

1. Khan, M.R. Optimality Criterion Techniques Applied to Frames Having General Cross-Sectional Relationships, pp. 669-676, AIAA Journal, Vol. 22, No. 5, May 1984.

2. Khan, M.R., Thornton, W.A. and Willmert, K.D. Optimality Criterion Techniques Applied to Mechanical Design, pp. 319-327, Journal of Mechanical Design, Transactions of the ASME, Vol. 100(2), 1978.

3. Venkayya, V.B., Khot, N.S., Tischler, V.A. and Taylor, R.F. Design of Optimum Structures for Dynamic Loads, 3rd

Conference on Matrix Methods in Structural Mechanics, Wright Patterson AFB, Ohio, 1971.

4. Venkayya, V.B., Khot, N.S. and Burke, L. Application of Optimality Criteria Approaches to Automated Design of Large Practical Structures, pp. 301-319, Second Symposium on Structural Optimization, Milan, Italy, AGARD Conference Proceeding No. 123, 1973.

5 Khot, N.S. Optimization of Structures with Multiple Frequency Constraints, pp. 869-876, Journal of Computers and Structures, Vol. 20, No. 5, 1985.

6. Khan, M.R. and Willmert, K.D. An Efficient Optimality Criterion Method for Natural Frequency Constrained Structures, pp. 501-507, Journal of Computers and Structures, Vol. 14, No. 5-6, 1981.

7. Haug Jr., E.J., Pan, K.C. and Streeter, T.C. A Computational Method for Optimal Structural Design. I. Piecewise Uniform Structures, pp. 171-184, International Journal for Numerical Methods in Engineering, Vol 5, 1972.

8. Mills-Curran, W.C. and Schmit, L.A. Structural Optimization with Dynamic Behavior Constraints, pp. 132-138, AIAA Journal, Vol. 23, No. 1, January 1985.

9 Cassis, J.H. and Schmit, L.A. Optimal Structures Design with Dynamic Constraints, pp. 2053-2071, Journal of Structural Division, ASCE, 102, No. ST10, 1976.

10. Yu, D. Optimization Design of Structures Subjected to Transient Dynamic Ground Motion, MS Dissertation, Mechanical and Industrial Engineering Department, Clarkson College of Technology, 1980.

11. Truman, K.Z. and Cheng, F.Y. Optimum Design of Steel and Reinforced Concrete 3-D Seismic Building Systems, pp. 475-482, Vol. V, Proceedings of the 8th World Conference on Earthquake Engineering, San Francisco, California, USA, 1984.

12. Jan, C.T. Optimum Structural Design with Dynamic Constraints, pp.730-738, Part 1, Proceedings of AIAA/ASME/ASCE/AHS 27th Structures, Structural Dynamics and Materials Conference, San Antonio, Texas, USA, 1986.

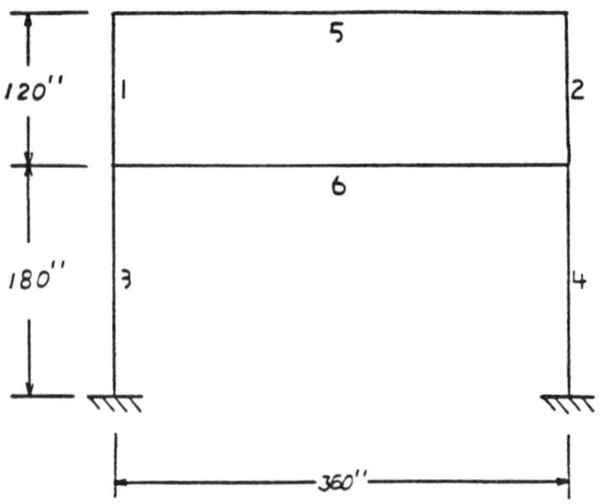

Figure 1. Two-Story One-Bay Frame

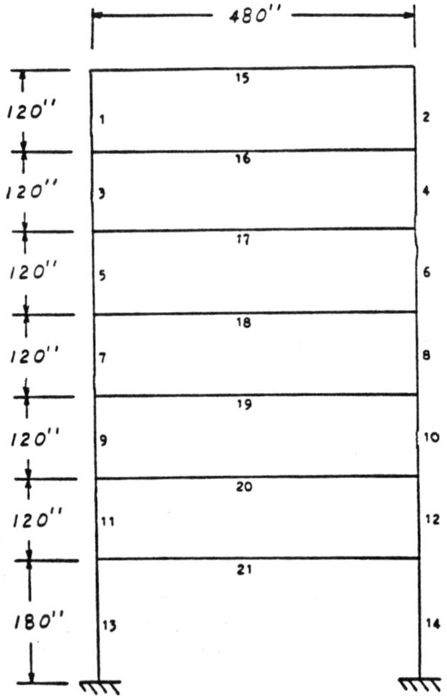

Figure 2. Seven-Story One-Bay Frame

Table 1. Two-Story, One-bay Frame

Approach	Khan & Willmert	Yu	OPFF
Analysis	28	8	20
Element Number	Area in^2	Area in^2	Area in^2
1	20.41	17.66	19.97
2	20.41	17.66	19.97
3	38.97	54.82	39.60
4	38.97	54.82	39.60
5	7.92	11.69	7.92
6	34.36	22.70	33.34

Table 2. Seven-Story, One-bay Frame

Approach	Cassis & Schmit	Khan & Willmert	OPFF
Analysis	-	8	4
Element Number	Area in^2	Area in^2	Area in^2
1 & 2	15.14	7.92	7.92
3 & 4	14.54	7.92	7.92
5 & 6	16.45	8.31	7.92
7 & 8	16.95	9.58	9.12
9 & 10	13.00	10.53	10.15
11 & 12	14.20	11.31	10.90
13 & 14	16.20	17.19	18.63
15	8.86	7.92	7.92
16	8.16	7.92	7.92
17	9.51	10.78	10.24
18	14.21	12.85	12.97
19	16.41	14.85	14.92
20	16.39	15.74	15.97
21	11.31	15.16	14.59
Weight (lbs)	19,106	16,918	16,861

Sequential Generator of Optimal Structures for Candidate Frequency Ranges

M. Ohsaki, T. Nakamura

Department of Architecture, Kyoto University, Sakyo, Kyoto 606, Japan

ABSTRACT

A computationally efficient method is presented of generating a set of optimum designs of structures based on the concept called *an ordered set of optimal structures for candidate frequency ranges*. The optimum designs of a structure, discretized by a finite element method, are swept out by the method of successive Taylor series expansion with respect to the prescribed eigenvalue. In the examples, ordered sets of optimal designs are generated for beams, arches and plates. The characteristics of the optimal structures are discussed in view of the optimality criteria.

INTRODUCTION

The significance of optimum design problem of distributed parameter structures has been well recognized in the field of mechanical and structural designs. There have been plenty of works on optimum designs of beams, arches and plates [1-12]. The numerical procedure in those papers may be classified into two groups. Almost any iterative improvement procedure for a structure discretized by means of a finite element method [1,2] requires computation of large sensitivity matrices and is not efficient for a large system. On the other hand, a numerical procedure for a set of governing nonlinear differential equations may be efficient only for a geometrically simple structure [3-12].

The authors have presented a computationally efficient method of generating an ordered set of optimal trusses for candidate frequency ranges [13]. That method has been successfully extended to the problem of finding earthquake-strain constrained designs of trusses [14]. In this paper, the authors' method is extended to optimum design problems of distributed parameter structures discretized by a finite element method. In the authors' method, the set of optimal trusses is swept out by successive Taylor series expansions with respect to the specified eigenvalue and various design charts of optimum structures may readily be presented for designers. It should be noted that the authors' method is far beyond the scope of optimum sensitivity analysis [15,16]. A similar method has been independently proposed on the basis of the concept of homotopy [17].

In the examples, ordered sets of optimum designs are generated for beams, arches and plates. The intrinsic relation of the optimum design to the prescribed eigenvalue is discussed in view of the optimality criteria. Karihaloo and Niordson [4] showed in several examples that the distribution of the optimal cross-sectional area of a cantilever is greatly affected by the ratio of the specified total mass to the nonstructural mass. The variation of the optimal stiffness with respect to the prescribed fundamental eigenvalue may easily be presented by the authors' method, since the optimum design is considered to be a function of the prescribed fundamental eigenvalue.

OPTIMUM DESIGN PROBLEM AND OPTIMALITY CRITERIA

Consider a distributed parameter structure such as a beam, an arch or a plate. The centerline (center surface) dimension of the structure is to be prescribed. The structure is discretized by a finite element method. Let $\mathbf{A} = \{A_i\}$ denote the design variable vector whose components are representative cross-sectional quantities. For a structure with an idealized sandwich cross-section, A_i denotes the thickness of an element flange. The structure is defined completely by $\mathbf{A}$. The stiffness matrix with respect to a global coordinate system is denoted by $\mathbf{K}(\mathbf{A})$. The mass matrix due to structural mass and that due to nonstructural mass are denoted, respectively, by $\mathbf{M}_D(\mathbf{A})$ and $\mathbf{M}_J$. The equation of eigenvibration may be reduced to the following form:

$$\mathbf{K}(\mathbf{A})\mathbf{\Phi}_r(\mathbf{A}) = \Omega_r(\mathbf{A})\{\mathbf{M}_D(\mathbf{A}) + \mathbf{M}_J\}\mathbf{\Phi}_r(\mathbf{A}) \qquad (r = 1, 2, \dots f), \tag{1}$$

where $\Omega_r(\mathbf{A})$ and $\mathbf{\Phi}_r(\mathbf{A})$ denote the rth eigenvalue and eigenvector, respectively, and where f denotes the degree of freedom of motion of the structure. The Rayleigh quotient is defined by

$$\Omega_r(\mathbf{A}) = \frac{\mathbf{\Phi}_r(\mathbf{A})^T\mathbf{K}(\mathbf{A})\mathbf{\Phi}_r(\mathbf{A})}{\mathbf{\Phi}_r(\mathbf{A})^T\{\mathbf{M}_D(\mathbf{A}) + \mathbf{M}_J\}\mathbf{\Phi}_r(\mathbf{A})} \qquad (r = 1, 2, \dots f), \tag{2}$$

where the superscript T denotes the transpose of a vector. The following normalization condition is adopted here:

$$\Phi_r(\mathbf{A})^T\{\mathbf{M}_D(\mathbf{A}) + \mathbf{M}_J\}\Phi_r(\mathbf{A}) = 1 \qquad (r = 1, 2, \dots f). \tag{3}$$

Let ρ and w_i denote the length (area) of the ith element and the mass density, respectively. The specified fundamental eigenvalue and the vector of minimum cross-sectional areas are denoted, respectively, by Ω_a and $\bar{\mathbf{A}}$. Then the problem of optimum design for specified fundamental eigenvalue may be stated as follows:

[Problem ODDF]

$$\text{Minimize} \qquad V(\mathbf{A}) \equiv \sum_{i=1}^{m} \rho A_i w_i, \tag{4}$$

$$\text{subject to} \qquad \Omega_a - \Omega_r(\mathbf{A}) \leq 0 \qquad (r = 1, 2, \dots f), \tag{5}$$

$$\bar{\mathbf{A}} - \mathbf{A} \leq \mathbf{0}, \tag{6}$$

where m denotes the number of the elments of the structure.

The multiplicity of the fundamental eigenvalue is denoted by s. Then the optimality criteria for ODDF may be written as

$$\sum_{r=1}^{s} \mu_r \zeta_i^r = w_i \quad \text{for} \quad A_i > \bar{A}_i \tag{7a}$$

$$\sum_{r=1}^{s} \mu_r \zeta_i^r < w_i \quad \text{for} \quad A_i = \bar{A}_i \qquad (i = 1, 2, \dots m), \tag{7b}$$

where μ_r is a nonnegative Lagrangian multiplier and ζ_i^r is defined by

$$\zeta_i^r(\mathbf{A}) = \Phi_r(\mathbf{A})^T \frac{\partial \mathbf{K}}{\partial A_i} \Phi_r(\mathbf{A}) - \Omega_a \Phi_r(\mathbf{A})^T \frac{\partial \mathbf{M}_D}{\partial A_i} \Phi_r(\mathbf{A}) \quad (r = 1, 2, \ldots s), \quad (i = 1, 2, \ldots m). \tag{8}$$

For a structure with an idealized sandwich cross-section, $\frac{\partial \mathbf{K}}{\partial A_i}$ and $\frac{\partial \mathbf{M}_D}{\partial A_i}$ do not depend on A_i and the conditions $(7a, b)$ are the necessary and sufficient conditions for optimality. It should be noted that the first and second terms of the right hand side of (8) are the specific strain energy and the specific kinetic energy of the ith element, respectively, of the corresponding eigenmode. It should also be noted that the second term is proportional to Ω_a. It is shown in the examples that the distributions of the design variables of an optimal structure are greatly affected by that of the following kinetic energy ratio:

$$e_i = \frac{\sum_{r=1}^{s} \left\{ \Omega_a \Phi_r(\mathbf{A})^T \frac{\partial \mathbf{M}_D}{\partial A_i} \Phi_r(\mathbf{A}) \right\}}{\sum_{r=1}^{s} \left\{ \Phi_r(\mathbf{A})^T \frac{\partial \mathbf{K}}{\partial A_i} \Phi_r(\mathbf{A}) + \Omega_a \Phi_r(\mathbf{A})^T \frac{\partial \mathbf{M}_D}{\partial A_i} \Phi_r(\mathbf{A}) \right\}} \quad (i = 1, 2, \ldots m), \tag{9}$$

which represents the ratio of the kinetic energy to the sum of the strain energy and the kinetic energy. Since the strain energy and the kinetic energy are large and the difference of those values is small for moderately large value of Ω_a, the ratio e_i represents the energy contents more clearly than the ratios of the strain energy and the kinetic energy to ζ_i^r.

AN ORDERED SET OF OPTIMAL STRUCTURES

The solution to ODDF may be found for each value of Ω_a. Therefore the optimum designs for a range of Ω_a may be regarded as a function of Ω_a which is denoted by $\hat{\mathbf{A}}(\Omega_a)$. Similar new functions, each denoted with a hat, are also defined for eigenvectors and Lagrangian multipliers, etc.. Differentiation of (1) with respect to Ω_a leads to the following equation:

$$\left\{ \sum_{i=1}^{m} \hat{A}_i' \frac{\partial \mathbf{K}}{\partial A_i} \right\} \hat{\mathbf{\Phi}}_r + \hat{\mathbf{K}} \hat{\mathbf{\Phi}}_r' = \{\hat{\mathbf{M}}_D + \mathbf{M}_J\} \hat{\mathbf{\Phi}}_r + \Omega_a \left\{ \sum_{i=1}^{m} \hat{A}_i' \frac{\partial \mathbf{M}_D}{\partial A_i} \right\} \hat{\mathbf{\Phi}}_r + \Omega_a \{\hat{\mathbf{M}}_D + \mathbf{M}_J\} \hat{\mathbf{\Phi}}_r' \quad (r = 1, 2, \ldots f), \tag{10}$$

where prime denotes differentiation with respect to Ω_a, and the argument Ω_a is omitted henceforth for brevity. From (3) and (8), the following equations may be derived:

$$2\hat{\mathbf{\Phi}}_r^T \{\hat{\mathbf{M}}_D + \mathbf{M}_J\} \hat{\mathbf{\Phi}}_r' + \hat{\mathbf{\Phi}}_r^T \left\{ \sum_{i=1}^{m} \hat{A}_i' \frac{\partial \mathbf{M}_D}{\partial A_i} \right\} \hat{\mathbf{\Phi}}_r = 0 \quad (r = 1, 2, \ldots f), \tag{11}$$

$$\hat{\zeta}_i^{r\prime} = 2\hat{\mathbf{\Phi}}_r^T \left\{ \frac{\partial \mathbf{K}}{\partial A_i} - \Omega_a \frac{\partial \mathbf{M}_D}{\partial A_i} \right\} \hat{\mathbf{\Phi}}_r' + \hat{\mathbf{\Phi}}_r^T \frac{\partial \mathbf{M}_D}{\partial A_i} \hat{\mathbf{\Phi}}_r \quad (r = 1, 2, \ldots s), \quad (i = 1, 2, \ldots m). \tag{12}$$

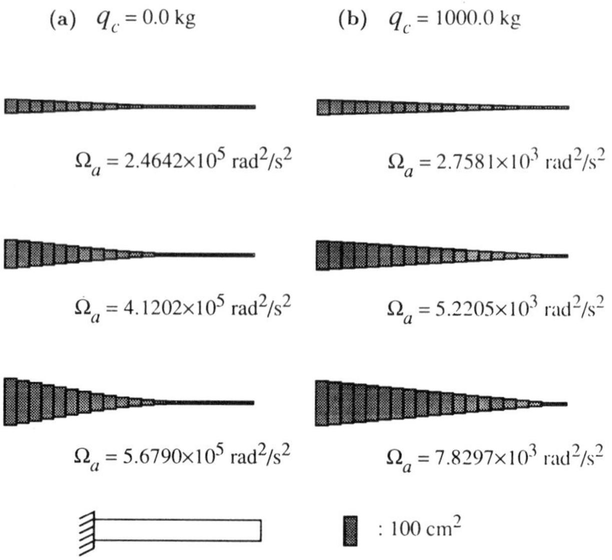

Fig.1 Two ordered sets of optimal cantilevers.

Differentiation of the optimality criteria leads to the following equations:

$$\left(\sum_{r=1}^{s} \hat{\mu}_r \hat{\zeta}_i^r\right)' = 0 \quad \text{for} \quad A_i > \bar{A}_i \tag{13a}$$

$$\hat{A}_i' = 0 \quad \text{for} \quad A_i = \bar{A}_i \qquad (i = 1, 2, \ldots m). \tag{13b}$$

These equations may further be differentiated to a desired order. The fundamental eigenvalue for the design with $\mathbf{A} = \bar{\mathbf{A}}$ is denoted by $\bar{\Omega}$. This design $\mathbf{A} = \bar{\mathbf{A}}$ is a trivial optimal solution of ODDF for $\Omega_a = \bar{\Omega}$. Then the ordered set of optimum designs can be swept out by successive Taylor series expansions with respect to Ω_a starting with the trivial initial optimal solution $\bar{\mathbf{A}}$ [13]. It should be noted here that the cross-sectional area of an element will be increased from its lower bound, as Ω_a is increased, after $\sum_{r=1}^{s} \mu_r \zeta_i^r$ of that element has reached w_i and the condition $(7a)$ is satisfied with equality. Since ζ_i^r in $(7a)$ is defined as the difference between the strain energy and the kinetic energy which is proportional to Ω_a, the cross-sectional area of that element with large specific strain energy and small specific kinetic energy increases as Ω_a is increased. The effect of the kinetic energy will be significant if Ω_a is large due to small value of the nonstructural mass compared to the total structural mass.

EXAMPLES

Ordered sets of optimum designs are generated for beams, arches and plates, each with idealized sandwich cross-sections. Both the stiffness and the structural mass are then proportional to the design variable. All the flanges consist of steel with elastic modulus E = 205.8 GPa and mass density ρ = 7.86 g/cm^3. The rotational inertia of the concentrated nonstructural mass is not considered.

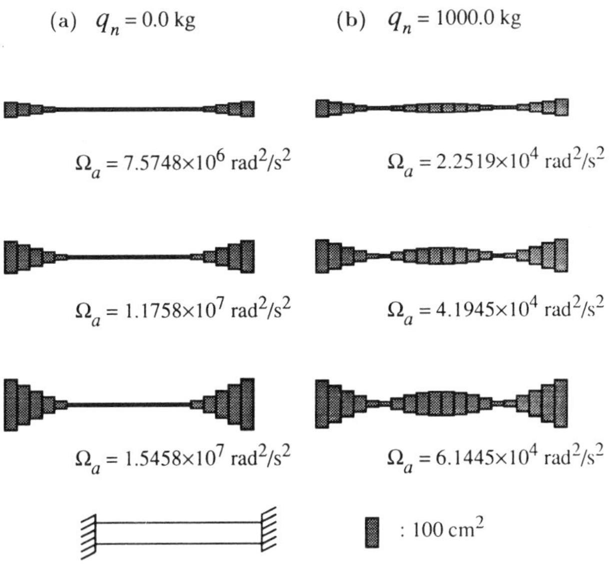

Fig.2 Two ordered sets of optimal clamped beams.

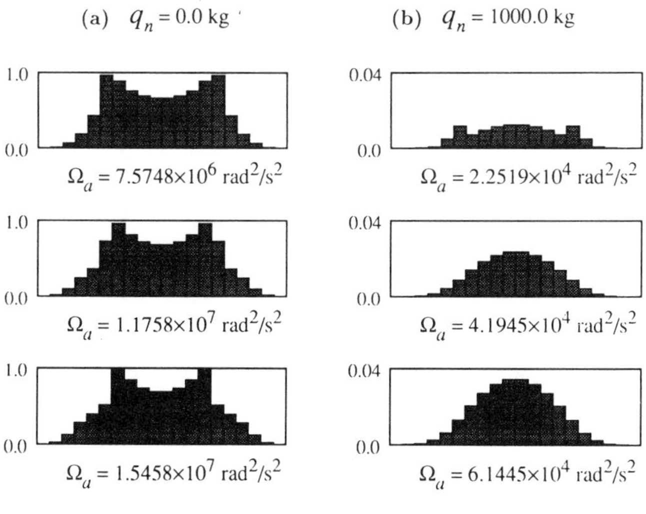

Fig.3 Kinetic energy ratio e_i of optimal clamped beams.

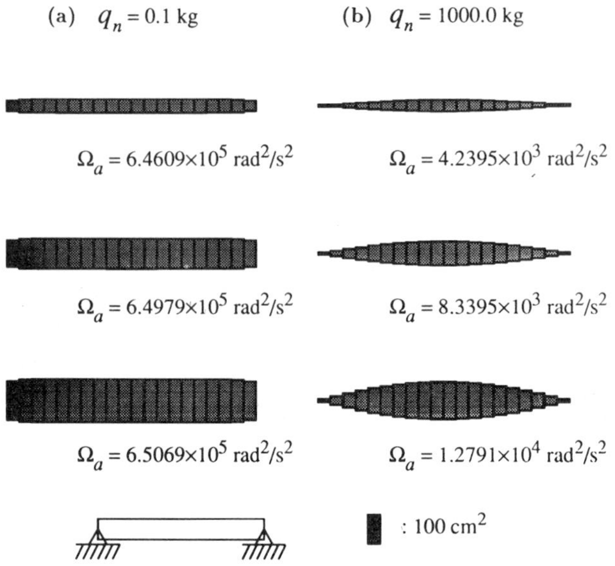

Fig.4 Two ordered sets of optimal simply supported beams.

Beams with various support conditions

The lateral displacement of each beam element is approximated with a cubic polynomial. The sum of the cross-sectional areas of a pair of the flanges of each element is chosen as the design variable. The distance between the two flanges is 50.0 cm. The minimum cross-sectional area $\bar{A}_i$ is equal to 10 cm^2 for all the elements. Consider first a cantilever with a tip mass q_c at the free end. The two ordered sets of optimal beams, one for $q_c = 0.0$ and the other for $q_c = 1000.0$ kg, are shown in Fig.1(a) and (b), respectively. It may be observed from Fig.1(a) that the cross-sectional areas of the elements near the support are increased drastically without accompanying any increase of A_i of the elements near the free end if there is no nonstructural mass at the free end. The distribution of the flange cross-sectional area of each example is quite similar to the result of Karihaloo and Niordson [4]. Consider next a clamped beam with nonstructural mass q_n at each node. The two ordered sets of clamped beams, one for $q_n = 0.0$ kg and the other for $q_n = 1000.0$ kg, are shown in Fig.2(a) and (b), respectively. It is observed from Fig.2(a) that the large eigenvalue due to small nonstructural mass leads to rapid increase of the cross-sectional areas of flanges of the elements near the supports. In the case of large nonstructural mass, the cross-sectional areas of flanges of the elements around the center also increases, because the fundamental eigenvalue is small and the effect of the second term in (8) is negligible. The kinetic energy ratio e_i defined by (9) is shown in Fig.3. It may be observed from Fig.3 that the second term is negligibly small for $q_n = 1000.0$ kg especially for the region of small eigenvalue. The two ordered sets of optimal simply supported beams, one for $q_n = 0.1$ kg and the other for $q_n = 1000.0$ kg, are shown in Fig.4(a) and (b), respectively. It may be observed that the cross-sectional area of the flange increases almost uniformly and rapidly for $q_n = 0.1$ kg, though Ω_a may be increased only slightly.

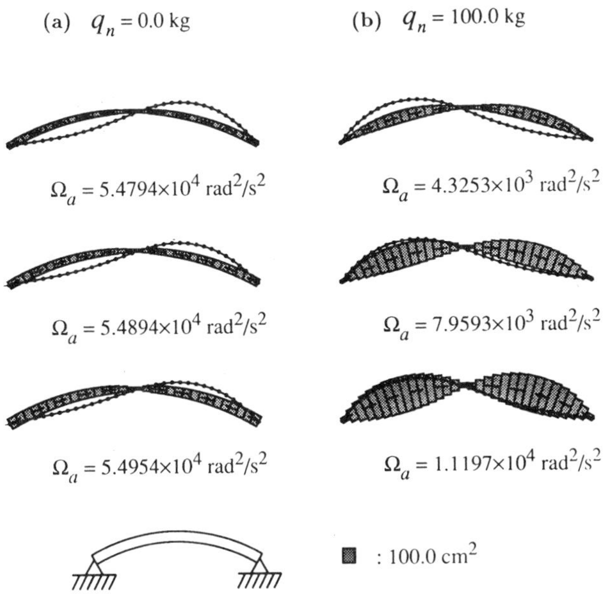

Fig.5 Two ordered sets of optimal simply supported arches.

Circular arches

Two ordered sets of optimal structures are generated for two circular arches with different support conditions. The cross-sectional area of the flange of the element is taken as the design variable. The shape function in Sabir and Ashwell [18] is used for discretization and the extensional strain is considered as well as the bending strain. The radius, the span and the flange distance of the arches are 8.0 m, 8.0 m and 30.0 cm, respectively. The minimum flange area is 10.0 cm^2 for all the elements. The two ordered sets of optimum designs of simply supported arches are shown in Fig.5(a) for $q_n = 0.0$ kg and in Fig.5(b) for $q_n = 100.0$ kg, where the fundamental eigenmodes have also been illustrated. It may be observed from Fig.5(a) and (b) that the flange cross-sectional area of the elements with large strain energy and small kinetic energy increases as the prescribed eigenvalue is increased. The optimal clamped arches, one for $q_n = 0.0$ kg and the other for $q_n = 100.0$ kg are shown in Fig.6(a) and (b), respectively. For $q_n = 0.0$ kg, flange cross-sectional areas of the elements near the supports increase rapidly. This tendency appears to be quite similar to that in the clamped beam. In the case of $q_n = 100.0$ kg, the fundamental eigenvalue is multiple for $\Omega_a \geq 1.2691 \times 10^4$ rad^2/s^2. Two independent fundamental eigenmodes are shown in Fig.5(b). The variation of the second eigenvalue is plotted in Fig.7 with respect to Ω_a. Eigenvalue analysis has been carried out only for the purpose of the verification for the arch with $q_n = 100.0$ kg and $\Omega_a = 2.0726 \times 10^4$ rad^2/s^2. Since the evaluated two lowest eigenvalues are both equal to 2.0726×10^4 rad^2/s^2, the truncation error due to Taylor series expansion is negligibly small. It is noted that no significant difference may be found in the distribution of the cross-sectional area of the two cases with single and multiple fundamental eigenvalue. The total flange mass is plotted in Fig.8 with respect to Ω_a. With the proposed method, such design charts as Fig.7 and 8 may easily be presented for designers.

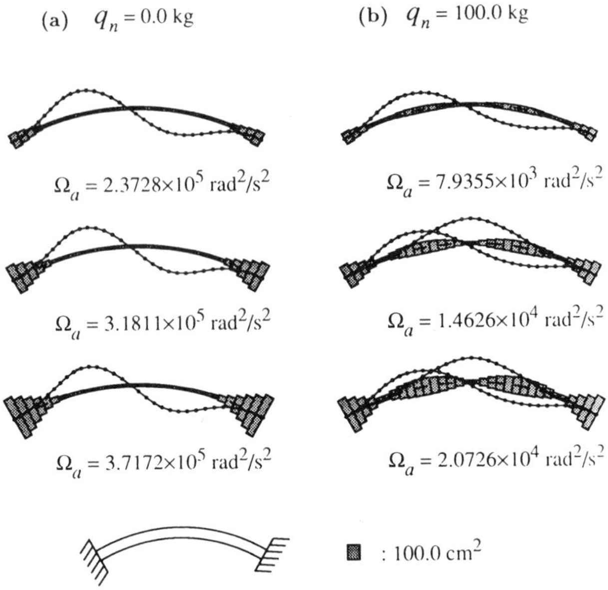

Fig.6 Two ordered sets of optimal clamped arches.

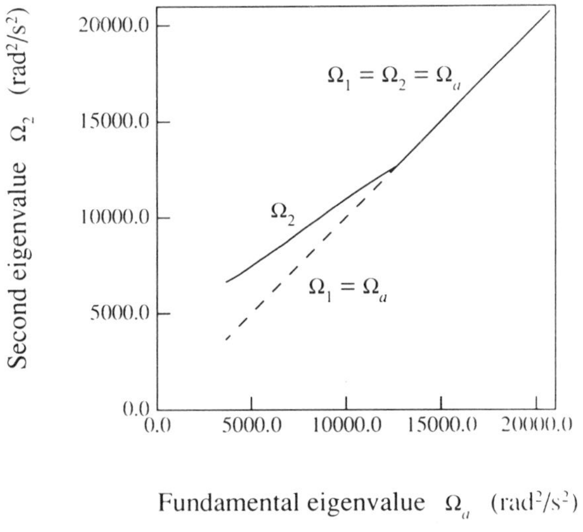

Fig.7 Variation of the second eigenvalue of the clamped arch with respect to Ω_a.

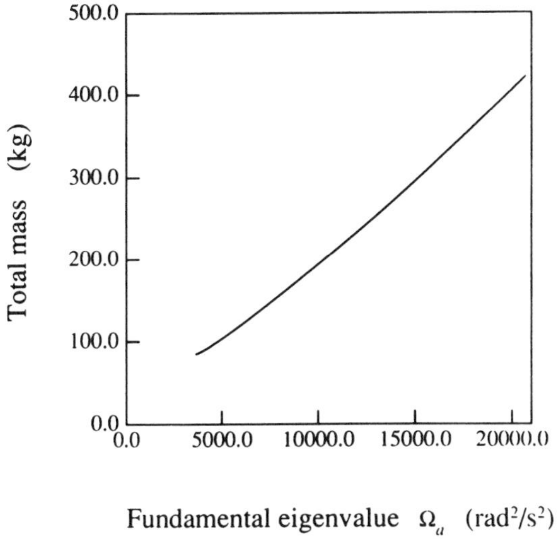

Fig.8 Variation of total mass of the clamped arch with respect to Ω_a.

Rectangular plates

A rectangular plate is treated as a system of equal rectangular elements. Cubic polynomial of twelve degree of freedom of displacements [19] is used for the shape function. The thickness of the flange is taken as the design variable. The distance between the two flanges is 40 cm. An ordered set of a simply supported plate is illustrated in Fig.9(a), where the distributed nonstructural mass is 500.0 kg/cm^2. It may be observed from Fig.9(a) that the thickness of the flanges of an element near the corner increase due to the large strain energy and the small kinetic energy in the fundamental eigenmode. It should be noted that the bending action is large in the elements near the corner of the simply supported plate. The thickness of the flanges of an element around the center also increases, especially for small value of Ω_a, because the second term in (8) is insignificant due to the large value of nonstructural mass. The set of optimal clamped plates is shown in Fig.9(b), where the distributed nonstructural mass is 100 kg/cm^2. It may be observed from Fig.9(b) that the thickness of the flange of an element along the two centerlines increases, as Ω_a is increased, due to the large strain energy and the small kinetic energy.

CONCLUSIONS

The authors' method of sequentially generating the set of optimal trusses by the Taylor series expansions with respect to the specified fundamental eigenvalue has been extended so as to be applicable to distributed parameter structures. With this method, two ordered sets of optimal structures have been generated for beams, arches and plates for specified frequency ranges. Various design charts may easily be presented for designers based on the optimum design theory. It has been shown in the examples that the proposed method is applicable also for the case where the fundamental eigenvalue is multiple. The characteristics of the optimal sandwich structures have been discussed in detail in view of the optimality criteria. This method can be extended further to any tapered or solid distributed parameter structures discretized by a finite element method.

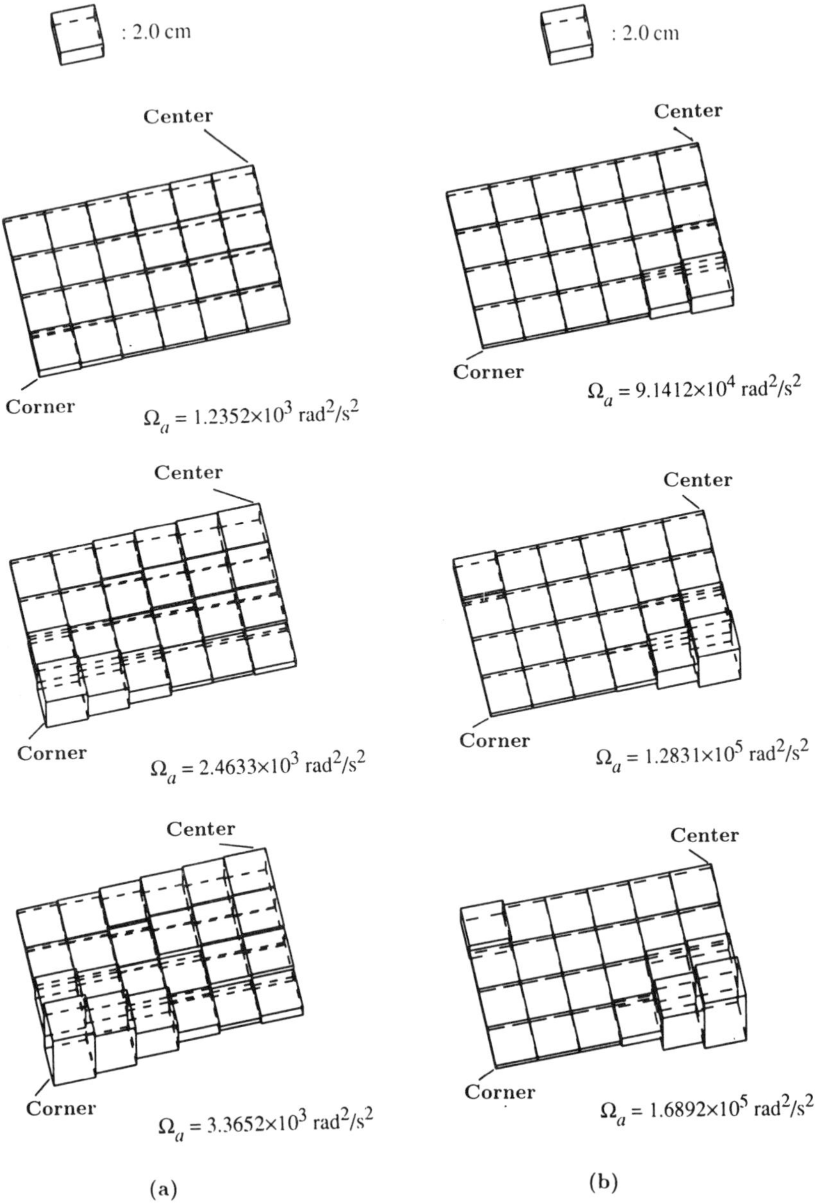

Fig.9 Two ordered sets of optimal rectangular plates, (a) simply supported plate, (b) clamped plate.

REFERENCES

1. Prasad, B and Haftka, R.T. Optimal Structural Design with Plate Finite Elements, J. Struct. Div., ASCE, Vol.105(11), pp2367-2382, 1979.

2. Domaszewski, M., Knopf-Lenoir, C., Batoz, J.L. and Touzot, G. Shape Optimization and Minimum Weight Limit Design of Arches, Eng. Opt., Vol.11, pp173-193, 1987.

3. Niordson, F.I. On the Optimal Design of a Vibrating Beam, Quart. Appl. Math., Vol.23, pp47-53, 1965.

4. Karihaloo, B.L. and Niordson, F.I. Optimum Design of Vibrating Cantilevers, J. Opt. Theory Appl., Vol.11(6), pp638-654, 1973.

5. Vepa, K. On the Existence of Solutions to Optimization Problems with Eigenvalue Constraints, Quart. Appl. Math., pp329-341, 1973.

6. Olhoff, N. Optimal Design of Vibrating Rectangular Plates, Int. J. Solids Struct., Vol.10, pp.93-109, 1974.

7. Plaut, R.H. and Olhoff, N. Optimal Forms of Shallow Arches with Respect to Vibration and Stability, J. Struct. Mech., Vol.11(1), pp81-100, 1983.

8. Blachat, J. and Gajewski, A. On Unimodal and Bimodal optimal Design of Funicular Arches, Int. J. Solids Struct., Vol.17, pp653-667, 1981.

9. Armand, J. Minimum-Mass Design of a Plate-Like Structure for Specified Fundamental Frequency, AIAA J., Vol.9(9), pp1739-1745, 1971.

10. Karihaloo, B.L. and Parbery, R.D. Minimum Weight Members for Given Lower Bounds on Eigenvalues, Eng. Opt. Vol.5, pp199-205, 1981.

11. Karihaloo, B.L. and Niordson, F.I. Optimum Design of Vibrating Beams under Axial Compression, Vol.24, pp1029-1037, 1972.

12. Olhoff, N. Optimal Design of Vibrating Circular Plates, Int. J. Solids Struct., Vol.6, pp139-156, 1970.

13. Nakamura, T. and Ohsaki, M. Sequential Optimal Truss Generator for Frequency Ranges, Comput. Meth. Appl. Mech. Engng, Vol.67, pp.189-209, 1988.

14. Nakamura, T. and Ohsaki, M. Sequential Generator of Earthquake-response Constrained Trusses for design strain Ranges, Int. J. Comp. & Struct., Vol.33(6), pp.1403-1416, 1989.

15. Sobieszczanski-Sobieski, J.S., Barthelemy, J. and Riley, K.M. Sensitivity of Optimum Solutions of Problem Parameter, AIAA J., Vol.20(9), pp1291-1299, 1982.

16. Schmit, L.A. and Chang, K.J. Optimum Design Sensitivity Based on Approximation Concepts and Dual Methods, Int. J. Numer. Meth. Engng, Vol.20, pp39-75, 1984.

17. Shin, Y.S., Haftka, R.T., Watson, L.T. and Plaut, R.H. Tracing Structural Optima as a Function of Available Resources by a Homotopy Method, Comp. Meth. Appl. Mech. Engng, Vol.70, pp151-164, 1988.

18. Sabir, A.B. and Ashwell, D.G. A Comparison of Curved Beam Finite Elements When Used in Vibration Problems, J. Sound Vib., Vol.18(4), pp.555-563, 1971.

19. Zienkiewicz, O.C. and Cheung, Y.K. The Finite Element Method in Structural and Continuum Mechanics, McGraw-Hill, New York, 1967.

Sequential Generator of Earthquake-Strain Constrained Curved Space Trusses for Design Strain Ranges

T. Nakamura, M. Ohsaki

Department of Architecture, Kyoto University, Sakyo, Kyoto 606, Japan

ABSTRACT

A practically efficient method is presented of constructing a design chart for a specified design strain set that represents n-dimensional locus or hypersurface of the evaluation function for earthquake-strain constrained designs (ESCD's) and optimum designs of a family of large trusses characterized with n geometrical parameters. The design chart data enable one to find the geometrically optimal truss of ESCD or of optimum design under the same earthquake-strain constraints. The efficiency of the proposed method is demonstrated through examples of large curved space trusses.

INTRODUCTIONS

The problems considered in this paper are (1) to find a sequence of earthquake-strain constrained designs (ESCD's) of a large curved space truss for candidate strain ranges[1] and a sequence of their optimum designs under the same constraints, and (2) to construct a locus and a design chart of ESCD's or of optimum designs of a family of curved space trusses characterized with different geometrical parameters for a design strain set. The mean maximum response strains to a set of design earthquake motions compatible with the prescribed design response spectrum are evaluated by the complete quadratic combination (CQC) method[2,3].

While various numerical methods have been proposed for the problems of finding response constrained designs and optimum designs of a truss with a fixed geometry for dynamic excitations[4-12], most of them require laborious computation of a great number of sensitivity coefficients even for finding just one such fully stressed design or optimum design. Any previous method of geometrical optimization of a large truss[13,14] requires far more laborious computation of a greater number of design sensitivity coefficients and geometrical sensitivity coefficients and does not seem to be efficient for large systems.

Nakamura and Ohsaki[1] presented a computationally efficient method of sweeping out the sequence of the earthquake-strain constrained designs (ESCD's) of large trusses of fixed geometry such that the mean maximum response strains due to the spectrum compatible earthquake motions would not exceed the specified upper bounds. The procedure is started with the trivial solution where the cross-sectional area of each member coincides with its specified lower bound, and ESCD's are swept out by successive Taylor series expansions with respect to the parameter which specifies the upper-bound strain. It should be noted that the eigenvalue analysis needs to be carried out only once for the initial design. This method of finding ESCD's is shown to be very efficient for the problem stated at the beginning.

The purpose of this paper is to present a practically efficient method of constructing a design chart for a specified design strain set that represents an $n-$dimensional locus or hypersurface of the evaluation function for ESCD's of a family of curved space trusses characterized with n geometrical parameters. If the ranges of the n geometrical parameters define a closed region, there must exist a minimal point on the hypersurface which represents the geometrically optimal truss of ESCD. The design chart data enable one to find the geometrically optimal truss of ESCD. Neither any ordinary design sensitivity analysis nor application of mathematical programming is required in the proposed method. It is shown that a sequence of ESCD's of a truss for a sequence of candidate design strain levels coincides at least partly with a sequence of optimum designs under the same constraints on earthquake-strains. A hypersurface of the evaluation function for ESCD's of a family of curved space trusses may therefore be shown to coincide as least partly with a hypersurface of the evaluation function for optimum designs of the same family of trusses. the design chart data also enable one to find the geometrically optimal truss among the optimally designed trusses under earthquake-strain constraints.

PROBLEM OF EARTHQUAKE-STRAIN CONSTRAINED DESIGNS FOR FIXED GEOMETRY

Consider a structure whose geometry is prescribed. The structure is discretized by a finite element method and is defined completely by the set of the design variable $\{A_i\}$ which is denoted by a vector $\mathbf{A}$. A set of lower bounds of the design variable is denoted by $\{\bar{A}_i\}$. The structure is to be designed so that each mean value of the maximum response representative strain of the element to a set of spectrum-compatible design earthquakes does not exceed the prescribed value. Since the structure is deformed considerably due to the self-weight, the initial static strain is taken into account in evaluating the maximum earthquake-strain.

Let $S_D(\Omega)$ denote a prescribed design displacement response spectrum with respect to the eigenvalue Ω. The stiffness matrix, the consistent mass matrix due to the structural mass and the mass matrix due to the nonstructural mass are denoted by $\mathbf{K}(\mathbf{A})$, $\mathbf{M}_D(\mathbf{A})$ and $\mathbf{M}_J$, respectively. The rth eigenvalue and the corresponding eigenvector are denoted by $\Omega_r(\mathbf{A})$ and $\mathbf{\Phi}_r(\mathbf{A})$, respectively. The equation of undamped free vibration in the rth eigenmode may be reduced to

$$\mathbf{K}(\mathbf{A})\mathbf{\Phi}_r(\mathbf{A}) = \Omega_r(\mathbf{A})\{\mathbf{M}_D(\mathbf{A}) + \mathbf{M}_J\}\mathbf{\Phi}_r(\mathbf{A}) \qquad (r = 1, 2, \ldots, s), \tag{1}$$

where the lowest s modes are taken into consideration in the following. The vector $\mathbf{\Phi}_r(\mathbf{A})$ may be normalized by

$$\mathbf{\Phi}_r(\mathbf{A})^T\{\mathbf{M}_D(\mathbf{A}) + \mathbf{M}_J\}\mathbf{\Phi}_r(\mathbf{A}) = 1 \qquad (r = 1, 2, \ldots, s), \tag{2}$$

where the superscript T denotes the transpose of a vector. The representative strain of the ith element corresponding to $\mathbf{\Phi}_r(\mathbf{A})$ is denoted by e_i^r. The relation between e_i^r and $\mathbf{\Phi}_r(\mathbf{A})$ may be written as

$$e_i^r(\mathbf{A}) = \mathbf{C}_i^T \mathbf{\Phi}_r(\mathbf{A}) \qquad (i = 1, 2, \ldots m), \qquad (r = 1, 2, \ldots s), \tag{3}$$

where m and $\mathbf{C}_i^T$ denote the number of the elements of the structure and the vector which defines the displacement-strain relation, respectively.

Let $\mathbf{I}$ denote a vector which defines the direction of the earthquake-motion. The participation factor $\eta_r(\mathbf{A})$ of the rth mode may be written, in terms of the normalized eigenvector $\mathbf{\Phi}_r(\mathbf{A})$ as follows:

$$\eta_r(\mathbf{A}) = \mathbf{\Phi}_r(\mathbf{A})^T[\mathbf{M}_D(\mathbf{A}) + \mathbf{M}_J(\mathbf{A})]\mathbf{I} \qquad (r = 1, 2, \ldots s). \tag{4}$$

Since the structure may have closely spaced natural frequencies, the mean maximum responses needs to be evaluated by the CQC method[2,3]. The modal damping ratio is considered to be equal to h throughout all the modes, for simpler representation of the method and for demonstrating the efficiency in a severer case. The cross-modal correlation coefficient $\rho_{pq}(\mathbf{A})$ between the pth and the qth modes is then given by

$$\rho_{pq}(\mathbf{A}) = \frac{8h^2\{1 + \alpha_{pq}(\mathbf{A})\}\alpha_{pq}(\mathbf{A})^{3/2}}{\{1 - \alpha_{pq}(\mathbf{A})^2\}^2 + 4h^2\alpha_{pq}(\mathbf{A})\{1 + \alpha_{pq}(\mathbf{A})^2\} + 8h^2\alpha_{pq}(\mathbf{A})^2} \tag{5}$$
$$(p, q = 1, 2, \ldots, s),$$

where

$$\alpha_{pq}(\mathbf{A}) = \sqrt{\{\Omega_q(\mathbf{A})/\Omega_p(\mathbf{A})\}}. \tag{6}$$

The mean maximum response strain $\varepsilon_i^v(\mathbf{A})$ in the ith element due to a set of earthquake motions may be evaluated in the following complete quadratic combination form:

$$\varepsilon_i^v(\mathbf{A}) = \sqrt{\sum_{p=1}^{s}\sum_{q=1}^{s}\{S_D(\Omega_p(\mathbf{A}))\eta_p(\mathbf{A})e_i^p(\mathbf{A})\}\rho_{pq}(\mathbf{A})\{S_D(\Omega_q(\mathbf{A}))\eta_q(\mathbf{A})e_i^q(\mathbf{A})\}} \tag{7}$$

Let $\mathbf{G}(\mathbf{A})$ denote a vector of static force due to the dead load. The static equilibrium equation may be written as

$$\mathbf{K}(\mathbf{A})\mathbf{u}(\mathbf{A}) = \mathbf{G}(\mathbf{A}), \tag{8}$$

where $\mathbf{u}(\mathbf{A})$ denotes the vector of static displacement. The initial static strain $\varepsilon_i^s(\mathbf{A})$ of the ith element may be written as

$$\varepsilon_i^s(\mathbf{A}) = \mathbf{C}_i^T\mathbf{u}(\mathbf{A}) \qquad (i = 1, 2, \ldots m). \tag{9}$$

The maximum strain $\varepsilon_i(\mathbf{A})$ of the ith element is defined as the sum of the nonnegative value $\varepsilon_i^v(\mathbf{A})$ and the absolute value of $\varepsilon_i^s(\mathbf{A})$ as follows:

$$\varepsilon_i(\mathbf{A}) = \varepsilon_i^v(\mathbf{A}) + |\varepsilon_i^s(\mathbf{A})| \qquad (i = 1, 2, \ldots m). \tag{10}$$

Let Λ_i denote the design earthquake-strain for $\varepsilon_i(\mathbf{A})$ specified by the designer's preference. Then the problem of finding an ESCD for a structure with a fixed geometry may be formulated as follows:

[Problem ECDF]

For a set of design earthquake-strains $\{\Lambda_i\}$, find an ESCD $\tilde{\mathbf{A}}$ which satisfies the following conditions:

$$\varepsilon_i(\tilde{\mathbf{A}}) = \Lambda_i \quad \text{for} \quad \tilde{A}_i > \bar{A}_i \tag{11a}$$
$$\varepsilon_i(\tilde{\mathbf{A}}) \leq \Lambda_i \quad \text{for} \quad \tilde{A}_i = \bar{A}_i \qquad (i = 1, 2, \dots m). \tag{11b}$$

AN ORDERED SET OF EARTHQUAKE-STRAIN CONSTRAINED DESIGNS OF A STRUCTURE WITH A FIXED GEOMETRY

The solution to Problem ECDF may be found for each set of $\{\Lambda_i\}$. Therefore the ESCD may be regarded as a function of $\{\Lambda_i\}$. A scaling parameter ξ is introduced, for brevity, as

$$\{\Lambda_i\} = \xi\{\Lambda_i^b\} \qquad (i = 1, 2, \dots m), \tag{12}$$

where $\{\Lambda_i^b\}$ is to be prescribed by the designer. Then $\tilde{\mathbf{A}}$ may be regarded as a function of ξ, and is denoted by $\hat{\mathbf{A}}(\xi)$. The corresponding response strains are also functions of ξ and denoted by

$$\hat{\varepsilon}_i(\xi) = \varepsilon_i(\hat{\mathbf{A}}(\xi)) \qquad (i = 1, 2, \dots m). \tag{13}$$

Hence the following problem of finding an ordered set of ESCD's of a structure with a fixed geometry may be stated:

[Problem OECDF]

For an ordered set $\{\Lambda_i\} = \xi\{\Lambda_i^b\}$ specified parametrically in terms of ξ, find an ordered set of ESCD's $\hat{\mathbf{A}}(\xi)$ which satisfy the following conditions:

$$\hat{\varepsilon}_i(\xi) = \xi\Lambda_i^b \quad \text{for} \quad \hat{A}_i(\xi) > \bar{A}_i \tag{14a}$$
$$\hat{\varepsilon}_i(\xi) \leq \xi\Lambda_i^b \quad \text{for} \quad \hat{A}_i(\xi) = \bar{A}_i \qquad (i = 1, 2, \dots m). \tag{14b}$$

Differentiation of $(14a, b)$ with respect to ξ leads to the following forms:

$$\frac{d\hat{\varepsilon}_i(\xi)}{d\xi} = \Lambda_i^b \quad \text{for} \quad \hat{A}_i(\xi) > \bar{A}_i \tag{15a}$$
$$\frac{d\hat{A}_i(\xi)}{d\xi} = 0 \quad \text{for} \quad \hat{A}_i(\xi) = \bar{A}_i \qquad (i = 1, 2, \dots m). \tag{15b}$$

If all the other governing equations are differentiated with respect to ξ, and if $\hat{\mathbf{A}}(\xi)$, $\hat{\Omega}_r(\xi)$,... are known for a certain value of ξ, then the ordered set of ESCD's may be swept out by successive Taylor series expansions with respect to ξ. The procedure is started with the trivial solution as $\hat{\mathbf{A}}(\xi) = \bar{\mathbf{A}}$. The details of the method have been shown in Nakamura and Ohsaki[1]. It should be noted that no successive improvement method utilizing ordinary design sensitivity analysis is needed. Furthermore, the eigenvalue analysis and the evaluation of the maximum earthquake-strain need to be carried out only once for the initial design.

OPTIMALITY CRITERIA FOR EARTHQUAKE-STRAIN CONSTRAINTS

The ESCD may also be regarded as a kind of fully stressed design. In this section, the corresponding problem of optimum design is formulated and the necessary conditions for optimality are derived. The optimum design problem under earthquake-strain constraints may be stated as follows:

[Problem ODPES]

Find the optimum design $\check{\mathbf{A}}$ which minimize the total volume $V(\mathbf{A}) = \sum_{i=1}^{m} A_i w_i$

under the constraints

$$\varepsilon_i(\check{\mathbf{A}}) \leq \Lambda_i \qquad (i = 1, 2, \ldots m), \tag{16a}$$
$$\check{A}_i \geq \bar{A}_i \qquad (i = 1, 2, \ldots m), \tag{16b}$$

where w_i denotes the structural volume per unit value of A_i.

The necessary conditions for optimality may be written as

$$w_i + \sum_{j=1}^{m} \nu_j \frac{\partial \varepsilon_j(\mathbf{A})}{\partial A_i} - \mu_i = 0 \qquad (i = 1, 2, \ldots m), \tag{17}$$
$$\nu_i(\varepsilon_i(\mathbf{A}) - \Lambda_i) = 0, \quad \nu_i \geq 0 \qquad (i = 1, 2, \ldots m), \tag{18}$$
$$\mu_i(\bar{A}_i - A_i) = 0, \quad \mu_i \geq 0 \qquad (i = 1, 2, \ldots m), \tag{19}$$

where $\{\nu_i\}$ and $\{\mu_i\}$ are Lagrangian multipliers.

It may be observed from $(16a, b)$ that both of the constraints for earthquake-strain and for the design variables may be satisfied in inequalities, whereas $\varepsilon_i(\mathbf{A})$ for the element with $A_i > \bar{A}_i$ of the ESCD must be equal to the prescribed value. Therefore the ESCD $\tilde{\mathbf{A}}$ may be regarded to be a trivial feasible solution of ODPES. As is well known, a fully stressed design is an optimum design if all the multipliers are nonnegative. Once $\check{\mathbf{A}}$ has been found through the procedure, design sensitivity coefficients of $\varepsilon_j(\mathbf{A})$ can be computed at $\check{\mathbf{A}}$ and the multipliers $\{\nu_j\}$ can be found from (17). It is shown in the following examples that ESCD's are also an optimum design for wide range of parameter ξ.

OPTIMAL GEOMETRY OF EARTHQUAKE-STRAIN CONSTRAINED TRUSSES WITH A FIXED TOPOLOGY

For every set of prescribed strains $\{\Lambda_i^b\}$ and prescribed geometrical parameters $\{b_i\}$, it is possible to find a sequence of ESCD's with respect to ξ. It is then convenient to conceive a set of ESCD's with one and the same value of ξ throughout the trusses with a fixed topology that can be described by exhausting all the possible combinations of $\{b_i\}$. The new function representing such a set is denoted by $\mathbf{A}^{\#}(\xi, \mathbf{b})$. The remaining parameters $\bar{A}_i$ need to be specified for a set $\mathbf{A}^{\#}(\xi, \mathbf{b})$. For instance, $\{\bar{A}_i(\mathbf{b})\}$ may be required to satisfy the following conditions:

$$\bar{A}_i(\mathbf{b}) = \bar{A}^*(\mathbf{b}) \qquad (i = 1, 2, \ldots, m), \tag{20}$$
$$V[\bar{A}^*(\mathbf{b})] = V^*. \tag{21}$$

The objective function $V[\mathbf{A}]$ for $\mathbf{A}^{\#}(\xi_k, \mathbf{b})$ specified by a fixed value of ξ, say ξ_k, may be represented by a hypersurface in an $(n+1)$–dimensional space whose cartesian coordinates are $\{b_i\}$ and V. V is now regarded as a function of $\mathbf{b}$ and an auxiliary variable ξ and denoted by $V^{\#}(\mathbf{b}, \xi)$. For a prescribed closed region B of $\{b_i\}$, there must exist a minimal point of $V^{\#}(\mathbf{b},\xi_k)$ in B. Therefore the following problem of geometrical optimization may be stated:

> [Problem GOESCD]
>
> For a prescribed set of $\bar{\mathbf{A}}^*(\mathbf{b})$, V^*, $\xi_k\{\Lambda_i^b\}$ and region B, find the optimum geometry $\mathbf{b}_m$ which minimizes $V^{\#}(\mathbf{b},\xi_k)$.

If $\mathbf{b}_m$ happened to be within a subregion of B such that $\mathbf{A}^{\#}(\xi_k,\mathbf{b})$ of Problem ECDF is also the optimal solution to Problem ODPES, then $V^{\#}(\mathbf{b}_m,\xi_k)$ is the geometrical optimum among the optimum earthquake-strain constrained designs.

Apparently a natural first impression on Problem GOESCD might be misleading if this problem were taken as a simple mathematical programming problem in n variables. Since $V^{\#}(\mathbf{b},\xi_k)$ has originally been defined in view of $V[\mathbf{A}^{\#}(\xi_k,\mathbf{b})]$, application of a nonlinear programming technique based upon ordinary sensitivity analysis would necessarily require laborious computation of a great number of sensitivity coefficients. It should also be noted that $V^{\#}(\mathbf{b},\xi_k)$ may have a number of ridge lines across which the first derivative of $V^{\#}(\mathbf{b},\xi_k)$ is discontinuous due to the piecewise continuous characteristics in previous sections.

It should be noted that a designer would welcome a set of $V^{\#}(\mathbf{b},\xi_k)$–surfaces (hypersurfaces) for a number of ξ_k values of interest as efficient design charts. In other words, the designer is interested not only in the single optimum geometry but also in the overall variation of a $V^{\#}(\mathbf{b},\xi_k)$–surface in the neighborhood of the optimal point. A primitive but practically simple second approach to Problem GOESCD is therefore to make use of the sequences of $A^{\#}(\xi,\mathbf{b})$, i.e. to compute $A^{\#}(\xi_k,\mathbf{b})$ for a set of N different $\mathbf{b}$'s and a set of several ξ_k values to draw several diagrams of $V^{\#}(\mathbf{b},\xi_k)$–surfaces. For each ξ_k value, a best-fit approximate surface with respect to the N points of $V^{\#}(\mathbf{b},\xi_k)$ may readily be defined and an approximate minimum of $V^{\#}(\mathbf{b},\xi_k)$ can be found. A set of those best-fit approximate surfaces may then be presented to the designer as the design charts just as illustrated in a later example. A better approximate optimal point in the n–space to a desired accuracy can be found if the foregoing pointwise procedure is repeated on a finer mesh near the previous approximate optimal point.

EXAMPLES

A sequence of ESCD's of a curved space truss and an approximate V–surface are illustrated here. The extensional strain of each member is taken as the representative strain. The response spectrum proposed by Newmark and Hall[17] is taken as the design response spectrum. The maximum values of the ground acceleration, velocity and displacement are 201.0 cm/s^2, 25.0 cm/s and 18.75 cm, respectively. The truss is composed of steel with elastic modulus 205.8 GPa and mass density 7.86 g/cm^3. The nonstructural mass at each node in the upper surface is 800.0 kg (400.0 kg for a side node and 200.0 kg for a corner node). The spans in x and y direction are 16 m and 12 m, respectively. The elastic spring attached to a support joint is to represent the stiffness of the support. The equal value of the strain ratios $\Lambda_i^b = 1$ is specified for all the members and h is taken to be equal to 0.02 throughout all the trusses.

A sequence of optimum design of a 504-bar cylindrical space truss

Consider a 504-bar cylindrical space truss as shown in Fig.1. The radii of the lower and the upper cylinders are 16 m and 17.5 m, respectively. The extensional

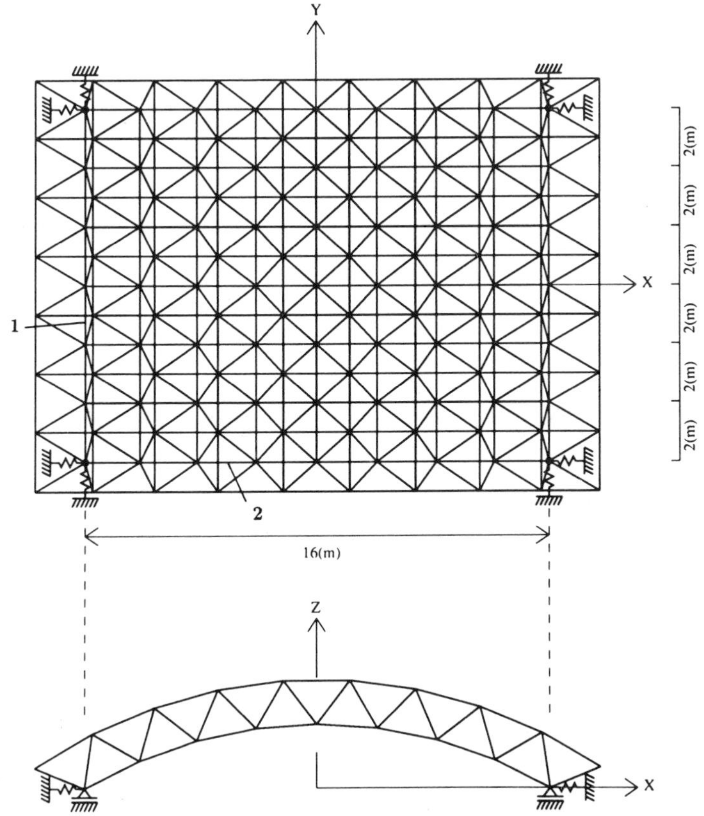

Fig.1 A 504-bar cylindrical space truss.

stiffness of the spring is 1.0290×10^5N/m. The minimum cross-sectional area is 5.0 cm^2 for all the members. The lowest four symmetric modes are taken into consideration, since the vertical excitations are considered and the participation factor for an antisymmetric mode vanishes. The value of ξ for the initial design $\mathbf{A} = \bar{\mathbf{A}}$ is 1.4648×10^{-3}, where the condition (11a) is satisfied in equality in member 1 in Fig.1. The parameter ξ has been decreased down to 2.0069×10^{-4}. The ESCD's for $\xi = 5.0762\times10^{-4}$, 3.2613×10^{-4}, 2.4711×10^{-4} and 2.0069×10^{-4} are shown in Fig.2(a)-(d) for one of the four symmetric parts. It may be observed from Fig.2 that the cross-sectional areas of the upper and lower chords between the supports and the diagonals near the supports increase as ξ is decreased. The maximum earthquake-strain for the truss with $\xi = 2.5081 \times 10^{-4}$ is 2.5055×10^{-4}. This result shows that the truncation error of the Taylor series expansions is sufficiently small. For $\xi > 1.0715 \times 10^{-3}$, the generated designs are the optimum designs for Problem ODPES, because all the Lagrangian multipliers are found to be nonnegative. At $\xi = 1.0715 \times 10^{-3}$, μ_i of member 2 turned negative and the generated designs for $\xi < 1.0715 \times 10^{-3}$ are not the optimum designs. The solid lines in Fig.3 show the variation of ν_2 with respect to ξ. The variation of ν_1 is also plotted with dashed lines in Fig.3. It may be observed from Fig.3 that ν_i is a discontinuous function of ξ. At those points where ν_i changes

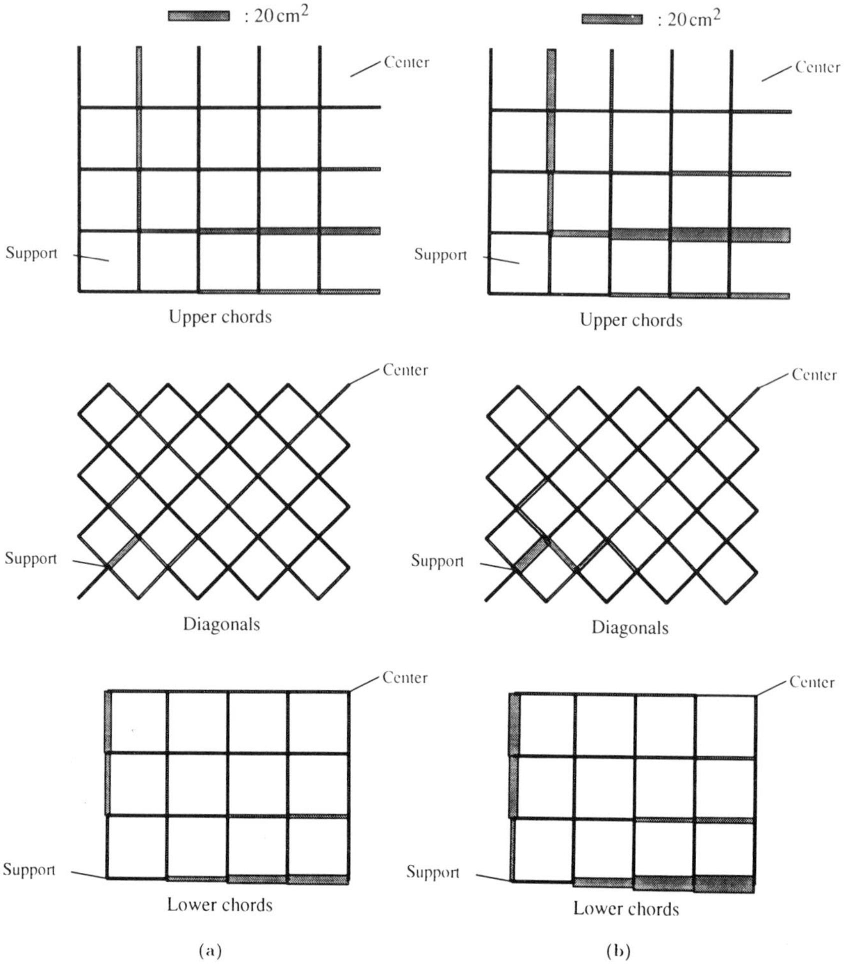

Fig.2 Cross-sectional area of the earthquake-strain constrained 504-bar cylindrical truss, (a) $\xi = 5.0762 \times 10^{-4}$ and (b) $\xi = 3.2613 \times 10^{-4}$.

discontinuously, the number of the members such that $A_i > \bar{A}_i$ increases. The total volume $\hat{V}$ of the earthquake-strain constrained design is plotted in the solid line in Fig.4. It may be observed from Fig.4 that $\hat{V}$ increases as ξ is decreased. Since Λ_i^b has been defined such that $\Lambda_i^b = 1$ for all the members, ξ is equal to the maximum earthquake-strain. The dashed line in Fig.4 shows the plots of the total volume V_c of a truss of uniform members with respect to the maximum earthquake-strain in all the members of the truss. It may be observed from Fig.4 that the ratio of $\hat{V}$ to V_c decreases rapidly as ξ is decreased, although the earthquake-strain constrained designs are no longer the optimum designs for $\xi \leq 1.0715 \times 10^{-3}$.

Design chart of ESCD's of a 504-bar doubly curved space truss and optimal geometry

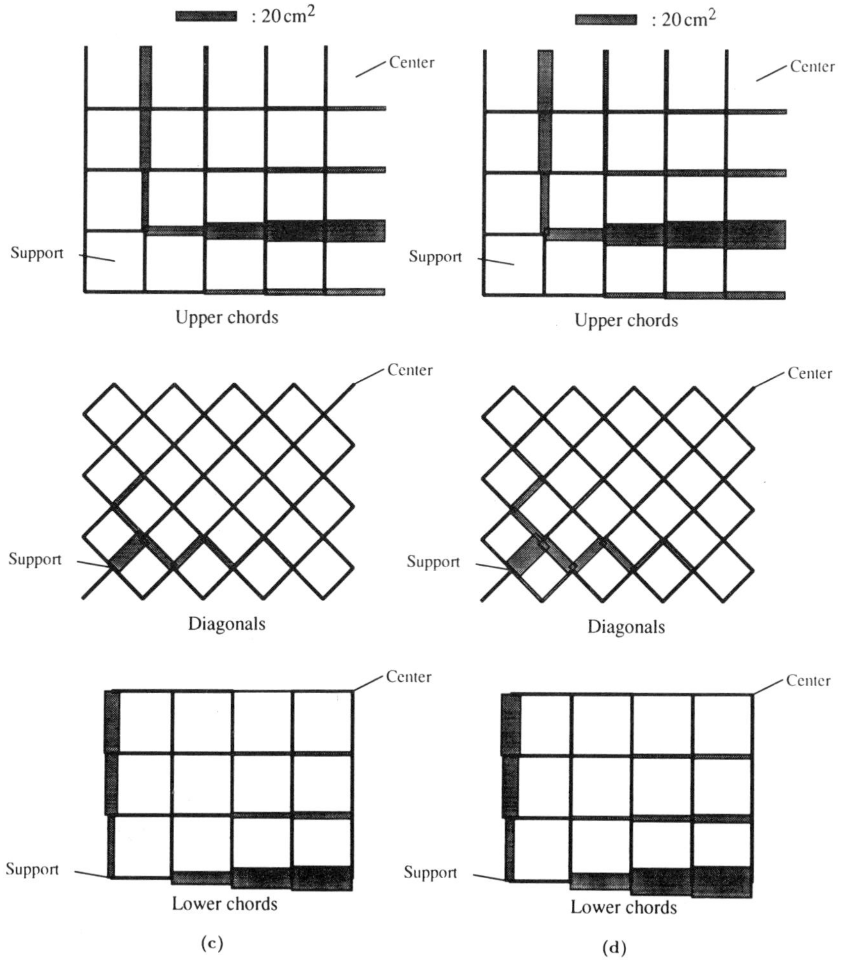

Fig.2 (Continued) Cross-sectional area of the earthquake-strain constrained 504-bar cylindrical truss, (c) $\xi = 2.4711 \times 10^{-4}$ and (d) $\xi = 2.0069 \times 10^{-4}$.

Consider a 504-bar doubly curved space truss whose topology is the same as that of the truss shown in Fig.1 and whose node configuration is defined completely by the open angles β_x and β_y in the x and y directions, respectively, of the lower surface. The radii in x and y directions of the upper surface are 17.5 m and 32.858 m, respectively. The extensional stiffness of the spring is 1.0290 N/m. The lowest four modes are taken into consideration for evaluation of the response strains. The earthquake motions in the direction of 45° from the $y-$axis in the $yz-$plane are considered. Twenty five ordered sets of ESCD's are generated for five different values of β_x and five different values of β_y with $V^* = 1.0\ \mathrm{m}^3$, $\xi_k = 7.0 \times 10^{-4}$ and $\Lambda_i^b \equiv 1$. The region B is defined by $20° \leq \beta_x \leq 140°$ $20° \leq \beta_y \leq 140°$. Several approximate contour lines of $V^{\#}(\beta_x, \beta_y, \xi_k)$ have been drawn in Fig.5 with reference to the five curves in β_x direction and five curves in β_y direction. An approximate optimal point $(\tilde{\beta}_x, \tilde{\beta}_y)$ and the variation of $V^{\#}(\beta_x, \beta_y, \xi_k)-$surface may readily be found visually.

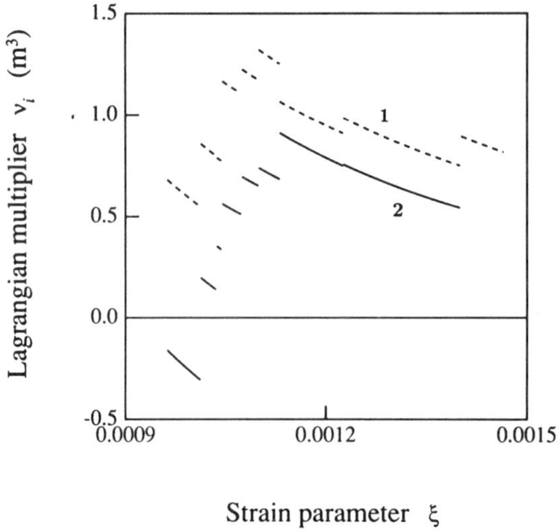

Fig.3 Variation of Lagrangian multiplier ν_i of the 504-bar cylindrical truss with respect to design strain parameter.

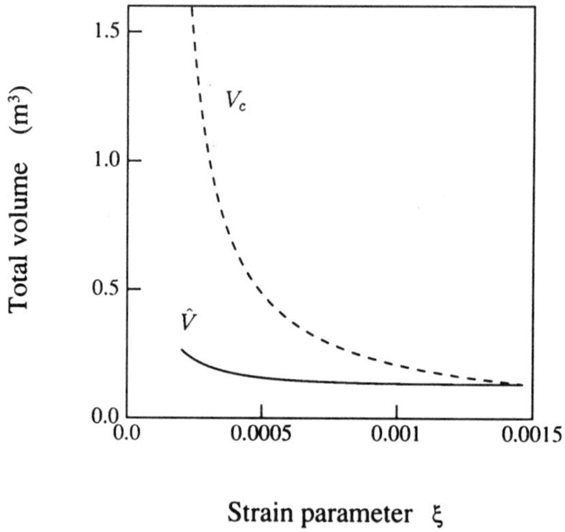

Fig.4 Variation of total volume of the 504-bar cylindrical truss with respect to design strain parameter.

CONCLUSIONS

It has been demonstrated through the examples that the authors' method of sequentially

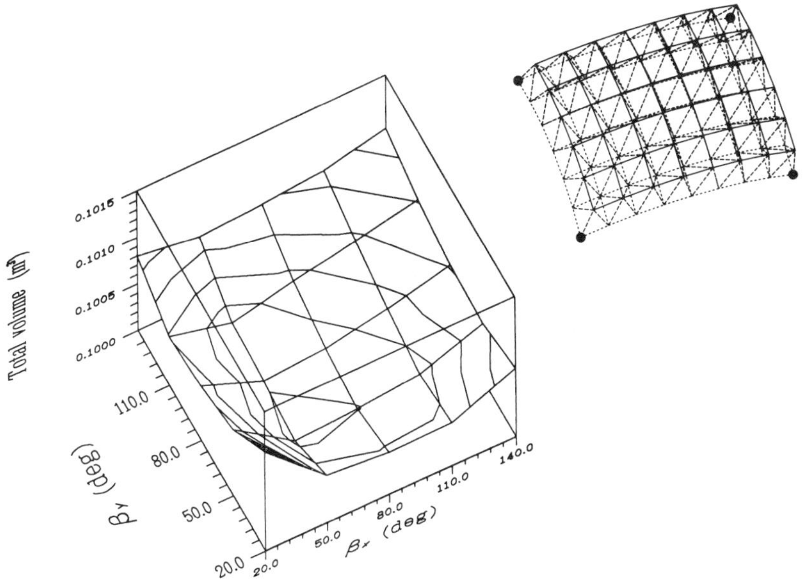

Fig.5 Hypersurface of total volume of earthquake-strain constrained 504-bar doubly curved space truss

generating earthquake-strain constrained designs (ESCD's) is very efficient for large curved space trusses. An ESCD is a fully stressed design such that the mean maximum response strains due to spectrum-compatible earthquake motions would either coincide with or be within the specified upper bound strains. A sequence of ESCD's for a design strain range has been examined within the framework of the optimum design under the same constraints and demonstrated to coincide at least partly with the sequence of the optimum designs under the same constraints on earthquake-strains. In view of this partial coincidence, it may be inferred that the remaining part of the sequence of ESCD's would be fairly close to the sequence of optimum designs and might practically be regarded as the latter.

The significance of a sequence of ESCD's has been demonstrated furthermore in the light of the optimum geometry design. It has been pointed out that the problem of optimum geometry design is not a simple mathematical programming problem in n geometrical parameters and would require laborious computation of a great number of sensitivity coefficients with respect to the original design variables if the optimal set of geometrical parameters were sought for by means of a mathematical programming technique. A diagram illustrating the overall variation of the objective function V with respect to the geometrical parameters among the ESCD's of a certain design strain level would be a far more useful information to a designer than the single information on the optimum geometry. An efficient method has been devised of drawing a design chart representing an approximate hypersurface (surface) of the objective function V in an $(n+1)$−dimensional space whose cartesian coordinates are the n geometrical parameters and V. An approximate optimal point in the n−space may readily be found from the design chart data and the geometrically optimal truss of ESCD may be found.

REFERENCES

1. Nakamura, T. and Ohsaki, M. Sequential Generator of Earthquake-response Constrained Trusses for Design strain Ranges, Int. J. Comput.& Struct., Vol.33(6), pp.1403-1416, 1989.

2. Wilson, E.L., Kiureghian, A.D. and Bayo, E.P. A Replacement for the SRSS Method in Seismic Analysis, Earthquake Engng Struct. Dyn., Vol.9, pp.187-194, 1981.

3. Kiureghian, A.D. On Response of Structures to Stationary Excitation, Report No. EERC 79-32, Earthquake Engineering Recearch Center, University of California, Berkeley, CA, 1979.

4. Haug, E.J. and Arora, J.S. Applied Optimal Design, John Wiley, New York, 1979.

5. Yamakawa, H. Optimum Structural Designs for Dynamic Response, New Directions in Optimum Structural Design, (Ed. Etres, E., Gallagher, R.H., Ragsdell, K.M. and Zienkiewicz, O.C.), pp.249-266, John Wiley, New York, 1984.

6. Hsieh, C.C. and Arora, J.S. Design Sensitivity Analysis and Optimization of Dynamic Response, Comput. Meth. Appl. Mech. Engng, Vol.43, pp.195-219, 1984.

7. Fox, R.L. and Kapoor, M.P. Structural optimization in the Dynamic response Regime; A Computational Approach, AIAA J., Vol.8, pp.1798-1804, 1970.

8. Arora, J.S. and Haug, E.J. Optimum Structural Design with Dynamic Constraints, J. Struct. Div., ASCE, Vol.103, pp.2071-2074, 1979.

9. Feng, T., Arora, J.S. and Haug, E.J. Optimal Structural Design under Dynamic Loads, Int. J. Numer. Meth. Engng, Vol.11, pp.39-52, 1977.

10. Bhatti, M.A. and Pister, K.S. A Dual Criteria Approach for Optimal Design of Earthquake-resistent Structural Systems, Earthquake Engng Struct. Dyn., Vol.9, pp.557-572, 1981.

11. Balling, R.J. Pister, K.S. and Ciampi, V. Optimal Seismic-resistent Design of a Planner Steel Frame, Earthquake Engng Struct. Dyn., Vol.11, pp.541-556, 1983.

12. Nakamura, T. and Yamane, T. Optimum Design and Earthquake-response Constrained Design of Elastic Shear Buildings, Earthquake Engng Struct. Dyn., Vol.14, pp.791-815, 1986.

13. Sadek, E.A. Dynamic Optimization of Framed Structures with Variable Layout, Int. J. Numer. Meth. Engng, Vol.23, pp.1273-1294, 1986.

14. Kirsch, U. optimal Topologies of Structures, Appl. Mech. Rev., Vol.42, pp.223-239, 1989.

15. Gunnlaugsson ,G.A. and Martin ,J.B. Optimality Condition for Fully Stressed Designs, SIAM. J. Appl. Math., Vol.25(3), pp.474-482, 1973.

16. Kicher, T.P. Optimum Design-Minimum Weight Versus Fully Stressed, J. Struct. Div., ASCE, Vol.92, pp.265-279, 1966.

17. Newmark, N.M. and Hall, W.J. Earthquake Spectra and Design, Earthquake Engineering Research Institute, Berkeley, CA, 1982.

18. Nakamura, T. and Ohsaki, M. Sequential Optimal Truss Generator for Frequency Ranges, Comput. Meth. Appl. Mech. Engng, Vol.67, pp.189-209, 1988.

Optimum Design of Thin Wall Structures Under Variable Amplitude Fatigue Loading

M.E.M. El-Sayed, E.H. Lund

Department of Mechanical and Aerospace Engineering, University of Missouri, Columbia, Missouri 65211, U.S.A.

ABSTRACT

This paper presents a method for considering fatigue life requirements in the optimal design of thin wall structures under variable amplitude load histories. The basic concept is to use the load histories combined with the stresses of the structure and the material fatigue properties to calculate the fatigue life during the optimisation process. The life requirement is considered as a side constraint and the structure weight as the objective function. The elements required for the stress and the fatigue life calculations are discussed. Some test cases to demonstrate the efficiency of the method are presented.

INTRODUCTION

With the increasing pressure of reducing the weight of many structures, such as aircraft and automotive structures the use of thin wall structures is increasing. In the design of these thin wall structures, under dynamic loading, the fatigue life requirement is one of the major design criteria for the safety of the structure.

Mechanical failure due to fatigue has caused over 50 percent of all mechanical failures. As a result, much work has been done in fatigue research in recent years. Criteria for fatigue design have evolved from infinite life to damage tolerance.

Infinite life design, simply put, means unlimited safety. This requires design stresses to be safely below the maximum fatigue limit. This is the oldest criterion and is only good for parts which undergo uniform cycles. This criterion uses the S-N curves and can not be applied to variable amplitude complex fatigue loading.

Safe-life design is the practice of designing for a finite life in a maximum load state. Safe-life designs include some statistical variations due to the scatter of fatigue results and other unknown factors. The fatigue calculations may be based on stress or strain life relations as well as crack growth. Safety factors may be calculated in terms of life, load, or both.

Fail-safe design is another criteria developed by the aerospace industry, due to the fact that the designers could not tolerate the extra weight due to large safety factors or the risk to life due to small ones. Fail-safe design basically operates under the premise that when cracks occur they will not result in failure of the structure until they can be identified and fixed.

Damage tolerant design is the most recent criteria developed of the ones discussed here and is a take-off of the fail-safe design. Damage tolerant design assumes cracks will exist and uses materials with slow crack growth and high fracture mechanics analysis to insure the cracks will not produce failure before they can be identified with routine inspections.

All of the above criteria suggest that it is necessary to keep the balance between the structural weight and durability. To achieve this balance a structural optimisation approach with fatigue life constraints was developed [1, 2]. In this paper we discuss this approach and its application to the structural optimisation of thin wall structures under variable amplitude fatigue loading.

OPTIMUM DESIGN OF FRAMES FOR A REQUIRED FATIGUE LIFE

The structural optimisation tasks with stresses, displacement, frequency and fatigue life constraints are discussed in [1-4]. Here, we will discuss the structural optimisation task, for thin wall structures, including fatigue life constraints. The basic problem is to minimise one function of the sizing variables of the structure, subject to limits on other functions. That is, find the set of design variables, X, which will

$$\text{Minimise} \quad W = F(X) \qquad \text{Objective Function} \qquad (1)$$

Subject to;

$$g_j(\underline{X}) \leq 0 \qquad j = 1, 2, .., m \qquad \text{Inequality Constraints} \qquad (2)$$

$$X_i^l \leq X_i \leq X_i^u \qquad i = 1, 2, .., n \qquad \text{Side constraints} \qquad (3)$$

Where n is the number of variables contained in $\underline{X}$ and m is the number of inequality constraints. For thin wall structures the physical design variables represent the cross-sectional dimensions or the thickness. The objective function to be minimised can be taken as the total weight of the structure. The side constraints of equation (3) are simple limits imposed on the design variables to provide practical limits on member sizes.

The inequality constraints of equation (2) are the constraints derived from performance requirements. These inequality constraints include limits on cross-sectional stresses, displacements, and vibration frequencies.

To include fatigue life constraints the entire load history of the required life span should be used in the optimum design of the structures. This can be accomplished by imposing a lower bound on the fatigue life, at the highest local stress area of the section, during the optimisation process. This constraint takes the form

$$1 - L_{ij} / \bar{L} \leq 0 \quad (4)$$

i = section number

j = loading condition

$\bar{L}$ = lower bound on fatigue life

The evaluation of the fatigue life constraints require the calculation of the fatigue life L_{ij} at every step of the optimisation. This requires a stress recovery routine and a fatigue life calculation routine to be connected to the optimisation routines. From the local stresses, the load histories and the material fatigue properties the fatigue life routine calculates L_{ij} in every optimisation step. In the following sections we discuss the basic requirements for local stresses and the fatigue life calculations.

LOCAL STRESSES

The most common fatigue failure in structures is at the thin wall locations. A detail discussion for different types of stress in thin wall structures is given in reference [5]. For fatigue life calculation the maximum distortion energy theory can be used to calculate the equivalent stress at any point in the section. For a combined normal and shear stress the equivalent stress takes the form,

$$S^2 = S_n^2 + 3\tau^2, \quad (5)$$

where S_n is the normal stress and τ is the shear stress. All critical points in the section should be considered. The maximum equivalent local stress value at the section should be used for the fatigue life calculation.

FATIGUE LIFE CALCULATION

Detailed discussions for the fatigue life calculation are given in [6, 7]. Here, we summarize the basic requirements for the fatigue life calculations.

Cyclic stress-strain relationship

An equation of the following form is generally used to express the cyclic stress-strain relationship

$$\frac{\Delta\varepsilon}{2} = \frac{\Delta\varepsilon_e}{2} + \frac{\Delta\varepsilon_p}{2} = \frac{\Delta\sigma}{2E} + \left(\frac{\Delta\sigma}{2K'}\right)^{1/n'} \quad (6)$$

The Δ, in the above equation, indicates completely reversed ranges of stresses and strains and subscripts e and p stand for elastic and plastic strain. The two material properties, n' and K' are the cyclic strain hardening exponent and cyclic strength coefficient, respectively.

The strain-life relationship

The cyclic plastic strain and fatigue life were shown by Coffin-Manson to be related by a simple power function, over the entire life range from only a half cycle to millions of cycles. The form of the Coffin-Manson relation is as follows

$$\frac{\Delta\varepsilon_p}{2} = \varepsilon_f' (2N_f)^c \qquad (7)$$

Where $2N_f$, indicate reversals or half cycles, while N_f means number of cycles to failure. Similarly, the stress-life relation can be represented as a power function of stress in a form similar to equation (7). This formulation was first suggested by Basquin and is of the following form

$$\frac{\Delta\sigma}{2} = \sigma_f' (2N_f)^b \qquad (8)$$

For the calculation of fatigue life, it is convenient to incorporate mean stress effects as an equivalent change in static strength. Equation (8) may then be modified as follows

$$\frac{\Delta\sigma}{2} = (\sigma'_f - \sigma_o)((2N_f)^b \qquad (9)$$

To obtain an expression relating total strain, mean stress and life, equation (9) is divided by the Young's modulus, E, to obtain the elastic strain and added to equation (7) for the plastic strain to yield

$$\frac{\Delta\varepsilon}{2} = \frac{\Delta\varepsilon_e}{2} + \frac{\Delta\varepsilon_p}{2} = \left(\frac{\sigma'_f - \sigma_o}{E}\right)(2N_f)^b + \varepsilon_f'(2N_f)^c \qquad (10)$$

The two exponents and the coefficients are regarded as fatigue properties of the metal, and they are designated as follows

b = Fatigue strength exponent
c = Fatigue ductility exponent
ε_f' = Fatigue ductility coefficient
σ'_f = Fatigue strength coefficient

By manipulating equations (7) and (8) to eliminate life and carrying the result to the form of equation (6), it can be shown that the cyclic strain hardening exponent, n', is determined by the fatigue strength and ductility exponents as follows

$$n' = \frac{b}{c} \qquad (11)$$

Similarly, it can also be shown that the cyclic strength coefficient can be determined from the fatigue properties as follows

$$K' = \frac{\sigma_f'}{(\varepsilon_f')^{n'}} \tag{12}$$

Cycle Counting

Some of the cycle counting methods in use today for fatigue analysis are peak, level crossing, range, range-mean, range-pair, and rainflow. Of these various methods, rainflow or its equivalent range-pair has been shown to yield superior fatigue life estimates [8]. The basic idea behind rainflow counting is to treat small events in the load history as interruptions over larger overall events and, in the simplest terms, to match the highest peak and deepest valley, then the next largest and smallest together, etc., until the peaks and valleys of the load history have been paired.

Simulation of the Stress-Strain Response

The purpose of simulating the stress-strain response is to determine the parameters necessary for cumulative damage fatigue analysis. Information such as stress amplitude, mean stress, elastic and plastic strain can be determined for each reversal in the load history. The most important feature of any simulation model is its ability to correctly describe the history dependence of cyclic deformation.

A model, was developed based on an "availability concept," in which the cyclic stress-strain curve equation (6) is approximated by a series of straight line segments or elements [9]. The number and size of the elements is arbitrary, depending on the manner in which elements are used.

Once the cycles have been defined and the stress-strain response determined, the appropriate fatigue parameters can be determined, so that a damage analysis can be performed.

Notch Analysis

In dealing with thin wall structures, it is necessary to relate the nominal loads or strains to the maximum stresses and strains at the critical location. Neuber derived a rule which applies when the material at the notch root deforms nonlinearly. The theoretical stress concentration, K_τ, is equal to the geometric mean of the actual stress and strain concentration factors, K_σ and K_ε

$$K_\tau = (K_\sigma K_\varepsilon)^{\frac{1}{2}} \tag{13}$$

Topper [10], modified Neuber's rule for use in cyclic loading applications by substituting the fatigue notch factor, K_f, for the stress concentration factor and rewriting equation (13) in the following form

$$K_\sigma = \frac{\Delta\sigma}{\Delta S}$$

$$K_\varepsilon = \frac{\Delta\varepsilon}{\Delta e} \qquad (14)$$

$$K_f = \left(\frac{\Delta\sigma\ \Delta\varepsilon}{\Delta S\ \Delta e}\right)^{\frac{1}{2}}$$

$\Delta\sigma$ = Stress range at notch root

ΔS = Nominal stress range

$\Delta\varepsilon$ = Strain range at notch root

Δe = Nominal strain range

This relationship is conveniently used in the following form

$$K_f^2\ \Delta S\ \Delta e = \Delta\sigma\ \Delta\varepsilon \qquad (15)$$

All terms on the left side are determinable for each reversal from the load history and cyclic stress-strain response of the material, and those terms on the right side represent the local stress-strain behavior of the material at the notch root. The terms on the left side are a determinable constant for each reversal and the result is an equation of the form, $xy = c$, which is a rectangular hyperbola. When the nominal strains are elastic, equation (15) may be used in the following form

$$\Delta\sigma\ \Delta\varepsilon = \frac{(K_f\ \Delta S)^2}{E} \qquad (16)$$

Combining this form with equation (6) , an expression for the notch stress becomes:

$$\frac{\Delta\sigma^2}{E} + \Delta\sigma \left(\frac{\Delta\sigma}{K'}\right)^{1/n'} = \frac{(K_f\ \Delta S)^2}{E} \qquad (17)$$

This equation is easily solved using the Newton-Raphson iteration technique. Once the notch stress is obtained, it can be used in equation (6) to solve for the elastic and plastic strains at the notch root. After each reversal, a new axis is defined and the right-hand side of equation (17) recalculated in order to solve the new stress and strain range.

Cumulative Damage Analysis

Cumulative damage fatigue analysis is usually based on the Palmgren-Miner linear damage rule. Fatigue damage is computed by linearly summing cycle ratios for the applied loading history, as indicated in the following equation

$$\text{Damage} = \sum \frac{n_i}{N_{fi}} \qquad (18)$$

n_i = Observed cycles at amplitude, i
N_{fi} = Fatigue life at constant amplitude, i

After the fatigue damage for a representative segment or block of load history has been determined, the fatigue life in blocks, L_{ij} of equation (4) , is calculated by taking the reciprocal.

The fatigue life for any cycle or reversal can be determined from equation (10). The mean stress, σ_o and cyclic strain range, $\Delta\varepsilon$, has been determined from the material response model. Equation (10) cannot be explicitly solved for life because of the negative fractional exponents involved, but can easily be solved using Newton-Raphson or interval halving iteration techniques.

TEST CASES

To demonstrate the application of the optimisation process and the effect of the fatigue life requirement, some test cases were performed using the keyhole specimen experimental test data of reference [6]. The minimum weight design of the SAE keyhole specimen shown in Figure (1), with fatigue life constraints was performed. The stress calculation was obtained using elementary beam theory. The optimisation was performed for different maximum loads to determine the thickness required for life equal to the actual tested fatigue life. The tests were made using data for RQC-100 steel with the Bracket, Suspension and Transmission load histories shown in Figure (2). The optimisation problem solved for each test case was:

Minimise $W = \rho * A * X$ (19)

Subject to $1 - L_i / \bar{L} \leq 0$ (20)

where
ρ = density
A = area
X = thickness
L_i = Life calculated with current thickness
$\bar{L}$ = Required fatigue life

An initial guess for the design variable X (thickness) was supplied to the optimisation routine as well as the bounds for the inequality constraints. For some load levels the upper and lower experimental fatigue life bounds were used to obtain upper and lower bounds for the optimum thickness.

The nominal stress is calculated by assuming that the key hole specimen acts as a simple beam subjected to bending and axial stresses.

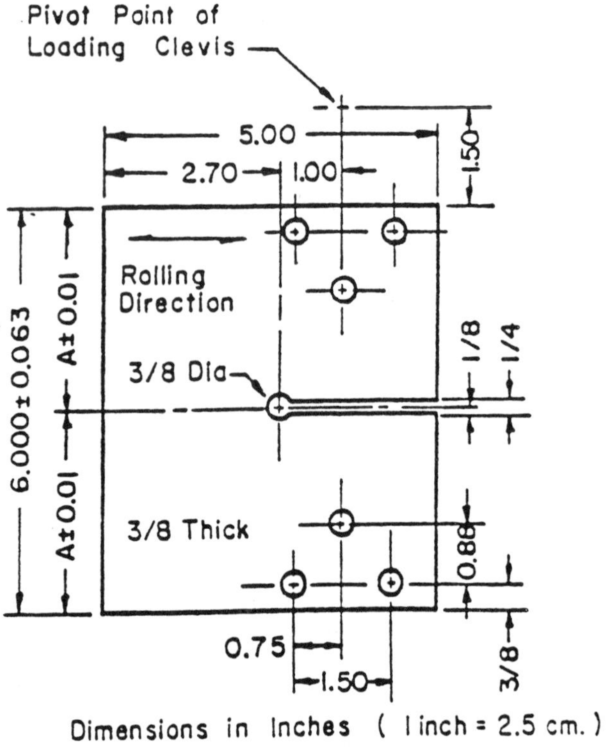

Figure (1) SAE Keyhole Specimen

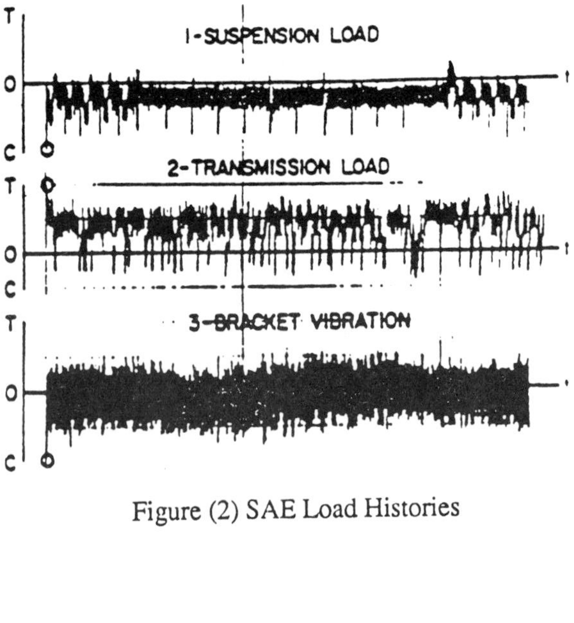

Figure (2) SAE Load Histories

From elementary beam theory:

$$S = \frac{Mc}{I} + \frac{P}{A} \tag{21}$$

where

S = Nominal Stress
I = Moment of Inertia
P = Applied Load
A = Area

For the keyhole specimen the theoretical stress concentration factor of equation (17) can be taken as:

$$K_f = 3.0 \tag{22}$$

The results of the numerical optimisation test cases are shown in tables (1), (2) and (3). In these tables the range of the obtained optimum thicknesses for each load level of the different histories are presented. In comparison with the actual Thickness of 0.375 in. the predicted optimum thicknesses are close and more conservative.

TABLE (1)

Predicted optimum Thicknesses for The SAE Keyhole specimen with RQC-100 steel and Bracket History (actual thickness = 0.375)

Max Load (lbs)	Blocks	Predicted Thickness
-16,000	3.3 - 5.1	0.5261 - 0.5733
-8000	47.0	0.4109

TABLE (2)

Predicted optimum Thicknesses for The SAE Keyhole specimen with RQC-100 steel and Suspension History (actual thickness = 0.375)

Max Load (lbs)	Blocks	Predicted Thickness
-16,000	19.9 - 64.0	0.3941 - 0.5342

TABLE (3)

Predicted optimum Thicknesses for The SAE Keyhole specimen with RQC-100 steel and Transmission History (actual thickness = 0.375)

Max Load (lbs)	Blocks	Predicted Thickness
16,000	22.2 -29.9	0.5421 - 0.5813
8000	269.0 - 460.0	0.4421 - 0.4816
3500	> 88,020.0	0.4132

CONCLUSION

A method for considering fatigue life requirements in the optimal design of thin wall structures under variable amplitude fatigue loading is presented. The method utilizes the local stresses, the entire load history of the required life span and the material properties directly in the optimisation process.

The numerical optimisation test cases performed demonstrated that reasonable designs can be obtained using this approach. It is clear from formulation that when the durability of the structure is of major concern more conservative designs could be achieved by introducing some factors of safety in the process. For light weight designs accurate methods such as finite element analysis should be used for the local stress recovery.

Due to the number of fatigue life evaluations and the time required for each evaluation, this approach needs a considerable amount of CPU time especially for large structures with complicated load histories. For such cases some approximation techniques should be developed to avoid the fatigue life calculation at every step of the optimisation process.

REFERENCES

1. EL-Sayed, M. and Lund, E. Optimum Design for a Required Fatigue Life Based On Nominal Stress, SAE Paper # 900832, 1990.

2. EL-Sayed, M. and Lund, E. Structural Optimization with Fatigue Life Constraints, to be published by International Journal of Engineering Fracture Mechanics, 1991.

3. Miuro, H., Lust, R. V. and Bennett, J. A. Integrated Panel and Skeleton Automotive Structural Optimisation, Proceedings of SAE 4th International conference on Vehicle Structural Mechanics, Detroit, MI, 1981.

4. Bennett, J. A. and Botkin, M. E. Automated Design for Automotive Structures,Journal of Mechanical Design,VOl.104, pp. 799-805, 1982.

5. Oden, J. T. and Ripperger, E. A. Mechanics of Elastic Structures, McGraw-Hill, 1981.

6. Socie, D. F. and Morrow, J. D. Review of Contemporary Approaches to Fatigue Damage Analysis, FCP Report No. 24, University of Illinois, 1976.

7. Fuchs, H. O. and Stephens, R. I. Metal Fatigue in Engineering, John Wiley and Sons, 1980.

8. Dowling, N. E. Fatigue Failure Predictions for Complicated Stress-Strain Histories, Journal of Materials, Vol.7 pp. 71-78, 1972.

9. Wetzel, R. M. A Method of Fatigue Damage Analysis, Ph. D. Thesis, University of Waterloo, Canada, 1971.

10. Topper, R. M., Wetzel, R. M. and Morrow, J. D. Nuber's Rule Applied to Fatigue of Notched Specimens, Journal of Materials, VOl.4, pp. 200-209, 1969.

Reliability-based Optimum Design with Random Seismic Input

F.Y. Cheng (*), C.-C. Chang (**)

(*) Department of Civil Engineering, University of Missouri-Rolla, Rolla, Missouri 65401, U.S.A.

(**) Delon Hampton & Associates, Rockville, Maryland 20852, U.S.A.

ABSTRACT

In this paper an algorithm for finding optimum parameters of a deterministic structure subjected to a stationary process with a seismic input spectrum for various failure probability expressions. The optimization algorithm involves considerations of the objective function of the construction cost and the expected failure cost, various constraints such as displacements of a system, yielding and buckling of constituent members, and the magnitude orders for different failure probability expressions and different system failure bounds. Numerical examples are provided and the magnitude orders of optimum solutions for different probability expressions are discussed.

INTRODUCTION

The problem of studying earthquakes has been very difficult because of its complicated characteristics. Recently, an optimum seismic structural design was proposed based on UBC seismic loading [1,2]. However, the problem of considering random loading expression has not received attention in the literature. In this paper, an approach to optimum structural design for random loading expression is presented with inclusion of reliability criterion.

A seismic stationary processes is a process whose statistics properties do not change with time and can be generally described by three classes of noise processes which are white noise, filter white noise and modified white noise processes. A stationary process with a constant power spectral density for all frequencies is known as a white noise process. The idea of using a white noise process to simulate a strong-motion earthquake was suggested by Bycroft [3] and was proven to be useful in examining the effects of earthquake. However, according to the existing strong motion accelerograms, their frequency spectra are not constant and may have peaks at one or several frequencies. Thus, a stationary filter white noise process provided by the transfer function characteristics could be more suitable to actual ground accelerograms. Previously, Kanai [4] and Tajimi [5] proposed a second order linear damped oscillator which includes the filter fundamental frequency ω_g and damping ratio ζ_g to simulate earthquake which has the form

$$G(\omega) = \frac{\omega_g^4 + 4\zeta_g^2\omega_g^2\omega^2}{(\omega_g^2-\omega^2)^2 + 4\zeta_g^2\omega_g^2\omega^2} G_0 \tag{1}$$

where G_0 is the constant value, ω = natural frequency. In general, these parameters are affected by the ground layer rigidity, the epicentral distance, and the earthquake magnitude.

The modified white noise process is a white noise process with a constant spectrum which is obtained by substituting the structural natural frequency into the Kanai-Tajimi seismic filter.

RELIABILITY AND PROBABILITY OF FAILURE

Although reliability is the possibility of structural survival during its lifetime, an alternative criterion is the probability of failure which is the opposite meaning of reliability and may also be adopted for the same purpose. In a classical formulation, the reliability and the probability of failure are defined to be

$$P_r(R \geq S) = \int_{-\infty}^{\infty} P_d(S) \left[\int_S^{\infty} P_d(R)\, dR \right] ds \tag{2}$$

$$P_f(R < S) = 1 - P_r(R \geq S) \tag{3}$$

where, R,S = structural resistance and response; $P_d(S)$, $P_d(R)$ = probability density function of structural response and resistance. Hereafter the reliability and probability of failure will also be represented as P_r and P_f.

Since in practice, complete knowledge of distributions of resistance and response are not possible to be obtained, two approximate expressions to the reliability or the failure probability have been used in many studies. One is the first-order second-moment expression that estimates the reliability and the probability of failure through the safety factor formulation. The other is the first passage expression that estimates the probability of crossing the specified barrier during the vibration time interval.

First-order second-moment expression

The first-order second-moment expression involves the evaluation of the mean and the variance of the structural resistance and response which represent the reliability and the probability of failure, respectively. Owing to the relationship of the structural resistance and response, normal and lognormal distributions are used.

normal distribution:

$$\textit{reliability,} \quad P_r = PN(\beta) \tag{4}$$

$$\textit{probability of failure,} \quad P_f = 1 - PN(\beta) \tag{5}$$

$$\textit{safety factor,} \quad \beta = (\bar{R} - \bar{S}) / (\sigma_R^2 + \sigma_S^2)^{1/2} \tag{6}$$

where PN() is the standard cumulative normal distribution and $\bar{R}$, $\bar{S}$, σ_R^2, σ_S^2 = the means and the variance of the structural resistance R and response S, respectively.

lognormal distribution:

$$\textit{reliability,} \quad P_r = PN(\beta) \tag{7}$$

$$\textit{probability failure,} \quad P_f = 1 - PN(\beta) \tag{8}$$

$$\textit{safety factor,} \quad \beta = \frac{\bar{Q}}{\sigma_Q} \tag{9}$$

where

$$\bar{Q} = \ell n\bar{R} - \ell n\bar{S} = \ell n\left[\left(\frac{\bar{R}}{\bar{S}}\right)\left(\frac{1+V_S^2}{1+V_R^2}\right)^{1/2}\right] \tag{10}$$

$$\sigma_Q^2 = \sigma_{\ell nR}^2 + \sigma_{\ell nS}^2 = \ell n\left[(1+V_R^2)(1+V_S^2)\right] \tag{11}$$

First passage expression

For a random vibration problem, the reliability is the probability of a stochastic response which falls in a prescribed bound during the system operating time. The probability of failure, which is called the first passage expression is defined as follows:

The reliability (P_r) and probability of failure (P_f) on time interval $0 < t < T_0$

$$P_f = P_f(T_0) = \int_0^{T_0} P_d(t)\,dt = 1 - \exp(-\nu_a T_0) \tag{12}$$

$$P_{r=Pr}(T_0) = 1 - P_f(T_0) = \exp(-\nu_a T_0) \tag{13}$$

where $\nu_a = \frac{\sigma_{\dot{s}}}{\pi\sigma_s}\exp(\frac{-\tilde{a}^2}{2\sigma_s^2})$, $\dot{s}$ = the derivative of s, σ_s and $\sigma_{\dot{s}}$ are the standard deviations of s and $\dot{s}$.

STRUCTURAL RESPONSE

Moments of structural responses

Because the ground acceleration is a zero-mean Gaussion process, the response process of a linear deterministic structural system to this excitation is also a zero-mean Gaussian process. Therefore, the statistical quantities involved in structural design are only second-order moments of structural responses, i.e., the variances of structural responses.

For lightly damped structural systems whose modal frequencies are well separated, the moments of structural displacements are obtained as [6]

$$\lambda_{u0} = \int_{-\infty}^{\infty} G_u(\omega)\,d\omega = \sum_n \{\Phi^2\}_n \Gamma_n^2 \lambda_{y0} \tag{14}$$

$$\lambda_{u1} = \int_{-\infty}^{\infty} \omega G_u(\omega)\,d\omega = \sum_n \{\Phi^2\}_n \Gamma_n^2 \lambda_{y1} \tag{15}$$

$$\lambda_{u2} = \int_{-\infty}^{\infty} \omega^2 G_u(\omega)\,d\omega = \sum_n \{\Phi^2\}_n \Gamma_n^2 \lambda_{y2} \tag{16}$$

where λ_{u0}, λ_{u1}, λ_{u2} = the zeroth, 1st, and 2nd moments of spectral displacements which can be seen and determined in Ref. 7; $\{\Phi\}_n$, Γ_n = the mode shape matrix and participation factor at nth mode; ω = structural frequency; G_u = the spectral density function of mean square displacement.

The moments of internal forces can be derived from the relationship

$$\{F\}_m = [S]_m [A]_m \{u\}_m \tag{17}$$

as

$$\lambda_{\{F0\}_m} = [S^2]_m [A^2]_m \lambda_{\{u0\}_m} \tag{18}$$

$$\lambda_{\{F1\}_m} = [S^2]_m [A^2]_m \lambda_{\{u1\}_m} \tag{19}$$

$$\lambda_{\{F2\}_m} = [S^2]_m [A^2]_m \lambda_{\{u2\}_m} \tag{20}$$

where [S] and [A] are stiffness and equilibrium matrix, respectively [2].

The statistics of peak responses

If the uncertainties of peak responses for random loads can be determined, the first-order second moment expression for the probability of failure is still applicable. Two expressions approximating to these uncertainties are used.

(1) Davenport [8] derived the mean and standard deviation of the maximum absolute value of a stationary Gaussian process, S(t), with zero mean over duration as

$$mean,\quad \overline{S}_{\max} = \left(\sqrt{(2\ell n\nu T_0)} + \frac{0.5772}{\sqrt{2\ell n\nu T}}\right)\sigma_S \tag{21}$$

$$standard\ deviation,\quad \sigma_{S_{\max}} = \left[\left(\frac{\pi}{\sqrt{6}}\right)\left(\frac{1}{\sqrt{2\ \ell n\ \nu\ T}}\right)\right]\sigma_S \tag{22}$$

where $\nu = (\lambda_{S2}/\lambda_{S0})^{1/2}$, $\lambda_{sk} = \int_{-\infty}^{\infty} \omega^k G_S(\omega)\,d\omega$, T = duration.

(2) An expression similar to Davenport's expression, Kirueghian [7] derived the following empirical expressions for the uncertainties of peak responses which are

$$mean,\quad \overline{S}_{\max} = \left(\sqrt{2\ell n\nu_e T} + \frac{0.5772}{\sqrt{2\ell n\nu_e T}}\right)\sigma_S \tag{23}$$

standard deviation,

$$\sigma_{S_{\max}} = \left[\frac{1.2}{\sqrt{2\ell n\nu_e T}} - \frac{5.4}{13 + (2\ell n\nu_e T)^{3.2}}\right]\sigma_S,\quad \nu_e T > 2.1 \tag{24}$$

$$= 0.65\sigma_S,\quad \nu_e T \leq 2.1 \tag{25}$$

where

$$\nu_e = (1.63q^{0.45} - 0.38)\nu, \quad q \le 0.69 \tag{26}$$

$$= \nu, \quad q \ge 0.69 \tag{27}$$

$$q = \sqrt{1 - \lambda_{S1}^2 / \lambda_{S0}\lambda_{S2}} \tag{28}$$

b = 0.2 is an empirically determined constant

OPTIMIZATION FORMULATIONS

In engineering optimum design, a best solution (optimum in some sense) provides an optimum value based on the designer's needs. An optimization problem for structural design is generally formulated as

minimize	objective function
subject to	constraints

Objective function
The objective function of structural design problem may be structural weight or structural cost which may be expressed as [2]

Weight (W) is consisted of structural member weights as

$$W = \sum_i r_{di} l_i A_i \tag{29}$$

where r_{di}, l_i, A_i = the mass density, length, and area of a member.

Total structural cost Total structural cost (C_T) consists of two parts: initial construction cost (C_I) and expected future failure loss ($L_f P_{fT}$); i.e.

$$C_T = C_I + L_f P_{fT} \tag{30}$$

in which L_f = expected failure cost; P_{fT} = system failure probability.

Initial construction cost C_I comprises of structural members cost and nonstructural members cost. Expected failure cost (L_f) is the total loss incurred in a structural failure state and includes additional replacement cost, damage to property, liability due to death and injury, business interruption.

Although initial construction cost and expected failure cost can be classified into many categories, these quantities are difficult to be estimated. Therefore two coefficients in the cost function, the ratio of initial cost to member cost (C_{in}) and the ratio of future failure cost to initial cost (C_{VL}), are used to represent the various magnitudes of initial construction cost and future failure cost. Based on the two above coefficients the initial cost and failure cost function are expressed as

$$C_I = C_{in} C_u \sum_i l_i A_i \tag{31}$$

$$L_f = C_{VL} C_I \tag{32}$$

where C_u = the unit volume cost of steel member,

By invoking these two coefficients, the influences of nonstructural cost and expected failure cost in cost optimum design problem can be observed.

Constraints

The constraints in reliability design consideration could be the failures of displacements, beams, and columns as shown in Eqs. 33, 34, and 35, respectively [2]. The failures can be expressed in terms of the safety factor or the probability of failure.

$$P_{f_u} - P_{f_{u0}} \leq 0 \tag{33}$$

$$P_{f_b} - P_{f_{b0}} \leq 0 \tag{34}$$

$$P_{f_{f1}} - P_{f_{f10}} \leq 0, \quad P_{f_{f2}} - P_{f_{f20}} \leq 0, \tag{35a}$$

$$P_{f_{f3}} - P_{f_{f30}} \leq 0, \quad P_{f_{f4}} - P_{f_{f40}} \leq 0. \tag{35b}$$

where $P_{f_{u0}}$, $P_{f_{b0}}$, $P_{f_{f10}}$, $P_{f_{f20}}$, $P_{f_{f30}}$, $P_{f_{f40}}$, = the given allowable bound values; where P_{f_u}, P_{f_b}, $P_{f_{f1}}$, $P_{f_{f2}}$, $P_{f_{f3}}$, $P_{f_{f4}}$, can be determined from Eqs. 3, 4, and 12.

The structural responses in Eqs. 3, 4, 12 are displacements for displacement failure; the applied moment for beam flexural failure; the interaction equations of yielding ($f1 = P/P_y + M/M_y$), instability of bending ($f2 = P/P_{cr} + C_m M/(M_y(1-P/P_E))$), lateral torsional buckling ($f3 = P/P_{cr} + C_m M/(M_{cr}(1-P/P_E))$), buckling about weak axis ($f4 = P/P_{cr_y}$), P,M = applied axial load or bending moment, P_y = yielding axial load, P_{cr} = critical axial load, P_E = Euler buckling load, M_y = yielding moment, M_{cr} = critical moment. The determination of the statistics of these quantities can be obtained in Ref. 2. However P_y, P_E, P_{cr}, M_y, M_{cr}, are considered as deterministic quantities.

The structural resistances are allowable displacement for displacement failure; the yielding moment for beam failure; the constant unit values of interaction equations for column failure.

NUMERICAL EXAMPLES AND OBSERVATIONS

To illustrate the optimum design procedure, an example of a ten-story steel shear-building shown in Figure 1 is implemented. The seismic stationary random load is used to investigate the comparisons of various random seismic inputs, various failure probability expressions, two cost ratios, and the differences of two system failure bounds and moment of inertia for weight and cost optimization. The parameters used in examples are: allowable displacement = 0.005 times corresponding the story height, yielding strength = 36 ksi, elastic modulus = 30000 ksi. Housner and Jennings [9] proposed values of ω_g = 15.6 rad/s, ζ_g = 0.64, and G_0 = 1.0 in^2/s^2 for firm ground conditions. The failure modes are displacement failures for each external degree of freedom and two column failure modes for each column member.

Comparison of various seismic input spectra in weight optimization

The stochastic seismic input spectra are chosen to be the white noise (WN) spectra, Kani-Tajimi filter white noise (FWN) spectra, or modified white noise (MWN) spectra. The results will demonstrate the differences of optimum solutions for the three spectra with the first passage equation.

Figure 2 shows that modified white noise yields the largest results over the other two. Nevertheless, the differences between the results due to the modified white noise and that due to the filter white noise are small. The results obtained for white noise is the smallest among all these three chosen spectra.

Comparison of various types of failure probability expressions in weight optimization

The magnitude orders of optimum weight for different types of probabilities of failure with modified white noise spectrum are shown in Fig. 3 from which the results may be summarized as follows:

(1) At P_{f0} = 10^{-1}, safety factor expression with the lognormal distribution and Davenport's equation (SLND) > safety factor expression with normal distribution and Davenport's equation (SND) > first passage equation (FP) > safety factor expression with lognormal distribution and Kiureghian's equation (SLNK) > safety factor expression with the normal distribution and Kiureghian's expression (SNK).

(2) At P_{f0} = 10^{-3}, SLND > SLNK > FP > SND > SNK

(3) At P_{f0} = 10^{-5}, 10^{-7}, SLND > FP > SLNK > SND > SNK

The ratio of initial cost to structural members cost, C_{in}, in cost optimization

The initial cost has two components which are structural members cost and nonstructural members cost. Since the terms involved in the nonstructural members cost are not clear, it is reasonable to assume that the structural members cost is a percentage of initial cost. It means that the initial cost is assumed to be the product of a ratio of initial cost to members cost (C_{in}) and members cost. Three ratio values of C_{in} (2., 5., and 20.) and one value of C_{VL} (1.0) are used to investigate the influence of nonstructural cost.

Figure 4 shows a difference between total cost and initial construction cost for three C_{in} values when allowable failure probability $< 10^{-3}$. When the opposite condition occurs, the total cost and initial construction cost are nearly the same, i.e., the influence of expected failure cost becomes small at high reliability levels. Although the optimum cost changes with various C_{in} values, the moments of inertia for column members do not change for all allowable failure probabilities. This indicates that the design sections are not affected by changes of nonstructural members cost.

The ratio of expected failure cost to initial cost, C_{VL}, in cost optimization

The expected failure cost consists of two components which are structural loss (repair costs) and nonstructural loss (business or human loss). However, this cost is also not clearly defined. So the expected failure costs considered here are all assumed to have relationship with initial cost. Three ratio values of C_{VL}, (0.5, 1.0, 10.0) and one value of C_{in} (5.0) are chosen to indicate the optimum design influence for magnitude of expected failure cost in Fig 5.

When allowable failure probability $> 10^{-3}$, there are differences between cost and design sections among three C_{VL} values.

When allowable failure probability $> 10^{-5}$, the cost and design sections are very close among three C_{VL} values. These conclude that at high safety criteria the influence of expected failure cost is very small.

comparison of system failure probability formulation in cost optimization

Although the exact system probability of failure is too complicated to obtain, the upper and lower bounds on this probability can be obtained respectively [10] by finding the sum of all individual failure modes and the maximum value among all individual failure modes used to investigate the range of exact system probability of failure may lie in.

Figure 6 shows the difference between maximum and minimum bounds at low reliability. The difference reduces when the allowable reliability increases.

CONCLUSIONS

The design for the modified white noise process requires the heaviest design among three seismic input spectra. Nevertheless, the difference between the modified white noise and filter white noise processes has been shown very small. The magnitude order for five failure probability expressions varies at low reliability criteria. For high reliability criteria, the order is SLND > FP > SLNK > SND > SNK.

For low reliability criteria, there are some differences between total cost and initial cost, but for high reliability level, the discrepancy between total cost and initial cost, however, is very small. The optimum moments of inertia of the constituent members are not affected by different C_{in} values at any reliability levels; the expected failure cost, however, affects the total cost and the member stiffness at low reliability design. There is not any noticeable difference in optimum costs and in member sizes for various C_{VL} values at high reliability criteria. At low reliability there is the difference between system maximum and minimum bounds. The difference reduces as reliability increases. At low reliability, there is a difference between cost and weight objective function. The difference reduces as reliability increases.

ACKNOWLEDGEMENTS

The paper is a partial result of the research project supported by the National Science Foundation under the Grant No. NSF CES 8403875.

REFERENCES

1. Cheng, F.Y., and Chang, C.C., Optimum Studies of Coefficient Variation of UBC Seismic Forces, Proceedings of 5th International Conference on Structural Safety and Reliability, International Association for Structural Safety and Reliability (LASSAR) and ASCE, Vol. III, pp. 1895-1902, 1989.

2. Cheng, F.Y. and Chang, C.C., Safety-Based Optimum Design of Nondeterministic Structures Subjected to Various Types of Seismic Loads, NSF Report, the U.S. Department of Commerce, National Technical Information Service, Virginia, NTIS No. PB90-133489/AS, (326 pages), 1988.

3. Bycroft, G.N., White Noise Representation of Earthquakes, J. of Eng. Mech. Div., ASCE, Vol. 86, No. EM2, April, 1960, pp. 1-16.

4. Kanai, K., Semi-empirical Formula for Seismic Characteristics of the Ground, Bull. of the Earthquake Research Institute, U. of Tokyo, 1957.

5. Tajimi, H., A Statistical Method of Determining the Maximum Response of a Building Structure During an Earthquake, Proc. 2nd World Conf. on Earthquake Engineering, Vol. 2, Tokyo, Japan, 1960.

6. Vanmarcke, E.H., Properties of Spectral Moments with Application to Random Vibration, J. of Eng. Mech. Div., ASCE, Vol. 98, No. EM2, April, 1972, pp. 425-446.

7. Kiureghian, A.H., Structural Response to Stationary Excitation, J. of Eng. Mech. Div., ASCE, Dec., 1980.

8. Davenport, A.G., Note on the Distribution of the Largest Value of a Random Value of a Random Function with Application to Gust Loading, Institution of Civil Engineers, Vol. 28, 1956.

9. Jennings, P.C., Housner, G.W., and Tsai, N.C., Simulated Earthquake Motions, Rept. Earthquake Eng. Res. Lab., CIT, April, 1968.

10. Cornell, C.A., Bounds on the Reliability of Structural Systems, J. Struct. Div., ASCE, Vol. 93, No. ST1, Feb. 1967, pp. 171-200.

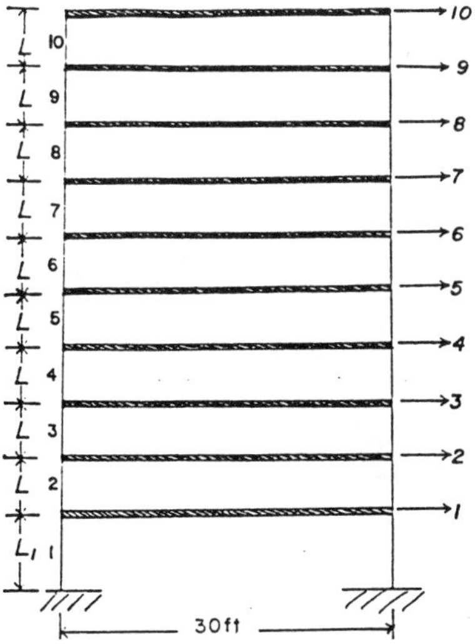

Figure 1. 10-Story Shear Building Structure (L_1 = 15 ft, L = 12 ft, 1 ft = 30.38 cm)

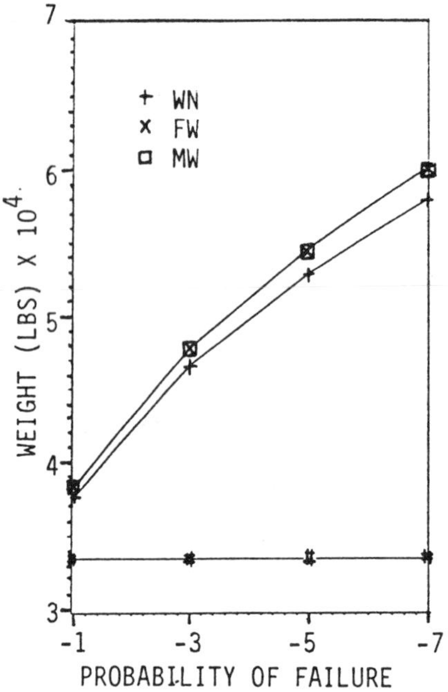

Figure 2. Optimum Weight for Three Seismic Input Spectra with FP and 10-Story Case (1 lb = 4.45 N)

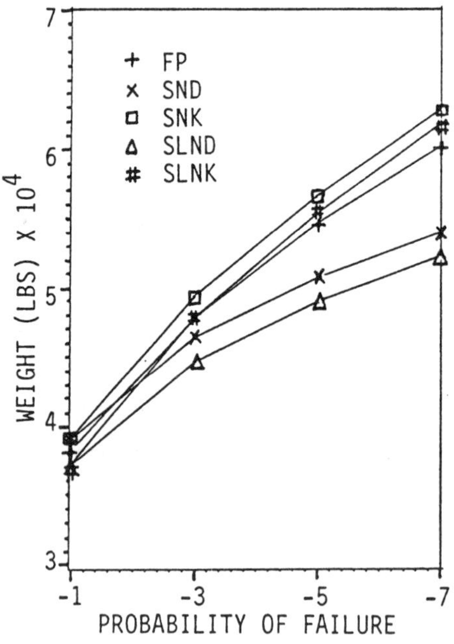

Figure 3. Optimum Weight for Various Failure Expressions with MWN Case (1 lb = 4.45 N)

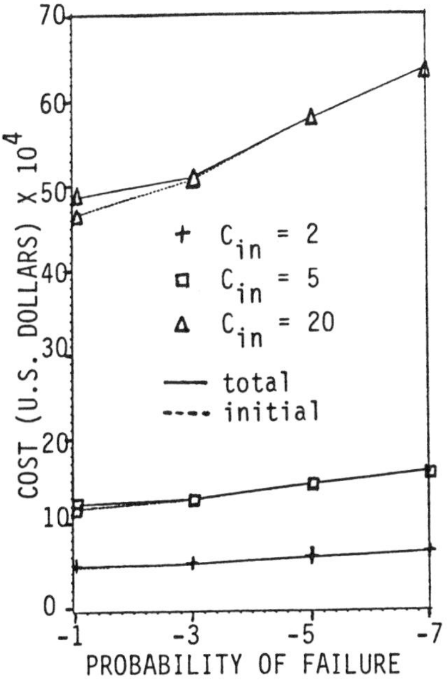

Figure 4. Optimum Cost for Various C_{in} with MW, FP, 10-Story Case

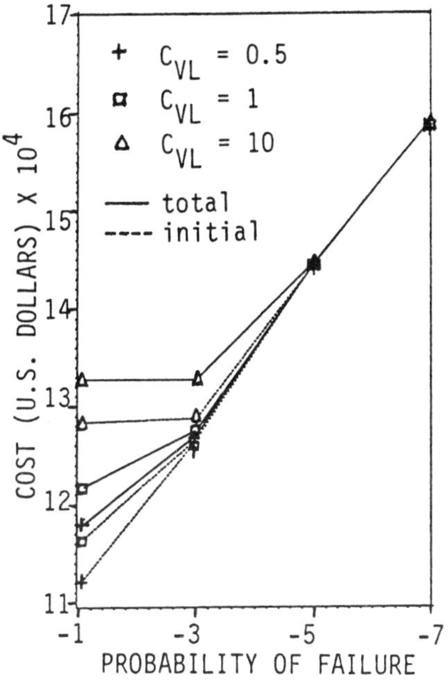

Figure 5. Optimum Cost for Various C_{VL} with MW, FP, and 10-Story Case

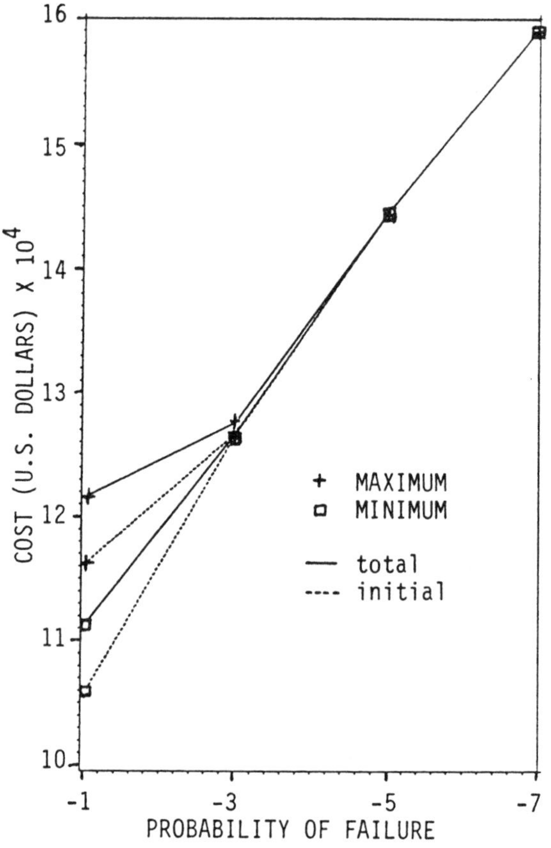

Figure 6. Cost for Maximum and Minimum System Failure with MW, FP, and 10-Story Case

Serviceability Design for Structural Vibration of Composite Beam and for Crack of Concrete Slab

K. Koyama (*), I. Yano (**)
(*) Department of Architecture and Civil Engineering, Shinshu University, 500 Wakasato, Nagano, Japan 381
(**) Structure and Planning Consultant CO., 3-34-1 Sugamo, Toshimaku, Tokyo, Japan 170

ABSTRACT

Reliability based serviceability index for dynamic vibration of composite beam and for crack of concrete slab is investigated in this study. The optimum serviceability indices which are first introduced by Reid and Turkstra[1] are obtained by setting the marginal psuedo limit state. The indices obtained in this paper are optimized to maximize the utility of structures.

INTRODUCTION

The basic limit states that are assigned to structures related to reliability are classified as (1) ultimate limit state(ULS),(2) serviceability limit state(SLS), and (3) fatigue limit state(FLS), generally. The design based on ULS is usually performed by using load factor or modified one, and the method seems to be practical in real design[4], while, the design based on SLS and FLS are not common despite having the same importance as ULS. The reason is considered that the failure boundary may not be able to define clearly, whether it is serviceable or not, for example[1].

In this paper, SLS design based on reliability

method for dynamic vibration of composite beam and for crack of concrete slab is investigated and the optimum serviceability indices for these problems are obtained. The serviceability parameters related to limit states are frequency and width of crack, in this papar, respectively.

SERVICEABILITY INDEX USING UTILITY FUNCTION

Using the concept of utility, Reid and Turkstra [1] first introduced the optimal serviceability index based upon reliability method. Setting the marginal pseudo limit state X^* and employing the trilinear utility function given in Eq.(1), they obtained the serviceability index β^*, equating the expected total utilities and the the utility for the marginal limit state, as follows:

$$u_X(x)=\begin{cases} u^*, & x > X_2 \\ (x-X_1)u^*/(X_2-X_1), & X_1 \leqq x \leqq X_2 \\ 0, & x < X_1 \end{cases} \qquad (1)$$

$$\beta^*=\Phi^{-1}\left(\frac{1}{\beta_2-\beta_1}\{\beta_2\Phi(\beta_2)+\phi(\beta_2)-\beta_1\Phi(\beta_1)-\phi(\beta_1)\}\right) \qquad (2)$$

in which, β^*=serviceability index for pseudo limit state X^*, β_1 and β_2 are standardized normal variates for normally distributed parameter X_1 and X_2, x=the value of serviceability parameter X, $\Phi(\ )$ =cumulative distribution function, $\phi(\ )$=pdf. of standardized normal variate, respectively.

In this paper, however, alternative serviceability index β^* is proposed. It is obtained from using an utility function given by Eq.3. It is considered to be a more adequate one to express the serviceability of structures, and it is shown in Fig.1.

$$u_X(x)=\begin{cases} 2u^*\{(x-X_1)/(X_2-X_1)\}^2, & x \leqq (X_1+X_2)/2 \\ -2u^*\{(x-X_1)/(X_2-X_1)\}^2+u^*, & x \geqq (X_1+X_2)/2 \end{cases} \qquad (3)$$

The index β^* is obtained, referring to Fig.1 and performing the Eq.4:

$$\int_{-\infty}^{\beta_1} 0 \cdot f_x(x)dx + \int_{\beta_1}^{(\beta_1+\beta_2)/2} u_x(x)f_x(x)dx + \int_{(\beta_1+\beta_2)/2}^{\beta_2} u_x(x)f_x(x)dx$$

$$+ \int_{\beta_2}^{\infty} u^* f_x(x)dx = \int_{\beta^*}^{\infty} u^* f_x(x)dx \qquad (4)$$

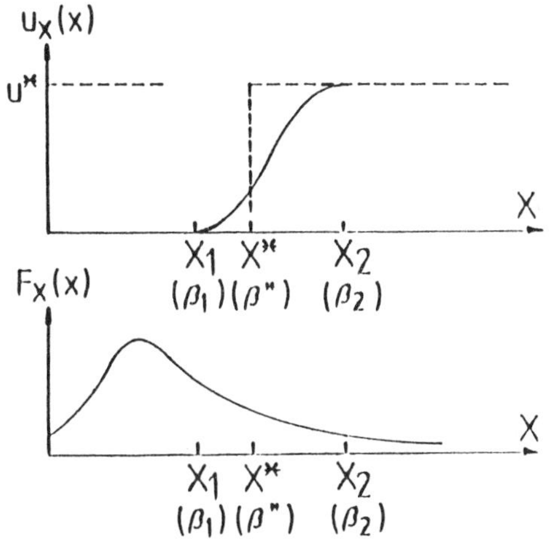

Fig.1 Utility and Density function of X

$$\beta^* = \Phi^{-1}\left(\frac{2}{(\beta_2-\beta_1)^2}\{A_1-B_1-B_2+A_2\}\right) \qquad (5)$$

in which,

$$A_1=(1+\beta_1^2)\Phi(\beta_1)+\beta_1\phi(\beta_1) \qquad (6)$$

$$A_2=(1+\beta_2^2)\Phi(\beta_2)+\beta_2\phi(\beta_2) \qquad (7)$$

$$B_1=\left(2+\frac{(\beta_1+\beta_2)^2}{2}\right)\Phi\left(\frac{\beta_1+\beta_2}{2}\right) \qquad (8)$$

$$B_2=(\beta_1+\beta_2)\phi\left(\frac{\beta_1+\beta_2}{2}\right) \qquad (9)$$

Serviceability is somewhat based on subjectivity. It would be, therefore, better to convert the utility, which expresses the quantity of serviceability, to numerical value like money. It is shown in the following section.

OPTIMIZATION OF SERVICEABILITY INDEX

(1) Dynamic Vibration Problem Of Composite Beam

The serviceability index is optimized to maximize the utility for serviceability of pseudo frequency limit state of composite beam shown in Fig 2.

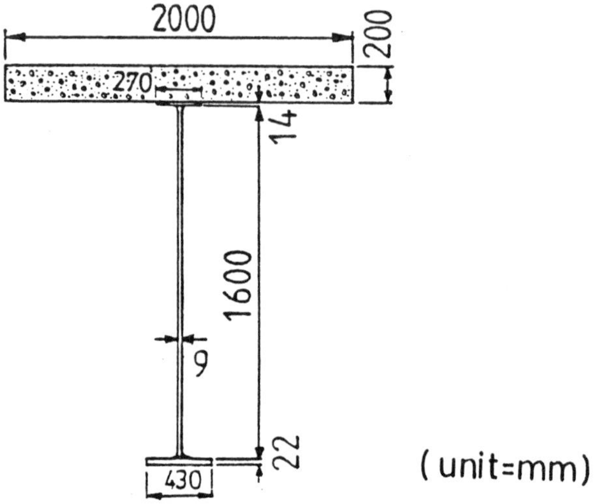

Fig.2 Girder of Beam with Concrete Slab

The utility function expressed by monetary scale is represented as[1]:

$$u_x(x)=B-C_1(x)-p_f(x)C_f \tag{10}$$

in which, B=benefit of the structures, $C_1(x)$=the initial cost of structures, $p_f(x)$=the probability that the structures lose the serviceability at $x=X^*$, C_f=the cost of unserviceability, respectively.

To maximize the utility, benefit B is being constant, the following equation has to be minimized, i.e.:

$$E(C_T)=C_1+p_fC_f \rightarrow \text{mini.} \tag{11}$$

The optimal serviceability index β_{opt} is, therefore, obtained solving $\partial E(CT)/\partial\beta=0$. The initial cost term is expressed as the function of the moment of inertia I of the beam which is shown in Fig.3 to be:

$$C_{i,ID}=C_{i,IO}+\frac{dC_i}{dI|_{I=I_0}}(I_D - I_0) \tag{12}$$

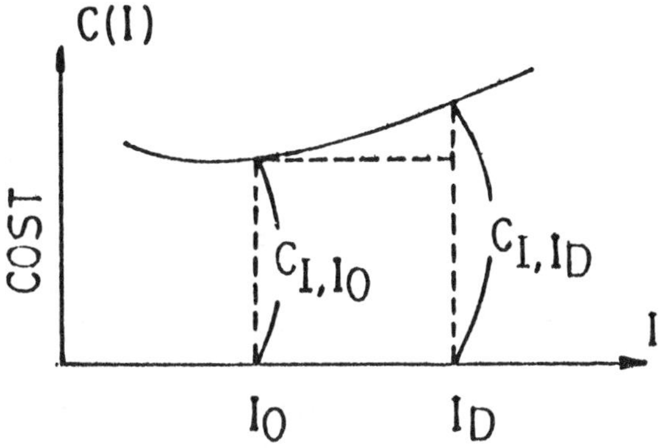

Fig.3 Cost Function of Beam with Inertia I

It is rewritten to be[1]:

$$C_{i,ID}=C_{i,I0}+\frac{a_2L}{a_1}(I_D-I_0) \tag{13}$$

in which $a_1=dI/dW|_{I=ID}$, $a_2=C_i/W\ L$, W=weight of beam per unit length, L=span length of beam.

Furthermore, the natural period of beam that is modeled by one degree of freedom system is expressed as:

$$T_D= 2\pi\sqrt{\frac{A_DL^3}{48E_SI_Dg}} \tag{14}$$

in which A_D=dead load of beam girder, E_S=elastic modulus of steel, $I_D=I_S-I_C$=the design variable, I_S=moment of inertia of steel girder, I_C=moment of inertia of concrete section about neutral axis of composite beam, g=gravity, respectively. The dead load A_D to be assumed normally distributed is given as follows:

$$A_D= m_A + \beta_D\sigma_A \tag{15}$$

in which m_A, σ_A are mean and standard deviation of

A_D, β_D=serviceability index to be optimized, respectively. It is considered here that the variation of A_D depends mainly on the variation of I_S of steel girder.

Solving I_D in Eq.14 and substituting it to Eq.12, and using Eq.11, the optimal index is obtained from $\partial E(CT)/\partial\beta=0$ as:

$$\beta_{D\ opt}=\sqrt{2\ln\{\frac{1}{(2\pi)^{5/2}}\ \frac{a_1C_F}{a_2}\ \frac{48ET_D{}^2g}{L^4\sigma_A}\}} \tag{16}$$

The index $\beta_{D\ opt}$ is,therefore, considered to be the most economic optimal index. While, β^* in Eq.2 is the serviceability index for pseudo limit state. If $\beta_{D\ opt}$ may be equal to β^*, then it is possible to get the most economic and serviceable design for pseudo limit state. To get this, $\beta_{D\ opt}$ is calculated by Eq.16, assuming the initial value of T_D. Then I_D is determined to substitute this $\beta_{D\ opt}$ to A_D in Eq.15 and Eq.14. When the mean and standard deviation of T_D is estimated, β^* can be calculated from Eq.5. The procedure is repeated until $\beta_{D\ opt}$ becomes equal to β^*, modifying I_D in Eq.14. The value T_D^* responds to β^* is the marginal natural period of serviceability for pseudo frequency limit state.

For example, a composite beam model shown in Fig.2 is adopted and investigated. In this example, the following data are used for calculation. The spanlength L is assumed to be 30m, a_1 and a_2 in Eq.13 are estimated about a_1=0.1m^4/tf/m and a_2=¥900,000/tf(≒$3000/kip), respectively. It is based on the construction cost build in our country and it includes, for example, lateral bracing members etc., besides the steel girder itself. Unserviceability cost $c_F(x)$ is estimated approximately to be ¥600,000/m^2, and E_S=2.1 x10^7tf/m^2[2,3].

Initial design value of I_D is 0.02566, while I_C=0.00729, and the width of influence area S=2.0m, respectively. From these,the mean value of dead load is estimated m_A=40.0tf. It includes members like lateral bracing overhead. To estimate the σ_A is difficult, though it seems to be small for dead load, so it is assumed to be 0.25, 0.50, 1.0tf.

The utility function that expresses the cost for unserviceability is shown in Fig.4.

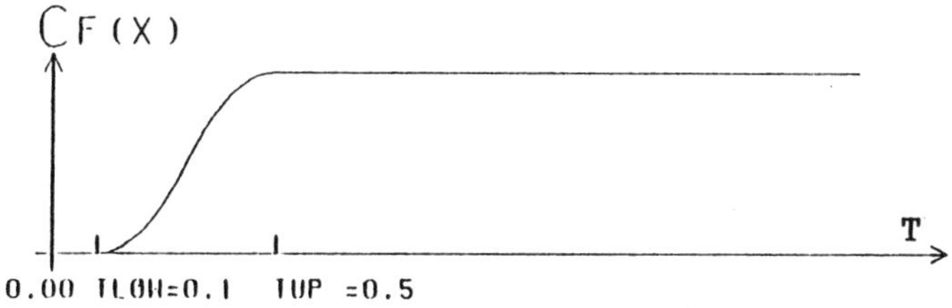

Fig.4 Cost function for Utility

In Fig.4, it shows that the composite beam begins to lose its serviceability from threshold value for the natural period TLOW=0.5 sec.,and reaches perfect unserviceability at TUP=1.0 sec. The results of pseudo natural period(=2π/frequency) limit state T_D^*, moment of inertia I_D, and optimal serviceability index $\beta_{D\ opt}$ for various TLOW, other things being equal, are shown in Fig.5. In Fig.5, solidline, dotted line and broken line are for $_A$=0.25, 0.50, 1.0, respectively. In this example,it is seen that T_D^* and I_D are sensitve to the change of TLOW while $\beta_{D\ opt}$ is not. It shows the same results as Reid et.al had just obtained. It is also found that the influence of other factors, for example the change of so called cost factor[1] $a_1C_F/\sqrt{2\pi}a_2$ in Eq.16, are small compare to TLOW. It means that the exact estimation of C_F or a_2 are not neccessarily needed for getting the optimal serviceability design for this problem. It is simply shown for only TD^* in Fig.6.

(2) Serviceability of Crack Width of Concrete Slab

It is required that the widths or depths of crack in concrete structures are small enough for the use of structures. Excessive widths of crack causes the

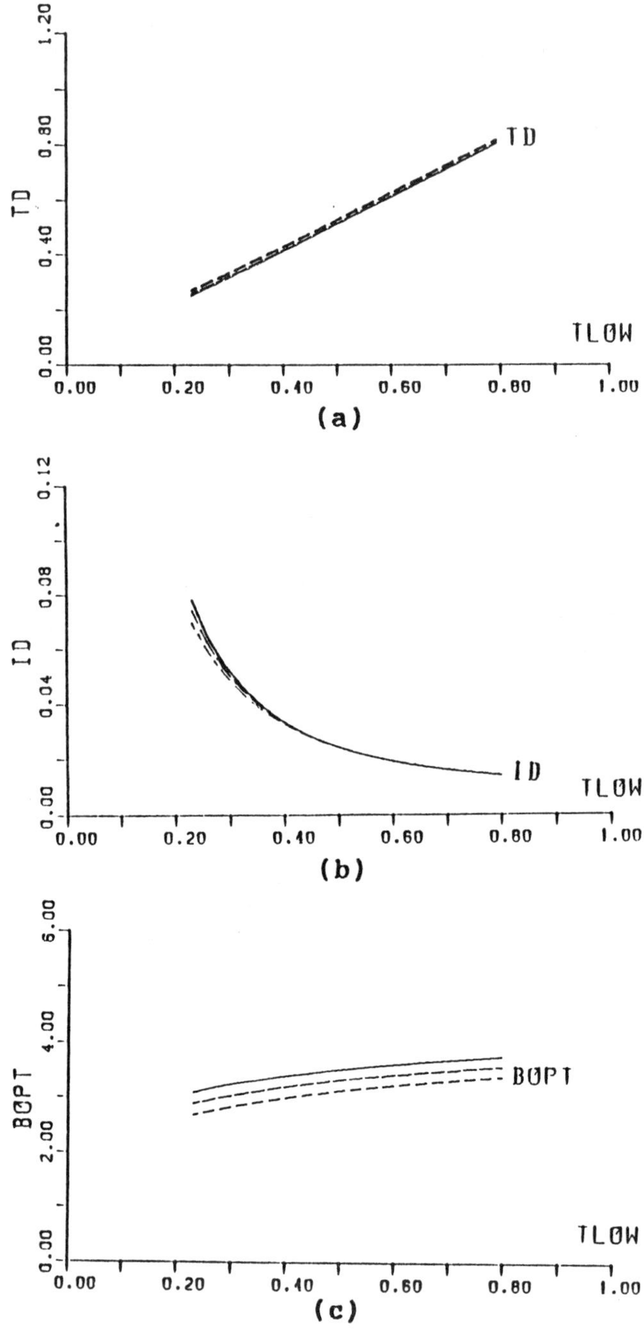

Fig-5 TD*,ID and $\beta_{D\,opt}$ for Composite Girder

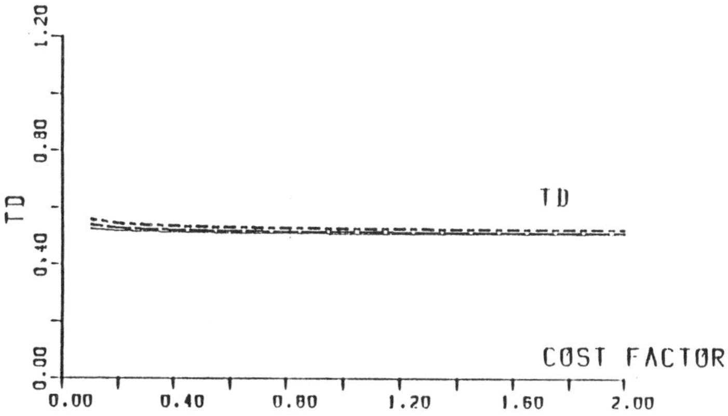

Fig.6 T_D^* for Cost Factor

unusability of rusting the steel in the concrete structures, and may give damage to the beauties of structures, also. Therefore, in our country, following limit states for crack of concrete structures are to be examined, for checking serviceability[4].

(1)limit state for occuring tensile stress, (2)limit state for occuring cracks, (3)limit state for crack width. In this section, type (3) limit state is adopted for SLS problem. The limit crack width of concrete structures with 30mm depth from concrete surface to reinforcement is given as[5]:

$$W_K=33.0x10^{-5}x\tau'_S+0.443x10^{-5}xSx\tau_S \qquad (17)$$

in which, W_K=limit width, τ'_S=average strength of steel, it means usually 137N/mm² in our country, τ_S=the actual stress in reinforcement of structures, S=the interval of reinforcement steel.

Assuming the interval S is random variable and its mean value μ_S and coefficient of variation δ_S are unknown, the optimal serviceability index for crack limit state is shown as:

$$\beta_{D\ opt}=\sqrt{2\ln(-0.443x10^{-5}\ \frac{a_1C_F}{\sqrt{2\pi}\,a_2}\ \frac{\tau_S}{L\sigma_{WK}})} \qquad (18)$$

in which, $a_1=dS/dW|_{S=SD}$, $a_2=C_i/WL$, σ_{WK}=standard deviation of W_K, respectively. To evaluate a_1, the following are used.

$$W=B(h\cdot\gamma_C+\phi_S\cdot\gamma_S/S) \qquad (19)$$

in which, B=width of slab, it is assumed to be 8.0m, h=thickness of concrete, γ_C, γ_S are weight density of concrete and steel, ϕ_S=sectional area of steel whose diameter is ϕ, respectively.

An example of SLS design for width of crack of slab, shown in Fig.7, is studied.

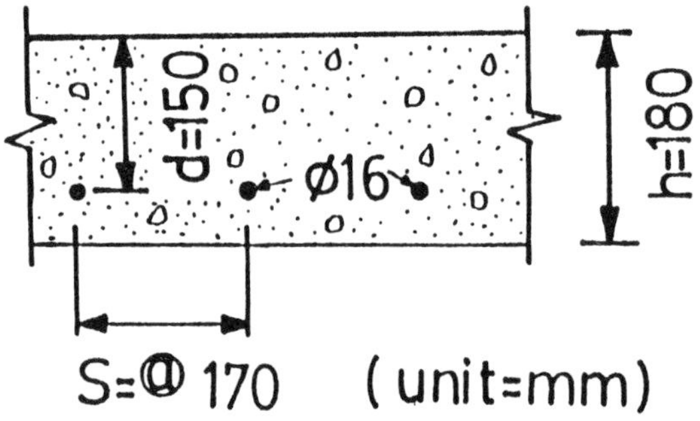

Fig.7 Reinforced Concrete Slab

The moment applied to slab is estimated as approximately 2.09tf cm. The required sectional area of steel is approximately 11.4cm² provided the tensile strength of steel is 137N/mm², therefore, if 16mm diameter of steel is used, the interval of steel is determined as 170mm. And from this, the required number of steel per 1m is assigned as 6, in ordinary design. In this case, the tensile stress in steel is calculated to be about 132N/mm² and the crack width is expected to be about 0.15mm.

The allowable limit widths of cracks are classified in three categories in our country, depending on where structures are to be built, as:
(1)0.2-0.3mm for normal circumstances, (2)0.1-0.2mm for bad circumstances, (3) 0.0-0.1mm for worst circumstances, like places very close to the seaside,

for example. It is assumed here that the WKLOW is 0.1mm and WKUP=0.5mm. For these values, the cost function is shown in Fig.8.

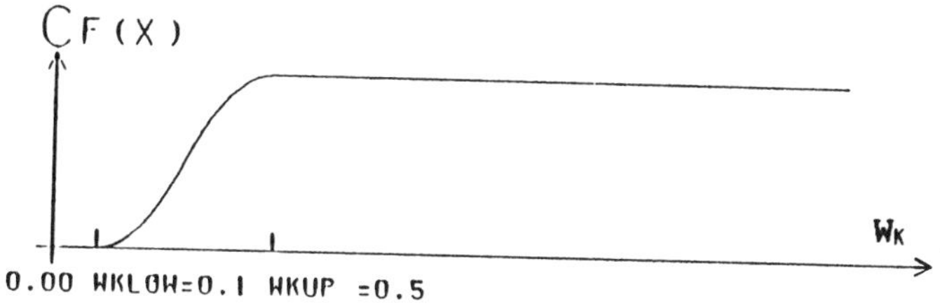

Fig.8 Cost Function for Utility of Slab

The SLS design for crack width based on Eq.18 is performed using the following data. γ_C=2.5tf/m³, γ_S=7.8tf/m³, a_1=185.0mm/N/mm, a_2=¥5.0/N. To estimate a_2, the construction cost for slab, in our country, is assumed to be ¥150,000/m³(=$1,100/m³), which includes the cost of concrete and material work overhead. In this example, the optimal serviceability index is calculated to be $\beta_{D\ opt}$=2.90. The optimal pseudo limit width of crack is obtained as W_K^*=0.112mm. In this case, the optimal interval of reinforcement steel is 143.11mm. If the interval is placed as 140mm, then the required number of steel per 1m is assigned to be 7, and therefore the stress in steel is calculated as 113N/mm. From this, the interval of reinforcement per unit length is needed to be rearranged from 170mm, in ordinary design, to 140mm for SLS one. It is shown that the required sectional area per unit length, therefore, should be increased for crack width SLS design.

Alternatively, if 13mm diameter steel is used in this problem, $\beta_{D\ opt}$ is calculated to be 2.98, the interval is obtained as S=115mm, the number of reinforcement per unit length is assigned as 9.

CONCLUSIONS

The SLS design for dynamic vibration of composite beam and for crack width of concrete slab is investigated. To obtain the optimal serviceability index, the utility expressed by monetary term is maximized based on the design that first proposed by Reid et.al, in this study. It is concluded that the serviceability pseudo limit state of a natural period for composite beam is sensitive to thresh-hold value TLOW, but rather not sensitive to so called cost factor $a_1C_F/\sqrt{2\pi}a_2$. It is also concluded that the optimal serviceability index $\beta_{D\,opt}$ is insensitive to both TLOW and cost factor. And for SLS design for crack width, it is further concluded that the sectional area of reinforcement per unit length should be increased more than ordinary design and the pseudo limit crack width is also sensitive to threshhold value from which the unserviceability begins. It is believed that if the more adequate utility function may be used, the more appropriate result may be obtained.

REFERENCES

1. Reid, S. G., and Turkstra, C. J. "PROBABILISTIC DESIGN FORSERVICEABILITY," 3rd International Conference on Structural Safety and Reliability, pp.583-592, 1981, Norway.
2. Society of Steel Construction of Japan, " A Manual for Design and Planning of Steel Girder Bridge", March, 1985(in Japanese).
3. Committee for Making Preliminary Data, "Collection and Arrangement of Preliminary Data for Structural design of Civil Engineering, No.2", Committee of Consultants Western Branch of Japan, June, 1984(in Japanese).
4. Hajime, Okamura, "Limit State Design of Concrete Structures, 2nd edition", Concrete Seminar 4, Kyoritsu Publishing, 5, 1984(in Japanese).
5. Shin, N., Yasuo, K., Michiaki, S.,"Maintenance and Repair of Concrete Structures", Kajima Publishing, 5, 1986(in Japanese).

SECTION 3: OPTIMUM CONTROL AND IDENTIFICATION OF STRUCTURES

Dynamic Structural Model Modification Using Mathematical Optimization Techniques

M.S. Ewing, R.M. Kolonay

Wright Laboratory (WL/FIBRA), Wright-Patterson Air Force Base, Ohio 45433-6553 U.S.A.

ABSTRACT

Mathematical optimization techniques are used within the framework of large structural analysis computer programs to update dynamic finite element models based on experimental measurements by modifying the physical variables used to define the finite elements. Error functions involving free vibration mode shapes and natural frequencies are minimized subject to separate error constraints on individual mode shapes and natural frequencies. Emphasis is placed on determining, or identifying, unknown stiffness and mass properties associated with structural support members. Numerical results are reported for a beam resembling an aircraft-borne missile supported flexibly at two intermediate points.

INTRODUCTION

In this study, the finite element model of a structure is updated in terms of physical variables. The basis of the update is a set of valid experimental measurements. The physical variables to be updated include straightforward quantities such as cross-sectional area, mass and mass moment of inertia, as well as joint and support flexibility - which are typically not modelled with great certainty. Other identifiable variables include "effective" cross-sectional area, which may be an engineering measure of damage in a structure.

The overall aim is to determine what model corrections, in terms of the physical variables, are required to obtain an "optimum" match between experimental and analytical natural frequencies and mode shapes. The class of model considered is the linear finite element model with "light damping" assumed, although damping is not included as a physical variable in the research reported here.

The long-term goal of this research is to incorporate structural model update algorithms into large-scale analysis programs. Increasingly, these analysis programs have the option to perform optimization tasks. Therefore, the algorithms reported here have been designed to be and have been incorporated into such programs.

BACKGROUND

Many able researchers have studied the problem of model updating and numerous surveys have been compiled. The recent survey article by Imregun and Visser[1] gives a very nice review as does an earlier survey by Caesar[2].

Years ago, Baruch and Itzhack[3] formulated techniques to update stiffness matrices by minimizing the change necessary to obtain agreement between analytical and experimental results. The basis of the method was enforcement of orthogonality and equilibrium constraints on the analytical mode shapes. The analytical mass matrix was assumed to be correct. Berman and Flannelly[4] proposed a method to change the mass matrix as well, by first enforcing orthogonality with respect to the experimental mode shapes. Numerous improvements to the basic techniques have been proposed, for instance by Berman and Nagy[5]. Berman has also published excellent commentaries on the limitations encountered when updating finite element models[6,7]. One of the drawbacks of the aforementioned techniques is that changes in the model occur which have no physical basis - terms which should be zero, for instance, are not. Kabe[8] proposed techniques to deal with this problem.

Numerous studies have addressed the structural identification problem in terms of assigning values to physical variables – as opposed to simply defining the elements of a mass or stiffness matrix. Chen and Garba[9] proposed a procedure in which the analytical model is updated iteratively so as to minimize the changes while enforcing increasingly better agreement with experimental mode shapes and frequencies. Wei and Janter[10] used non-linear programming optimization techniques to match experimental and analytical natural frequencies. Most recently, Ojalvo[11,12] has used least-squares ("pseudo-inverse") techniques to match analytical and experimental frequencies *and* mode shapes.

It is of academic interest to note that the model update problem posed here is mathematically identical to the design problem in which specific frequencies and mode shapes are the design objectives. The problem of designing a structure to have a specific fundamental natural frequency has been posed as an optimization problem and solved by Venkayya and Tischler[13]. Grandhi and Venkayya[14] solved the same problem for multiple frequencies. Design optimization to mode shape constraints has not yet been tackled.

DIFFICULTIES WITH MODEL UPDATING

The difficulties associated with model modification have been discussed elsewhere (most completely in references 1,6,7 and 10), but the most pathological ones warrant mention in any treatment on the subject.

The most troublesome limitation on model update schemes is the requirement to determine many design parameters based on a relatively small number of measurement data. This presents a difficult scientific problem, yet at the same time suggests the best situations in which to attempt an automated model update.

With relatively few measurement data and many physical parameters which must be determined, the problem is very much underspecified[6,7] and an infinite number of solutions are possible. More data tends to alleviate this problem in the sense that equal numbers of data and physical variables *may* allow a unique, and presumeably nearly exact, determination of the physical parameters. The incorporation of mode shape data in addition to frequency data into the algorithm tends to alleviate this problem.

The underspecificaton problem also tends to be alleviated by reducing the number of physical variables one seeks to update. As a result, the greatest success in identification may occur in the class of problems in which only a small number of physical varibles, for instance support structure, are identified. A two-phase process in other instances is also suggested: during the first phase the "areas" of uncertainty are identified, and then, in a second model update, the uncertain physical parameters are quantified. Unfortunately, in some cases, the most "sensitive" properties tend to change during identification, rather than the ones which have been modelled incorrectly.[1] In a variant process, an "experienced" engineer determines which properties are the most uncertain and then only these properties are updated.

Another associated problem encountered in identification is the lack of one-to-one correspondance between analytical degrees of freedom and measurement degrees of freedom. If mode shapes are to be matched only where measurement points and analytical node points coincide, some uncertainty exists in the agreement of the mode shapes at other points. One possibility is to expand the experimental modes, while another is to reduce the analytical modes.[1]

PROBLEM STATEMENT

Consider a structural model with physical parameters, r_i, as well as analytically determined mode shapes, or eigenvectors, $\underset{\sim}{\phi}_j$ (with individual elements,ϕ_j^k), and natural frequencies, ω_j. For notational convenience, we shall speak of the frequencies in terms of the eigenvalues, $\lambda = \omega^2$. The actual structure has physical parameters, $\bar{r}_i$, mode shapes, $\underset{\sim}{\bar{\phi}}_j$, (with individual elements,$\bar{\phi}_j^k$), and natural frequencies, $\bar{\omega}_j$. To identify the structure in terms of the model's physical parameters, the following optimization problem may be solved.

Minimize either:

$$\sum_{j=1}^{J} \|\underset{\sim}{\phi}_j - \underset{\sim}{\bar{\phi}}_j\|^2 \tag{1}$$

or:

$$\sum_{j=1}^{J} (\lambda_j - \bar{\lambda}_j)^2 \tag{2}$$

subject to the constraints:

$$\left| \frac{\lambda_j}{\bar{\lambda}_j} - 1 \right| < a_j \quad for\ j = 1, 2, ...J' \tag{3}$$

$$\|\phi_j - \bar{\phi}_j\|^2 < b_j \quad for\ j = 1, 2, ...J" \tag{4}$$

$$\sum_{i=1}^{I} m_i = M \tag{5}$$

where: J is the number of modes used for the identification task (some inappropriate modes can be neglected); both a_j and b_j are "small" numbers chosen to quantify how close to target is "close enough"; J' is the number of modes which are to have particularly accurate frequency matching; J'' is the number of modes which are to have particulary accurate mode shape matching; m_i is the mass of an individual finite element or concentrated mass; and M is the total mass of the physical structure.

The objective function minimized in equation 1 is a measure of the deviation of the analytical and experimental mode shapes. Note that a term-by-term difference between these mode shapes, $(\phi_j^k - \bar{\phi}_j^k)$, is squared, then summed for each mode (giving the "norm" of the difference vector), and finally summed for all modes considered. In general, of course, the number of elements in ϕ_j^k is less than the number in $\bar{\phi}_j^k$, so either ϕ_j^k must be reduced or $\bar{\phi}_j^k$ must be expanded. The objective function in equation 2 is the sum of the squared errors between measured and analytical eigenvalues. Equation 3 is a set of frequency constraints for a selected set of modes. Equation 4 is a set of individual mode shape constraints – a constraint is violated if the analytical mode shape is not close in value to the measured mode shape. Finally, equation 5 enforces the total mass of the structure to be equal to the physical mass.

SOLUTION

Mathematical optimization techniques are used to solve the problem posed in the previous section. Numerous non-linear programming techniques have been implemented into a wide variety of commercially available software codes. Many different codes should be investigated; however, Schittkowski's[15] sequential quadratic programming technique, NLPQL, available in the popular IMSL mathematics library, has been used successfully in this research.

The solution of the free vibration problem within the optimization algorithm was accomplished with a planar frame analysis program developed for optimization research by Kolonay[16]. This program was written with the same architectural attributes of modern, large-scale analysis and optimization programs such as NASTRAN and ASTROS[17].

Gradient Calculation

As with any solution of an optimization problem, the gradients, or sensitivities, of the objective function (equation 1 or 2) and the constraints (equations 3-5) need to be calculated. In this study, exact gradients were formulated in all cases. To find these sensitivities, one must first find the gradients of the analytical model's natural frequencies and mode shapes. Techniques, due to Nelson[18], for calculating the eigenvalue and eigenvector gradients are well-known (see for instance Ojalvo[19]). Nelson's technique is quite expensive, computationally, but has been used in this study due to a high confidence in the technique. Other more efficient techniques are available, for instance, Ojalvo's[20].

Nelson's formula for eigenvalue gradients is:

$$\lambda_{j,i} = \underset{\sim}{\phi}_j^T \underset{\sim}{K}_{,i} \underset{\sim}{\phi}_j - \lambda_j \underset{\sim}{\phi}_j^T \underset{\sim}{M}_{,i} \underset{\sim}{\phi}_j \tag{6}$$

where subscripts, j, denote the jth mode number and subscripts, ",i", denote differentiation with respect to the ith design variable. Also, $\underset{\sim}{K}$ is the analytical stiffness matrix and $\underset{\sim}{M}$ is the analytical mass matrix. Eigenvectors are ortho-normalized with respect to the analytical mass matrix. The experimentally determined eigenvectors are normalized with respect to the analytical eigenvectors.

To find the eigenvector gradients, $\underset{\sim}{\phi}_{j,i}$, the following set of equations must be solved:

$$(\underset{\sim}{K} - \lambda_j \underset{\sim}{M})\underset{\sim}{\phi}_{j,i} = \lambda_{j,i}\underset{\sim}{M}\underset{\sim}{\phi}_j - \underset{\sim}{K}_{,i}\underset{\sim}{\phi}_j + \lambda_j \underset{\sim}{M}_{,i}\underset{\sim}{\phi}_j \quad (7)$$

Unfortunately, the nth order matrix, $(\underset{\sim}{K} - \lambda_j \underset{\sim}{M})$, is singular. However, the desired gradients may be found from:

$$\underset{\sim}{\phi}_{j,i} = \underset{\sim}{V}_{ji} + \alpha_{ji}\underset{\sim}{\phi}_j \quad (8)$$

Here, $\underset{\sim}{V}_{ji}$ is the solution for $\underset{\sim}{\phi}_{j,i}$ from equations (7) with the following modification. First, the element of $\underset{\sim}{\phi}_j$ with the greatest absolute value is found. Then, the associated element of $\underset{\sim}{V}_{ji}$ is set to zero. Finally, the remaining elements of $\underset{\sim}{V}_{ji}$ are found from the set of equations *now* of order $(n-1)$:

$$(\underset{\sim}{K} - \lambda_j \underset{\sim}{M})\underset{\sim}{V}_{ji} = \lambda_{j,i}\underset{\sim}{M}\underset{\sim}{\phi}_j - \underset{\sim}{K}_{,i}\underset{\sim}{\phi}_j + \lambda_j \underset{\sim}{M}_{,i}\underset{\sim}{\phi}_j \quad (9)$$

In equation (8):

$$\alpha_{ji} = -\underset{\sim}{\phi}_j^T \underset{\sim}{M}\underset{\sim}{V}_{ji} - \frac{1}{2}\underset{\sim}{\phi}_j^T \underset{\sim}{M}_{,i}\underset{\sim}{\phi}_j \quad (10)$$

Consideration of Computational Requirements

The computational requirements of the model updating task using mathematical optimization must be carefully considered when applying the method to "large" problems. The following discussion highlights the computational "expenses" involved, which may form the basis for comparison to other techniques. In this discussion, the following parameters are used: n is the number of degrees of freedom of the analytical model; b is the bandwidth of the problem; m is the number of modes for which eigenvalue and eigenvector gradients are calculated (m is the maximum of $J, J' and J''$); and, p is the number of variable physical parameters. Note that m, being dependent on the number of valid, experimental modes known, is typically small compared with n, and that p can be greatly reduced using parameter linking.

For each iteration, the eigenvalue and eigenvector problems must be solved. This amounts to performing computations in proportion to between $n \cdot (b+m)^2$ and n^3 if a "transformation" method is used; and, in proportion to between $n \cdot b^2 \cdot m$ and $n^2 \cdot m$ if a "tracking" method is used. (see, for instance, reference 21.) A tracking method is ued in this study. Whenever a model reduction technique, such as Guyan reduction, is appropriate, its use substantially reduces the computation required. However, the sensitivities with respect to physical variables become more difficult to calculate. Chen and Garba[9] have addressed this problem; but, their solution – essentially, the calculation of finite difference gradients – is also computationally intensive.

For most structures, unless physical parameter linking is used, the major effort in the model update task using optimization is the calculation of the eigenvalue and eigenvector gradients. The frequency derivatives for each analytical frequency with respect to each physical parameter requires the following to be done $m \cdot p$ times:

calculate the derivatives of the mass and stiffness matrix with respect to the physical variables

$2 \cdot n \cdot (n+1)$ multiplications to find $\lambda_{j,i}$ from equation (6)

The much more time-consuming process of finding the mode shape derivatives requires the following to be done $n \cdot m \cdot p$ times:

n^2 multiplications to form the right-hand side of equation (9)

solve the $(n-1)$ linear, algebraic equations (9) for $\underset{\sim}{V}$

$2 \cdot n^2$ multiplcations to find α from equation (10)

For a structure with a large number of design variables, p, substantial reductions in computation are accrued by either choosing not to update certain physical parameters (due to high confidence in their values) or by enforcing some form of linkage between the parameters. A 10-fold decrease in the value of p translates into a 10-fold decrease in the number of algebraic equations (9) which must be solved.

Subtleties of the Updating Algorithm

Numerous subtleties of implimentation of the constrained minimization problem are worthy of note since they influence the architecture of the software. Some highly desireable characteristics of a structural identification algorithm will now be described.

As discussed earler, the identification process may, in some cases be a two-step process: first the location of poorly modelled areas of a model are identified; then they are quantified more accurately. As a result, software must be able to assign only a subset of physical quantities as variables. (This capability is available in the software developed in the current research effort.)

Many times large sections or repeated substructures share a common physical parameter (such as sheet thickness). Rather than expend computational resources to identify all such parameters, it is expedient to allow such parameters to be "linked." This technique, known as variable linking, is a well-established capability of most structural optimization codes (and has been used in this work).

A much more demanding requirement is that the identification routine must be able to handle "mode switching." As a structure's physical parameters are varied, the numerical ordering of mode shapes will vary. For instance, in a cantilevered structure, the alteration of a physical parameter can cause the frequency of a bending mode of vibration to increase in magnitude so as to exceed that of a torsion mode which originally had a higher value. One possiblity is to allow for mode shape tracking, that is, to provide a way to always compare "like" mode shapes in the objective and constraint functions.

EXAMPLE PROBLEM

To illustrate the application of optimization techniques in model updating, an example problem has been formulated: a straight beam supported at two intermediate points by a support structure with a stiffness which is not known with great accuracy. This arrangement is typical of a military missile attached to an aircraft. A sketch of the beam is shown in figure 1. The beam is modelled as 27 planar, rectangular cross-section, beam bending elelments (for a total of 84 degrees of freedom). The cross-sectional area is uniform. The supports are modelled as short, rectangular cross-section beams with one tenth of the length and cross-sectional area of the primary beam. Table 1 lists the physical parameters associated with each node and element. For the missile structure, the modulus of elasticity $E = 30\ Msi$ and the density $\rho = 0.10\ lb_m/in^3$. The flexural rigidity of the beams is assumed to be a cubic function of the cross-sectional area. (Other relationships are possible, and decoupling of the cross-sectional area and the flexural rigidity is actually desirable for general applicability.) .

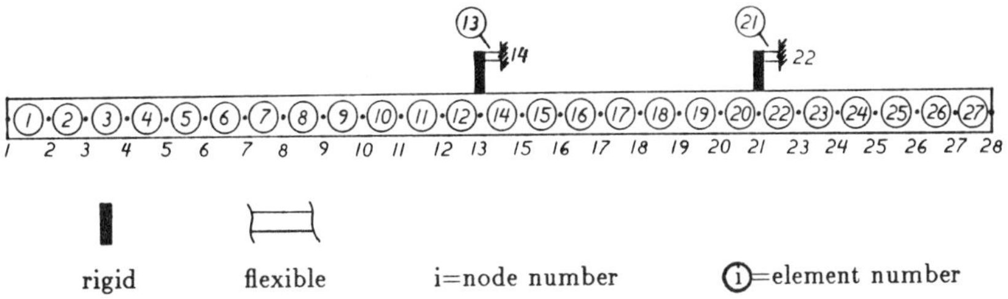

Figure 1. Beam supported at two intermediate points.

The support beams are attached to the primary beam at one end and rigidly fixed at the other end, as shown. "Point" masses were stationed at each node piont in a fashion typical of a tactical missile. For this study, the masses were not subject to change. The first five modes of free vibration were calculated, then tabulated at 6 evenly spaced locations along the beam. These are assumed to be the "experimental" mode shapes. In descriptive terms, the first two modes are, in order of increasing frequency, a forward section bending and a rear section bending mode. The rest are more difficult to describe, but involve substantial involvement of the support beams.

Table 1. Values of physical variables for the beam.

elements	area (in^2)	length (in)	nodes	lumped mass (lb_m)
1-12	2.0	10.0	1-10	.3
13	0.2	1.0	11-13	.6
14-20	2.0	10.0	15-21	.6
21	0.2	1.0	23-28	.6
22-27	2.0	10.0		

With the "experimental" natural frequencies and mode shapes known, initial "guesses" as to the value of the support cross-sectional area were made, with varying levels of deviation from the actual areas. Finally, using the free vibration properties of the altered structure as target properties, the "unknown" areas were determined through minimization of either the eigenvalue or eigenvector errors. As expected, the further from target the support flexibility was assumed, the greater the probability the algorithm would fail. However, excellent identification was possible for initial guesses for the cross-sectional area of as much as 25% high. (This corresponds to flexural rigidity being nearly twice the actual.)

Table 2 shows the identified support cross-sectional areas, using both of the objective functions with and without physical variable linking. The linking involved keeping the two support cross-sectional as variables and linking the other cross-sections with three variables: one each for the left, middle and right sections of the beam. In table 2, three cross-sectional areas areas are reported: two for the supports and one for the area most poorly identified. The identified areas of the left and right support are different as a result of using a numerical search technique for the optimization. The number of iterations required to identify the cross-sectional areas is also given. In all cases, there were 6 measurement degrees of freedom – corresponding to the lateral motions. There were 26 cross-sectional areas identified without linking and only 5, as described above, with linking. In all cases, $J = J' = J" = 5$ and $a_j = b_j = 0.000001$.

Table 2. Identified cross-sectional areas (in^2) as a function of objective function and parameter linking scheme used.

objective function	linked?	cross-sectional area left	right	other	iterations
eqn. 1	yes	.20085	.19910	2.0044	5
eqn. 2	yes	.20043	.19933	1.9977	5
eqn. 1	no	.23538	.20239	1.8549	7
eqn. 2	no	.19976	.19941	1.9893	4

Table 3. Free vibration eigenvalues (sec^{-2}) as a function of objective function and parameter linking scheme used.

objective function	linked?	mode number 1	2	3	4	5
eqn. 1	yes	352.91	2015.7	15645	48758	109140
eqn. 2	yes	352.91	2015.4	15643	48746	109140
eqn. 1	no	351.45	2014.3	15638	48701	109140
eqn. 2	no	353.13	2012.6	15630	48476	108450
target		352.91	2015.0	15644	48761	109140

In an earlier, preliminary examination of the use of mathematical optimization to identify physical variables[22], an attempt was made to determine the best combination of objective functions and constraints to use. For the simple structures studied (a 9 degree of freedom springs-in-series problem and a 12 degree of freedom beam problem) it was apparent that both eigenvalue and eigenvector constraints were helpful. However, there was no conclusive evidence supporting one objective function over another.

For this example problem, it seems the eigenvalue objective function is superior in the case of no physical variable linking. The deviation of the identified cross-sectional areas from the target when eigenvalue errors formed the objective function never exceeded 0.5%. Using the eigenvector error objective function resulted in the same level of deviation when linking was used (an easier problem), but nearly 18% deviation without linking. However, without further study, general conclusions are not warranted on the use of alternative objective functions.

Concerning the use of variable linking, common sense suggests and the facts support the notion that whenever possible – that is, whenever there is a clear justificaton – to link variables, one should. As discussed earlier, there is a significant computational time savings. As shown above, the easier optimization with fewer varibles is nearly certain to be superior in practical problems.

CONCLUSIONS

The use of optimization techniques within standard structural analysis computer programs for model updating has been demonstrated. This article is intended to show that the techniques work for yet another simple structure. Continuing efforts are focused on expanding the complexity of the problems being solved, particularly with problems in which a limited number of valid measurement points severely taxes the identification algorithm.

Future work also includes adding new capabilities to the model updating algorithm. Planned additions include mode shape tracking and allowance for separate axial and bending stiffness property identification within a finite element. Efforts are also planned for investigating alternative optimization schemes, in particular, a recently-developed compound scaling algorithm.[23,24]

REFERENCES

1. Imregun, M. and Visser, W. J., "A Review of Model Updating Techinques", The Shock and Vibration Digest, vol. 13, Jan 1991, pp 9-20.
2. Caesar, Bernd, "Updating System Matrices Using Modal Test Data", Proceedings, 5th International Modal Analysis Conference, London, 1987, pp 453-459.
3. Baruch, M. and Bar Itzhack, I. Y., "Optimal Weighted Orthogonalization of Measured Modes", AIAA Journal, vol. 16, April 1978, pp 346-351.
4. Berman, A. and Flannelly, W. G., "Theory of Incomplete Models of Dynamic Structures", AIAA Journal, vol. 9, Aug 1971, pp 1481-1487.
5. Berman, A. and Nagy E. J., "Improvement of a Large Analytical Model Using Test Data", AIAA Journal, vol. 21, Aug 1983, pp 1168-1173.
6. Berman, A., "System Identification of Structural Dynamic Models – Theoretical and Practical Bounds", Proceedings, AIAA 25th Structures, Structural Dynamics and Materials Conference, Palm Springs, CA, 14-16 May, 1984. (AIAA paper 84-0929)
7. Berman, A.. "Nonunique Structural System Identification", Proceedings, 8th International Modal Analysis Conference, 1990, pp 355-359.
8. Kabe, A. M., "Stiffness Matrix Adjustment Using Mode Data", AIAA Journal, vol. 23, Sept 1985, pp 1431-1436.

9. Chen, J. C. and Garba, J. A., "Analytical Model Improvement Using Modal Test Results," AIAA Journal, vol. 18., June 1980, pp 684-690.

10. Wei, M.L. and Janter, T., "Optimization of Mathematical Model Via Selected Physical Parameters," Proceedings, 6th International Modal Analysis Conference, Kissimmee, FL, Feb 1988, pp 73-79.

11. Ojalvo, I. U., and Pilon, D., "Diagnostics for Geometrically Locating Structural Math Model Errors From Modal Test Data", Proceedings, AIAA 29th Structures, Structural Dynamics and Materials Conference, Orlando, FL, Feb 1988, pp 1174-1186. (AIAA paper 88-2358)

12. Ojalvo, I. U., Ting, T., and Pilon, D., "Practical Suggestions for Modifying Math Models to Correlate with Actual Modal Test Results", Proceedings, 7th International Modal Analysis Conference, Las Vegas, NV, Jan 1989, pp 347-354.

13. Venkayya, V. B., and Tischler, V. A., "Optimization of Structures with Frequency Constraints", Computer Methods for Nonlinear Solids and Structural Mechanics, 1983, pp 239-259.

14. Grandhi, R. V., and Venkayya, V. B., "Structural Optimization with Frequency Constraints", Proceedings, 28th Structures, Structural Dynamics and Materials Conference, Monterey, CA, 1987, pp 322-333.

15. Schittkowski, K., "NLPQL: A FORTRAN Subroutine Solving Constrained Nonlinear Programming Problems", Annals of Operations Research, vol. 5, 1986, pp 485-500.

16. Nelson, R. B., "Simplified Calculation of Eigenvector Derivatives", AIAA Journal, vol. 14, Sept. 1976, pp 1201-1205.

17. Ojalvo, I. U., "Gradients for Large Structural Models with Repeated Frequencies", SAE Aerospace Technology Conference, Long Beach, CA, 13-16 Oct 1986. (SAE paper 861789)

18. Ojalvo, I. U., "Efficient Computation of Mode-Shape Derivatives for Large Dynamic Systems", AIAA Journal, vol. 25, Oct 1987, pp 1386-1390.

19. Kolonay, R. M., Venkayya, V. B., Tischler, V. A. and Canfield, R. A., "Structural Optimization of Framed Structures with Generalized Optimality Criteria", Proceedings, Recent Advances in Multidisciplinary Analysis and Optimization, Hampton, VA, 18-30 Sep 1988, pp 955-970. (NASA Conference Publication 3031, part II)

20. Johnson, E. H. and Venkayya, V. B., "Automated Structural Optimization System (ASTROS)", AFWAL-TR-88-3028, Dec 1988.

21. NASTRAN Theoretical Manual, NASA SP-221(06), Jan 1981, pp 10.1-1 to 10.1-3.

22. Ewing, M. S. and Venkayya, V. B., Structural Identification Using Mathematical Optimization Techniques, Proceedings, AIAA 32nd Structures, Structural Dynamics and Materials Conference, Baltimore, MD, April 1991. (to appear)

23. Venkayya, V. B., "Optimality Criteria: A Basis of Multidisciplinary Design Optimization", Journal of Computational Mechanics, vol. 5, 1989, pp 1-21.

24. Venkayya, V. B. and Tischler, V. A., "Generalization of Compound Scaling for Mathematical Optimization", Wright Laboratory Technical Report, WL-TR-91-00xx, to appear.

Optimum Design of Truss Structures with Active and Passive Damping Augmentation

R.A. Manning

TRW Space and Technology Group, Redondo Beach, California, U.S.A.

ABSTRACT

This work is concerned with the development of an optimum design capability for truss structures augmented with active and passive members. The active members consist of piezoelectric sensors and actuators embedded within the layup of graphite epoxy composite members. The passive members consist of cocured viscoelastic material/graphite epoxy composite tubes. A two stage design optimization procedure is described which avoids the solution of the combinatoric optimization problem associated with active/passive member placement. A numerical example of a large complex space mission is included to demonstrate the feasibility of the design optimization procedure.

INTRODUCTION

New and innovative structural/control hardware and design techniques are necessary for the success of future space missions. The synergistic design application of passive damping will also aid in the elimination of harmful controls/structure interactions and model uncertainties.

Viscoelastically damped passive members [1] and active members with embedded piezoelectric sensors and actuators [2,3] have recently been fabricated and tested at the component level. These technologies suppress structural vibrations with minimal mass, complexity, and power consumption impact. The next step in the development of these technologies is the design and demonstration of the active and passive members at the system level. In order to do this, a simultaneous design procedure is needed.

Bronowicki and Diaz presented analysis and optimization techniques for constrained layer, viscoelastically damped members in Ref. 1. The techniques were applied to a single component which was then successfully fabricated and tested. A system level design optimization technique for viscoelastic damping treatments based on the modal strain energy method was presented in Ref. 4. The method utilized a pre-packaged finite element method where behavior sensitivities were obtained semi-analytically.

Regarding active members, a number of previous studies have demonstrated integrated structure/control design methods using member actuators. An integrated optimization methodology for member actuators using direct output feedback was examined by Lust and Schmit [5]. Thomas and Schmit [6] advanced this idea and added dynamic stability constraints to insure convergence to a stable design for non-colocated sensors and actuators. McLaren and Slater [7] included various control compensators in their covariance approach to the integrated optimization problem. A useful finding of this latter work was the decrease in the value of the objective function as the order of the compensator is increased (see Figures 4 and 8, Ref. 7).

The work presented herein is concerned with the development of an integrated design optimization methodology where structural, passive member, active member, and control compensator design variables are treated simultaneously. Analytical mass and stiffness matrices, as well as mass and stiffness sensitivities, are obtained analytically by deriving these quantities at the element level. The integrated design process is intended for large complex structures where purely structural solutions or even sequential design solutions have little or no potential for success in meeting the stringent performance requirements. By including the compensator parameters in the optimization procedure, better performance can be obtained while meeting stability constraints than for direct output feedback.

SYSTEM DESCRIPTION

Many future space systems will consist of baseline truss structures from which reflectors, antennae, communications equipment, and electronics equipment are mounted. The work considered herein is concerned with truss structures augmented with passive and active members. It has been assumed that overall system performance can be derived from motions of specific points on the truss (e.g., sensitive optical components are hard-mounted to certain points on the structure).

The structural dynamic equations of motion can be written as

$$M\ddot{z} + [(1 + i\gamma^s)K^s + i\gamma^p K^p]z = R + bu \qquad (1)$$

where M and K^s are the real mass and stiffness matrices, z is the vector of physical displacements, and R is the vector of externally applied nodal loads. Inherent damping in the inert truss is modeled as structural damping and enters the equations through the γ^s parameter. The passive damping members contribute to the standard real mass and stiffness matrices, but also add the complex damping term $i\gamma^p K^p$ [8]. Active members contribute to the standard real mass and stiffness matrices, but also add the final term on the right hand side of equation (1). This final term consists of the b matrix locating actuators on the structure and the vector of actuator forces, u. Equation (1) can be written in state space form as

$$\dot{Z} = AZ + Bu \tag{2}$$

where undamped normal modes have been used to reduce the size of the matrices. The state vector, Z, is the stacked vector of modal displacements and velocities

$$Z = \left\{ \begin{array}{c} q \\ \dot{q} \end{array} \right\} \tag{3}$$

and the plant matrix, A, and input matrix, B, are given by

$$A = \left[\begin{array}{cc} 0 & I \\ -\omega^2 & -\phi^T \gamma^p K^p \phi - \phi^T \gamma^s K^s \phi \end{array} \right] \tag{4}$$

and

$$B = \left\{ \begin{array}{c} 0 \\ \phi^T b \end{array} \right\} \tag{5}$$

and ϕ is the matrix of normalized undamped modes.

A block diagram for the systems under study is shown in Figure 1. The plant (Box A) consists of inert truss members, concentrated masses, viscoelastically damped members, and the unenergized active members. Sensor measurements are available and are given by (Box B)

$$y = C_s Z + D_s u \tag{6}$$

where the C_s matrix locates sensors on the structure. Closing the local loops around the active members activates the compensator (Box D in Figure 1) which consists of the A_c, B_c, and C_c matrices. Compensator voltage degrees-of-freedom are given by

$$\dot{V} = A_c V + B_c y \tag{7}$$

The specific elements of each of the A_c and B_c matrices depend on the control law being used. In this work, either a third order Strain Rate Feedback (SRF) or a fourth or sixth order Positive Position Feedback (PPF) compensator was tied around each active member.

Closing the loops around the active members results from combining equations (2), (6), (7), and the feedback relation

$$u = -C_c V \tag{8}$$

to give

$$\dot{X} = \tilde{A} X + \tilde{B} u \tag{9}$$

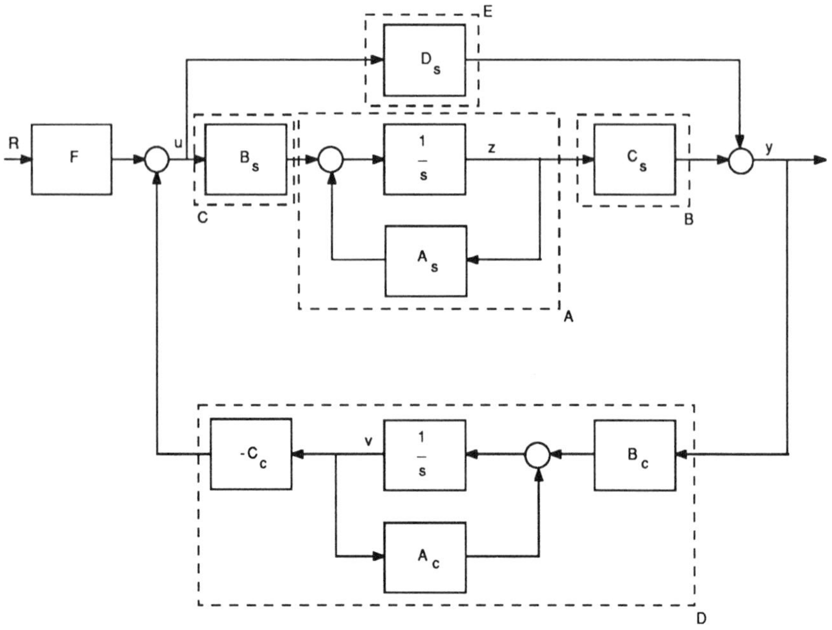

Figure 1: System Block Diagram

where the vector X includes both modal coordinates and compensator voltages. Solution of equations (9) in the frequency (or time) domain yields the closed loop system response.

The design variables and optimization variables for each type of element in the system are shown in Figure 2. For the inert truss members, the tube diameter and wall thickness were taken as design variables while the reciprocal of the cross sectional area was used as the optimization variables. For the passive members, the inside tube diameter and wall thickness, viscoelastic material thickness, and constraining layer thickness were the design variables. The optimization variables were chosen as the reciprocal of the areas of the inside tube, the viscoelastic material, and the constraining layer. For the active members, the inside dimension and wall thickness of the member were taken as design variables whereas the reciprocal of the area was used as the optimization variable. Design and optimization variables for the local compensator wrapped around each active member were the filter frequency and damping ratio and the overall compensator gain.

A mass penalty of 100% of the structural mass of the passive member was added to account for thermal control hardware. In addition, a mass penalty of 200% of the structural mass of the active member was added to

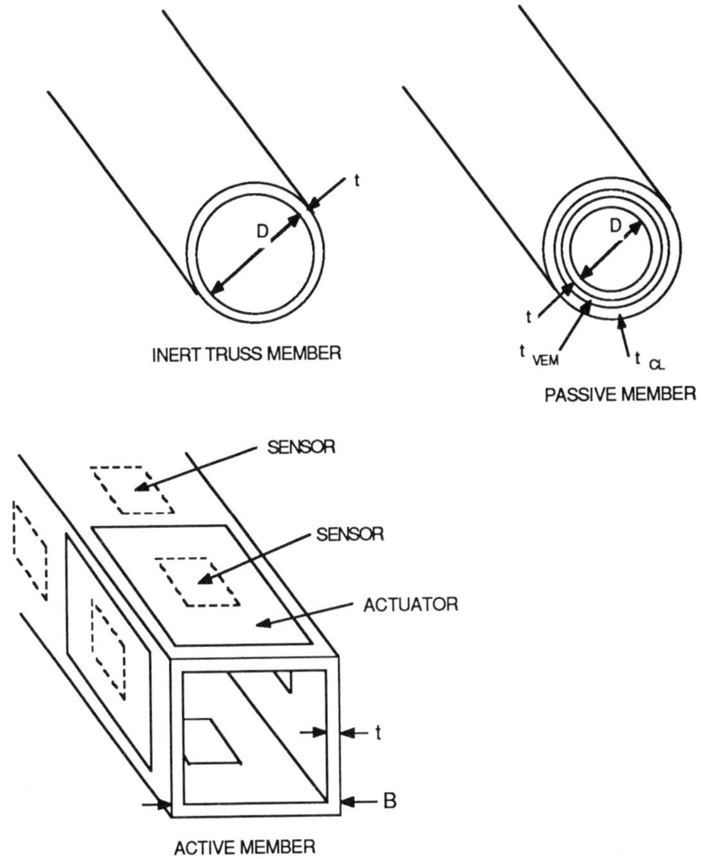

Figure 2: Design Element Details

account for thermal control, electronics, and power consumption hardware.

OPTIMUM DESIGN PROBLEM STATEMENT

For many space missions, a single performance index, such as line-of-sight (LOS) or pointing error, is the critical parameter for mission success. Minimization of this quantity maximizes system performance. However, additional constraints must be imposed on other quantities within the system. For example, line-of-sight may depend on the relative displacement and rotation of a number of optical elements within the system. The goal of the optimization process is to minimize LOS without allowing any of the optical elements to exceed their range of motion or hit their mounting stops.

For such missions, the optimization problem can be stated as

$$\min \ \mathrm{LOS}(\mathrm{d}, \mathrm{f}) \tag{10}$$

subject to

$$g(d, f) \leq 0 \tag{11}$$

along with the side constraints

$$d^l \leq d \leq d^u \tag{12}$$

where it is understood that d is the vector of design variables for the inert truss members, passive damping members, and active members as discussed in the previous section.

In general, the design problem stated in equations (10) through (12) is an implicit nonlinear mathematical programming problem. Furthermore, embedded within the optimum design problem statement is the placement of the passive and active members on the structure. The placement aspect of the design problem requires a computationally burdensome combinatoric optimization solution technique. It can be stated from a computational experience viewpoint that the design problem posed in equations (10) through (12) defies attempts at a direct solution.

SOLUTION METHODOLOGY

In this work, the solution procedure for the optimum design problem stated in equations (10) through (12) is broken into a heuristic subproblem and a formal subproblem. A pictorial description of the complete solution sequence to the original optimum design problem is shown in Figure 3. In the heuristic subproblem, locations for the passive and active members are determined by examining regions of high strain energy for those modes which contribute most to the objective function and the constraints. Once the locations of the passive and active members have been found, a formal subproblem based on equations (10) through (12) is solved. The formal subproblem replaces the implicit problem posed in equations (10) through (12) with the explicit approximate problem

$$\min \ \tilde{\mathrm{LOS}}(\mathrm{d}, \mathrm{f}) \tag{13}$$

subject to

$$\tilde{g}(d, f) \leq 0 \tag{14}$$

along with the side constraints

$$d^l \leq d \leq d^u \tag{15}$$

The $\tilde{LOS}$ and $\tilde{g}$ represent explicit first order Taylor series approximations for the objective function and constraints, respectively.

Solution of the implicit optimum design problem posed in equations (10) through (12) proceeds by solving a sequence of heuristic and formal subproblems. Each formal subproblem involves solving a sequence of approximate problems (stated in equations (13) through (15)).

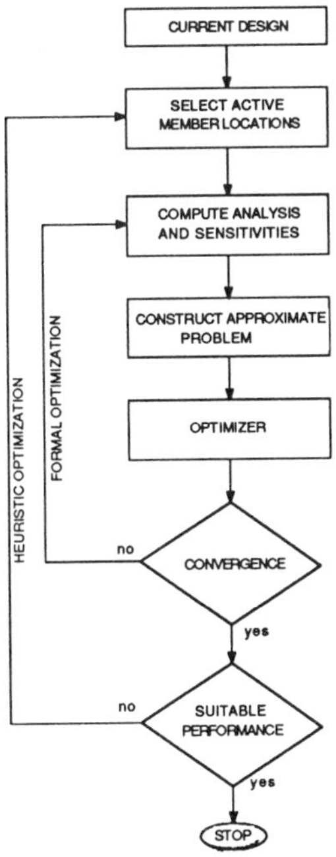

Figure 3: Solution Procedure Flow Diagram

EXAMPLE PROBLEM

An example problem is used to demonstrate the feasibility of the optimum design procedure. Figure 4 shows a potential concept for a Space Based Interferometer (SBI) [10]. The interferometer consists of an 11 meter tower with a telescope running down the center of it. Two 13 meter arms are attached at the base of the tower and support collecting telescopes at their tips. The 13 meter arms yield a baseline optical path length of 26 meters. Internal measurements are available from the laser metrology equipment mounted at the end of an 11 meter truss.

Maximum performance from the SBI is obtained when optical path length excursions from 26 meters are minimized during normal operations. Limited field of view of the collecting telescopes require tip and tilt of the telescopes to be below 5 μradians. A wide-band disturbance is applied at the central structure of the SBI to simulate reaction wheel out-of-balance

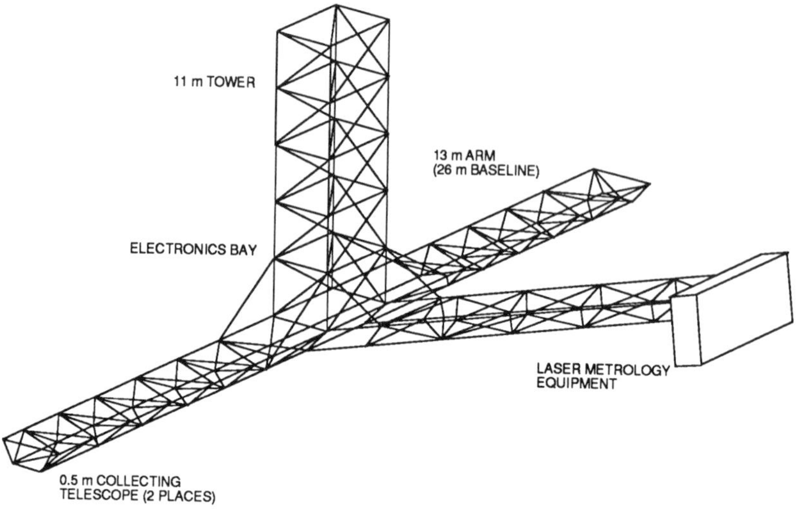

Figure 4: Space Based Interferometer Concept

and noise. The optimum design problem is to minimize optical path length excursions from 26 meters with upper bound constraints of 5 μradians on the relative tilt and tip of the collecting telescopes. An upper bound cap on the total system mass is also imposed to insure that the SBI is within the launch vehicle's capability for a given orbit.

The initial design of the SBI (with no active or passive damping augmentation) has a mass of 252 kgs. Strength and geometric considerations were used to arrive at sizes for the members. The initial mass of 252 kgs also serves as the upper bound mass cap during the optimization. The top traces of Figure 5 show the performance of the interferometer at the initial design. The modes at 7.2, 8.4, 16.4, 18.9, 27.7, and 36.8 Hz contribute to unacceptable optical lengths as well as relative tilt and tip motions of the collecting telescopes exceeding their bounds.

The heuristic subproblem for active and passive member locations was solved by placing these members in regions of high strain energy for the troublesome modes. Optimum values for the design variables were then found by solving the formal subproblem. Four to five approximate problem solutions were typically needed to achieve acceptable designs.

Table 1 gives the levels of damping that were obtained in each of the complex structural modes below 40 Hz as well as the values of the constraints and objective function. Fourteen active members and ten passive members with a total mass of 15 kgs were utilized in achieving these results. The bottom trace in Figure 5 shows the performance of the interferometer at the optimum design. The integrated design optimization methodology achieved a factor of 40 improvement in SBI baseline while regaining feasibility with respect to telescope tilt and tip responses.

Table 1: Initial and Optimum Frequencies and Damping Ratios

	Initial Design		Optimum Design	
Mode Number	Frequency (Hz)	$\zeta(\%)$	Frequency (Hz)	$\zeta(\%)$
1-6	0.0	0.0	0.0	0.0
7	4.4	0.2	4.3	10.7
8	6.5	0.2	6.3	2.9
9	7.2	0.2	6.9	6.9
10	8.4	0.2	8.0	3.6
11	8.5	0.2	8.3	4.1
12	12.9	0.2	12.4	11.9
13	16.4	0.2	15.3	13.1
14	18.9	0.2	17.7	10.1
15	19.1	0.2	18.1	20.1
16	21.7	0.2	20.7	0.5
17	24.5	0.2	23.8	0.5
18	27.7	0.2	26.8	11.7
19	29.0	0.2	27.2	1.8
20	36.8	0.2	33.4	11.7
Mass (kgs)	252		252	
Tip (μrad)	21		4.8	
Tilt (μrad)	48		4.2	
Baseline(μm)	3.2		.08	

CONCLUDING REMARKS

A two stage heuristic/formal optimization procedure has been described for the design of trusses with active and passive damping augmentation. The heuristic subproblem is concerned with the placement of the active and passive members in efficient locations. The formal subproblem then "sizes" all of the design variables with the fixed active and passive member locations. Approximation concepts are used to yield optimum designs in relatively few complete structural dynamic analyses. The design methodology is an integrated procedure which treats all design variables simultaneously.

The simultaneous treatment of the design variables takes advantage of the synergy that exists between the disciplines. By designing damping into the system early in the design process, superior performance can be achieved when compared to retrofit damping and/or purely structural solutions. Damping is designed into the lower modes without destabilizing the higher modes by designing the roll-off characteristics of the compensator at the same time as the structural characteristics.

REFERENCES

[1] Bronowicki, A.J. and Diaz, H.P., Analysis, Optimization, Fabrication and Test of Composite Shells with Embedded Viscoelastic Layers, Proceedings of Damping '89, West Palm Beach, Florida, February 8-10, 1989, pp. GCA-1-GCA-21.

[2] Bronowicki, A.J., Manning, R.A., and Mendenhall, T.L., TRW's Approach to Intelligent Space Structures, presented at the ASME Winter Annual Meeting, San Fransisco, California, December 13-15, 1989.

[3] Fanson, J.L., Blackwood, G.H., and Chu, C-C., Active-Member Control of Precision Structures, Proceedings of the AIAA/ASME/ASCE/AHS/ASC 30th Structures, Structural Dynamics, and Materials Conference, Mobile, Alabama, April 3-5, 1989, pp. 1480-1494.

[4] Gibson, W.C. and Johnson, C.D., Optimized Designs of Viscoelastic Damping Treatments, Proceedings of Damping '89, West Palm Beach, Florida, February 8-10, 1989, pp. DBD-1-DBD-23.

[5] Lust, R.V. and Schmit, L.A., Control-Augmented Structural Synthesis, AIAA Journal, Vol. 26, Jan. 1988, pp. 86-94.

[6] Thomas, H.L. and Schmit, L.A., Control Augmented Structural Synthesis with Dynamic Stability Constraints, Proceedings of the AIAA/ASME/ ASCE/AHS/ASC 30th Structures, Structural Dynamics, and Materials Conference, Mobile, Alabama, April 3-5, 1989, pp. 521-531.

[7] McLaren, M.D. and Slater, G.L., A Covariance Approach to Integrated Control/Structure Optimization, Proceedings of the 31st AIAA Dynamics Specialists Conference, Long Beach, California, April 5-6, 1990, pp. 189-205.

[8] Hedgepeth, J.M. and Mobren, M., Investigation of Passive Damping of Large Space Truss Structures, Proceedings of Damping '86.

[9] Fanson, J.L. and Caughey, T.K., Positive Position Feedback Control for Large Space Structures, Proceedings of the 28th AIAA Dynamics Specialists Conference, Monterey, California, April 9-10, 1987, pp. 588-598.

[10] Laskin, R.A., A Spaceborne Optical Interferometer: The JPL CSI Mission Focus, Proceedings of the NASA/DoD Controls-Structures Interaction Technology Conference, NASA CP 3041, San Diego, California, January 29-February 2, 1989, pp. 1-16.

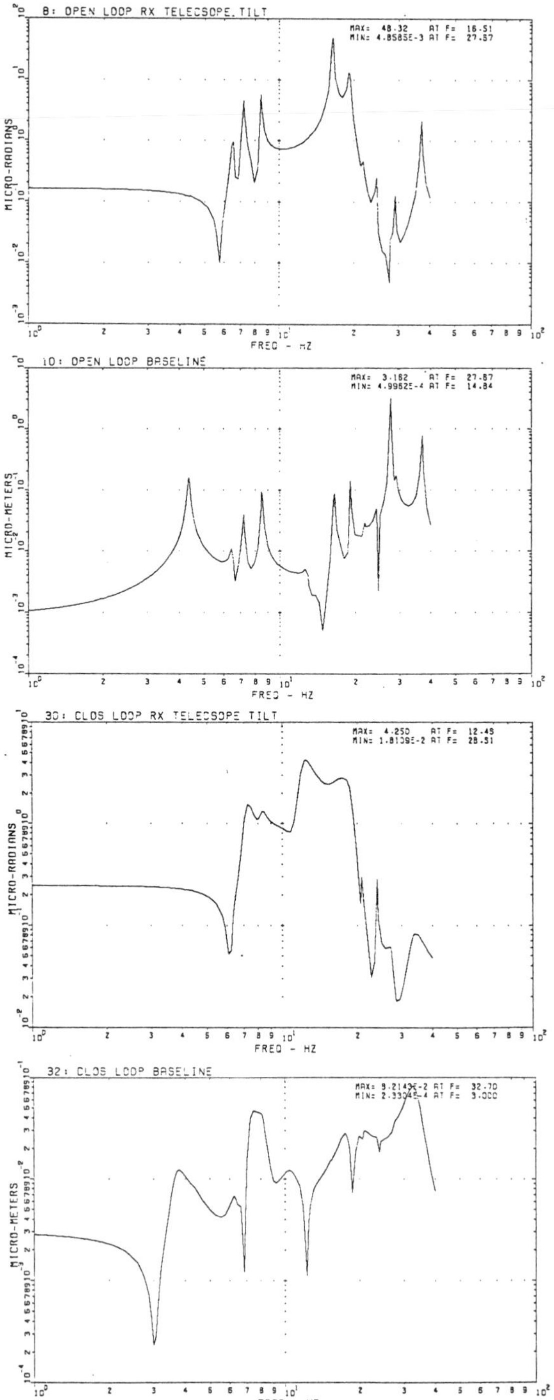

Figure 5: Initial and Final Baseline Response

Pareto Solution of an Inverse Problem in Elastoplasticity

L.M.C. Simões

Departamento de Engenharia Civil, Faculdade de Ciências e Tecnologia, Universidade de Coimbra, 3049 Coimbra, Portugal

ABSTRACT

An elastic-perfectly plastic discretized structure subjected to given proportional loads, undergoes displacements, some of which are measured. On the basis of this experimental data the yield limits and the hardening coefficients are sought, whereas the elastic properties are known. A number of possible ways of tackling this inverse problem are outlined and discussed. The present paper contains results on the sensitivity analysis for elastoplastic problems in the case of discrete structures and structures modelled by finite elements. This formulation covers situations where inaccuracies of practical significance with known statistical properties affect both the measurements and the modeling of the real system.

INTRODUCTION

Little attention has been paid to inverse problems in quasi-static elastoplasticity. However, there are practical reasons for the indirect identification (through in situ measurements on the overall response) of parameters which characterize the local resistance of complex systems, such as rock masses and other geotechnical formations. The calibration in this sense of an elastoplastic mathematical model embodying the Mohr-Coulomb yield criterion has been investigated[1]. A quite natural measure of discrepancy between the measured and the theoretical displacements is provided by the Euclidean norm of the difference vector. The minimization of this error with respect to the parameters appears to provide a way of identifying these parameters. Direct search techniques have been applied but involve by its very nature the solution of a sequence of analysis problems of the system for given parameters and, hence, turns out to be generally time consuming and costly. This special problem of identification under suitable hypothesis of piecewise linear yield surfaces and no local unstressing under increasing loads is amenable to the minimization of a convex quadratic function under linear and complementary constraints. Exploiting this circumstance, the resistance identification was reduced to a particular problem in nonconvex quadratic programming[2]. However, this algorithm is limited to structures where the hardening coefficients are assumed as constants.

The formulation proposed here treats the hardening moduli as parameters to be identified (together with the yield limits). In the inverse problem, it is convenient to have a suitable method for obtaining the sensitivity of the elastoplastic deformation, ie: directional derivatives of response with respect to variation of design parameters. The sensitivity result can be used in its own right for the solution of the inverse problem and a sequential quadratic programming algorithm is suggested.

Alternatively, this mathematical program is set as a multicriteria optimization and a Pareto solution is sought. By using the maximum entropy formalism a solution may be found indirectly by the unconstrained minimization of a scalar function which is both continuous and differentiable and thus considerably easier to solve. The post-optimality analysis also shows the sensitivity of the parameters to identify with respect to perturbations of the measured displacements. The procedure developed is tested by means of a 50 element elastoplastic beam on a elastoplastic unilateral foundation.

THE ANALYSIS PROBLEM

The problem of elastoplastic analysis of structures modelled by finite elements can under the usual assumptions of small displacements and deformations be formulated as quadratic programming problems. For the sake of simplicity, reference will be made to truss-like structures. The matrix relations which govern the elastoplastic response of these structures are known to cover implicitly, just by re-interpretation of symbols, a broader category of discrete structural models of continua with piecewise yield locci. For the ith element, a reversible (path independent or holonomic) stress-strain relation of the type depicted in Fig.1 is fully described analytically by the equations:

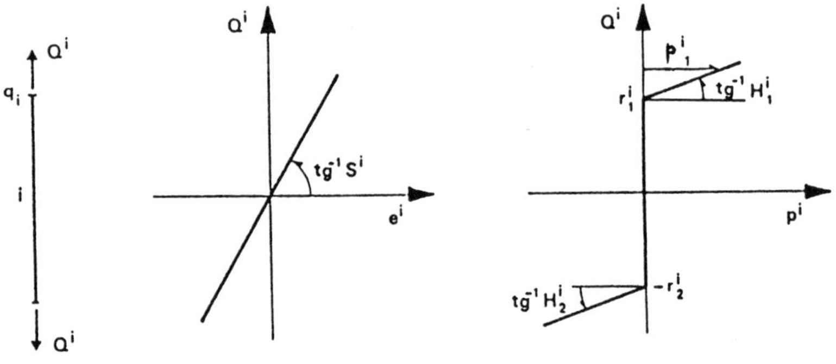

Figure 1

$$q^i = e^i + p^i = (S^i)^{-1} Q^i + p_1{}^i - p_2{}^i \qquad (1a)$$
$$\phi_1{}^i = - Q^i + (r_1{}^i + H_1{}^i p_1{}^i) \geq 0 \qquad (1b)$$
$$\phi_2{}^i = Q^i + (r_2{}^i + H_2{}^i p_2{}^i) \geq 0 \qquad (1c)$$
$$p_1{}^i, p_2{}^i \geq 0 \quad ; \quad \phi_1{}^i p_1{}^i = 0 \quad ; \quad \phi_2{}^i p_2{}^i = 0 \qquad (1d)$$

where q^i is the total generalized strain, S^i denotes the elastic stiffness, Q^i the generalized stress, $p_j{}^i$ are non-negative measures of the inelastic strain termed plastic multipliers. $\phi_j{}^i$ are auxiliary variables, referred to as yield functions. $r_j{}^i$ and $H_j{}^i$ are nonnegative constants, which define the yield limits and the hardening modulus for each i and j=1,2, respectively. Plastic flow implies energy dissipation and hence nonnegative $p_j{}^i$. It can occur in a yield mode if and only if the corresponding yield limit is reached.

Holonomic elastoplastic deformability laws which are represented, for one-component two-yield-mode elements, by the relation set (1) will be considered here in finite (not incremental) quantities. Such laws cover both truly reversible nonlinear elastic cases and irreversible elastoplastic situations susceptible to the non-local-unstressing hypothesis under proportional loads. The relations for i=1,..,m will be assembled for convenience in the following matrix relations,

$$q = (S)^{-1} Q + N p \tag{2a}$$

$$\phi = - N^t Q + H p + r \geq 0 \tag{2b}$$

$$p \geq 0 \quad ; \quad \phi^t p = 0 \tag{2c}$$

where S and H are diagonal matrices and N is a diagonal [I -I] matrix. Let u and F denote respectively vectors of the displacements of the free nodes (n degrees of freedom) and of the corresponding given independent nodal loads; Q and q will represent the m-vectors of generalized stresses and strains, respectively. The geometric compatibility and equilibrium equations read:

$$q = C u \tag{3}$$

$$C^t Q = \alpha F \tag{4}$$

where C is a m by n matrix which depends only on the given layout of the structure and α is the load factor. For structures described by an elastoplastic stress-strain law with workhardening and for a given α, the resulting stress vector Q is given by the minimizer of,

$$\text{Min } 1/2\, Q^t S^{-1} Q + 1/2\, p^t H p \tag{5a}$$

subject to,

$$C^t Q = \alpha F \tag{5b}$$

$$\phi = - N^t Q + H p + r \geq 0 \tag{5c}$$

$$Q \text{ real} \; ; \; p \geq 0 \tag{5d}$$

This problem has a solution if the design makes the structure capable of carrying the given loads. The dual of (5) is the convex quadratic program,

$$\text{Min } 1/2\, [u^t \;\; p^t] \begin{vmatrix} C^t S C & - C^t S N \\ - N^t S C & N^t S N + H \end{vmatrix} \begin{vmatrix} u \\ p \end{vmatrix} + [\alpha F^t \;\; r^t] \begin{vmatrix} u \\ p \end{vmatrix} \tag{6a}$$

subject to, u real ; $p \geq 0$ (6b)

By substitution of vectors q and Q, the relationship (2a), (3) and (4) lead to the following expression for the displacements:

$$u = \alpha u^e + G p \tag{7a}$$

where,

$$u^e = K^{-1} F \; , \; K = C^t S C \;\text{ and }\; G = K^{-1} C^t S N \tag{7b}$$

The vector u^e represents the elastic displacement vector and matrix G transforms vector p into the vector of plastic displacements. By setting,

$$Q^e = S C K^{-1} F \tag{8a}$$

$$A = H + N^t S N - G^t K G \tag{8b}$$

one obtains the quadratic program,

$$\text{Min } 1/2\, p^t A p - p^t (\alpha N^t Q^e - r) p \tag{9a}$$

subject to, $p \geq 0$ (9b)

that is equivalent to (6). At the optimum solution of the elastoplastic analysis problems (8) and (11) all the matrices and vectors involved are differentiable. Moreover the active constraints columns are linearly independent.

SENSITIVITY ANALYSIS

Elastoplastic analysis problems formulated as quadratic programming problems involve energy functionals, equilibrium and yield constraints that depend on the structural data and loading. The problem of determining the variation of structural response subject to variation of these parameters is considered in this section. A general result for discrete structures is presented and implications for the elastoplastic inverse problem are discussed. The general parametric quadratic program in the form,

$$\text{Min } \psi(x,\varepsilon) = 1/2\, x^t Q(\varepsilon)\, x - f(\varepsilon)^t x \tag{10a}$$

subject to,

$$A_1(\varepsilon)^t x - b(\varepsilon) = 0 \tag{10b}$$

$$A_2(\varepsilon)^t x - b(\varepsilon) \leq 0 \tag{10c}$$

will be considered, where ε is a real positive parameter, x are real and the matrices $Q(\varepsilon)$ are symmetric and positive definite. Also, it is assumed that the matrices $Q(\varepsilon)$, $A(\varepsilon)^t = [A_1(\varepsilon)^t \; A_2(\varepsilon)^t]$ and the vectors $f(\varepsilon)$, $b(\varepsilon)$ are differentiable at $\varepsilon = 0$, with derivatives Q', A' and f', b'. The Lagrange multipliers for problem (10) are given by the dual problem,

$$\text{Min } \psi(x,\varepsilon) = 1/2\, \mu^t P(\varepsilon)\, \mu - g(\varepsilon)^t \mu \tag{11a}$$

with,

$$\mu_1 \text{ real} \quad ; \quad \mu_2 \geq 0 \tag{11b}$$

where the matrix $P(\varepsilon)$ and the vector $g(\varepsilon)$ are differentiable with respect to ε and given by,

$$P(\varepsilon) = A(\varepsilon)^t Q(\varepsilon)^{-1} A(\varepsilon) \quad ; \quad g(\varepsilon) = A(\varepsilon)^t Q(\varepsilon)^{-1} f(\varepsilon) - b(\varepsilon) \tag{12}$$

In the dual problem $P(\varepsilon)$ and $g(\varepsilon)$ have derivatives P' and g', respectively. For $\varepsilon \geq 0$ small enough, the right-derivative (sensitivity) μ' of the Lagrange multipliers is given as the unique solution of,

$$\text{Min } 1/2\, v^t P(o)\, v - [g' - P' \mu(o)]^t v \tag{13a}$$

subject to,

$$v_i \text{ real} \quad \text{for} \quad \mu_1(o)_i \text{ real} \tag{13b}$$

$$v_i > 0 \quad \text{for} \quad a(o)_i^t x - b(o)_i = 0 \text{ and } \mu_2(o)_i > 0 \tag{13c}$$

$$v_i \geq 0 \quad \text{for} \quad a(o)_i^t x - b(o)_i = 0 \text{ and } \mu_2(o)_i = 0 \tag{13d}$$

$$v_i = 0 \quad \text{for} \quad a(o)_i^t x - b(o)_i < 0 \tag{13e}$$

The stationarity conditions can be used to find sensitivities of the primal variables x',

$$x' = Q(o)^{-1} [- Q' x(o) + f' - A' \mu(o) - A(o) \mu'] \tag{14}$$

It should be emphasized that this procedure gives the right-derivatives. The left derivatives are the symmetric solutions if the set of active constraints with zero Lagrange multipliers is empty. In elastoplastic analysis of structures with positive strainhardening, the solution is unique and the active yield constraints are associated with positive plastic multipliers. The optimization procedure for the elastoplastic analysis problem can be used to provide the sensitivities as well.

NUMERICAL EXAMPLE

As an example, the plastic deformations and associated sensitivities are computed for the elastoplastic beam on elastoplastic foundation represented in Fig.2. The model has 26 degrees of freedom and consists of 50 deformable elements. Precisely 24 hinges where the flexural deformability of the beam is lumped, and 26 springs, account for the foundation deformability. The structural model is subjected to live loads only.

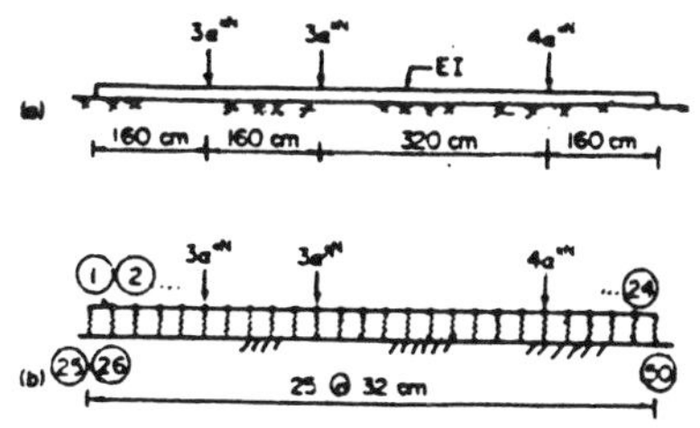

Figure 2

For the elastic hardening behaviour characteristic of the elements indicated in Fig.3, the hardening stiffness equals 5% of the elastic stiffness specified for each element.

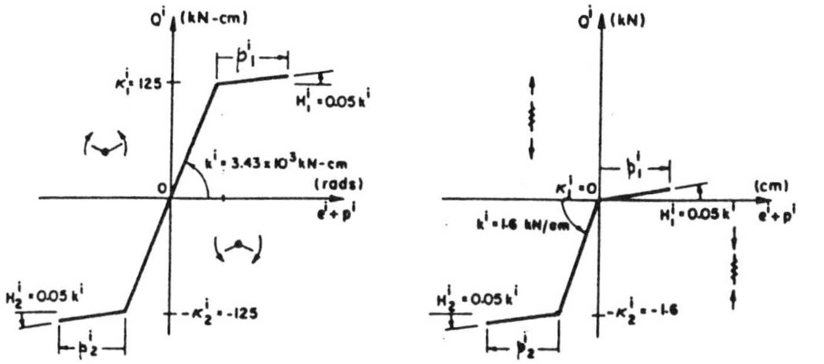

Figure 3

The deformation profile at $\alpha = 2.13$ for the beam foundation is shown in Fig.4, where the springs undergoing plastic strains are indicated as straight lines. It is worth noting that only the 20th hinge has been activated (in sagging bending), whereas the upper yield limit is just reached only in the 17th hinge, but no plastic rotation has developed here.

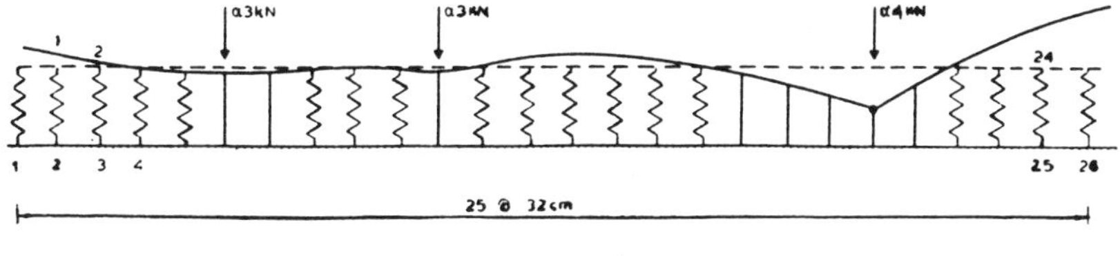

Figure 4

Fig.5 represents the nodal displacements obtained at different locations by using the sensitivity information for variations of the beam sagging moment, foundation yield limit and hardening moduli.

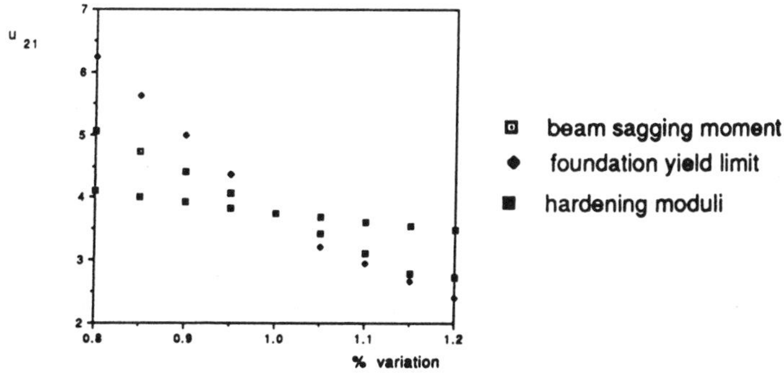

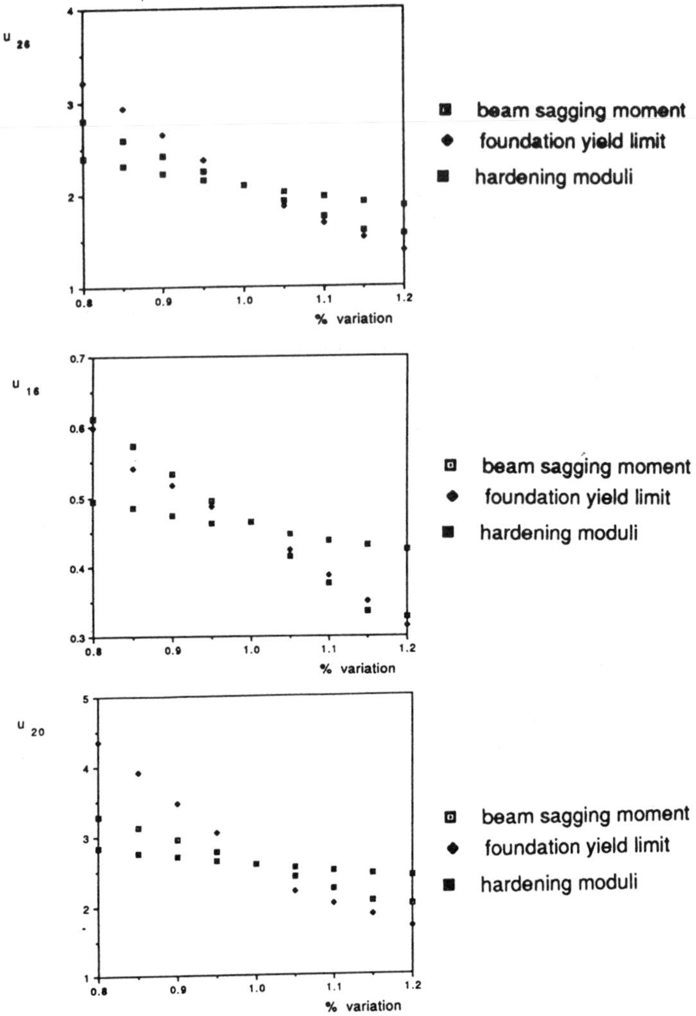

Figure 5

The approximations agree actual results given that the derivatives of the beam deformations do not change significantly. Since the quadratic coefficients for the sensitivity problem are the same as for the analysis problem, it is necessary to assemble only one stiffness matrix.

THE INVERSE ELASTOPLASTIC PROBLEM

The inverse elastoplastic problem can be described as follows. Whereas the elastic stiffnesses are all known, the strainhardening coefficients defining the diagonal matrix H and the element resistances r will be the parameters to identify. The yield limits depend linearly on some unknown parameters gathered in vector R,

$$r = B\,R \qquad (15)$$

For instance, the m structural members might be "a priori" subdivided in g groups of equal members, in each of which the 2 g resistances are unknown; then R becomes an identification or colocation matrix of order 2 m x 2 g with binary entries. The diagonal elements of matrix

A in (8b) are directly related to the hardening matrix H. Some (say d) displacements u_{hk}^{M} are measured in t tests performed on the structure along a loading process, say at levels $\alpha_1,...,\alpha_t$. This experimental information (u_{hk}^{m};h=1,...,d;k=1,...,t) on the overall response of the system to be exploited to determine the unknown parameters R governing the local elemental strength. Let u_{hk}^{c} indicate the calculated displacements, ie: the values which would be supplied by the governing relations set (mathematical model) under the same loads for the same displacement components subjected to measurements; generally depends on the values of parameters fed in those relations.

A quite natural measure of discrepancy between the measured and the theoretical displacements is provided by the Euclidean norm of the difference vector or its square (should measures exhibit different levels of confidence, then different weighting coefficients would be appropriate).

$$F = \Sigma_{h=1,d}\, \Sigma_{k=1,t}\, (u_{hk}^{m} - u_{hk}^{c})^2 \tag{16}$$

The minimization of this error with respect to the parameters appears to provide a way of identifying these parameters.

Entropy-based weighting coefficients

The maximum entropy formalism is a fundamental concept in information theory[3]. In general terms it is concerned with establishing what logical, unbiased inferences can be drawn from available information. On the basis of the information available, there is no logical justification for a criterion which unduly favours one specific coefficient rather than another. In view of Shannon's intrepretation of the entropy function it is entirely logical to maximize the entropy of the weighting coefficients w_{hk}. These are obtained by solving the maximum entropy mathematical problem:

$$\text{Max} \qquad S/K = -\Sigma_{h=1,d}\, \Sigma_{k=1,t}\, w_{hk} \ln w_{hk} \tag{17a}$$

$$\text{subject to,} \qquad \Sigma_{h=1,d}\, \Sigma_{k=1,t}\, w_{hk} = 1 \tag{17b}$$

$$\Sigma_{h=1,d}\, \Sigma_{k=1,t}\, w_{hk}\, g_{hk} = \varepsilon \tag{17c}$$

$$w_{hk} \geq 0 \tag{17d}$$

S is the Shannon entropy, K is a positive constant and the g_{hk} represent the levels of confidence (or the square of the discrepancy between measured and theoretical displacements).

Equation (17c) has an expected value of zero. The entropy maximization problem has an algebraic solution for w_{hk}:

$$w_{hk} = \frac{\exp[\beta\, g_{hk}]}{\Sigma_{h=1,d}\, \Sigma_{k=1,t}\, \exp[\beta\, g_{hk}]} \tag{18}$$

in which β, the Lagrange multiplier for Eq. (17c) is closely related to ε and can be found by substituting result (18) into Eq. (17c). It may be deduced that for ε to approach zero from above β must be chosen positive.

VARIOUS APPROACHES TO THE INVERSE PROBLEM

In this section a number of possible ways of tackling the identification problem formulated earlier is envisaged and briefly outlined. Let D denote a binary dxn matrix which selects, among the n displacement components, those subjected to measurements; thus, through (7a) the d vector of calculated values for the kth test can be expressed in the form,

$$u_k^c = D(\alpha_k u^e + G p_k) \tag{19}$$

Making use of eq.(19) in (16), the function to minimize becomes a quadratic form in the plastic multipliers only.

$$\text{Min } F = \Sigma_{k=1,t} (p_k^t M p_k + b_k^t p_k + c_k) \tag{20}$$

where,

$$M = G^t D^t D G \; ; \; b_k = 2 G^t D^t (\alpha_k D u^e - u_k^m) \; ; \; c_k = (u_k^{m\,t} - \alpha_k u^{et} D^t)(\alpha_k D u^e - u_k^m)$$

It is worth noting that F is convex (as matrix M is positive semidefinite) and does not depend on the parameters R. These intervene in the minimization constraints, which are directly supplied by the formulation (2) on the analysis problem and read:

$$\phi_k = -\alpha_k N^t Q^e + A(H) p_k + B R \geq 0 \tag{21a}$$

$$p_k \geq 0 \quad ; \quad \phi_k^t p_k = 0 \tag{21b}$$

The constrained optimization problem to which the identification of the local resistance parameters has been cast, is characterized by a complementary constraint requiring that between a certain pair of corresponding variables at least one component must vanish. Its peculiarity rests on the fact that the constraints are all linear except the complementary condition. Besides the nonlinearity steming from the product A(H) p_k in (21a), the complementary condition makes the parameter identification problem nonlinear and nonconvex. For simplicity sake, the experimental data is assumed to be derived from a single loading condition in the following sections.

Nonconvex Parametric Quadratic Program

If the hardening coefficients in matrix H are assumed as given constants and the identification problem is concieved as a constrained minimization with respect to vectors p and R, the only real source of mathematical and numerical difficulties is represented by the complementary requirement. At first, a quite natural way of trying to circumvent this difficulty is to augment the objective function with the complementary condition, removing it from the constraints:

$$\text{Min } [p^t \;\; R^t] \begin{bmatrix} (M + \rho A) & \rho/2\, B \\ \rho/2\, B^t & 0 \end{bmatrix} \begin{bmatrix} p \\ R \end{bmatrix} + (b - \rho N^t Q^e)^t\, p + c \tag{22a}$$

$$\text{st,} \quad A p + B R \geq N^t Q^e \tag{22b}$$

$$p \geq 0 \tag{22c}$$

where p is a real positive constant. It can be stated that problem (20) subject to (21) is equivalent to the nonconvex QP problem (22) in the sense that both minimum points coincide. Ref.[2] provides a two-phase method for obtaining the global optimum. Now, if the hardening coefficients are regarded as parameters to be identified, then inequality constraints (22b) become nonlinear and the nonlinearity of the objective function increases in view of the parametric nature of A(H). The sensitivity analysis procedure described previously for the general convex parametric QP can be adapted here to identify a (at least) local solution.

Sequential Quadratic Programming

The minimization of the quadratic error function F in (16) with respect to both R and H can be done numerically. The objective function F(R,H) is given implicitly via the elastoplastic analysis programs. $u(R^o,H^o)^c$ is the displacement vector of the structure characterized by the yield limits R and hardening moduli H in the dual program (6) (or in (9)). Therefore, F is a continuously differentiable function of u^c. The derivatives of F with respect to the parameters can be computed by using the results of the parametric quadratic program (13).

$$\text{Min } F = \Sigma_{h=1,d} \left[u_h^m - u_h(R^o,H^o)^c \; \Sigma_{j=1,g_1} \left(\frac{\partial u_h^c}{\partial R_j}\right)_o \Delta R_j + \Sigma_{j=1,g_2} \left(\frac{\partial u_h^c}{\partial H_j}\right)_o \Delta H_j\right]^2 \quad (23)$$

For this problem where there are only simple range constraints imposed on $(\Delta R_j, \Delta H_j)$, computational results can be obtained by use of standard sequential quadratic programming. By letting $x^t = [R^t H^t]$, (23) can be written,

$$\text{Min } F = \Sigma_{i=1,n} \Sigma_{j=1,n} \Sigma_{h=1,d} \left(\frac{\partial u_h^c}{\partial x_i}\right)_o \left(\frac{\partial u_h^c}{\partial x_j}\right)_o \Delta x_i \, \Delta x_j \; +$$

$$+ \; \Sigma_{i=1,n} \Sigma_{h=1,d} \left(\frac{\partial u_h^c}{\partial x_i}\right)_o [\, u_h^m - u_h(x^o)^c \,] \;\; \Delta x_i \; + \; \Sigma_{h=1,d} \, [\, u_h^m - u_h(x^o)^c \,]^2 \quad (24)$$

Solving (24) for particular numerical values of $u_h(x^o)^c$ and $(\partial u_h^c / \partial x_i)_{x^o}$ forms only an iteration. The solution vector x^1 of such an iteration represents a new set of parameters which must be analysed and gives new values for $u_h(x^1)^c$ and $(\partial u_h^c / \partial x_i)_{x^1}$ to replace those corresponding to x^o in (24). Iterations continue until changes in the design variables become small.

Minimax Formulation

The information provided by the plastic multipliers and yield functions reduces the dependence on the stipulated bounds for the parameter changes in each iteration. The mathematical programming algorithm described in this section consists of solving a mini-max problem that is found by rewriting the objective function (16) and the constraints (21) as goals in a normalized form. If $\underline{F}$ represents a reference error, and $\underline{p}$, $\underline{\phi}$ the corresponding plastic multipliers and yield functions, relation (16) becomes,

$$\Sigma_{h=1,d} \, (u_h^m - u_h^c)^2 \leq \underline{F} \quad \Rightarrow \quad g_1(x) = \frac{(u_h^m - u_h(x)^c)^2}{\underline{F}} - 1 \leq 0 \qquad (25a)$$

The sign constraints which impose limits on the variations of nonzero multipliers $\underline{p}$ and yield functions $\underline{\phi}$ lead to,

$$g_2(x) = -\frac{\Delta p(x)}{\underline{p}} - 1 \leq 0 \qquad (25b)$$

$$g_3(x) = -\frac{\Delta \phi(x)}{\underline{\phi}} - 1 \leq 0 \qquad (25c)$$

where ϕ is given by (2b). The sensitivity result is obtained by considering the primal as well as its dual problem and sensitivities are given for both the primal and dual variables. The components of $\partial p/\partial x$ and $\partial \phi/\partial x$ are obtained from the piecewise solution of the quadratic program and hence are discontinuous piecewise functions of x. The complementary constraint is checked after each iteration.

The problem of finding values for the cross sectional areas which minimize the maximum of the goals has the form,

$$\min_x \max_{k=1,c} (g_1, \ldots, g_k \ldots g_c) \tag{26}$$

and belongs to the class of minimax optimization. The procedure used to solve this problem is a recently developed entropy-based appoach. The minimax problem (26) is discontinuous and non-differentiable, of which both attributes make its numerical solution by direct means difficult. In ref.[4] it is shown that the minimax solution may be found indirectly by the unconstrained minimization of a scalar function which is both continuous and differentiable, and is thus considerably easier to solve:

$$\mathrm{Min}_x \ \mathrm{Max}_{k\in K} < g_k(x) > \ = \ \mathrm{Min}\,(1/\rho)\log\{\Sigma_{k=1,c}\exp[\rho\ g_k(x)]\} \tag{27}$$

over variable x with a sequence of values of increasingly large positive $\rho \geq 1$.The scalar function minimization allows the use of algorithms for convex optimization. The strategy adopted was to solve the implicit optimization problem by means of an iterative sequence of explicit approximation models. An explicit approximation can be formulated by taking Taylor series expansions of all the goal functions in problem (27), truncated after the linear term for $g_2(x)$ and $g_3(x)$ and the quadratic term in the case of $g_1(x)$:

$$\mathrm{Min}\,(1/\rho)\log\{\Sigma_{k=2,c}\exp\rho\,[g_k(x^o)+\Sigma_{i=1,N}\left(\frac{\partial g_k}{\partial x_i}\right)_o \Delta x_i +$$

$$\exp\rho\,[g_1(x^o)+\Sigma_{i=1,N}\left(\frac{\partial g_1}{\partial x_i}\right)_o \Delta x_i + \frac{1}{2}\ \Sigma_{i=1,n}\ \Sigma_{j=1,n}\left(\frac{\partial g_1}{\partial x_i}\right)_o\left(\frac{\partial g_1}{\partial x_j}\right)_o \Delta x_i\,\Delta x_j \tag{28}$$

that is solved iteratively until changes in the design parameters become small.

Discussion

In order to identify R+H parameters, one obviously needs at least (R+H) t independent measured values. This condition is not sufficient: if in the experiments the plastic properties are not activated, the yield limits are nowhere reached and lower bounds on the parameters are provided by the elastic stresses related to the solution.Practically, the solution is highly dependent upon the number and positions of the measurements.Measurements should be made where the discrepancies are potentially higher: in the neighborhood of the most critical points both in compression and tension in an unsymmetrical way.

Normally the number R+H of independent parameters to identify is much less than the number (2 m) t of the variables p which characterize the dimensions of the nonconvex quadratic parametric program. Clearly the size of this identification problem increases almost proportionally with the number of t different tests and should be ruled out. This is in contrast with both the sequential quadratic programming technique and the minimax formulation which are rather insensitive to the test number. In the sequential quadratic programming approach, the nodal displacements u are approximated by first order Taylor series at the current parameters (R^o,H^o) yielding a quadratic program in the parameter changes $(\Delta R, \Delta H)$. Since the solution largely depends on the assumed bounds, move limits should be small to avoid an erratic behaviour. The minimax problem is less dependent on the bounds stipulated for the parameter changes, providing a smoother convergence. Hence the number of cycles of analysis/optimization in SQP is potentially greater than in the case of the the minimmax approach. On the other hand, the quasi-Newton algorithm used to solve (28) is less efficient than the routine used for quadratic programming. The algorithm used to minimize quadratic functions is subject only to upper and lower bounds employed for the elastoplastic analysis, sensitivity analysis and SQP uses partial LDL^t factorization. The computational times are comparable to those required for the factorization of the quadratic coefficients matrix.

INFLUENCE OF INACCURACIES IN MEASUREMENTS

As long as it is 'a priori' known that there is no inaccuracies in measurements and modelling of the identification problem treated in purely deterministic terms, the minimum of the error function is zero. In real-life situations inaccuracies of practical significance affect both the measurements and the modelling of the real system. One of the procedures for filtering such 'noises' is the post-optimality analysis of the quadratic program giving the error function (24) with respect to each parameter in turn. If the coefficients a_h represent such sensitivities, the parameter changes can be obtained by the linear approximation:

$$\Delta x = \Sigma_{h=1,d}\, a_h\, \Delta u_h \qquad (29)$$

If the inaccuracies have known statistical properties, the mean and variance of the parameter changes are given by,

$$\mu_{\Delta x} = \Sigma_{h=1,d}\, a_h\, \mu_{\Delta x_h} \qquad (30a)$$

$$\sigma_{\Delta x}^2 = \Sigma_{h=1,d}\, a_h^2\, \sigma_{\Delta x_h}^2 + \Sigma_{h=1,d}\, \Sigma_{i=1,d;i\neq h}\, a_h\, a_i\, \rho_{hi}\, \sigma_{\Delta x_h}\, \sigma_{\Delta x_i} \qquad (30b)$$

where ρ_{ij} is the correlation coefficient between Δu_h and Δu_i. Inaccuracies due to other instrumental errors and modeling can be treated in a similar way.

NUMERICAL EXAMPLE

The vertical displacements of the hinge points of the beam on elastoplastic foundation represented in Fig.2 are assumed to be measurable and the two yield limits of the hinges, the compressive strength of the springs and the hardening coefficients are to be identified on the basis of those measurements. In principle it should be possible to identify the yield limits, if the corresponding yield modes are activated and if the number of measured displacements is not less than the number of unknowns. Different starting points were used and both the sequential quadratic programming and the minimax formulation gave results in 2-3 iterations. As stated before, the quality of these solutions depends on the location and number of the measurements and this can be seen in the following Table:

Mesured Displacements	Foundation yield limit	Beam sagging moment	hardening moduli
All	0.0%	0.1%	0.1%
17,19,21,23,24,26	0.0	0.1	0.1
14,15,16,17	0.4	1.0	5.4
17,18,19,21	0.1	0.3	1.6
19,20,21,22	0.0	0.1	0.1
19,20,21	0.0	0.1	0.1
17,18,21	0.1	0.3	1.8

It should be noted that the hogging yield limit of the beam cannot be identified correctly on the basis of the available information, since there are no positive plastic rotations in the simulated experiment.

Since the parameter identification process depends on the precision of the measures, it is appropriate to test the sensitivity of the method with respect to possible measurement errors. An investigation has been carried out by giving a 5% increment to every measurement. The resulting errors in the parameters are 0.6, 1.3 and 3.4%, respectively. It can be seen that the effect on the parameter identification is much less than the order of magnitude of the measurement errors in the case of the foundation yield limit and the sagging limit moment and smaller in the case of the hardening coefficient. For some other combinations of measurement discrepancies, the error involving this parameter might be higher, because the displacements are rather insensitive with respect to hardening moduli changes (as can be seen in Fig.5). In order to partially simulate measurement errors the generated 'measured' displacements were rounded off at the first decimal. Also in this case the disturbances on the identified parameters proved to be negligible (0.7, 0.1 and 0.1%, respectively). The procedure has also worked in a subsequent numerical test, where the number of parameters to be identified is increased by also assuming the zero tensile yield limit of the spring to be unknown.

CONCLUSIONS

The present paper contains results on the sensitivity analysis for elastoplastic problems in the case of discrete structures and structures modelled by finite elements. The result shows that determination of the sensitivities can be based on the solution of an associated quadratic programming with unchanged quadratic term but with changed linear terms and constraints which are given by the derivatives of the matrices involved as well as by the solution of the primal analysis problem.The inverse problem of identifying yield limits on the basis of information on displacement response to given loads has been tackled here in the context of discrete structural models with holonomic elasto-plastic piecewise-linear laws governing the local deformability. As the dual variables in the elastoplastic analysis problems have a physical interpretation (eg, displacements and plastic strains) the sensitivities for the dual variables are also of interest in the parameter identification context. Various solution procedures resting on mathematical programming methods, all capable of exploiting the peculiar mathematical features of the proposed formulation have been devised and discussed. Inaccuracies primarily due to the approximations embodied in the model and instrumental'noises' affecting the measures have been considered by the procedure employed for the sensitivity analysis. The examples reported show that the yield limits and hardening moduli are relatively insensitive with respect to perturbation of the measured displacements, provided these are chosen in location and number such that they are affected by the local yielding processes.

ACKNOWLEDGEMENTS

The author wish to thank the financial support given by JNICT (Junta Nacional de Investigação Cientifica e Tecnológica, Proj. 87 230) and Calouste Gulbenkian Foundation.

REFERENCES

1. Gioda, G. e Maier, G. "Direct Search Solution of an Inverse Problem in Elastoplasticity: Identification of Cohesion, Friction Angle and "in situ" Stress by Pressure Tunnel Tests", Int. J. Num. Meth. Eng., 15, 1823-1848, 1980.
2. Maier, G. e Gianessi, F. "Indirect Identification of Yeld Limits by Mathematical Programming", Eng. Struct., 4, 86-98, 1982.
3. Shannon, C.E. "A Mathematical Theory of Communication", Bell System Technical Journal, 27, 379-428, 1948.
4. L.M.C. Simões and A.B.Templeman "Entropy-based Synthesis of Pretensioned Cable Net Structures", Eng. Optimization, Vol.15, 121-140, 1989.

SECTION 4: SENSITIVITY ANALYSIS AND SHAPE OPTIMIZATION

A General Related Variational Approach to Shape Design Sensitivity Analysis

J. Unzueta, E. Schaeidt, A. Longo, J.J. Anza

Analysis & Design Department, LABEIN, Cuesta de Olabeaga, 16, 48013, BILBAO, Spain

ABSTRACT

A continuous approach with boundary representation for 2D + Axisymmetric potential problems is presented to develop shape sensitivities. The funcions to be derived are boundary functionals involving the gradient of the state variable. BEM is chosen as analysis tool for solving both the original problem and the one derived (a direct method is followed in calculating sensitivities). A numerical example is presented, illustrating the capabilities of the proposed solution procedure.

INTRODUCTION

In optimum design methodologies, the sensitivity analysis is defined as the effect of a change in the current design on a response functional. This calculus is a central step in formulating the mathematical model of an optimization problem; it can be used for approaching the cost function and constraints in terms of the design variables, providing the essential gradient information to compute a search direction in the optimization process.

The gradients are also important in their own right as they represent a trend for the problem performance. Thus, the sensitivity analysis has recently emerged as a powerful design tool, not only from the standpoint of the role of derivatives in optimization, but also from the point of view of using sensitivity information in a computer-aided engineering environment for interactive design.

Many papers have come up over the past few years in the field of shape design sensitivity analysis (see Haftka [1] and references quoted therein). There have been two common approaches contributing to this topic, namely the discrete approach and the continuous approach.

The first approach, a natural adaptation of sizing design sensitivity, derives the implicit discretized equilibrium equation. Since there is not an explicit relation of shape variables for the terms appearing in that equation, generally speaking, the calculus of each derivative can be more expensive than the original analysis. Moreover, the derivatives obtained have a spurious component, which reflects the changing accuracy of the solution when the mesh is distorted.

Using the BEM as analysis tool (Defourny [2], Kane[3], Saigal [4]), this technique requires the numerical integration of a new class of sensitivity Kernels, which exhibit a singular behaviour similar to that of the analysis fundamental solution. Nevertheless, the implementation of this approach requires a significant programming effort to develop the numerical integration.

The second approach uses a continuous model of the problem and the material derivative method of continuum mechanics to obtain computable expressions for the effect of shape variation on the functionals arising in the shape design problem.

Two equivalent formulations, either on the boundary or in the domain, can be obtained for the continuous approach depending upon the way of calculating the derivatives of the functionals. The expressions reached through the boundary approach are functions of the partial shape derivatives of state variables. The domain approach, on the other hand, includes the total shape derivatives.

To evaluate the sensitivity equation derived by the boundary approach, only the design velocity field along the varied boundary is required. This represents a considerable computational saving compared to the domain approach, in which the design velocity field over the entire domain needs to be specified.

As it is done in this paper, the boundary approach is usually employed for shape sensistivity analysis in conjunction with the boundary element method for analysis. Thus, inherent numerical difficulties in MEF regarding lack of accuracy for the response over the boundaries can all be avoided.

The continuous shape sensitivity is obtained in terms of state variables, their shape derivatives and geometric changes (velocity field). The evaluation of those derivatives requires the response of an associated problem, for each shape variable by directly differentiating the governing equations - direct method -, or an adjoint problem for each functional which removes the shape derivatives from the formulation - adjoint method.

Since it has been concluded that the adjoint method is more efficient than the direct method if the number of functionals is less than the number of design variables, and active constraints selection algorithms are included in most optimization codes, developments in shape sensitivities analysis were oriented towards the adjoint method. Examples of that can be found in references Choi [5], Haug [6], Meric [7], Mròz [8], Kwak [9].

Although the formal adaptation of the adjoint technique is conceptually straightforward, major computational difficulties arise when evaluating sensitivities of functionals at discrete boundary points, or sensitivities of boundary functionals involving spatial gradients of state variables. The problem stems from the fact that the adjoint solution can not be expressed in terms of the boundary element formulation, because they give rise to infinite integrals (i.e., in elasticity the adjoint solution would correspond to a concentrated force and moment for displacement and stress sensitivities, respectively).

Nevertheless, the direct method, which does not present these drawbacks because it calculates sensitivities of the entire response field regardless of the functionals, has been mostly ignored and few references (Dems [10-12]) using this approach can be found.

In this paper, shape sensitivity analysis of boundary functionals defined as either integrals involving the gradient of the state variable or gradient functional applied on isolated points of the boundaries, is presented. The main claim of the developments reached is, besides treatment of the above mentioned functionals, its generality. The domain can show C^k regular boundaries, because the tangential velocity is considered, there are not restrictions on the movements of the boundaries (Dirichlet, Neumann and interfaces), functionals can be applied over whichever varied boundary and a closed or open piece of it containing non-singluar corners can be used to define the integrals. Punctual functionals can also be placed on non-singular corners. Finally, the formulation takes into account the possibility of dealing with orthotropic materials.

MATERIAL DERIVATIVE FOR SHAPE SENSITIVITY ANALYSIS

Since the domain Ω is to be varied, it is treated as a continuum moving with a "time like" parameter t in the material derivative formulation of shape design sensitivity analysis. The variations of a point $\underline{x}$ in the domain are expressed in terms of a velocity field $\underline{V}$, which defines the direction of movement of $\underline{x}$ on the nominal domain Ω to points $\underline{x}_t$ on a deformed domain Ω_t, given by the transformation,

$$\underline{x}_t = \underline{x} + t\underline{V}(\underline{x})$$

Then, the variation of a general differentiable function Ø with respect to a shape variation is expressed as the material derivative or total derivative of Ø at t = 0,

$$\dot{\phi} = \frac{d\phi_t(\underline{x}_t)}{dt}\Big|_{t=0} = \lim_{t\to 0}\frac{\phi_t(\underline{x}_t) - \phi(\underline{x})}{t} = \phi'(\underline{x}) + \underline{\nabla}\phi\underline{V}(\underline{x})$$

where,

ϕ_t $(\underline{x}_t)$ is the solution of the boundary value problem on Ω_t, evaluated at a point $\underline{x}_t$ that moves with t, and

$$\phi'(\underline{x}) = \lim_{t\to 0}\frac{\phi_t(\underline{x}) - \phi(\underline{x})}{t}$$

namely local derivative, is the variation of ϕ at point $\underline{x}$.

Material derivatives of the unit normal to a boundary

Let $\underline{n}$ be a normal vector to a boundary Γ. Its material derivative is (Zolésio [13]),

$$\dot{\underline{n}} = (\underline{n}D\underline{V}\underline{n})\underline{n} - D\underline{V}^T\underline{n}$$

where

$$D\underline{V} = \left(\frac{\partial V_i}{\partial x_j}\right)$$

and by simplicity, it will be understood that

$$\underline{a}B\underline{c} = \underline{a}^T.B.\underline{c}$$

Let $\underline{N}_o$ and $\underline{S}_o$ be two unitary extensions of the fields $\underline{n}$ and $\underline{s}$ respectively, then

$$\underline{V} = u\underline{N}_o + v\underline{S}_o \quad , \quad u = V_n, \ v = V_s$$

$$D\underline{V}^T\underline{n} = \underline{\nabla}u + vD\underline{S}_o^T\underline{n}, \quad \text{because } D\underline{N}_o^T\underline{n} = 0$$

$$(\underline{n}D\underline{V}\underline{n})\underline{n} = (\underline{n}\underline{\nabla}u)\underline{n} + v(\underline{n}D\underline{S}_o\underline{n})\underline{n}, \text{ because } D\underline{S}_o^T\underline{s} = 0$$

Therefore,

$$\dot{\underline{n}} = -\frac{\partial V_n}{\partial s}\underline{s} + V_s \underline{D}\underline{N}_o\underline{s} \qquad (1)$$

and,

$$\underline{n}' = -\frac{\partial V_n}{\partial s}\underline{s} - V_n \underline{D}\underline{N}_o\underline{n} \qquad (2)$$

Material derivatives of integral functionals

Let ψ_t be a functional defined over an open or closed boundary Γ_{At} piece-wise C^{k+1} regular on the varied domain Ω_t in R^2.

$$\psi_t = \int_{\Gamma_{At}} g_t(\underline{x}_t) d\Gamma_t \qquad (3)$$

Defining χ as a characteristic function,

$$\chi(\underline{x})|_\Gamma = \begin{cases} 1, \underline{x} \in \Gamma_A \\ 0, \underline{x} \notin \Gamma_A \end{cases}$$

the expression (3) has the form

$$\psi_t = \int_{\Gamma_t} \chi_t(\underline{x}_t) g_t(\underline{x}_t) d\Gamma \qquad (4)$$

where Γ_t is $\partial\Omega_t$.

The material derivative of ψ at Γ is,

$$\psi' = \int_{\Gamma_A} [g' + (\nabla g \underline{n} + gH) V_n] d\Gamma + \sum^{corners \Gamma_A} [\![g V_s]\!] + g V_s |_A^B \qquad (5)$$

H being the curvature of Γ, $[\![\ ]\!]$ a jump operator and A, B, the initial and final points of Γ_A.

In many cases, it will be interesting to formulate functionals weighted with the measure m_A of Γ_A,

$$\psi_t = \frac{1}{m_{At}} \int_{\Gamma_{At}} g_t(\underline{x}_t) d\Gamma_t \qquad (6)$$

also defined over an open or closed boundary Γ_{At} piece-wise C^{k+1} regular in R^2. The material derivative of ψ at Γ is,

$$\psi' = \frac{1}{m_A}[\int_{\Gamma_A} [g' + (\underline{\nabla} g \underline{n} + (g - \psi)H)V_n] d\Gamma +$$

$$+ \sum^{corners\Gamma_A} [\![(g - \psi)V_s]\!] + (g - \psi)V_s|_A^B] \qquad (7)$$

SHAPE SENSITIVITY IN ELECTROSTATIC PROBLEMS

Let Ω be an heterogeneous domain in R^2 over which an electrostatic phenomenon is defined, with Dirichlet, Neumann and interface boundary conditions. The governing equations are:

$$\begin{aligned}
&\nabla^2_k \phi = 0 && \text{in } \Omega_i \\
&\phi = \phi_o && \text{on } \Gamma^i_\phi \qquad (8) \\
&\underline{\nabla}_k \phi \underline{n} = q_o && \text{on } \Gamma^i_q \\
&\phi^i = \phi^j && \\
& && \text{on } \Gamma^{i/j}_I \qquad i, j = 1 \ldots\ldots n \\
&q^i + q^j = 0 &&
\end{aligned}$$

being,

n = number of subregions of Ω

$\Gamma^i = \partial\Omega^i$

$\Gamma^i_\phi \cup \Gamma^i_q \cup \Gamma^i_I = \Gamma^i$

$\underline{\nabla}_k = (K_x\, \partial/\partial x,\ K_y\, \partial/\partial y)$

K_x, K_y = orthotropic permitivities in Ω^i.

The functionals determining a design in electrostatics (and in most physical phenomenons) are functions of the state variable gradient on the boundaries, that is, punctual functionals,

$$\psi = g(\underline{\nabla}\phi) \qquad \textit{on whichever point of } \Gamma \quad (9)$$

which material derivative is,

$$\psi' = g_{\underline{\nabla}\phi}\,[\underline{\nabla}\phi' + D^2\phi\underline{V}] \qquad (10)$$

and also integral functionals,

$$\psi = \int_{\Gamma_A} g(\underline{\nabla}\phi)d\Gamma \qquad (11)$$

which has the form of eq.(4). Therefore, its material derivative is like eq.(5), where

$$g' = g_{\underline{\nabla}\phi}\underline{\nabla}\phi' \qquad (12)$$

For calculating the material derivatives of eq. (9) and (11), that is eq.(10) and (7) with (12), it is needed the direct differentation of the governing equations in order to get the response derived.

- In Ω,

  $$\nabla_k^2\,\phi = 0$$

  Taking the material derivative at both hands of the equality,

  $$\nabla_k^2\,\phi' = 0 \qquad (13)$$

  where the nullity of the Laplacian has been impossed.

- On $\Gamma\phi$,

  $$\phi = \phi_o$$

  $$\phi' = -\underline{\nabla}\phi\underline{V} \qquad (14)$$

  where ϕ_o has been supossed constant.

- On Γq

$$q = q_o$$

$$q' = -\underline{\nabla} q \underline{V} \qquad \text{because } q_o = \text{constant}$$

$$\underline{\nabla}_k \phi' \underline{n} = - \underline{\nabla}_k \phi \underline{n}' - \underline{\nabla} q \underline{V}$$

Taking into account eq.(2), after operating

$$\underline{\nabla}_k \phi' \underline{n} = \frac{\partial V_n}{\partial s} \underline{\nabla}_k \phi \underline{s} - \underline{n}_k D^2 \phi \underline{V} - V_s \underline{\nabla}_k \phi \frac{\partial \underline{n}}{\partial s} \tag{15}$$

where

$$D^2 = [\frac{\partial^2}{\partial x_i \partial x_j}]$$

- On $\Gamma_I^{i/j}$,

$$\phi^i = \phi^j$$

$$\phi^{i'} = \phi^{j'} + (\underline{\nabla}\phi^j - \underline{\nabla}\phi^i)\underline{V} \tag{16}$$

$$q^i = - q^j$$

Differentiating both hands of this equation, similar to (15),

$$\underline{\nabla}_k \phi^{i'} \underline{n}^i = - \underline{\nabla}_k \phi^{j'} \underline{n}^j - \underline{V} \sum^2 [D^2 \phi \underline{n}_k] - \sum^2 [\underline{\nabla}_k \phi \dot{\underline{n}}] \tag{17}$$

where $\dot{n}$ is given in eq.(1).

The problem governed by eq. (13-17) allows the calculus of the unknown term $\underline{\nabla}\phi'$ in eq.(10) and (7) with (12) and therefore, the desired sensitivities.

EXAMPLE

Figures 1 to 3 show a real life design application example. Figure 1 shows the design models, both the initial geometry (a) and final geometry (b) (regarding tidiness the boundary elements analysis models have not been presented). It is an axisymmetric insulator, of which only half has been modelled, surrounded by air. The electrode is conected to 75 Kv. Dimensions selected as design variables are displayed. The optimization problem consists in determining V1, V2, V3, V4 and V5 in order to minimize,

$$\frac{1}{m_c}\int_{\Gamma_c}(\|\underline{\nabla}\phi\|)^2 d\Gamma$$

where Γ_c is the curved interface boundary close to ground. There are two behaviour constraints at interface points A and B,

$$\|\underline{\nabla}\phi\| \le 400 \ v/mm \quad at \quad A$$

$$\|\underline{\nabla}\phi\| \le 350 \ v/mm \quad at \quad B$$

Figure 2 shows sensitivities at initial geometry obtained with the direct method (a) compared to finite differences (b).

Figure 3 shows objective function (a) and maximum constraint violation (b) history. Notice that even though one constraint (A) is violated at initial shape, there is an objective function decrease during the optimization process. Constraint A is active at the optimum.

REFERENCES

1. Haftka R.T. and Adelman H.M. Recent Developments in Structural Sensitivity Analysis. Third International Conference on CAD/CAM, Robotics and Factories of the Future, Southfield, Michigan, August 14-17, 1988.
2. Defourny M. Optimization Techniques and Boundary Element Method. Proc. Boundary Elements X. Vol 1: Mathematical and Computational Aspects. Springer-Verlag, 1988.
3. Kane J.H., Saigal S. Design Sensitivity Analysis of Solids Using BEM. Journal of Engineering Mechanics. Vol 114, pp 1703-1722, 1988.
4. Saigal S., Borggaard J.T., Kane J.H. Boundary Element Implicit Differentiation Equations for Design Sensitivity of Axisymetric Structures. International Journal for Solids and Structures. Vol 25, No 5, pp. 527-538, 1989.
5. Choi K.K., Seong H.G. A Domain Method for Shape Design Sensitivity of Built-up Structures. Computer Methods in Applied Mechanics and Engineering, 57, pp. 1-15, 1986.
6. Haug E.J., Choi K.K. Komkov V. Design Sensitivity Analysis of Structural Systems. Academic Press. New York, 1986.
7. Meric R.A. Boundary Elements in Shape Design Sensitivity Analysis of Thermoelastic Solids. Computer Aided Optimal Design: Structural and Mechanical Systems. Ed: C.A. Mota Soares. Springer-Verlag, Berlin, Heidelberg, 1987.
8. Mróz Z. Sensitivity Analysis and Optimal Design with Account for Varying Shape and Support Conditions. Computer Aided Optimal Design: Structural and Mechanical Systems. Ed: C.A. Mota Soares, Springer-Verlag, 1987.
9. Kwak B.H., Choi J.H. Design Sensitivity Analysis Based on Boundary Integral Equation Method Considering General Shape Variations. Part I: For Self-Adjoint Elliptic Operator Problems. KSME Journal. Vol 1, No 1, pp. 70-73, 1987.
10. Dems K. Sensitivity Analysis in Thermal Problems-II: Structure Shape Variation. Journal of Thermal Stresses, Vol 10, pp. 1-16, 1987.
11. Dems K., Mroz Z. Variational Approach to Sensitivity Analysis in Thermoelasticity Journal of Thermal Stresses, Vol 10, pp. 283-306, 1987.
12. Dems K., Haftka R.T. Two Approaches to Sensitivity Analysis for Shape Variation of Structures. Mechanics of Structures and Machines, 16 (4), 501-522, 1989.
13. Zolesio J.P. The material Derivative (or Speed) Method for Shape Optimization. Optimization of Distributed Parameter Structures (Eds. E.J. Haug, J. Cea), Sijthoff & Noordhoff, Alphen aan den Rijn, Netherlands, pp. 1089-1153, 1981.

Acknowledgement: This research was carried out under project OCIDE 132135.

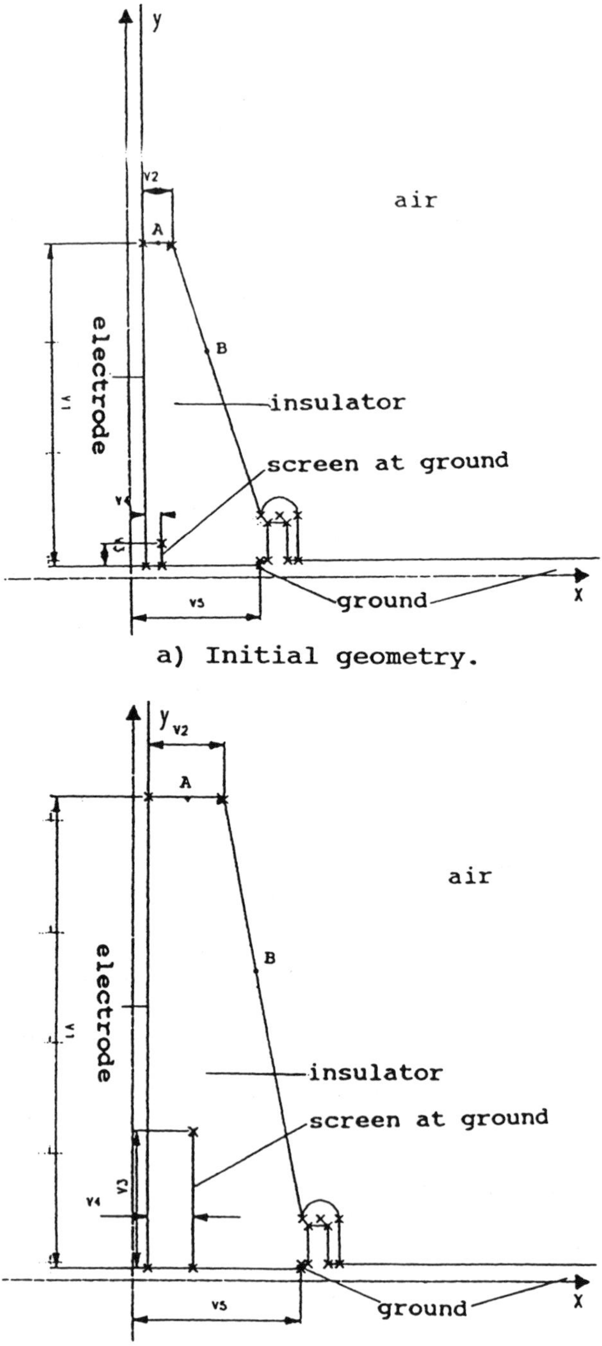

a) Initial geometry.

b) Final geometry.

Figure 1. Insulator: Design model.

	OBJECT. FUNCT.	CONSTRAINT 1	CONSTRAINT 2
VARIABLE 1	207.342631	-0.002277	-0.004620
VARIABLE 2	592.113125	-0.020854	-0.009060
VARIABLE 3	-1235.737593	0.000991	0.002243
VARIABLE 4	-817.864734	0.000548	0.001227
VARIABLE 5	-1682.823067	0.002543	0.002316

a) Direct Method.

	OBJECT. FUNCT.	CONSTRAINT 1	CONSTRAINT 2
VARIABLE 1	213.917890	-0.003756	-0.004525
VARIABLE 2	627.283961	-0.020767	-0.009000
VARIABLE 3	-1209.870590	0.000993	0.002247
VARIABLE 4	-779.077121	0.000547	0.001220
VARIABLE 5	-1973.503866	0.002191	0.001726

b) Finite differences.

Figure 2. Insulator: Sensitivity values.

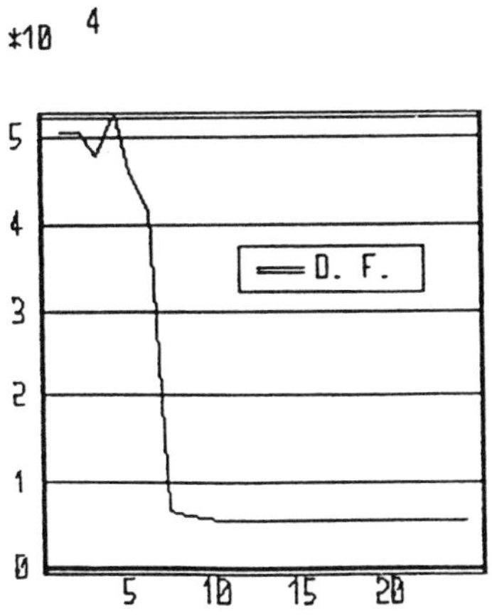

a) Objective function v. optimization iterations.

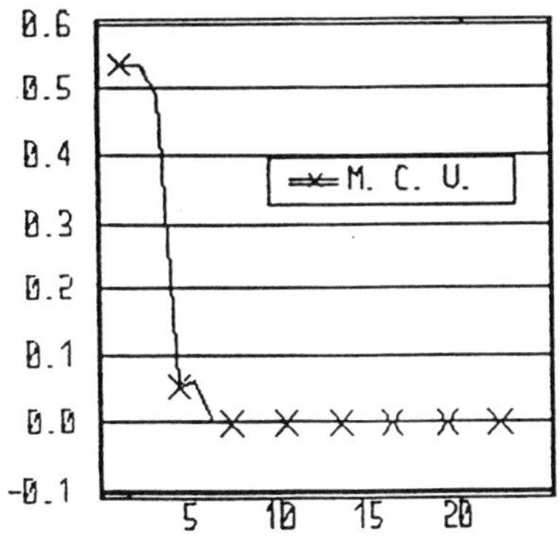

b) Maximum constraint violation v. optimization iterations.

Figure 3. Insulator: Optimization history.

Disk Shape Optimization Using the Three Unknowns Method

P. Cizmas

Department of Aerospace Engineering, Polytechnic Institute Bucharest, 77206 Bucharest, Romania

ABSTRACT

The Three Unknowns Method (T.U.M.) is applied to the optimization of a disk shape so as to minimize the stresses. The axisymmetric disk is subjected to centrifugal forces, uniformly distributed radial or axial loads, shearing forces and temperature variation along the radius. Considering a hyperbolic radial thickness variation one obtains similar governing equations for disk and plate. The disk to be optimized is divided into tracks with hyperbolic thickness variation. This enables the use of a single algorithm to compute the stresses due to disk and plate loads, using the T.U.M. For given geometrical restrictions one finds the disk shape for which the maximum stress is less than an imposed value. The use of the T.U.M. instead of the F.E.M. as the basic algorithm in the optimization process sharply reduces the computing time.

NOMENCLATURE

D - bending stiffness, $D = \dfrac{Eh^3}{12(1+\nu)}$

E - Young modulus

F_i - sum of forces inside the circle of radius r_i

h_i^R - thickness of disk/plate on the right side of the i node

h_i^L - thickness of disk/plate on the left side of the i node

p_i - uniformly distributed load on the track between the radii r_{i-1} and r_i

Q - shearing force

u - radial displacement

w_i - weighting function

α - thermal expansion coefficient

Φ - slope of the deflection surface

ν - Poisson ratio

ρ - density

σ_i - equivalent stress at radius i

σ_{max} - imposed maximum stress

σ^* - medium stress

θ - strain due to the temperature gradient, $\theta=\int_{t_0}^{t} \alpha \, dt$

INTRODUCTION

Optimization calculations usually involve repeated computations of the object to be optimized. Therefore to obtain an eficient optimization process it is necessary to improve the basic algorithm.

In this paper one intends to find the optimum shape of an axisymmetric disk subjected, in addition to the centrifugal forces, to uniformly distributed radial or axial loads, shearing forces (acting on circles of given radii) and also to the temperature variation along the radius. The disk may be subject to different geometric restrictions.

In the process of optimization, using the stress analysis, one tries to obtain a reduction of the maximum stress and, if possible, a quasi constant value of the equivalent stress along the radius.

The computation of the disk stresses is usually carried out using the F.E.M. which is a very general but rather time-consuming method. For improving the speed in the basic algorithm and obtaining an eficient optimization process the author proposes the use of the T.U.M. with a hyperbolic function thickness approximation for the disk tracks.

GOVERNING EQUATIONS

The governing equations for disks and plates are derived for the case of hyperbolic thickness variation along the radius.

Disk governing equation

For a disk which rotates with the angular speed ω and which has a known radial temperature variation, the equilibrium differential equation, e.g. Ponomariov [5], is:

$$\frac{d^2u}{dr^2}+\frac{1}{r}(1+\frac{r}{h}\frac{dh}{dr})\frac{du}{dr}-(1-\nu\frac{r}{h}\frac{dh}{dr})\frac{u}{r^2}+\frac{1-\nu^2}{E}\rho\omega^2 r-$$
$$-(1+\nu)(\frac{d\theta}{dr}+\frac{\theta}{h}\frac{dh}{dr})=0 \qquad (1)$$

Considering a hyperbolic radial thickness variation

$$h = h_o\, r^{-\beta} \qquad (2)$$

in equation (1) one obtains:

$$\frac{d^2u}{dr^2}+\frac{1-\beta}{r}\frac{du}{dr}-\frac{1+\nu\beta}{r^2}u = f_1(r,\omega,\theta) \qquad (3)$$

where for the linear variation of the strain due to the temperature gradient

$$\theta = \theta_1 + \lambda(r - r_1) \qquad (4)$$

the right hand side is:

$$f_1(r,\omega,\theta) = \lambda(1+\nu)(1-\beta) - \beta(1+\nu)\frac{\theta_1-\lambda r_1}{r} - \frac{1-\nu^2}{E}\rho\omega^2 r \qquad (5)$$

Circular plate governing equation

One considers a circular plate with axisymetric load and a certain radial thickness variation. The governinng equation for an element of the plate is, e.g. Conway [4]:

$$D\frac{d}{dr}(\frac{d\Phi}{dr}+\frac{\Phi}{r})+\frac{dD}{dr}(\frac{d\Phi}{dr}+\nu\frac{\Phi}{r}) = -Q \qquad (6)$$

Considering a hyperbolic bending stiffness

$$D = D_0\, r^{-\beta} \qquad (7)$$

the equation (6) becomes:

$$\frac{d^2\Phi}{dr^2}+\frac{1-\beta}{r}\frac{d\Phi}{dr}-\frac{1+\nu\beta}{r^2}\Phi = f_2(r,Q) \qquad (8)$$

where
$$f_2(r,Q) = -\frac{Q}{D_0 r^{-\beta}} \qquad (9)$$

General solution

The governing equations (3) and (8) differ only in the terms f_1, f_2, so the solutions of the homogeneous equations will be similar:

$$u_0 = A_u r^m + B_u r^{-n} \qquad (10)$$

$$\Phi_0 = A_\Phi r^m + B_\Phi r^{-n} \qquad (11)$$

where the arbitrary constants A_u, B_u, A_Φ, B_Φ are derived from the boundary conditions and the exponents are:

$$m = \frac{\beta + \sqrt{4+\beta(4\nu+\beta)}}{2} \tag{12}$$

$$n = \frac{-\beta + \sqrt{4+\beta(4\nu+\beta)}}{2} \tag{13}$$

The particular solutions for equations (3) and (8) are:

$$u_p = (\theta_1 - \lambda r_1)r + \frac{\lambda(1+\nu)(1-\beta)}{3-\beta(2+\nu)}r^2 - \rho\frac{\omega^2(1-\nu^2)}{E[8-\beta(3+\nu)]}r^3 \tag{14}$$

$$\Phi_p = -\frac{F_i}{2\pi D_0\beta r^{-\beta-1}(1-\nu)} - \frac{p_i}{2D_0 r^{-\beta-1}}\left[\frac{r^2}{8+\beta(3-\nu)} - \frac{r_i^2}{\beta(1-\nu)}\right] \tag{15}$$

where the definition of the shearing force Q for a circular plate:

$$2\pi r Q = F_i + \pi(r^2 - r_i^2)p_i \tag{16}$$

was used.

Using the equations (10-15) one obtains the solutions of the governing equations:

$$u = A_u r^m + B_u r^{-n} + (\theta_1 - \lambda r_1)r + \frac{\lambda(1+\nu)(1-\beta)}{3-\beta(2+\nu)}r^2 - \rho\frac{\omega^2(1-\nu^2)}{E[8-\beta(3+\nu)]}r^3 \tag{17}$$

$$\Phi = A_\Phi r^m + B_\Phi r^{-n} - \frac{F_i}{2\pi D_0\beta r^{-\beta-1}(1-\nu)} - \frac{p_i}{2D_0 r^{-\beta-1}}\left[\frac{r^2}{8+\beta(3-\nu)} - \frac{r_i^2}{\beta(1-\nu)}\right] \tag{18}$$

Disk stresses

The disk stresses result from the superposition of the stresses due to disk load and plate load. With the values of the displacement given by the equation (17) one can find the stresses due to disk load using the well known equations, e.g. Ponomariov [5], for a rotating disk with given temperature variation:

$$\sigma_\theta = \frac{E}{1-\nu^2}\left[\frac{u}{r} + \nu\frac{du}{dr} - \theta(1+\nu)\right] \tag{19}$$

$$\sigma_r = \frac{E}{1-\nu^2}\left[\frac{du}{dr} + \nu\frac{u}{r} - \theta(1+\nu)\right] \tag{20}$$

The stresses due to the plate load are obtained using the equations for a circular plate with constant temperature, e.g. Buzdugan [2]:

$$\sigma_\theta = \frac{Ez}{1-\nu^2}\left(\frac{\Phi}{r} + \nu\frac{d\Phi}{dr}\right) \tag{21}$$

$$\sigma_r = \frac{Ez}{1-\nu^2}\left(\frac{d\Phi}{dr} + \nu\frac{\Phi}{r}\right) \tag{22}$$

Using the expressions of the displacement $u(r)$ and of the slope of the deflection surface $\Phi(r)$ one obtains the stresses presented in Table 1.

	Disk	Plate
Variable Parameter	$h = h_0 r^{-\beta}$	$D = D_0 r^{-\beta}$
Differential equation	$\frac{d^2u}{dr^2} + \frac{1-\beta}{r}\frac{du}{dr} - \frac{1+\nu\beta}{r^2}u = f_1(r,\omega,\theta)$	$\frac{d^2\Phi}{dr^2} + \frac{1-\beta}{r}\frac{d\Phi}{dr} - \frac{1+\nu\beta}{r^2}\Phi = f_2(r,Q)$
Constants	$\beta = -\frac{\ln h_1/h_2}{\ln r_1/r_2}$ $h_0 = r_1^{\beta} h_1$	$\beta = -3\frac{\ln h_1/h_2}{\ln r_1/r_2}$ $D_0 = r_1^{\beta} D_1$
Solution of Differential Equation	$u = A_u r^m + B_u r^{-n} + u_p$	$\Phi = A_\Phi r^m + B_\Phi r^{-n} + \Phi_p$
	$m = \frac{\beta + \sqrt{4+\beta(4\nu+\beta)}}{2}$ $n = \frac{-\beta + \sqrt{4+\beta(4\nu+\beta)}}{2}$	
Stresses σ_θ	$\frac{E}{1-\nu^2}[A_u[r^{m-1}(1+\nu m)] + B_u[r^{-n-1}(1-\nu n)] + \frac{u_p}{r} + \nu u'_p - \theta(1+\nu)]$	$\frac{Ez}{1-\nu^2}[A_\Phi[r^{m-1}(1+\nu m)] + B_\Phi[r^{-n-1}(1-\nu n)] + \frac{\Phi_p}{r} + \nu\Phi'_p]$
Stresses σ_r	$\frac{E}{1-\nu^2}[A_u[r^{m-1}(m+\nu)] + B_u[r^{-n-1}(\nu-n)] + u'_p + \nu\frac{u_p}{r} - \theta(1+\nu)]$	$\frac{Ez}{1-\nu^2}[A_\Phi[r^{m-1}(m+\nu)] + B_\Phi[r^{-n-1}(\nu-n)] + \Phi'_p + \nu\frac{\Phi_p}{r}]$

Table 1. Stresses in disk and circular plate.

To obtain the value of the stresses one must compute the constants A_u, B_u, A_Φ, B_Φ on each track of the disk.

THE THREE UNKNOWNS METHOD

The T.U.M. presented by Blumenfeld [1] enables us to derive the integration constants obtained in the solution of the differential equation with constant coeficients, provided that one can produce two functions which depend on the integration constants. The well known "equation of the three moments" is a particular case of the T.U.M.

One divides the element to be calculated into groups of two intervals. At the end sections of these two intervals three unknowns are chosen; these unknowns are introduced in a relation which may be applied to each group of two intervals, into which the element has been divided. In this way an algebraic equation system is obtained, the roots of which are the unknowns previously chosen.

One considers that the two functions obtained in the integration process are:

$$\varphi = A\,a(r) + B\,b(r) + c(r) \tag{23}$$

$$\psi = A\,d(r) + B\,e(r) + f(r) \tag{24}$$

where A, B are the integration constants and $a(r)$, $b(r)$, $c(r)$, $d(r)$, $e(r)$ and $f(r)$ are known continuous functions.

Supose the element (i.e. the disk) is divided into n intervals (i.e. the tracks) and consider the group of two intervals between the nodes i-1 and i+1. Consider the variation of the functions φ and ψ given in figure 1, where the functions could have discontinuities in the nodes. The value of the functions on the left side of the nodes have the superscript L while on the right side have the superscript R.

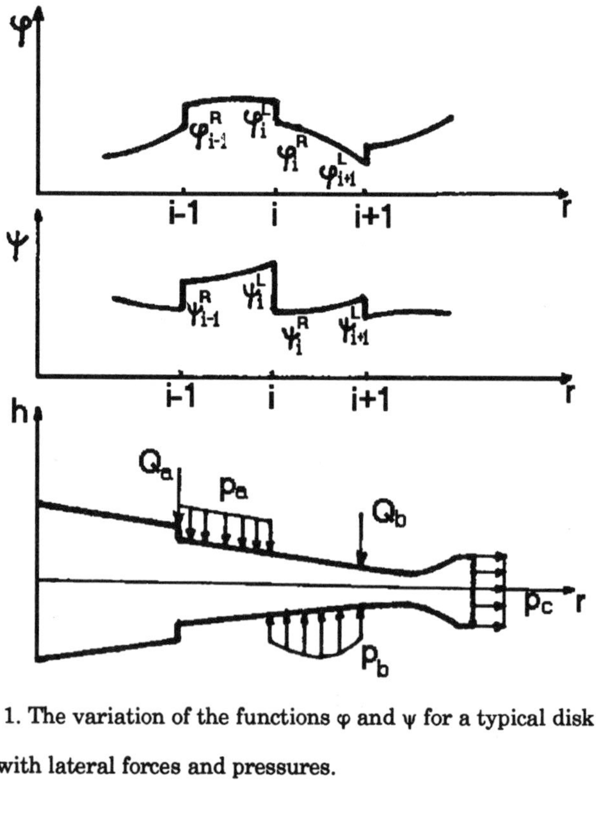

Figure 1. The variation of the functions φ and ψ for a typical disk case,

with lateral forces and pressures.

One suposes that between the functions φ and ψ in the i and i+1 nodes the following relations hold:

$$\varphi_i^L = k_1 \varphi_i^R \tag{25}$$

$$\varphi_{i+1}^L = k_4 \varphi_{i+1}^R \tag{26}$$

$$\psi_i^L = k_2 \psi_i^R + k_3 \varphi_i^R \tag{27}$$

where the constants k_1, k_2, k_3 and k_4 are known. In the case of the disk these are

$$k_1 = \frac{h_i^L}{h_i^R}, \qquad k_2 = 1, \qquad k_3 = \nu(k_1 - 1), \qquad k_4 = \frac{h_{i+1}^L}{h_{i+1}^R} \tag{28}$$

while in the case of the plate

$$k_1 = 1, \qquad k_2 = \left(\frac{h_i^R}{h_i^L}\right)^3, \qquad k_3 = \nu(1 - k_2), \qquad k_4 = 1 \tag{29}$$

To establish the recurrence relation for the T.U.M. one writes the equations (23) and (24) for two intervals:

- the interval i-1,i

$$\varphi = A_{i-1,i}\, a^I(r) + B_{i-1,i}\, b^I(r) + c^I(r) \tag{30}$$

$$\psi = A_{i-1,i}\, d^I(r) + B_{i-1,i}\, e^I(r) + f^I(r) \tag{31}$$

- the interval i,i+1

$$\varphi = A_{i,i+1}\, a^{II}(r) + B_{i,i+1}\, b^{II}(r) + c^{II}(r) \tag{32}$$

$$\psi = A_{i,i+1}\, d^{II}(r) + B_{i,i+1}\, e^{II}(r) + f^{II}(r) \tag{33}$$

Using the equations (30) and (32) one expresses the values of the φ function in the nodes $(i\text{-}1)^R$, i^L, i^R, $(i+1)^L$ and using the equation (27) one obtains five linear equations for the unknowns $A_{i-1,i}$, $B_{i-1,i}$, $A_{i,i+1}$, $B_{i,i+1}$. The compatibility condition provides the recurrence relation:

$$-\varphi_{i-1}\frac{P}{M} + \varphi_i\left(k_2\frac{S}{N} + k_1\frac{R}{M} - k_3\right) - \varphi_{i+1}\frac{T}{N}k_2k_4 =$$

$$= -c_{i-1}^I\frac{P}{M} + c_i^I\frac{R}{M} + c_i^{II}k_2\frac{S}{N} - c_{i+1}^{II}k_2\frac{T}{N} - f_i^I + f_i^{II}k_2 \tag{34}$$

where one drops the superscript R for φ. The constants M, N, P, R, S and T are:

$$M = a_{i-1}^I b_i^I - a_i^I b_{i-1}^I \qquad N = a_i^{II} b_{i+1}^{II} - a_{i+1}^{II} b_i^{II}$$

$$P = a_i^I e_i^I - d_i^I b_i^I \qquad R = a_{i-1}^I e_i^I - d_i^I b_{i-1}^I \tag{35}$$

$$S = e_i^{II} a_{i+1}^{II} - b_{i-1}^{II} d_i^{II} \qquad T = e_i^{II} a_i^{II} - b_i^{II} d_i^{II}$$

where $a_{i-1}^I = a^I(r_{i-1}), \ldots, f_i^{II} = f^{II}(r_i)$.

For an element divided into n intervals one can obtain only n-2 recurrence relations (34). To obtain the n unknowns φ_i one must use the boundary conditions.

The solution of the system of linear algebraic equations is carried out recursively as the system is tridiagonal. Knowing the values of the φ function in the nodes i, one can compute the constants A and B and then the functions ψ:

$$\psi(r) = \frac{(b_{i-1}^I c_i^I - b_i^I c_{i-1}^I)d^I(r) + (c_{i-1}^I a_i^I - c_i^I a_{i-1}^I)e^I(r)}{a_{i-1}^I b_i^I - a_i^I b_{i-1}^I} + f^I(r) +$$

$$+ \varphi_{i-1} \frac{b_i^I d^I(r) - a_i^I e^I(r)}{a_{i-1}^I b_i^I - a_i^I b_{i-1}^I} + \varphi_i \frac{a_{i-1}^I e^I(r) - b_{i-1}^I d^I(r)}{a_{i-1}^I b_i^I - a_i^I b_{i-1}^I} \tag{36}$$

When the T.U.M. is applied to the disk stress analysis, the φ and ψ functions are σ_r and σ_θ.

THE OPTIMIZATION PROCESS

One considers a part of the disk with a given shape and/or given restrictions of this geometry. The disk shape on the area which is optimized could be introduced by the analyst or could be computed step by step. In the later case the thickness is:

$$H(r) = H_1 e^{x_1 r} + H_2 e^{x_2 r} \tag{37}$$

where the constants x_1 and x_2 are imposed in the iterative process and the constants H_1 and H_2 result from the boundary conditions. In the generation of the thickness as a sum of exponentials, some shapes are a priori avoided using geometric criterions. A mixed kind of shape generation is also possible.

For the disk shape optimization the objective function is:

$$F(x) = \sum_i (\sigma_i - \sigma^*)^2 w_i \tag{38}$$

with the restriction $\sigma_1 < \sigma_{max}$. One finds the optimum values for x, where $x = (x_1 \; x_2)^T$ usinng the gradient method:

$$x^{k+1} = x^k - \lambda \nabla F(x^k) \tag{39}$$

To illustrate the optimization procedure, consider the compressor disk presented in figure 2. The speed of revolution is 10500 rpm; the disk is subjected to the lateral presure p_1=0.1 MPa between the radii 100 and 175 mm and p_2=0.08 MPa between 185 and 210 mm. The pressure exerted by the blades at the external diameter is 32 MPa. The temperature distribution and the Young modulus variation are presented in figure 2.

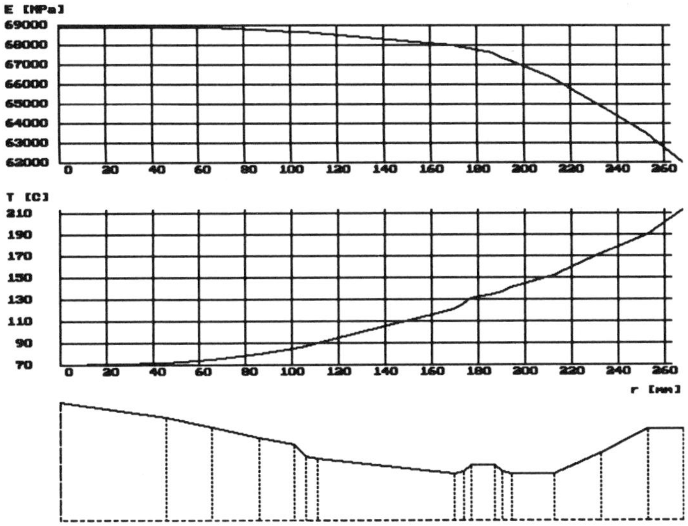

Figure 2. The temperature and Young modulus variation of the disk to be optimized.

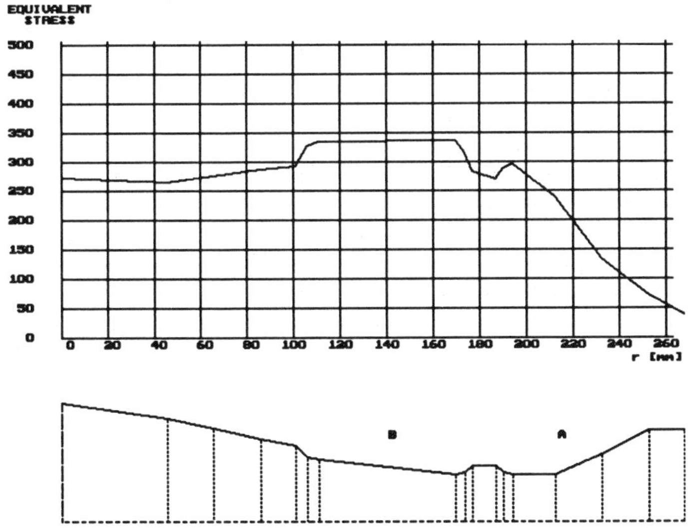

Figure 3. The stresses of the initial disk.

Given the shape of the disk between the radii 0- 100 mm, 175-185 mm and 250-265 mm, one must find the optimum thickness variation in the areas *A* and *B*.

In this case the optimization is made succesively. For a given shape of the *B* area one finds the optimum *A* zone; then with the fixed *A* area one finds the optimum *B* area. Iterations continue until the shape disk variation is less than an imposed value. Results are presented in figures 3,4,5 and 6.

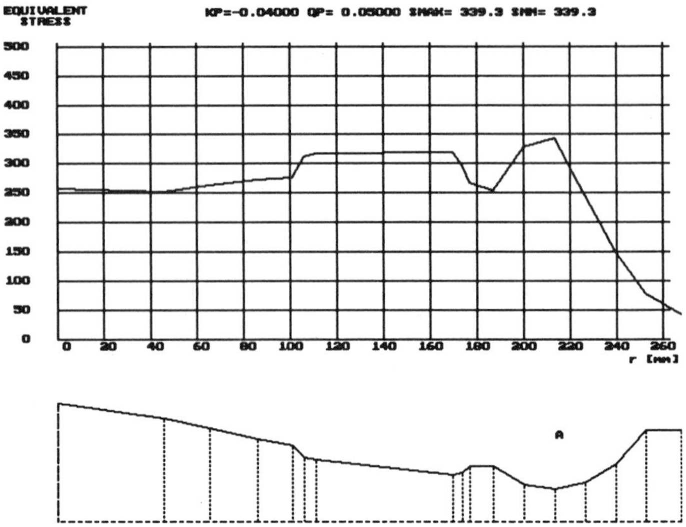

Figure 4. The optimization of the A area.

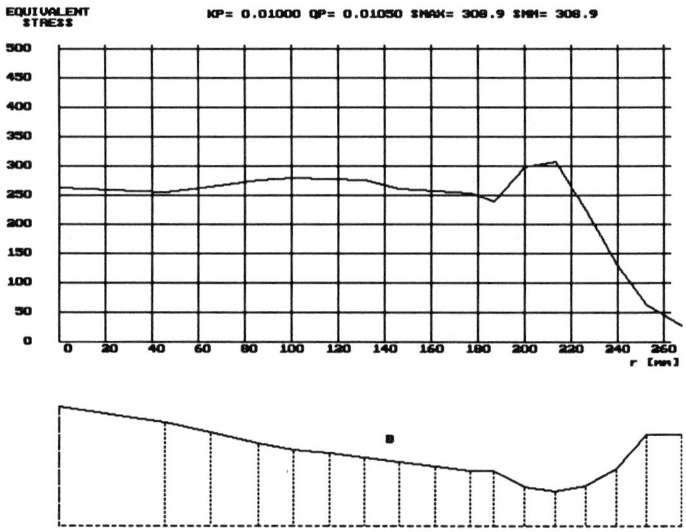

Figure 5. The optimization of the B area.

CONCLUSIONS

The similarity of the equilibrium equations for disks and circular plates with hyperbolic variation enables one, using the recurrence relations of the T.U.M., to obtain a common code, which represents the basic step of the optimization.

The use of analytic relations permits one to obtain the exact solution if the thickness of the tracks, in which the disk is divided, varies as a hyperbolic function. Hence for a disk/plate of a given shape the only source of error of the algorithm is the approximation of the disk contour.

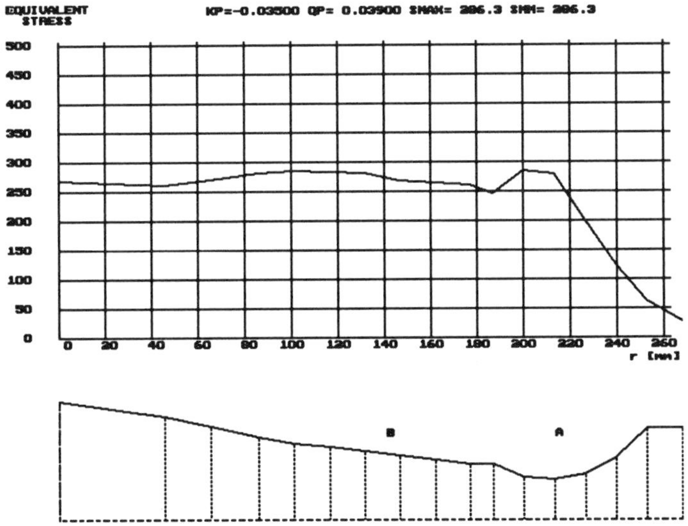

Figure 6. The optimized shape of the disk.

The algorithm based on the T.U.M. is much faster than the F.E.M. In a comparison made by the author in [3], the time necessary for a computation using the F.E.M. was about 50 times longer than that for the T.U.M., the results being in good agreement.

In addition to the speed and accuracy of the T.U.M., the effort made in the discretization and data input is much less than for the F.E.M. However the F.E.M. has the great advantage of generality; the F.E.M. can be used in a broader domain than the T.U.M. which has restrictions for the shape and the load of the disks. In the case when the disk to be optimized satisfies the restrictions imposed by the T.U.M. it is advisable, due to his accuracy and rapidity, to use it instead of the F.E.M.

BIBLIOGRAPHY

1. Blumenfeld, M. Relatii generalizate ale metodei de recurenta a celor trei necunoscute, pentru studiul sistemelor elastice, St. cerc. mec. apl., Vol 14, pp. 715-726, 1963.

2. Buzdugan, Gh. Rezistenta materialelor, Edit. Tehnica, Bucuresti, 1980.

3. Cizmas, P. Calculul discurilor si placilor circulare incarcate axial simetric prin metoda de recurenta a celor trei necunoscute, St. cerc. mec. apl., Tom 49, pp. 71-84, 1990.

4. Conway, H.D. Note on Bending of Circular Plates of Variable Thickness, Journal of Applied Mechanics, Vol.16, 1949.

5. Ponomariov, S.D., Biderman, V.L., Liharev, C.C., Macusin, V.M., Malinin, N.N., Fedosiev, V.I., Osnovî sovremionnah metodov rasciota na procnosti v masinostoienii, Masgiz, 1952.

Optimal Design of Thin Elastic Annular Plate Under Strength and Stiffness Constraints

H.R. Yu, F. Wang

Department of Mechanics, Lanzhou University, Lanzhou, GanSu, China

ABSTRACT

This paper presents the optimal design of thin elastic annular plate subject to strength and stiffness constraints. By means of the stepped reduction method suggested by Yeh Kaiyan in 1965, the function of optimal thickness distribution can be approached by stepped function. Explicit expressions of the design–variables can also be obtained. Further, the problem can be posed as a standard non–liner pragramming problem with non–liner constraints. With two methods, we calculate several cases; the results show that our methods have many advantages.

INTRODUCTION

During recent decades, the optimal design of elastic continuum has aroused scientists great interest, because of studying deeply in the optimal design of frame structure. Thus it has become one of the important subjects of modern optimization. Although there are many articles about optimal design, there are only a few about optimal design of elastic continuum under synthetic constraint conditions. Karihaloo [1] used the splines method to optimal design of beams in his article, but it did not concern optimal design of plates and shells under synthetic constraint conditions. So, this problem must be studied farther and the method of optimization needs to be developed further. One of the problems of optimal design for elastic thin plates is to find the thickness distribution function which allows the volume of the elastic thin plate to be minimal, and the maximal deflection and stress to be less than the permissible valve assigned. In this problem, the explicit formulations of both objective function and constrained conditions can not be given by traditional methods, so, it leads to many computational difficulties. The stepped reduction method can offer the solution of deflection and stress of plate with variable thickness; further, the explicit expression of the objective function and constrained condition can be obtained. The expression is suitable for arbitrary boundary conditions and arbitrary distributed loads. This problem is reduced to a non–linear programming problem with non–linear constraints.

FUNDAMENTAL EQUATIONS

We consider an annular elastic plate with a variable thickness of which outer and inner radii, are a and b respectively (Fig.1). We suppose that the elastic modules E and Poisson ration v of the plate are all constant. The annular plate is acted upon by an axisymmetric, arbitrarily distributed load and has an axisymmetric boundary condition. The deflection is thus axisymmetric.

We take the middle plane of the annular plate as (r,θ) coordinate and let the center of the annular plate be located at the origin.

Now our optimal design problem is to determine the thickness function H(r) of the plate and make the volume of the plate be minimal

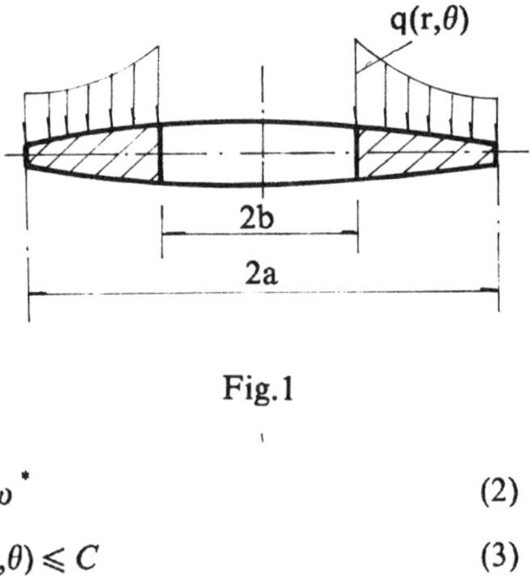

Fig.1

subjected to

$$\omega_{max} \leqslant \omega^{*} \tag{2}$$

$$G(r,Z^{*},h(r),\theta) \leqslant C \tag{3}$$

$$H_{min} \leqslant H(r) \leqslant H_{max} \tag{4}$$

Formula (2) is a deflection constrained condition; (3) is a strength constrained condition; and (4) is a geometry constrained condition.

Under those conditions, the differential equation of the plate deflection can be written as:

$$\nabla^2\nabla^2\omega + \frac{1}{D}\frac{d^2D}{dr^2}[\frac{\partial^2\omega}{\partial r^2} + v(\frac{1}{r}\frac{\partial\omega}{\partial r} + \frac{1}{r^2}\frac{\partial^2\omega}{\partial\theta^2})] + \frac{1}{D}\frac{dD}{dr}[2\frac{\partial^3\omega}{\partial r^3} + (2$$

$$+ v)\frac{1}{r}\frac{\partial^2 \omega}{\partial r^2} - \frac{1}{r^2}\frac{\partial \omega}{\partial r} + \frac{2}{r^2}\frac{\partial^2 \omega}{\partial r \partial \theta} - \frac{3}{r^3}\frac{\partial^2 \omega}{\partial \theta^2} = \frac{q(r,\theta)}{D} \tag{5}$$

where

$$\nabla^2 = \frac{\partial^2}{\partial r^2} + \frac{1}{r}\frac{\partial}{\partial r} + \frac{1}{r^2}\frac{\partial^2}{\partial \theta^2} \tag{6}$$

$$D = D(r) = E(h(r))^3 / 12(1 - v^2)$$

$$\lambda_1 = 1 - v; \qquad \lambda_2 = 1 + v \tag{7}$$

The relation between deflection and bending moment, shear force, reaction force respectively can be described by the following equations;

$$\left.\begin{aligned}
M_r &= -D(r)(\lambda_1 \frac{\partial^2 \omega}{\partial r^2} + v\nabla^2 \omega) \\
M_\theta &= -D(r)(\nabla^2 \omega - \lambda \frac{\partial^2 \omega}{\partial r^2}) \\
M_{r\theta} &= -D(r)\lambda_1 \frac{\partial}{\partial r}(\frac{1}{r}\frac{\partial \omega}{\partial \theta}) \\
Q_r &= -D(r)\frac{\partial}{\partial r}\nabla^2 \omega, \quad Q_\theta = -D(r)\frac{\partial}{\partial \theta}\nabla^2 \omega \\
V_r &= -D(r)[\frac{\partial}{\partial r}\nabla^2 \omega + \frac{\lambda_1}{r}\frac{\partial}{\partial r}(\frac{1}{r}\frac{\partial^2 \omega}{\partial \theta^2})]
\end{aligned}\right\} \tag{8}$$

Three usually used boundary conditions are:

(1) simply supported $\omega = 0$, $Mr = 0$;

(2) Fixed $\omega = 0$, $(\frac{\partial \omega}{\partial r}) = 0$; (9)

(3) Free $Mr = 0$, $Vr = 0$.

First, we expand the load function $q(r,\theta)$ into Fourier series

$$q(r,\theta) = q_0(r) + \sum_{m=1}^{\infty} (q_m(r)cosm\theta + \bar{q}_m(r)sinm\theta] \tag{10}$$

Then, ω, Mr, Vr can be also formuate

$$\omega = \sum_{m=0}^{\infty} \omega_m(r)cosm\theta + \sum_{m=1}^{\infty} \bar{\omega}_m(r)sinm\theta$$

$$M_r = \sum_{m=0}^{\infty} M_{rm}(r)cosm\theta + \sum_{m=1}^{\infty} \overline{M}_{rm} sinm\theta \tag{11}$$

$$V_r = \sum_{m=0}^{\infty} V_{rm}(r)cosm\theta + \sum_{m=1}^{\infty} \overline{V}_{rm} sinm\theta$$

Substituting (10), (11) into (5), we can obtain the differential equation of ω_m.

$$V_m^2 V_m^2 \omega_m + \frac{1}{D}\frac{d^2 D}{dr^2}[\frac{\partial^2 \omega_m}{\partial r^2} + v(\frac{1}{r}\frac{\partial \omega_m}{\partial r} - \frac{m^2}{r^2}\omega_m)] + \frac{1}{D}\frac{dD}{dr}[2\frac{\partial^3 \omega_m}{\partial r^3}$$

$$+ (2+v)\frac{1}{r}\frac{\partial^2 \omega_m}{\partial r^2} - \frac{1}{r^2}\frac{\partial \omega_m}{\partial r} - \frac{2m^2}{r^2}\frac{\partial \omega_m}{\partial r} + \frac{2m^2}{r^3}\omega_m] = \frac{q_m(r)}{D} \quad (12)$$

where

$$V_m^2 = \frac{\partial^2}{\partial r^2} + \frac{1}{r}\frac{\partial}{\partial r} - \frac{m^2}{r^2}$$

The boundary conditions (9) become

$$\text{(1) Fixed} \qquad \omega_m = 0 \quad \frac{d\omega_m}{dr} = 0;$$

$$\text{(2) Simply supported} \quad \omega_m = 0 \quad M_{rm} = 0; \qquad (13)$$

$$\text{(3) Free} \qquad M_{rm} = 0 \quad V_{rm} = 0.$$

In order to use the stepped reduction method, we divide the whole plate into n annular segments at constant thicknesses; h(r) is approximated by a segmentwise constant function. The radii of the boundary segments are $b = r_0$, $r_1 \cdots\cdots r_n = a$ (Fig.2). This means that h(r) is replaced with n variables H_1, H_2, H_3, $\cdots$, H_n.

The differential equation of i–th annular segment with a thickness of Hi, a inner radius of r_{1-i}, a outer radius of r_i and a stiffness of D_i, $D_i = E_i^3 / 12(1-v)$, can be written as:

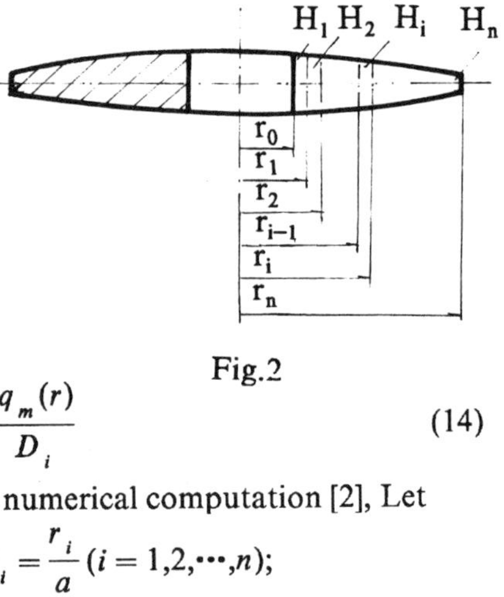

Fig.2

$$V_m^2 V_m^2 \omega_m = \frac{q_m(r)}{D_i} \qquad (14)$$

For convenience of deducing and numerical computation [2], Let

$$W = \frac{\omega}{a}; \quad x = \frac{r}{a}; \quad \beta_0 = \frac{r_0}{a}; \quad \beta_i = \frac{r_i}{a}(i = 1,2,\cdots,n);$$

$$P_m(x) = \frac{q_m(r)}{D_0}a^3; \quad M_m(x) = \frac{M_{rm}(r)}{D_0}a; \quad V_m(x) = \frac{V_{rm}(r)}{D_0}a^2;$$

$$M_\theta^m = \frac{M_{\theta m}(r)}{D_0}a \quad D_0 = \frac{EH_0^3}{12(1-v^2)}; \quad D_i = \frac{EH_i^3}{12(1-v^2)}; \quad \delta_i = \frac{D_0}{D_i}$$

H_0 is the thickness of a uniform plate which has the same inner radius, outer radius, load, stiffness and strength–constrained conditions as the whole annular plate.

Thus, equation (14) becomes

$$\nabla^2_m \nabla^2_m W_m = P_m \tag{15}$$

where

$$\nabla^2_m = \frac{d^2}{dx^2} + \frac{1}{x}\frac{d}{dx} - \frac{m^2}{x^2}$$

According to Ref [3], the exact solution of the equation (15) can be written as:

$$\left.\begin{aligned}
W_m(x,\delta_i) &= W_m(\beta_{i-1})f_1 + W'_m(\beta_{i-1})f_2 \\
&+ M_m(\beta_{i-1})f_3 + V_m(\beta_{i-1})f_4 + f_5 \\
\theta_m(x,\delta_i) &= W_m(\beta_{i-1})f'_1 + W'_m(\beta_{i-1})f'_2 \\
&+ M_m(\beta_{i-1})f'_3 + V_m(\beta_{i-1})f'_4 + f'_5 \\
M_m(x,\delta_i) &= W_m(\beta_{i-1})F_1^{(\mathrm{II})} + W'_m(\beta_{i-1})F_2^{(\mathrm{II})} \\
&+ M_m(\beta_{i-1})F_3^{(\mathrm{II})} + V_m(\beta_{i-1})F_4^{(\mathrm{II})} + F_5^{(\mathrm{II})} \\
V_m(x,\delta_i) &= W_m(\beta_{i-1})F_1^{(\mathrm{III})} + W'_m(\beta_{i-1})F_2^{(\mathrm{III})} + \\
&M_m(\beta_{i-1})F_3^{(\mathrm{III})} + V_m(\beta_{i-1})F_4^{(\mathrm{III})} + F_6^{(\mathrm{III})}
\end{aligned}\right\} \tag{16}$$

where $W_m(\beta_{i-1})$, $W'_m(\beta_{i-1})$, $M_m(\beta_{i-1})$,$V_m(\beta_{i-1})$ is dimensionless deflection, slope, bending monent and reaction force at $(x=\beta_{i-1})$ respectively.

$f_i(i=1,2,3,4,5)$ are the functions of $x(\beta_{i-1}<x<\beta_i)$,m,δ_i and P_m.

For different m. fi has different formulations. Ref [3] gives the formulation of fi(i=1,2,3,4,5).

Let

$$S^i_m(x) = \begin{bmatrix} W_m(x) \\ \theta_m(x) \\ M_m(x) \\ V_m(x) \end{bmatrix} \qquad B^i_m(x,\delta_i) = \begin{bmatrix} f_5 \\ f'_5 \\ F_5^{(\mathrm{II})} \\ F_5^{(\mathrm{III})} \end{bmatrix} \tag{17-A}$$

$$A^i_m(x,\delta_i) = \begin{bmatrix} f_1 & f_2 & f_3 & f_4 \\ f'_1 & f'_2 & f'_3 & f'_4 \\ F_1^{(\mathrm{II})} & F_2^{(\mathrm{II})} & F_3^{(\mathrm{II})} & F_4^{(\mathrm{II})} \\ F_1^{(\mathrm{III})} & F_2^{(\mathrm{III})} & F_3^{(\mathrm{III})} & F_4^{(\mathrm{III})} \end{bmatrix} \tag{17-B}$$

So, equations (16) can be written as

$$S^i_m(x) = A^i_m(x,\delta_i) \cdot S^i_m(\beta_{i-1}) + B^i_m(x,\delta_i) \tag{18}$$

$(\beta_{i-1} \leqslant x \leqslant \beta_i) \quad i = 1,2,3,4,\cdots \quad m = 0,1,2,3,\cdots$

In order to make the relation between the subject problem and the variables $(\delta_1\cdots\delta_n)$ more clearly, we let $\delta = (\delta_1\cdots\delta_n)$.

Because continuity conditions at the junction of two neighboring elements must be satisfied, we can get the following equation

$$S^i_m(\beta_i,\delta) = S^{(i+1)}_m(\beta_i,\delta) \tag{19}$$

Thus, we have

$$S^i_m(x,\delta) = AA^i_m \cdot S^1_m(\beta^0, \ \delta) + BB^i_m \tag{20}$$

where

$$AA^i_m = A^i_m(x,\delta_i) \cdot \prod_{j=1}^{i-1} A^j_m(\beta_j,\delta_j) \tag{21}$$

i = 1,2,3,⋯,n

$$BB^i_m = B^i_m(x,\delta_i) + A^i_m(x,\delta_i)[B^{i-1}_m(\beta_{i-1},\delta_{i-1}) + \sum_{k=1}^{i-2} B^k_m(\beta_k,\delta_k) \prod_{l=k+1}^{i-1} A^l_m(\beta_l,\delta_l)] \tag{22}$$

i = 1,2,3,⋯,n

When i = n, x = β_n we get the relation formulation between inner boundary's deflection, slope, bending moment, shear force and outer boundary's for whole plate

$$S^n_m(\beta_n\delta) = A \cdot A^n_m(\beta_n,\delta)S'_m(\beta_0,\delta) + BB^n_m(\beta_n,\delta) \tag{23}$$

Letting

$$A \cdot A^n_m = AO, \qquad BB^n_m = BO$$

The formula (23) becomes

$$S^n_m(\beta_n\delta) = AO \cdot S'_m(\beta_0\delta) + BO \tag{24}$$

If the boundary conditions of the plates are given, the explicit expression of S'_m (β_0,δ) can be obtained.

Further the explicit expression of deflection, slope, bending moment and shear force of the whole plate at any point can also be obtained.

$$S^i_m(x,\delta) = AA^i_m S'_m(\beta_0,\delta) + BB^i_m \tag{25}$$

After W_m has been determined, i–th annular segment's M^m_θ can be written as

$$M^m_\theta = -\frac{1}{\delta_i}[\nabla^2_m W_m(x) - \lambda_1 \frac{d^2 W_m(x)}{dx^2}] \tag{26}$$

The form of solution about $\overline{W}_m, \overline{\theta}_m, \overline{M}_m, \overline{V}_m$ is similar with formula (25).

OBJECTIVE FUNCTION AND CONSTRAINED CONDITIONS

1. objective function

Let

$$V_0 = (\pi a^2 - \pi b^2)H_0; \qquad \overline{V} = \frac{V}{V_0}$$

V_0 is the volume of a uniform plate with an inner radius of b, an outer radius of a and a thickness of H_0.

So, after dividing the whole plate into n annular segments, the dimensionless objective function is

$$\overline{V} = \sum_{i=1}^{\infty} (\delta_i)^{-\frac{1}{3}} \frac{(\beta_i^2 - \beta_{i-1}^2)}{(1 - \beta_0^2)} \tag{27}$$

2. constrained conditions

(1). Deflection Constrained Conditions

From formula (2), we know that the deflection constrained condition is

$$W_{max} \leqslant W^*$$

For the sake of explaining the problem, we let $q(r,\theta) = q_0 \cos m\theta$

From (11), we have the following equation

$$W(x) = W_m(x) \cdot cosm\theta \tag{28}$$

Maximizing to (28), we know that when $\theta = 0$, $x = x^*$ W(x) reaches maximum.

That is

$$W_{max} = W_m(x^*) = [W_m(x)]_{max} \tag{29}$$

Since we have the explicit expression of $W_m(x)$ and $\overline{W}_m(x)$, we can obtain the maximum point of deflection under an abitrary distribution load.

In order to compare conveniently, we make the allowed value of deflection equal to or less than maximum deflection of uniform plate.

(2). Strength Constrained Condition

From formula (3), we know that strength condition is

$$G(r,z^*,h(r),\theta) = C$$

G stands for function relation of strength condition, c is a constant.

We adopt Mises' strength condition [5], it's formulation can be written as:

$$(\sigma_r - \sigma_\theta)^2 + (\sigma_\theta - \sigma_z)^2 + (\sigma_z - \sigma_r)^2 + 6(\tau_{r\theta}^2 + \tau_{\theta z}^2 + \tau_{zr}^2) = 2\sigma_s^2 \quad (30\text{-}a)$$

Where σ_s is yield stress.

In general points, shearing stress and press commpared with bending stress is small; therefore, we only considered the contribution of bending stress.

Not considering shearing stress, the formula (30–a) can be written as:

$$\sigma_r^2 + \sigma_\theta^2 - \sigma_r\sigma_\theta = \sigma_s^2 \quad (30-b)$$

From ref [6], we know that

$$\sigma_r = \frac{12Mr}{b^3}Z, \quad \sigma_\theta = \frac{12M_\theta}{b^3}Z \quad (31)$$

The maximum value being at the upper and the bottom surface, we can choose $Z^* = \pm\frac{h}{2}$. Substituting (31) into (30–b), Mises's strength condition can be written as

$$(M_r^2 + M_\theta^2 - M_r M_\theta)\frac{36}{H(r)^4} = \sigma_s^2 \quad (32)$$

After dividing the whole plate into n annular segments whose thickness is Hi (i = 1,2,3,⋯,n) and having dimensionless transformation, the formula (32) becomes

$$cos^2 m\theta[(M_m^i)^2 + (M_\theta^{mi})^2 + M_m^i M_\theta^{mi}] \cdot \delta_i^{\frac{4}{3}} = \frac{\sigma_s^2 a^2 H_0^4}{36D_0^2} \quad (33)$$

Let $R_m^i = [(M_m^i)^2 + (M_\theta^{mi})^2 + M_m^i M_\theta^{mi}] \cdot \delta^{\frac{4}{3}}$

$$G_{max} = \frac{\sigma_s^2 a^2 H_0^4}{36D_0^2}$$

We call R_m^i the bending moment of equal effect for R_m concerned with Mr, M_θ.

G_{max}'s meaning is maximum bending moment of equal effect on Mises' strength condition under the condition of uniform plate.

Therefore a strength constrained condition can be written as:

$$(R_m^i)_{max} \leqslant G_{max} \quad (34)$$

(3). Final Expression of the Problem

Determine distributed variables of thickness for whole elastic annular plate, and minimize volume of whole plate

$$\overline{V} = \sum_{i=1}^{n} (\delta_i)^{\frac{1}{3}} \frac{(\beta_i^2 - \beta_{i-1}^2)}{(1-\beta_0^2)} \tag{35}$$

subjected to

$$\left. \begin{array}{l} W_{max} \leqslant W_0 \\ (R_m^i)_{max} \leqslant G_{max} \\ \delta_{min} \leqslant \delta_i \leqslant \delta_{max} \end{array} \right\} \tag{36}$$

THE METHOD OF OPTIMAL DESIGN

From the above analysis for the problem, we can see that the problem is a non–linear programming problem with multi–constraint conditions. There are many methods for non–linear programming problems, we adopted the synthesized constrained dual–densent method [7] and flexible tolerance method [8].

Now, we deduce the formulation of the gradient of objective function and constraned conditions.

(1). Objective Function

From (35), the partial differental of $\overline{V}$ can be written as

$$\frac{\partial \overline{V}}{\partial \delta_i} = -\frac{1}{3}(\delta_i)^{\frac{-4}{3}} \frac{(\beta_i^2 - \beta_{i-1}^2)}{(1-\beta_0^2)} \tag{37}$$

$$(i = 1,2,3,\cdots,n)$$

(2). Constrained Conditions

Letting $q(r,\theta) = q_0 \cos m\theta$, in this case, both Wmax and $(R_m^i)_{max}$ can be obtained when $\theta = 0$. So deducing $\frac{\partial w}{\partial \delta_i}, \frac{\partial R_m^i}{\partial \delta_i}$ we do not consider $\cos m\theta$. That is

$$S^i(x,\delta) = S_m^i(x,\delta) \tag{38}$$

$$(i = 1,2,3,\cdots,n)$$

Substituting (20) into (38), we can get

$$S^i = (x,\delta) = AA_m^i S'_m(\beta_0) + BB_m^i \tag{39}$$

We have got the explicit expression of formula (39) about design variables, so we can obtain the formulation of the partial differential for (39) by directly differentiating with respect to δ_i. In order to avoid long deductions, it is not necessary to give detail.

EXAMPLES AND DISCUSSION

During computation, we select $\beta = \frac{b}{a} = 0.2$, $v = 0.3$ [4].

We computed the minimum volume of plates which subjected to un–uniform, distributed loads and uniform distributed loads with deflection, strength and geometrically constrained conditions.

Example 1: both edge clamped $q(r,\theta) = q_0 \cos 4\theta$.

The volume optimized is 67.27 percent of the volume of uniform plate, that is.

$$\overline{V} = V / V_0 = 0.6727$$

(In all the folowing figures, solid lines represent optimal results. dotted lines represent results of uniform plate. (a): thickness distribution; (b): deflection distribution; (c): equivalent bending moment distribution.)

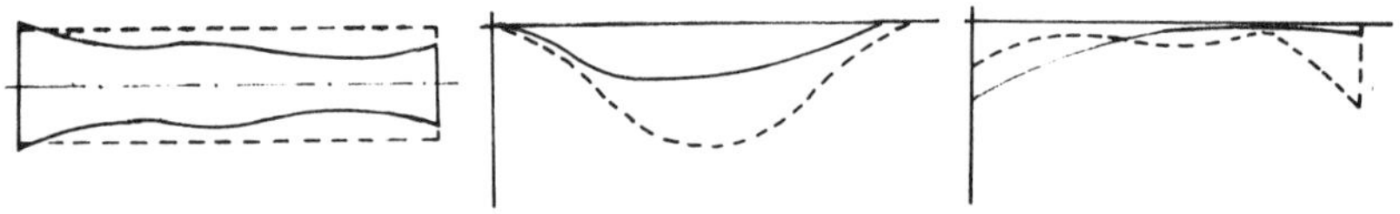

(a) (b) (c)

Fig.3

Example 2: inner edge clamped, outer edge simply supported: $q(r,\theta) = q_0 \cos 4\theta$; $\overline{V} = 0.883$

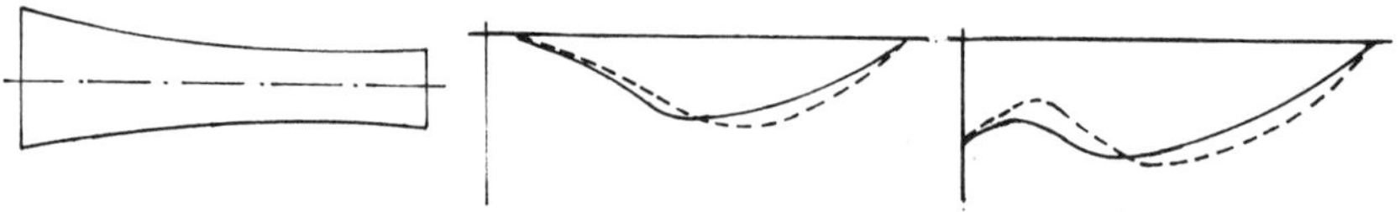

(a) (b) (c)

Fig.4

Example 3: inner edge clamped, outer edge free; $q(r,\theta)=q_0\cos4\theta$; $\overline{V}=0.983$

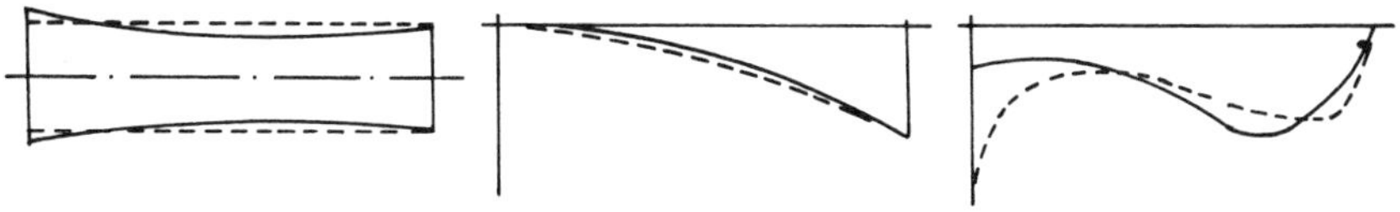

(a) (b) (c)

Fig.5

Example 4: both edges clamped. $q(r,\theta)=q_0$; $\overline{V}=0.5878$

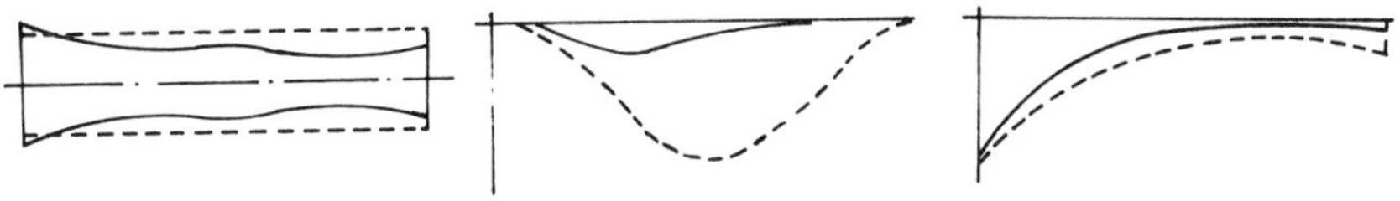

(a) (b) (c)

Fig.6

Example 5: inner edge clamped, outer edge simply supported; $q(r,\theta)=q_0$; $\overline{V}=0.843$

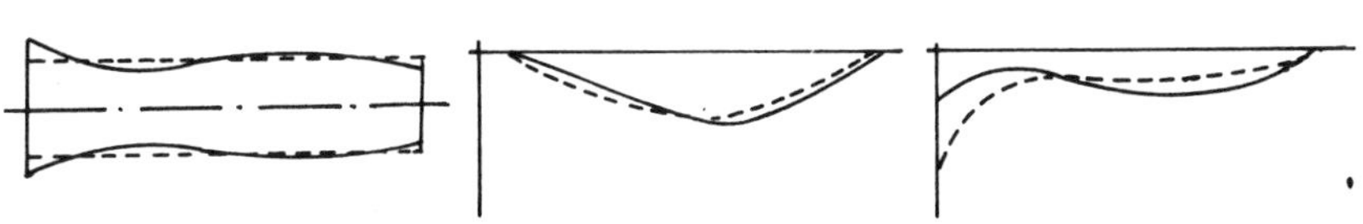

(a) (b) (c)

Fig.7

Example 6: inner edge clamped. outer edge free; $q(r,\theta)=q_0$; $\overline{V}=0.967$

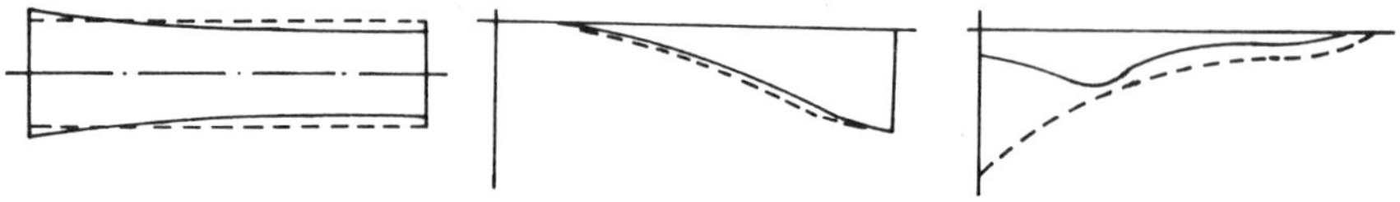

(a) (b) (c)

Fig.8

(1). In presenting the problem, we empolyed geometry constrained conditions($\delta_{min}<\delta_i<\delta_{max}$). but during optimal computing, we relaxed the geometry constraint by only ensuring ($\delta_i>0$); and the results are still reasonable. In the case of the above discussion, there is neither very thin concentration of mass whose height is infinite, nor is there a case in which local thickness is too thin.

(2). From above results, we can see that the results of both edges clamped are best; the results of inner edge clamped and outer edge simply suported are better; for inner edge clamped and outer edge free, the volume optimized only descends a little compared with the volume of uniform plate. So, we can conclude that the more constraints on the boundary, the better the optimal results.

(3). Having analyzed the curve of bending moment and deflection, we find that the point of maximum deflection moves towards the inner boundary for either both edges clamped or inner edge clamped and outer edge simply supported; the point of maximum bending moment moves towards the outer boundary for either inner edge clamped and outer edge simply supported or inner edge clamped and outer edge free.

REFERNCES

[1] B. L. Karihaloo and S. Kanagasundaram "Optimum Structures under strength and stiffness constraints".Computers and Structure Vol. 28. No. 5. P641–661

[2] Yu. H. R and K. Y. Yeh "Optimal design of thin elastic plate under

arbitrarily load". Acta Mech. Sinica. Proceedings of China Researcher of the 16–th ICTAM, (1986). P348–358.

[3] Yeh, K. Y. and J. H. Kue, Bending of arbitrary axisymmetrically non–homogeneous and variable thickness circular plate with holes at centers under arbitrary steady temperature", J. of Lanzhou University. special Number of Mech, 19 (1979) P75–114.

[4] Yu Huan–ran and Yeh Kai–Yuan, "Optimal Design of Minimax Deflection of an Annular Plate "Applied Mathematics and Mechanics. Vol 9. No. 1

[5] Xu bing–ye and Chen Sheng–can, General Treatise of Plasticity Theory, Tsing Hua Vniuersity Press. 1981.

[6] Xu, zhi–lun. Elasticity, People Education Press, 1980.

[7] Wan. Yao–qing, Liang, geng–rong and Chen, zhi–qiang. Commom using programming collection of optimal design method. Worker Press. 1983.

[8] Himmelblan D. M. Appiled Nonlinear Programming.

[9] Timoshenkos's and S wrionwsky–krieger. Theory of plates and shells. McGraw–Hill Book. Com. Inc. 1959.

Gradientless Computer Methods in Shape Designing

E. Schnack, G. Iancu

Institute of Solid Mechanics, Karlsruhe University, Kaiserstrasse 12, 7500 Karlsruhe 1, Germany

INTRODUCTION

The finite element formulation has been used by many researchers for shape optimization. Nonlinear programming with sensitivities obtained by implicitly differentiating the discretized equations has been used by Zienkiewicz and Campbell [46], Francavilla, Ramakrishnan and Zienkiewicz [07], Ramakrishnan and Francavilla [23] and Kristensen and Madsen [18] to solve this problem in two dimensions. The papers of Pedersen and Laursen [22], Zhang and Beckers [44] and Trompette and Marcelin [41] treat shape optimization of axisymmetric structures in a similar manner. Aspects associated with three-dimensional structures are discussed in this context by Botkin, Yang and Benett [03], Imam [14], and Kodiyalam and Vanderplaats [17]. A detailed description about computation of structural response using the FE based discrete approach and numerical problems associated with this are presented by Haftka [08], Wang, Sun and Gallagher [42] and Haftka and Barthelemy [09].

Variational equations as such derived in the papers of Choi and Haug [04], Chun and Haug [05,06], Zolesio [45] and in the book of Haugh, Choi and Komkov [10] take advantage of the variational character of FE formulation. A comparison between the discrete and continuum approach can be found for instance in [44].

Beside the classical gradient methods of mathematical programming, we have the possibility to develop a nongradient strategy of the feasible direction type for minimization of stress concentrations. The research work has shown that this is possible for two-dimensional, axisymmetric and general three-dimensional problems. For previous works and actual state of research on this topic, see Schnack [26-34], Schnack and Iancu [35-37], Schnack and Spörl [38], Schnack, Spörl and Iancu [39], Iancu and Schnack [12-13], Iancu [11] and Spörl [40].

Attention has also been paid to shape optimization with the boundary element approach. Mota Soares et al. [20,21], Rodriguez and Mota Soares [24] and Rodriguez [25] use a variational approach to optimize the shape of shafts. The paper of Kane and Saigal [15] is devoted to the sensititivity analysis by differentiating the discretized BE-equations for two-dimensional problems. Shape sensitivity using Analytical differentiation of the boundary integral equation was formulated for three-dimensional linearly elastic structures by Barone and Yang [01] and by Zhang and Mukherjee [43] for the plane case.

Contributions about adaptive meshing in context of shape optimization can be found in the works of Benett and Botkin [02], Kikuchi et al. [16] and Leal [19].

ANALYSIS

Several problems of engineering are formulated for weight and cost minimization of structures involving beam, truss and plate elements with constraints on the design variables, displacements, stresses or eigenvalues. These are called sizing problems because the design variables: cross sectional dimensions, area's moments of inertia, moments of resistance as well as the physical constraints are defined on a given domain. In the linear case one can derive analytical expressions for partial derivatives of the problem functions without any difficulties.

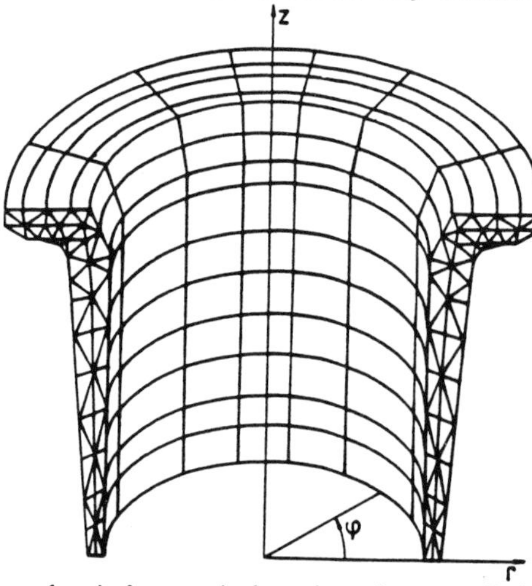

Figure 1. Axisymmetric structure - Rotor mast.

Because shape optimization problems with objective function or constraints depending on state of system do not fall into this category, they require a more complex treatement. An example is shape optimization with the objective minimization of stress concentration of the rotor mast of a helicopter shown in Figure 1. An axial section of this axisymmetric structural component together with the boundary conditions, design boundary and variation domain are given in Figure 2. Here Γ denotes design boundary and Γ^* variation domain. In Figure 3 is shown a typical FE mesh for the starting design. This problem has been treated in the case of axisymmetric loading, also in [44]. In the present lecture we compare two optimal solutions for nonaxisymmetric loading. They have been obtained in [41] using the augumented Lagrangian multiplier together with the DFP method of nonlinear programming and in [13] by a nongradient method.

The optimal solution with the nongradient strategy is shown in Figure 4(a), while the optimal design with the DFP-Method from [41] is demonstrated in Figure 4(b). A comparison of the stress distributions of the starting design and of the solutions from [41] and [13] is shown in Fig. 5.

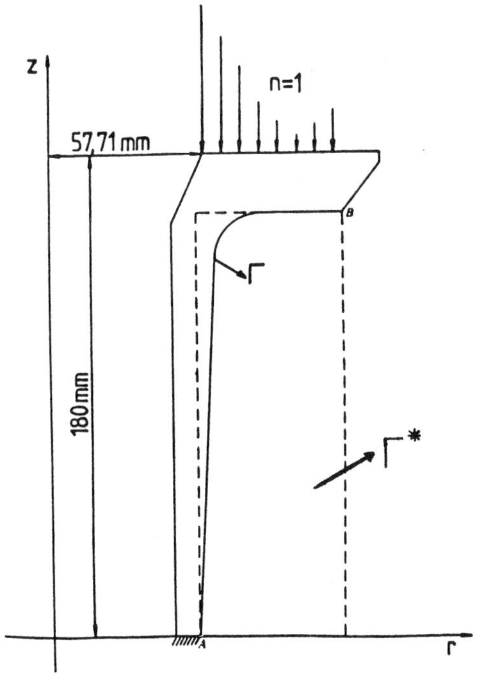

Figure 2. Axial section of the rotor mast.

It can be seen that in the critical area, the stress peak is rapidly reduced by the nongradient strategy. The design variable of such a problem is the shape. This will be described in the following by the vector of design variables b. The vector b appears in the functions which describe the physical state of structure such as the stress components B in two ways: explicitly and implicitly through the displacement vector u(b):

$$B(b,\ u(b)) \qquad\qquad b \in \mathbb{R}^n \qquad (1)$$

The minimization of stress concentrations can be written as follows:

$$\min \quad f(B(b,\ u(b))) \qquad b \in \mathbb{R}^n \tag{2}$$

$$g_i(b) \le 0 \tag{3}$$

$$h_j(B(b,\ u(b))) \le 0 \tag{4}$$

with $i = 1(1)k,\ j = 1(1)\ell$.

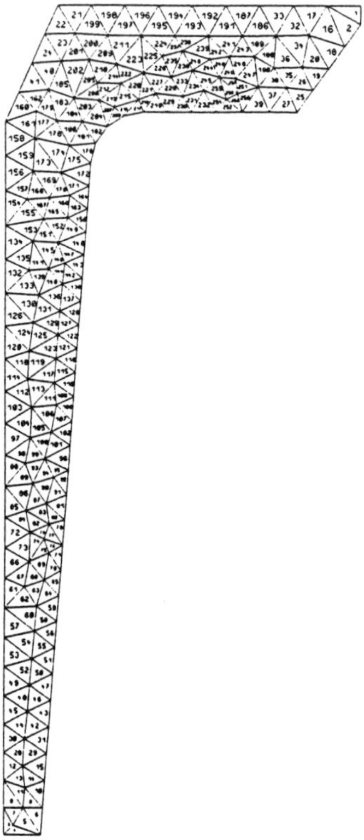

Figure 3. Starting Design.

The function f is defined as the maximum von Mises stress value $\bar{\sigma}_\mu$ of all M loading cases in a subdomain of the boundary value problem Ω^*:

$$f: \text{Max } \bar{\sigma} \qquad \text{in } \Omega^* \subset \Omega \text{ and for } \mu = 1(1)M \tag{5}$$

The geometrical and physical constraints denoted by g and h respectively mean:

$$g: \qquad \Gamma \subset \Gamma^* \tag{6}$$

$$h: \qquad \bar{\sigma}_\mu \le \sigma \quad \text{in } \Omega \tag{7}$$

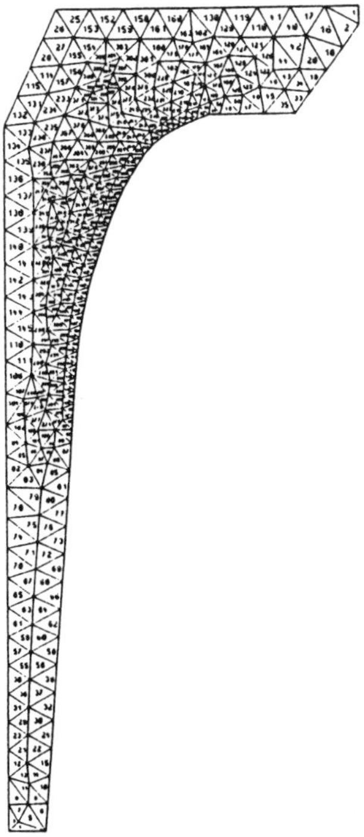

Figure 4(a): Optimal design with the nongradient method from [13]

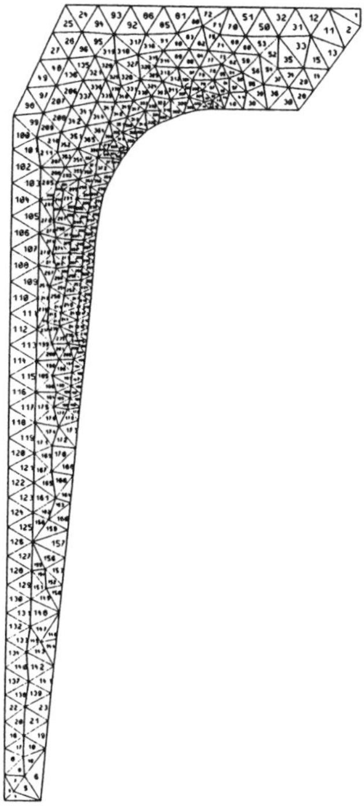

Fig. 4(b): Optimal design with the DFP-method from [41]

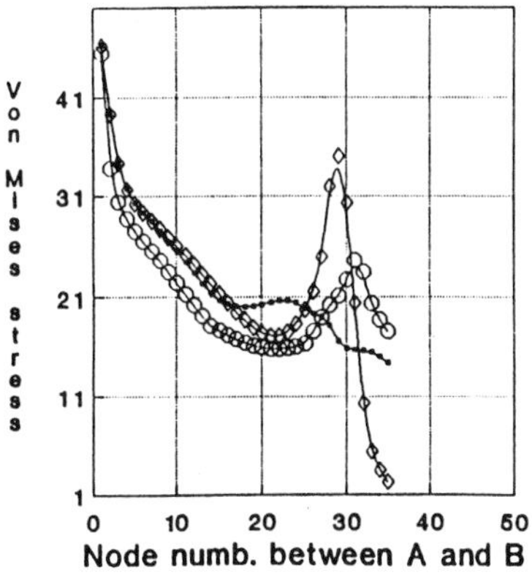

Figure 5. Stress distributions on Γ for: --◇-starting design

- -⊖- DFP-method from [41]

- -•- nongradient method from [13]

In the Inequality (7) denotes σ an upper stress bound. Because the shape optimization problem (5-7) is in general nondifferentiable, we have to transform it, if we want to use mathematical programming procedures for the solution. This can be done by many methods. In all these cases we have to compute the sensitivities of functions of type B.

We can also predict pointwise the stress response to shape variation at a traction free boundary using the monotonicity theorem. The nongradient procedure for minimization of stress concentration has the following form:

$$b^T = \left[x_1,\ldots,x_{NB},\ y_1,\ldots,y_{NB},\ z_1,\ldots,z_{NB}\right] \tag{8}$$

$$\min\max\left(\bar{\sigma}^{\mu}_{1}(b),\ldots,\bar{\sigma}^{\mu}_{NB}(b)\right) \tag{9}$$

$$\bar{\sigma}^{\mu}_{i}(b) - \sigma \leq 0 \tag{10}$$

$$Ab - B \leq 0 \tag{11}$$

$$\text{with } \mu = 1(1)M \quad \text{and} \quad i = 1(1)N$$

with NB number of nodal points on the design boundary.

The transition function f_j which describes the iteration rule for changing the nodal point coordinates on the optimizing boundary is defined on a physical basis:

- A geometrical perturbation on the design boundary Γ produces a rapid fade away of the von Mises stress $\bar{\sigma}$ in the neighbourhood of the perturbation.

- From the monotonicity relation of the two principal stresses with respect to the corresponding normal curvatures in the principal stress directions we can derive a relation for the control of the von Mises stress on Γ. We have pointwise:

$$\delta\bar{\sigma}^{\mu} = K^{s}_{w}\delta h_{n} \tag{12}$$

  with K^{s}_{w} as weighted surface curvature which depends generally on stress state and geometrical data on the surface, and δh_n as the perturbation in the normal direction of the boundary Γ.

- By increasing the minimum effective stress, the maximum effective stress in the direct neighbourhood can also be reduced:

$$b_j = f_j(b_{j-1}, v_j) \quad \text{explicit}$$

$$\text{(for tetrahedron elements)} \tag{13}$$

$$x_j^i = x_{j-1}^i + v_j^i \cdot \frac{1}{N_i} \sum_{k=1}^{N_i} \bar{a}_{j-1}^{i,k} \tag{14}$$

$$y_j^i = y_{j-1}^i + v_j^i \cdot \frac{1}{N_i} \sum_{k=1}^{N_i} \bar{b}_{j-1}^{i,k} \tag{15}$$

$$z_j^i = z_{j-1}^i + v_j^i \cdot \frac{1}{N_i} \sum_{k=1}^{N_i} \bar{c}_{j-1}^{i,k} \tag{16}$$

for i = 1(1)NB

with $\bar{a}_{j-1}^{i,k}$, $\bar{b}_{j-1}^{i,k}$ and $\bar{c}_{j-1}^{i,k}$ coefficients of the Hessian form for each of the N_i surface triangles abutting the node i. Because the magnitude of shifting v_j is controlled by an arithmetical smoothing algorithm, we have at the nodal point i:

$$v_j^i = v_j^{i^*} \xi_{j-1}^i \left(1 - \frac{s^i}{M_s}\right), \quad s^i = 1(1)M_s \tag{17}$$

with M_s = number of smoothing zones

$v_j^{i^*}$ = magnitude of shifting at a point with maximum or minimum stress

Making the approximation that each of two principal stresses is a linear function of the normal curvature we get for the sign of shifting ξ_{j-1}^i:

$$\xi_{j-1}^i = \operatorname{sgn}(\sigma_{1,j-1}^i + \sigma_{2,j-1}^i)\ \operatorname{sgn}(\bar{\sigma}_{j-1}^i - \tilde{\sigma}_{j-1}) \tag{18}$$

for i = 1(1)NB

where σ_1^i and σ_2^i denote the two principal stresses at the nodal point i and $\tilde{\sigma}$ the average von Mises stress on the design boundary.

As a result, we have a discrete, dynamic optimization problem with the following 'cost function' g_j:

$$g_j := (\bar{\sigma})_j^{max} - (\bar{\sigma})_{j-1}^{max} \tag{19}$$

$$(\bar{\sigma})_j^{max} = \text{Max}\ ((\bar{\sigma}_1^{\mu})_j, \ldots, (\bar{\sigma}_{NB}^{\mu})_j) \qquad \mu = 1(1)M \tag{20}$$

Problem:

$$\min \sum_{j=1}^{l} g_j\left(b_{j-1}, v_j\right) \tag{21}$$

$$b \in \Xi$$

$$v_j \in \Omega_j\left(b_{j-1}\right) \tag{22}$$

with $\Omega_j(b_{j-1})$ as feasible control space

$$\left(x_j^i, y_j^i, z_j^i\right) \in \bar{\Omega}_j^* \subset \mathbb{R}^n \tag{23}$$

with $n = 3$, $i = 1(1)NB$, $j = 1(1)\ell$

$\bar{\Omega}^*$ closed set: $\bar{\Omega}^* = \Omega \cup \Gamma_- \cup \Gamma_+$

State space Ξ: $\Xi \subset \mathbb{R}^{3NB}$

Ξ is compact, as a product of compact sets from geometrical constraints:

$$v_j^i \in \left[-d^i\left(\Gamma_{d,j-1}, \Gamma_-\right),\ d^i\left(\Gamma_{d,j-1}, \Gamma_+\right)\right] = \Omega_j^i\left(b_{j-1}\right) \tag{24}$$

For the decision space it follows:

$$\Omega_j\left(b_{j-1}\right) = \prod_{i=1}^{NB} \Omega_j^i\ \left(b_{j-1}\right) \subset \mathbb{R}^{NB} \tag{25}$$

with the theorems of Tychonoff (Ω_j^i is compact) and Weierstrass and making the supposition that the objective function from the dynamic optimization problem is continuous, we have the existence of the solution.

3D-TEST PROBLEM

An example of application of the nongradient procedure for three-dimensional structures is the stress optimal shape of a cavity in a large elastic domain (see [11] and Figure 6). The starting geometry for this problem (see Figure 7) was a spherical hole:

$$a : b = a : c = 1 \tag{26}$$

with the maximum stress 3.36 σ_{0x}.

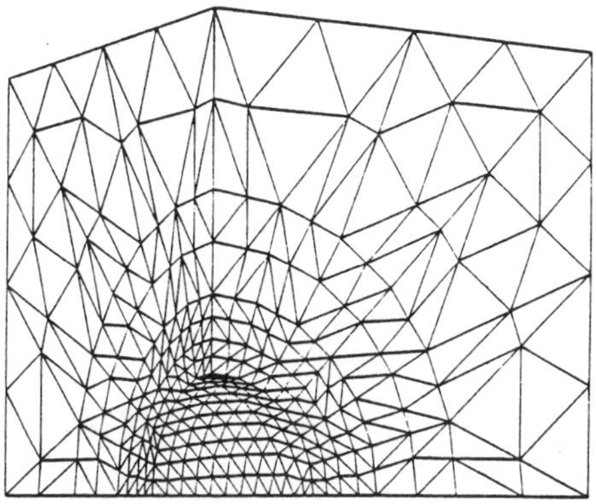

Figure 6. FE-discretization of body with spherical hole.

The loading case is:

$$\sigma_{0x} : \sigma_{0y} : \sigma_{0z} = 1 : 2 : 2. \tag{27}$$

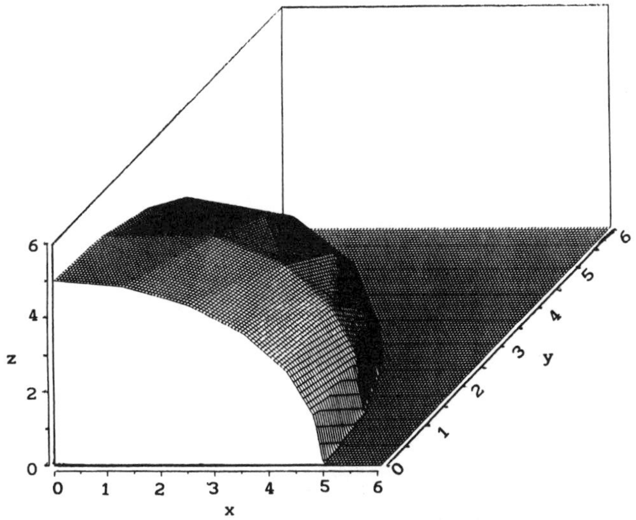

Figure 7. Shape of the starting surface.

After 24 iteration steps we obtain an axisymmetric ellipsoid with:

$$b : a = c : a = 2.160 \tag{28}$$

The maximum stress value is reduced here to 2.70 σ_{0x}. This shape is shown in Figure 8. The proof of optimality for the cavity problem can be given analytically with the maximum priciple. The result is:

$$b : a = c : a = 2.614 \tag{29}$$

and a constant von Mises stress on Γ of 2.5 σ_{0x}.

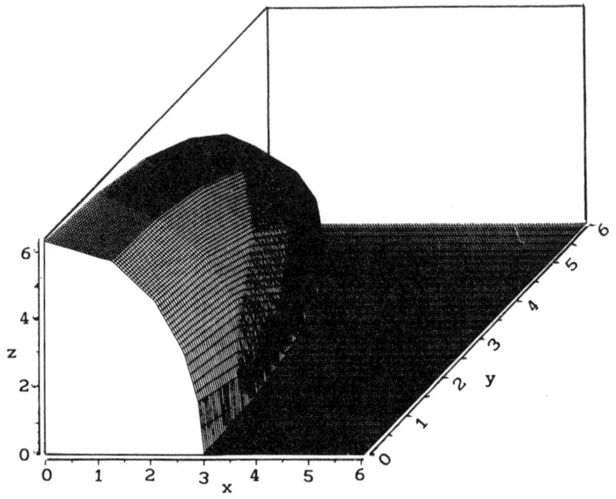

Figure 8. Optimal shape of cavity in a large elastic domain.

The examples of the last chapter show the high performance of the nongradient strategies in optimizing notch problems of elasticity.

REFERENCES

01. Barone, M.R. and Yang, R.J. A Boundary Element Approach for Recovery of Shape Sensitivities in Three-Dimensional Elastic Solids, Computer Methods in Applied Mechanics and Engng., Vol. 74, pp. 69-82, 1989.
02. Benett, J.A. and Botkin, M.E. Shape Optimization of Two-Dimensonal Structures with Geometric Problem Description and Adaptive Mesh Refinement, AIAA Journal, 1983.
03. Botkin, M.E., Yang, R.J. and Benett, J.A. Shape Optimization of Three-Dimensional Stamped and Solid Automotive Components. Paper presented at the International Symposium on Optimum Shape, General Motors Research Labs, Warren, Michigan, 1985.
04. Choi, K.K. and Haug, E.J. Shape Design Sensitivity Analysis of Elastic Structures, J. of Structural Mech., Vol. 11, No. 2, pp. 231-269, 1983.
05. Chun, Y.W. and Haug, E.J. Two-Dimensional Shape Optimal Design, Int. J. Num. Meth. Engng., Vol. 13, pp. 311-336, 1978.

06. Chun, Y.W. and Haug, E.J. Shape Optimization of a Solid of Revolution. J. of Engng. Mech., Vol. 109, No. 1, pp. 30-46, 1983.
07. Francavilla, A., Ramakrishnan, C.V. and Zienkiewicz, O.C. Optimization of Shape to Minimize Stress Concentration, J. of Strain Analysis, Vol. 10/2, pp. 63-69, 1975.
08. Haftka, R.T. Finite Elements in Optimal Structural Design. Computer Aided Optimal Design: Structural and Mechanical Systems, (Ed. Mota Soares, C.A.), pp. 271-297, Springer-Verlag, Berlin and New York, 1987.
09. Haftka, R.T. and Barthelemy, B. On the Accuracy of Shape Sensitivity. Computer Aided Optimum Design of Structures: Recent Advances, (Ed. Brebbia, C.A. and Hernandez, S.), pp. 327-336, Springer-Verlag, Berlin, Heidelberg and New York, 1989.
10. Haug, E.J., Choi, R.K. and Komkov, V. Design Sensitivity Analysis of Structural Systems. Mathematical Science and Engineering, (Ed. Ames, W.F.), Academic Press, Orlando, San Diego and New York, 1986.
11. Iancu, G. Optimierung von Spannungskonzentrationen bei dreidimensionalen elastischen Strukturen, Doctoral Thesis, Karlsruhe University, 1991.
12. Iancu, G. and Schnack, E. Knowledge-Based Shape Optimization, in Computer Aided Optimum Design of Structures: Recent Advances (Ed. Brebbia C.A. and Hernandez, S), pp 71-83, Proceedings of the First Conference on Computer Aided Optimum Design of Structures (CAOD) - OPTI 89, Southampton, United Kingdom, 1989. Springer-Verlag, Berlin, Heidelberg and New York, 1989.
13. Iancu, G. and Schnack, E. Shape Optimization Scheme for Large Scale Structures. Proceedings of the Second World Congress on Computational Mechanics, 27 - 31 August 1990, Stuttgart, Germany. to appear
14. Imam, M.H. Three-Dimensional Shape Optimization, Int. J. of Num. Meth. Engng., Vol. 18, pp. 661-673, 1982.
15. Kane, J. and Saigal, S. Design-Sensitivity Analysis of Solids Using BEM, J. of Engng. Mech., Vol. 114, No. 10, pp. 1703-1722, 1988.
16. Kikuchi, N., Chung, K.Y., Torigaki, T. and Taylor, J.E. Adaptive Finite Element Methods for Shape Optimization of Linearly Elastic Structures, Comp. Meth. in Appl. Mech. and Engng., Vol. 57, pp. 67-89, 1986.
17. Kodiyalam, S. and Vanderplaats, G.N. Shape Optimization of Three-Dimensional Continuum Structures via Force Approximation Techniques, AIAA Journal, Vol. 27 No. 9, pp. 1256-1263, 1989.
18. Kristensen, E.S. and Madsen, N.F. On the Optimum Shape of Fillets in Plates Subjected to Multiple In-Plane Loading Cases, Int. J. Num. Meth. Engng., Vol. 10, pp. 1007-1019, 1976.
19. Leal, R.P. Boundary Elements in Bidimensional Elasticity, Master Sc. Thesis, Technical University of Lisbon, 1985.
20. Mota Soares, C.A, Rodrigues, H.C., Oliveira Faria, L.M. and Haug, E.J. Optimization of the Geometry of Shafts Using

Boundary Elements, ASME J. of Mechanisms, Transmissions and Automation in Design, Vol. 106, pp. 199-203, 1984.
21. Mota Soares, C.A., Rodriguez, H.C., Oliviera Faria, L.M. and Haug, E.J. Boundary Elements in Shape Optimal design of Shafts. Optimization in Computer Aided Design, (Ed. J.S. Gero, J.S.), pp. 155-175, North-Holland, 1985.
22. Pedersen, P. and Laursen, L.L. Design for Minimum Stress Concentration by Finite Element Elements and Linear Programming, J. of Struct. Mech., Vol. 10/4, pp. 375-391, 1982-83.
23. Ramakrishnan, C.V. and Francavilla, A. Structural Shape Optimization Using Penalty Functions, J. of Struct. Mech., Vol. 3/4, pp. 403-422, 1974-1975.
24. Rodriguez, H.C. and Mota Soares, C.A. Shape Optimization of Shafts. Third National Congress of Theoretical and Applied Mechanics, Lisbon, Portugal, 1983.
25. Rodriguez, H.C. Shape Optimization of Shafts Using Boundary Elements, Master Sc. Thesis, Technical University of Lisbon, 1984.
26. Schnack, E. Ein Iterationsverfahren zur Optimierung von Spannungskonzentrationen, Habilitationsschrift, Univ. Kaiserslautern, 1977.
27. Schnack, E. Ein Iterationsverfahren zur Optimierung von Kerboberflächen, VDI-Forschungsheft, Nr. 589, VDI-Verlag, Düsseldorf, 1978.
28. Schnack, E. An Optimization Procedure for Stress Concentrations by the Finite Element Technique, Int. J. Num. Meth. Engng., Vol. 14, No. 1, pp. 115-124, 1979.
29. Schnack, E. Optimierung von Spannungskonzentrationen bei Viellastbeanspruchung, ZAMM, Vol. 60, T151-T152, 1980.
30. Schnack, E. Optimal Designing of Notched Structures without Gradient Computation, in Control of Distributed Parameter Systems (Ed. Barbary J.P. and Le Letty, L.), pp 365-369, Proceedings of the 3rd IFAC-Symposium, Toulouse, France, 1982. Pergamon Press, Oxford, New York, Toronto, Sydney, Paris and Frankfurt, 1982.
31. Schnack, E. Computer Simulation of an Experimental Method for Notch-Shape-Optimization, in Simulation in Engineering Sciences (Ed. Burger J. and Janny, Y), Tome 2, pp. 269-275, Proceedings of the Int. Symp. of IMACS, Nantes, France, 1983. Elsevier Science Publishers B.V. (North Holland), Amsterdam, The Netherlands, 1985.
32. Schnack, E. Local Effects of Geometry Variation in the Analysis of Structures. Studies in Applied Mechanics 12: Local Effects in the Analysis of Structures, (Ed. Ladevèze, P.), pp. 325-342, Elsevier Science Publisher, Amsterdam, 1985.
33. Schnack, E. Free Boundary Value Problems in Elastostatics, in Innovative Numerical Methods in Engineering (Ed. Shaw, R.P., Periaux, J., Chaudouet, A., Wu, J., Marino, C., Brebbia, C.A.), pp. 435-440, Proceedings of the 4th Int. Symp. on Numerical Methods in Engineering, Atlanta, Georgia/USA, 1986. Computational Mechanics Publications Southampton, Springer-Verlag, Berlin, Heidelberg and New York, 1986.
34. Schnack, E. A Method of Feasible Direction with FEM for Shape Optimization, in Structural Optimization (Ed. Rozvany,

G.I.N., Karihaloo, B.L.), pp. 299-306, invited lecture: Proceedings of the IUTAM-Symp. on Structural Optimization, Melbourne, Australia, 1988. Kluwer Academic Publishers, Dordrecht, Boston and London, 1988.

35. Schnack, E. and Iancu, G. Control of the von Mises Stress with Dynamic Programming, in Proceedings of the GAMM-Seminar, (Ed. Eschenauer, H.A. and Thierauf, G.), Vol. 43, pp. 154-161, GAMM-Seminar on Discretization Methods and Structural Optimization - Procedures and Applications, Siegen, Federal Republic of Germany, 1988. Springer-Verlag, Berlin and Heidelberg 1989.

36. Schnack, E. and Iancu. G. Shape Design of Elastostatics Structures Based on Local Perturbation Analysis, Structural Optimization, Vol. 1,pp. 117-125, 1989.

37. Schnack, E. and Iancu, G. Non-Linear Programming Applicable for the Control of Elastic Structures, in Preprints of the 5th IFAC Symposium on Control of Distributed Parameter Systems (Ed. El Jai, A. and Amouroux., M.), pp. 163-168, Proceedings of the 5th IFAC Symposium on Control of Distributed Parameters, Perpignan, France, 1989. Institut de Science et de Génie des Matériaux et Procédés (CNRS), Groupe d'Automatique, Université de Perpignan, 1989.

38. Schnack, E. and Spörl, U. A Mechanical Dynamic Programming Algorithm for Structure Optimization, Int. J. Num. Meth. Engng., Vol. 23, No. 11, pp. 1985-2004, 1986.

39. Schnack, E. Spörl, U. and Iancu, G. Gradientless Shape Optimization with FEM, VDI Forschungsheft 647/88, pp. 1-44, 1988.

40. Spörl, U. Spannungsoptimale Auslegung elastischer Strukturen, Docto- ral Thesis, Karlsruhe University, 1985.

41. Trompette, Ph. and Marcelin, J.L. On the Choice of the Objectives in Shape Optimization. Computer Aided Optimal Design: Structural and Mechanical Systems, (Ed. Mota Soares, C.A.), pp. 247-261,Springer-Verlag, Berlin and New York, 1986.

42. Wang, S.-Y., Sun, Y., Gallagher, R.H. Sensitivity Analysis in Shape Optimization of Continuum Structures, Computer & Structures, Vol. 20, No. 5, pp. 855-867, 1985.

43. Zhang, Q., Mukherjee, S. Design Sensitivity Coefficients for Linear Elasticity Problems by Boundary Element Methods, (Ed. Kuhn, G. and Mang, H.), pp. 283-289, Proceedings of the IUTAM/IACM Symposium on Discretized Methods in Structural Mechanics, Vienna,Austria, 1989. Springer-Verlag, Berlin and Heidelberg, 1990.

44. Zhang, W.H. and Beckers, P. Comparison of Different Sensitivity Analysis Approaches for Structural Shape Optimization. Computer Aided Optimum Design of Structures: Recent Advances (Ed. Brebbia, C.A. and Hernandez, S.), pp. 346-356, Springer-Verlag, Berlin, Heidelberg and New York, 1989.

45. Zolesio, J.-P. The Material Derivative (or speed) Method for Shape Optimization of Distributed Parameter Structures. Eds.: Haug E.J., J. Cea. Sijthoff and Noordhoff, Alphen aan den Rhijn (1981), pp. 1089-1151.

46. Zienkiewicz, O.C. and Campbell, J.S. Shape Optimization and

Sequential Linear Programming. Optimum Structural Design, (Ed. Gallagher R.H. and Zienkiewicz, O.C.), John Wiley & Sons, London, New-York, and Sydney, 1973.

2D- and 3D-Shape-Optimization Based on Biological Growth

C. Mattheck (*), M. Beller (**), J. Schäfer (*)

(*) Kernforschungszentrum Karlsruhe GmbH, Institute for Material- and Solid State Research IV, P.O.Box 3640, 7500 Karlsruhe 1, Germany

(**) PREUSSAG Anlagenbau GmbH, Pipeline-Service, Breslauerstr.56b, 7500 Karlsruhe 1, Germany

ABSTRACT

Biological load-carriers (e.g. wood, bone etc.) selfoptimize by adaptive growth in order to achieve a more homogeneous surface stress distribution. The selfoptimization is carried out for the most important natural loading condition encountered by the structure concerned. This natural procedure has been copied and implemented into an optimization procedure developed at the Karlsruhe Nuclear Research Center (KfK). Termed CAO (Computer Aided Optimization), the method can be used in the fields of biomechanics and engineering and is intended as an easy-to-use, straight forward optimization tool in conjunction with commercial FE-codes.

This paper introduces further examples illustrating the 2D- and 3D-application of the method. An example taken from nature is used to investigate the consequence of specialization due to optimization. T-joint structures (2D and 3D) are investigated in order to illustrate the use of the method for a technical application.

INTRODUCTION

A lot is to be learnt by carefully observing nature. In recent years engineers and scientists have become increasingly aware of ingeniuous solutions provided by nature for a number of mechanical problems [1], [2], [3] encountered by 'living structures'. It is well known that biological load carriers are subject to a hard competition for energy and living space. Only the best mechanical construction has a chance to survive. This optimum in nature is characterized by minimized weight and sufficient mechanical strength. Preliminary studies by Mattheck [4] , [5] have shown that biological structures adapting their growth with respect to external loads will grow into a state of constant mechanical stress at their surface. Adaptive growth thus leads to a contour where no point along the surface is either under- or overloaded. A homogeneous surface stress distribution is aimed for and notch stresses are eliminated. Especially the latter is a good insurance to have in order to ensure a long fatigue life of a component.

CAO: COMPUTER AIDED OPTIMIZATION

Observing and learning from nature

Nature has had ample time in order to mechanically optimize biological structures. An 'optimized' design shall always imply structural shape optimization in the context of this paper. It is the aim of the procedure used to reduce or even eliminate stress concentrations induced at the surface of a structure for a given set of loading and boundary conditions. Achieving a homogeneous stress distribution at the surface of a component, be it biological or technical, will increase its fatigue endurance. Even when limiting observation of nature just to the case of trees, it can clearly be seen that the results achieved through the mechanism of adaptive growth in surface stress reduction are very impressive indeed. All this therefore suggests to observe and learn from nature and to simulate this natural mechanism in order to improve engineering design proposals.

Different Optimization Techniques

Optimized transition contour lines for beams with a narrowing cross-section have been evaluated by Baud [6] as early as 1934. Baud used an experimental approach in order to obtain transition lines without notch stresses. More recently many attempts of structural optimization using numerical methods are reported in the literature. For reasons of space only two of those shall be mentioned here. Umetanu and Hirai [7] also decided to copy biological growth for structural optimization purposes. Huiskens et.al. [8] reported a similar method. A more detailed comparison of these individual methods is given in [4].

Simulating adaptive growth

The CAO method simulates the natural mechanism of adaptive growth by special application of a commercial FE-code on a computer. The procedure consists of the following steps as shown in Fig.1. Initially a reasonable design proposal has to be made. This is followed by a standard elastic FE-analysis. The FE-run will calculate the Mises stress distribution present in the structure due to the external loading applied. In a subsequent step only the stress values obtained for the surface are used. The main principle of the technique is to set the Mises equivalent stresses formally equal to a fictitious temperature field. This implies that the problem of stress homogenization is solved by transformation into a problem of thermal strains. The resulting temperature field is applied in a further FE-run as the only applied loading, using the thermal expansion or thermal swelling routine provided by most commercial FE-codes. The structure will then "grow" by stress controlled thermal expansion which simulates the mechanism of adaptive growth and will lead to a better structural design regarding the initially applied load. A subsequent standard FE-run will reveal the Mises stress distribution present in the "swollen" structure. The procedure can be repeated until a satisfactory stress distribution is obtained. Alternatively CAO provides the option to remove material from the region of a structure which is underloaded. In that case a minimized weight design can be obtained for a given stress distribution. The main advantages of CAO are:

- Any standard FE-code including a thermal expansion option can be used.
- There is no need for costly postprocessing routines adapted to individual problems.
- A very much improved surface stress distribution with drastically reduced stress concentrations can be obtained within only a few computing cycles, which is a major bonus considering costs.

EXAMPLES

Branch Joint or Optimization Is Specialization

Fig.2a shows the contour and loading conditions of a branch joint taken from nature. A newly developed mesh generator [9] allows a 2D-mesh to be modelled directly from a photograph, drawing or actual specimen. A branch joint of a tree was modelled using this technique and analysed using the FE-code ABAQUS [10] . The natural loading conditions for such a joint is bending as indicated in Fig.2a. The Mises stress distribution along the contour investigated is shown in Fig.2c. It can clearly be seen that no notch stresses exist along the contour line s. In other words, a homogeneous stress distribution exists for this specific loading condition.

The same structure (same shape and mesh) was then subjected to a different, non-natural loading situation (Fig.2b). It can clearly be seen that this biological load carrier is not optimized for the tensile load applied. This is just one example in order to illustrate that optimization implies specialization. A structure which is shape optimized for a given set of loading and boundary conditions will only be an optimum for just that set of conditions [11]. In addition it was found [12] that the optimization path followed also plays an important role in the optimizing process. Wrong assumptions as to actual loading and boundary conditions can easily lead to erroneous results.

T-Joints

The following examples were chosen in order to verify the use of the CAO-method on hollow cylindrical structures. Especially T- and Y-joints are commonly used in the field of pipeline-engineering and especially in marine and offshore applications. Two previous investgiations have shown that stress concentrations due to the redirectioning of force lines in joint-structures can be reduced considerably using CAO [13], [14]. Stress peaks are usually to be found in the intersection region between brace and chord. FE-computations have shown that, depending on loading conditions, the maximum stress is situated on the inside of the brace within the intersection region. However the notch stress intensity in the weld region is the more critical as the material properties of the heat affected zone (HAZ) are usually inferior to those of the base metal.

CAO aims at reducing notch stresses present at the surface of a component by shape optimization of the component in order to achieve a homogeneous stress distribution at the surface. It was the aim of the developer of this approach to supply an optimization tool which is easy to handle, can be applied with most commercial FE-codes and uses a minimum of CPU time in addition to the FE-analysis. Although initially designed for the field of biomechanics, engineering components can easily be improved (i.e.shape optimized) within very few computing cycles.

Reducing the notch intensity factor for a given component under a static load will also increase the fatigue endurance of the component, provided the loading spectrum does fall within a certain frequency band. Very high frequencies tend to shift the location of maximum stress peaks. This aspect, however, is of no concern for the case of T- and Y-joints considered here. The procedure used does not require a sensitivity analysis as the surface stress distribution is the sole parameter considered. Multiparameter conditions present in global optimization approaches do hinder an easy to use straight forward tool for engineering application. On the other hand it must be pointed out of course that CAO, as far as engineering application is concerned, is limited to stress reduction by shape optimization and is not intended as a general optimization tool. The variety of application for CAO in biomechanics and arboricultural science is shown in [4].

Fig.3 shows the 2D-structure used to model a T-joint consisting of two small diameter pipes. In order to use a simple structure, a joint was modelled between a small diameter pipe and a sphere. Using axi-symmetric 8-node elements, the FE-code ABAQUS [10] was used to investigate the stress distribution induced due to an applied internal pressure, resulting in hoop stresses within the pipe wall. Following this initial investigation the CAO-method was used to reduce the peak notch stresses present in the structure. It should be noted that an optimization is carried out for specific loading and boundary conditions and that inaccurate or wrong assumptions used for these conditions can not only lead to an non-optimized structure but can even result in a detrimental effect [11].

Fig.3 shows the 2D-structure optimized as well as the non-optimized and optimized contour superimposed onto each other. Fig.4 shows the von Mises stress distribution along the contour line s. The stress reduction obtained was 39% within only 4 computing cycles.

Fig.5 shows the FE-mesh used to model a 3D-T-Joint. Only half the structure is modelled for reasons of symmetry. A joint has been modelled by a pipe-plate connection assuming a small diameter and larger diameter T-joint. The plate was subjected to a distributed load representing the hoop stresses and the brace was loaded by a bending moment. Again the initial and final contour are shown. A cut through the location of largest surface stresses (along line s) has been chosen to show the initial and final stress distribution along the surface of the structure. The stress reduction obtained for these loading conditions was 20% obtained after only 4 computing cycles, see Fig.6. Further computations will be concerned with modelling entire T-, Y- and K-joints and improving their shape in order to increase their fatigue resistance.

CONCLUSIONS

It was the aim of this paper to show further applications of the CAO-method. CAO can be used in the fields of biomechanics and engineering and has been developed as an easy-to-use, practical tool which allows the shape optimization of 2D- and 3D-structures. A homogeneous surface stress distribution is aimed for, including a marked reduction or even elimination of any stress concentrations (notch stresses) present. The method can be applied in conjunction with any commercial FE-code incorporating a thermal expansion option. It is ideally suited for an engineer in industry who is aiming for a light weight and fatigue resistant design.

REFERENCES

[1] Gordon, J.E. Structures or Why Things Don't Fall Down, Penguin Books, England, 1978.

[2] Nachtigall, W. Funktionen des Lebens, Hoffmann und Campe, Hamburg, 1977.

[3] Nachtigall, W. Konstruktionen - Biologie und Technik, VDI-Verlag, Düsseldorf, 1987.

[4] Mattheck, C. Engineering Components Grow Like Trees, Mat.-wiss. u. Werkstofftech., 21, pp.143-168, 1990.

[5] Mattheck, C. Design And Growth Rules For Biological Structures And Their Application, Fatigue Fract.Engng Mater.Struct., Vol.13, No.5, pp.535-550, 1990.

[6] Baud, R. Beiträge zur Kenntnis der Spannungsverteilung in prismatischen und keilförmigen Konstruktionselementen mit Querschnittsübergängen, Report 29, Schw. Verb. für Materialprüfung i.d. Technik, Zürich, 1934.

[7] Umetani, Y., Hirai, S. An adaptive shape optimization method for structural material using growing-reforming procedure, Proc. of 1975 Joint ISME-ASME, Applied Mechanics Western Conference, pp.359-365, 1975.

[8] Huiskens, R., Weinans, H., Grootenboer, H., Dalstra, M., Fudala, B., Sloof, T. Adaptive bone remodelling theory applied to prosthetic design analysis, J. Biomechanics, 20, pp.1135-1150, 1987.

[9] Polyphem User's Manual, Vers.1.0, Science and Computing, Tübingen, 1990.

[10] Hibbit, H.D., Karlsson, B.J., Sorensen, E.P. ABAQUS User's Manual, Vers.4.9, Providence, R.I., 1990.

[11] Mattheck, C., Harzheim, L. Effect Of Loading On The Optimized Shape Of A Kinked Bar, entered for publication in Int.J. for Num.Meth. in Eng., 1991.

[12] Harzheim, L., Mattheck, C. 3D-Shape Optimization: Different Ways To An Optimum Design, Proc. of Int. Conf. on Eng. Optim. in Design Proc., eds.H.A. Eschenauer, C. Mattheck, N. Olhoff, Springer-Verlag, 1991.

[13] Mattheck, C., Beller, M., Bethge, K., Erb., D. Computer Aided Optimization of T-Joint Structures By Simulation of Biological Growth, Proc.First Offshore Mechanics Symposium, ISOPE, Trondheim, 1990.

[14] Beller, M., Mattheck, C., Schäfer, J. Shape Optimization on the Basis of Biological Growth with Special Regard to Slanted Joints, in Comp. Meth. in Marine and Offshore Application, Ed. T.K.S. Murthy, Computational Mech. Publ., pp.357-365, 1991.

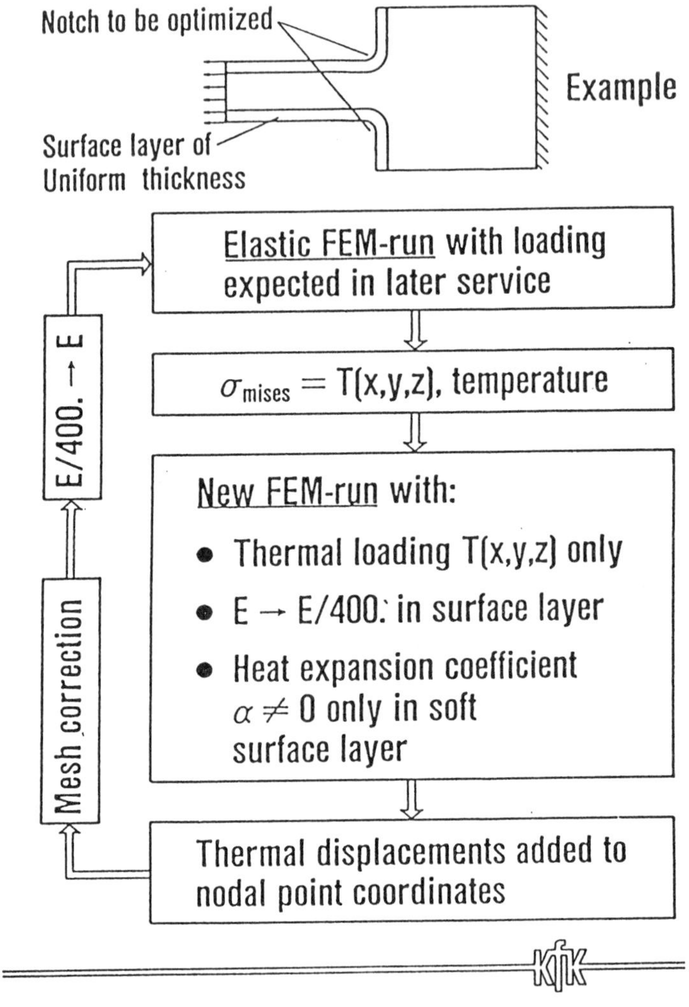

Figure 1. The Procedure of CAO [14]

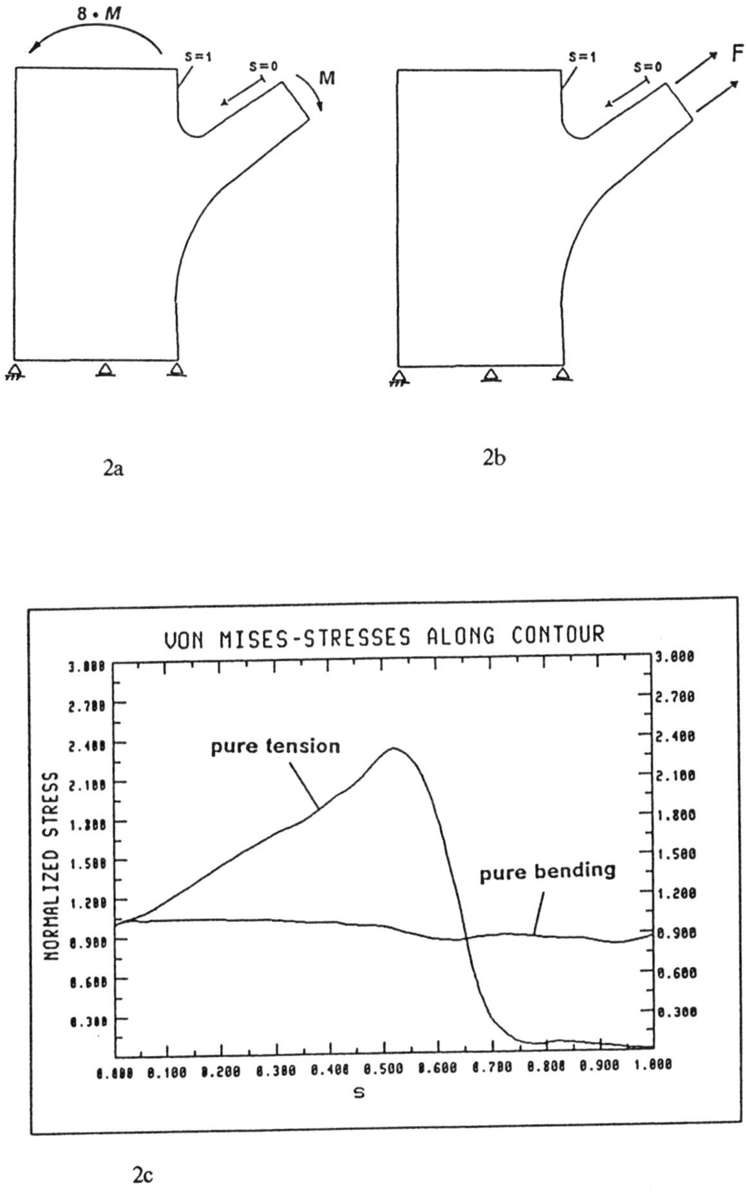

Figure 2. Branch Joint

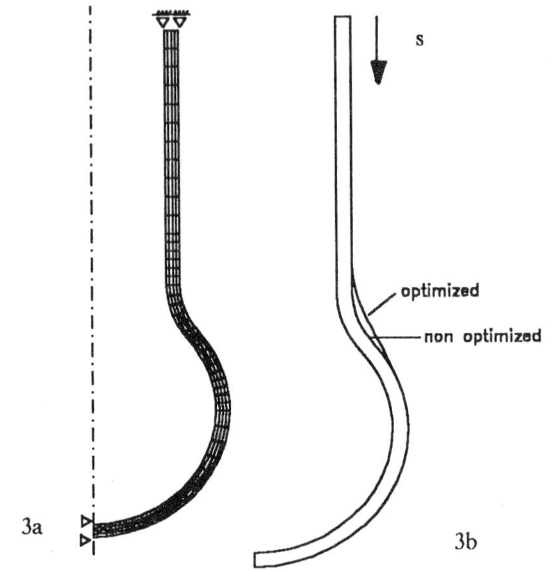

Figure 3. 2D model: a. mesh and boundary conditions b. non-optimized and optimized contour.

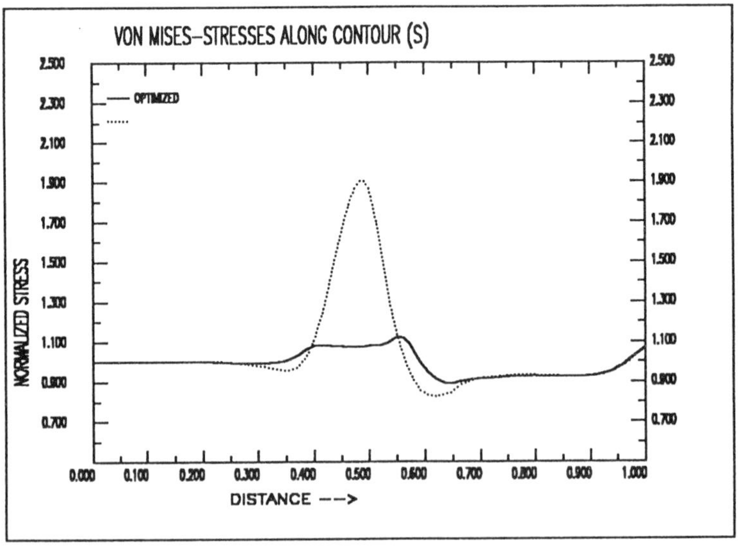

Figure 4. Mises stress distribution: surface stresses along contour line s

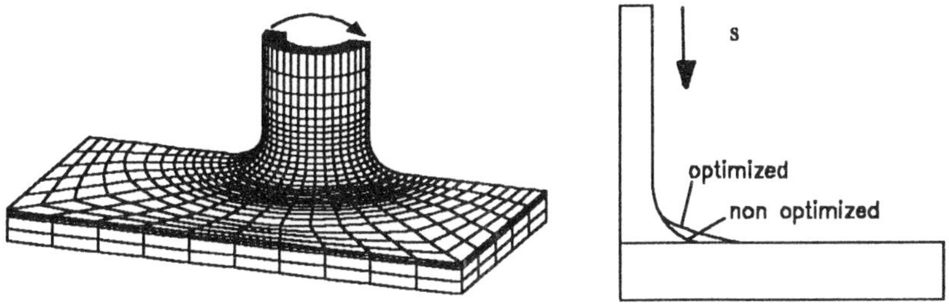

5a 5b

Figure 5. 3D model: a. mesh and boundary conditions b. non-optimized and optimized contour.

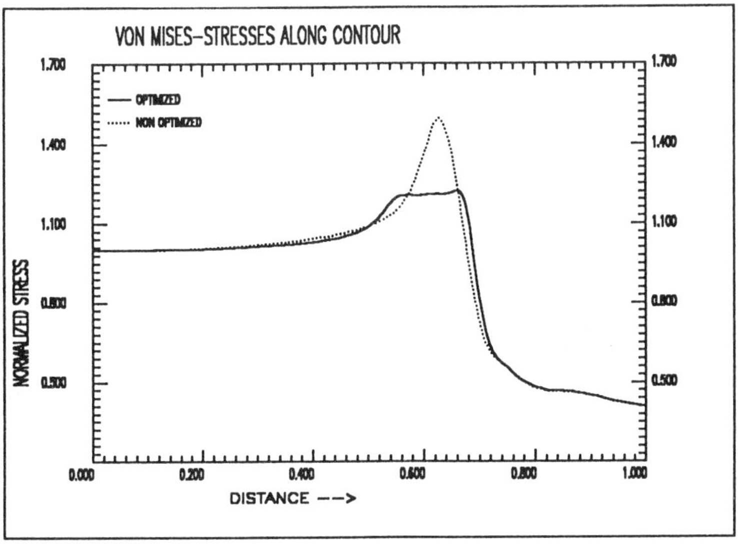

Figure 6. Mises stress distribution: surface stresses along contour line s

Super-Elliptic Geometry as a Design Tool for the Optimization of Dome Structures

P. Huybers

Civil Engineering Department, Delft University of Technology, 1 Stevinweg, 2628 CN, DELFT, The Netherlands

ABSTRACT

Dome building structures are often constructed from small elements, struts or plates. For the determination of the dimensions of these structural elements the surface of the dome has to be subdivided: a pattern based on one of the three triangular Platonic solids (tetrahedron, octahedron or icosahedron) is usual. An approach is suggested in this paper to compare these three basic patterns and to evaluate two different methods of further triangulation. This is done from the point of view of a distribution of the nodal points as equal as possible on the surface. This aspect is of importance to its economy, as it limits the variety in size of the constituting parts.

The shape of domical structures is in most cases spherical, which means that all points on its surface are at equal distances from the system centre. Spherical domes have often a great amount of waste space, as most building purposes do not need all the height that is available in the middle of such a dome. It is however possible to squeeze or stretch the sphere in one or two directions in order to make it fit better to the required space contours.

The influence of the various aspects, mentioned in the foregoing, were studied and are demonstrated with the help of the interactive computer-program CORELLI. This is being developed at the Delft University of Technology and is meant for the study and the design of structural forms.

SUBDIVISION METHODS

Shapes like these can be subdivided by the superposition of grids, that are generated on the basis of one of the regular polyhedra with triangular faces: the tetrahedron (4 faces), the octahedron (8 faces) or the icosahedron (20 faces).

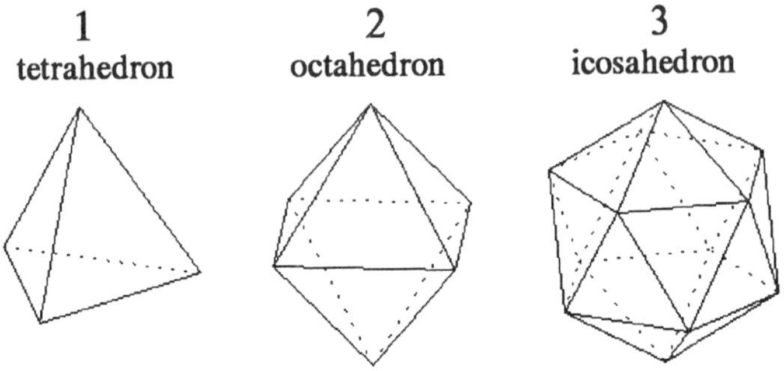

Figure 1. The three triangular Platonic solids

The original polyhedron triangle has to be subdivided in a suitable frequency, so that elements are produced of the required maximum or minimum size. This can be done by a number of different methods, each of which has its own specific advantages and drawbacks.

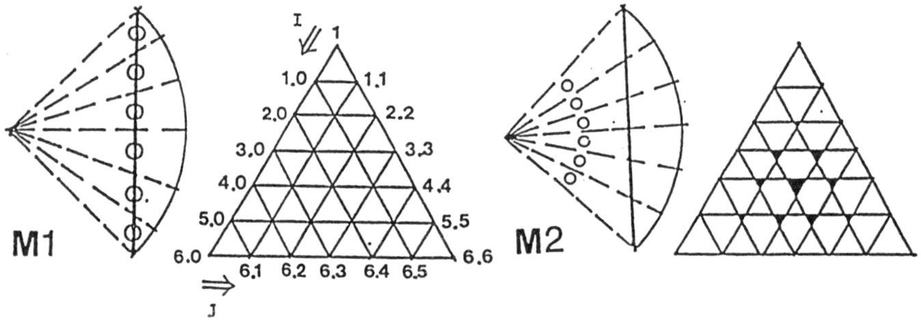

Figure 2. Two main methods of subdivision.

For the purpose of this study two main methods were considered:

1. Subdivision of the polyhedron edge in equal parts and successive interconnection of corresponding points on opposite edges of the triangular polyhedron face, so that a pattern of regular small triangles is found. This pattern is then projected from the centre onto the sphere.
2. Subdivision of the polyhedron edge in equal parts of the spherical angle under which this edge is seen from the centre, so that in the case of the

sphere equal chords are found. The parts into which the edge is subdivided are no longer equal. If in this case opposite points are interconnected, the connection lines do not intersect in points but they form small triangular 'windows' - as J.D. Clinton calls them [Ref. 1]. The centres of these windows are also projected onto the envelope and form the inner points of the spherical triangle. The basic subdivision pattern according one of the 3 Platonic solids with triangular faces is called 'Class' by Clinton. This terminology is adopted here also. With the help of the computer programme 'CORELLI' these 3 classes of spherical subdivision are worked out in Fig. 3, using both methods and for a frequency of 8.

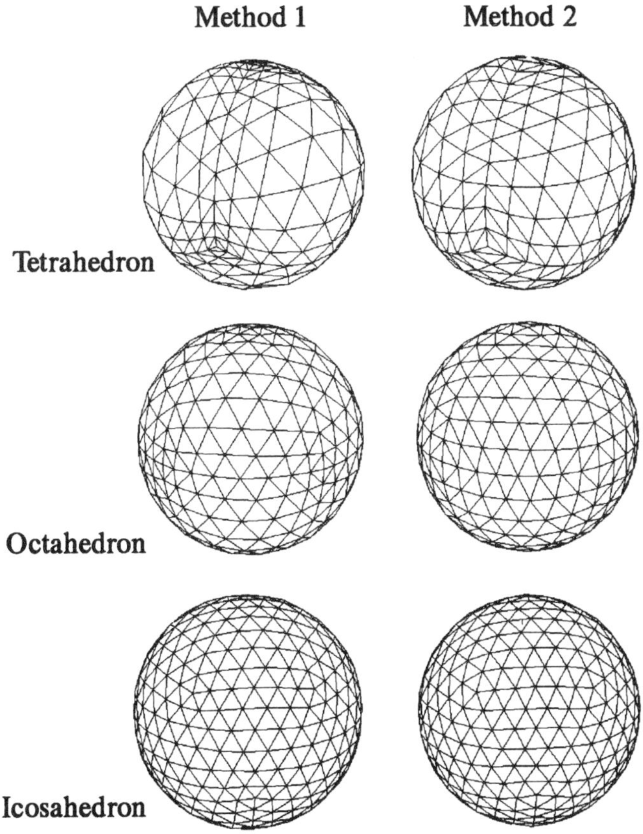

Figure 3. Subdivision patterns for the three main Classes and two methods.

This programme has been developed in its present state by G. van der Ende together with the author. It is written in GFA-basic for an Atari configuration and it works on an inter-active basis. The development of an MS-DOS version is being considered. For each of the Classes in Figure 3 a specific frequency for the triangular face has been chosen in order to obtain a more or less similar overall subdivision in all cases. These frequencies are based on the angles under which the original polyhedron edges are seen from the centre:

1. Tetrahedron: 109.47121950° -> frequency 10
2. Octahedron : 90.00000000° -> frequency 8
3. Icosahedron: 63.43494854° -> frequency 6

The data found with this program can at wish also be obtained in alphanumeric form. These data can be compared for the various cases on a statistical basis. This is done in table 1 for the specimens in Figure 3 in relation to the areas of their triangular faces. A radius=70 was chosen, in order to make these data comparable to those derived in a later stage (see table 3, No. 12).

	Tetrahedron		Octahedron		Icosahedron	
Method	1	2	1	2	1	2
Min. area	24.96	77.90	49.24	94.14	61.48	78.95
Max. area	471.50	274.69	192.88	144.18	99.17	89.81
Ave. area	150.43	151.34	118.70	118.81	84.79	84.80
Stand. dev.	106.50	51.80	37.70	12.60	10.88	3.18
Var. coeff.	70.79	34.22	31.76	10.61	12.83	3.75

Table 1. Variation in area for the different methods of subdivision

It appears from this table, that a greater evenness of distribution is found for the higher Class numbers (i.e. a greater number of basic polyhedron triangles) and that Method 2 is generally more economical than the other.

DESCRIPTION AS ELLIPSOID

If the dome is a true sphere, its vertical cross-section answers the equation

$$x^2 + y^2 = a^2 \tag{1}$$

In this case the radius R = a, but if it is an ellipse, the equation turns into

$$x^2/a^2 + y^2/b^2 = 1 , \tag{2}$$

or if a = 1 and b = E (Expansion):

$$x^2 + y^2/E^2 = 1 \qquad \{3\}$$

The radius varies between 1 and E. This equation can be written in polar notation as (see Figure 4):

$$R_p{}^2(\sin^2\theta + \text{cosin}^2\theta/E^2) = 1 \qquad \{4\}$$

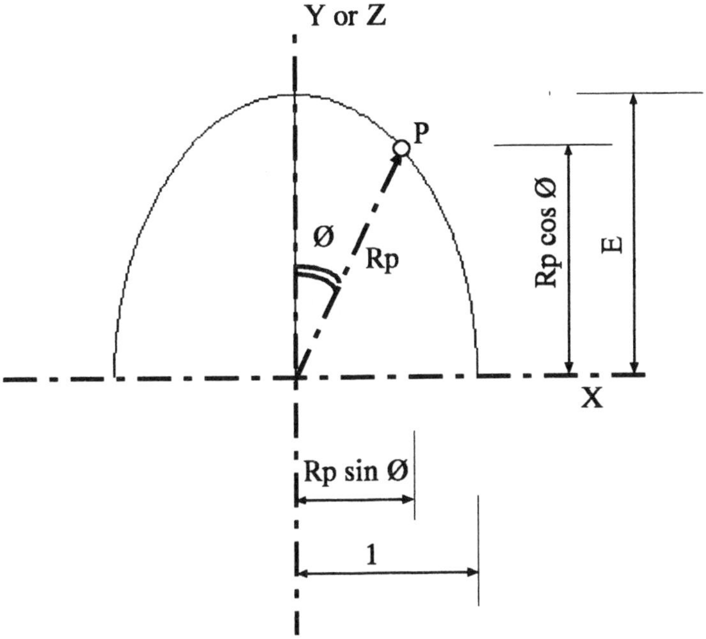

Figure 4. Basic form of ellipse

If the exponent of this equation is made n instead of 2, where n can have any positive value - even broken numbers - the equation obtains a more general character.

$$R_p = E/(E^n\sin^n\theta + \cos^n\theta)^{1/n} \qquad \{5\}$$

Kenner [Ref. 2] gives very valuable and interesting suggestions in this respect. The next section contains a further elaboration of these ideas.

Figure 5 shows the effect of varying this value of n. The curvature is elliptical for n=2 but if n is raised a form is found, which approximates the circumscribed rectangle. If n is decreased, the curvature is flattened until n=1 and the ellipse then has the form of a pure rhombus with straight sides, connecting the maxima on the co-ordinate axes. For n<1 the curvature becomes concave and obtains a shape, reminiscing a hyperbola. For n=0 the figure coincides completely with the X-, and Y-axes.

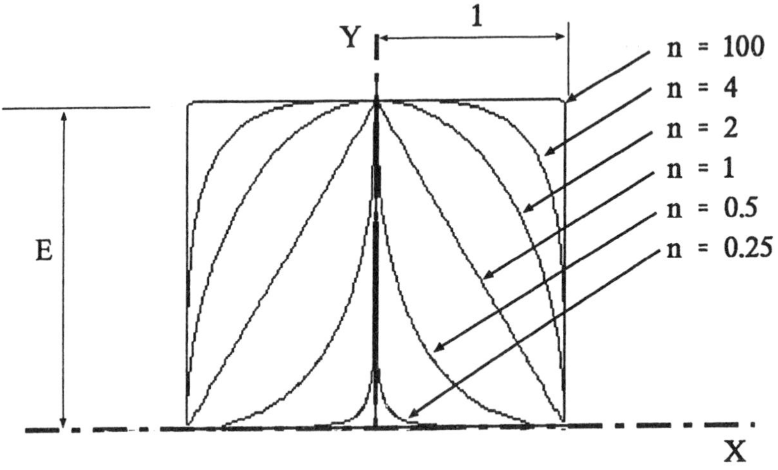

Figure 5. Change in form of an ellipse by variation of exponent

MATHEMATICAL DESCRIPTION OF ELLIPSOIDAL SHAPES

The ellipse forming the horizontal cross-section of the hemisphere can be expressed in its radius R_1, the rotation angle $\emptyset$, the ratio of its axes E_1 and its exponent n1 as:

$$R_1 = E_1/(E_1^{n1}\sin^{n1}\emptyset + \cos^{n1}\emptyset)^{1/n1} \qquad \{6\}$$

E_2/R_1 is the ratio of the axes of the vertical ellipse. If equation {5} is considered to represent the general form of the vertical ellipse and if substitutions are done so that $R_P = R_2$, $E = E_2/R_1$ and n = n2, it changes into:

$$R_2 = R_1E_2/(E_2^{n2}\sin^{n2}\emptyset + R_1^{n2}\cos^{n2}\emptyset)^{1/n2} \qquad \{7\}$$

Equation {6} and {7} form a coherent pair and if used together, not only the ratios E_1 and E_2 of the horizontal and of the vertical ellipse can be chosen independently but also their exponents n1 and n2. The latter fact means, that the visual appearance of the hemispherical shape can be altered accordingly. The pure sphere forms in fact only one specific representative out of a great number of possible shapes that are formed by a combination of different horizontal and vertical ellipses. Some of these are not even reminiscent of the original convex ellipsoidal shape, yet are very familiar such as the pyramid, the cone, the cylinder, the cube, etc.

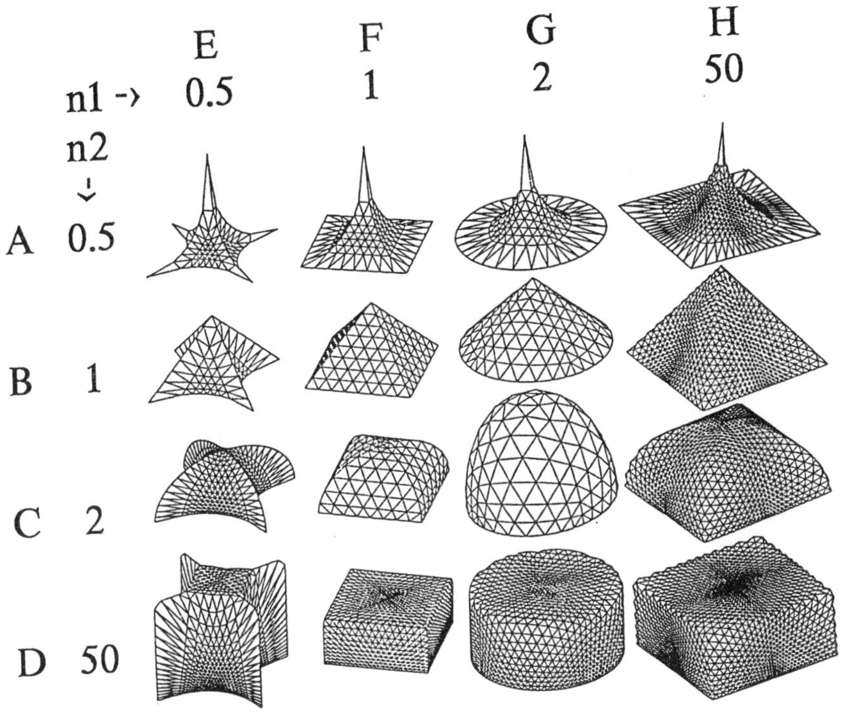

Figure 6. Ellipsoidal shapes formed by different combinations of horizontal and vertical ellipses with varying exponents

A PRACTICAL EXAMPLE

The apparent elasticity in the combinatory sense of the horizontal and vertical cross-sections of such 'pseudo-hemispherical' structures offers the opportunity to follow the outlines of the required space more closely than pure spheres do. Apart from the exponent there is still the ratio of the axes that may be varied. In order to demonstrate the potentials of the optimization procedure, that the program offers, a practical example has been worked out. It was assumed that a rectangular space with a width x length x height of 40 x 60 x 60 units length had to be enclosed.

The largest dimension of the spatial prism is equal to R_p, at the angle ∅ in the horizontal plane and the angle θ in the vertical plane. R_p, θ and ∅ can be derived from Figure 7:

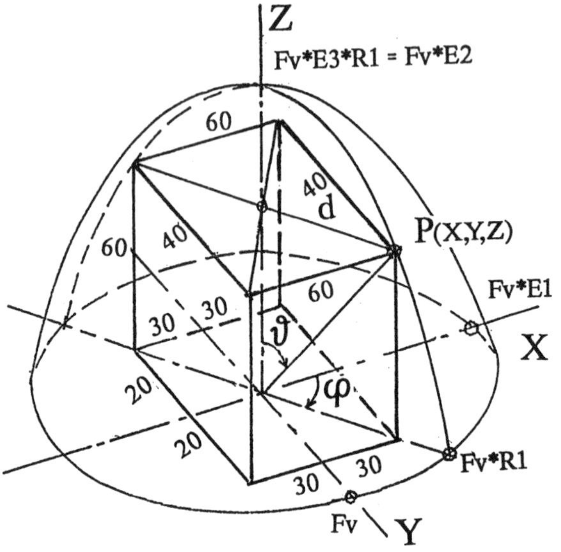

Figure 7. Scheme of the space to be enclosed by a hemi-ellipsoid

R_p = sqrt(20^2 + 30^2 + 60^2) = 70

θ = arctan (sqrt(30^2 + 20^2))/60 = 31.002719°

∅ = arctan 20/30 = 33.690067°,

with d = sqrt(30^2 + 20^2) = 36.055513 and E_3 = E_2/R_1,

of which the latter represents the ratio of the axes of the vertical ellipse passing through point P (X,Y,Z) = P(30,20,60). For a start this ratio was chosen here as

E_3 = 60/d = 1.6641006.

In further approximations this value may be varied as well. A number of ellipses passing through the point P with different values of the exponent is shown in Figure 8. The most characteristic values are gathered in table 2, where

$F_v * R_1 = R_p/k = 70/k$,

$k = E_3/(E_3^{n2} \sin^{n2}\theta + \cos^{n2}\theta)^{1/n2}$ and

$F_v * E_3 * R_1 = F_v * E_2$

N_2	E_3^{n2}	k	F_v*R_1	F_v*E_2
0	1.0000	0.0000	∞	∞
0.5	1.2900	0.4854	144.2221	240.0000
1	1.6641	0.9707	72.1110	120.0000
2	2.7692	1.3728	50.9902	84.8528
2.5	3.5723	1.4713	47.5755	79.1705
10	162.8530	1.8114	38.6433	64.3064
100	1.31E+22	1.9280	36.3063	60.4173
∞	∞	1.9415	36.0555	60.0000

Table 2. Characteristic values of ellipses through point P

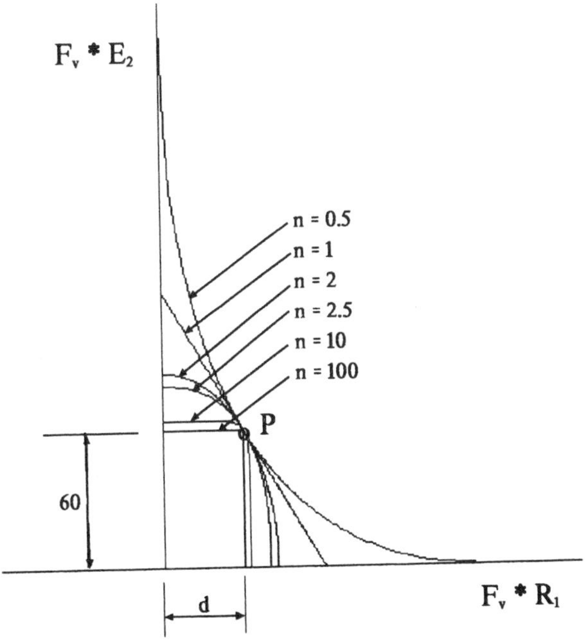

Figure 8. Ellipses of varying exponent through point P

Similar exercizes as with the ellipse in one plane can be made in the third dimension. A number of possible shapes, all of them touching the space prism to be enclosed in its outer most corner, has been worked out in Table 3 and shown in Figure 9 and 10. The table gathers some physical magnitudes of the thus found shapes also: the covered floor area, the surface area of the dome and its volume. These data give information, that might be used in further evaluation phases.

No	n1	n2	E1	E2	Factor Fv	Floor Area	Dome Area	Volume
1	2	2	1.0	1.6641	50.9902	8110	23548	449383
2	2	2.5	1.0	1.6641	47.5755	7097	22267	409860
3	2	10	1.0	1.6641	38.6433	4827	18255	281158
4	2	100	1.0	1.6641	36.3063	4320	16656	237969
5	2	1	1.0	1.6641	72.1110	16219	31433	640340
6	2	0.5	1.0	1.6641	144.2221	64877	82485	1162252
7	2	2.5	1.5	2.1213	37.3213	6547	21764	377038
8	2.5	2.5	1.5	2.2736	34.8220	6144	20950	352184
9	2	10	1.5	2.1213	30.3143	4459	17806	258580
10	2.5	10	1.5	2.2736	28.2843	4243	17137	241634
11	100	100	1.5	2.9793	20.2792	2724	12725	136312
12	2	2	1.0	1.0000	70.0000	15284	30389	701618
13	1	1	1.0	1.0000	103.3542	21364	37004	736026
14	1	2	1.0	0.7071	98.4886	19400	38539	888628
15	100	100	1.0	0.9931	30.2087	3863	10174	105929
16	2	2	1.5	2.1213	40.0000	7482	23024	413446

Table 3. Data of ellipsoidal shapes passing through point P in Figure 7 according Class 2, Method 1

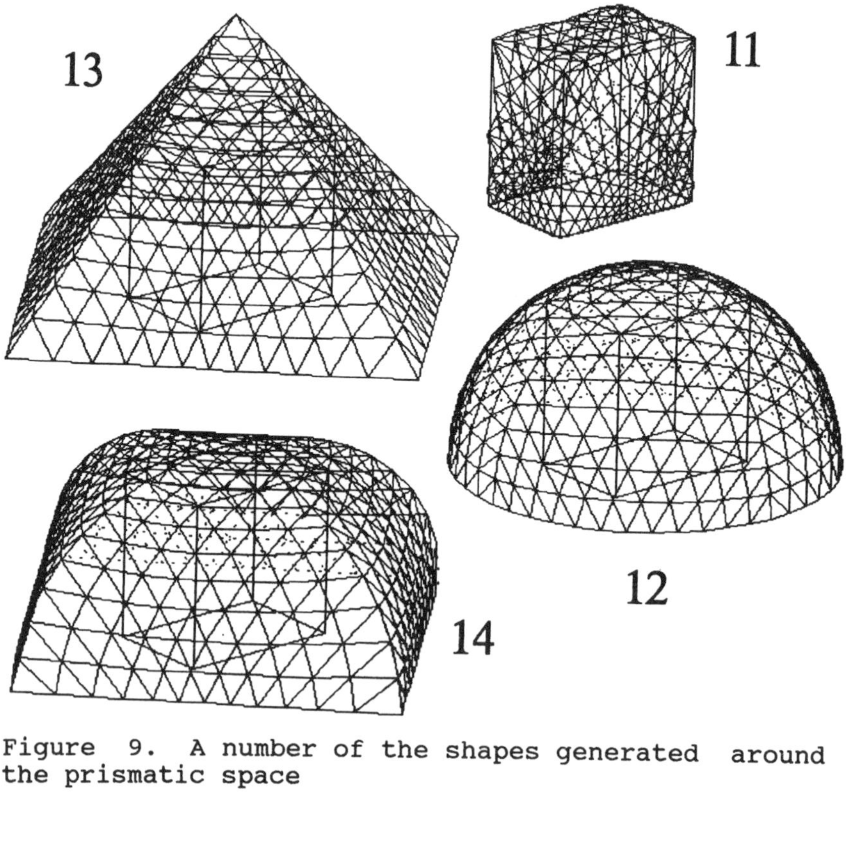

Figure 9. A number of the shapes generated around the prismatic space

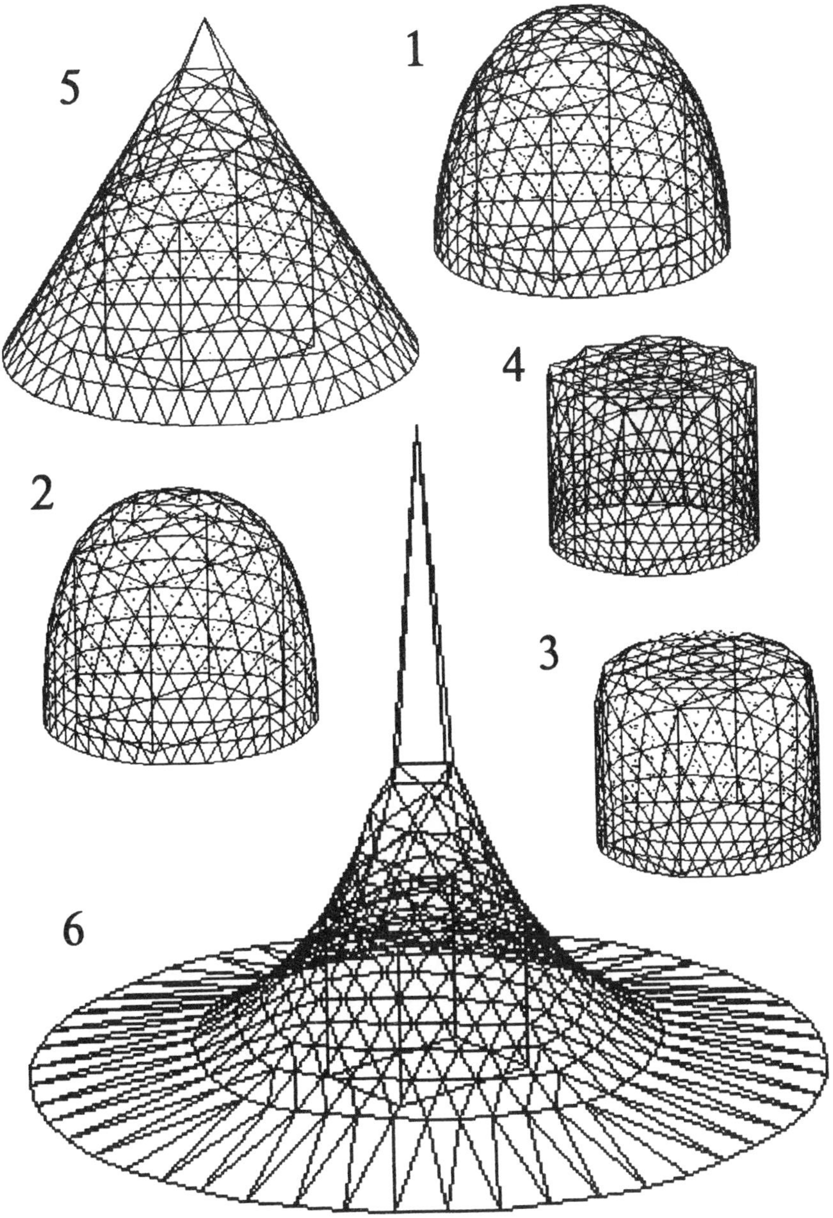

Figure 10. Another group of the ellipsoidal approximations

DISTRIBUTION OF THE NODAL POINTS

The subdivision of the surface of such an ellipsoidal shape may be based on the same Methods and Classes as previously described. The thus formed pattern is projected onto the surface of the ellipsoid from the inside, using origo as the projection centre. However, when the ellipsoid has a very much distorted shape, for instance if large values of E are used, in that case a correspondingly uneven distribution of the nodal points is found. This is demonstrated in Table 2 and Figure 7 for three ellipsoids of revolution with E2=1.6641006 and subdivided according Class 2 and Method 2. In the second and third case a Correction Factor is introduced. This changes the value of θ to θ':

$$\theta' = \arctan((\tan\theta)/F_{corr}) \qquad \{8\}$$

A value for F_{corr} of E2 or of sqrt(E2) gives good results. Their effect has in Table 4 and Figure 11 been compared for dome No. 1 in Table 3.

	(1)	(2)	(3)
Correction Factor	1	1.6641006	1.2900002
Minimum area	50.36	50.79	64.70
Maximum area	149.76	117.13	115.88
Average area	91.90	92.31	92.20
Standard deviation	30.67	15.53	13.59
Largest difference	99.3	66.34	51.19
Variation coeff.	33.37	16.83	14.74

Table 4. Area compensation of ellipsoid (with E1=1, E2=1.6641006, n1=n2=2, Method=2, Frequency=8) by the use of the correction factors E2 and sqrt(E2)

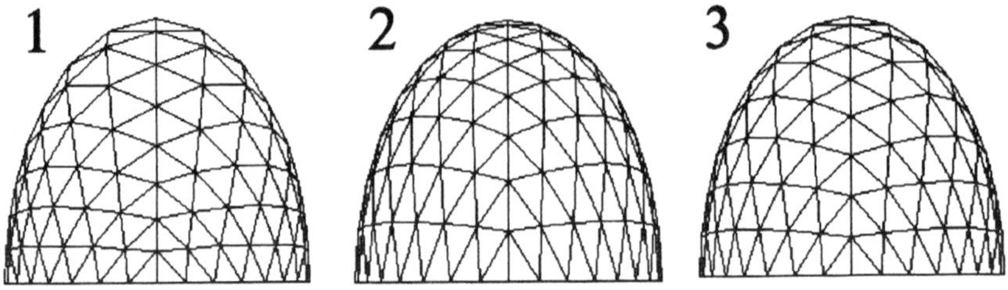

Figure 11. The effect of the correction factor on the pattern

CONCLUSIONS

This study contains a number of aspects that allow evaluation procedures of structural shapes. The data in Table 3 offer f.i. the opportunity to express the designed shape in material properties, such as weight, cost, strength, thermal insulation, energy requirement, a.s.o. It would therefore be interesting to incorporate suitable extensions in the existing program.

REFERENCES

1 Clinton, J.D., Advanced structural geometry studies. NASA Contract Report CR.1735

2 Kenner, H. Geodesic math, and how to use it. University of California Press Ltd., London, 1976

3 Huybers, P. The use of polyhedra for building structures, Structural Topology,(6), 1982, p.33-42

4 Huybers, P. and G.J. Arends, Double-elliptic vaults. Chapter 4 in "Analysis, Design and Construction of braced Barrel Vaults", Z.S. Makowski ed., Elsevier Applied Science Publ. Ltd., London, 1985, p. 66-75

5 Huybers, P. Polyhedral shapes visualized with CAD/CAM. IASS-Conference on 11-15th Sept. 1989, Madrid, 14 pp.

A Parametric Model of Structures Representation and its Integration in a Shape Optimal Design Procedure

A. Arias (*), J. Canales (**), J.A. Tárrago (**)

(*) Dpto. de Proyectos y Expresión Gráfica en Ingeniería

(**) Dpto. de Ingeniería Mecánica

Escuela Superior de Ingenieros Industriales y de Ingenieros de Telecomunicación, Alda. Urquijo s/n, 48013-Bilbao, Spain

ABSTRACT

In this paper we present a structural model suitable for a geometric representation orientated towards an Automatic Shape Optimal Design, based on the conceptual difference between the geometric model (CAD model) and the analysis model (FEM model), using the well known design elements technique. By means of two partition levels, it passes naturally from the geometric model to the analysis model, obtaining an interrelation between both models. The first partition level divides the geometric model into subregions -design elements and subregions with fixed geometry- obtaining the design model. Linked to the design elements, the design variables are defined. A design elements library for two-dimensional or axisymmetric space structures has been created. The second partition level divides each of the subregions resultant from the previous partition into finite elements, obtaining the analysis model. In this way, through the design model which is used as a link between the geometric model and the analysis model, it has been defined what could be named as a Parametric geometric model (Fig. 1), where the parameters are a few scalars which control the structure geometry and which will be suitable to be incorporated as design variables in the optimization problem.

INTRODUCTION

This work is concerned with Structural and Mechanical Systems Computer Optimal Design. We are talking about an investigation area which has an increasing attention in all industrialized countries [1].

The application of computers to mechanic systems design can be considered as the result of two fundamental factors confluence: on one hand, the progressive development of Computer Aided Design Systems (CAD) and its application in the Mechanical Engineering, and on the other hand, the wide acceptance of theFinite Element Method (FEM) as the most powerful tool of structural analysis available today. Nevertheless, and because there still exists an important gap between both technologies, its integration into a system that would allow a maximum automatization of the complete design cycle has not been possible (Fig. 2).

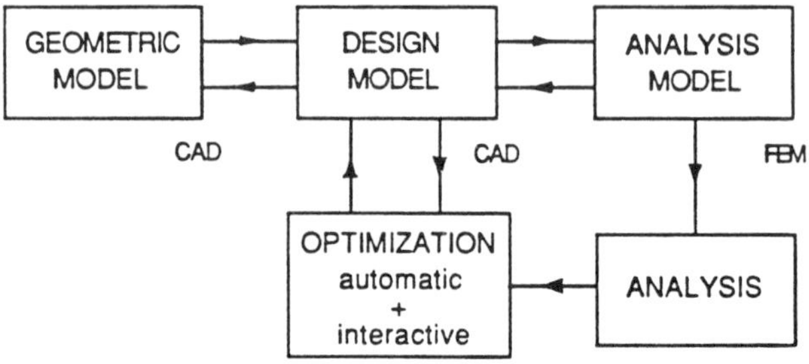

Figure 1 Parametric geometric model scheme

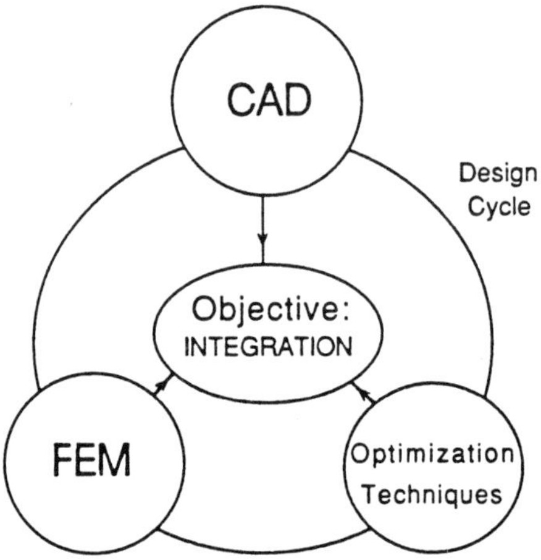

Figure 2 Design cycle scheme

Firstly, the CAD technology deals mainly with the objects geometric definition, having as first task their manufacturing without hardly contemplating the specific aspect of design, infravalorating the graphic and interactive possibilities which it obviously has. Secondly, the mechanical systems design remains highly dependant on structural analysis methods which generally have poor graphic capabilities, as its main purpose is numeric results.

Nowadays, the situation is developing fast and the difference between both systems tend to shorten. The surface representation and the solid modeling require the inclusion of mathematic formulations and very complex algorithms in the CAD systems. Likewise, the spectre of analysis programmes packages' users, particularly the FEM, is on continuous increase, and so they tend to use them as a black box for a great diversity of applications. Because of all this, the analysis programmes need efficient data processing systems, besides good graphic and interactive capabilities. From what has been said, the conclusion can be reached that both technologies need the same kind of tools, so that its integration is not only interesting or convenient, but it is also possible.

The scheme which must be developed for the integration of these techonologies, must be based on the introduction of an optimization interactive process which automatizes the design cycle, allowing the user's interaction (fig. 3). The possibility of performing the sensibility analysis in a FEM system, and transferring the optimizer's solutions to the structure geometric model in a CAD system, is the basis for the automatic design cycle's closure.

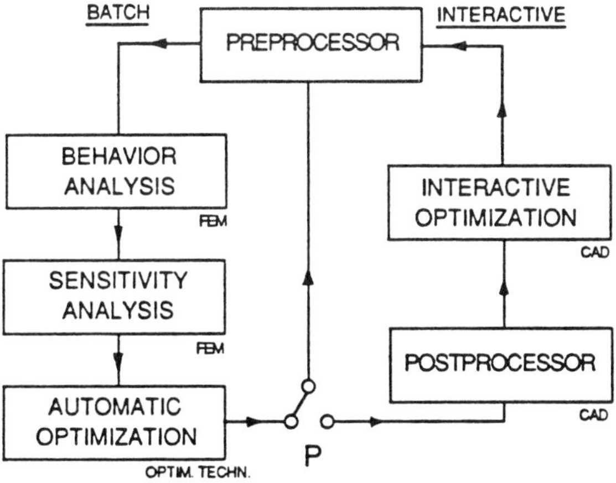

Figure 3 Interactive optimization cycle

GEOMETRIC REPRESENTATION AND DESIGN VARIABLES ELECTION

The process for the geometric representation chosen in this work is based on the design elements technique, which was first developed by the scientist Botkin and Iman in the General Motors Research Laboratories in 1982. This technique has incorporated the modifications suggested by Braibant and Fleury, which are based on the use of the blending functions and usual procedures in CAD [2,3,4,5] for the description of the boundaries and the surfaces.

Under these premises, a formulation has been proposed for geometric representation which distinguish three models: the proper structure geometric representation model and the design and analysis models. For the description of the geometric model -which confines itself to the definition of the piece geometry- some functions and usual techniques in CAD for the curves interactive generation, are used.

Design Model

Assuming that the geometric model to obtain the analysis model has been defined, two partition levels are proposed (Fig. 4).

The first one divides the geometric model into an assembly of subregions, some of them of simple geometry (i.e. quadrangular or triangular) whose shape can be modified (the ones that hold the moving boundaries) and which are named as design elements. Others called fixed elements have a geometry that will remain unalterable during the optimization process. They both define the so called design model, taking it as the ensamble of the necessary data for the definition of a structure geometric model set for its automatic optimization, that is, with an educated definition of its moving boundaries, whose shape must be controlled by convenient design variables.

The second partition level divides, in a natural way, each of the previous subregions into finite element meshes, generating the precise model for the structural analysis by a FEM system, and both partitions try to keep using the usual concepts in CAD.

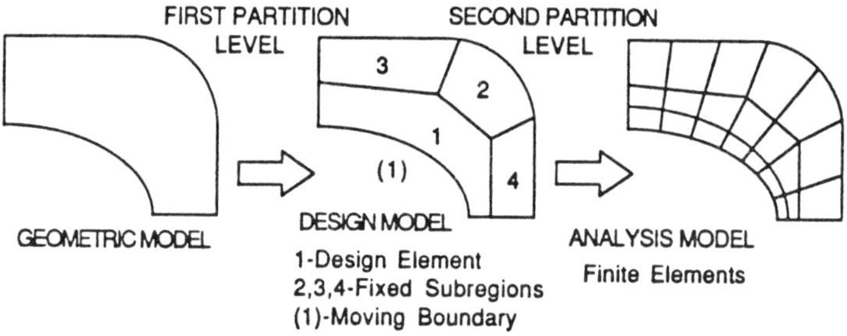

Figure 4 Partition levels

The design model proposed and implemented in a pre-processor is basically composed of the following fundamental components: the boundary-lines, the design elements, transfer elements and design variables.

Boundary Lines

For the description of the boundaries of the design elements implemented in the pre-processor, we have used parametric functions normally used in computer graphics representations for the curves interactive generation. Therefore, for instance, the following type of boundary-lines are used: line segment, conics arc (particularly circumference arc) Bèzier arc, B-spline arc (non-periodic and periodic), curves composed of the previous types, such as polilineal made by a circumference arcs, etc..., expressed -all of them- in a normalized parametric way (Fig.5).

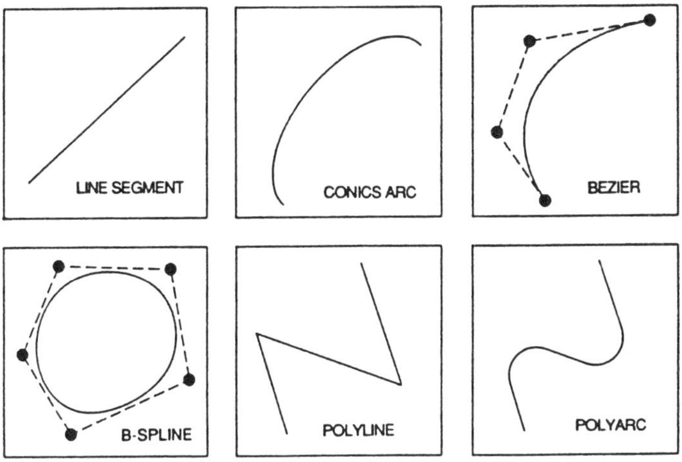

Figure 5 Boundary lines

<u>Design Elements</u>
Once the boundaries have been defined, the same has to be done with the design elements, which due to being superficial elements, require the use of specific procedures for the generation of surfaces. The design elements implemented in this work can be completely defined by its boundary-lines, or they could also need an internal set of points to the element for its complete definition.

To define the first ones, the interpolation techniques between its boundary-lines (Coons techniques) have been used. These elements have different parametric expressions, depending on the result of being completely defined by two, three or four boundary-lines, and the functions used for its interpolation.

To define the second ones, the cartesian technique has been employed. The elements defined by this technique are obtained by the cartesian product of two families of coordinate curves onto the surface. The assemblage of curves chosen are Bèziers or B-Splines.

It has been proved that these techniques, originally developed for Computer Graphic Representation, can be very well adapted to the problems in Shape Optimal Design. In this sense, they represent a very important feature: they provide the use of a simple procedure to up-date the finite elements meshes when the structure geometry changes during the optimization process.

Through these techniques of surface definitions and with the previously quoted boundary-lines, a design elements library has been defined, general and flexible enough to allow to perform, without difficulty, the first partion level onto the geometric model, of a large number of two-dimensional or axisymmetric space structures (Fig. 6 and 7).

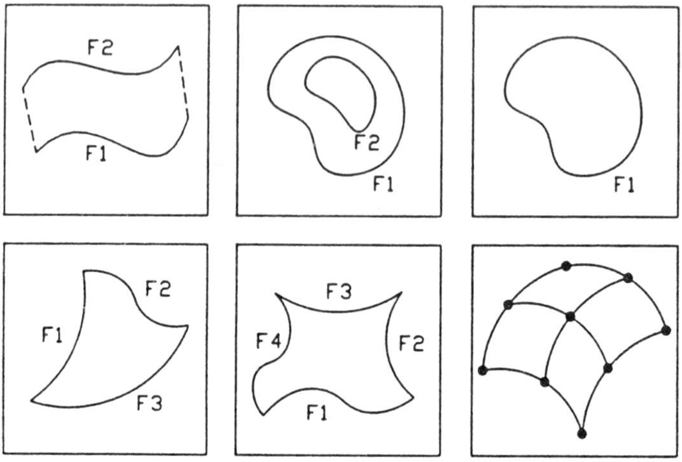

Figure 6 Design elements

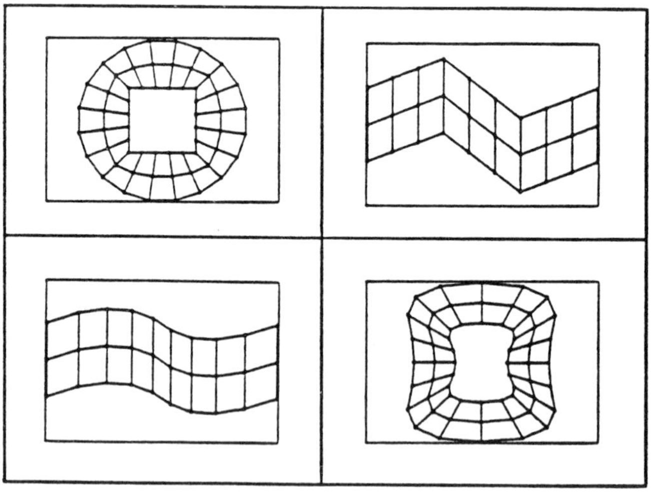

Figure 7 Discretized design elements

Design Variables
To complete the design model, the design variables connected with it must also be defined. The design variables chosen are scalars linked to the "control points" or "master points" which control the shape of the moving boundaries and, therefore, the shape of the design elements. These scalars are the modules of the vectors whose up-most point is on a moving control point "P" and its initial points are on a previously defined reference point "R", so that the "RP" vector defines the movement direction of point "P" (Fig. 8).

These moving points have been divided into two groups: free and link groups. The movement of the first ones is directly specified by the design variables. The movement of the second ones (as its own name shows) is linked to the first ones. Various types of links have been defined (homothey, similarity, symmetry, alignment, tangents, etc...) so that the moving boundaries shape can be controlled, not being necessary to impose new geometric restrictions to the optimization problem, which would make its resolution even more difficult. It has been also tried to completely maintain the independence of the variables linked to the same moving control point. To reach this aim, some moving or sliding references which will be used when a control point is linked to two or more design variables, have been defined.

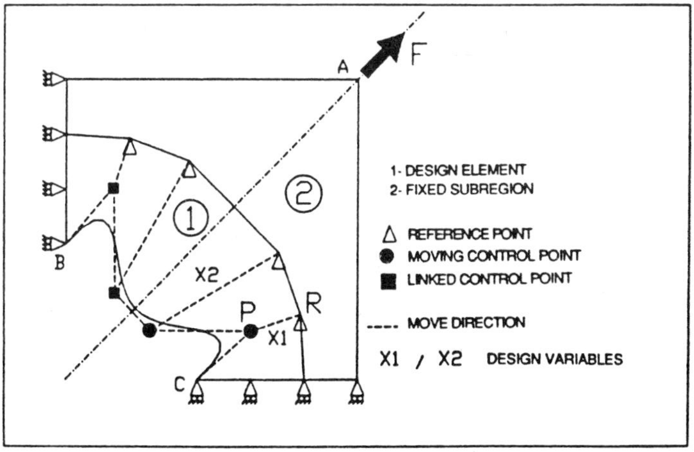

Figure 8 Design model

MESH GENERATION AND ITS UP-DATING

Analysis Model
Once the first partition level has been performed and the design model has been obtained, this last one becomes fractionated, as it has already been explained, into two types of subregions: the fixed ones, which remain unaltered during the optimization process and the design elements, whose shape can change during the course of the process. Now, a new partition that allows to discretize

the structure to perform its analysis by the FEM, is needed. Therefore, it is obliged to define the analysis model of the structure.

To generate the mesh of finite elements from the design model, once different procedures have been studied and compared, the transfinite interpolation techniques have been chosen, because they can be understood as a natural extension of the methods used for the definition of superficial elements [8, 9]. Through this technique a mesh of points is created. On this mesh, the finite elements -previously chosen from its library- are defined, getting in, this way, the second partition level on the geometric model and obtaining the finite elements analysis model, after imposing the applied loads and the boundaries conditions. These transformations set up an explicit relation which permits the determining of the coordinates of any point (a finite element node) inside the design element or in its boundary-lines, depending on the control points actual position, providing -in a natural way- with a generation and automatic up-dating of the finite elements mesh procedure. In the design elements, this mesh is up-dated in each interaction, according to the results obtained in the optimization. Inside of the fixed subregions, the user can define the mesh which he thinks as more suitable and that will remain unaltered during the whole process.

Transition Elements

Finally, a transition elements library has also been defined. They are composed of various design elements of quadrangular topology degenerated into another triangular one, and they are used to link design elements of different mesh densities, an aim which they acomplished by interrupting the course of meshing in one point or in one side (Fig. 9).

One important particularity of the design model proposed is the possibility of performing an adaptative mesh, which allows actuation on two different levels for the enrichment of the mesh: In the first partition level, it actuates over the design elements, adjusting the longitudinal and transverse densities of the mesh. In the second partition level, it actuates on the finite elements. As regards the Sensibility Analysis and Optimization Algorithms, aspects of particular importance in the general process of Shape Optimal Design, have been used in techniques proposed in references[10, 11, 12].

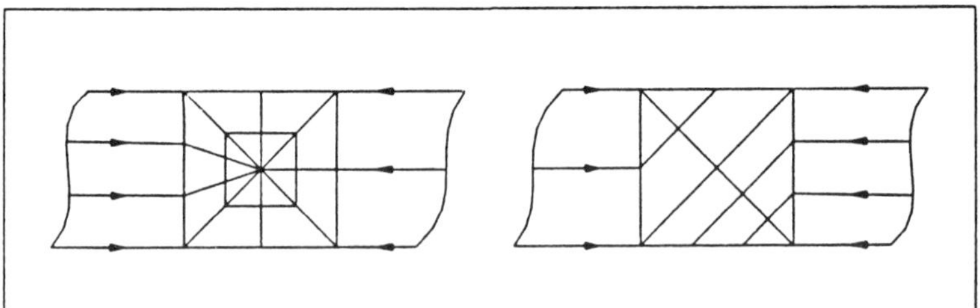

Figure 9 Transition elements

APPLICATIONS

A typical application in the bibliography relating to the Shape Optimal Design, which can be used as a test to verify the efficiency of the concepts and methods shown, is the connecting-rod optimization. The rod of the figure 10a, which must be made of steel and of constant thickness, is the structure to be optimized. The weight of the rod must be minimized and if the material density is considered as constant, this can be replaced by the area.

The internal and external boundaries of the rod are designed by circumference arcs and by rectilineal segments. Thus, the "shape" of the boundary's curves is fixed, but its position and magnitude are not, which will determine the final optimal shape of the rod. In the figure 10b, the moving boundary-line curves are shown. The internal circumferences, which will be determined by the cylindrical shape and by the spindle and crankshaft diameters, are considered fixed. The boundary- lines F1 and F2 are circumference arcs which can only be transformed by radial expansion (homothety) with its own center point -the center of their respective circumferences- that is to say, only their radius will be able to modified. The boundary-lines F3 and F4 are circumference arcs as well, and they can be transformed by homothety with its own center point, the center of their circumferences, and by a homothety with improper center the improper point on the X axis. That is, boundary-lines F3 and F4 will be able to move in the same direction of the X axis and change their radius. Finally, the F5,F6,F7 and F8 boundary-lines are line segments that link

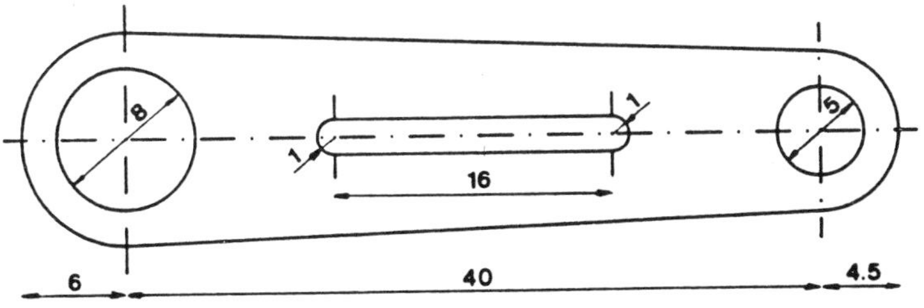

Figure 10a Geometric model

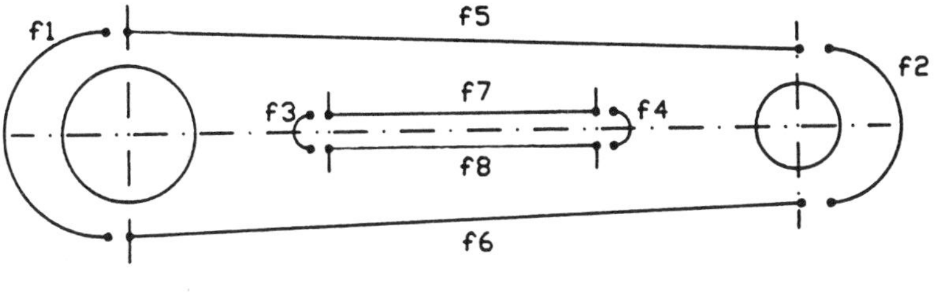

Figure 10b Moving boundary-line curves

the previous arcs. This way, the rod's shape is controlled by the design variables shown in Fig.10c.

The first partition level was carried out on the geometric model of Fig. 10a, dividing it into six subregions which are modeled through superficial elements that belong to the design elements library (Fig. 10d). Once the mesh densities for the different subregions have been chosen and the second partition level has been performed, the initial finite elements model of the complete model is obtained. (Fig. 10g)

The rod is attached and loaded as shown in Fig. 10e. The tension of Von-Mises is limited to specified finite elements of the model. Restrictions are also imposed to the boundary-lines which define the variance intervals possible for the design variables.

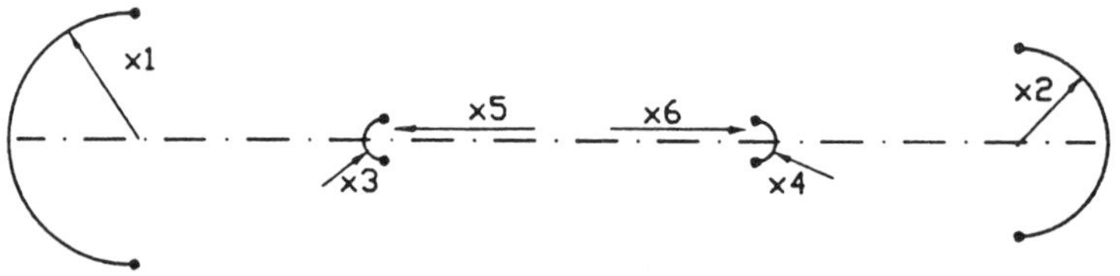

Figure 10c Design variables

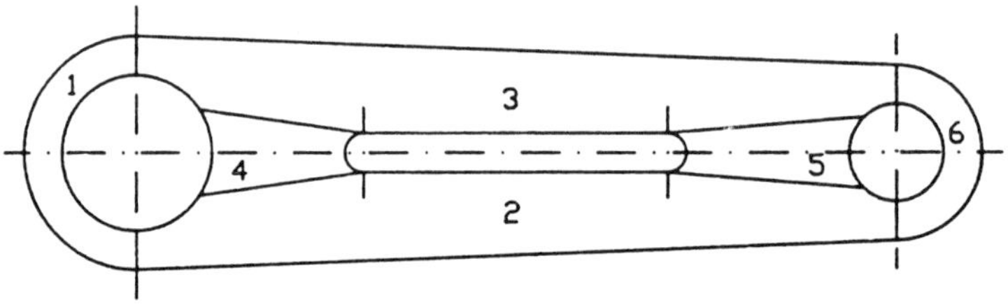

Figure 10d Subregions in the geometric model

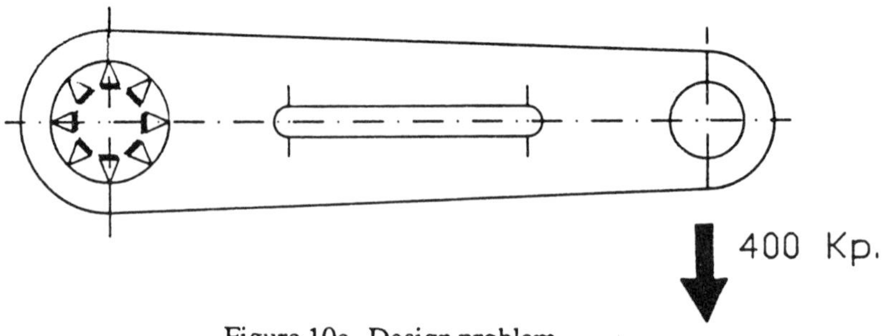

Figure 10e Design problem

The problem has been solved directly and using the approaching techniques (Fig.10f shows the history of the process). In Fig.10h final shape obtained in the optimization process is shown.

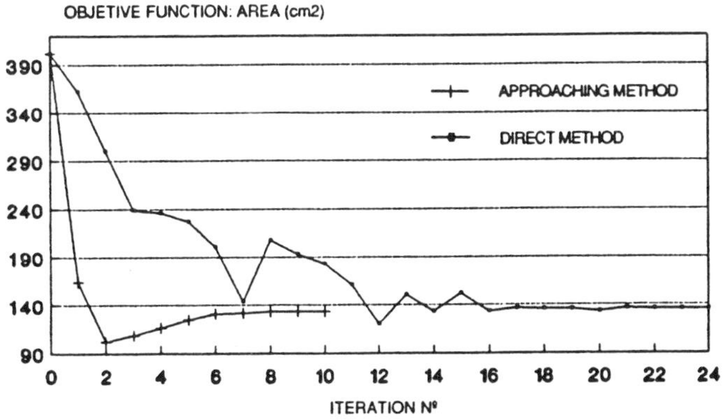

Figure 10f Optimization history

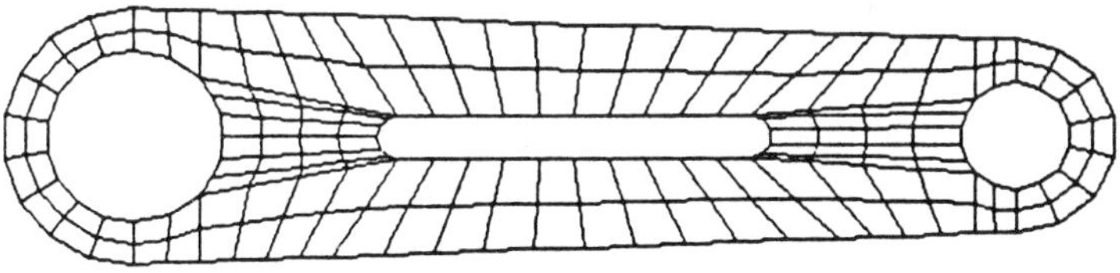

Figure 10g Analysis model (initial shape)

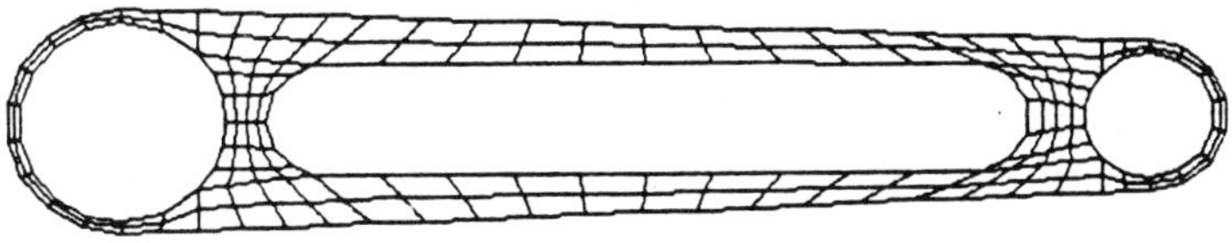

Figure 10h Final shape

CONCLUSIONS

In this paper, a procedure for bidimensional or axisymmetric structures geometric representations particularly suitable to Automatic Shape Optimal Design of elastic structures discretized by the FEM has been described. This process has been implemented in a set of subroutines which are part of the Optimal Design Programme developed in the Mechanical Engineering Department at the Basque Country University.

REFERENCES

1. Botkin,M.E., Gupta,A. and Yen,A. Considerations in using Interactive Graphics for Structural Optimization Design Modeling, Computers in Engineering, Vol.3, Book no. GO431C, V.A. Tipnis and E.M. Patton, 1988.

2. Botkin,M.E., Shape Optimization of Plate and Shell Structures, AIAA Journal, Vol. 20, No. 2,pp.268-273, 1982.

3. Bennett, J.A., Structural Optimization in the Industrial Design Process, General Motors Research Laboratories, GMR-4 353, 1983.

4. Braivant,V. and Fleury,C., Shape Optimal Design using B-Splines, Computer Methods in Applied Mechanics and Engineering, Vol.44, pp.247-267, 1984.

5. Braivant,V. and Fleury,C., Shape Optimal Design - A Performing CAD Oriented Formulation, Proc, AIAA/ASME/ASCE/AHS 25th. Structures, Structural Dynamics and Materials Conference, Palm Springs, California, 1984.

6. Yamaguchi, F., Cures and Surfaces in Computer Aided Geometric Design, Springer-Verlag, Berlín-Heidelberg, 1988.

7. Coons,S.A., Modification of the shape of Piecewise Curves, Computer-Aided Design, 9, pp.178-180, 1977.

8. Heber,R., Shepard,M.S., Abel,J.F., Gallagher,R.H. and Greenberg,D.P., A General Two-Dimensional, Graphicall Finite Element Preprocesor utilizing Discrete Transfinite Mappings, International Journal for Numerical Methods in Engineering, Vol.17, pp.1015-1044, 1981.

9. Hall,C.A., Transfinite interpolation and applications to engineering problems, Theory of Aproximation, Academic Press, 1976.

10. Braivant,V. and Fleury,C., Sensitivity Analysis in Shape Optimal Design, Paper presented at NATO/NASA NSF/USAF ASI on Computer Aided Optimal Design: Structural and Mechanical Systems, Troia, Portugal, July 1986.

11. Canales,J., No,M. and Tarrago, J.A., Un Algoritmo eficaz para Diseño Optimo de Estructuras basado en Aproximaciones Explícitas y Métodos Duales, VII Congreso Nacional de Ingeniería Mecánica, Valencia, Diciembre de 1988.

12. Tarrago,J.A. and Aviles,R., Diseño Optimo de Forma de Sistemas Resistentes. Tecnologías Avanzadas de Diseño y Fabricación, Ed. por Servicio Central de Publicaciones del Gobierno Vasco, pp.503-515, 1988.

SECTION 5: STRUCTURAL OPTIMIZATION IN MECHANICAL AND AIRCRAFT INDUSTRIES

Computer Analysis and Optimization of the Automotive Valve Gear

C. Chiorescu (*), M. Oprean (*), P. Jebelean (**)

(*) Politechnic Institute of Bucharest, Faculty of Transports, Chair of Road Motor Vehicles, str. Independentei 313, 77206, Bucharest, Romania

(**) Politechnic Institute of Timisoara, Department of Mathematics, Piata Horatiu, nr. 1, 1900, Timisoara

ABSTRACT

Described here are computer methods for analysis and optimization of valve gear train which includes a conventional mechanical tappet or a hydraulic one. In this sense, is conceived a multi-mass valve train model for overhead camshaft, which takes into account of the effects of the linking elements stiffness and damping under dynamic working conditions. The computer model supplies valuable information about displacements, velocities and accelerations of the component elements, allowing the accurate representation of detailed aspects of valve train dynamical behaviour. In order to assess the engine volumetric efficiency for a given cam profile a computer model is used for the engine cycle simulation. An important asset of our work is the achievement of a nonstandard cam profile processing method. This provides to design an optimum profile from both, dynamical behaviour and volumetric efficiency points of view.

INTRODUCTION

The valve gear, at the I.C.Engines, is conceived to provide the opening and closing of the valve at the defined moment of the engine cycle.

The valve, the most important part of the valve mechanism, is located at the end of a long train of connected elements. Due to the elasticity of these elements the motion commanded by the cam is reproduced by the valve with considerable alteration. It must be remarked the great influence of the inertia forces and of the gas pressure force, varying during the valve lifting. The valve gear train is found to oscilate during the engine operating event because of alternative compression and decompression of its elements.

The vibratory behaviour of the valve gear train causes high accelerations, much higher than those that are cam-controlled [4], [10], [11], [13]. These accelerations lead to very high dynamical stresses in the system as well as the increase of the contact pressure and of the cam wear [6], [14].

The correct evaluation by mathematical calculations of the dynamical behaviour of the valve gear train is dependent upon the characteristics of the valve operating conditions and upon the accuracy evaluation of the masses and stiffness of the component elements. These aspects as well as those refering to the wear are well documented in the literature [1], [5], [9], [11].

This paper presents a multi-mass valve train model on a four-cylinder in- line S.O.H.C. engine with usual tappet or with a hydraulic one. The model takes into account the effects of the elasticity of the elements as well as of the damping and the frictional losses under the dynamical working conditions. This allows to predict the displacements, velocities and accelerations of all elements of the model.

It is well known that the optimization of the valve gear system is far from being easy. Taking into account the fact that the cam profile considerably influences the dynamical behaviour of the valve mechanism as well as the gas flow capability, we focused our attention on it. In the present paper the dynamical behaviour of the system and the volumetric efficiency of the engine are criteria for the optimization of the cam profile. The testing of the optimized cam profile from the volumetric efficiency point of view is made by a package of computer programs [7], [10], which simulate the engine cycle, in view to determine, on the one hand, the instantaneous pressure differences acting on the valve disk, and on the other hand, the engine performances. A good agreement was found between the analytical results and the experimental data. The work methodology is presented in figure 1.

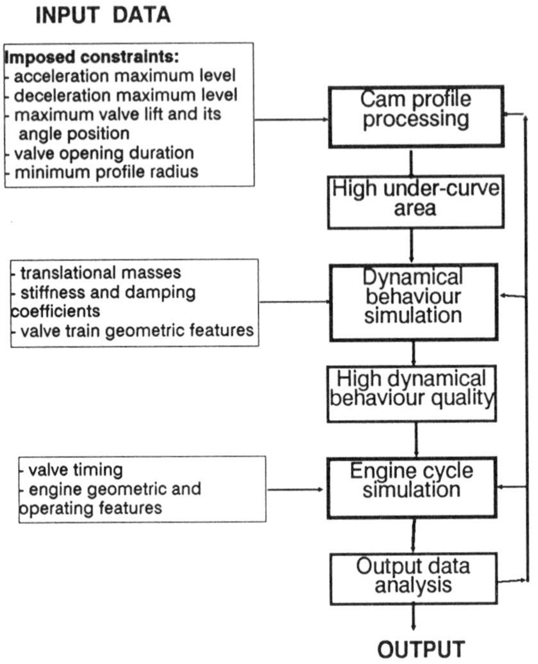

Fig. 1 Work methodology

MULTI-MASS VALVE TRAIN MODEL

Experimental approaches [5] revealed that a high percentage of the overall vibration level is induced by the valve train vibrations. Emphasis is laid on the control of the valve train elements' movements and also on the assessment of the dynamic loads present in the mechanism.

In this respect we develop a multi-mass valve train numerical model for single overhead camshaft. Mechanical and hydraulic tappets are considered. In a similar manner other valve train configurations, including cam and rocker or finger follower and push-rod-rocker arrangments can be simulated. The computer model is in such a manner structured that a simple valve train modification is allowed. The figures 2 and 3 present the schematic layout of the valve gear model with conventional mechanical and hydraulic tappets. The valve train is considered to be a system of concentrated translational mass elements (M_i), for $i = 1,N$, associated with massless springs of representative stiffness (K_i) and massless viscous dampers (C_i). In addition to these elements the frictional losses and the external viscous damping due to the motion of some elements in their guides are included. Special attention was paid to design a submodel to simulate the complex phenomena concerning the hydraulic tappet dynamical behaviour. The angular camshaft displacement due to the drive flexibility is neglected in this stage of our work. The spring-mass model adopted for the camshaft-bearing system represents the bending degree of freedom along the valve axis (K_1, C_1). The valve spring is represented using a multi-mass model which comprises translational elements corresponding to the free coils of the valve spring (M_i, K_i, C_i, $i = 4,N$).

The above simplifications have been found to be acceptable in relation with the complexity of the phenomena occurring in the system.

A certain element of the valve gear train is modeled as a concentrated mass which linked with other elements by means of massless stiffnesses (K_i, K_{i+1}) and viscous dampers (C_i, C_{i+1}) as it is shown in figure 4. Other possible interactions acting to this element are represented as concentrated forces of representative varying law during the valve cycle. This modular representation enables the user to easily include another element, if it is necessary, for a new valve train configuration. The new element can be assigned by mentioning its mass, linkages with other elements, the other concentrated forces acting on it and initial position. In order to account the separation possibility of the components it is used the only-compression elements (K_c, C_c). These specificates the linkeage elements action transmiting only compression forces (see figure 4).

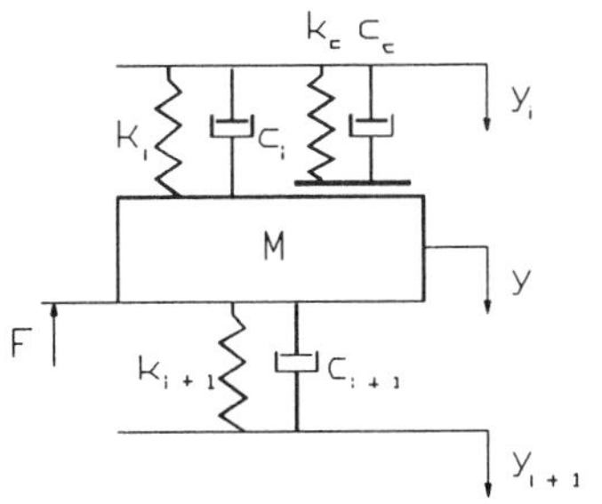

Fig. 4 Concentrated mass element representation

The symbols used in the valve gear models depicted in the figures 2 and 3 have the following significances:

M_1, reduced camshaft mass determined by assumption that the camshaft is a simply supported beam on two bearings; M_2, tappet mass; M_3, total mass of the valve and of the components moving with it; M_4 to M_N, masses of the free coils of the valve spring.

The below mentioned stiffnesses and dampers model the following:

K_2, C_2, the cam lobe-tappet interface; K_t, C_t, the valve steam including the elasticity of the spring retainer; K_{ss}, C_{ss}, the valve-seat contact which acts during valve seating and takes into account the elasticity of the valve seat and of the valve disk; C_s, external viscous damping due to the motion of the tappet in its guide.

In addition : e , excentricity of the cam-tappet contact point; x_t, valve lift; θ, angular camshaft torsional displacement; α, rotation of camshaft - degree.

The current displacements, Y_i, are defined as positive in the direction shown in the figures 2 and 3 and they are considered to be zero at the initial moment (valve closed).

The model of the valve train configuration including hydraulic tappet needs additional specifications. In accordance with the figures 3 and 5, the dynamical tappet model contains : m_1, mass of the tappet body; m_2, piston mass including the mass of the check valve ball (m_4); m_3, mass of the barrel; M_2, global tappet mass, for this configuration.

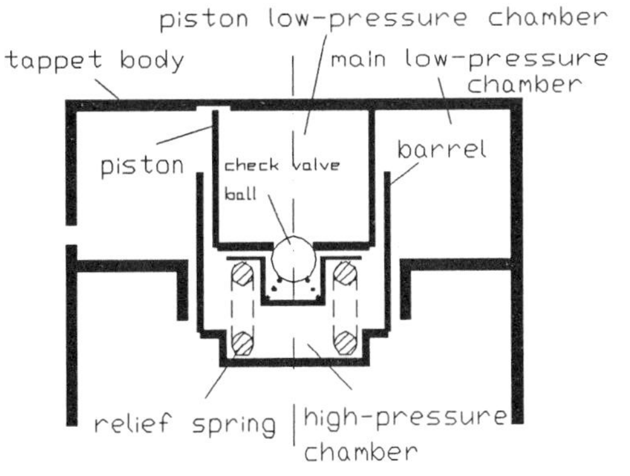

Fig. 5 Hydraulic tappet

The pairs of stiffnesses and dampers (K_{12},C_{12}), (K_{13},C_{13}), (K_{23},C_{23}), respectively, simulate the mechanical contact between piston and tappet body, the barrel-tappet body contact and the relief spring action. In order to predict the current position of the ball check valve relatively to its seat, which in turn defines the oil flow to or from the high-pressure chamber (h. p. c.), a simple dynamical check valve submodel is conceived. This, takes into account the ball-seat contact (K_{sb}, C_{sb}), the actions of the ball spring (K_{42},C_{42}) and also the forces determined by the oil pressures in both low and high pressure chambers. The oil effect is simulated by means of the simple stiffness (K_u) placed in parallel with a viscous damper (C_u). During the circle base period of the cam (valve on its seat), the oil is fed through the inlet passage to the main low-pressure chamber (m.l.p.c.) and to the piston low-pressure chamber (p.l.p.c.),successively. Hence, the oil passes through the orifice controlled by the check valve ball fills the h.p.c. and compensates the clearance existing in the valve train. Subsequently, during the valve lift event the cam acts on the tappet body, which in turn, transmit t s the forces to the valve by means of the piston and of the oil trapped into the h.p.c.. In the shut-off engine condition the oil in the m.l.p.c. and in the h.p.c. is drained. A sufficient quantity of oil for a quick h.p.c. refilling during a new engine start is trapped in the p.l.p.c..

A difficult problem is to estimate the stiffness and damping values of the model elements . In this respect, the stiffnesses of the camshaft and of the valve spring are experimentally determined. The others stiffnesses, excluding the oil equivalent stiffness are determined using the Finite Element Method.

Special attention is given to the phenomena occurring inside the hydraulic tappet mainly with regard to the h.p.c.. For a given constant pressure supply, the problem consists in assessing the equivalent oil stiffness and the effect of the oil flow to or from the h.p.c.. To compute the oil stiffness, we use [1]:

$$K_u = \frac{1}{\overline{\beta}} \times \frac{A_p^2}{V_p} = \frac{1}{\overline{\beta}} \times \frac{A_p}{L} \quad [\text{N/m}^2]$$

where: $\overline{\beta}$, oil-gas mixture compressibility coefficient; A_p, cross section area of the h.p.c.; V_p, h.p.c. instantaneous volume; L, distance between the piston and the barrel;

The equivalent compressibility coefficient of the oil and gaseous inclusions is obtained by [1] :

$$\overline{\beta} = (1-a)\beta + a\beta' \quad [\text{m}^2/\text{N}] \qquad (1)$$

where: a, percentage of gas present in oil-gas mixture; β', gas compressibility in isothermic evolution; β, compressibility coefficient for mineral oil without gaseous inclusions.

The β coefficient is calculated using the following empirical relation:

$$\beta = \beta_o\,(1 + A\,t + B\,t^2)\ [\text{m}^2/\text{N}]$$

where:

$\beta_o = 6 \cdot 10^{-10}$ [m²/N]
$A = 6 \cdot 10^{-3}$ [1/°C]
$B = 2 \cdot 10^{-5}$ [1/°C]
t : oil temperature in °C.

On the other hand, the quantity of oil contained in the h.p.c. varies during the valve cycle, due to the oil transfer through the check valve and due to the oil seepage through the gap between the piston and the barrel. The flow rate of oil passing through the check valve is [12]:

$$Q_b = C_d \times A_o \sqrt{(2/\rho)\,\Delta p}$$

where : A_o, minimum cross section of passage; ρ, oil density; Δp, pressure differential between h.p.c. and l.p.c.; C_d, discharge coefficient which in turn is calculated by (see fig. 6) :

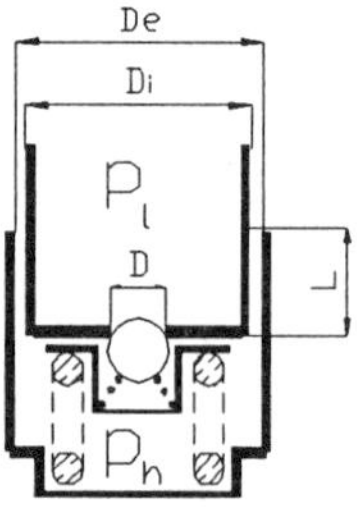

Fig. 6 Tappet detail

$$C_d = \frac{1}{\sqrt{0.5 + 0.066 R^2/x^2}}$$

where x is the ball displacement along the seat axis and $R = D/2$.

The volume of the oil transferred through the check valve is :

$$\Delta V = Q_b \cdot \Delta t$$

where Δt is the computation time step.

The flow rate of oil passing through the annular slot between piston and barrel is given by [12]:

$$Q_j = \frac{\pi D_m J^3 \alpha}{96 \eta L} \Delta p$$

where : Δp, pressure differential between h.p.c.(p_h) and l.p.c.(p_l); $D_m = (D_i + D_e)/2$ average diameter of the annular slot; L, length of the annular slot; $J = D_e - D_i$, diametral gap; α, ($\in [1, 2.5]$) coefficient of excentricity; η, absolute viscosity of oil, dependent on temperature and pressure.

The forces acting on the elements of the models are : F_{ft}, tappet-guide friction force; F_t, steam valve-guide friction force; F_r, F_{ss}, valve disk-seat friction force; F_p, force of the gas pressure in the engine cylinder; F_{pu}, oil pressure force due to the oil supply pressure; F_{ot}, preloading force of the piston relief spring; F_{pj}, force of the oil pressure in the p.l.p.c.; F_{pi}, force of the oil pressure in the h.p.c..

NUMERICAL GENERATION OF THE CAM PROFILE

Overall engine performances including aspects of reliability and gas flow capability are mainly determined by the camshaft design. In this sense a general purpose method for cam profile processing should be of primary concern.

The cam profile is designed to assure the required timing with regard to the valve opening advance and the valve closing delay and, also, a proper valve lift law in order to enhance volumetric efficiency of the engine. For a certain engine geometry the instantaneous geometric flow areas are determined by the momentary valve lift, thereby gas flow efficiency is proportional to the under-curve area of the cam profile.On the other hand, the cam profile aspects involved in the dynamical behaviour of the valve gear train is of great interest for our work.

Thesetworequirements, gas flow capability and dynamical behaviour are approached in different ways, giving rise to the proper investigation tools. Our work is focused on this two main directions, valve lift curve being accordingly analysed on angular intervals. Each of them induces particular effects on the above mentioned engine performances. Figure 7 depicts valve lift curve and respectively the velocity and the acceleration resulting by differentiation. The above mentioned angular intervals are:

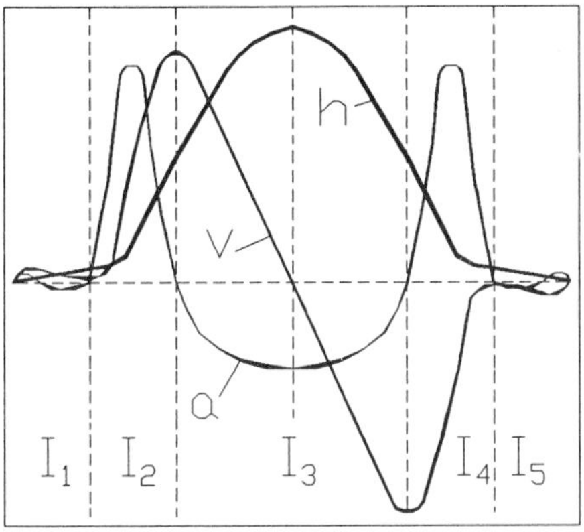

Fig. 7 Lift (h), velocity (v) and acceleration (a) laws of the cam profile

The *impact cam ramp interval* (I_1) including the moment of the tappet-cam impact after the valve train clearance is diminished to zero; hence the valve opening occurs. The volumetric efficiency influence is less significantly because of the low valve lift, but the dynamical behaviour is greatly influenced.

The *ascending flank interval* (I_2) is important from both dynamical behaviour (acceleration reaches maximum value) and volumetric efficiency (high valve lift) points of view.

The *cam nose interval* (I_3) decisively influences the volumetric efficiency because of the maximum effective gas flow area (highest valve lift).

If the deceleration load exceeds the valve spring force, the cam-tappet separation occurs and hence high impact forces apear. In addition, the local radii are small and so, high Hertzian pressure is induced.

The *descending flank interval* (I_4) is similar to I_2.

The *impact cam slope interval* (I_5) governs the valve-seat contact. Therefore, the level of the impact forces may be controlled in this phase. The valve lift is low so, the volumetric efficiency is less affected.

From the mathematical point of view, a good dynamical behaviour means low level of the variation of the second derivative and a high gas flow efficiency means a large under-curve area. To this aim, usually, there are used elementary functions under imposed constraints concerning the level of the acceleration and of the deceleration, the duration of the valve opening and the maximum valve lift. So, in the case of the classical Kurtz cam, trigonometric and polynomial functions are used. In [8] the presented method use high degree (50) polynomials.

The start point of our method is the fact that it is difficult to process a shape using curves of function type and the efficient geometric intuition is better served by techniques of computational geometry. So, the parameter curve technique is advisable. In view of the computation of the cam profile characteristics, the obtained parameter curve is approximated by an interpolating cubic spline function.

In order to generate the cam profile, we succesively use the technique of *Bezier curves* and the technique of *B-spline curve* as it follows.

Let $P_1, P_2, \ldots, P_n$ be distinct points in plane ($\mathbb{R}^2$). The *Bezier curve* is defined by the vectorial form of the (n-1)-th degree *Bernstein's operator*:

$$B_{n-1}(P_1,\ldots,P_n) = \sum_{j=1}^{n} \binom{n-1}{j-1} t^{j-1}(1-t)^{n-j} P_j \,, \quad t \in [\,0, 1\,]$$

To improve the approximation of the polygonal line $P_1 \ldots P_n$, we proceed to inserting new points on the segments $P_i P_{i+1}$, $i = \overline{1, n-1}$. The computation of the curve points is made using their well known geometric interpretation [2] and so, large degree of the above operator may be obtained (≅ 150 on CORAL 4021).

To define the *B-spline curve* which we have used, the *Bernstein's operator* is replaced by the *Schöenberg's operator*. The variation diminishing *B-spline* is obtained.

Shortly, for $k \in N$, $k \geq 2$, we define:

$$t_1 = 0 \qquad t_j = \frac{\sum_{i=1}^{j-1} \| P_{i+1} - P_i \|}{\sum_{i=1}^{n-1} \| P_{i+1} - P_i \|}$$

for $j = \overline{2,n}$ and let

$$s_1 = \ldots = s_k = t_1 \qquad s_{m+1} = \ldots = s_{m+k} = t_n$$

$$s_{k+j} = t_{j+1}$$

for $j = \overline{1,n-2}$, with $m = n+k-2$. For $i = \overline{1,m}$, we put:

$$s_i^* = \frac{1}{k-1}(s_{i+1} + \ldots + s_{i+k-1})$$

Let $f : [0, 1] \to \mathbb{R}^2$ be defined by :

$$f(t) = \frac{1}{t_{j+1} - t_j}[(t_{j+1} - t)P_j + (t - t_j)P_{j+1}]$$

for $t \in [t_j, t_{j+1}]$, $j = \overline{1,n-1}$

The *B-spline curve* is given by the vectorial form of the (k-1)-th degree *Schöenberg's operator* :

$$V_k f(s) = \sum_{i=1}^{m} f(s_i^*) B_{i,k}(s)$$

where the $B_{i,k}$ functions are defined by :

$$B_{i,1}(s) = \begin{cases} 1 & \text{for } s_i \leq s < s_{i+1} \\ 0 & \text{otherwise} \end{cases}$$

$$B_{i,j}(s) = \frac{s - s_i}{s_{i+j-1} - s_i} B_{i,j-1}(s) + \frac{s_{i+j} - s}{s_{i+j} - s_{i+1}} B_{i+1,j-1}(s)$$

for $j = \overline{2,k}$, $i = \overline{1,m}$. Details on *Schöenberg's operator* can be found in [3, pp.154-161] . In our computations, the k degree was at most 9.

To obtain a cam profile we start with a polygonal line $P_1 \ldots P_n$ which is approximated using *Bezier curves.*

The cam profile processing means acting on the position of the vertex of the polygonal line and on the degree of the operators. The *Bezier curve* allows us only a global control of the shape but good properties of tangency to the base circle and smoothness are obtained without exceeding a maximum imposed acceleration level. The *Bezier curve* sert as input data for the *B-spline curve.* The aim of this " conversion " is to refine the regions of high acceleration .

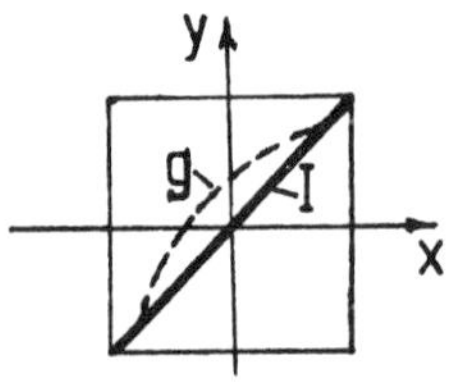

Fig. 8 Identity map (I) and spline function (g)

The entrance on the cam profile and the exit from the cam profile (the impact intervals)are, usually refined in the same manner. So the quality of local control of the B-spline curve is exploited.

Regarding the engine volumetric efficiency it is known that the gasflow depends on the depression in the cylinder, which is induced by the piston movement law. As, the angle of the maximum piston speed is between 70 and 77 crankshaft-deg. A.T.D.C., a gain in volumetric efficiency would be obtained if the angle of maximum valve lift is more closely to the maximum piston speed angle. An asymmetric cam profile is needed in consequence, with positive results mainly for the high speed engines. To obtain an asymmetrical cam profile, we start from an optimized symmetrical profile seen as a function $f : [-1,1] \rightarrow \mathbb{R}$. Obviously, $f = f \circ I$ where $I : [-1,1] \rightarrow [-1,1]$ is the identity map. Now, if I is replaced by a function $g : [-1,1] \rightarrow [-1,1]$ (usually a spline function), which has the graph as in fig. 8, then an asymmetrical profile is obtained. So the function $f \circ g$ gives the cam profile. This ideea to start with an optimized symmetrical profile offer results better than those obtained working directely on an asymmetrical profile.

To design only the impact cam intervals is conceived in addition to a "constant-velocity" method. It consist of the junction between two intervals:one of them providing a linear lift variation, i.e. a constant velocity, represent the main impact interval; the other one is a polynomial shape and makes the linkeage between the constant-velocity interval and the cam circle base.

RESULTS

In seeking to increase the accuracy of the predicted results, the dynamical model program is coupled with an engine cycle model program [7]. The cam profile lift is the input to this complex model and it varies in different increments of time in accordance with the instant frequency of the oscillations. The highest oscillation frequences is reached when an impact between two elements of the valve train occurs. During the engine cycle, the camshaft angular velocity is asumed to be constant. Valuable information about displacements, velocities and accelerations of the component elements througtout cam profile duration are available as outputs. The engine cycle simulation supplies informations concerning volumetric efficiency, mean indicated pressure and other engine performances. In addition to this, it is possible to assess the effect of the predicted cylinder gas pressure on the valve disk.

The experimental results, in accordance with the computed ones, performed for the initial Dacia 1300 car engine with mechanical valve train, indicate that valve acceleration reaches 8000 m/s^2 on its flank intervals and over 7000 m/s^2 during valve-seat impact, at 5000 rpm engine speed. Some of the typical predicted results obtained for optimized cam profile for the valve gear train with mechanical tappet are plotted in the figures 9a and 9b. These show a reduced acceleration level even for a larger under-curve area. The figure 9c, shows an asymmetric valve law processed by asymmetrization method previously presented. The ascending flank acceleration is higher than the descending one, therefore, while the first peak valve acceleration reaches 9000 m/s^2 the second one remain up to 5500 m/s^2 at 5000 rpm engine speed. This provides low acceleration level during valve seat impact.

The analysis of the contact force between cam and tappet (figure 10), is important in revealing the possible separation between the valve train components. These occur when the contact force reaches zero.

As we have already defined our target in enhancing the dynamical behaviour by a proper design of the impact cam ramp interval and of the impact cam lope interval, our inquiry indicates that a small value of the profile velocity in this Interval(up to $5 \cdot 10^{-3}$ mm/deg.), would be able to accomplish the dynamical requirements. The figure 11 shows a detailed representation of the impact cam lope interval velocity and of the valve velocity, for a cam profile obtained using *Bezier curves* technique. To assure a low force of impact, valve seating must occur within the vicinity of the minimum absolute velocity (e.g. in figure 11, from 167 to 175 camshaft-deg.). Turning to figure 9c, it can be noticed that the impact cam intervals are designed, in this case, using constant-velocity method, above mentioned. This provides low velocity but, consequently, long angular durations of this intervals. We must emphasize that an optimum design from the dynamical point of view needs low velocities and also, negative accelerations in the impact moments. If it is possible to keep accurately the clearance in the system, the impact intervals can be designed with a common angular durations because of the narrow lift limits.

Refering now to the gas flow efficiency appro aches a set of different angular duration cam profile are designed (e.g. 128,132,136,140 camshaft-deg. efective angular duration). Because of the impact intervals the total angular durations become very long, respectively 188,142,196,190 camshaft-deg. The under-curve areas for the above mentioned profiles are 673, 698, 720 and 725 mm·deg. Using the engine cycle simulation program, the volumetric efficiencies of this profiles are predicted, considering also different values for the angular valve opening advance. The results reveal. that, in fact, the beginning of the gas cylinder filling occurs in the vicinity of the T.D.C., because of the piston movement. So, the opening advance of the valve is more important for the exhaust residuals influence than for the volumetric efficiency gain. In addition, an evident increase in the under-curve area of the cam profile, e.g. from 673 to 720 mm·deg, for the above examples, does not necessary induce a volumetric efficiency gain in the same measure : from 0.79 to 0.81 at

5250 rpm engine speed and 100 % engine load or, from 0.73 to 0.74 at 3500 rpm- 50 % load. This is because that the gas flow through the inlet port is essentially determined by the relationships of the valve lift curve to the piston movement law. On the other hand, it is noticebly that a long angular duration of the impact intervals does not produce, in fact, significant disturbances to the volumetric efficiency, because of the low valve lift. The inverse gas flow ussually occurs in this case, but the losses in volumetric efficiency are low (e.g. up to 1 % for a valve closing delay of 50 crankshaft-deg.). In addition, the low valve lift in the impact cam ramp interval leads to a high gas flow velocity through the valve port inducing thereafter, an adequate turbulence in the cylinder. So, this profiles provides to enhance the gas flow velocity with 8 % to 17 % comparison with the initial engine cam profile. The increase in volumetric efficiency is evidently when the angular position of the maximum valve lift is advanced (asymmetric profile), but in the usual engine speed range, the gain is small (e.g. for 6 crankshaft-deg. advance the gain is about 2 %, for the engine speed up to 5500 rpm). However the maximum acceleration of the ascending flank interval increases at the asymmetric cam profile (see figure 9c). For the above example the 6 deg.-advance induces a 13 % increase in maximum valve acceleration yet, a 16 deg.-advance induces a 32 % increase. So, the profile asymmetrization must be carefully applied.

The valve train configuration including a hydraulic tappet requires a new cam profile to be optimized (figure 12b). So, the results obtained using an optimized cam profile for mechanical tappet are unsatisfactory, from the dynamical behaviour point of view, when it is used for a hydraulic one (figure 12a). In order to relieve the influence induced by the quantity of oil escaped from the h.p.c. while the valve is open, different values of the diametral gap (J) between the piston and the barrel are considered. Some of the results can be viewed in figure 13. It can be noticed that a diametral gap up to 0.02 mm has a less significantly influence to the loss of valve lift (up to 50 μm) and hence of volumetric efficiency (1-2%). A larger gap induces important malfunctions e.g. 6% loss of volumetric efficiency for $J = 0.05$ mm and even 24% for $J = 0.1$ mm, wich is an accidental case. In addition due to the significant loss of lift, valve-seat impact occurs in advance to the impact slope interval, to a point where the velocity and acceleration of the profile is higher, according to the kinematic profile. Because of the high impact velocity, can be noticed a pronounced valve bounce after the first contact with the seat, evident source of noise and stress. The influence of the engine speed on the pressure in the h.p.c. and on the leakage of oil from the h.p.c. (for a normal $J = 0.01$mm) can be viewed in the figures 14 and 15. The parameter exerting the greatest influence on the oil elasticity is the percentage of the gaseous inclusions (see figure 16 and equation (1)). Even for a small quantity of gas in the oil the dynamical behaviour is compromised due to the separation occurring between the tappet and the cam.

CONCLUSIONS

Using the above computer system and the cam profile numerical generation method, optimized valve gear train were obtained. So, the original Dacia 1300 valve gear train was improved from the dynamical behaviour and from the volumetric efficiency points of view.

REFERENCES

1. Bakaj, L.F., Rechnerische Simulation der Ventiltriebsdynamik eines OHC-Schlepphebelventiltriebes mit Hydraulischem Ventilspielausgleich, Dissertation, Universität Karlsruhe, 1988

2. Bezier, P. (translated by Forrest, A.R.), Numerical Control - Mathematics and Applications, Wiley, New York, 1972.

3. Boor, C.de, A Practical Guide to Splines, Springer Verlag, New York, Heidelberg, Berlin, 1982.

4. Colombo, T., Pagliarulo, V. and Virgilio, U., Computer Simulation of the Dynamic Behaviour of the Overhead Camshafts Valve Train Assembly with Hydraulic Lash Adjuster, Proceedings of XXIII FISITA Congress, 435 - 450, 1990.

5. Eichhorn, V. and Schonfeld, H., The Valve-Train of I.C.E. as a Source of Vibrations Experimental Results and a Method of Calculation, Proceedings of XXIII FISITA Congress

6. Fessler, H. and Ham, R., Lubrication and Stress Analysis as a Basis for Camshaft Optimization, Proceendings of XXIII Congress, 565 - 579, 1990.

7. Grunwald, B. and Gheorghiu, V., Gas Flow Computation in I.C. Engines,Polytechnic Institute of Bucharest, 1982 (in roumanian).

8. Herrmann, R., Delange, J. and Louradour, G., L'evolution du trace des cames en fonction des possibilites offertes par l'ordinateur, Ingenieurs de l'Automobile, 11, 655 - 665, 1969.

9. Morel, T., Flemming, M.F. and Buuck, B.A., Evaluation of Variable Camshaft Effects on Performance of a High Output, 4-Valve SI Engine, Proceedings of XXIII FISITA Congress, 397 - 433, 1990.

10. Oprean I.M., Study of the Interaction Cam-Valve Spring in the High Speed S.I. Engines, Thesis - Polytechnic Institute of Bucharest, 1983.

11. Roskilly, M., Feran, W.H Windsor, C.,Valve Gear Analysis, Proceedings of XXII FISITA Congress : 1.193 - 1.200, 1988

12. Vasiliu, N., Catana, I., Hydraulic and Electrohydraulic Power Systems, Bucharest, 1988

13. Valve Train Compliance Measurement Method Developed, Automotive Engineering, Vol. 90, June, 1982

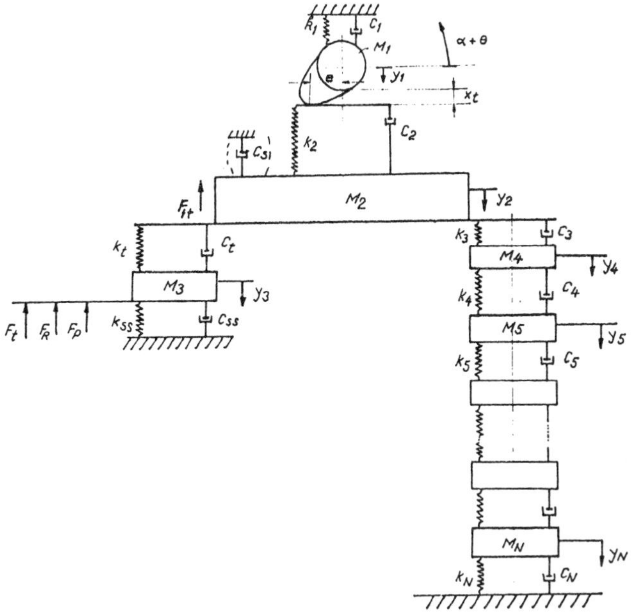

Fig. 2 Valve train model with mechanical tappet

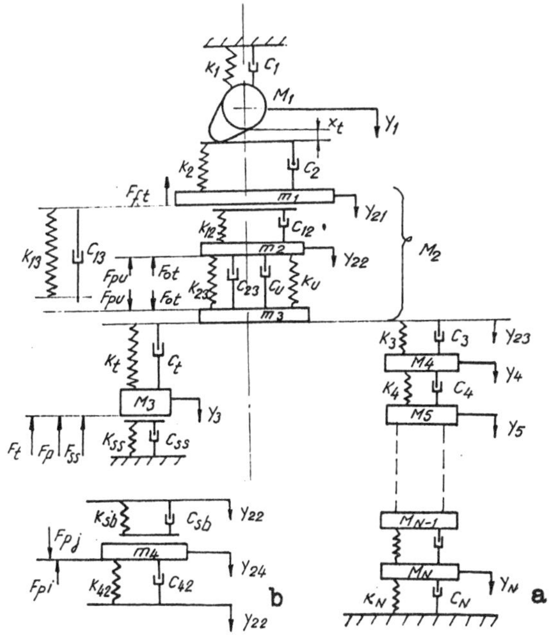

Fig. 3 Valve train model with hydraulic tappet (a), check valve model (b)

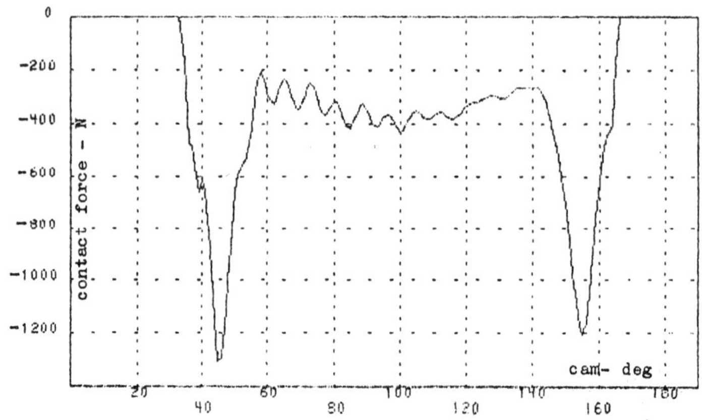

Fig. 10 Cam - tappet contact force

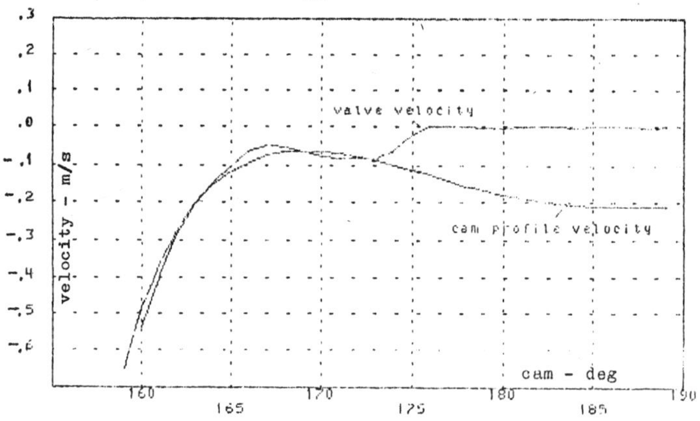

Fig. 11 Valve and cam profile velocities in the impact lope intervals

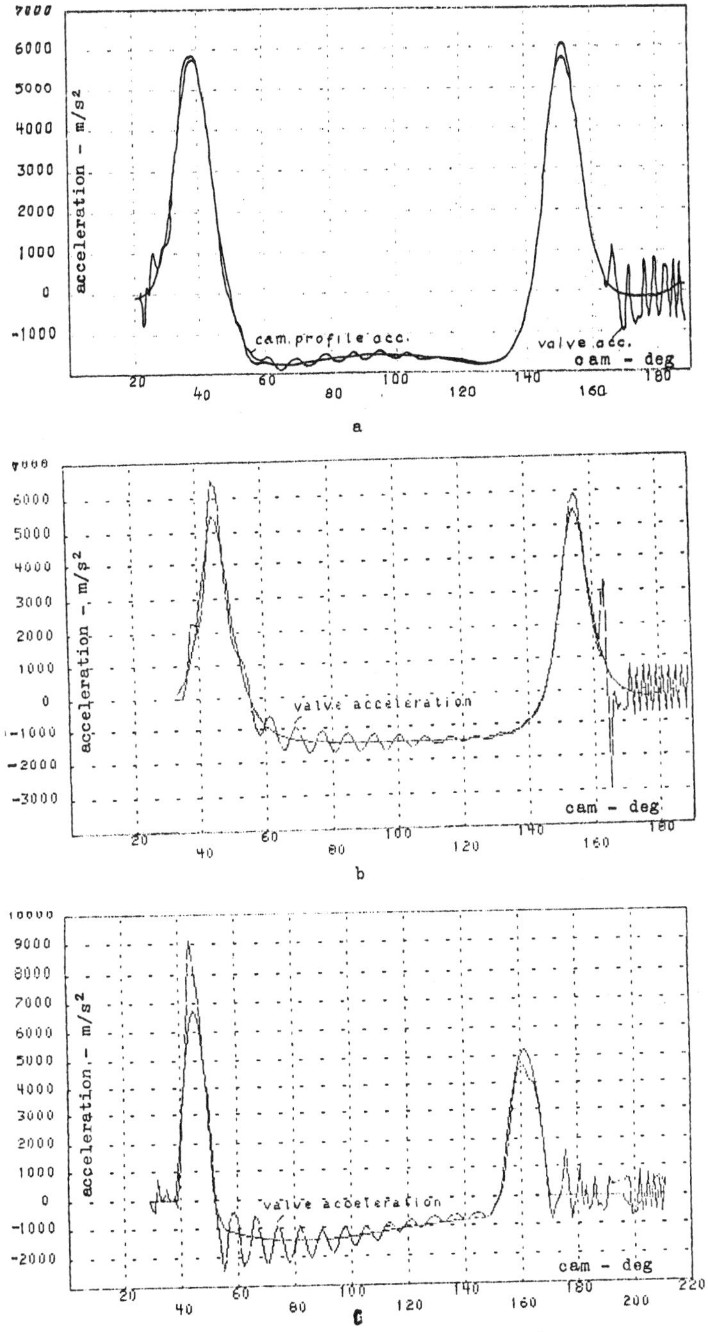

Fig. 9 Valve and cam profile accelerations

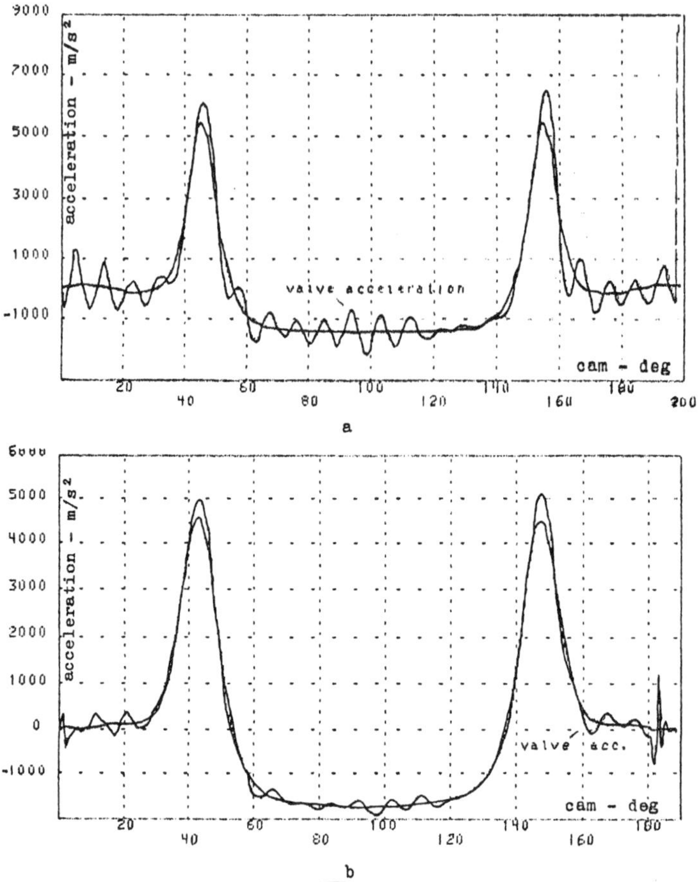

Fig. 12 Valve and cam profile accelerations without (a) and with (b) optimized cam profile

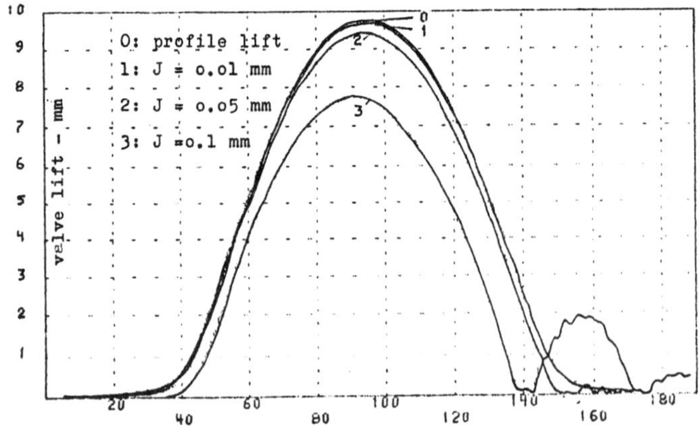

Fig. 13 Valve lift for different values of the piston - barrel gap (J)

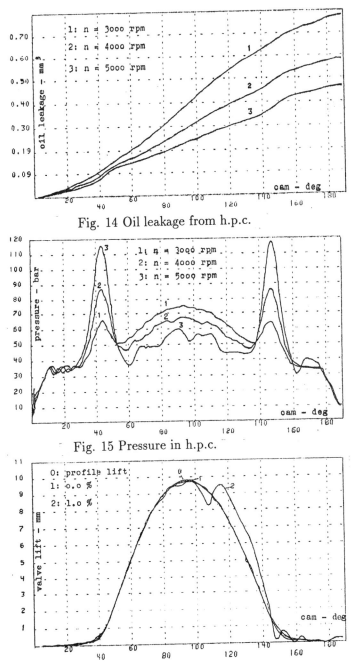

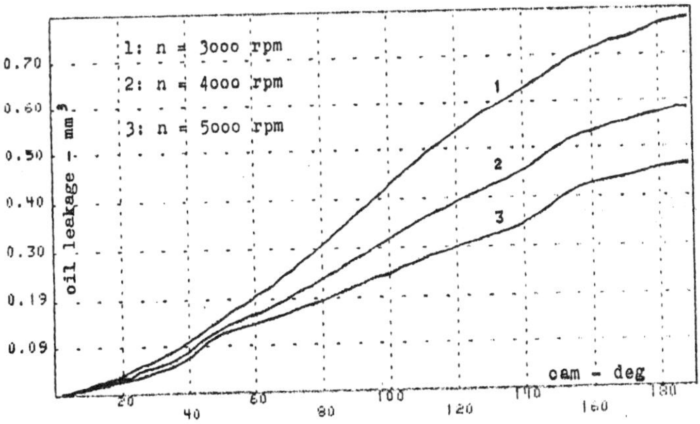

Fig. 14 Oil leakage from h.p.c.

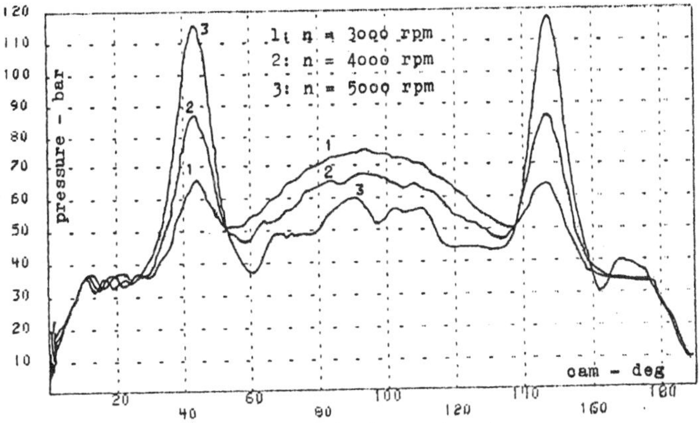

Fig. 15 Pressure in h.p.c.

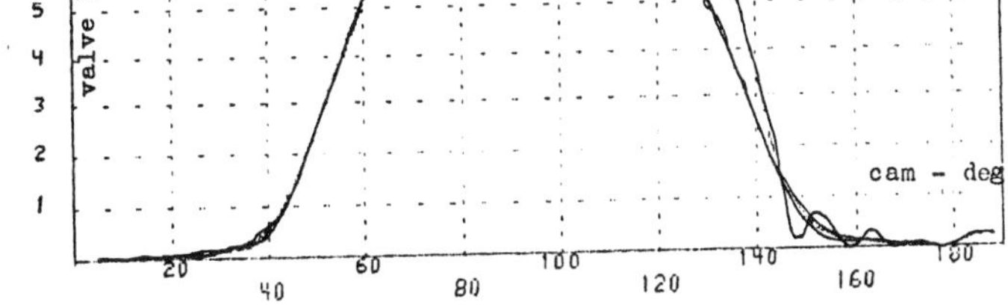

Fig. 16 Influence of the percentage ot gaseous inclusions on the valve lift

Design of a Minimum Weight Wing Enhancing Roll Beyond Reversal

M. French (*), F.E. Eastep (**)
(*) Aerospace Engineer, Wright Laboratories, Wright-Patterson AFB, OH, U.S.A.
(**) Head, Graduate Aerospace Engineering Department, University of Dayton, Dayton, OH, U.S.A.

Abstract

Fighter wing design is often driven by the need to maintain good roll performance at high dynamic pressures where control surface effectiveness is poor. A finite element based optimization and analysis code is used to design the structure of a typical fighter wing for minimum weight using stress constraints applied during a 9G pullup and a 180 deg/sec roll at very high dynamic pressure. Significant weight savings result from using leading edge control surfaces and trailing edge control surfaces beyond the reversal dynamic pressure.

Introduction

The design of fighter wings is often conditioned by the need for the aircraft to maintain a given roll rate capability throughout the flight envelope. At high dynamic pressure conditions, this requirement can be difficult to meet due to a serious decrease in the effectiveness of control surfaces due to flexibility of the wing. Here, control surface effectiveness, e, is defined as the ratio of flexible to rigid rolling moment coefficients due to control surface deflection.

$$e = \frac{\left[C_{l_\delta}\right]_{flexible}}{\left[C_{l_\delta}\right]_{rigid}}$$

Designers have used various approaches to deal with this decrease in roll effectiveness. Typical solutions are to stiffen the wing beyond what is needed to meet the other design constraints and to allow differential horizontal tail deflections (often called a rolling tail or tailerons). Another solution has been proposed in which the wing is not stiffened and wing control surfaces are used at dynamic pressures beyond those at which reversal occurs. This approach can be further improved through the use of leading edge control surfaces for which reversal is not a problem. Leading edge control surfaces on flexible wings have the benefit of getting more effective as dynamic pressure increases rather than less effective as do surfaces at the trailing edge. The benefits of using leading edge surfaces and trailing edge surfaces beyond reversal potentially include improved roll performance at critical parts of the flight envelope and decreased structural weight of the vehicle.

The practicality of using leading edge control surfaces and trailing edge control surfaces beyond reversal has been addressed both analytically and experimentally[1,2,3]. The availability of general purpose finite element based optimization and analysis codes offers the possibility of considering the influence of different control surface configurations on structural weight. This work applies a general purpose finite element based design code called ASTROS[4] to a generic fighter wing with the objective of minimizing weight subject to stress constraints and a roll rate constraint.

Wing Modelling

The wing is a low aspect ratio clipped delta. There is an inboard trailing edge control surface, an outboard trailing edge control surface and a single leading edge control surface. The aerodynamic planform is presented in Figure 1. ASTROS uses a linear panel method for both subsonic and supersonic steady aerodynamic modelling. The aerodynamic model of the wing was kept simple so that run times did not become prohibitively long; the aerodynamic representation of the wing used a 6 x 6 panel grid. The wing section was assumed to be symmetric, so no camber was included in the model. Thickness has very little effect on on aeroelastic results using linear aerodynamics, so thickness was not modelled either. The wing has a conventional built up structure made of aluminum with 12 spars and five ribs. The structural model was also kept relatively simple. A picture of the finite element model is presented in Figure 2. The results of this study are meant to be largely qualitative, so the ability to make large numbers of runs quickly was considered more important than that of resolving small details in the structure. Most runs were made on a Sun SPARCstation 1, with some others being done on a VAX 8650 and a VAXstation III. Typical run times on the SPARCstation were 10 to 20 minutes for a design run.

The results of any optimization effort are strongly dependent on the number and type of design variables chosen. There are 235 finite elements in this wing model, thicknesses or areas of which are all potential design variables. Using each element as a design variable would result in extremely long run times and would be inconsistent with the nature of the aerodynamic and structural models. The number of design variables was reduced to 5 using very heavy linking of the element sizes. This linking certainly adds weight to the final designs. In return, though, the models run relatively quickly and extended series of runs can be made in a short time. It was assumed that although the design variable linking added weight to the final design, it did not change the basic phenomena involved and therefore, the results presented here could be qualitatively applied to other design problems.

Analysis

ASTROS computes both rigid and flexible stablility derivatives from the structural and aerodyanmic models defined in the input file. It then uses these values to solve the rolling moment equation

$$M = qSb\left[\sum_{i=1}^{n} C_{l_{\delta_i}}\delta_i + C_{l_p}\frac{pb}{2V}\right]$$

where

δ_i= deflection of ith control surface
b = reference semi-span
M = rolling moment
n = number of control surfaces
p = roll rate
q = dynamic pressure
S = reference area
V = velocity

For a steady roll the moment, M, must be zero. The equation is then solved for either δ_i or p using user input values for the other parameters and ASTROS calcuated values for the stability derivatives. In this work, p was set to be 3.14159 rad/sec (180 deg/sec). In cases involving only one control surface, the deflection, δ, of that control surface was used as the dependent variable. In cases involving more than one control surface. The deflection of one was used as a dependent variable and the deflection of the others were set to 5 degrees.

A nominal model was developed to serve as an initial design using conservative estimates of the design variable values. The weight of the nominal model is 497 lbs. Figure 3 shows the effectiveness of the three control surfaces for a range of dynamic pressures at Mach 0.85. The purpose of showing results for flight conditions in which the mach numbers and dynamic pressures are not consistent is to remove the Mach number effects from the results and present information which can be readily interpreted in the context of subsonic aerodynamics.

The nominal model was used as the starting point for a design case in which the wing was optimized for minimum mass using stress constraints during a 9G symmetric pullup maneuver. The flight condition for the pullup was Mach .85 and a dynamic pressure, q, of 44.9 psi. This dynamic pressure was chosen since it was the reversal dynamic pressure of the inboard trailing edge surface for the nominal wing. The weight of the final model was 242 lb. Figure 4 shows a plot of the control surface effectiveness of this model for various dynamic pressures at Mach 0.85. The effectiveness of the trailing edge surface decreased significantly and the leading edge surface effectiveness increased. This is due to a large decrease in torsional stiffness of the wing as the sizes of the elements decreased.

It seems intuitive that a weight penalty is imposed as more constraints are included. To define the weight penalty which results from imposing a roll rate requirement, a roll rate constraint and stress constraints during roll as well as stress constraints during a symmetric pullup were applied for the next design case. The roll constraint required a roll rate of 180 degrees per second at Mach .85 and q=44.9 psi given a maximum control surface deflection (positive or negative) of no more than 10 degrees. Maximum allowable stresses were 30 ksi in tension and compression and 20 ksi in shear. The final weight for this wing was 452 lbs with a control surface deflection of 6.8 degrees. Figure 5 shows the control surface effectiveness for the range of dynamic pressures. The figure shows that effectiveness of the control surfaces is greatly increased throughout the dynamic pressure range in question, but at the price of a 210 lb (87%) increase in structural weight over that of the lift optimized wing.

An attempt was then made to use the other control surfaces to meet the roll rate requirements without such a large weight increase. The trailing edge outboard surface proved not to be very useful by itself or in conjunction with the inboard trailing edge surface. It reversed at a relatively low dynamic pressure, which does not preclude its use, but it is also small enough that it could not satisfy the roll rate requirement by itself. When it was used along with the trailing edge control surface only, the two surfaces were deflected in opposite directions. The result was a mediocre roll rate and an unusual stress condition which incurred a large weight penalty.

A more promising approach was to use the leading edge control surface along with one or both of the other two. Most fighters have leading edge surfaces which are used symmetrically for low speed or high angle of attack flight, and it is reasonable to assume that they could be deployed antisymmetrically for roll control as well. It is likely that, along with control system modifications, higher bandwidth actuators would be required. This may incur a weight penalty, but it is assumed that this additional weight is small compared to the potential savings in wing structural weight.

An attempt was made to use the leading edge control surface by itself to satisfy the roll rate requirement. However, there was not enough control power to get the required roll rate without using an unrealistic deflection. Deflections of leading edge control surfaces should be limited to relatively small values for thin wings, since the combination of a small leading edge radius and a large angle of attack at the leading edge can easily cause flow separation. For this reason, leading edge control surface deflections were limited to 10 degrees positive or negative deflection.

The next step was to use both the inboard trailing edge surface and the leading edge surface to satisfy the roll rate requirement. There can be only one dependent variable in the roll equation as solved by ASTROS, so the deflection of the trailing edge surface was fixed at 5 degrees and that of the leading edge surface was left as the dependent variable. Again, a limit of 10 degrees deflection plus or minus was applied to the surface. For this case, the structural weight of the wing was 427 lb. Figure 6 shows the control surface effectiveness for this design. Control surface effectiveness values indicate this wing has higher torsional stiffness than the nominal model even though the structural weight is lower.

Finally, the deflections of the leading edge surface and the inboard trailing edge surface were both fixed at 5 degrees and the deflection of the outboard trailing edge surface was left as the dependent variable. For this case, the final structural weight was 358 lb. The trailing edge surface was used beyond reversal and went to -3.2 degrees deflection. It is interesting to note that the use of the relatively small outboard trailing edge surface cut the resulting structural weight significantly.

The results of all the design runs are summarized in Table 1. In the notation used for the first column of the table, TEI refers to the trailing edge inboard surface, TEO, the trailing edge outboard surface and LE the leading edge surface.

It is apparent that the roll rate constraint is a critical one in the sizing of the wing structure and a heavy weight penalty results from using only the trailing edge inboard surface to satisfy the roll rate constraint. This is consistent with experience gained from past fighter designs which have been driven by high dynamic pressure roll rate constraints. By using the leading edge control surface, the weight penalty due to the roll rate constraint was greatly reduced. The weight savings due to using the trailing edge inboard surface in conjuction with the leading edge surface was small and would probably be eaten up by the extra local structure and actuators needed for the extra control surface. A much more significant weight savings was realized when all three surfaces were used for roll.

As stated earlier, the results presented here are intended for qualitative use. For quantitative results, several modifications should be made to the approach used. The heavy design variable linking greatly reduced run times, but limited the degree to which the final design could reflect changes in the loading conditions. A model with 20 or 30 well chosen design variables would almost certainly result in lower weights than those presented here. The accuracy of stresses computed from a finite element model is a strong function of the aspect ratios of elements in the model. Refining the structural representation would be required before detailed stress information could be reliably computed. The final modification would be to increase the number of ribs. Rib sizes got large for some of the design cases, and an increase in the number of ribs would create more potential load paths through the wing.

References

1. Miller, G; "Active Flexible Wing (AFW) Technology"; AFWAL-TR-87-3096; February 1988.

2. Pendleton, E.W., Lee, M.R. and Wasserman, L.; "An application of the Active Flexible Wing Concept to an F-16 Derivative Wing Model"; AIAA-91-0987.

3. Pendleton, E.W., Lee, M.R. and Wasserman, L.; "A Low Speed Model Simulating an F-16 Derivative Wing Design", WRDC-TR-90-3083, January 1991.

4. Neill, D.J., Johnson, E.H. and Herendeen, D.L; "Automated Structural Optimization System (ASTROS)"; AFWAL-TR-88-3028. April 1988.

Table 1
Summary of Optimization Results

Control Surfaces	Constraints	Weight (lb)
TEI	N/A - Nominal Model	497
TEI	Stress - 9G Pullup	242
TEI	Stress - 9G Pullup Stress - Roll TEI Deflection	452
TEI, LE	Stress - 9G Pullup Stress - Roll LE Deflection TEI - 5 Deg	427
TEI, TEO, LE	Stress - 9G Pullup Stress - Roll TEO Deflection TEI - 5 deg LE - 5 deg	358

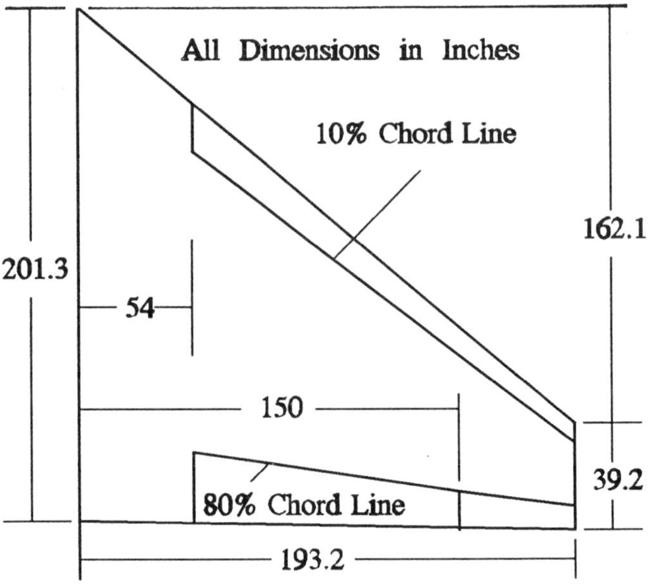

Figure 1 - Wing Aerodynamic Planform

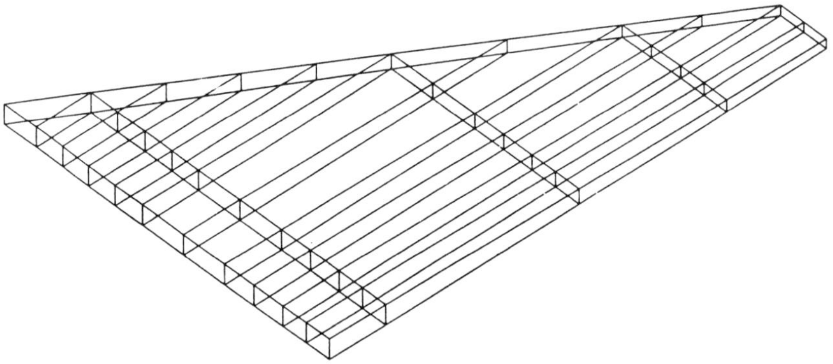

Figure 2 - Wing Finite Element Model

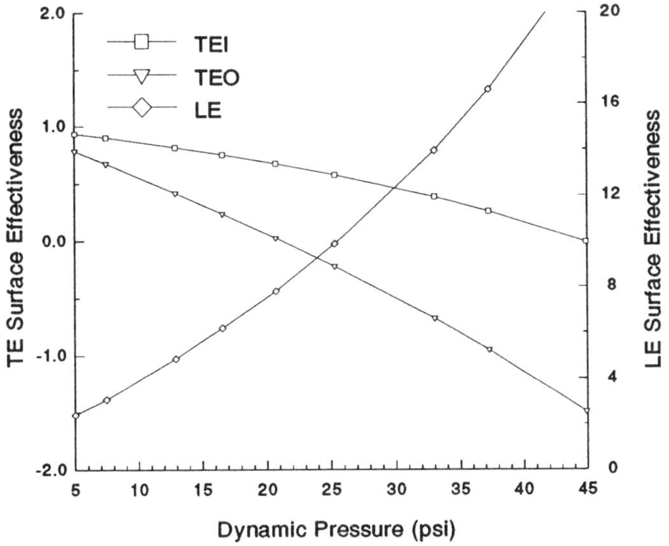

Figure 3 - Control Surface Effectiveness of Nominal Model

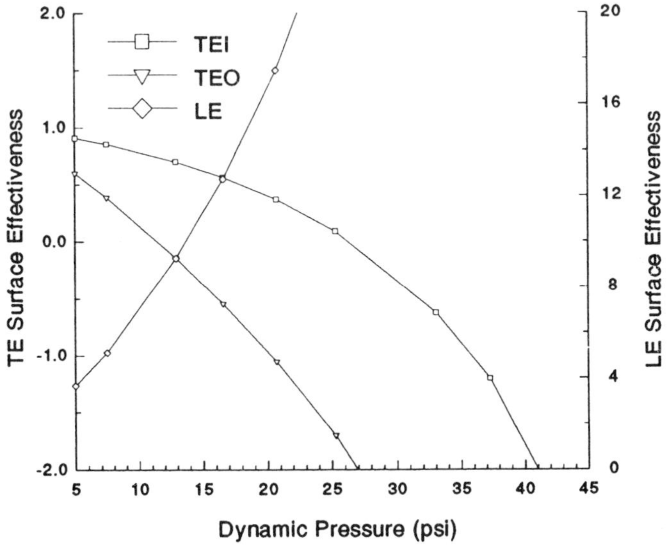

Figure 4 - Control Surface Effectiveness of Lift Optimized Model

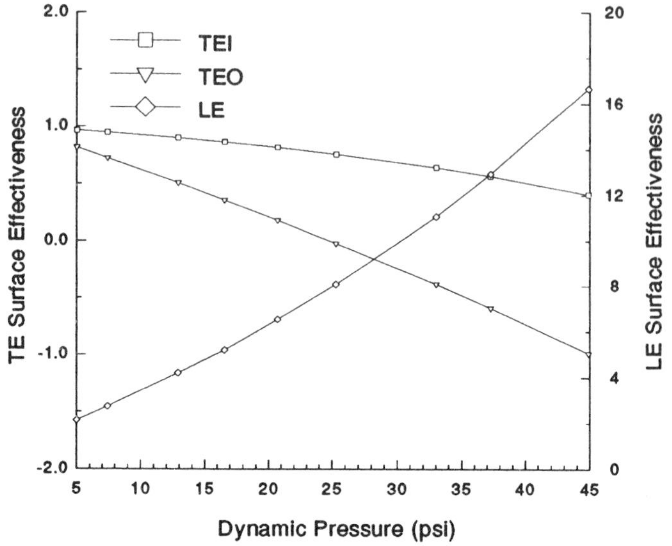

Figure 5 - Control Surface Effectiveness of Wing With TEI Surface

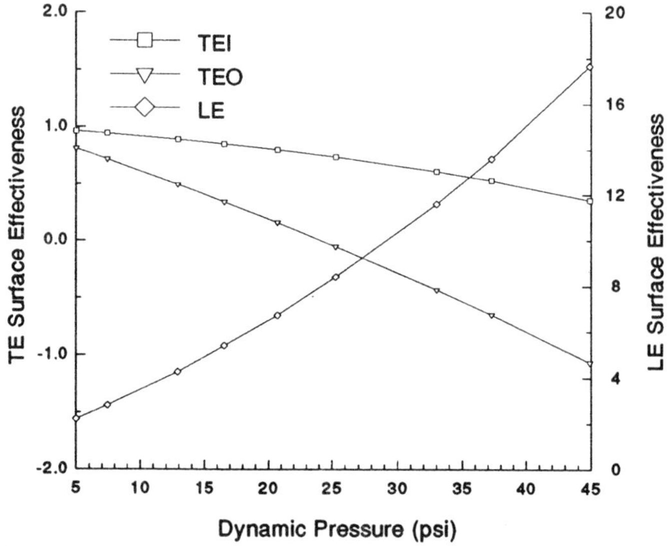

Figure 6 - Control Surface Effectiveness of Wing With TEI and LE Surfaces

Optimization Applications for Aircraft Engine Design and Manufacture

T.K. Pratt (*), L.H. Seitelman (*), R.R. Zampano (*), C.E. Murphy (*), F. Landis (**)

(*) Management Information Systems, Pratt & Whitney, United Technologies Corporation, East Hartford, CT 06108, U.S.A.

(**) Department of Mechanical Engineering, University of Wisconsin-Milwaukee, Milwaukee, WI 53201, U.S.A.

ABSTRACT

Numerical optimization is an important tool in many aspects of jet engine design, development, and test. Early optimization work centered on structural optimization projects, while more recent applications are multidisciplinary in nature. A variety of constrained and unconstrained minimization techniques have been applied to the solution of these problems. Some projects resulted in substantial cost savings relative to previously used methods; others provided automated solutions not previously attainable when an engineer was part of the iteration process. A compressor/turbine performance optimization effort produced an automated technique for preliminary design of advanced gas turbine engines. By varying flowpath design parameters, component efficiencies were maximized, subject to both structural and aerodynamic constraints. In a manufacturing problem, a technique was developed to inspect drill hole patterns and tolerances automatically, and to compare them with corresponding template geometries. An interesting hybrid solution process combined penalty function and constrained optimization techniques. An acoustic data reduction project provided an automated procedure to match analytical models to engine data.

INTRODUCTION

Engineering and manufacturing applications of optimization techniques have dramatically expanded at Pratt & Whitney (P&W) during the past decade, with support from two NASA-Lewis Research Center projects. The Structural Tailoring of Engine Blades (STAEBL) project [1,2] designed composite and hollow fan blades to minimize direct operating cost (DOC), subject to constraints on resonance margins, stresses, and flutter. The

Structural Tailoring of Advanced Turboprops (STAT) project [3] applied numerical optimization and finite element analyses to highly-swept propfan blades to minimize either DOC or aeroelastic differences between a physical model and a corresponding computational model, subject to structural, aerodynamic, and acoustic constraints.

Other minimum weight structural optimization projects included compressor and turbine blade, compressor disk, rotor shaft, engine case, and burner liner designs [4]. Some of the novel, non-intuitive designs provided feasible solutions not previously attainable by design engineers. The optimization procedures considered a multiplicity of complex design criteria, and produced final designs that were not only feasible, but optimally feasible as well. The principal savings were obtained in the automation and compression of the design iteration cycle, rather than computer time reductions, so that high quality designs were produced in shorter elapsed times.

As optimization experience spread beyond structural design, applications appeared in other areas, some of which are illustrated in this paper. Efforts are currently underway to automate the entire engine design procedure using rules-based design techniques. By collecting and quantifying design rules for common parts used in all engines, techniques from optimization and artificial intelligence will be used to establish more sophisticated design procedures in the future [5].

DISCUSSION AND RESULTS

The general nonlinear optimization problem is the determination of design variables $\underline{x} = (x_1, \ldots, x_n)$ for the objective function $f(\underline{x})$ that will

$$\text{minimize } f(\underline{x}) \, , \tag{1a}$$

subject to behavior constraints

$$g_i(\underline{x}) \leqslant 0 \quad , \quad i = 1, \ldots m \, , \tag{1b}$$

and side constraints on the design variables

$$L_i \leqslant x_i \leqslant U_i \quad , \quad i = 1, \ldots n \, . \tag{1c}$$

In some applications with more than one objective to be optimized, the scalar objective function $f(\underline{x})$ represents a weighted combination of several individual quantities.

Several computer programs are available in software libraries for the solution of (1). Because of their versatility in solving a variety of engineering problems, the Control Program for Engineering Synthesis/Constrained Minimization (COPES/CONMIN) [6,7] and the Control Program for Engineering

Synthesis/Automated Design Synthesis (COPES/ADS) [8] programs developed by Dr. G. N. Vanderplaats have been utilized at P&W during the past decade. The widely used CONMIN program is based on the method of feasible directions (MFD), while ADS includes the modified method of feasible directions (MMFD), sequential linear programming (SLP), and the Broyden-Fletcher-Goldfarb-Shanno (BFGS) method [9]. COPES is a preprocessor program that interfaces the optimizers with user subroutines for the objective function and constraints.

The optimizations were performed using a sequence of constrained one-dimensional line searches. The search directions were established using the gradients of the objective function and active constraints. For most engineering applications, one-sided finite difference approximations to the gradients were sufficiently accurate. Global optima and multiple local minima were demonstrated by varying the initial guesses.

Subsequent sections discuss examples from the wide variety of fields where optimization procedures are currently in use at P&W. Representing the entire engine development cycle, these applications include optimization of flowpath designs in engineering, an optimized bolt hole inspection program in manufacturing, and an acoustic data reduction program in engine test.

Compressor and Turbine Design Optimization

The Advanced Engine Group at P&W receives thrust and weight requirements for specific missions from airframe manufacturers. Once preliminary aerodynamic specifications and flowpaths have been established, structures models are used to design prototype engine cross-sections. Rotor dynamics, vibration, and flutter analyses are studied for these models. Finally, economic analyses estimate weight and DOC for each particular design. This project coupled the flowpath simulation program with the COPES/ADS optimization program so that individual components, namely compressors and turbines, could be evaluated.

The compressor flowpath design system specified over 160 input parameters (dimensions, losses, etc.). For a realistic prototype problem, however, only four basic design variables (inlet flow, exit Mach number, and inlet and exit root/tip ratios) were selected. Compressor efficiency was the objective function. Twenty-seven constraints (surge margin, rotor speed, etc.) provided realistic limitations commonly encountered by performance design engineers. In particular, surge margin determined the limit in engine pressure ratio above which large scale flow instability and flow reversal could occur. Because there was a trade-off between efficiency and surge margin, this constraint was nearly always active at the optimum design.

The turbine design system, shown in Figure 1, involved more than 190 input parameters. Three design variables were determined to maximize efficiency, subject to twenty-two constraints.

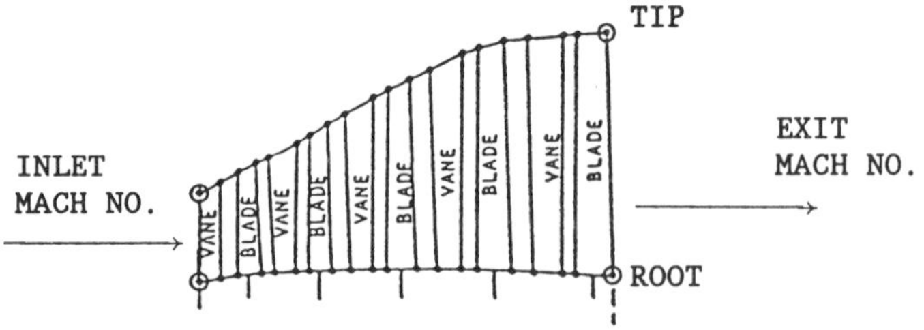

Figure 1: Turbine flowpath design permitted specification of over 190 input parameters.

As shown in Table 1, the high pressure compressor (HPC) flowpath design was optimized using MMFD, MFD, SLP, and a Taylor series approximation method (APPROX). Starting from an initially feasible base case, all methods achieved nearly a 1% gain in efficiency, with three constraints active at convergence. For mature engine designs, efficiency gains of even 0.1% were considered significant. Gradient-based methods (MFD, MMFD) generally required more function calls than sequential programming strategies. Apparently, these problems were only slightly nonlinear, so that the linear approximation subproblems in SLP and APPROX were able to identify an optimum relatively early in the iterative process.

Table 1: High pressure compressor optimization produced 1% efficiency gain over base design

	Base Case	MMFD	MFD	SLP	APPROX
Design Variables*					
Inlet Flow	.775	.084	.176	.241	.084
Exit Mach No.	.460	.336	.180	.000	.336
Inlet Root/Tip Ratio	.587	.287	.293	.283	.290
Exit Root/Tip Ratio	.985	1.000^{A}	1.000^{A}	1.000^{A}	1.000^{A}
Objective Function					
Efficiency Increase (%)	.0	1.03	1.03	1.05	1.03
Selected Constraints*					
Exit Wheel Speed	.909	$.951_{A}$	$.966_{A}$	$.983_{A}$	$.951_{A}$
Diffusion Factor	.931	1.000^{A}_{A}	1.000^{A}_{A}	1.002^{A}_{A}	1.000^{A}_{A}
Surge Margin	.029	$.000^{A}$	$.000^{A}$	$-.001^{A}$	$.001^{A}$
No. of Function Calls		71	52	23	14

*All design variables and constraints were normalized between 0. and 1.   A = Active

The high pressure turbine (HPT) flowpath design was also optimized using a variety of techniques from COPES/ADS. Starting from initially infeasible designs, efficiency increases of about 0.7% were realized in all cases. The final design shown in Table 2 was fully constrained, with three active constraints.

Table 2: Optimized high pressure turbine produced 0.7% gain in efficiency starting from initially infeasible design.

	Base Case	MMFD	MFD	SLP	APPROX
Design Variables*					
Velocity Ratio	.000^A	1.000^A	1.000^A	1.000^A	1.000^A
Exit Mach No.	.867	.931	.927	.927	.938
Inlet Mach No.	.105	.000^A	.000^A	.000^A	.000^A
Objective Function					
Efficiency Increase (%)	.0	.74	.73	.74	.74
Selected Constraints*					
Blade Stress	1.099^V	.996^A	.999^A	1.000^A	.990^A
Exit Flow Angle	-.007^V	.042	.042	.042	.044
Exit Mach No.	.992	.973	.972	.972	.974
Flow Turning	1.136^V	.905	.905	.906	.904
No. of Function Calls		42	24	15	14

*All design variables and constraints were normalized between 0. and 1.

A - Active
V - Violated

For both the HPC and HPT optimization problems, each function call required only 1-2 CPU seconds on an IBM 3090 computer, and an entire optimization required 1-2 CPU minutes, permitting sophisticated trade-off studies that uncovered potential problem areas early in the design process.

Bolt Hole Inspection

The inspection of complex and accurately machined and drilled components to determine which are in tolerance and which can be reworked is an extremely time-consuming manual effort. In our application, bolts were supposed to fasten two essentially planar components together. Due to required clearances between holes and bolt diameters, there was a tolerance associated with each hole. To pass inspection, nominal bolt hole locations and part diameters were checked against corresponding locations and tolerances on an exact template design for the component. Since the inspection table permitted 2-D planar motion, the problem was to determine the best orientation of the part so that as many bolt holes as possible could be lined up with their respective template locations.

The primary difficulty in the inspection problem was reconciling nominal and template data measurements taken from two different coordinate systems. As shown in Figure 2, there were three independent design variables given by $\underline{x} = (\Delta x, \Delta y, \Delta\theta)$ representing the rigid body translations Δx and Δy along the x- and y-axes, respectively, and a rigid body rotation $\Delta\theta$ about the origin that could be used to transform the nominal data for comparison with the template data. The distances between nominal bolt hole centers and their corresponding template hole centers determined "goodness of fit" between the component and the template. For the ith bolt hole center located at (x_i, y_i) and its corresponding template center at (x_i', y_i'), a suitable objective function $f(\underline{x})$ was given by

$$f(\underline{x}) = \sum_{i=1}^{n} d_i^{\,2} \quad , \tag{2}$$

where d_i was the distance between the $i^{\underline{th}}$ bolt center and $i^{\underline{th}}$ template hole center,

$$d_i = [(x_i - x_i')^2 + (y_i - y_i')^2]^{1/2} \; . \tag{3}$$

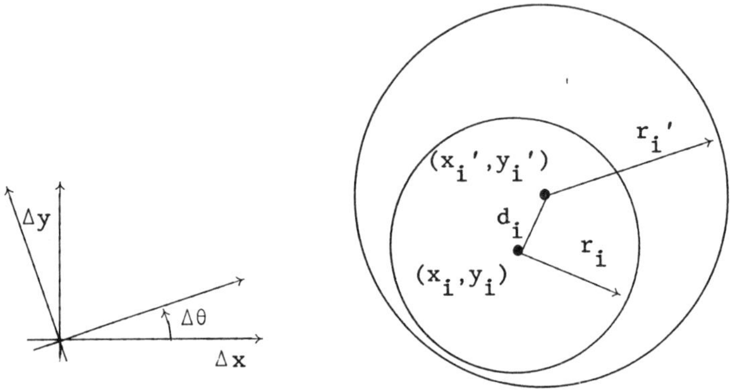

Figure 2: Translation and rotation applied to nominal coordinates (x_i, y_i) for comparison with template coordinates (x_i', y_i').

To minimize rework, as many holes as possible were brought within tolerance. To achieve this goal, a penalty term on the number of out-of-tolerance bolt holes was appropriate. Denoting the radius of the $i^{\underline{th}}$ nominal hole as r_i and the radius of the corresponding template hole as r_i', the tolerance was determined by the difference $r_i' - r_i$. A nominal hole was contained within its corresponding template hole, provided $d_i \leqslant r_i' - r_i$. Consequently, the appropriate penalty function to be appended to the objective function was

$$P(\underline{x}) = N \sum_{i=1}^{n} \Big\langle (r_i + d_i) - r_i' \Big\rangle_+^2 \; , \tag{4}$$

where N was the number of holes out of tolerance, and the bracket operator was defined as

$$\Big\langle g(\underline{x}) \Big\rangle_+ = \begin{cases} g(\underline{x}) & , \quad g(\underline{x}) > 0 \\ 0 & , \quad g(\underline{x}) \leqslant 0 \; . \end{cases} \tag{5}$$

The augmented objective function was given by

$$F(\underline{x};R) = f(\underline{x}) + R\,P(\underline{x}) \; , \tag{6}$$

where the penalty parameter R (typically 10^6) was used to accelerate convergence by rapidly decreasing the number of holes that were out of tolerance.

In practice, the penalty function approach was still inefficient and ineffective for placing bolt holes in tolerance. Consequently, a more practical technique involving constraints was developed. For a manual inspection, if bolts could be placed through two relatively good nominal hole locations, the part would be "pinned" to the template so that only minor adjustments would be necessary to qualify the remaining bolt hole locations. In fact, if the two selected holes were very accurate relative to the template, then most other holes would likely to be within tolerance once these holes were pinned.

To determine the "best hole", the locations of each bolt hole relative to all other holes were compared to corresponding relative distances on the template. As shown in Figure 3, the distance between the $i^{\underline{th}}$ and $j^{\underline{th}}$ nominal holes was

$$d_{ij} = [(x_i - x_j)^2 + (y_i - y_j)^2]^{1/2} . \qquad (7)$$

The corresponding distance between the $i^{\underline{th}}$ and $j^{\underline{th}}$ template holes was denoted by d_{ij}' and was defined analogously. The sum of the squares of all the deviations between the nominal and corresponding template pattern distances for the $i^{\underline{th}}$ hole was

$$G_i = \sum_{i=1}^{n} (d_{ij} - d_{ij}')^2 . \qquad (8)$$

The "best hole" was taken to be the one with the smallest value of G_i, namely G_{min}, representing the smallest overall deviation between corresponding nominal and template distances. Similarly, G_{max} represented the "worst" hole.

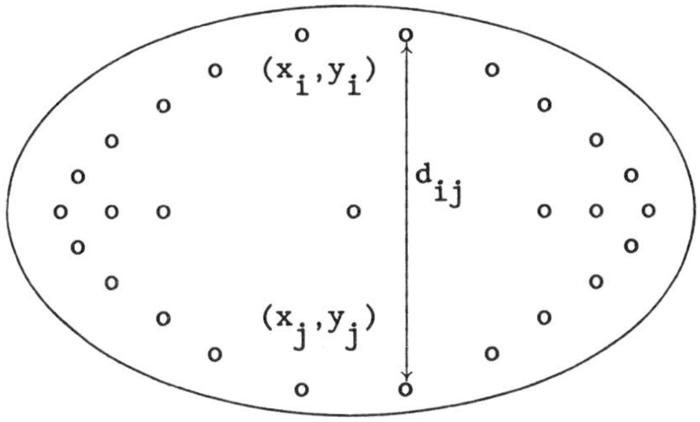

Figure 3: Accuracy of bolt hole location determined by relative deviations from template distances.

In addition to being another "accurate" hole with a low value of G_i, the second hole to be constrained was selected to be relatively far from the best hole to minimize small angle errors when both holes were "pinned". Consequently, the "second best" hole was selected to maximize the term

$$\max_{\text{All } i} \left\{ \left(\frac{D_i}{D_{max}} \right) \left(\frac{G_{max} - G_i}{G_{max} - G_{min}} \right) \right\}, \qquad (9)$$

where D_i was the distance from the best hole to the $i^{\underline{th}}$ hole, and D_{max} was the maximum distance from the best hole to any hole. The constraints

$$g_i(\underline{x}) = (r_i + d_i) - r_i' \leqslant 0 \ , \qquad (10)$$

were applied for the two indices corresponding to the two "best" bolt holes. Ironically, when such constraints were applied to all the bolt holes, the optimizer locked up. Once a constraint was satisfied, even for a "bad" hole, it could never be violated subsequently in order to satisfy other constraints.

The nonlinear constrained optimization problem was solved using the MFD in the COPES/CONMIN program. For dimensional consistency, the design variable $\Delta\theta$ was scaled using a characteristic length L so that $L\Delta\theta$ was approximately the same order of magnitude as Δx and Δy. Constraining the two "best" holes was successful provided the maximum number of remaining holes could be brought into tolerance by varying the "pinned" holes within their allowed tolerances. The strategy of selecting these "best" holes was critical since if, by chance, a "bad" hole were "pinned", then many potentially "good" holes could not be brought within tolerance.

The program was checked using a component with 100 randomly located bolt holes. To model typical manufacturing situations, the hole center locations were perturbed away from their corresponding template values using a Gaussian process. Adjusting the variance controlled the approximate number of holes expected to violate the tolerance band. After an initial displacement of the inspection table in the Δx, Δy, and $\Delta\theta$ directions, the optimizer determined shifts to place as many holes as possible within tolerance when compared to the template.

The results of this program checkout were summarized in Table 3. In each case small initial displacements were applied to the nominal hole patterns. The optimizer determined subsequent translations and rotations to bring as many holes as possible within tolerance (.001 in.). Perturbations with successively larger variances were applied to the hole center locations. In all cases, the final shifts were close to the exact template locations (Δx, Δy, $\Delta\theta$) = (0, 0, 0). The procedure was considered robust enough for manufacturing applications, and is currently being used for on-site inspection.

Table 3: Optimization of shifted templates with perturbed hole patterns.

Variance of Perturbation (in.)	Holes Out of Tolerance After: Perturb Only	Perturb & Shift	Optimize	Final Shifts Δx (in.)	Δy (in.)	$\Delta\theta$ (rad.)
.0002	0	92	0	.0000	.0000	-.0001
.0004	3	89	1	-.0003	.0000	.0000
.0006	8	89	9	.0000	.0001	.0002
.0008	16	87	15	-.0004	.0000	.0012
.0010	24	86	24	-.0002	.0000	.0004

Automated Noise Separation Optimization

Identifying acoustic components that contribute to aircraft engine noise is very important in understanding how noise is produced, how to measure it, and how to control it. Jet noise is caused by turbulent mixing of the engine exhaust with the atmosphere; core noise is due primarily to combustion; turbo-machinery noise refers to noise from the fan, compressor, and turbine components. During an actual engine test, acoustic data were received by an array of microphones distributed between 0° and 180° sufficiently far from the engine. As illustrated in Figure 4, the noise signal from each microphone was then processed using a fast Fourier transform (FFT) and converted into a sound pressure level (SPL) frequency spectrum.

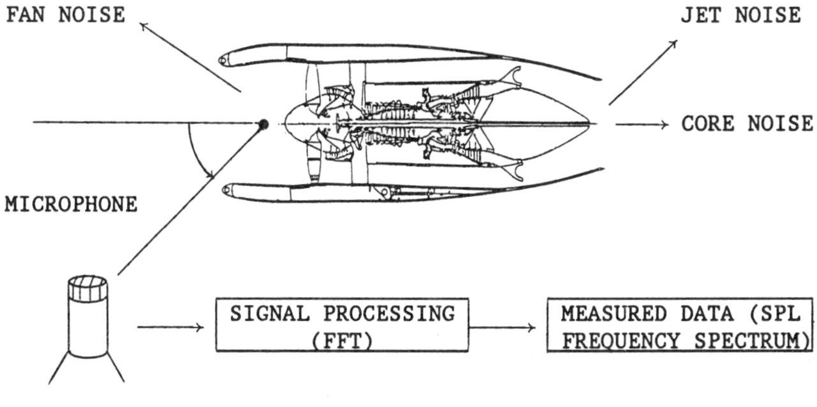

Figure 4: Measured data determined sound pressure level frequency spectrum for components of engine noise.

A typical SPL curve is shown as "measured data" in Figure 5. The corresponding jet and core noise spectral shapes were predicted using an existing data base or from an empirical prediction system. The predicted jet and core noise curves were assumed to be the two main components of low frequency engine noise. SPL's from two different source components were combined according to a standard logarithmic addition formula:

$$PT_i = 10 \ LOG_{10} \ (10^{PJ_i/10} + 10^{PC_i/10}) \ . \tag{11}$$

where PJ_i, PC_i, PT_i were the SPL values of the predicted jet, core, and total, respectively, at the $i^{\underline{th}}$ frequency level.

Each of the jet and core curves could be shifted in either amplitude or frequency, for a total of four design variables. Prior to this project, these adjustments were made manually until agreement with measured data was achieved, and jet and core components were identified. This time-consuming procedure, whose output was both user-dependent and largely unrepeatable, was automated and optimized. The objective function to be minimized for this program was the standard measure,

$$ERROR = \sum_{i=1}^{N} [(M_i - PT_i)^2/N \ ]^{1/2} \ , \tag{12}$$

where M_i was the measured data at the $i^{\underline{th}}$ frequency, and N was the number of data points in the SPL spectrum.

In order to solve the unconstrained (except for side constraints) minimization problem, the BFGS method was used. A test case for a typical aircraft engine is illustrated in Figure 5; here the initial locations for the jet and core curves were quite far from the measured data. After optimal shifts were determined, the predicted total curve was indistinguishable from the measured data, with an average error of less than 1/2 decibel (dB), including the core noise "bump."

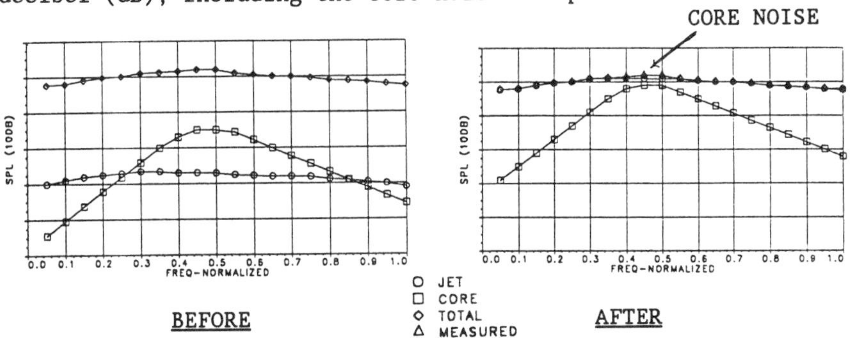

Figure 5: Automated system accurately modeled engine noise within limits of experimental error.

This program successfully identified noise components for several different P&W engine types for a wide range of operating speeds and microphone angle locations. Tasks that previously required elapsed times of four weeks were performed in four days. In addition, the variability of human interaction was removed. Average SPL errors in amplitude were generally less than 1 dB; this level was considered acceptable when compared to typical levels of experimental error.

CONCLUSIONS AND FUTURE DIRECTIONS

In the aircraft engine industry, compressors, turbines, and rotors have been re-designed many times for derivative engine applications. For years, standard computer-aided design (CAD) procedures have been applied in an ad hoc manner. Sets of rules and numerical optimization procedures have been established for many specific part types. Using parametric design techniques, prototypes have been rapidly generated for trade-off studies. Currently, a large engineering-wide effort is underway at P&W to incorporate optimization, expert system, artificial intelligence, and "intelligent" CAD techniques into rules-based design procedures for generic part types.

Historically, optimization techniques have made design improvements based upon search directions determined by gradient calculations. The key feature of this approach was coupling the analytic description of the design procedure with the optimizer. The major disadvantages were the inability to make discontinuous changes in design variables (e.g., changing materials, or adding or deleting design features) and the premature termination of the search procedure after encountering local optima or binding constraints. By incorporating rules-based design techniques into the search procedure, some subjective and intuitive decisions could be used to complement analytical criteria in determining improved designs.

The importance of optimization techniques in the design, manufacture, and testing of future aircraft engines is increasing. Even modest parallel processing capabilities will drastically reduce the computer time required by the gradient search techniques, since finite difference function evaluations will be performed simultaneously. The availability and speed of workstations will permit the rapid analysis and trade-off studies required by minimization algorithms.

ACKNOWLEDGMENTS

The authors would like to express their gratitude to Dr. G. N. Vanderplaats, VMA Engineering, Inc., for many helpful suggestions in implementing his software for these applications, and to P&W employees C. Cook and F. Pinney (Manufacturing), A. Peracchio (Acoustics), B. Robinson (Advanced Engines), and S. Tanrikut (Turbine) for many helpful technical comments.

REFERENCES

1. Platt, C.E., Pratt, T.K., and Brown, K.W., Structural Tailoring of Engine Blades (STAEBL), NASA CR-167949, June, 1982.

2. Brown, K.W., Pratt, T.K., and Chamis, C.C., Structural Tailoring of Engine Blades (STAEBL), Proc. AIAA/ASME/ASCE/AHS 24th Structures, Structural Dynamics, and Materials Conference, Lake Tahoe, NV, pp. 79-88, May, 1982.

3. Brown, K.W., Structural Tailoring of Advanced Turboprops (STAT), NASA CR-180861, August, 1988.

4. Pratt, T.K., Optimization Applications in Aircraft Engine Design and Test, pp. 815-832, Proceedings of Recent Experiences in Multidisciplinary Analysis and Optimization, NASA Conference Publication 2327, April, 1984.

5. Tong, S.S., and Gregory, B.A., Turbine Preliminary Design Using Artificial Intelligence and Numerical Optimization Techniques, ASME Gas Turbine Congress and Exposition, Brussels, Belgium, June 11-14, 1990.

6. Vanderplaats, G.N., CONMIN - A FORTRAN Program for Constrained Function Minimization, NASP TM X-62, 282, August, 1973.

7. Madsen, L.E., and Vanderplaats, G.N., COPES - A FORTRAN Control Program for Engineering Synthesis, Report NPS69-81-003, Naval Postgraduate School, Monterey, CA, March, 1982.

8. Vanderplaats, G.N., COPES/ADS - A FORTRAN Program for Engineering Synthesis Using the ADS Optimization Program, Dept. of Mechanical Engineering Report, Naval Postgraduate School, Monterey, CA, October, 1984.

9. Vanderplaats, G.N., Numerical Optimization Techniques for Engineering Design: With Applications, McGraw-Hill, NY, 1984.

Multiobjective Design Optimization of Helicopter Rotor Blades with Multidisciplinary Couplings

A. Chattopadhyay, T.R. McCarthy

Department of Mechanical and Aerospace Engineering, Arizona State University, Tempe, Arizona, 85287 - 6106, U.S.A.

ABSTRACT

The application of multiobjective optimization is investigated in a practical design environment. A procedure is developed for the design of helicopter rotor blades with multiple design objectives and involving multidisciplinary couplings. The objective is to reduce blade vibration without degrading certain rotor performance and structural requirements. The 4/rev vertical shear and the 3/rev inplane shear forces, at the blade root, are selected as objective functions. Constraints are imposed on the remaining critical forces and moments, frequencies, centrifugal stresses, autorotational inertia, and rotor thrust. Solutions to this two-objective design problem are obtained using three different multiple objective approaches. These approaches are used in conjunction with nonlinear programming technique and an approximate analysis procedure. Substantial reductions are obtained in the vibratory root forces and moments without degrading rotor performance and structural integrity. Comparative results are presented for all three multiobjective formulations.

NOMENCLATURE

c	chord, ft
$f_3\ f_4, f_5, f_6$	natural frequencies of first four coupled elastic modes, per rev (/rev)
f_r	3/rev radial shear, lb
f_x	3/rev inplane shear, lb
f_z	4/rev vertical shear, lb
$\bar{f}_x, \bar{f}_z$	prescribed values of f_x and f_x, respectively
f_{x_0}, f_{z_0}	values of f_x and f_z, respectively, at the beginning of an iteration
g_1, g_2, g_3	constraint functions
k_r	principle radius of gyration at blade root, ft
m_c	3/rev torsional moment, lb-ft
m_x	3/rev flapping moment, lb-ft
m_z	4/rev lagging moment, lb-ft
w_j	nonstructural weight at j^{th} node, lb
x, y, z	reference axes
AI	autorotational inertia, lb-ft^2
C_T	thrust coefficient
EI_{xx}, EI_{zz}	bending stiffnesses, lb-ft^2
F_k	k^{th} objective functions

GJ	torsional stiffness, lb-ft^2
K	sum of number of constraints and objective functions
N	number of blade nodes
NCON	number of constraints
NDV	number of design variables
NOBJ	number of objective functions
NSEG	number of blade segments
R	blade radius, ft
T	thrust, lb
W	total blade weight, lb
β_1, β_2	pseudo design variables
Ω	rotor angular velocity, rad/sec
φ_i	i^{th} design variable
λ	taper ratio
μ	advance ratio
ρ	K-S function multiplier
σ	area solidity
σ_i	centrifugal stress in i^{th} segment, lb/ft^2

Subscripts

max	maximum value
r	value at the blade root
ref	reference blade value
t	value at the blade tip
L	lower bound
U	upper bound

INTRODUCTION

In recent years formal optimization methods have become an important tool which can often expedite design. A significant amount of work has been done to bring the state of the art to a very high level [1,2]. Practical engineering structures, however, are often too complicated to be defined as a single objective design problem. The quality of the design of these complex engineering systems generally depends upon the improvements of several objectives, which are often conflicting in nature. Lately, systematic decomposition and design synthesis techniques have been proposed to decompose a complex problem into parent systems and subsystems [3]. However, within these various systems, it is still difficult to decide on a single objective function as the critical design criterion. In spite of this commonly recognized fact, most optimization applications have been limited to single objective function with multiple design constraints. Although this reduces the complexity of the problem, the solution of an optimum design problem obtained using multiple objectives differs significantly from that obtained by formulating it with a single objective function and introducing the remaining objectives as constraints. The promising alternative is to consider all the different criteria in the framework of a multicriteria optimization.

Multicriteria optimization refers to problems where the objective function comprises of a set of distinct criteria, e.g., in a structural design problem these may be stresses, displacements, weight, etc. Lately, as structural optimization is emerging as a practical design tool in the industry, the need for multiobjective decision making is being recognized. Therefore, there is a renewed interest in multicriteria programming for application to such design problems. The concept of multiobjective optimization, perhaps, dates back to Pareto [4] who introduced the concept within the framework of welfare economics. Most applications of these problems in structural and mechanical designs have been based on an ordering of the objective functions prior to optimization. The optimal solution to such problems is an element of the set of Pareto optimal designs. Stadler [5] provides an extensive survey describing the mathematical basis for such methods. The application of these methods to structural design problems is recognized. Koski [6, 7] provides a description of variational problem statements and their applications to truss design problems. Osyczka [8, 9] and Rao [10, 11] have addressed numerous applications of these techniques in mechanical design problems. Bendsoe et al. [12] used a variational approach to address multiobjective design problems for application to beam problems. Hajela et al. [13] addressed the use of multiple objectives in the presence of mixed integer and discrete design variables. It appears, however, that the applications of such techniques have been limited to the design of basic structural elements like beams and trusses and limited literature exist on the application of these methods in a practical design environment. This paper addresses the use of multiobjective optimization procedures for applications to optimum helicopter rotor blade design.

The helicopter rotor design problem is multidisciplinary in nature and requires the coupling of several disciplines such as dynamics, aerodynamics, structures and controls. For example, the blade must be designed for reduced vibration, it must be aeroelastically stable, it should satisfy specified performance criteria and must be structurally safe. A true optimum design of a rotor blade, therefore, would require multiobjective formulation with the coupling of all of these disciplines. Vibration has been a major source of problems in helicopters and its alleviation plays a major role in the rotor blade design process. In the past, conventional design methods mainly used the designer's experience and trial and error methods. However today with efficient optimization schemes available and improvements towards understanding helicopter analysis, attempts are being made to apply design optimization techniques to the problem. Due to the importance of the problem, there has been a considerable amount of research aimed at reducing vibration, primarily at the blade level, as shown in Reference 22. However, in most of the previous work, dealing with optimum blade designs for reduced vibration, optimizations were either performed with single objective functions and/or without the couplings of the necessary disciplines [14-16]. Weller formulated the problem of reducing vibration of a rotor blade with two objective functions using a "Minimum Sum β" formulation [17]. However, the aerodynamic loads on the blade were prescribed and the effects of the design changes, during optimization, on changes in the blade airloads were not included thereby eliminating the discipline couplings. Chopra [19] addressed a similar problem using a more simpler approach of using the sum of the squares of the individual objectives to form a single objective function. Chattopadhyay et al. [21, 22] addressed the problem of vibration reduction in rotor blade with the coupling of aerodynamic loads and dynamics. A modified "Global Criterion" approach [21] was used to formulate the optimization problem consisting of two objective functions. The blade root 4/rev vertical shear and the blade weight were minimized, in Reference [21], with constraints on coupled lead-lag and flapping frequencies, blade autorotational inertia and centrifugal stress. The 4/rev vertical root shear and the 3/rev inplane root shear were used as design objectives with constraints on stresses, frequencies, autorotational inertia and rotor thrust in Reference 22. However, the emphasis in these research efforts was on the formal integration of the two disciplines, blade aerodynamics and dynamics, inside a closed-loop optimization environment to obtain a truly coupled rotor blade design. The efficiency of the multiobjective formulation procedure was not investigated.

In the present paper, the rotor blade design problem of Reference 22 is addressed using three different multiple objective function formulation techniques. The results obtained in Reference [22] using a modified Global Criterion approach are compared to those obtained using a Minimum Sum β formulation and a "Kreisselmeier-Steinhauser function" approach [23, 24]. The Minimum Sum β formulation is a further modification of the Global Criteria approach. The comprehensive helicopter analysis code program CAMRAD [25] is used for the blade dynamic and aerodynamic calculations. Its presence in the closed loop optimization procedure allows the inclusion of the effects due to changes in the airloads with changes in design variables. The nonlinear programming method of feasible directions, as implemented in the program CONMIN [26], is used along with a linear Taylor series based approximation technique [21]. The critical vibratory shear forces and moments, at the blade root, primarily responsible for vibration at the rotor hub are used as design objectives and/or constraints. The natural frequencies are also separated from values which are integer multiples of the rotor speed. The blade must also have sufficient autorotational inertia to autorotate in case of engine failure. The vibration reduction is achieved without incorporating penalties on blade weight and centrifugal stresses, while maintaining a required amount of rotor lifting capability.

BLADE MODEL

The blade planform is modelled with linear taper (Figure 1), with the assumption that the blade stiffnesses are contributed entirely by the blade structural components. The stiffnesses contributed by nonstructural masses, the skin, the honeycomb, etc., are assumed to be negligible. The linear chord distribution is given as

$$c(y) = c_r \left[\frac{y}{r}\left(\frac{1}{\lambda} - 1\right) + 1\right] \tag{1}$$

where c_r is the root chord and λ is the taper ratio. Based on the above chord distribution, the stiffness EI_{xx} is represented as follows:

$$EI_{xx}(y) = EI_{xx_r}\left[\frac{y}{r}\left(\frac{c(y)}{c_r} - 1\right) + 1\right]^4 \tag{2}$$

Further details of the blade modelling can be found in Reference 21.

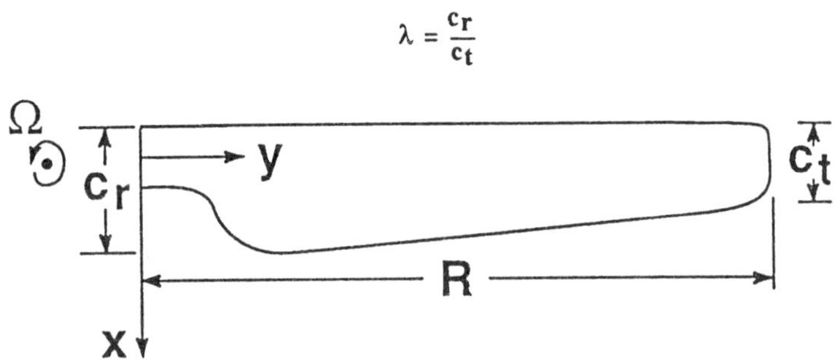

Figure 1 Simplified rotor blade model with linear taper

OPTIMIZATION FORMULATION

The optimization problem can be stated as follows

Minimize	$F_k(\varphi)$	$k = 1, 2,..., NOBJ$	(objective functions)
subject to			
	$g_{1_j}(\varphi) \leq 0$	$j = 1, 2,..., NCON$	(inequality constraints)
	$\varphi_{i_L} \leq \varphi_i \leq \varphi_{i_U}$	$i = 1, 2,..., NDV$	(side constraints)

In the present paper, the two most critical forces at the blade root, the 4/rev vertical shear (f_z) and the 3/rev inplane shear (f_x) are used as objective functions. Upper bound constraints are imposed on the 3/rev radial shear (f_r), the 3/rev flapping moment (m_x), the 3/rev torsional moment (m_c) and the 4/rev lagging moment (m_z). All of these forces and moments are critical for a four-bladed rotor (the magnitudes of the 5/rev forces and moments are small and are ignored) [27]. Upper and lower bounds are imposed on the first four coupled lead-lag natural frequencies. Constraints are also imposed on the blade weight (W), the autorotational inertia (AI), the centrifugal stresses (σ_i, i = 1, 2,..., NSEG), and the rotor thrust (T). Side constraints are also imposed on the design variables. The design variables are the blade stiffnesses at the root, EI_{xx_r}, EI_{zz_r}, GJ_r, the taper ratio, λ, the root chord, c_r, the radius of gyration at the blade root, k_r, and the nonstructural weights w_j, j= 1,2,... N, where N is the total number of blade nodes. The blade preassigned properties and the rotor performance parameters are used from an existing rotor blade's data [21], which will hereafter be referred to as the "reference" blade. The optimization problem is solved using three different techniques. Brief descriptions of each of these techniques follows.

Modified Global Criteria Approach

This method was used by Chattopadhyay et al. [21, 22] for rotor blade optimization and is presented here for the sake of comparison. Using this method, a global objective function is formulated using the values of the single objective function optimizations, as follows

$$F_1(\varphi) = \left(\left[\frac{f_z(\varphi) - f_z(\varphi_1^*)}{f_z(\varphi_1^*)} \right]^2 + \left[\frac{f_x(\varphi) - f_x(\varphi_2^*)}{f_x(\varphi_2^*)} \right]^2 \right)^{1/2} \tag{3}$$

The design variable vectors φ_1^* and φ_2^* are obtained by individually minimizing $f_z(\varphi)$ and $f_x(\varphi)$, respectively. The optimization reduces to minimizing $F_1(\varphi)$ subject to the set of constraints $g_{1_j}(\varphi) \le 0$ ($j = 1, 2, ..., NCON$). A description of this method is presented in Reference [21].

Minimum Sum β (Min∑β) Approach

This method is a further modification of the Global Criteria approach in which the individually optimized values $f_z(\varphi_1^*)$ and $f_x(\varphi_2^*)$ are replaced by specified target values. The objective function, $F_2(\varphi)$, is also linear and is defined as

$$F_2(\varphi) = \beta_1 + \beta_2 \tag{4}$$

where β_1 and β_2 are two pseudo design variables with properties such that the original objective functions f_x and f_z remain within $\pm\beta_i$ (i=1,2) percent tolerance of some prescribed values. This requirement introduces two new constraints which assume the following form

$$\frac{f_x - \bar{f}_x}{\bar{f}_x} \le \beta_1 \tag{5}$$

$$\frac{f_z - \bar{f}_z}{\bar{f}_z} \le \beta_2 \tag{6}$$

The quantities $\bar{f}_x$ and $\bar{f}_z$ are the prescribed target values of the objective functions f_x and f_z, respectively. In this paper, these values are estimated from the optimum values of f_x and f_z obtained in the Global Criteria approach. From the above formulation, as the objective function reduces and therefore the values of β_1 and β_2, the values of f_x and f_z are driven to their prescribed values. The design variables for the Min∑β formulation comprise the original set of design variables and the two pseudo design variables β_1 and β_2. The new constraint vector, $g_{2_j}(\varphi)$, k = 1, 2,..., K, comprises the original constraints and the two new constraints presented in Equations (5) and (6), i.e., K = NCON + NOBJ.

Kreisselmeier-Steinhauser (K-S) Function Approach

The first step in formulating the objective function in this approach involves transformation of the original objective functions into reduced objective functions. These reduced objective functions are of the form

$$f_x^*(\varphi) = \frac{f_x(\varphi)}{f_{x_0}} - 1 - g_{max} \le 0 \tag{7}$$

$$f_z^*(\varphi) = \frac{f_z(\varphi)}{f_{z_0}} - 1 - g_{max} \le 0 \tag{8}$$

where f_{x_0} and f_{z_0} are the values of f_x and f_z, respectively, calculated at the beginning of each iteration. The quantity g_{max} is the value of the largest constraint corresponding to the design variable vector φ and taken to be constant for this iteration [24]. Because these reduced objective functions are analogous to the previous constraints, a new constraint vector $g_{3_j}(\varphi)$, k = 1, 2,..., K, is introduced, where K = NCON + NOBJ. The new objective function to be minimized is then defined, using the K-S function as follows [23]

$$F3(\varphi) = g_{3_{max}} + \frac{1}{\rho}\log_e \sum_{k=1}^{K} e^{\rho(g_{3_k}(\varphi) - g_{3_{max}})} \tag{9}$$

where the multiplier ρ can be considered analogous to a draw-down factor with ρ controlling the distance from the surface of the K-S objective function to the surface of the maximum function value. The design variable vector φ is identical to that used in the Global Criteria approach.

ANALYSIS

The program CAMRAD is used for both blade dynamic and aerodynamic analyses. The blade is trimmed at each cycle (with changes in design variables); therefore each intermediate design is a trimmed design. A wind tunnel trim option is selected (as the reference blade is a wind tunnel model). The basic optimization algorithm used is the method of feasible directions as implemented in the program CONMIN. Evaluations of the objective functions and the constraints using full analyses, during each iteration of CONMIN, is computationally prohibitive. Therefore, a linear Taylor series based approximation procedure is used. Using this procedure, the objective function and the constraints are approximated using first order linear Taylor series expansions based on the design variable values obtained from CONMIN and the sensitivity information obtained from the full analysis [21]. The assumption of linearity is valid over small increments in the design variable values and does not introduce large errors if the increments are small. A 'move limit', defined as the maximum fractional change of each design variable value, is imposed as an upper and lower bound on each design variable φ_i. A variable move limit procedure is used in this paper. Move limits of the order of 0.1 - 0.2 are used initially to bring the design close to a local minimum. Smaller move limits (0.01 - 0.001) are then used to converge smoothly to the optimum design.

RESULTS

The optimization procedure developed is applied to a reference blade which is a modified version of the wind tunnel model of an existing articulated rotor blade [21]. The reference blade has a radius, R = 4.685 ft and rotating speed, Ω = 639.5 rad/sec. The blade is discretized into six segments (i.e. NSEG = 6). Optimization is performed under forward flight conditions, with an advance ratio $\mu = 0.3$. The rotor is trimmed at each design step to maintain the same C_T/σ as the reference rotor (C_T being the thrust coefficient and σ the blade solidity). This, together with the constraint on the rotor thrust, forces the blade solidity (σ) to remain unchanged during optimization.

Optimum results (using all three formulations) are obtained within 20 - 25 cycles. Due to the large amount of nonlinearities present in the objective functions and/or constraints, the move limits used in the approximation procedure are carefully monitored. Tables 1 and 2 present summaries of the constraints used. Table 1 indicates that the second elastic flapping dominated mode (f_4) is active for both Min$\sum\beta$ and K-S formulations and the third elastic flapping (f_5) and the first elastic lead-lag (f_6) dominated modes are active in the K-S approach. The autorotational inertia constraint (AI) is also at its prescribed upper bound in the K-S case. The weight constraint is active for the Min$\sum\beta$ and the K-S approaches, but reduces by 0.6 percent in the Global Criteria approach. The thrust constraint is active in all three cases.

Figure 2 presents the distribution of the centrifugal stresses along the blade span (prescribed σ_{max} = 25 x 10^6 lb/in^2), and indicates a significant reduction of these stresses from the reference blade values. The most significant reduction in these stresses occurs in the K-S case.

Table 2 presents a summary of the reference blade and the optimized blade design variables (except for the nonstructural masses). All of the design variables, with the exception of k_r and nonstructural masses, remain unchanged in the Global Criteria approach. Substantial changes occur in both Min$\sum\beta$ and K-S cases. For example, the root chord c_r is reduced by 2.9 percent in the Min$\sum\beta$ case and by 21.8 percent in the K-S case. The planform remains uniform before and after optimization (λ = 1.0) using the Global Criteria approach, whereas optimization produces an "inverse taper" (i.e. larger tip chord relative to the root chord) of λ = 0.97 in the Min$\sum\beta$ case and λ = 0.75 in the K-S case. The significant decrease in the value of the root stiffnesses, e.g. EI_{xx_r}, in the K-S case is due to the reduction in the value of c_r from the reference blade value.

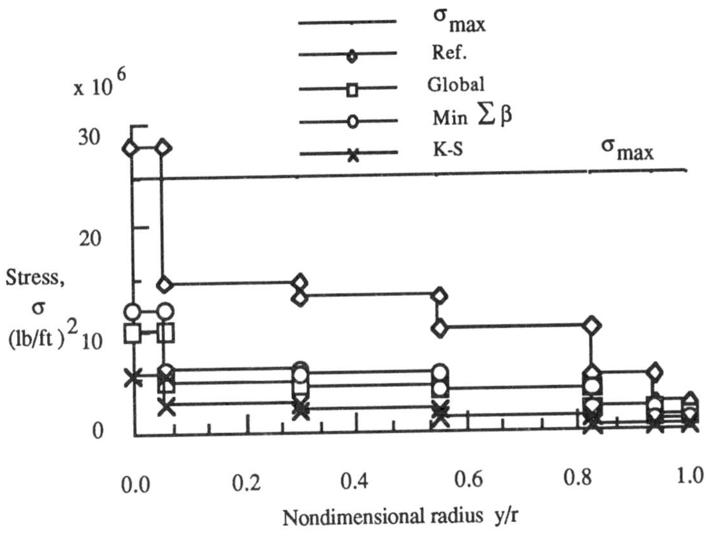

Figure 2 Centrifugal stress distribution

Figure 3 presents the nonstructural weight distributions of the reference and the optimum blades. There is a similarity in the distribution pattern for the Global Criteria and the Min$\sum\beta$ cases, with reduced values (from reference blade values) towards the inboard section of the blade and increased values towards the outboard section. However, the reductions (inboard) are more significant in the Min$\sum\beta$ case, which also exhibits a tip value lower than that of the reference. The optimized blade from the K-S approach results in a significantly different distribution, with large increases towards the root, and large decreases towards the tip. A possible explanation is the relative increase in the blade structural mass, towards outboard, due to increased value of c_t causing a reduction in the nonstructural masses at those locations in order to satisfy the constraint on the total weight. The satisfaction of the autorotational inertia constraint also requires the total mass to be sufficient towards blade outboard. The flapwise bending stiffness EI_{xx} is plotted, along the blade radius, for the reference and the optimum blades in Figure 4. The figure shows a significant increase in the EI_{xx} value towards blade tip for the optimum blade in the K-S case due to increased c_t.

Table 1 Constraint values

		Reference blade	Prescribed bounds lower	Prescribed bounds upper	Optimum Global	Optimum Min$\sum\beta$	Optimum K-S
f_3 (per rev)	(flap)	3.07	3.05	3.50	3.13	3.12	3.05
f_4 (per rev	(flap)	6.76	6.50	6.90	6.87	6.90	6.90
f_5 (per rev)	(flap)	9.28	9.25	9.50	9.38	9.48	9.50
f_6 (per rev)	(lead-lag)	12.63	12.50	12.75	12.75	12.7	12.75
AI (lb-ft^2)		19.71	19.71	-	20.26	21.8	19.71
W (lb)		3.41	-	3.41	3.39	3.41	3.41
3/rev f_r (lb)		2.71	-	2.81	2.65	2.43	2.35
3/rev m_x (lb -ft)		0.69	-	0.69	0.69	0.62	0.43
3/rev m_c (lb -ft)		0.24	-	0.24	0.24	0.22	0.22
4/rev m_z (lb -ft)		0.63	-	0.63	0.58	0.54	0.42
Thrust, T (lb-ft)		297.10	297.10	-	297.10	297.10	297.10
β_1		0.10	0.0005	0.1050	-	0.0606	-
β_2		0.10	0.0005	0.1050	-	0.0980	-

Figures 5 and 6 present comparisons of the objective functions of the reference and the optimum blades. The most significant reductions in both the 4/rev vertical shear f_z (16.8 percent), and the 3/rev inplane shear f_x (16.5 percent) are achieved by using the K-S approach. The reduction in f_z is 10.9 percent in the Global Criteria case and 3.3 percent in the Min$\sum\beta$ case. The situation is reversed with f_x, the reduction is 4.10 percent in the Global Criteria case and 8.69 percent in the Min$\sum\beta$.

The convergence history of the two new approaches presented in this paper, the Min$\sum\beta$ and the K-S function approach, is presented in Figure 7. The figure indicates a smoother convergence of the objective function, $F_3(\varphi)$, used in the K-S approach , although the objective function, $F_2(\varphi)$, used in the Min$\sum\beta$ formulation is strictly linear ($F_2(\varphi) = \beta_1 + \beta_2$).

Table 2 Optimization results

	Reference blade	Optimum Global	Optimum Min$\sum\beta$	Optimum K-S
EI_{xx_r} (lb-ft^2)	10277.0	10277.0	10605.9	8563.0
EI_{zz_r} (lb-ft^2)	354.0	354.0	329.8	290.7
GJ_r (lb-ft^2)	261.0	261.0	333.1	299.6
k_r	0.27	0.16	0.17	0.11
λ	1.00	1.00	0.97	0.75
c_r (ft)	0.45	0.45	0.44	0.35

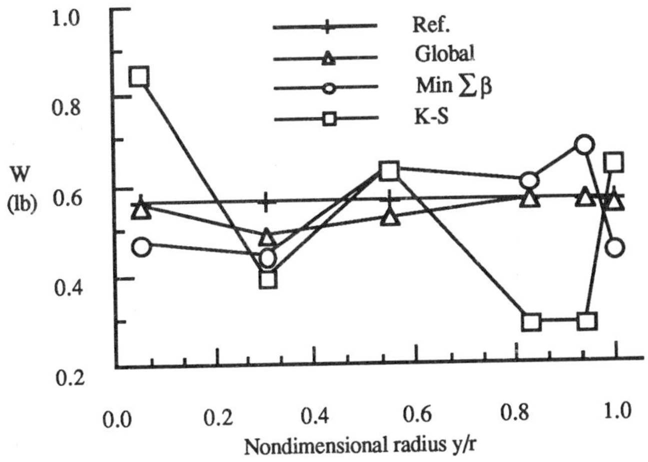

Figure 3 Blade nonstructural weight distribution

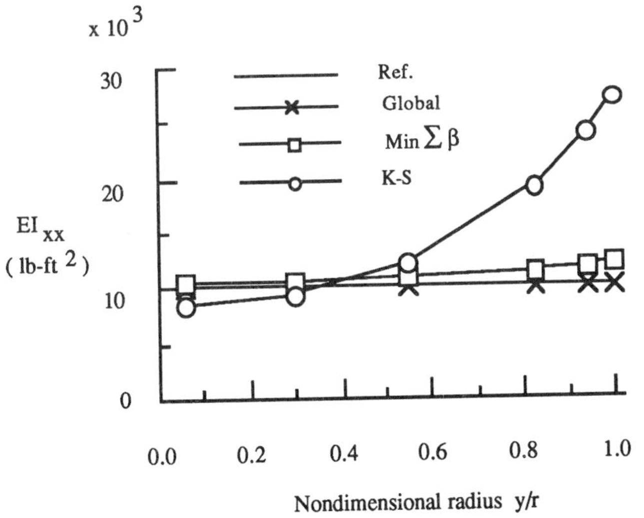

Figure 4 Blade Stiffness (EI_{xx}) distribution

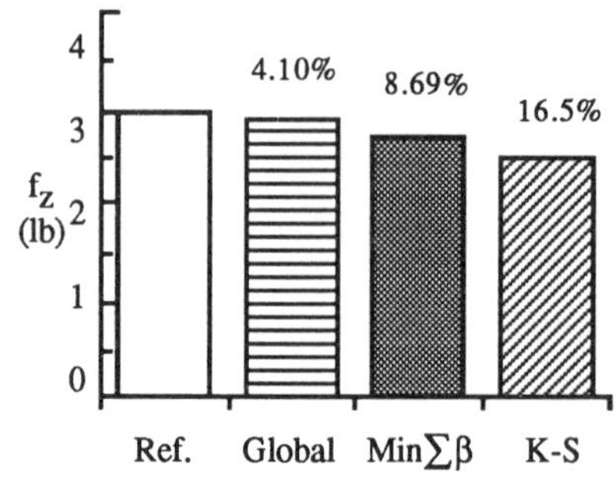

Figure 5 4/rev Vertical shear

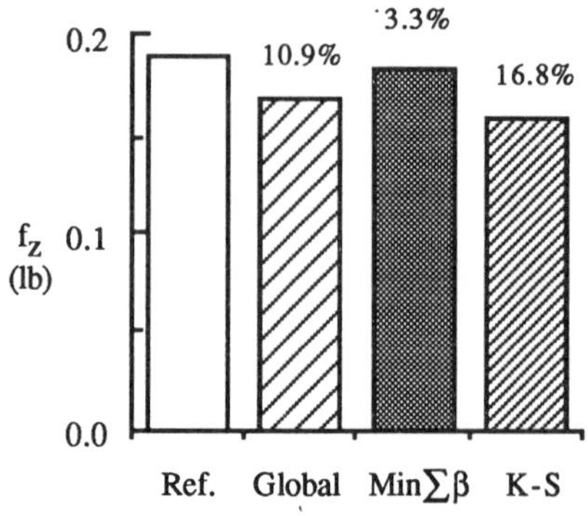

Figure 6 3/rev Inplane shear

The results obtained indicate the convergence to three different local minimum using the three approaches which is expected in a nonlinear programming approach. However, the K-S function approach proved to be the least judgmental of the three approaches. It did not require single objective optimizations as required by the Global Criteria approach, or specific target values of the objectives, as required by the Min$\sum\beta$ approach. The convergence was smoother and the reductions in the objective function values are more significant in the K-S function. A K-S factor of $\rho = 200$ proved to be effective in obtaining convergence.

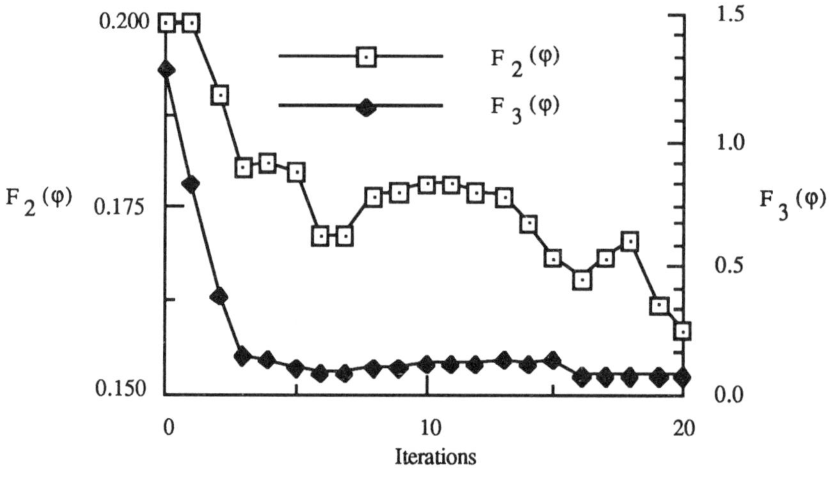

Figure 7 Objective function iteration history

CONCLUDING REMARKS

The application of three different multiple objective optimization procedures are investigated for optimum design of helicopter rotor blades with the couplings of aerodynamics and dynamics. The 4/rev vertical and the 3/rev inplane root shears are minimized with constraints on remaining critical vibratory forces and moments, frequencies, autorotational inertia, and rotor thrust. The results obtained using the modified Global Criteria approach, the Min$\sum\beta$ approach, and the K-S function approach are compared with a reference blade design. The following conclusions are made from this study:

1) All three approaches were effective in solving the optimization problem. However, the Min$\sum\beta$ and the K-S function approaches were computationally more efficient since they did not require optimization of the individual objective functions.

2) The K-S function approach provided smoother convergence, and was less judgmental.

3) All of the three approaches used, provided significant reductions in the objective function values, with the maximum reductions obtained by using the K-S function approach.

4) The three approaches converged to three different local minima. The optimum blade was closer to the reference blade design in the Global Criteria approach, and differed most significantly from it in the K-S function approach.

ACKNOWLEDGEMENTS

The authors appreciate the computational help extended by Dr. Richard M. Casey, Manager Supercomputing Services, Arizona State University.

REFERENCES

1. Vanderplaats, G. N., "Structural Optimization - Past, Present, and Future," *AIAA Journal* 20, July 1982, pp. 992 - 1000.

2. Schmit, L. A., "Structural Synthesis - It's Genesis and Development," *AIAA Journal* 19 (10), October 1981, pp. 1249 - 1263.

3. Sobiesczczanski-Sobieski, J. and Barthelemy, J. F., "Improving Engineering System Design by Formal Decomposition, Sensitivity Analysis, and Optimization," NASA TM 86377, February, 1985.

4. Pareto, V., "Cours D'economie Politque," F. Rouge, Lausanne, 1896.

5. Stadler, W., "A Survey of Multicriteria Optimization or the Vector Maximum Problem," Part I: 1776 - 1960, *J. Optimization Theory and Applications* 29 (1): 1 - 52 (1979).

6. Koski, J., "Truss Optimization with Vector Criterion," Tampare University of Technology, Publications 6, Tampare, 1979.

7. Koski, J., "Multicriteria Optimization in Structural Deign," *Proc.* of the International Symposium on Optimum Structural Design, 11th ONR Naval Structural Mechanics Symposium, Tucson, Arizona, October 19 - 22, 1981.

8. Osyczka, A., "An Approach to Multicriteria Optimization Problems for Engineering Design," *Comp. Methods Appl. Mech. Eng.*, Vol. 15, pp. 309 - 333, 1978.

9. Osyczka, A., "An Approach to Multicriteria Optimization for Structural Design," *Proc.* of the International Symposium on Optimum Structural Design, 11th ONR Naval Structural Mechanics Symposium, Tucson, Arizona, October 19 - 22, 1981.

10. Rao, S. S., and Hati, S. K., "Game Theory Approach in Multicriteria Optimization of Function Generating Mechanisms," *J. Mech. Design*, Trans. ASME, 101, pp. 348 - 405, 1979.

11. Rao, S. S., "Multiobjective Optimization in Structural Design with Uncertain Parameters and Stochastic Processes," *AIAA Journal*, Vol. 22 (11), November 1984, pp. 1670 - 1678.

12. Bendsoe, M. P., Olhoff, N. and Taylor, J. E., "a Variational Formulation for Multicriteria Structural Optimization," *J. Struct. Mech.*, 11 (4), 1983 - 84, pp. 523 - 544.

13. Hajela, P. and Shih, C. J., "Multiobjective Design in Mixed Integer and Discrete Design Variable Problems," *Proc.* AIAA/ASME/AHS/ASC 30th Struct. Dynamics and Materials Conference, Mobile, Alabama, April 3 -5, 1989.

14. Taylor, R. B., "Helicopter Vibration Reduction by Rotor Blade Modal Shaping", *Proc.* of the 38th Annual Forum of the AHS, May 4-7, 1982, Anaheim, California.

15. Chattopadhyay, A. and Walsh, J. L., "Minimum Weight Design of Rotorcraft Blades with Multiple Frequency and Stress Constraints," *Proc.* AIAA/ASME/ASCE/AHS 29th Structures, Structural Dynamics and Materials Conference, Williamsburg, Virginia, April 18-20, 1988, *AIAA Journal*, March, 1990.

16. Peters, D. A. Rossow, M. P., Korn, A., and Ko, T., "Design of Helicopter Rotor Blades for Optimum Dynamic Characteristics," Computers & Mathematics with Applications Vol 12A, No. 1, 1986, pp 85-109.

17. Hanagud, S., Chattopadhyay, A., Yillikci, Y. K., Schrage, D., and Reichert, G., "Optimum Design of a Helicopter Rotor Blade," Paper No. 12, *Proc.* of the 12th European Rotorcraft Forum, September 22-25, 1986, Garmisch-Partenkirchen, West Germany.

18. Celi, R. and Friedmann, P. P., "Efficient Structural Optimization of Rotor Blades with Straight and Swept Tips," *Proc.* of the 13th European Rotorcraft Forum, Arles, France, September 1987. Paper No. 3-1.

19. Lim, J., W. and Chopra, I., "Aeroelastic Optimization of a Helicopter Rotor," *Proc.* of the 44th Annual Forum of the AHS, June 16-18, 1988, Washington, D. C.

20. Weller, W. H. and Davis M. W., "Applications of Design Optimization Technique to Rotor Dynamics Problems," *Proc.* of the 42nd Annual Forum of the AHS, June 2-4, 1986, pp. 27 - 44.

21. Chattopadhyay, A., Walsh, J. L. and Riley, M. F., "Integrated Aerodynamic/Dynamic Optimization of Helicopter Blades," *Proc.* AIAA/ASME/ASCE/AHS 30th Structures, Structural Dynamics and Materials Conference, Mobile, Alabama, April 3-5, 1989. AIAA Paper No. 89-1269. Also available as NASA TM-101553, February, 1989.

22. Chattopadhyay, A. and Chiu, Y. D., "Details of an Enhanced Aerodynamic/Dynamic Optimization Procedure for Helicopter Rotor Blades," NASA CR - 181899.

23. Sobiesczczanski-Sobieski, J., Dovi, A., and Wrenn, G., "A New Algorithm for General Multiobjective Optimization," NASA TM - 100536, March, 1988.

24. Wrenn, G. A., "An Indirect Method for Numerical Optimization Using the Kreisselmeier-Steinhauser Function," NASA CR - 4220, March 1989.

25. Johnson, W., "A Comprehensive Analytical Model of Rotorcraft Aerodynamics and Dynamics," Part II: User's Manual, NASA TM 81183, June 1980.

26. Vanderplaats, G. N., "CONMIN - A Fortran Program for Constrained Function Minimization," User's Manual, NASA TMX-62282, August 1973.

27. Gessow, A. and Myers, G. C., Jr., Aerodynamics of the HELICOPTER, College Park Press, USA, 1985.

A Review of Analysis and Optimization Methods for Controlled Flexible Mechanisms

P.J. Woytowitz, T.K. Hight

Department of Mechanical Engineering, Santa Clara University, Santa Clara, California, U.S.A.

1 Introduction

For purposes of this paper, a flexible mechanism is defined as a machine or structure consisting of mechanically flexible components. Further, a mechanism is differentiated from most common structures in that it is typically required to, or is capable of undergoing kinematically large rotations and displacements. Controlling the mechanism refers to the act of placing either driving forces or some type of other control (such as prescribed motions) on the mechanism.

Of course it is known that all mechanical systems consist of flexible components. However, the extent to which the flexibility affects the performance of the system is normally the important feature. Many mechanisms can be adequately designed and analyzed as if they were *rigid.* If the mechanism can be considered rigid, then there are many well established design procedures, analysis methods and computational tools at the designers disposal. On the other hand, flexibility effects in mechanism design have only been more recently addressed. Despite recent advances, rigorous analysis of such systems is often lacking and ad hoc methods for including flexibilities are prevalent. Optimal design of such systems is even less developed despite their ever increasing technological importance.

As strains on natural resources and the drive for more efficient designs increases, so will the importance of considering both the flexibilities and optimization of such systems. Examples of such systems can already be found in many fields including; deployable solar arrays and other satellite appendages; robotic arms; helicopter rotor blades; airplane control surfaces; track positioning servo's for computer disk drives and hi-rise construction cranes. In addition to these commom applications, the emerging technologies such as *biomechanics* and *adaptive structures* [1] will place new demands on rigorous and robust methodologies for designing and analyzing controlled flexible mechanisms.

The traditional design and analysis methodologies applied to controlled mechanism problems is as follows. Often the mechanism is considered as rigid, or, it is designed based upon analysis or judgement so that it is *essentially* rigid. A rigid-body type dynamics analysis is performed which determines inertia loads which are then checked against the strength of the various components. Inclusion of control complicates the matter further in that capabilities and strengths of actuators need to be considered along with control system performance requirements. As can be seen, the above approach involves several disciplines including structural analysis, rigid-body dynamics and control system design. The chance of ever optimizing such a system with regards to all these disciplines is quite bleak using the above described design approach.

In the following we first review methodologies for dynamic analysis of flexible systems undergoing kinematically large motions. This discussion includes considerations of control system/structural interaction. Next, a review of general optimization and techniques applicable to the present class of problems are presented. Finally a proposed methodology for optimal design of controlled flexible mechanisms is presented.

2 Analysis of Controlled Flexible Mechanisms

Shabana [2] presents fundamentals and analysis procedures for flexible structures subjected to large arbitrary motions. The key component to the methodology is the definition of a set of *body reference* coordinates. Body reference coordinates are chosen that remain fixed with respect to the orientation of the body. A generalized displacement vector is formed consisting of the vector quantities $\mathbf{R}$, $\boldsymbol{\theta}$ and $\mathbf{q}_f$ for each body. Here $\mathbf{R}$ is the vector locating the origin of the moving body reference frame, $\boldsymbol{\theta}$ are the angles (or parameters) describing the orientation of the moving reference frame and $\mathbf{q}_f$ are the generalized coordinates describing the flexible deformation of the body. Using these generalized coordinates, the displacement and velocity of every point on each of the bodies can be developed. Formation of the total kinetic energy by integrating over the total volume of the system and substitution into Lagranges equations of motion results in the equations of motion for the system.

Applications of this method have appeared in numerous journal articles. Reference [3] describes analysis of an aircraft landing gear and suspension system. In this reference, the flexible deformations are considered elastic and are transformed to modal coordinates. The effect of using different numbers of modes in the analysis is presented. Reference [4] discusses the selection of the moving body reference frames, which is not a unique process, for different situations. Results are then presented for a slider-crank mechanism, a six-bar mechanism and a vehicle with a flexible chassis. Chen [5] presents an analysis of an initially curved flexible slider-crank mechanism and discusses the effect of the initial curvature.

Kane, Ryan and Banerjee [6] developed a dynamics approach for a cantilever beam subjected to arbitrarily large base motions. In their approach fundamental kinematic considerations are applied to a typical cross-section of a cantilever beam. The beam considered is quite general.

After developing the velocities and angular velocities, partial velocities and partial angular velocities (confer Kane [7]) are formed and are subsequently linearized. The equations are only linearized with respect to the generalized coordinates governing the elastic motion of the beam. Full nonlinear terms are retained for the base motion and therefore the beam can undergo large rigid-body motion. This approach, as pointed out by Kane, leads to the same linearized equations which would be obtained by retaining all nonlinear terms, forming the equations of motion, then linearizing them. The linearized partial velocities and partial angular velocities can then be differentiated to form the accelerations and angular accelerations and substituted into Kane's equations of motion [7].

Banerjee and Kane [8] extended this approach to a problem involving spin-up of a plate imbedded in a rigid plane. The beam theory developed in Reference [6] is analogously extended to a small displacement plate theory, and, results are presented that appear to be good.

Belytschko [9] has developed a nonlinear dynamics approach using a finite element formulation based upon convected coordinates (also known as corotational coordinates). Convective coordinate systems are defined by the deformation of the body and therefore change with time. In a two-dimensional beam element for example, the convected coordinates can be defined by the vector connecting the two end nodes of the element. Convective coordinates are not uniquely defined in this manner, and as pointed out by Belytschko and Glaum [10], results are approximately independent of the convective coordinate for small or moderate rotations of the element.

Belytschko [9] develops the approach for a class of small-strain large-displacement problems. Belytschko develops explicit forms for the mass matrix and out-of-balance forces for a simplex element and for a two-dimensional beam element. Another feature of the method is that global stiffness matrices need not be formed, therefore, savings in storage are large. The approach was extended by Belytschko et. al. in Reference [11] to allow for moderate variations of rotations within the element. This involves the use of corotational formulations for strain measures. A corotational strain includes terms in the strain-displacement relations which approximately account for rotations of the material precluded in linearized forms. Belytschko and Glaum [10] apply the methodology to nonlinear statics involving curved beam elements which undergo moderate rotation variations within elements.

Simo and Vu-Quoc [12, 13] have presented a formulation for analyzing both finite strain and rotation of a two-dimensional beam. The analysis includes axial, shear and bending deformations. The work is based upon a more general three-dimensional beam formulation presented by Simo [14] which in turn is based upon formulations due to Reissner [15] and Antman [16]. In Reference [12] the preferred treatment uses an inertial (fixed) reference frame to develop the appropriate displacement fields, strain measures and conjugate stress measures. Expressions for the kinetic and potential energies of the beam are developed and Hamilton's principle is then invoked to arrive at the partial differential equation for the beam. The advantages of the formulation are that the resulting nonlinearities occur only in the stiffness terms of the resulting partial differential equation. The inertia terms are all linear in the time derivatives of the displacement measures.

Simo [12] also demonstrates development of a beam formulation which uses a moving (or floating) reference frame similar to that used by Shabana. This development produces inertia terms which are nonlinear in the derivatives of the kinematic variables as does Shabana's development, however, the resulting stiffness terms do not contain nonlinearities. That the method is easily implemented is demonstrated in Reference [13] where explicit forms for the element mass and stiffness matrices, load vectors and numerical integration schemes are presented. Also, several problems are solved showing good performance with both very large flexible deformations and rigid body deformations combined.

Simo's two-dimensional formulation has also been used by Eischen [17] for the solution of multi-rigid-body kinematics. In Reference [17] Eishen and Sun analyze a four-bar linkage using both standard rigid-body kinematics and Simo's two-dimensional formulation. It is then demonstrated that the kinematics can be determined by solving a nonlinear system of equations with only the velocities unknown.

The "standard" nonlinear finite element method presented in such references as Bathe [18] can also be used to solve the present class of problems. Using this approach, one can develop the applicable matrix equations starting from the principle of virtual work, which is written as :

$$\int_V \mathbf{S} \cdot \delta\mathbf{E}\, dV = \mathcal{R} \tag{1}$$

where $\mathbf{S}$ is the second Piola-Kirchhoff stress tensor, $\mathbf{E}$ is the Green-Lagrange strain tensor, $\mathcal{R}$ is the virtual work of the applied forces, V is the volume of the body and δ denotes an arbitrary (virtual) variation of the quantity.

The objective of the nonlinear finite element procedure is to solve the "mixed" boundary value problem of nonlinear continuum mechanics [19]. This boundary value problem consists of simultaneously solving the equations of momentum balance, compatibility and the constitutive relations, along with satisfaction of imposed displacements and/or surface traction boundary conditions. All previously described methods use a form of this approach for at least the flexible degrees of freedom of the system.

As stated by Bathe[18], the virtual work principle expressed in terms of second Piola-Kirchhoff stresses and Green-Lagrange strains is fully general and does not involve any assumptions regarding kinematics or material response. With reference to large kinematical motions of a nominally small strain linear elastic material, the virtual work equations, if solved correctly, should produce theoretically correct results.

The vast majority of work in control of *flexible* structures, has been limited to analysis of linear structures. For complex systems the structural modeling is almost exclusively carried out using the finite element method. Many commercial finite element codes have at least rudimentary capability for modeling of control systems [20, 21, 22], however, these finite element systems typically do not allow for control system design. On the other hand, many commercially available control system design programs exist [23, 24] but typically do not allow for determination of the required structural matrices.

Optimal control techniques have also been applied to structural control. Soong [25] indicates how linear optimal control techniques can be applied to structural systems. This consists of a rewriting of the second order structural equations in state space form, and, application of optimal control techniques. State, output, instantaneous and bounded feedback control laws are discussed.

Another popular form of control used for stuctural applications is *independent modal space control*, most notably advocated by Meirovitch and co-workers [26, 27, 28]. Independent modal space control is based upon first applying a standard modal transformation to the equations of motion. The resulting state equations are uncoupled, the modal gains may be chosen readily using either classical design techniques, optimal control techniques, or standard modern techniques. The modal reduction greatly simplifies the control system design.

Output feedback control eliminates the need of estimating modal states and potentially provides a solution to distributed control using a finite number of actuators and sensors. Balas [29] has shown that the method is particularly well suited to distributed systems. Normally the method is used by colocating sensors and actuators and making a given actuator force a function only of its colocated sensor reading. The method however does suffer from some theoretical questions concerning arbitrary pole assignment, and therefore, the stability of the resulting system.

Many practical considerations of designing control systems for flexible structures may be considered constraints on (or objectives of) the combined analysis/control optimization process. These include control/observer spillover effects; time delay and discrete time control effects; nonlinearities such as friction and backlash; structural parameter uncertainties and system stability. Additionally, specifications for such systems can often be inconsistent, and, normally a design based upon certain compromises is obtained. Robustness of computationally developed "optimized" designs can also be a problem.

3 Methodologies for Design Optimization

In this section we will review current methodologies used in design optimization. We do not present a thorough discussion of general optimization procedures which can be found in standard optimization texts [30, 31, 32, 33]. We will first briefly present the general optimization problem in discrete form. Next, we will discuss more specialized optimization techniques applicable to the present class of problems.

General Optimization Procedures

In discussing discipline specific optimization procedures the same general form of an optimization problem occurs. The general optimization problem of interest is:

$$\begin{aligned} \textbf{Minimize} \quad & F(\mathbf{x}) \\ \textbf{Subject to} \quad & g_i(\mathbf{x}) \geq 0 \quad i = 1, \ldots, n_g \\ & h_j(\mathbf{x}) = 0 \quad j = 1, \ldots, n_h \end{aligned} \tag{2}$$

where F is the objective function and $\mathbf{x}$ is the vector of design variables. The functions g and h are inequality and equality constraints imposed on the optimization procedure. In the present research, both the objective function and constraints are, in general, nonlinear functions of the design variables, and therefore, the current optimization problem is a nonlinear constrained type.

Although we are interested in solving a constrained optimization problem, a large number of constrained optimization algorithms are based upon unconstrained methods. Any unconstrained optimization technique can be used for constrained optimization by including penalty or barrier functions in the objective function. On the other hand, some constrained optimization procedures are distinct from the unconstrained algorithms. A partial list of popular *unconstrained* and *constrained* optimization algorithms, and a somewhat non-unique classification, is presented in Table 1.

The list of constrained optimization techniques of Table 1 includes these distinct algorithms (the "primal" methods), techniques for using unconstrained algorithms on constrained problems (penalty and barrier methods), and, mixtures thereof. It has been pointed out by Leunberger [30] that each of the major classifications given above correspond roughly to the space in which the algorithm operates.

Structural Optimization

One approach to structural optimization would be to select an appropriate multidimensional optimization procedure from the list previously discussed (i.e. Table 1). These methods assume that the objective function and constraints are evaluated, as needed, by the optimization procedure. Due to the large number of evaluations usually required, coupled with the expense of each evaluation, these techniques have not seen extensive direct usage in large structural analyses.

A recent overview of structural optimization techniques is presented by Haftka, Gurdal and Kamat [34]. This survey indicates that the majority of recent work in structural optimization is concentrated in *sequential approximate optimization* methods pioneered by Schmit and Farshi [35]. These techniques attempt to minimize the number of structural analyses performed in the optimization procedure by using approximations to the structural response due to design variables perturbations. Limits are placed on the maximum deviation that a design variable can undergo and re-analysis is performed once the design variables have deviated a specified amount.

The approximations to the objective function are based upon *sensitivity derivatives* associated with the structural response, and, these are typically calculated during the structural analysis. These sensitivity derivatives are derivatives of the objective function and constraints with respect to the design variables. By using a truncated Taylor expansion about the analysis configuration, the objective function for perturbed values of the design variables can be approximated without resort to complete re-analysis.

Various methods are used for evaluation of the sensitivity derivatives used in *approximation* methods. These include, finite difference approximations, direct method, adjoint method and variational sensitivity techniques. The reader is referred to Haftka [34] for a discussion of the various techniques used to evaluate sensitivity derivatives, and especially to Haug, Choi and Komkov [36] for a thorough treatment of variational sensitivity techniques.

Not all structural optimization *approximation methods* rely on explicit evaluation of sensitivity derivatives. Fast re-analysis approximations [34] use known displacement solutions from a previous step in the optimization procedure to produce new displacement solutions for a perturbed stiffness matrix. Once the displacements for the perturbed problem are evaluated, any other quantity needed in either the objective function, or, constraint relations, can be directly evaluated. However, sensitivity derivatives are not computed so that optimization procedures which need this information can not be used.

Another class of structural optimization techniques applicable to both static and dynamic design are *optimality criteria* methods [34, 37]. Optimality criteria methods have come under attack due to their occasional lack of rigor and intuitive nature, however, advances in recent years have addressed some of these criticisms and the techniques can prove to be effective.

The structural dynamic optimization problem differs from the static primarily in that both the objective and constraint functions may now be functions of time. There are several ways to handle the time dependence of these functions. The first way is to choose a set of discrete times, $t_i, i = 1, \ldots, n_t$, at which the objective function and constraints will be evaluated. Another approach is to replace either the objective function and/or the constraint functions by approximate equivalent measures. For example, suppose the objective function, F, to be minimized is the weight of the system. We may then consider a new objective function, the average weight of the system, as,

$$\overline{F}(\mathbf{x}) = \frac{1}{t_f} \int_0^{t_f} F(\mathbf{x}, t)\, dt \tag{3}$$

which removes the explicit dependence of the objective function on time. Constraints for the structural dynamic optimization problem are handled similarly.

Sensitivity derivatives of these integral objective functions and constraints for dynamic problems can be developed. Analogous to the static problem, these sensitivity derivatives are used in either a sequential linear or sequential nonlinear programming technique, where, linearization or approximation of some, or all, of the constraints and/or objective function would be advantageous.

Finite difference techniques for sensitivity derivatives are completely analogous to those of static problems. A finite difference approximation is used to approximate the term $\partial \mathbf{d}/\partial dx_i$. The finite difference technique normally requires a complete reanalysis of the transient equations of motion for each design variable of interest.

Adjoint methods compute sensitivity derivatives assuming that the system dynamics may be written in state space form as

$$\dot{\mathbf{z}} = \mathbf{f}(\mathbf{z}, \mathbf{u}, \mathbf{x}, t) \tag{4}$$

where $\mathbf{u}$ is the input or forcing function. By differentiating Equation 4 with respect to a typical design variable, x_i, and assuming that the input is not a function of the design variables we arrive at a first order vector differential equation which can be solved for $\partial \mathbf{z}(t)/\partial x_i$ for the initial condition $\partial \mathbf{z}(0)/\partial x_i = 0$. Note that a system of differential equations needs to be solved for each design variable, x_i.

Fast reanalysis techniques have received most attention with regards to linear structural systems which can be solved using modal superposition methods. In the present research, the system is nonlinear and modal superposition is not primarily of interest. The reader is referred to a recent monogram [38] for a presentation of this topic.

Optimal Control Techniques

The optimal control problem is concerned with design of a feedback control law for a dynamic system described in terms of state variables as follows

$$\dot{\mathbf{z}} = \mathbf{f}(\mathbf{z}, \mathbf{u}, t) \tag{5}$$

where $\mathbf{z}$ is the vector of state variables, and $\mathbf{u}$ is the control vector, or, inputs to the system. The function $\mathbf{f}$ is in general nonlinear.

The typical optimal control problem is concerned with minimizing an objective function (also called a performance index) which is typically expressed in integral form typically in order to eliminate the explicit time dependence. To be specific, we wish to find an extremum of the performance index given by

$$\phi(\mathbf{z}, t_f) + \int_{t_o}^{t_f} L(\mathbf{z}, \mathbf{u}, t)\, dt \tag{6}$$

with respect to the input vector $\mathbf{u}$ and subject to the constraints imposed by Equation 5. The optimization is with respect to $\mathbf{u}$, since, given $\mathbf{u}$, the state variables $\mathbf{z}$ are completely determined by Equation 5.

In order to minimize Equation 6 subject to the constraints of Equation 5 we employ the Lagrange multiplier method and append the constraints to Equation 6 to obtain the modified performance index J.

The necessary conditions for an extremum are that the variation of J with respect to the variables $\mathbf{z}$, $\mathbf{u}$ and $\boldsymbol{\lambda}$ be zero. After integrating by parts and rearranging, the necessary conditions for the optimum performance, subject to the given constraints results in a coupled set of first order differential equations for the Lagrange multipliers $\boldsymbol{\lambda}(t)$. Many different problems of optimal dynamics may be solved by either specializations or extensions of these equations. Bryson and Ho [39] is an authoritative reference on these topics.

Optimal Structural Control

Optimization methods applied to structural control problems have only recently appeared in the literature and we present some of these results here. Methods discussed include control and structural parameters being simultaneously optimized, and optimization of the control system parameters alone.

Soong [25] presents an optimization method which addresses the concerns discussed under control of flexible structures. The concern is that in applying discrete control to continuous parameter systems, observation spillover can lead to instability of the physical system. This problem occurs due to the fact that only a finite number of modes can be controlled using a finite dimensional controller. The procedure assumes that a finite number of modes are to be controlled, but, that a model is available which predicts a larger (although not infinite) number of modes. The goal is to place the controlled poles at desired locations while keeping the residual poles stable. Soong [25] reports success using this approach in conjunction with a Lagrange multiplier method for the optimization algorithm.

Haftka et al [40, 41] have shown that simultaneously optimized structural and control systems can result in significant improvements in the overall system performance. They develop sensitivity measurements which indicate that actuator requirements can be significantly reduced by careful modification of stiffnesses and/or masses.

A procedure for simultaneous optimization of control and structural parameters is presented by Soong [25]. This procedure uses optimal control concepts in conjunction with standard numerical optimization techniques to determine the optimum design parameters. The first variation of the augmented performance index is taken considering variations in the state variables $\mathbf{z}$, the structural parameters $\mathbf{x}$ and the Lagrange multipliers $\boldsymbol{\lambda}$. Setting the first variation to zero, integrating various terms by parts and rearranging, yields a set of nonlinear differential equations as the necessary conditions for stationary values of J^*. Soong presents results using this approach for several simple structural control problems and indicates good performance.

4 A Proposed Methodology for Optimal Design

As can be seen, an integrated methodology for optimal design of controlled flexible mechanisms is currently not available. The goal of this section is to outline a plan for developing a software system which provides this capability.

The following features are considered key requirements of the software system

1. the ability to model and analyze flexible mechanisms undergoing kinematically large displacements and rotations,
2. the ability to include control systems into the analysis and
3. the capability to simultaneously optimize structural and control system parameters.

The system should be as general as possible and not limited to certain mechanism geometries, control laws or classes of optimization problems. Numerous checks must be present in the software to detect anomalous behavior caused by such things as unstable control systems, non-converging optimization iterations, singular mechanism geometries, etc.

The ability to model and analyze flexible mechanisms undergoing large displacements and rotations will be accommodated using a nonlinear finite element formulation. As has been discussed, these formulations have been used extensively in the past for related problems, but to a lesser degree for analysis of mechanisms. The reason for selecting this approach in the present case is that the methodology is well established with many efficient algorithms and software design techniques publically available. The method enjoys a firm theoretical basis and no reasonable lack of rigor can be identified. Finally, the method is very general and can model a wide range of systems.

Incorporation of control systems into the design and analysis procedure will extend currently available methods. The capability for the structure and control system to be analyzed simultaneously is most important for developing an optimized configuration. Incorporation of control systems into the nonlinear finite element formulation presents no obvious problems, however, it is noted that the normal symmetry enjoyed by structural matrices is typically destroyed by such additions. The resulting discretized dynamics equations to be solved will be of a more general class than normally obtained in traditional nonlinear finite element formulations.

Finally we envision at least three potential approaches for the optimization algorithm. The possible approaches include techniques of general (or mathematical) optimization, structural optimization and optimal control. General or mathematical optimization techniques, as pointed out earlier, have not been used directly in many structural optimization systems. However, mathematical programming techniques are used extensively in conjunction with various structural optimization approaches such as *approximation* methods. It is anticipated that this will also be the case with the current approach. Therefore, the approaches we intend to pursue are:

1. combined structural *approximation* and mathematical programming methods;

2. optimal control techniques in conjunction with mathematical programming techniques; and

3. combined structural approximation, optimal control and mathematical programming methods.

The goal is to devise a software system that allows the user to control the combinations of optimization methods to be used. As a result of this research we will propose various combinations of optimization procedures and present their merits relative to a set of test cases.

The class of test cases currently envisioned is exemplified by the following. Given the two link robot arm and geometry shown in Figure 1, determine optimal structural parameters (say area's and area moments of inertia), and control gain parameters (given the general form of the control law).

The objective function might be the total cost of the system, or the weight. The weight to be considered could include, for example, both the actuators and the basic structural elements. The structural constraints might include strength considerations, vibration frequencies or the weight of individual members. The controls related constraints could include such requirements as trajectory tracking error, steady state error or settling time. It is judged that development of such a software system will be of great utility in design and analysis of optimal controlled flexible mechanisms.

5 Conclusions and Summary

A brief review of techniques for analysis and optimal design of controlled flexible mechanisms has been presented. It is noted that a comprehensive integrated methodology for analysis and optimization of such systems is currently unavailable. A methodology has been proposed which shows good promise of providing such a system. Future research will implement this system and identify combined analysis and optimization procedures which can be used to optimized controlled flexible mechanisms.

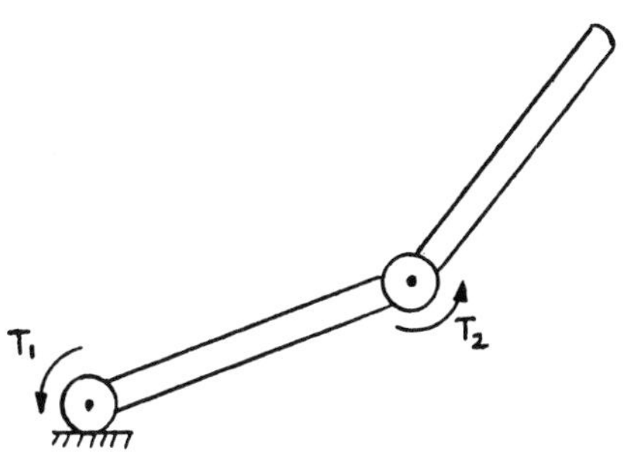

Figure 1: Typical Test Problem for Proposed Methodology

Unconstrained	Constrained
• Basic Descent Methods – Steepest Descent by Coordinate Steps – Steepest Descent by Line Searches – Newton's Method • Conjugate Direction Methods – Powells Conjugate Direction Method – Parallel Tangents – Conjugate Gradient (Fletcher-Reeves) • Quasi-Newton Methods – Modified Newton – David-Fletcher-Powell (DFP) – Broyden-Fletcher-Goldfarb-Shanno (BFGS) • Pattern Search Methods – Hooke and Jeeves – Rosenbrock Method – Nelder-Mead Simplex • Miscellaneous – Total Search – Random Search – Sectioning Method – Area Elimination – Sequential Nonlinear Programming	• Primal Methods – Feasible Direction Methods – Active Set Methods – Gradient Projection – Reduced Gradient – Sequential Linear Programming • Penalty and Barrier Methods – Exterior Penalty Function – Interior (or Barrier) Function – Quadratic Extended Penalty – Mischke Quadratic Form – Variable Penalty Fucntions – Logarithmic Barrier Function • Dual and Cutting Plane Methods – Augmented Lagrangian – Kelly's Cutting Plane • Lagrange Methods – Direct Methods – Sequential Quadratic Programming – Multiplier Update Methods – Powell's Projected Lagrangian

Table 1: General Optimization Procedures

References

[1] Ben K. Wada, James I. Fanson, and Edward F. Crawley. Adaptive structures. *Mechanical Engineering*, 41–46, 1990. American Society of Mechanical Engineers.

[2] Ahmed A. Shabana. *Dynamics of Multibody Systems.* John Wiley and Sons Inc., 1989.

[3] Koorosh Changizi, Yehia A. Khulief, and Ahmed A. Shabana. Transient analysis of flexible multi-body systems, Part II: application to aircraft landing. *Computer Methods in Applied Mechanics and Engineering*, 54:93–110, 1986.

[4] O. P. Agrawal and A. A. Shabana. Application of deformable-body mean axis to flexible multibody system dynamics. *Computer Methods in Applied Mechanics and Engineering*, 56:217–245, 1986.

[5] Da-chih Chen and A. A. Shabana. Effect of the coupling between the longitudinal and transverse displacements on the dynamics of rotating curved beams. *Journal of Applied Mechanics, Brief Notes*, 56:979–981, 1989.

[6] T. R. Kane, Ryan R. R., and Banerjee A. K. Dynamics of a cantilever beam attached to a moving base. *Journal of Guidance, Control, and Dynamics*, 10(2):139–151, 1987.

[7] Thomas R. Kane and David A. Levinson. *Dynamics: Theory and Practice.* McGraw Hill Book Company, 1985.

[8] A. K. Banerjee and T. R. Kane. Dynamics of a plate in large overall motion. *Journal of Applied Mechanics*, 56:887–892, 1989.

[9] T. Belytschko and B. J. Hsieh. Non-linear transient finite element analysis with convected coordinates. *International Journal for Numerical Methods in Engineering*, 7:255–271, 1973.

[10] T. Belytschko and L. W. Glaum. Applications of higher order corotational stretch theories to nonlinear finite element analysis. *Computers and Structures*, 10:175–182, 1979.

[11] T. Belytschko, E. Welch, and R. Bruce. Large displacement nonlinear transient analysis by finite elements. In *Proc. Int. Conf. Vechicle Struct. Mech.*, pages 188–197, SAE, New York, 1974. also in; SAE Transations, 1461-1468 (1974).

[12] J. C. Simo and L. Vu-Quoc. On the dynamics of flexible beams under large overall motions – the plane case:Part I. *Journal of Applied Mechanics*, 53:849–854, 1986.

[13] J. C. Simo and L. Vu-Quoc. On the dynamics of flexible beams under large overall motions – the plane case:Part II. *Journal of Applied Mechanics*, 53:855–863, 1986.

[14] J. C. Simo. A finite strain beam formulation. The three-dimensional dynamic problem. Part I. *Computer Methods in Applied Mechanics and Engineering*, 49:55–70, 1985.

[15] E. Reissner. On one-dimensional finite strain beam theory : The plane problem. *J. Appl. Math. Phys.*, 23:795–804, 1972.

[16] S. S. Antman. Kirchhoff's problem for nonlinearly elastic rods. *Quart. J. Appl. Math.*, 32:221–240, 1974.

[17] J. W. Eishen and H. K. Sun. Multi-rigid-body kinematic analysis with elastic finite elements. In M. Joshi, S., L. Silverberg, and E. Alberts, T., editors, *Dynamics and control of Multibody/robotic Systems with space applications, DSC-Vol. 14*, pages 7–17, ASME, 1989.

[18] Klaus-Jurgin Bathe. *Finite Element Procedures in Engineering Analysis.* Prentice-Hall Inc., 1982.

[19] Thomas J. R. Hughes. *The Finite Element Method, Linear Static and Dynamic Finite Element Analysis.* Prentice-Hall Inc., 1987. Section 2.7.

[20] *NASTRAN User's Manual.* NASA, June 1986. NASA SP-222(08).

[21] *MSC/NASTRAN User's Manual.* The MacNeal-Schwendler Corp., Los Angeles, Ca., msr-39 edition, November 1985.

[22] *ANSYS User's Manual.* Swanson Analysis Systems, Inc., Houston, Pa., revision 4.4 edition, May 1989.

[23] *CNTL-C.* System Control Technology Inc., Palo Alto, Ca.

[24] *PRO-MATLAB.* Math Works Inc., Sherborn, Ma., 1987. Version 3.2.

[25] T. T. Soong. *Active structural control : theory and practice.* Longman Scientific and Technical, 1990.

[26] L. Meirovitch and H. Oz. Active control of structures by modal synthesis. In H. H. E. Leipholz, editor, *Structural Control*, pages 505–521, North Holland, 1980.

[27] L. Meirovitch and H. Oz. Modal space control of distributed gyroscopic systems. *Journal of Guidance, Control, and Dynamics*, (3):140–150, 1980.

[28] Meirovitch L. and H. Baruh. Control of self-adjoint distributed parameter systems. *Journal of Guidance, Control, and Dynamics*, (5):60–66, 1982.

[29] M. J. Balas. Direct velocity feedback control of large space structures. *Journal of Guidance and Control*, 2(3):157–180, 1979.

[30] David G. Luenberger. *Linear and Nonlinear Programming.* Addison-Wesley, second edition, 1984.

[31] G. N. Vanderplaats. *Numerical Optimization Techniques for Engineering Design.* McGraw-Hill, 1984.

[32] Philip E. Gill, Walter Murray, and Margaret H. Wright. *Practical Optimization.* Academic Press, Inc., 1981.

[33] Terry E. Shoup and Farrokh Mistree. *Optimization Methods with Applications for Personal Computers.* Prentice-Hall, Inc., 1987.

[34] R. T. Haftka, Z. Gurdal, and M. P. Kamat. *Elements of Structural Optimization.* Kluwer Academic Publishers, 1990.

[35] L. A. Schmit and B. Farshi. Some approximation concepts for structural synthesis. *AIAA Journal*, 12(5):692–699, 1974.

[36] E. J. Haug, K. K. Choi, and V. Komkov. *Design Sensitivity of Structural Systems.* Academic Press, 1986.

[37] L. Berke and B. Venkayya, V. Review of optimality criteria approaches structural optimization. In *Structural Optimization Symposium, AMD, Volume 7*, pages 23–34, American Society of Mechanical Engineers, New York, 1974.

[38] B. P. Wang, editor. *Reanalysis of Structural Dynamic Models, AMD-Vol. 76.* The American Society of Mechanical Engineers, 1986.

[39] Arthur E. Bryson and Yu-Chi Ho. *Applied Optimal Control.* Hemisphere Publishing Corporation, 1975.

[40] R. T. Haftka, Z. N. Martinovic, and Hallauer W. L. Enhanced vibration controllability by minor structural modifications. *AIAA Journal*, 23:1260–1266, 1985.

[41] R. T. et al Haftka. Sensitivity of optimized control systems to minor structural modifications. In *AIAA/ASME/ASCE/AHS 26 th Structures, Structural Dynamics, and Materials Conference*, 1986. Paper No. 85-0801-CP.

SECTION 6: STRUCTURAL OPTIMIZATION IN BUILDING AND CIVIL ENGINEERING

Optimization in Building Design

R. Gilsanz, A. Carlson
Gilsanz Murray Steficek, New York, N.Y., U.S.A.

ABSTRACT

A joint effort of SOM/IBM resulted in the AES Architecture & Engineering Series software. This software contains an optimization routine that minimizes the volume of material in structures subjected to displacement limitations. The optimality criterion is based on the strain energy distribution in the structure. The enforcement of geometric and strength constraints is demonstrated. A detailed explanation of how the algorithm is implemented and three examples are presented. The first one is a direct application of an improved algorithm. The second one shows the magnitude of expected savings in high rise design. The third one explores the design process using optimization.

INTRODUCTION

A joint effort by SOM/IBM to develop an integrated software package for architects and engineers has resulted in the AES Architecture & Engineering Series software. The software is available in the commercial market based on the IBM RISC/6000 machine. One of the features of this integrated package is an optimization routine contained in the design unit of the structural modelling program. The routine reduces the total material volume of a structure subjected to displacement constraints.

The design of tall buildings is primarily controlled by the wind load. Once the geometry and location of a building is established, the response to the wind is a function of the stiffness, mass and damping. The deflection of the last occupied floor is related to the issues of comfort and serviceability. The deflection of the last occupied floor is a measure of the stiffness.

The routine uses a criterion that minimizes the volume of material to satisfy a given deflection. The criterion states that the contribution of a given volume of material to the tip deflection of a building has to be constant throughout the structure. If this is not the case, there are members in the structure where the contribution per volume is different from the average. If the contribution per volume is less than the average, the member is over designed and the size will be reduced. If the contribution per volume is more than the average, the member is under designed and the size will be increased. When the structure is optimal the contribution per volume will be the same for all the members.

An optimization criterion that designs large structures based on the deflection of one point only is somewhat lacking. A better optimization criterion for buildings is one based on the first period of the building. The criterion then states that the contribution of a given volume of material to the fundamental period of a building has to be constant throughout the structure. Material quantity is only one factor of the cost equation in building design. Other factors left to the judgment of the designer are material combinations, type of construction, ease of erection, etc.

ALGORITHM

The routine first calculates the contribution of each member to the specified node deflection using the virtual work method. The method consists of placing a virtual load at the node where we want to compute the deflection and analyzing the structure. The contribution of each member to the node deflection is the strain energy of each member divided by the virtual load. The strain energy of a member may have six components. The components are two moments, two shears, one torsion and one axial. The strain energy of each component is calculated by integrating the product of the real and virtual forces along the member. The strain energy of any component is a function of the forces at the end of the member. Loads distributed along the member are neglected in this calculation.

If the virtual load is a unit load, the method becomes the dummy load method. The strain energy is then equivalent to the deflection contribution.

The contribution per unit volume of a member is the member contribution divided by the member volume. A structure is optimal when the contribution per unit volume is the same for all the members. If the virtual and real load cases are equal, the contribution becomes the strain energy, and the contribution per unit volume is the strain energy density.

By specifying different virtual loads the user can use two optimization criteria. If the virtual load is the same as the real load, constant strain energy density is the goal. In a loaded truss, this criterion will produce constant stress on all the members. If the virtual load is a dummy load at the node with a specified deflection, the goal is a constant contribution per unit volume throughout the structure.

Member size selection starts at this stage. The routine selects member sizes from a table of available sections. If it is a steel structure, the available sections can be mill rolled steel sections or any built-up sections the designer wants to include. If it is a concrete structure, the designer can incorporate any reasonable sections. By limiting the available sections the designer imposes design constraints. The constraints relate to the strength or to the geometry of the member. The architect, mechanical engineer, and fabricator generate geometric constraints on the member size. Checking all the available sections, the routine selects the most efficient member size that satisfies the geometric constraints for each member.

The strength constraints are generated from the analysis and design of the original structure. They represent the minimum size that each member requires for strength. The strength constraints are enforced by comparing the selected section to the strength required section. The strength constraints are kept in a separate file called the constraint file.

To select the most economical section we followed an optimality criterion based on constant strain energy density. Energy density is the ratio of work to volume. The goal is to find the structure with the least volume that has the desired target work. The target work is the product of the real deflection at the specified node times the virtual load. When all the members have the highest possible strain energy density, the structure has the least weight. The first target density is the average strain energy density of the original structure scaled up or down to meet the specified target work.

For every member we check the difference between the target density and the strain energy density of all the sections available for that member. We will find that several are close to the target density but have different volumes. To reach the least weight structure the section chosen is the one that weighs least and produces the least work. We choose the lighter section instead of the heavier section to insure the highest possible value for the final strain energy density of the structure. The routine selects the section that minimizes the following expression:

$$[1+\mathrm{abs}\{(\text{target\_density} - \text{section\_density})/\text{target\_density}\}]\mathrm{x}(\text{volume\_of\_member}^{2})$$

The next step is to compare the section selected against the constraint size. If the constraint size is stronger than the selected section, the constraint size replaces the selected size. Otherwise, we keep the selected size. Typically, stronger is the same as bigger or heavier.

After selecting a new set of sizes the routine calculates the new target density and repeats the procedure. Because some of the sizes selected are bigger than required for stiffness, the other sizes have to decrease to meet the specified deflection or target work. The reverse can also happen, where some sizes are smaller than optimum because of geometric constraints. This causes an increase in the other sizes to meet the specified deflection. The procedure is iterative, in each iteration the target strain energy is increased until the deflection is met.

After achieving convergence, the structure is reanalyzed and the members are checked for strength. This procedure insures that the redistribution of forces in the structure does not over stress some members. The final set of sizes can be used as the constraint sizes for the next deflection constraint to be implemented. The additional deflection constraints are enforced sequentially, one deflection constraint after another. For each displacement constraint a virtual loading case has to be analyzed. This procedure does not produce the absolute minimum structural weight, but it improves upon standard practice. The alternative is to solve for all the constraints simultaneously which involves an intensive computational effort. The order in which the deflection constraints are enforced affects the results, but is the writers experience that this is not a significant factor.

Future releases of the routine will work with the algorithm described above. The following enhancements can be included:

More than one material: Changing the criterion to work density per unit cost of material instead of per unit volume will handle structures composed of several materials.
Member Grouping: The designer can constrain several members to have the same size. This reflects the standards of practice in fabrication of structures.
Joint Effects: Estimation of the joint size effect in beam column connections from a computer analysis that uses center line dimensions.
Dynamic Optimization: Implementing an optimization criterion that optimizes for a specified period may result in better designed high-rise buildings.
Simultaneous Solutions of Different Directions: The present algorithm implements deflection constraints sequentially. A simultaneous implementation would produce bigger material savings.

The following three examples are representative of the method. The first shows how the algorithm works, the second shows the savings obtained and the third presents a different use of the optimization program. The second and third examples were run with the program that is available in the market.

EXAMPLE I

This is an example of a vierendeel frame that spans fifty feet. Figure 1 shows the geometry and loading of the frame. The frame supports a curtain wall with a maximum allowable deflection at the center of 0.75 inches under the loading shown. The columns of the frame have slotted connections and carry no axial load. There are three different member types in the structure: column, main span, and side span. The structure is indeterminant to the first degree. The starting point for both optimizations was a very stiff structure composed of a W36x300 girder and a W14x730 column. The original deflection was 0.12 inches. Table 1 shows all the results of the original structure.

Two different cases were calculated. In the first case there is no restriction in the depth of the column or beam. In the second case the column is limited to a W14 and the girder to W33. Graph 1 and table 2 show the iteration steps for both cases. Both cases converged after four iterations. In the constrained case the main span

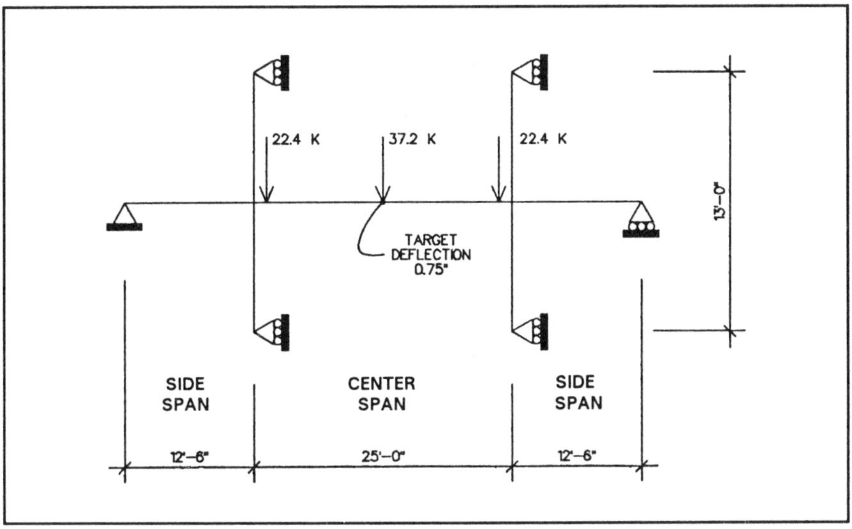

Figure 1. Computer model example I

	UNITS	SIDE SPAN	COLUMN	CENTER SPAN
SIZE		W36x300	W14x730	W36x300
LENGTH	IN.	150	78	150
Ax	SQ. IN.	88.3	215	88.3
Ay	SQ. IN.	34.719	68.829	34.719
Iz	$INCH^4$	20300	14300	20300
YOUNG MODULUS	KSI	29000	29000	29000
SHEAR MODULUS	KSI	111530	111530	111530
REAL CASE ANALYSIS RESULTS				
LOADING CONDITION: 37.2 KIP @ CENTER OF MAIN SPAN AND 22.4 KIP @ EA. COLUMN				
SHEAR	KIP	41.0	41.8	18.6
MOMENT-I	KIP-IN	0.0	0.0	377.9
MOMENT-J	KIP-IN	6150.0	3263.9	2412.1
VIRTUAL CASE ANALYSIS RESULTS				
LOADING CONDITION: 100 KIP @ CENTER OF MAIN SPAN				
SHEAR	KIP	50.0	62.4	50.0
MOMENT-I	KIP-IN	0.0	0.0	2233.4
MOMENT-J	KIP-IN	7500.0	4866.7	5266.6
MEMBER RESULTS				
SHEAR WORK	LB-IN	794	265	360
MOM. WORK	LB-IN	3918	996	837
TOTAL WORK	LB-IN	9423	5045	2395
VOLUME	$INCH^3$	26490	67080	26490
DENSITY	$LB\text{-}IN/IN^3$	0.356	0.075	0.090
STRUCTURE RESULTS				
TOTAL WORK	LB-IN	16863		
TOTAL VOL.	$INCH^3$	120060		
DENSITY	$LB\text{-}IN/IN^3$	0.140		
DISPLACEMENT	INCH	0.169		
TARGET WORK	LB-IN	0.75x100,000.	=75000	
TARGET DENSITY	$LB\text{-}IN/IN^3$	75000/120060	=0.625	

TABLE 1. ORIGINAL STRUCTURE PROPERTIES AND TARGET DENSITY

converges after the second iteration. The algorithm then continues with the side span and column member. Reanalysis showed displacements of 0.75 and 0.74 inches, respectively. The structures achieved by the algorithm have no guarantee of being the lightest possible. For the constrained case a w14x132 column, a w33x130 side span, a w33x118 main span has a 0.75 inch deflection at main span. This structure weighs 0.4% less than the computer solution presented.

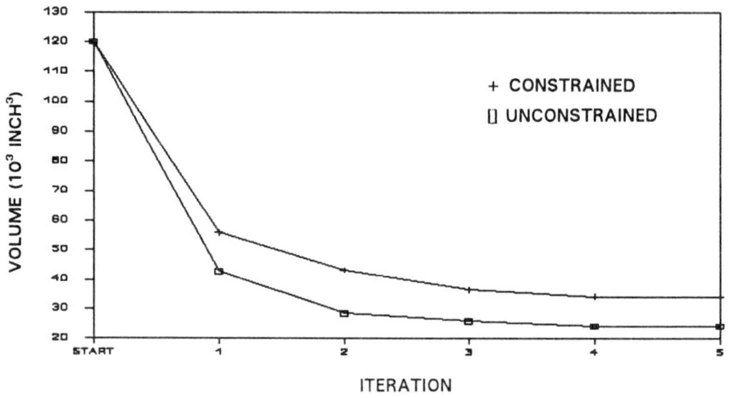

GRAPH 1. VOLUME vs. ITERATIONS FOR CASES 1 AND 2

	COLUMN	CENTER GIRDER	SIDE GIRDER	COLUMN DENSITY	CENTER DENSITY	SIDE DENSITY	TOTAL DENSITY
CASE 1: UNCONSTRAINED OPTIMIZATION							
START	W14X730	W36X300	W36X300	0.08	0.09	0.36	0.14
1	W36X150	W33X118	W36X210	0.69	0.69	0.75	0.72
2	W30X99	W27X84	W36X135	2.09	1.88	1.93	1.97
3	W27X94	W24X76	W33X118	2.66	2.76	2.88	2.78
4	W27X84	W24X68	W33X118	3.35	3.50	2.88	3.19
5	W27X84	W24X68	W33X118	3.35	3.50	2.88	3.19
CASE 2: CONSTRAINED OPTIMIZATION							
START	W14X730	W36X300	W36X300	0.08	0.09	0.36	0.14
1	W14X283	W33X118	W33X221	0.70	0.69	0.74	0.71
2	W14X211	W33X118	W33X152	1.35	0.69	1.65	1.28
3	W14X159	W33X118	W33X130	2.51	0.69	2.31	1.93
4	W14X145	W33X118	W33X118	3.05	0.69	2.88	2.28
5	W14X145	W33X118	W33X118	3.05	0.69	2.88	2.28

	CASE 1: UNCONSTRAINED TOTALS			CASE 2: CONSTRAINED TOTALS			
	WORK	VOLUME	TARGET DENSITY	WORK	VOLUME	INTERIM* DENSITY	TARGET DENSITY
START	16863	120060	0.62	16863	120060	-	0.62
1	30568	42740	1.75	39888	55900	-	1.34
2	55957	28429	2.64	55390	43164	1.47	2.07
3	71534	25772	2.91	70253	36470	2.42	2.60
4	77061	24178	3.10	77739	34142	2.97	2.86
5	77061	24178	3.10	77739	34142	2.97	2.86

* The density of the members which are not constrained.

TABLE 2. OPTIMIZATION ITERATIONS FOR EXAMPLE I

EXAMPLE II

This example describes the use of the optimization program as a tool helping the designer select the final beam and column sizes of complex structures. Nineteen ninety saw the completion of a steel building located at 1540 Broadway, New York (Fig. 2). The mixed use office and retail building has forty-two stories and five basements. The lateral load resisting system consists of moment frames and a braced core. The marketing demands, the different building uses, and the architectural design made the lateral system complex (Fig. 4&5). Figure 3 shows the computer analysis model.

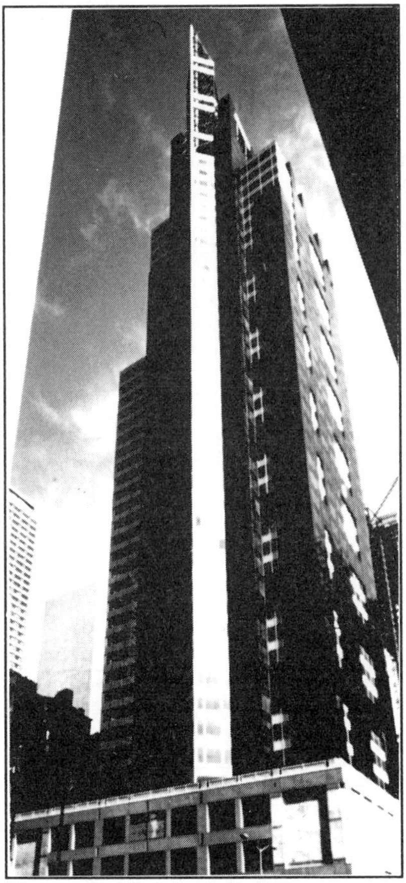

Figure 2. Finished Building.

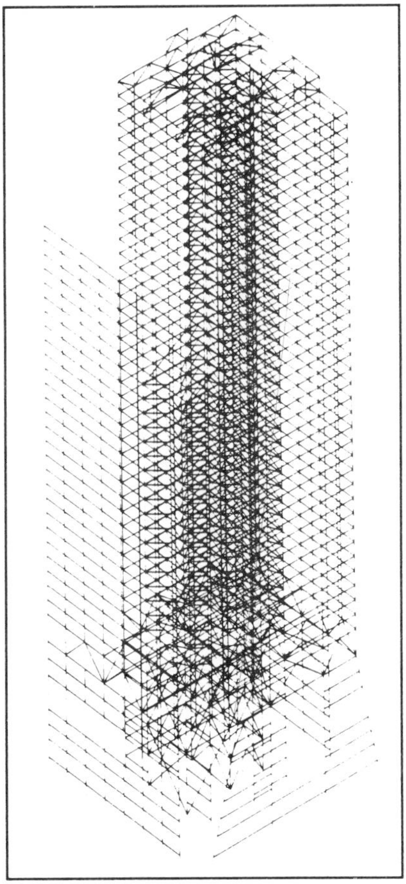

Figure 3. Computer Model

The building was designed using accepted standards of practice. A computer optimization performed at the end of the design gave an estimate of the savings that can be accomplished. Table 3 shows the results of the optimization. The target movement for the last occupied floor is 14.5 inches in the X direction and 13.0

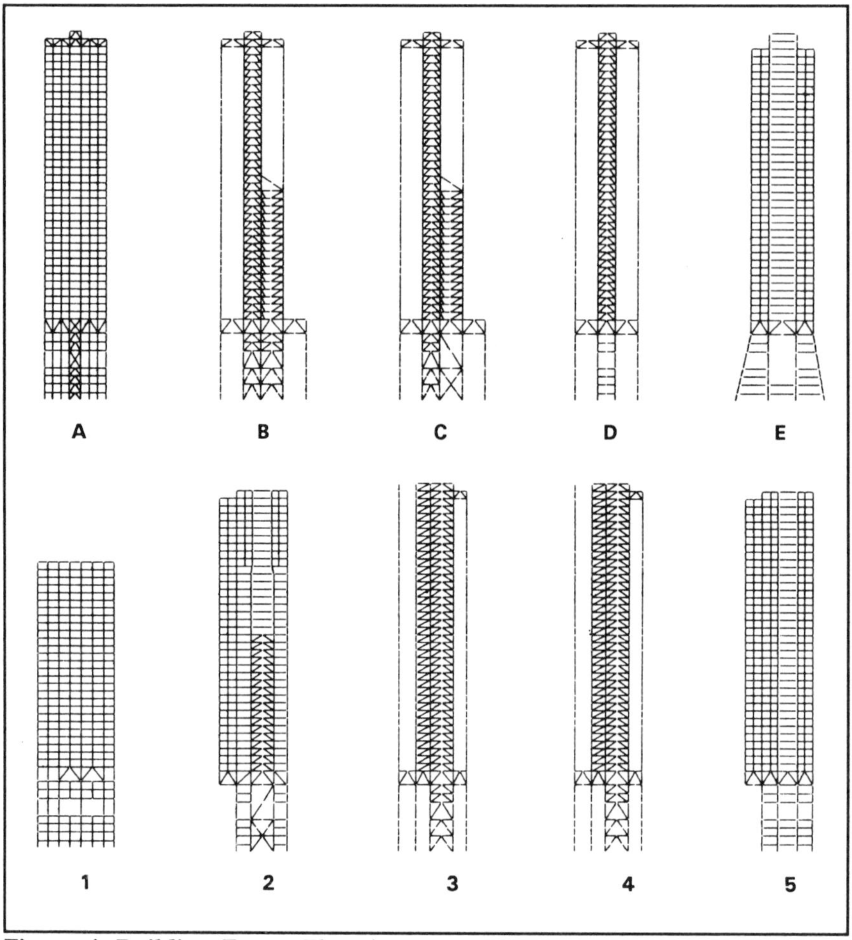

Figure 4. Building Frame Elevations

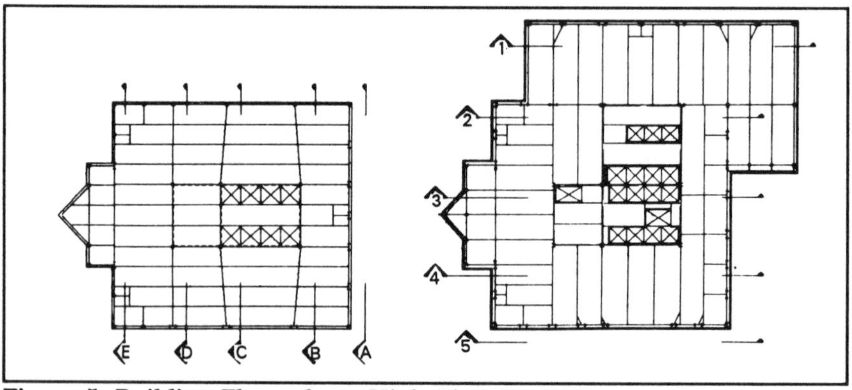

Figure 5. Building Floor plans: High Rise, Low Rise

inches for the Y direction. The original design moved 14.8 and 11.1 inches in the X and Y directions respectively. The program stiffened the X direction and weakened the Y direction producing a design closer to the target and 450 tons lighter. This represents a 5.5% savings in the weight of the lateral load resisting structure.

	X - DIRECTION		Y - DIRECTION		
	Top Displacement (inches)	Maximum Interstory Drift @ Floor	Top Displacement (inches)	Maximum Interstory Drift @ Floor	Total Weight (kips)
1st Run	14.8	L/321 @ 31	11.1	L/544 @ 29	17100
1st OPT	14.5		12.7		16500
2nd OPT	14.5		13.0		16200
2nd Run	14.3	L/334 @ 35	12.3	L/482 @ 33	16200

TABLE 3. OPTIMIZATION RESULTS FOR EXAMPLE II

EXAMPLE III

This example describes the use of the optimization program as a design tool. A vaulted space 80 feet wide and 280 feet long was proposed as a new extension of an existing airport. Repetition of a forty foot module seven times in the longitudinal direction achieved the required length. The designers declared interest in expressing a skewed grid. Placing the tie-rods and beams in a skewed pattern created a tied arch vaulted space with the intended grid.

The computer model reflected the geometry of an eighty foot wide by forty foot long slice of vault. For clarity the figures do not show the tie-rods. The geometry of the vault came from the funicular shape of the load. All the members had the same properties in the first model. This first model was optimized using the real load as a virtual load. Using the real load as a virtual load produces a constant strain energy. Figure 6 shows in plan the result of the first optimization. The thickness of the line expresses the optimal member size. In the next step (Figure 7) the smaller members were removed creating a new model with all the members having the same size. By removing members the force distribution and the behavior of the structure changes.

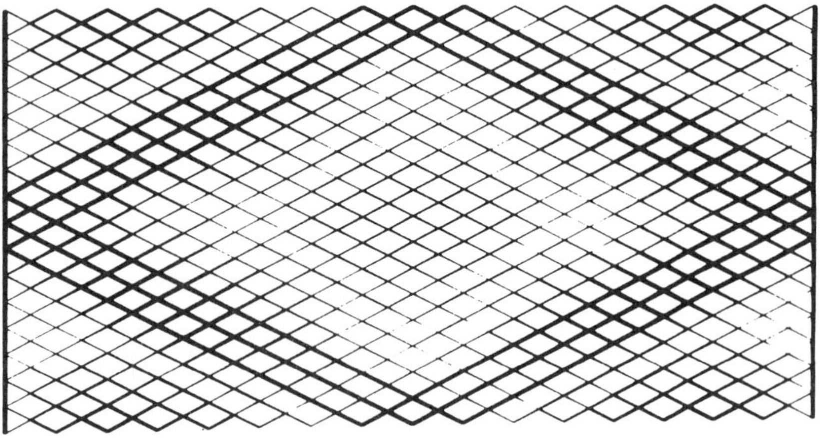

Figure 6. Results of the optimization on the first uniform model

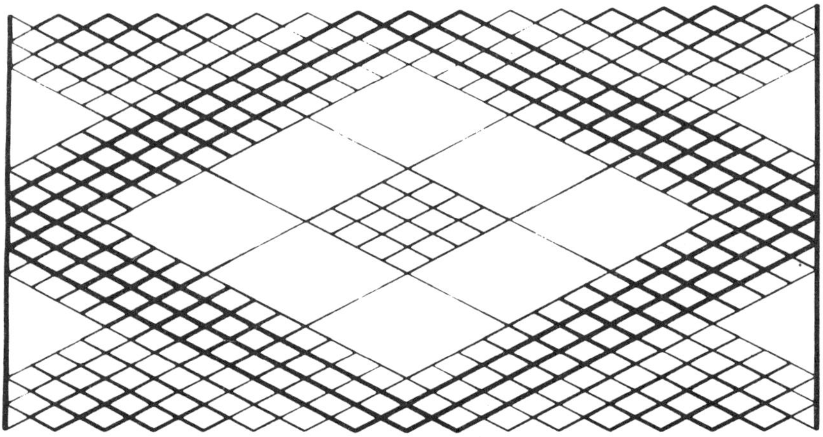

Figure 7. Revised model of the first iteration with deleted members

A new analysis of the computer model was performed before proceeding to the next optimization. The final structure resulted from repeating the procedure two more times. Figures 8 and 9 show all the iterations of the modelled slice in plan and in isometric view. Figure 10 shows the resulting structure with the webs and flanges drawn. The command GEOM>DRAW BEAM SHAPE of the structural modelling program generated figure 10. The structure is economical because the final disposition of the members relates to the structural behavior. Another proposal was a conventional steel framed scheme. It consisted of seven foot deep trusses spanning 80 feet and placed forty feet on center. Wide flange rolled beams spanned between the trusses. A cost evaluation of both schemes was done. The optimized scheme was as inexpensive as the conventional scheme. The use of the optimization program in the manner described accomplishes a rational design.

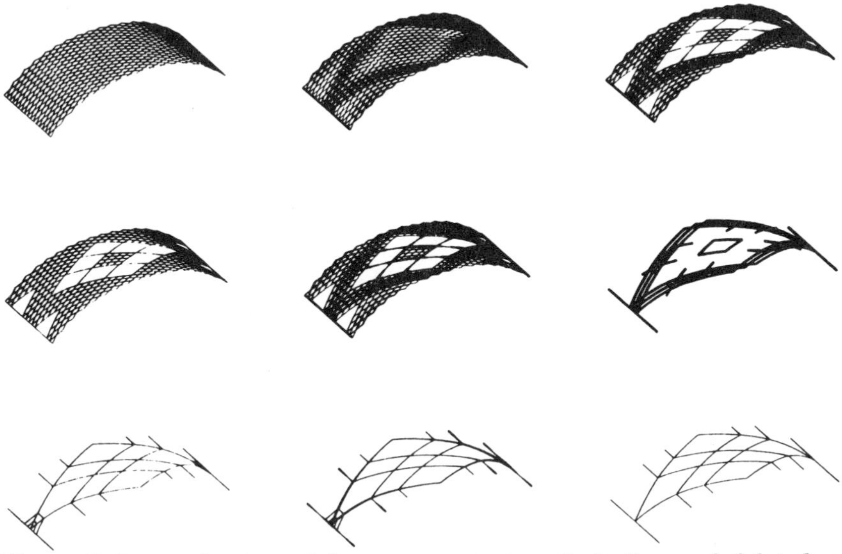

Figure 8. Isometric view of three steps: start, optimization and deleted members for each iteration.

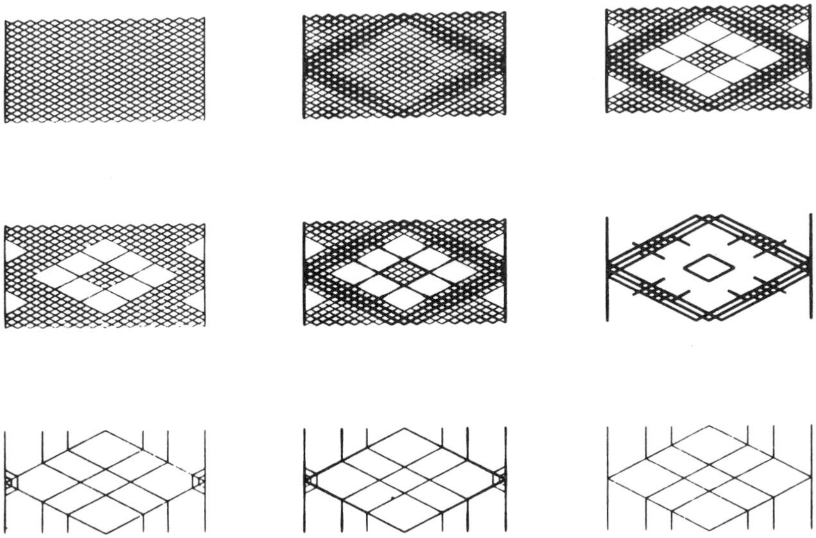

Figure 9. Plan view of three steps: start, optimization and deleted members for each iteration.

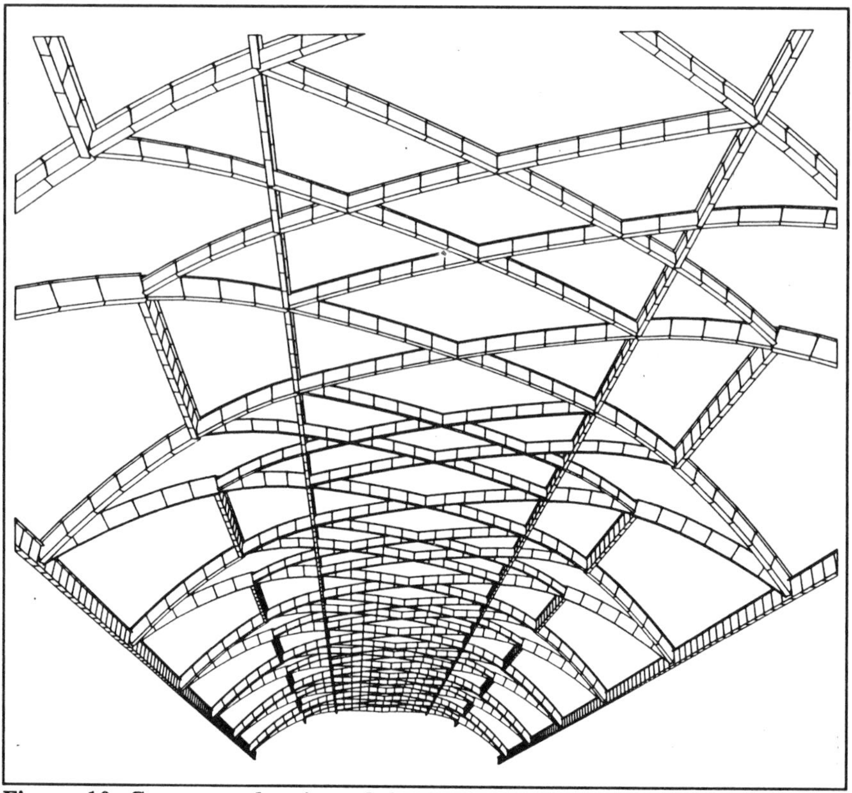

Figure 10. Computer drawing of the final structure

ACKNOWLEDGEMENTS

The algorithm developed, the examples presented, and the coding of the source code was done during the stay of both authors at the New York office of Skidmore Owings & Merrill. Robert A. Halvorson was the structural partner responsible for this project. The experience of R. Gilsanz with F. Sinclair and R. Henige among others at the office W. LeMessurier provided the foundation for this effort.

REFERENCES

1.Velisakis, E., and DeScenza, R. Design Optimization of Lateral Load Resisting Frameworks., American Society of Civil Engineers, Proceedings of the Eighth Conference on Electronic Computation, New York, N.Y., 1983.

METRIC CONVERSIONS

1 inch (in)	=25.4 mm;	1 foot (ft)	=304.8 mm;		
1 pound (lb)	=4.45 N;	1 kip	=4.45 kN;	1 ksi	=6.89 MPa;

Optimum Design of Earthquake Resistant Structures Subject to Seismic Provisions

C.T. Jan (*), K.Z. Truman (**)
(*) DRC Consultants, Inc., Flushing, New York, U.S.A.
(**) Department of Civil Engineering, Washington University, St. Louis, Missouri, U.S.A.

ABSTRACT

This paper presents an optimality criterion method for the design of structural systems under pseudo-dynamic displacement and story drift constraints. The optimization and seismic analyses are performed according to the criteria within the ATC-3, UBC, and BOCA provisions. For simplicity in computation, it has long been common practice to follow code formulae to estimate earthquake-induced forces and have design shears and moments determined from static analysis. In response to the need of a better seismic code, these building code formulae for generating base shear and lateral forces have gone through changes every few years. In this study, the application of structural optimization and seismic code provisions are implemented into a finite element program by the authors, and are used to solve several benchmark problems. The optimal structures provide a system which satisfies all drift, displacement, and stability criteria for the appropriate seismic provisions used in the analysis. The use of these seismic provisions also provides a means of comparing the effects of the different provisions on the optimal design.

INTRODUCTION

In general, there are three major specifications provided to design and analyze buildings and their components under the impacts of earthquakes: the ATC-3 Tentative Provisions [1], Uniform Building Code [2], and, BOCA Basic/National Building Code [3]. This paper is presented to use an optimality criterion method to analyze and compare optimal designs for different building codes. It is utilized to investigate structures located in different seismic zones, and study the effect of displacement and drift constraints on different structural configurations using the pseudo-dynamic loads specified in the ATC-3 provisions, UBC code, and BOCA code.

This optimization program, on the weight minimization of plane framed structures subjected to pseudo-dynamic displacement and story drift constraints, consists of efficient schemes for iterative scaling, the linking of design variables, the calculation of Lagrange multipliers, the formulations of recurrence relationships, and the choosing of active, passive and side constraints. The design procedures include nonstructural masses and the AISC Manual of Steel Construction [5] wide-flanged shapes using nonlinear cross-sectional relationships among area, the moments of inertia with respect to major axis, and section modulus. The optimization algorithm presented provides rapid convergence. This paper demonstrates the practicality of structural optimization in satisfying all drift, displacement, and stability criteria for the appropriate seismic provisions used in the analysis.

SEISMIC PROVISIONS

The ATC-3 Tentative Provisions

The ATC-3 provisions [1], prepared by Applied Technology Council associated with the Structural Engineers Association of California in 1978, give tentative provisions for the development of seismic regulations for buildings. In the ATC-3 code, there are specific requirements to select the seismic analysis and design of buildings and their components. The designing seismic forces, and their vertical distributions over the height of the buildings, shall be established in accordance with one of two procedures:

Equivalent Lateral Force Procedure The Equivalent Lateral Force procedure is a method that replaces seismic lateral force by equivalent static lateral force for simplicity in computation. It had long been common practice to assume lateral forces of a constant K times the weight of each element of the structure. More recently there has been a move to use the concept of seismic base shear, whereby the structure is designed to resist a force applied at the ground which is equal to a constant C_s times the total weight of the structure and which is transmitted to each story of the structure. In the Equivalent Lateral Force procedure it will be seen that C_s varies between 0.05 and 0.2 and depends on regional and geological conditions, importance, natural period, ductility and stiffness distribution of the structure, and other factors.

The Equivalent Lateral Force procedure is based strictly upon the distribution of weight and story heights. Almost all building codes adopt the static method of analysis because of the simplicity of its application to seismic design. The building, considered to be fixed at the base, shall be designed to resist the lateral seismic base shear. The lateral seismic base shear, V, is determined in accordance with the following formula:

$$V = C_s W \tag{1}$$

where W is the total weight, and C_s is the seismic coefficient.

P-delta effects on story shears and moments, the resulting member forces and moments, and the story drifts induced by these effects need not be considered when the stability coefficient, θ, as determined in accordance with Equation (8), is equal to or less than 0.10. It is defined as the ratio of the P-delta moment to the story moment due to lateral loading:

$$\theta = \frac{P_x \Delta}{V_x h_{sx} C_d} \tag{2}$$

where Δ = design story drift.
V_x = seismic shear force between level x and x-1.
h_{sx} = story height below level x.
P_x = total gravity load at and above level x.

For seismic design of medium-size structures, the Equivalent Lateral Force procedure is generally used. This approach requires very little in terms of computations since these formulae allow the use of an approximate fundamental period.

<u>**Modal Analysis Procedure**</u> Since the lateral force acting on a structure during an earthquake cannot be evaluated precisely by the Equivalent Lateral Force procedure, a dynamic analysis, as proposed in the Modal Analysis procedure, is adopted when a more accurate evaluation of seismic force and structural behavior is required. Dynamic analysis allows the response of a statically designed structure under dynamic force to be determined and a judgment to be made on the safety of the structure's response. If the response is unsafe, the design is modified to satisfy the required performance of the structure. In this case, the first step of the static design plays an important role.

The Modal Analysis procedure should be used for buildings

with vertical irregularities. The required periods and mode shapes of the building in the direction under consideration shall be calculated by established methods for the fixed base condition using the masses and elastic stiffness of the seismic resistant system. The seismic base shear for m^{th} mode, V_m, is given as

$$V_m = C_{sm} \bar{W}_m \tag{3}$$

where the modal seismic design coefficient C_{sm} is determined similar to coefficient C_s in the Equivalent Lateral Force procedure, and $\bar{W}_m$ is the effective modal gravity load determined in accordance with the following formula:

$$\bar{W}_m = \frac{\left[\sum_{i=1}^{n} W_i \phi_{im} \right]^2}{\sum_{i=1}^{n} W_i \phi_{im}^2} \tag{4}$$

where ϕ_{im} is the displacement amplitude at the ith level of the building when vibrating in its mth mode.

The provisions of the Equivalent Lateral Force procedure for story seismic shear distribution, torsion, P-delta effects and story drifts apply to Modal Analysis procedure. Modal Analysis procedure is provided for comparatively large and important structures. For structures with nonuniform vertical distribution of stiffness or mass, so that modes are superimposed to obtain the appropriate vibrational response, Modal Analysis procedure should also be used. In design practice, modal analysis is employed in conjunction with the response spectra. The total response, that is, the story shears, overturning moments, drift quantities, and deflection at each level, all of which are to be used in design, can be computed by the root-sum-square method, using the above-obtained modal values.

1985 Uniform Building Code

The key elements of the earthquake regulations in the Uniform Building Code [2] for design of buildings are the formulae for base shear and distribution of lateral forces over the height of the building. Many design codes, such as the Uniform

Building Code, define the seismic-design coefficient as the product of several other factors. The design base shear is to be determined from the formula:

$$V = Z\ I\ K\ C\ S\ W \tag{5}$$

where Z = Numerical coefficient depending on the seismic zone,
I = Occupancy Importance Factor;
K = 0.67 to 2.5, depending on the structural system.
C = $1/15\sqrt{T}$, but need not exceed 0.12.
T = Fundamental natural period of vibration of the building in seconds.
S = 1.0 to 1.5, the characteristic period of the site.
W = Total dead load.

Buildings having setbacks wherein the plan dimension of the tower in each direction is at least 75 percent of the corresponding plan dimension of the lower part may be considered as uniform buildings without setbacks, provided other irregularities as defined do not exist. The distribution of the lateral forces in structures which have highly irregular shapes, large differences in lateral resistance or stiffness between adjacent stories, or other unusual structural features, shall be determined considering the dynamic characteristics of the structure.

1984 BOCA Basic/National Building Code

The typical provisions of earthquake loads in the BOCA Code [3] for design of buildings are the formulas for base shear and distribution of lateral forces over the height of the building. The design base shear is to be determined from the formula:

$$V = Z\ K\ C\ W \tag{6}$$

where Z = Numerical coefficient depending on the seismic zone.
K = Horizontal force factor,
C = $0.05 / \sqrt[3]{T}$,
T = Fundamental natural period of vibration of the building in seconds.
W = Total dead load.

Buildings more than 160 feet in height shall have ductile moment-resisting space frames which (including connections) are capable of resisting not less than 25 percent of the required seismic force for the structure as a whole. All buildings designed with a horizontal force factor K of 0.67 or 0.80 shall be ductile moment-resisting space frames. Connections and panel joints shall allow for a relative movement between stories of not less than two times story drift caused by wind or seismic forces, or 1/4 inch, whichever is greater.

STRUCTURAL OPTIMIZATION

Objective functions are the functions to be optimized. The total weight of the structure

$$W = \sum_{i=1}^{N} \rho_i A_i(I_i) L_i + W_{non} \tag{7}$$

in which W is the total weight, ρ_i the density for element i, L_i the length of the ith element, W_{non} the non-structural weight, and N the number of structural elements.

The virtual load technique [5] is used to find the the gradient of displacement constraint

$$\frac{\partial u_{jl}}{\partial I_i} = - \{u_j\}^T \frac{\partial [K]}{\partial I_i} \{u_l\} \tag{8}$$

With the use of the binomial theorem it provides a linear recurrence relationship such that

$$I_i^{k+1} = \left[1 + \frac{1}{r} \left[\frac{- \sum_{j=1}^{M} \mu_j \dfrac{\partial h_j}{\partial I_i}}{\partial W / \partial I_i} - 1 \right] \right] I_i^k \qquad = 1,..,N \tag{9}$$

By using the linear recurrence relationship for design variables , a set of simultaneous equations for solving Lagrange multipliers are expressed as

$$\sum_{p=1}^{M} \mu_p \sum_{i=1}^{N} \frac{\dfrac{\partial h_j}{\partial I_i} \dfrac{\partial h_p}{\partial I_i} \Delta I_i}{\partial W / \partial I_i} = r\, h_j^k - \sum_{i=1}^{N} \left[\frac{\partial h_j}{\partial I_i} \right]_k \Delta I_i \qquad j = 1,...M \tag{10}$$

If there are N1 active elements and (N-N1) passive elements, Equation (10) becomes

$$\sum_{p=1}^{M} \mu_p \sum_{i=1}^{N} \frac{\frac{\partial h_j}{\partial I_i} \frac{\partial h_p}{\partial I_i} \Delta I_i}{\partial W / \partial I_i} = r\, h_j^k - \sum_{i=1}^{N1} \left[\frac{\partial h_j}{\partial I_i} \right]_k \Delta I_i$$

$$+ \; r \sum_{i=N1+1}^{N} \frac{\partial h_j}{\partial I_i} (I_i^p - I_i) \qquad j = 1,\ldots,M \qquad (11)$$

where I_i^p represents a member size which becomes passive, and N1 is the total number of active elements. Note that as long as a passive element remains at a constant value, the last term is zero. If any Lagrange multiplier is found to be zero or negative after solving Equation (10) or (11), the associated terms with respect to this constraint should be removed from the simultaneous equations and a new set of Lagrange multipliers must be found. This is an iterative procedure until all the Lagrange multipliers are found to be positive or zero.

NUMERICAL EXAMPLE

This numerical example is used to illustrate the application of the ATC-3, UBC, and BOCA seismic provisions. By using the AISC Manual, A36 [4] wide-flange compact sections with cross-sectional areas ranging from 3.83 in^2 to 88.3 in^2, a fixed relationship generated by Truman and Cheng [5] is represented in Equation (12) where Equation (12) is used to describe a variety of wide-flanged sections through statistical means.

$$A = 0.5008\, I_x^{0.487} \qquad (12)$$

By a similar argument, the properties of the steel bracing members defined herein are calculated from appropriate tables for A36 double-angled compact sections from the AISC Manual with cross-sectional areas ranging from 8.72 in^2 to 22.0 in^2, the relationship can be formulated as shown in Equation (13).

$$A = 0.2954\, I_x \qquad (13)$$

Each beam or column member is considered to be a wide flange steel section with Young's modulus of 29,000 ksi and a density of 7.34×10^{-4} $lb\text{-}s^2/in^4$. Each bracing member is taken to be a double angled section. It also has imposed design variable linking on all columns, beams, and braces in the same floor level having identical cross sections. The initial cross-sectional areas are 46.0 in^2 for each member.

This example is used to find the optimal design for the seven story, one-bay frames with uniform X-bracing as shown in Figure 1. First, the structures were subjected to the ATC-3 seismic provisions, the lateral displacements u_j were constrained to 0.315 (inch) multiplying the story number j for the jth floor, and the story drifts were 0.351 (inch) for all seven floors. Secondly, the structures were subjected to the UBC and BOCA seismic provisions, the lateral displacements u_j were constrained to 0.07 (inch) multiplying the story number j for the jth floor, and the story drifts were 0.078 (inch) for all seven floors. A uniformly distributed weight of 208.3 lb/in has been treated as non-structural weight on the floors.

Table 1, Figures 2 and 3 are obtained from optimal designs by the ATC-3 two procedures and UBC and BOCA seismic provisions. In both ATC-3 analysis procedures, values of A_a, A_v were based upon map area 7, Soil Profile Group II, a regular configuration, a framing coefficient of 0.035, a response modification factor, R, of 5.0, and a deflection amplification factor, C_d, of 4.5 were used. The results of the two ATC-3 analysis procedures show a similar distribution of weight and stiffness throughout the entire structure. The Equivalent Lateral Force procedure produces a heavier system, while the Modal Analysis procedure takes more computing time since it requires the actual eigenvalues and eigenvectors. As observed by the comparison, the optimal structural weight produced by the 1984 BOCA Code is much less than the other three approaches. The design base shear determined by the 1984 BOCA Code is nearly one fourth of the design base shears from the other three provisions. It is impractical compared to other provisions. Some of the parameters used in the formulae of the BOCA Code are smaller than the parameters used in the formulae of the UBC Code [6].

CONCLUSIONS

A variety of buildings subjected to the ATC-3, UBC, and BOCA seismic provisions for different structural parameters are presented in this paper. The following conclusions can be drawn from the present study:

Being able to constrain displacements and story drifts of structures under seismic loadings specified in the ATC-3 Equivalent Lateral Force and Modal Analysis procedures, the effects of code analysis techniques can be evaluated. The

Modal Analysis procedure seems to give lighter structures with nearly identical stiffness distributions as the Equivalent Lateral Force procedure at the cost of computational effort [6].

It is apparent from the comparison among the ATC-3, UBC, and BOCA seismic provisions to observe the poor performance of 1984 BOCA code. The optimal solutions indicate that the seismic provisions in 1984 BOCA are inappropriate. It is most economical to use the ATC-3 seismic provisions. At the cost of computational effort, the Modal Analysis procedure is suggested. Through the comparative study presented, it is sufficient to incorporate the ATC-3 two procedure in existing seismic codes.

ACKNOWLEDGEMENTS

This research was performed with the financial support of the Department of Civil Engineering at Washington University, their assistance is gratefully acknowledged.

REFERENCES

1. Applied Technology Council, Tentative Provisions for the Development of Seismic Regulations for Buildings, ATC-3-06, National Bureau of Standards, Special Publication 510, 1978.

2. International Conference of Building Officials, Uniform Building Code, 1985 Edition, Whittier, California, USA.

3. Building Officials and Code Administrators International, Inc., The BOCA Basic/National Building Code/1984, Country Club Hills, Illinois, USA.

4. American Institute of Steel Construction, Manual of Steel Construction, 8th Edition, USA, 1980.

5. Truman, K.Z. and Cheng, F.Y. Optimum Design of Steel and Reinforced Concrete 3-D Seismic Building Systems, pp. 475-482, Vol. V, Proceedings of the 8th World Conference on Earthquake Engineering, San Francisco, California, USA, 1984.

6. Jan, C.T. and Truman, K.Z. Optimum-Based Evaluation of Building Code Formulae for Earthquake Loads, pp.37-69, Proceedings of the 50th Regional Conference on Tall Buildings in Seismic Regions, Los Angeles, California, USA, 1988. Council on Tall Buildings and Urban Habitat, Lehigh University, Bethlehem, Pennsylvania, USA, 1990.

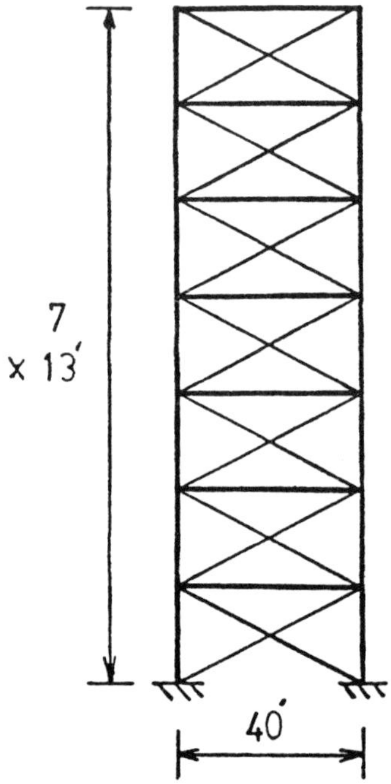

Figure 1. Seven-Story One-Bay Diagonal Braced Frame

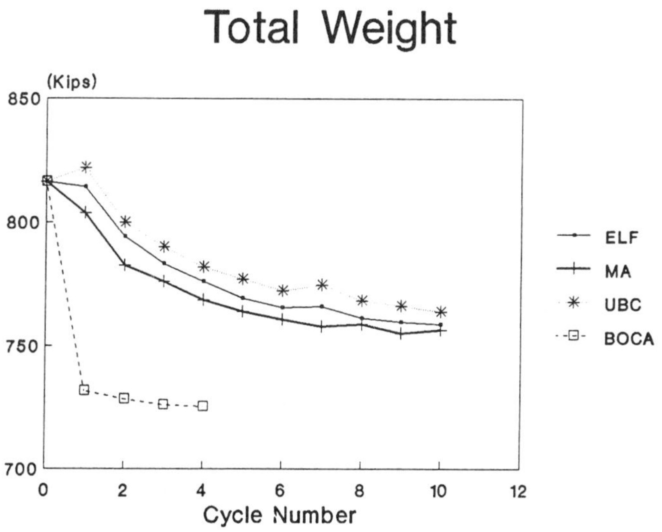

Figure 2. Seven-Story One-Bay Diagonal Braced Frames

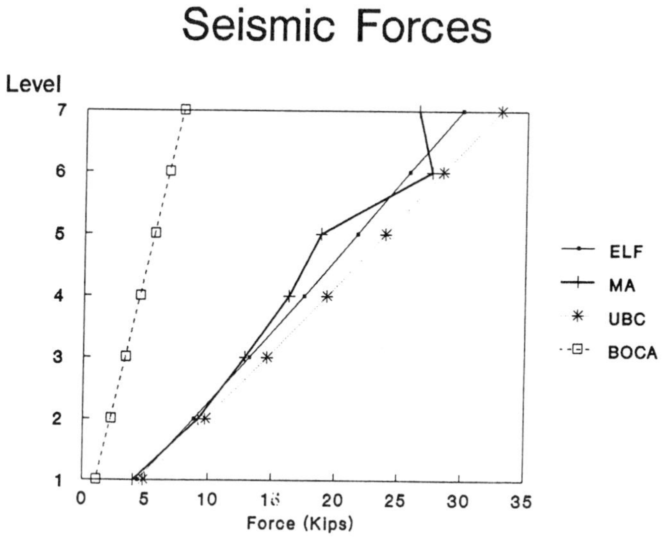

Figure 3. Seven-Story One-Bay Diagonal Braced Frames

Table 1. Seven-story One-bay Diagonal Braced Frames

Provisions	ELF	MA	UBC	BOCA
Cycles	10	9	10	4
Total V (kips)	121.2	115.1	133.8	31.1
Initial W_s (kips)	114.3	103.8	121.8	31.5
Optimum W_s (kips)	58.8	55.1	63.8	25.1
Reduction (%)	48.6	46.9	47.6	20.3
Max. Drift (inch)at Story No.	0.36 7	0.35 4	0.08 7	0.08 5

Shape Optimization in Lateral Load Resistant Systems

G. Castro (*), C.T. Jan (**)

(*) Weidlinger Associates, New York, New York, U.S.A.

(**) DRC Consultants, Inc., Flushing, New York, U.S.A.

ABSTRACT

The design of lateral load resistant systems is a very sophisticated process in which the configuration of the structure is of vital importance. A simplified shape optimization procedure for the design of lateral load resistant systems is investigated through several benchmark problems. The shape optimization procedure is developed based on the principle of virtual work and the assumption of redistribution in strain energy density due to one displacement constraint. The proposed effort shows fast rate of convergence in finding its optimal solution. As it has been noted before, it usually starts with initial sizes required accordingly to gravity loads. The question that arises is that of what initial sizes should be used for structural optimization. The observed structure is found to be adequately optimized by the proposed method. The shape optimization can be carried out by considering structures of different materials. An outrigger structure with inside core is investigated for its performance under shape optimization. It is interesting to note that the role of shape optimization in choosing lateral load resistant systems provides an efficient tool in schematic designs. It is hoped that this work will contribute in some way to a better understanding of the lateral load resistant systems and to the development of better conceptual guidelines using shape optimization.

INTRODUCTION

In building industries, with the growing trend in high-rise buildings, there are numerous progresses in their design technologies. Most likely, the emphasis on schematic design plays an important role in the early phase of the entire project. Finding an optimal framing system, especially a proper lateral load resistant system, is a complicated issue involving professionals from different trades.

Nevertheless, from the architectural point of view, the application of shape optimization helps a great deal in determining an optimal framing scheme that is structurally reliable.

The application of structural optimization in lateral load resistant systems has offered great advantage in easing off the complexity of drift control. A few mathematical programming techniques and optimality criterion methods are proven to be very favorable. However, they ought to be optimized by means of powerful computer facilities. The proposed approximate approach is rather simple. The structural optimization procedures are consisted of multiple levels. It is first treated as primary level in shape optimization, and then optimized as secondary level with lateral displacement constraint. In essence, its formulations are based on Kuhn-Tucker conditions [1] from non-linear programming. It is primarily formulated by virtual strain energy, and provides an approximate solution to design lateral load resistant systems under one displacement constraint. In fact, Lagrange multipliers are closely related with respect to strain energy and structural weight.

The application of virtual strain energy performs its optimization process by keeping constant strain energy density re-distributed in each member equally throughout the whole structure. The use of such approach reveals the efficiency of each individual member's capability in resisting lateral loads. The percentage of one member's strain energy divided by the gross strain energy indicates its influence. Thus, such percentages are used as indicators. For members having much smaller percentages, it is feasible to remove those members from the lateral load resistant system. Under the same circumstance of identical applied forces, the deletion of unnecessary members is an iterative procedure. It becomes a simple form of shape optimization. After the completion of shape optimization, it is then constrained to satisfy displacement in the targeted directions.

Based on the proposed optimization approximation, several benchmark problems have been solved to verify its feasibility. So far, it has been proven to be very accurate. To perform shape optimization to design lateral load resistant systems, this approach certainly has an advantage over other techniques. However, the optimization on joint coordinates are not considered in this paper. It will be carried out in the future development.

OPTIMIZATION APPROXIMATION

It was first presented by Venkayya, Khot and Burke [2] while introducing optimality criterion methods. Researches and developments were mostly concentrated on different iterative and recursive schemes in order to improve the rate of convergence. In the meantime, the use of principle of virtual work was adopted. For structures having identical material density, the approximate optimization keeps constant strain energy density, one member's strain energy divided by its volume, distributed evenly over the whole structure. This basis sometimes gives misconception for optimizing composite structures. In fact, for composite structures of one displacement constraint, it is to achieve constant ratio of strain energy versus structural weight for each member. For truss structures of member $i = 1, 2, 3, \ldots, n$, the total structural weight, W, is

$$W = \sum_{i=1}^{n} W_i = \sum_{i=1}^{n} \rho_i A_i L_i \qquad (1)$$

where n is the total number of members, ρ_i is the density of member i, A_i is the cross sectional area of member i, and L_i is the longitudinal length of member i. In addition, $W_{NON\text{-}STRUCTURAL}$ is the total weight of non-structural elements.

In static analysis of truss members, the strain energy of member i subjected to axial force P_i, U_i, can be expressed as

$$U_i = \frac{P_i^2 L_i}{2 E_i A_i} \qquad (2)$$

where E_i is the modulus of elasticity for member i.

As for the gross strain energy of the whole structure, U_E can be represented as

$$U_E = \sum_{i=1}^{n} U_i = \sum_{i=1}^{n} \frac{P_i^2 L_i}{2 E_i A_i} \qquad (3)$$

By means of the principle of virtual work, displacement at joint j can be expressed as

$$\Delta_j = \frac{\partial U_E}{\partial P_j} = \sum_{i=1}^{n} \frac{P_i L_i \, \partial P_i / \partial P_j}{E_i A_i} \qquad (4)$$

Thus, based on Kuhn-Tucker conditions, at the optimum, it yields

$$\frac{U_E^*}{W^*} = \frac{U_i^*}{W_i^*} = \frac{P_i}{\lambda^*} \qquad (5)$$

where λ^* is the Lagrange multiplier.

In the light of framed structures, in addition to axial force P_i (or F_{Xi}), there are more actions attributed to flexural moments M_{Xi}, M_{Yi}, M_{Zi}, and shear forces F_{Yi}, F_{Zi}. Accordingly, the total strain energy U_E for framed structures is expressed as

$$U_E = \sum_{i=1}^{n} \left(\frac{F_{Xi}^2 L_i}{2 E_i A_{Xi}} + \frac{F_{Yi}^2 L_i}{2 G_i A_{Yi}} + \frac{F_{Zi}^2 L_i}{2 G_i A_{Zi}} + \frac{M_{Xi}^2 L_i}{2 G_i I_{Xi}} + \frac{M_{Yi}^2 L_i}{2 E_i I_{Yi}} + \frac{M_{Zi}^2 L_i}{2 E_i I_{Zi}} \right) \qquad (6)$$

$\Delta_j = \partial U_E / \partial P_j$ holds true for framed structures, but involves with much complicated forms than Equation (4).

The derivation of $\partial \Delta_j / \partial A_{Xi}$ leads us to the degree of approximation for framed structures.The basis of formulations for framed structures is similar to the ones shown in Equations (1) to (6). Note that the relations shown in Equation (6) hold true, including framed structures, for one displacement constrained problems [2].

ILLUSTRATIVE EXAMPLES

A computer program OPTIMA was developed to serve as a post-processor of a commercial finite element computer package. The structures are analyzed accordingly to the input format of this finite element package. However, in the loading cases, there is an additional concentrated point load applied at the location and along the direction of the constrained displacement. The optimal solutions are divided into two portions. First, the input member sizes are considered as side constraints, and the optimal member sizes are greater than or equal to their initial sizes. Secondly, it is treated as no size constraint. Thus, some member sizes may turn out to be very small or zero. It implies that those members can be taken out without affecting the structure's performance.

Based on the optimal solutions provided by OPTIMA, the efficiency of each member in satisfying displacement constraint can be easily identified. If the value of one member's strain energy divided by its own structural weight is much smaller than others, it implies that this member does not contribute significant resistance to comply with the desired displacement. Then, such member is a good candidate to be deleted in shape optimization. Two benchmark problems were solved to demonstrate its application.

Example 1. Optimal Shape Design of Truss Structure

The truss structure shown in Figure 1(a) was optimized for the design subjected to vertical loads 33.72 Kips acting at joints 6 and 7. There were total of 33 members. The modulus of elasticity was 29000 ksi. The design requirements also limited the vertical deflections at joints to 0.6 inches and the horizontal deflections at joints to 0.2 inches. The initial maximum deflections of the truss structure shown in Figure 1 were 0.096 inches vertically and 0.081 inches horizontally. In addition, there were requirements on allowable stresses of 16.8 ksi for each member. The initial volume was 1.53 cubic feet.

This truss structure was first designed by Majid and Elliott [3], and then re-solved by Saka [4]. The member removal during each shape optimization iterations are given in Table 1. Other information obtained by shape optimization are also included. The original requirements on allowable stresses were ignored during those iterations. This member removal continues until no further topological improvements are possible. The intermediate and final shapes of the truss during iterations are shown from Figures 1(a) to 1(c). The final solution based on the approximate shape optimization reduces the initial structure to 15 members. Finally, the remaining members were optimized to satisfy the stress constraints. The final volume of the optimal structure was 0.51 cubic feet. This example compares well with the solutions obtained in References 3 and 4. It serves as a test of the proposed methodology.

Example 2. Fourteen-Story K-Braced Frame with Outrigger

A fourteen-story and one-bay K-braced frame was first studied by Velivasakis and others [5] based on the principle of virtual work and applying the concept of strain energy density with different optimization approach. With outrigger members added to their model as shown in Figures 2(a) and 2(b), it is utilized to study its importance. The displacement at the top floor was constrained to 4.5 inches. The above discussed methodology has been used to study this K-braced frame with and without outriggers. The initial member sizes of fourteen-story K-braced frame with and without outrigger are summarized in Table 2. Table 3 outlines the comparison of all six schemes.

Model 1 corresponds to the case of a K-braced frame without outriggers, Models 2 through 5 include the outriggers, and Model 6, in addition to the outrigger, includes two diagonals from the core to the perimeter columns. The comparison between the different systems can be done by looking at the weight of the core. Models 2 through 5 show the benefit obtained by the inclusion of the outriggers, but at the same time reveal the importance of the choice of initial sizes for the analysis optimization. Model 2 uses heavy girders from the core to the perimeter columns, but small floor beams in the core. The solution obtained in Model 2 revealed no improvement in resisting lateral forces and raised the doubt in its modeling. Model 3 uses floor beam sizes in the core that are as large as the outside girders, and Models 4 and 5 are similar to Model 3 but the girder's moment of inertia has been decreased and increased, respectively, by 100 times in order to consider the effects of different initial member sizes on structural optimization. The results of Table 3 suggest that the choice of initial moment of inertia affects the final solution. This motivates Model 6, where the top level stiffness has been increased by adding a diagonal. As expected, the result is beneficial.

CONCLUSIONS

The application of an approximate optimization to find the optimal shapes of lateral load resistant structures is described in this paper.

It is observed that the shape optimization alone can be achieved without taking constrained conditions into account. In essence, the ratio of strain energy over weight indicates the importance of the member lateral load resistance. The iterative process of member removal converged fast. After the completion of shape optimization, other considerations such as displacement and stress constraints are treated.

However, for structural optimization using the concept of strain energy, the choosing of initial member sizes is very important. Examples showed that the accuracy of structural optimization is very much dependent on initial member sizes. Traditionally, algorithms that perform structural optimization bring the energy density of each member to the same value by modifying the cross-sectional area. What is suggested by the results of this study is that an improved method can be reached by altering the energy density of certain members through the modification of their moment of inertia. For members having noticeable flexural strain energy under lateral forces, it is recommended to assign larger values of moment of inertia in those members and re-run the structural optimization. Nevertheless, the ease of its application and its fast rate of convergence make this approach very appealing. The method described herein serves as an efficient tool in determining the optimal shapes of lateral load resistant structures.

ACKNOWLEDGEMENTS

The authors wish to acknowledge the supports of Weidlinger Associates during the preparation of this paper.

REFERENCES

1. Kirsch, U., Optimum Structural Design, McGraw-Hill, Inc., New York, 1981.

2. Venkayya, V.B., Khot, N.S. and Burke, L. Application of Optimality Criteria Approaches to Automated Design of Large Practical Structures, pp. 301-319, Second Symposium on Structural Optimization, Milan, Italy, AGARD Conference Proceeding No. 123, 1973.

3. Majid, K.I. and Elliott, D.W.C., Topological Design of Pin Jointed Structures by Non-Linear Programming, Proceedings, Vol. 55, Part 2, Institute of Civil Engineers, London, England, 1973.

4. Saka, M.P., Shape Optimization of Trusses, pp. 1155-1174, Journal of the Structural Division, Proceedings of the American Society of Civil Engineers, Vol. 106, No. ST5, May, 1980.

5. Velivasakis, E.E., Joseph, L., Scarangello, T.Z. and DeScenza, R., Design Optimization of Tall Steel Structures for Lateral Loads, pp. 104-120, Special Structures, ASCE Metropolitan Section - The Structures Group, Spring Seminar, 1989.

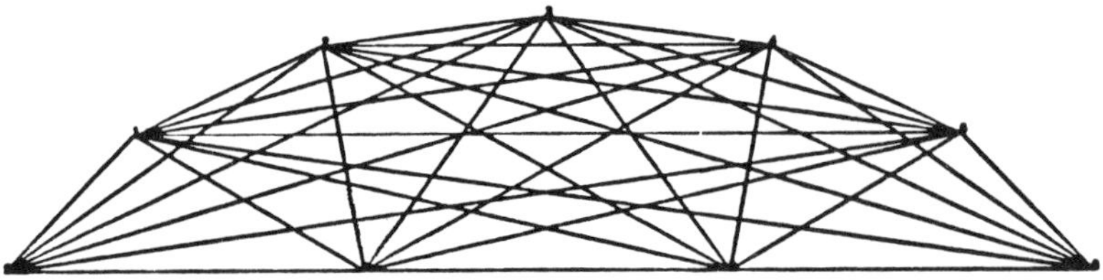

a) Initial Configuration

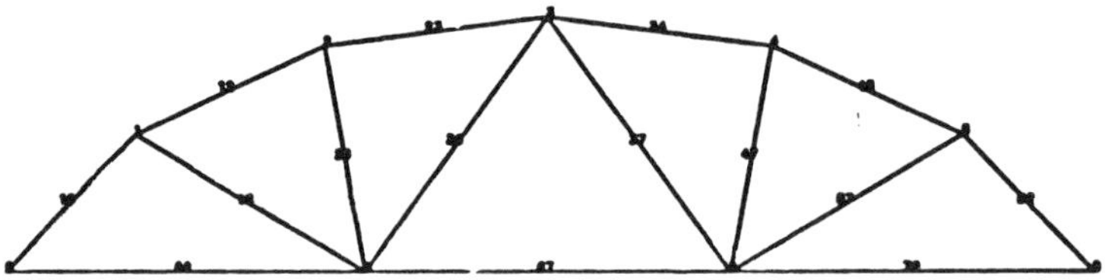

b) Final Configuration

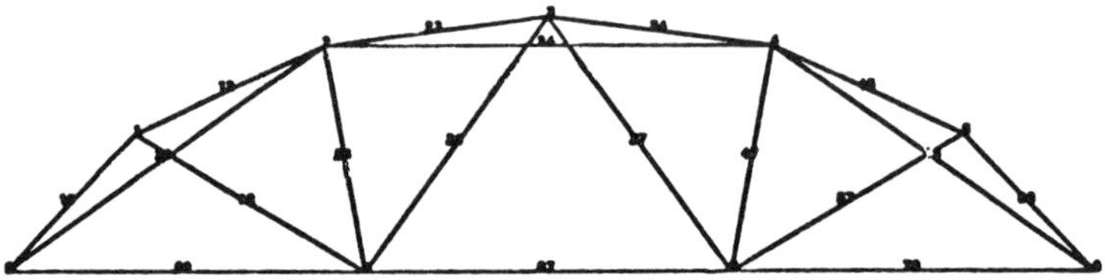

c) Intermediate Configuration

FIGURE 1. TRUSS STRUCTURE

Table 1. A Truss Structure

ITERATION NUMBER	CONSTRAINED DIRECTION	DISPLACEMENT (INCH)	MEMBER(S) REMOVAL
1	Vertical	0.096	5
2	Vertical	0.1	2
3	Vertical	0.104	8
4	Vertical	0.14	0
5	Horizontal	0.115	3
6	Horizontal	0.1	0

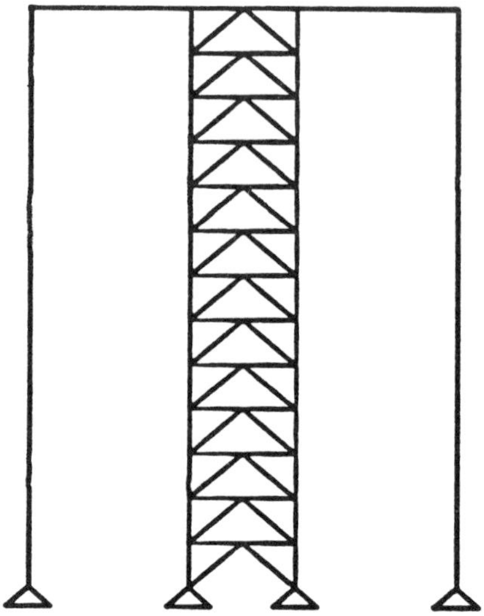

a) Models 2 thru 5

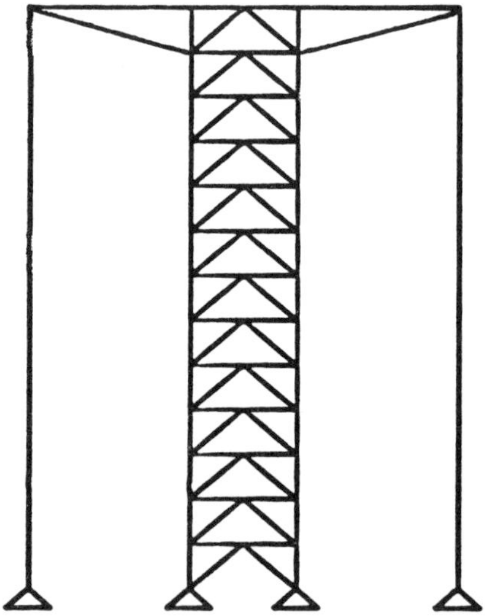

b) Model 6

FIGURE 2. K-BRACED FRAMES WITH OUTRIGGERS

Table 2. Initial Model of K-Braced Frames

		MODELS 03, 03A, 03B, 03C, 03D					
	LAT. LOAD	CORE			OUTRIGGER		
LEVEL	(K)	COL	BEAM	BRACE	COL	BEAM	DIAG
14	7.5		W14X34*			W14X730**	
13	15.	W6X20	W14X34	W6X20	W14X120		***
12	15.	W6X20	W14X34	W6X20	W14X120		
11	15.	W6X20	W14X34	W8X24	W14X120		
10	15.	W10X33	W14X34	W8X24	W14X120		
9	15.	W10X33	W14X34	W8X24	W14X120		
8	15.	W14X68	W14X34	W8X24	W14X120		
7	15.	W14X68	W14X34	W8X24	W14X120		
6	15.	W14X68	W14X34	W8X28	W14X120		
5	15.	W14X90	W14X34	W8X28	W14X120		
4	15.	W14X90	W14X34	W8X28	W14X120		
3	15.	W14X90	W14X34	W8X31	W14X120		
2	15.	W14X120	W14X34	W8X31	W14X120		
1	15.	W14X120		W8X31	W14X120		
		MODEL 01					

NOTE:

* - W14X730 for 03A; 1/100 I_z (W14X730) for 03B; 100 I_z (W14X730) for 03C.

** - 1/100 I_z (W14X730) for 03B; 100 I_z (W14x730) for 03C; 1/2 A_x (W14X730) for 03D.

*** - 1/2 A_x (W14X730) for 03D.

Table 3. Final Results of K-Braced Frames

System No.	1	2	3	4	5	6
Model Name	01	03	03A	03B	03C	03D
Initial Core W (K)	39.05	39.05	39.05	39.05	39.05	39.05
Initial Out. W (K)	——	81.38	95.33	95.33	95.33	83.07
Optimal Core W (K)	54.89	52.84	46.58	54.56	40.21	40.06
Initial Displ. (")	7.44	7.00	5.78	7.33	4.89	4.79
Optimal Displ. (")	4.50	4.50	4.50	4.50	4.50	4.50

Optimum Design of Radio Telescope Backup Trusses

V.D. Wagle

Department of Civil Engineering, California State University Sacramento, 6000 J Street, Sacramento, California, 95819-6029, U.S.A.

ABSTRACT

A hybrid method of optimum design is developed to minimize the weight of radio telescope backup trusses. The design variables consist of member areas and node locations of some selected joints of the backup truss. The method of optimization treats the member areas and geometry variables in two separate design spaces and alternates between the two phases. The member areas are updated by a recurrence relation based on the optimality criteria approach, whereas the node locations are modified by the gradient projection method of mathematical programming. The design is subjected to constraints on performance (RMS error) and geometry of the backup truss. The loads on the backup truss consist of combinations of gravity and wind. The numerical results show that significant reduction in the weight can be achieved by changing both the member areas and the node locations. Also, the geometry constraints are likely to become active at the optimum design. The study shows that a hybrid method of optimization can be successfully applied for minimizing the weight of large telescope backup trusses.

INTRODUCTION

A radio telescope essentially consists of a huge reflector surface, paraboloidal in shape, which collects the radio waves that are emitted from a celestial object and redirects them to a receiver placed at the focus. The reflector surface consists of metal panels is supported by a backup truss which is usually fabricated from structural steel. A significant engineering problem in operating a radio telescope is to maintain the required accuracy of the reflector surface. If the surface deviates more than a fraction of a wavelength the waves arriving at the focus are out of phase and the strength of the signal is reduced.

The accuracy of a radio telescope is measured by the root mean square path length error (RMS error) of the incident radio waves and is related to the exactness of the paraboloidal surface. In practice, to work efficiently at a given wave length λ, the surface must not deviate more than $\lambda/16$ from the exact paraboloidal shape. The RMS error does not depend on the absolute magnitude of the distortions, but rather depends on the relative distortions between the deviated surface and a best fit paraboloid that can be made to fit the distorted surface. Moreover, the gravity load vector is not constant but changes as the telescope is rotated about the horizontal axis. The RMS error is obviously a function of the stiffness related properties of the backup truss, namely the geometry, the topology and the member areas.

Von Hoerner [9] did pioneering work on the design of radio telescope backup trusses. He developed a procedure based on the theory of homology wherein it is possible to proportion a backup truss such that it deforms from one paraboloid into another. Biswas (23) extended the homology theory and proportioned several backup trusses. Levy [4, 5] observed that a homologous design may not always be achievable. Initially Levy based his work on redistributing a fixed amount of a material to obtain improved values of the RMS error, but later he converted the problem to a standard problem in optimum structural design, where he sought the minimum weight of a backup truss by changing member areas. The present research [10], extends Levy's work by minimizing the weight of the backup truss by changing both its member areas and geometry of the truss joints.

RMS ERROR AND HOMOLOGY PARAMETERS

The half path length error is derived by considering the geometry of the deformed surface. The distortion vector is composed of two parts: the displacements of the surface, and the displacements due to the rigid body motion of the backup truss. This rigid body motion defined by a maximum of six fitting parameters, called "homology parameters" [2, 9], consists of three translational shifts of the vertex parallel to the coordinate axes, two rotations about the X and Y axes, and one focal length change parameter. Each of the six parameters contribute to the half path length error at each surface joint, and the displacements due to these rigid body modes need to be eliminated from the displacements of the surface joints. Reference [3, 7 and 10] explain the mathematics in detail.

The Root Mean Square error is defined as,

$$\text{RMS error} = \text{SQRT}\ (\ SS/s\ ) \tag{1}$$

where
$$SS = (\ \{R\}^T\ [W]\ \{R\}\ ) \tag{2}$$

and
$$\{R\} = [A]\{r\} + [B]\{h\} \tag{3}$$

where,
- s = the number of surface joints
- SS = the sum of squares of the path-length errors
- {R} = the vector of half pathlength changes or the residuals
- {r} = the vector of the distortion components of the surface joints
- {h} = the vector of the best-fit homology parameters
- [A} = an invariant matrix
- [B] = an invariant matrix
- [W] = weighting matrix (invariant)

Matrices [A] and [B] contain the geometric information used to transform {r} and {h}. Individual terms in these matrices are composed of the coordinates of the surface joints, the direction cosines of the normals to the undeformed paraboloid and the original focal length [10]. The best-fit paraboloid is found by minimizing the sum of squares (SS) of the residuals of the half pathlength changes. The minimization is accomplished by setting the partial derivatives of SS with respect to the six homology parameters as zero and solving the resulting set of the six linear normal equations. The solution to these normal equations are the values of the best-fit homology parameters.

$$\{h\} = -[\ [B]^T\ [W]\ [B]\ ]^{-1}\ [B]^T[W]\ \{r\} \tag{4}$$

As indicated by the above equation, the vector of the best-fit parameters is given as a product of the displacement vector {r} and the invariant matrices, [A], [B] and [W].

Equation (4) is substituted in Equation (3) to express the residuals in the following form,

$$\{R\} = [P]\ \{r\} \qquad (5)$$

where,

$$[P] = [A]-[B][\ \ [B]^T[W][B]\ ]^{-1}[B]^T[W][A]\ \ ] \qquad (6)$$

[P] of size sx3s is also an invariant matrix. Once the residuals are known, the RMS error can be computed by Equation (1). The path length error, the residuals and the homology parameters are all functions of the displacements of the 's' surface joints.

AREA VARIABLES

The area variables are resized or updated using the optimality criteria approach. This approach is an iterative technique for obtaining a design which satisfies the definition of an optimum point as per the Kuhn-Tucker conditions of nonlinear programming. The application of this approach to articulated structures has been presented by various researchers [1, 8]. Levy [4, 5] extended this approach to radio telescope backup trusses and derived an optimality criteria. Wagle [10] recast the recurrence formula in a different form. The weight (or the volume) of the backup truss can be expressed as follows:

$$V(A) = \sum A_i\ L_i \qquad (7)$$

The RMS error constraint (a displacement constraint) can now be expressed as follows :

$$\sum F_i^a F_i^v L_i /\ A_i E \quad - \quad D \ = \ 0 \qquad (8)$$

where;

A_i = cross-section area of truss member i
D = limiting value of the RMS error
E = modulus of elasticity of the truss member
F_i^a = axial force in member i under actual loads
F_i^v = axial force in member i under virtual loads
L_i = the length of the truss member

The virtual load vector is a specially constructed vector such that the summation of the virtual strain energies of the individual members equal the squares of the half path length errors or the residuals. The recurrence relation to update the member areas is obtained by setting up the Lagrangian and differentiating with respect to zero.

$$A_i = [\sum (F_i^a F_i^v / E)^{1/2} L_i] [(F_i^a F_i^v / E)^{1/2}] / D \qquad (9)$$

The member areas obtained by the recurrence relation given above considers only the RMS error constraint and need to be updated to account for the side constraints on allowable stresses and minimum sizes. The updated area for the member i will be given by the maximum value obtained to satisfy all the different constraints (envelope method). Berke and Khot [1] observed that the solution by the envelope method is close to the solution obtained by optimality criteria for multiple constraints.

$$A_i = \max.(A_{io}, A_{it}, A_{ic}, A_{ie}, A_{lb}) \qquad (10)$$

where,
A_{io} = area obtained from optimality criteria
A_{it} = required area to satisfy constraint on tensile stress.
A_{ic} = required area to satisfy constraint on max. compressive stress
A_{ie} = required area to satisfy constraint on Euler buckling.
A_{lb} = minimum required area (lower bound)

The individual expression for the constraint functions are given in detail in reference [10].

GEOMETRY VARIABLES

The RMS error is an implicit function of the design variables, i.e., the member areas and the coordinates of the movable joints. Using a Taylor series expansion, an approximate but explicit expression for the RMS error can be obtained in terms of the geometry variables.

Let m = number of geometry variables
{X} = vector of geometry variables (size:mx1)
{r} = vector of displacements of surface joints (size : 3sx1)

m is the actual number of the movable coordinates and is different from the number of movable joints, since a joint can have more than one movable coordinate. Combining Equations (2), (3), (4) and (5) gives the following expression,

$$\{r\}^T\ [Q]\ \{r\} \quad \leq (D^2)(s) \tag{11}$$

where $[Q] = [P]^T\ [P]$ is in invariant matrix, and D is limiting value of the RMS error in inches, which is normally $\lambda/16$. Equation (11) expresses the RMS error constraint in terms of the displacement vector $\{r\}$. The RMS error can now be expressed in terms of the geometry variables by considering the Taylor series expansion of $\{r\}$ in terms of $\{X\}$, about a given point $\{X*\}$. Expanding up to the linear term gives,

$$\{r\} = \{r*\} \tag{12}$$

$$\text{or } \{r\} = \{r*\} + [\nabla_x r^*]\{\ X - X*\ \} \tag{13}$$

where $\{r*\}$ = displacement vector at the given point X*

and $[\nabla_x r^*]$ = matrix of the partial derivatives of the displacements with respect to the node coordinates (gradient matrix).

The approximated RMS error constraint can now be written as (Reference [10]),

$$e + \{G4\}^T\ \{x\} + \{x\}^T\ [G3]\{x\} - (D^2)(s) \ \leq 0 \tag{14}$$

where, the scalar term e, the vector {G3} and the matrix [G4] are all functions of the gradient matrix $[\nabla_x r^*]$, the invariant matrix [Q] and displacements and coordinates at the known point, i.e., {r*} and {X*}. This expression is an estimate of the RMS error at {X}, based on a first order Taylor series expansion about {X*}. The error due to the truncation of the higher order terms could be reduced by considering second order terms, but the computational cost of obtaining and storing the Hessian matrix of second derivatives makes that approach unattractive. Moreover, at the end of a cycle of geometry optimization, the design is cycled through at least one additional iteration of area optimization, where an exact analysis is involved. For these reasons, using an approximated RMS error for geometry optimization seemed to be a reasonable approach, and proved to be as was seen in the design examples.

The RMS error constraint is the main behavioral constraint on the minimum weight design of the backup truss, and through Equation (14) is expressed explicitly in terms of the geometry variables. Thus it is possible to obtain a particular solution vector {X} that satisfies Equation (14) and also gives a low weight design. However, it may not always be possible to locate the joints at the positions given by the above solution vector. Every backup truss will have some restrictions on the movement of the joints, so that they do not interfere with the overall framing of the backup truss. Hence, in addition to the RMS error constraint, it is necessary to include constraints on the position (geometry) of these joints. These constraints on the geometry could be in the form of individual upper and lower bounds on the coordinates, or could also be in the form of a relationship linking the coordinates of two or more joints. The upper and lower bounds on the joint coordinates are similar to the upper and lower bounds on the area variables, the upper and lower bounds basically guarantee a practical solution vector (i.e., a solution of the vector {A} with non-negative values as well as not too large values of the member areas) and thus essentially the bounds act as side constraints. However, in the case of geometry variables, the upper and lower bounds together with the RMS error constraint, form the boundaries of the feasible region. Thus they have a potential of becoming active constraints (i.e., where the solution vector {X} is governed by the magnitudes of the upper or the lower bounds).

The optimum design problem with respect to the geometry variables can now be written as follows:

Find a vector $\{X\}$ (size: mx1)

to minimize, $$V = \sum A_i L_i \tag{15}$$

such that,

$$e + \{G4\}^T \{X\} + \{X\}^T [G3]\{X\} - (D^2)(s) \leq 0 \tag{14}$$

$$\{X\} - \{X^{UB}\} \leq 0 \tag{16}$$

and $$\{X^{LB}\} - \{X\} \leq 0 \tag{17}$$

Equations (16 & 17) are the geometry

constraints, i.e., the upper and lower bounds on the movement of the joints. The member lengths, L_i, are functions of the design variables {X}. Equations (14, 16 and 17) form a set of (2m+1) constraints for the solution of the m values of the design vector {X}.

The mathematical programming problem formulated above was solved using the gradient projection algorithm (Rosen [6]). The matrix of partial derivatives of displacements with respect to the node coordinates was computed by implicit differentiation of the equilibrium equations (Reference [10]).

DESIGN EXAMPLES

Reference [10] presents results of a number of different studies on the design optimization of three different backup trusses. The results from two studies on one of those telescopes (Telescope B) is presented below. Figure 1 shows the isometric view of the backup truss. The initial structure of this backup truss was taken from an earlier study based on homology [2]. The backup truss has 12 joints of which 7 are surface joints. There are 41 main members. The number of movable degrees of freedom considered for this study is 2; joints 11 and 12 can move along the Z or the focal axis. Figure 2 shows the iteration history under gravity load. The RMS error constraint has a limiting value of $\lambda/16$. Figure 3 shows the initial and final geometry and weight. The changes in the geometry contributed about 21% to the overall reduction in the weight.

Figure 4 shows the iteration history for the same backup truss under combinations of gravity and wind loads. There are a total of three separate loading cases; (i) gravity load alone with a RMS error constraint of $\lambda/40$, (ii) gravity load together with a normal operating wind condition (30 mph wind) with RMS value of $\lambda/16$, and (iii) gravity load plus a survival wind load (70 mph) with no stipulation on RMS error. In this case the net reduction in weight due to geometry changes is about 13%

SUMMARY AND CONCLUSIONS

A hybrid method for optimum design is presented

which treats the design variables in two separate design phases. The member areas are resized by the optimality criteria approach using a recurrence relation considering the RMS error as a primary constraint. The node coordinates are updated by the gradient projection method of mathematical programming. The RMS error is expressed explicitly in terms of the node coordinates by a first order Taylor series expansion. The feasible design space for the gradient projection method is bounded by the constraints on the geometry and by the explicit expression of the RMS error constraint. All the partial derivatives in the Taylor series expansion are computed by closed form analytical expressions.

It is observed that changing member areas and geometry leads to significant savings in the weight of the backup truss. The percentage of weight reduction due to geometry changes alone depends on the framing of the backup truss. The geometry constraints have a potential of becoming active at the optimum point and hence caution must be exercised in specifying their limiting values. The member stresses remain passive throughout the iteration cycles and can be considered as secondary constraints. Using a second order Taylor series expansion and subdividing the feasible space into smaller regions may be necessary for backup trusses with a large number of movable degrees of freedom.

ACKNOWLEDGEMENTS

The author would like to thank the following for their advice and encouragement during various stages of this research; Dr. Roy Levy formerly of the Jet Propulsion Laboratory, Professor Moy Biswas of Duke University and Professor Harry Jones of Texas A&M University.

REFERENCES

1. Berke, L. and Khot, N.S., Use of Optimality Criteria Methods for Large Scale Systems, AGARD Lecture Series No. 70, Structural Optimization, 1974.

2. Biswas, M., Homology in Structures, thesis presented to the University of Virginia in 1970, in partial fulfillment of the requirements for the degree of doctor of philosophy.

3. Katow, M.S., Antenna Structures : Evaluation of Reflector Surface Distortions, Technical Report 32-1526, Vol I, Jet Propulsion Laboratory, California, 1971.

4. Levy, R., Computer-Aided Design of Antenna Structures and Components, Computers and Structures, Pergammon Press, Vol. 6, pp. 419-428, 1976.

5. Levy, R., Optimization of Antenna Structure Design, Proceedings of the Eighth Conference on Electronic Computation, ASCE, February 1983.

6. Rosen, J. B., The Gradient Projection Method of Nonlinear Programming, Part II : Nonlinear Constraints, Journal of the Society of Industrial And Applied Mathematics, Vol. 9, 1961.

7. Utku, S. and Barondess, S.M., Computation of Weighted Root Mean Square of Path Length Changes Caused by the Deformations and Imperfections of Rotational Paraboloidal Antennas, Technical Memorandum 33-118, Jet Propulsion Laboratory, California, 1963.

8. Venkayya, V.B., Khot, N.S. and Berke, L., Application of Optimality Criteria Approaches to the Automated Design of Large Practical Structures, Proceedings of the Second Symposium on Structural Optimization, AGARD-CP-123, Italy, 1973.

9. Von Hoerner, Sebastian, Homologous Deformations of Tiltable Telescopes, Journal of the Structural Division, ASCE, Vol. 93, pp 461-485, 1957.

10. Wagle, V.D., Optimum Design of Radio Telescope Backup Trusses, thesis presented to Texas A&M University in 1984, in partial fulfillment of the requirements for the degree of Doctor of Philisophy.

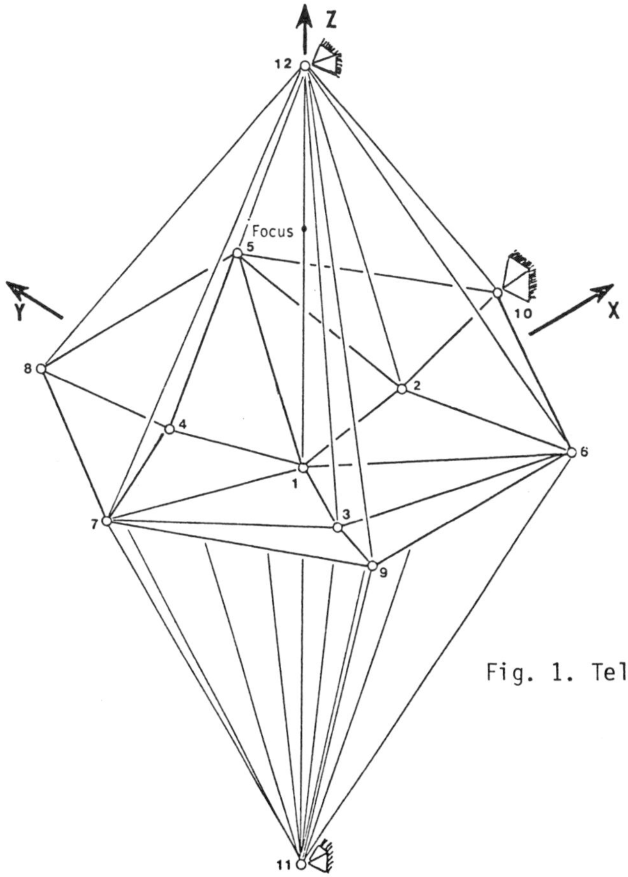

Fig. 1. Telescope B

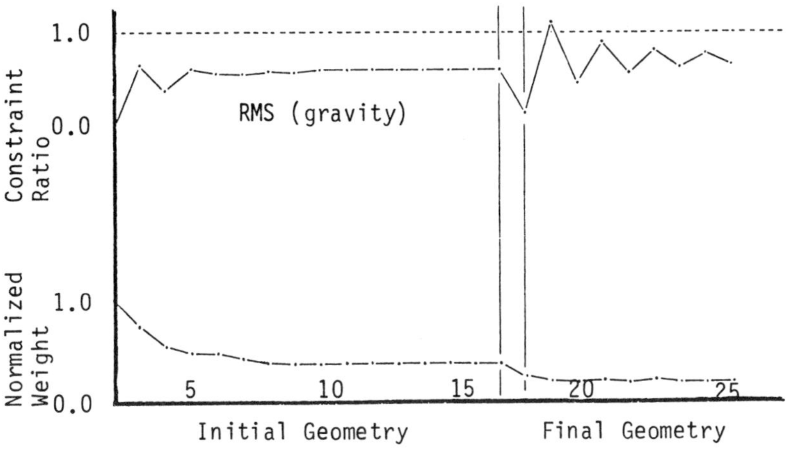

Fig. 2 Iteration History - Gravity Load

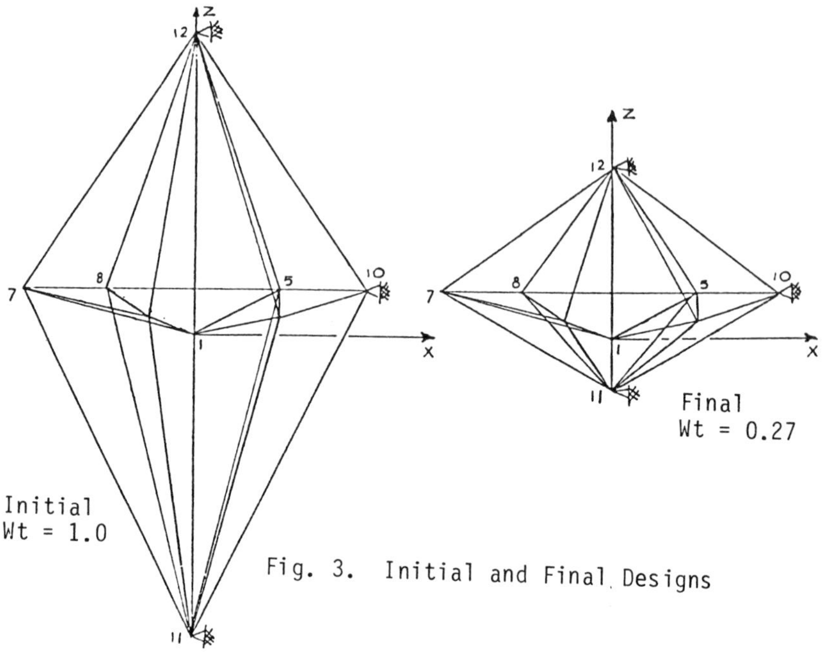

Fig. 3. Initial and Final Designs

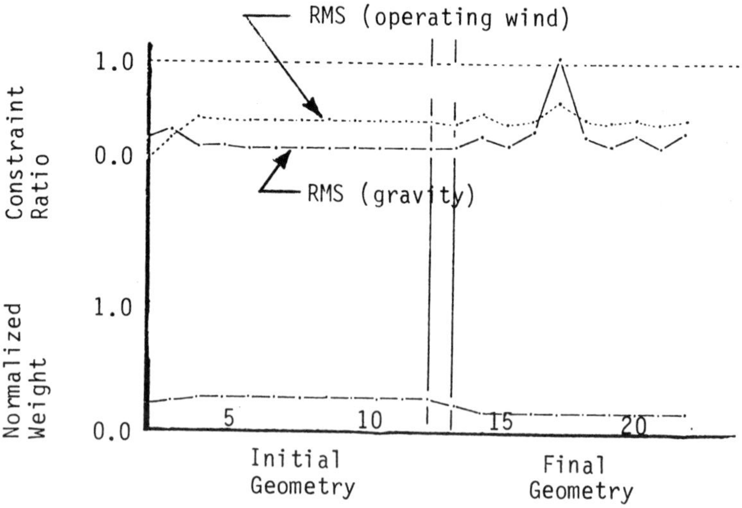

Fig. 4. Iteration History - Gravity & Wind Loads

Shape and Sizing Optimization of Steel Cable-Stayed Bridges

S. Ohkubo, K. Taniwaki

Department of Civil Engineering, Ehime University, Matsuyama, Ehime, Japan

ABSTRACT

A minimum-cost design method for steel cable-stayed bridges is developed, with the cable anchor positions on the main girder and pylon, and the sectional dimensions of the member elements dealt with as the design variables. The cost minimization problem under stress constraints is solved by using the dual method with mixed direct/reciprocal design variables. The optimum solutions can be obtained quite rigorously and efficiently by the proposed design method. The significance of dealing with cable anchor positions as the design variables is also emphasized.

INTRODUCTION

The cable-stayed bridge is one of the most attractive types of bridges by its ability to overcome large spans, and economical and aesthetic characteristics. Because the cable-stayed bridge is a highly statically indeterminate structure, its structural behaviors and total cost are greatly affected by the cable arrangement and stiffness distribution in the cables, pylon and main girder. Furthermore, the aesthetic view of the bridge is also affected by these design variables. Therefore, the establishment of a rational and efficient computer-aided design method for cable-stayed bridges which can determine the optimum values of the design variables mentioned above theoretically and automatically has been awaited expectantly.

The study of the optimum design of cable-stayed bridges was begun in the late 1970s. Yamada and Daiguji studied an optimum design method based on the optimality criteria.[1] Kobayasi et al. presented a multilevel optimal design method by using the SLP algorithm and applied it to three types of cable-stayed bridges with different supporting systems.[2] Gimsing[3] and Nagai et al.[4] investigated the rational cable arrangement of cable-stayed bridges from the structural system analysis viewpoint.

In this paper, a rigorous and efficient optimum design method for steel cable-stayed bridges is developed. In this method, not only the sectional dimensions of the cables, main girder and pylon elements, but also the cable anchor positions on the main girder and pylon are considered as the design variables. The design problem is formulated as a minimum-cost design problem subjected to the stress constraints taken from the Japanese Specifications for Highway Bridges.[5] The cost-minimization problem is solved by using the approximate concept and dual method with mixed direct/reciprocal design variables developed by Fleury et al.[6)-9)] The proposed optimum design method has been applied to the minimum-cost design problems of practical-scale steel cable-stayed bridges, and the theoretical rigorousness and the efficiency of the proposed optimum design method are demonstrated by giving numerical results of three-span steel cable-stayed bridges under various design conditions. The significance of dealing with cable anchor positions on the main girder and pylon as the design variables is also emphasized for the minimum-cost designs of cable-stayed bridges.

FORMULATION OF OPTIMUM DESIGN PROBLEM

Design variables

In this study, the shapes of cross sections of the main girder and pylon are assumed as box types depicted in Figs.1 (a) and (b) respectively, and the span lengths, number of cables, nc, height and width of the elements of the main girder and pylon, material types to be used for each structural element are the preassigned constant design parameters.

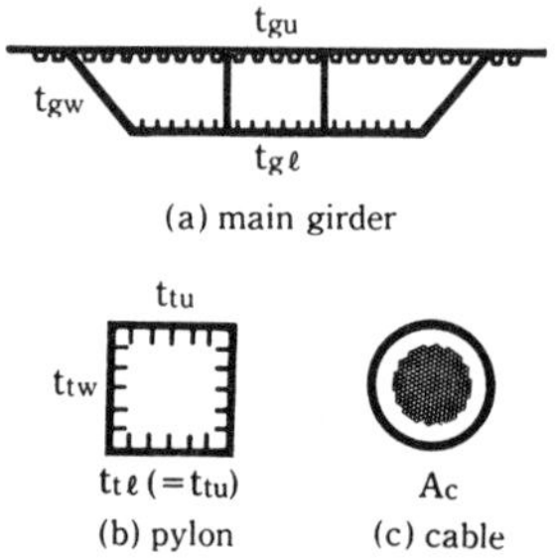

Fig.1 Cross sections and sizing variables t_{gu}, $t_{g\ell}$, t_{tu}, $t_{t\ell}$ and Ac in main girder, pylon and cable elements

The design variables selected with respect to the sizes of the cross sections are the cross-sectional area of each cable, A_c, and thicknesses of

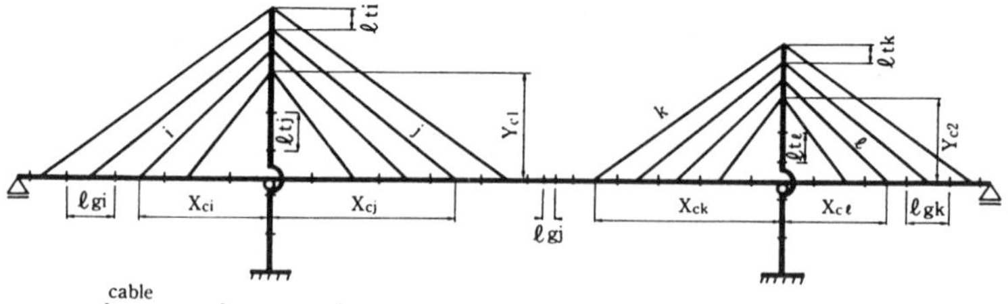

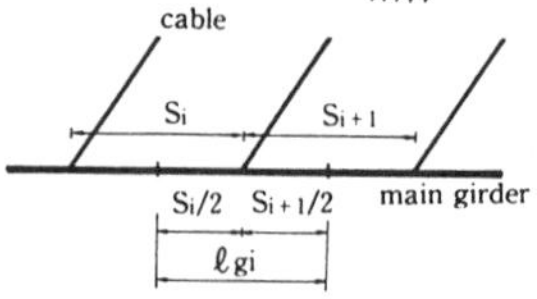

Fig.2 Design variables Xc and Yc and element lengths ℓ_{gi} and ℓ_{tj} of main girder and pylon element

upper and lower flange plates of each main girder element, t_{gu}, t_{gl}, and pylon element, t_{tu}, t_{tl}, as shown in Fig.1, where t_{tu} and t_{tl} are assumed to be the same. The thicknesses of these flange plates are dealt with as the converted thicknesses which include the contributions of the longitudinal stiffeners. Z is the vector of these design variables with respect to the sizes of the cross sections.

$$Z=[Z_1^T..,Z_i^T,..,Z_n^T]^T \quad (i=1,...,n) \tag{1}$$

if i denotes the element of main girder, $Z_i=[t_{gui}, t_{gli}]^T$
if i denotes the element of pylon, $Z_i=[t_{tui}, t_{tli}]^T$ where $t_{tui}=t_{tli}$
if i denotes the cable element, $Z_i=A_{Ci}$

The distance from the pylon to each cable anchor position in the main girder, X_{Ck}, and the height of the lowest cable in the pylon from the axis of main girder, Y_{Cl}, in Fig.2, are dealt with as the design variables with respect to the cable anchor positions. The distances of each cable in the pylon, ℓ_{ti} and ℓ_{tk} are assumed as the preassigned constant values. X_C and Y_C are the vectors with the elements X_{Ck} and Y_{Cl} respectively.

Design constraints

In this study, the constraints on stress are considered as the behavioral constraints. The working stresses at cables, elements of the main girder and pylon are summarized as the stresses due to the dead load acting in the cantilever system at the erection closing stage and the stresses due to the traffic load and a part of the dead load acting in the continuous girder system at the service stage.(see Fig.3) The dead load and traffic load, impact factor, effective widths of the flange plates and effective lengths of the pylon elements for bucklings are taken from the Japanese Specifications for Highway Bridges.[5)]

The following constraints related to the stresses at each cable and elements of the main girder and pylon, slenderness ratios of the pylon elements, and upper and lower limits of the design variables are considered in the optimization process:

(a) The stress at the main girder element:

$$g\sigma_{gi}(Z,X_C,Y_C)=\sigma_i(Z,X_C,Y_C)-\sigma_{ai}(Z)\leqq 0 \; (i=1,...,m_g) \tag{2}$$

$\sigma_i(Z,X_C,Y_C)$: The stress due to design loads.
$\sigma_{ai}(Z)$: The allowable stress.

(b) The stress at the pylon element:

$$g\sigma_{t1j}(Z,X_C,Y_C)=\frac{\sigma_{cj}(Z,X_C,Y_C)}{\sigma_{cazj}(Z,Y_C)}+\frac{\sigma_{bcyj}(Z,X_C,Y_C)}{\sigma_{cagyj}\left(1-\dfrac{\sigma_{cj}(Z,X_C,Y_C)}{\sigma_{eayj}(Z,Y_C)}\right)}-1\leqq 0 \; (j=1,...,m_t) \tag{3}$$

$$g\sigma_{t2j}(Z,X_C,Y_C)=\sigma_{cj}(Z,X_C,Y_C)+\frac{\sigma_{bcyj}(Z,X_C,Y_C)}{\left(1-\dfrac{\sigma_{cj}(Z,X_C,Y_C)}{\sigma_{eayj}(Z,Y_C)}\right)}-\sigma_{calj}(Z)\leqq 0 \; (j=1,...,m_t) \tag{4}$$

$\sigma_{cj}(Z,X_C,Y_C)$: The axial compressive stress due to design loads.
$\sigma_{bcyj}(Z,X_C,Y_C)$: The bending stress due to design loads.
$\sigma_{cazj}(Z,Y_C)$: The allowable axial compressive stress.
σ_{cagyj} : The allowable bending stress without local buckling.
$\sigma_{eayj}(Z,Y_C)$: The allowable stress for Euler's buckling.

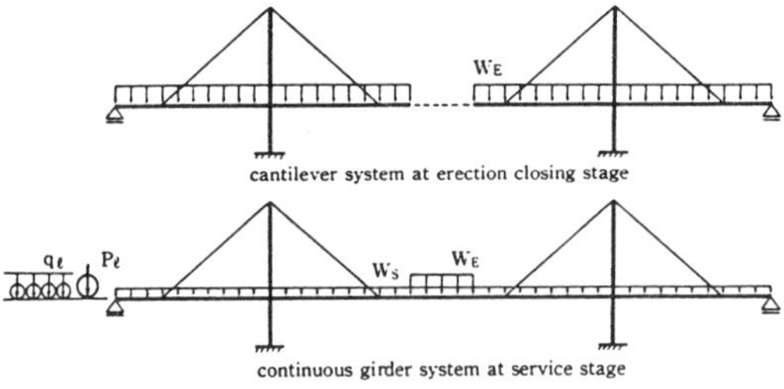

Fig.3 Structure-load systems at erection closing and service stages

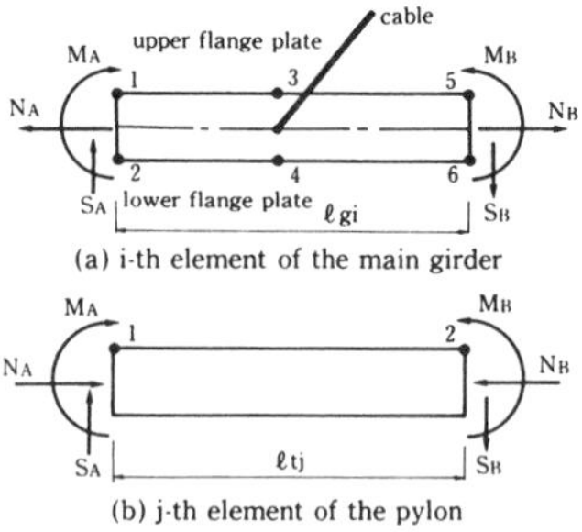

Fig.4 Stress inspection points in the elements of the main girder and pylon

$\sigma_{calj}(Z)$: The allowable bending stress for local buckling.

(c) The slenderness ratio of the pylon element:

$$g\ell_j(Z, Y_C) = \ell_j(Y_C)/r_j(Z) - 120 \leqq 0 \; (j = 1, \ldots, nt) \quad (5)$$

$\ell_j(Y_C)/r_j(Z)$: The slenderness ratio.

(d) The stress at the cable element:

$$g\sigma_{ck}(Z, X_C, Y_C) = \sigma_k(Z, X_C, Y_C) - \sigma_{ak} \leqq 0 \; (k = 1, \ldots, nc) \quad (6)$$

$\sigma_k(Z, X_C, Y_C)$: The tensile stress due to design loads.

σ_{ak} : The allowable tensile stress.

(e) The upper and lower limits of the design variables:

$$Z^{(l)} \leqq Z \leqq Z^{(u)}, \quad X_C^{(l)} \leqq X_C \leqq X_C^{(u)}, \quad Y_C^{(l)} \leqq Y_C \leqq Y_C^{(u)}. \quad (7)$$

After the determination of Z, X_C and Y_C, the minimum web plate thickness of each element of the main girder and pylon is determined so as to satisfy the composite stress criteria on the web plate.

Stress analysis and critical stress condition

In this paper, the cable-stayed bridge is analysed by the finite element method as a 2-dimensional plane frame structure. The critical stresses at the upper and lower flange plates in each element of the main girder and pylon are inspected at the points shown in Fig.4 for the determination of these plate thicknesses. The maximum and minimum values of N, S, and M at the stress inspection points due to traffic and impact loads are calculated by using corresponding influence lines.

Formulation of optimum design problem

The minimum-cost design problem of the steel cable-stayed bridge can then be formulated as follows:

find $Z, X_C, Y_C,$ which

minimize $\mathrm{TCOST}(Z, X_C, Y_C) = \sum_{j=l}^{n} W_j A_j$ (8)

subject to

$$g\sigma_{gj}(Z, X_C, Y_C) \leqq 0 \quad (j = 1, \ldots, m_g) \quad (9)$$

$$g\sigma_{t1j}(Z, X_C, Y_C) \leqq 0 \quad (j = 1, \ldots, m_t) \quad (10)$$

$$g\sigma_{t2j}(Z, X_C, Y_C) \leqq 0 \quad (j = 1, \ldots, m_t) \quad (11)$$

$$g\ell_j(Z, Y_C) \leqq 0 \quad (j = 1, \ldots, n_t) \quad (12)$$

$$g\sigma_{cj}(Z, X_C, Y_C) \leqq 0 \quad (j = 1, \ldots, n_c) \quad (13)$$

$$Z^{(l)} \leqq Z \leqq Z^{(u)}, \quad X_C^{(l)} \leqq X_C \leqq X_C^{(u)}, \quad Y_C^{(l)} \leqq Y_C \leqq Y_C^{(u)} \tag{14}$$

where

$j = 1 \sim n_g$: $W_j = \rho_{gj} \cdot \ell_{gj}(X_C)$, $A_j = A_{gj}(Z_j)$

$j = n_g + 1 \sim n_g + n_t$: $W_j = \rho_{t(j-ng)} \cdot \ell_{t(j-ng)}(Y_C)$, $A_j = A_{t(j-ng)}(Z_j)$

$j = n_g + n_t + 1 \sim n_g + n_t + n_C$: $W_j = \rho_{C(j-ng-nt)} \cdot \ell_{C(j-ng-nt)}(X_C, Y_C)$, $A_j = A_{C(j-ng-nt)}(Z_j)$

$m = m_g + m_t + m_t + n_t + n_c$, $n = n_g + n_t + n_C$, $\rho_{gi}, \rho_{ti}, \rho_{Ci}$: The price per unit volume of each member. A_{gi}, A_{ti}, A_{Ci} : The cross-sectional area of each member.

OPTIMIZATION METHOD

Approximate convex design problem

Utilizing the convex and linear approximation concept developed by Fleury et al.[6)-9)], the objective function, eq.(8), and the behavior constraints, eqs.(9)~(14), are approximated by using first-order partial derivatives and Z, X_C, Y_C and the reciprocal variables $1/Z, 1/X_C, 1/Y_C$. In the objective function, the constant term can be neglected in the optimization process, and only the change in the objective function, $\Delta\mathrm{TCOST}(Z, X_C, Y_C)$, is dealt with as the objective in place of the objective function, $\mathrm{TCOST}(Z, X_C, Y_C)$, then the following approximate subproblem can be derived:

find $\quad Z, X_C, Y_C,$ which

$$\text{minimize} \quad \Delta\mathrm{TCOST}(Z, X_C, Y_C) = \sum_{i=1}^{n} W_i \cdot \frac{\partial A_i}{\partial Z_i} \cdot Z_i + \sum_{k=1}^{K} \underset{(+)}{\left(\sum_{i=1}^{n} W_{Xik} \cdot A_i(Z_i^o)\right)} \cdot X_{Ck}$$
$$- \sum_{k=1}^{k} \underset{(-)}{\left(\sum_{i=1}^{n} W_{Xik} \cdot A_i(Z_i^o)\right)} \cdot (X_{Ck}^o)^2 \cdot \frac{1}{X_{Ck}} + \sum_{l=1}^{L} \underset{(+)}{\left(\sum_{i=1}^{n} W_{yil} \cdot A_i(Z_i^o)\right)} \cdot Y_{Cl} \tag{15}$$
$$- \sum_{l=1}^{L} \underset{(-)}{\left(\sum_{i=1}^{n} W_{yil} \cdot A_i(Z_i^o)\right)} \cdot (Y_{Cl}^o)^2 \cdot \frac{1}{Y_{Cl}}$$

$$\text{subject to} \quad g_j(Z, X_C, Y_C) = \sum_{i=1}^{n} \left(\underset{(+)}{a_{ij}} \cdot Z_i - \underset{(-)}{a_{ij}} \cdot (Z_i^o)^2 \cdot \frac{1}{Z_i}\right) + \sum_{k=1}^{K} \left(\underset{(+)}{b_{kj}} \cdot X_{Ck}\right. \tag{16}$$
$$\left. - \underset{(-)}{b_{kj}} \cdot (X_{Ck}^o)^2 \cdot \frac{1}{X_{Ck}}\right) + \sum_{l=1}^{L} \left(\underset{(+)}{c_{lj}} \cdot Y_{Cl} - \underset{(-)}{c_{lj}} \cdot (Y_{Cl}^o)^2 \cdot \frac{1}{Y_{Cl}}\right) + \overline{U}_j \leqq 0 \quad (j = 1, \ldots, m)$$

$$Z^{(l)} \leqq Z \leqq Z^{(u)}, \quad X_C^{(l)} \leqq X_C \leqq X_C^{(u)}, \quad Y_C^{(l)} \leqq Y_C \leqq Y_C^{(u)}$$

where $W_i = \rho_i \cdot \ell_i$, $W_{Xik} = \dfrac{\partial W_i}{\partial X_{Ck}}$, $W_{yil} = \dfrac{\partial W_i}{\partial Y_{Cl}}$, $a_{ij} = \dfrac{\partial g_j}{\partial Z_i}$, $b_{kj} = \dfrac{\partial g_j}{\partial X_{Ck}}$, $c_{lj} = \dfrac{\partial g_j}{\partial Y_{Cl}}$

$$\overline{U}_j = g_j(Z^o, X_C^o, Y_C^o) + \sum_{i=1}^{n} \left(\underset{(-)}{a_{ij}} \cdot Z_i^o - \underset{(+)}{a_{ij}} \cdot Z_i^o\right) + \sum_{k=1}^{K} \left(\underset{(-)}{b_{kj}} \cdot X_{Ck}^o - \underset{(+)}{b_{kj}} \cdot X_{Ck}^o\right) + \sum_{l=1}^{L} \left(\underset{(-)}{c_{lj}} \cdot Y_{Cl}^o - \underset{(+)}{c_{lj}} \cdot Y_{Cl}^o\right)$$

$$g = [g_1, .., g_m]^T = [g\sigma_{g(1)}, .., g\sigma_{g(mg)}, g\sigma_{t1(1)}, .., g\sigma_{t1(mt)}, g\sigma_{t2(1)}, .., g\sigma_{t2(mt)}, g_{l(1)}, .., g_{l(nt)}, g\sigma_{C(1)}, .., g\sigma_{C(nC)}]^T$$

In the above expressions, the symbols (+) and (-) express the signs of the first-order partical derivatives.

Optimization algorithms by dual method

The above approximate subproblem is solved by the dual method. The optimization algorithm is as follows.

① Derive the following separable Lagrangian function for each approximate subproblem:

$$L(Z, X_C, Y_C, \lambda) = \Delta\mathrm{TCOST}(Z, X_C, Y_C) + \sum_{j=1}^{m} \lambda_j \cdot g_j(Z, X_C, Y_C) \tag{17}$$
$$= \sum_{i=1}^{n} L_i(Z_i, \lambda) + \sum_{k=1}^{K} L_k(X_{Ck}, \lambda) + \sum_{l=1}^{L} L_l(Y_{Cl}, \lambda) + \sum_{j=1}^{m} \lambda_j \cdot \overline{U}_j, \quad \lambda_j \leqq 0 \; (j = 1, \ldots, m)$$

where λ_j is the Lagrange multiplier for the j-th behavior constraint and $\overline{U}_j$ is a constant term. L_i, L_k and L_l are in turn given by the following expressions:

$$L_i(Z_i,\boldsymbol{\lambda})=W_i\cdot\frac{\partial A_i}{\partial Z_i}\cdot Z_i+\sum_{j=1}^{m}\lambda_j\cdot(\underset{(+)}{a_{ij}}\cdot Z_i-\underset{(-)}{a_{ij}}\cdot(Z_i^o)^2\cdot\frac{1}{Z_i}) \tag{18}$$

$$L_k(X_{Ck},\boldsymbol{\lambda})=\underset{(+)}{(\sum_{i=1}^{n}W_{Xik}\cdot A_i(Z_i^o))}\cdot X_{Ck}-\underset{(-)}{(\sum_{i=1}^{n}W_{Xik}\cdot A_i(Z_i^o))}\cdot(X_{Ck}^o)^2\cdot\frac{1}{X_{Ck}}$$
$$+\sum_{j=1}^{m}\lambda_j\cdot(\underset{(+)}{b_{kj}}\cdot X_{Ck}-\underset{(-)}{b_{kj}}\cdot(X_{Ck}^o)^2\cdot\frac{1}{X_{Ck}}) \tag{19}$$

$$L_l(Y_{Cl},\boldsymbol{\lambda})=\underset{(+)}{(\sum_{i=1}^{n}W_{yil}\cdot A_i(Z_i^o))}\cdot Y_{Cl}-\underset{(-)}{(\sum_{i=1}^{n}Wy_{il}\cdot A_i(Z_i^o))}\cdot(Y_{Cl}^o)^2\cdot\frac{1}{Y_{Cl}}$$
$$+\sum_{j=1}^{m}\lambda_j\cdot(\underset{(+)}{c_{lj}}\cdot Y_{Cl}-\underset{(-)}{c_{lj}}\cdot(Y_{Cl}^o)^2\cdot\frac{1}{Y_{Cl}}) \tag{20}$$

② The solution of the dual problem $\boldsymbol{Z}^*$, $\boldsymbol{X}_C^*$, $\boldsymbol{Y}_C^*$ and $\boldsymbol{\lambda}^*$ can be obtained by minimizing $\boldsymbol{L(Z, X_C, Y_C, \lambda)}$ with respect to $\boldsymbol{Z}$, $\boldsymbol{X_C}$ and $\boldsymbol{Y_C}$, and maximizing it with respect to $\boldsymbol{\lambda}$. $\boldsymbol{D}^*=[\boldsymbol{Z}^{*T}, \boldsymbol{X}_C^{*T}, \boldsymbol{Y}_C^{*T}]^T$ which minimize $\boldsymbol{L(Z, X_C, Y_C, \lambda)}$ can be obtained analytically by the following expressions derived from the necessary conditions for the minimums of L_i, L_k, L_l, namely $\partial L_i/\partial Z_i=0$, $\partial L_k/\partial X_{Ck}=0$ and $\partial L_\ell/\partial Y_{C\ell}=0$ and the side constraints:

(a) $\omega_i=\partial\mathrm{TCOST}/\partial D_i\ \geqq 0$

$$\begin{aligned}
D_i^*&=D_i^{(l)} &&\text{if}\quad R_i-\omega_i=0\quad\text{or}\quad D_i\leqq D_i^{(l)}\\
D_i^*&=D_i^{(u)} &&\text{if}\quad (\omega_i+V_i)=0\quad\text{or}\quad D_i\geqq D_i^{(u)}\text{, where}\quad D_i=\sqrt{\frac{R_i}{\omega_i+V_i}}\\
D_i^*&=D_i &&\text{if}\quad D_i^{(l)}<D_i<D_i^{(u)}
\end{aligned} \tag{21}$$

(b) $\omega_i=\partial\mathrm{TCOST}/\partial D_i<0$

$$\begin{aligned}
D_i^*&=D_i^{(l)} &&\text{if}\quad R_i=0\quad\text{or}\quad D_i\leqq D_i^{(l)}\\
D_i^*&=D_i^{(u)} &&\text{if}\quad V_i=0\quad\text{or}\quad D_i\geqq D_i^{(u)}\text{, where}\quad D_i=\sqrt{\frac{R_i-\omega_i}{V_i}}\\
D_i^*&=D_i &&\text{if}\quad D_i^{(l)}<D_i<D_i^{(u)}
\end{aligned} \tag{22}$$

where $R_i=-\sum_{j=1}^{m}\lambda_j\cdot\underset{(-)}{T_{ij}}\cdot(D_i^o)^2$, $\quad V_i=\sum_{j=1}^{m}\lambda_j\cdot\underset{(+)}{T_{ij}}$

T_{ij} and ω_i in the above expressions take the following values with respect to $\boldsymbol{Z}$, $\boldsymbol{X_C}$ and $\boldsymbol{Y_C}$ respectively:

(a) $D_i=[t_{ui}, t_{li}, A_{ci}]^T$: $\qquad T_{ij}=a_{ij}$ (23)

(b) $D_i=X_{Ci}$: $\qquad T_{ij}=b_{ij}$

$$\omega_i=\underset{(-)}{(\sum_{j=1}^{n}W_{xji}\cdot A_j(Z_j^o))}\cdot(X_{Ci}^o)^2\ \text{if}\ \omega_i<0,\quad \omega_i=\underset{(+)}{(\sum_{j=1}^{n}W_{xji}\cdot A_j(Z_j^o))}\ \text{if}\ \omega_i\geqq 0 \tag{24}$$

(c) $D_i=Y_{Ci}$: $\qquad T_{ij}=c_{ij}$

$$\omega_i=\underset{(-)}{(\sum_{j=1}^{n}W_{yji}\cdot A_j(Z_j^o))}\cdot(Y_{Ci}^o)^2\ \text{if}\ \omega_i<0,\quad \omega_i=\underset{(+)}{(\sum_{j=1}^{n}W_{yji}\cdot A_j(Z_j^o))}\ \text{if}\ \omega_i\geqq 0 \tag{25}$$

The minimized Lagrangian function with respect to $\boldsymbol{Z}$, X_C and Y_C by the above expressions is denoted as $\ell(\boldsymbol{\lambda})$:

$$\ell(\boldsymbol{\lambda})=\min_{\boldsymbol{Z},X_C,Y_C} L(\boldsymbol{Z},X_C,Y_C,\boldsymbol{\lambda}) \tag{26}$$

Following the minimization process with respect to $\boldsymbol{Z}$, X_C and Y_C, the minimized Lagrangian function $\ell(\boldsymbol{\lambda})$ is maximized with respect to λ by using a Newton-type algorithm. In the algorithm, the Lagrange multipliers $\bar{\lambda}$ corresponding to the active behavior constraints at the current stage is modified iteratively as

$$\bar{\lambda}^{(t+1)} = \bar{\lambda}^{(t)} + \alpha^{(t)} \cdot S^{(t)} \tag{27}$$

or in a scalar form

$$\bar{\lambda}_j^{(t+1)} = \bar{\lambda}_j^{(t)} + \alpha_{max}^{(t)} \cdot S_j^{(t)} \qquad (j \in S_{AG}) \tag{28}$$

wher·· $S^{(t)}$ is the search direction of $\bar{\lambda}^{(t)}$ for active constraints, S_{AG} s the set of active behavior constraints and $\alpha^{(t)}$ is the step length parameter. $S^{(t)}$ is given by

$$S^{(t)} = -[H(\bar{\lambda}^{(t)})]^{-1} \cdot \nabla \ell(\bar{\lambda}^{(t)}) \tag{29}$$

where $\nabla \ell(\bar{\lambda}^{(t)})$ is the vector of first derivatives of $\ell(\lambda)$ with respect to $\bar{\lambda}$. $H(\bar{\lambda}^{(t)})$ is the Hessian matrix of $\ell(\lambda)$ with respect to $\bar{\lambda}_j (j \in S_{AG})$ and the jk-th component of the matrix is given by the following expressions:

$$H_{jk} = \sum_{i=1}^{n+K+L} B_i \ , \quad \text{where} \quad \begin{array}{ll} B_i = Q_i \cdot (D_i^o)^4 / D_i^3 & \text{if } T_{ij} < 0,\ T_{ik} < 0 \\ B_i = Q_i \cdot (D_i^o)^2 / D_i & \text{if } T_{ij} \geqq 0,\ T_{ik} < 0 \\ B_i = Q_i \cdot (D_i^o)^2 / D_i & \text{if } T_{ij} < 0,\ T_{ik} \geqq 0 \\ B_i = Q_i \cdot D_i & \text{if } T_{ij} \geqq 0,\ T_{ik} \geqq 0 \end{array} \tag{30}$$

Q_i is given by

$$\text{if } \omega_i < 0: Q_i = -\frac{1}{2} \cdot T_{ij} \cdot T_{ik} / V_i \ , \quad \text{if } \omega_i \geqq 0: Q_i = -\frac{1}{2} \cdot T_{ij} \cdot T_{ik} / (\omega_i + V_i) \tag{31}$$

The step length $\alpha^{(t)}$ is first set as 1.0; however, its maximum value must be restricted to ensure the nonnegativity of $\bar{\lambda}$ when $S^{(t)}$ includes negative components. When $\ell(\bar{\lambda}^{(t+1)})$ exceeds the maximum point, $\alpha^{(t)}$ is reduced and the search is continued until $\ell(\bar{\lambda}^{(t+1)})$ is maximized. According to the modification of λ, the primary variables Z, X_C and Y_C are improved by eqs.(21)~(22) and the set of S_{AG} must also be updated. The min.-max. process described above is iterated until Z, X_C, Y_C and λ converge to constant values Z^*, X_C^*, Y_C^* and λ^*.

③ By using Z^*, X_C^*, Y_C^*, the minimum web thicknesses of the main girder and pylon elements t_w are improved so as to satisfy the corresponding stress constraints on t_w.

④ Z^*, X_C^* and Y_C^* are assumed as new initial values of the design variables, and a new approximate subproblem is derived. The final optimum solution can be determined by iterating the above improvement processes ①~③ until Z, X_C, Y_C and λ converge to constant values.

In the above optimization process, it should be noted that if the changing rates of X_C, Y_C by eqs.(21)~(22) are too large in one iteration of the improvement process, the successive solutions may oscillate and smooth convergence may not be obtained in some cases. For this reason, the adaptive move limit constraints, maximum 10%, are imposed on the changing rates of X_C and Y_C. It is also important to note that if two or more stress constraints given by eq.(16) become active with respect to one flange plate in a main girder element and a_{ij}, b_{kj} and c_{ij} in the constraints are almost same, these constraints become linearly dependent on each other and in consequence the Hessian matrix $H(\lambda)$ becomes singular. In this case, λ can be successfully improved by considering that only the most critical stress constraint is active and deleting the other stress constraints on the same flange plate from the set of active constraints S_{AG}.

DESIGN EXAMPLES

Various minimum-cost design problems of steel cable-stayed bridges have been solved by the proposed method. In this section, the numerical results of the three-span steel cable-stayed bridge shown in Fig.5 under various conditions are presented. The design constants used in the numerical examples are tabulated in Table 1. and 2. Since the skeleton of this design model is symmetric, the numbers of the independent design variables and constraints to be dealt with in the optimization process are 67 and 158, respectively. In the design problems, the lower limits of the cross-sectional areas of cables are set as 0.1cm^2, and the objective is to determine the optimum topological cable arrangement.

To investigate the significance of dealing with X_C and Y_C as the design variables, the optimum solutions for the cases in which Z only and Z, X_C and Y_C are dealt with as the design variables are compared. The global optimality of optimum solutions obtained by the proposed design method is confirmed by comparing the optimum solutions which are obtained by assuming the initial values of Y_C^0 , as 75m and 45m.

Table 3 summarizes the optimum solutions for the above four cases, namely, two for Y_C^0 =75m and another two for Y_C^0 =45m. The cable arrangement, maximum and minimum bending moments and axial forces distributions, Mmax, Mmin, Nmax, Nmin, and upper and lower flange plate thicknesses distributions, t_{gu} , t_{gl} , at the optimum solutions in which Y_C^0 is assumed to be 75m are shown in Fig.6. The distributions of Mmax, Mmin, Nmax, Nmin and the cross-sectional areas in the pylon at the above optimum solutions are depicted in Fig.7. The values of bending moments and axial forces shown in Fig.6 express the magnitudes of these acting on the half of the cross-sectional area, depicted in Fig.1, of the main girder, while the total cost expresses the cost in the whole system.

As clearly seen from Table 3, the optimum solutions can be obtained efficiently after 15 and 13 iterations respectively for the cases of Y_C^0 =75m and 45m even if the adaptive move limit constraints, maximum 10%, are imposed on X_C and Y_C. The optimum heights of the lowest cable, Y_{Copt} are determined as 57.52m and 57.45m respectively for Y_C^0 =75m and 45m. The optimum cable anchor positions on the main girder, X_{Copt} are also almost the same for the two cases, although a slight difference of 7.6m is observed in X_{C3}. The difference of the minimum total costs for both cases is 0.12%. From these results, it can be said that quite similar optimum solutions, the global optimum solutions, can be obtained by the proposed design method even if the optimization process is started from extremely different initial values of Z and Y_C.

At the optimum cable arrangement, the top two cables in the side span are parallel and are anchored at the end support. Their cross-sectional areas are 3.6~1.4 times larger than those of the middle cables. In the center span, all cables are distributed as the geometric series and the cross-sectional areas of the top two cables are also 2.6~1.7 times larger than those of the middle cables. The cross-sectional areas of the lowest cables in the side and center spans are determined as 0.1 cm² by the lower limit constraint, which indicates that the lowest cables in the side and center span are unnecessary from the static optimization view point.

In the optimum solutions in which Z, X_C and Y_C are dealt with as the design variables, the cable arrangements are determined so as to decrease the critical local peaks of the maximum and minimum bending moments at the main girder and pylon. As clearly seen from Fig.6, the local peaks of the minimum bending

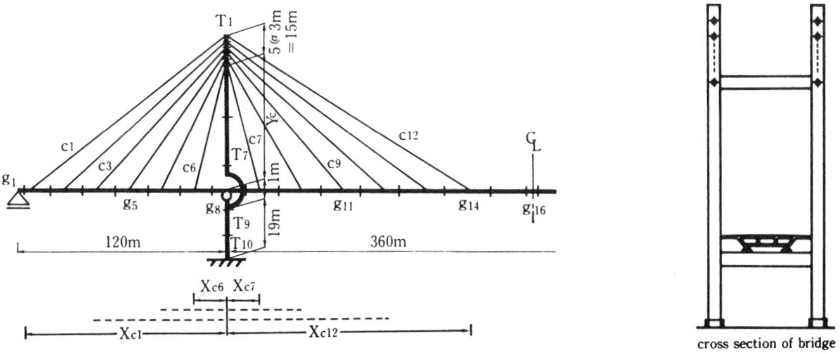

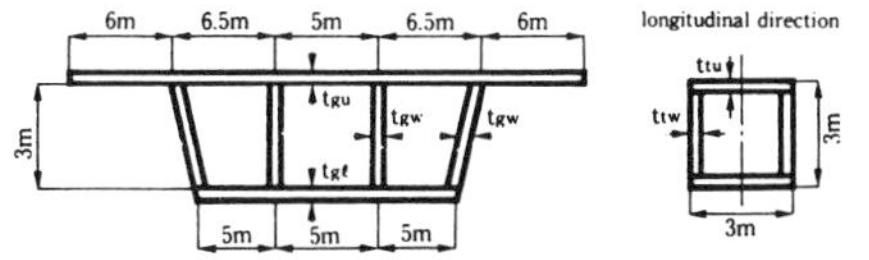

Fig.5 Three-span steel cable-stayed bridge with 48 cable stays

Table 1 Dead load and traffic load at erection closing and service stages

erection closing stage	D. Load[1]	main girder[3]	4.0 tf/m
		pylon[3]	2.0 tf/m
		steel weight	7.85 tf/m³
service stage	D. Load[1]	main girder[4]	3.4 tf/m
	T. Load[2]	uniform	2.25 tf·m
		line	54.1 tf
		impact	0.11

1) Dead load 2) Traffic load
3) Dead load due to cable anchors etc.
4) Dead load due to asphalt pavement etc.

Table 2 Material properties and minimum plate thicknesses used for the elements of the main girder, pylon and cables

No. of element	E(kgf/cm²)[1]	ρ (1/m³)[2]	material kind	t_u^a(mm)[3]	t_l^a(mm)[4]	t_w^a(mm)[5]
g_1~g_6 g_7 g_8 g_9 g_{10}~g_{16}	2.1×10⁶	500	SS41	18.5	15.3	12.0
T_1~T_5 T_6~T_8 T_9~T_{10}	2.1×10⁶	700	SM50 SM58 SM58	26.0	26.0	28.0
C_1~C_{12}	2.0×10⁶	900	5100 [6]	A_C^a = 0.00001 (m²)		

1) Modulus of elasticity 2) Price per unit volume
3) Converted minimum upper flange thicknesses including longitudinal stiffeners 4) Converted minimum lower flange thicknesses including longitudinal stiffeners 5) Converted minimum web thicknesses including longitudinal stiffeners 6) Allowable tensile stress of cable member element

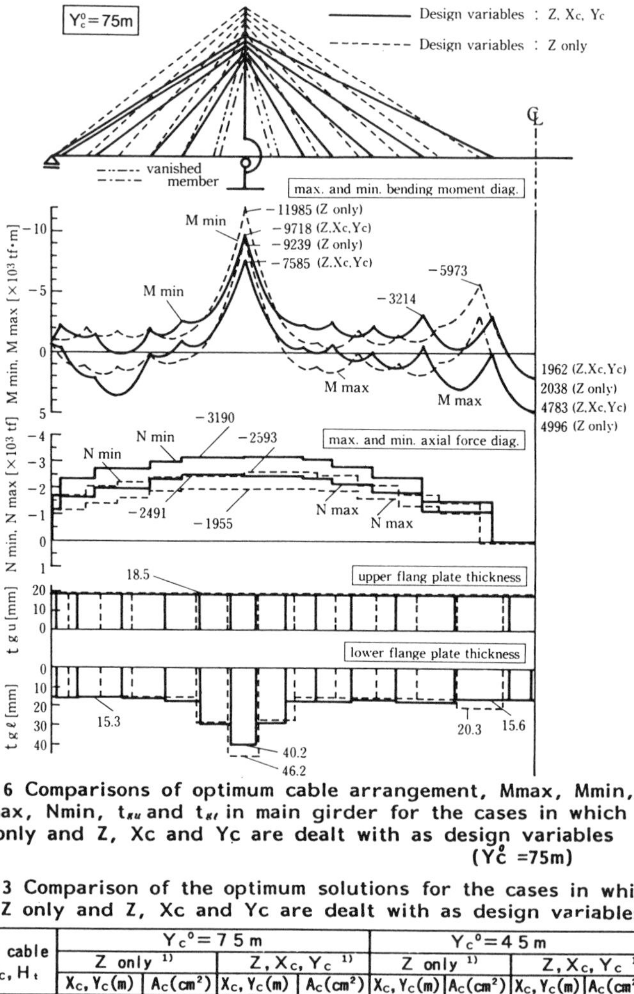

Fig. 6 Comparisons of optimum cable arrangement, Mmax, Mmin, Nmax, Nmin, t_{gu} and $t_{g\ell}$ in main girder for the cases in which Z only and Z, Xc and Yc are dealt with as design variables (Y_c^0 =75m)

Table 3 Comparison of the optimum solutions for the cases in which Z only and Z, Xc and Yc are dealt with as design variables

No. of cable C_i, Y_c, H_t	Y_c^0=75m				Y_c^0=45m			
	Z only [1]		Z, X_c, Y_c [1]		Z only [1]		Z, X_c, Y_c [1]	
	X_c, Y_c(m)	A_c(cm²)	X_c, Y_c(m)	A_c(cm²)	X_c, Y_c(m)	A_c(cm²)	X_c, Y_c(m)	A_c(cm²)
C_1	119.00	422	119.00	391	119.00	553	119.00	379
C_2	99.00	94	114.07	157	99.00	141	114.06	186
C_3	79.00	77	93.28	102	79.00	19	85.70	107
C_4	59.00	122	58.22	116	59.00	178	56.35	100
C_5	39.00	64	38.87	133	39.00	162	38.23	135
C_6	19.00	0	15.57	0	19.00	0	16.02	0
C_7	20.00	0	16.15	0	20.00	0	16.46	0
C_8	45.00	66	36.56	89	45.00	154	36.10	92
C_9	70.00	128	54.10	127	70.00	143	53.56	119
C_{10}	95.00	95	78.83	121	95.00	115	77.19	122
C_{11}	120.00	77	109.33	214	120.00	122	107.06	215
C_{12}	145.00	324	152.40	326	145.00	430	152.38	334
Y_c	75.00	—	57.52	—	45.00	—	57.45	—
[2] H_t	90.00	—	72.52	—	60.00	—	72.45	—
[3] ITE	9		15		9		13	
TCOST	529132.8		483890.7		496760.3		483312.9	
[4] R.TCOST	1.000		0.914		1.000		0.973	

1) Design variables 2) Total pylon height: total length from main girder to the top of pylon 3) Number of iterations 4) Ratio between TCOST (Z only) and TCOST (Z, X_c, Y_c)

moments at the middle support and near the center point in the main girder are reduced to 81% and 54%, respectively, compared with the ones for the case in which Y_c^0 =75m and Z only is dealt with as the design variable. At the center point of the main girder, it seems that a large maximum bending moment still acts on the cross section, although the cable arrangement is optimized; however, the upper and lower flange plate thicknesses at this point are determined to be the same as the lower limit plate thicknesses. This means that it is not necessary to decrease the maximum bending moment at this point. This result also shows the reliability and rigorousness of the proposed design method.

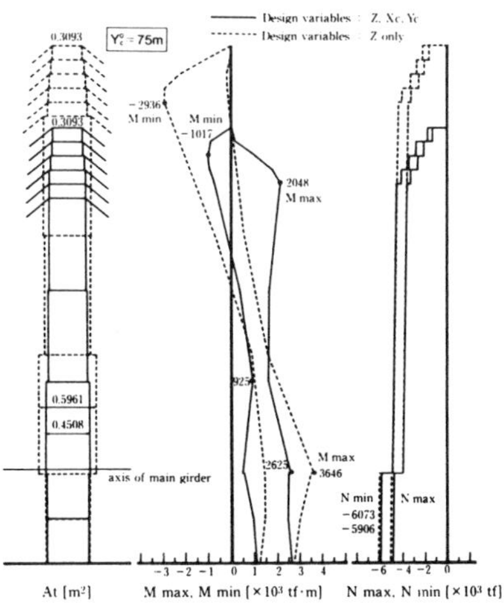

Fig.7 Comparisons of At, Mmax, Mmin, Nmax and Nmin for the cases in which Z only and Z, Xc and Yc are dealt with as design variables (Y_c^0 = 75m)

In the optimum solutions in which Z, X_c and Y_c are dealt with as the design variables, the horizontal components of the tensions due to dead load in the left and right cables at each set in the pylon are well balanced, and the magnitudes of the maximum and minimum bending moments in the pylon are reduced drastically as shown in Fig.7, at most, with 28% reduction in the maximum bending moment, and are well averaged through the pylon compared with the optimum bending moment distributions in which Z only is dealt with as the design variable.

As the consequence of the improvements mentioned above due to the changes in X_c and Y_c, the total cost of the bridge decreases as much as 8.6% compared with that for the case in which Z only is dealt with as the design variable.

Similar comparisons can be made for the two optimum solutions obtained from Y_c^0 =45m which are given in Table 3. In this case, 2.7% reduction in the total cost of the bridge is observed when Z, X_c and Y_c are dealt with as the design variables.

From the investigation of the optimum solutions for various design conditions, it is clear that the proposed optimum design method can determine the optimum cable arrangement and anchor positions on the main girder and pylon and the optimum cross-sectional dimensions of all structural elements of a steel cable-stayed bridge superstructure efficiently and rigorously. Furthermore, the significance of dealing with cable anchor positions on the main girder and pylon as the design variables is clarified for the minimum cost design of steel cable-stayed bridges.

CONCLUSIONS

The following conclusions can be drawn from this study.

(1) The global optimum solutions can be determined quite rigorously and efficiently by the proposed optimum design method.

(2) The optimum values of pylon height, cable anchor positions, steel plate thicknesses of each main girder and pylon element, cross-sectional area of each cable obtained are quite reasonable and well balanced.

(3) The total cost of the steel cable-stayed bridge is greatly affected by the cable anchor positions on the main girder and pylon. Therefore, the treatment of the cable anchor positions and the height of the pylon as the design variables is extremely significant in the optimum design problem of steel cable-stayed bridge.

(4) With regard to the optimum cable arrangement in the numerical examples, the top two cables are parallel and are anchored at the end support in the side span; on the other hand, the cables are distributed as the geometric series in the center span. The sectional areas of the top two cables in the side and center spans are determined to be 3.6~1.4 times larger than those of the middle cables. The cross-sectional areas of the unnecessary cables at the optimum solution can be reduced automatically to the imposed lower limit by the proposed optimum design method.

REFERENCES

1. Yamada, Y. and Daiguji, H. Optimum Design of Cable Stayed Bridges using Optimality-Criteria, Proc. of JSCE, No.253, pp.1-12, 1976. (in Japanese)
2. Kobayashi, I. et al. Multilevel Optimal Design of Cable-Stayed Bridges with Various Types of Anchorages, Proc. of JSCE, No. 392/I-9, pp.317-325, 1988. (in Japanese)
3. Gimsing, N. J. Cable Supported Bridges, Concept and Design, John Wiley & Sons, Ltd., 1983.
4. Nagai, M. et al. A study on the Determination of the Basic Configuration of the Three Span Continuous Cable-Stayed Girder Bridge with Multiple Cables, Proc. of JSCE, No.362/I-4, pp.343-352, 1985. (in Japanese)
5. Japanese Road Association, Specifications for Highway Bridges, Steel Bridges, 1980. (in Japanese)
6. Fleury, C. and Braibant, V. Structural Optimization: A New Dual Method using Mixed Variables, Int. J. for Num. Meth. in Engng., Vol.23, pp.409-428, 1986.
7. Starnes, J. H. and Haftka, R. T. Preliminary Design of Composite Wings for Buckling, Strength, and Displacement Constraints, J. Air craft, Vol.16, No.8, pp.564-570, 1979.
8. Prasad, B. Explicit Constraint Approximation Forms in Structural Optimization-Part I: Analyses and Projections, Comp. Meth. Appl. Mech. Eng., Vol.40, pp.1-26, 1983.
9. Ohkubo, S., Nakajima, T. and Asai, K. Total Optimization of Truss Considering Shape, Material and Sizing Variables, Computer Utilization in Structural Engineering, ASCE, pp.247-256, 1989.

Design Optimization of Tall Steel Buildings

D.E. Grierson, C.-M. Chan

Department of Civil Engineering, University of Waterloo, Ontario, Canada N2L 3E5

ABSTRACT

An efficient computer-based method is developed for the optimum design of tall steel building frameworks. Specifically, an Optimality Criteria method is applied to minimize the weight of a lateral load-resisting structural system of fixed topology subject to constraints on overall and interstorey drift. By exploiting the fact for building frameworks that member forces are relatively insensitive to changes in member sizes, rigorously-derived optimality criteria are shown to be readily satisfied through an iterative redesign procedure that converges in but a few cycles. A steel framework example is presented from a variety of viewpoints to illustrate the features of the design optimization method.

INTRODUCTION

A primary concern for the design of a tall slender building is to adequately control lateral deflections induced by dynamic loadings due to wind and earthquake. Typically, in accordance with the provisions of most current building codes, the design is carried out by transforming the dynamic loads to equivalent static loads applied horizontally on the structure. Moreover, while there is no universally established drift criteria for tall building design, the allowable value for overall building drift or interstorey drift is usually specified as some ratio of the building or storey height (e.g., 1/400).

Once the topology and support conditions have been established for a framework, the main design effort involves sizing the individual beam, column, and bracing members. For tall slender buildings, the design of structural members is generally governed by lateral drift criteria rather than by stress criteria. Moreover, as such structures usually consist of thousands of members, it is generally a complex task to size the members so that the drift criteria is satisfied.

There are a number of powerful computer software packages available today for the response analysis of tall slender buildings, e.g. [1]. On the other hand, there is presently no such computer-based methodology available that efficiently sizes members of tall structures so as to economically satisfy lateral drift criteria. In fact, in the absence of an optimization tool, the design of a tall building typically involves an educated trial and error process, which is generally highly iterative and quite time consuming. Moreover, while feasible, the final design so found is often not optimal in the sense of being of minimum structural weight or cost.

This paper presents a computer-based Optimality Criteria (OC) method for the optimum design of tall slender lateral load-resisting steel frameworks of fixed topology subject to constraints on overall and interstorey drift. The design optimization problem is first formulated, and then the details of the OC algorithm are developed. Finally, a worked example is presented to illustrate the features of the design method.

DESIGN OPTIMIZATION PROBLEM

The problem to find the minimum weight (or cost) design of a lateral load-resisting building framework composed of i=1,2,...,n members, to satisfy j=1,2,...,m lateral drift constraints, can be generally stated as

Minimize: $$W(A_i) = \sum_{i=1}^{n} \rho_i L_i A_i \qquad (1a)$$

subject to : $$(\delta_{j+1} - \delta_j) / h_j^{j+1} \le d_j^U \qquad (j=1,2,\ldots,m-1) \qquad (1b)$$

$$\delta_m / H \le d_m^U \qquad (1c)$$

$$A_i^L \le A_i \le A_i^U \qquad (i=1,2,\ldots,n) \qquad (1d)$$

where: eq. (1a) is the weight function for the structure, in which ρ_i, L_i, A_i are the material density, length and cross-section area for member i, respectively; eqs. (1b) define the interstorey drift constraints for the structure, in which δ_{j+1} and δ_j are the lateral deflections at the (j+1)th and jth floor levels, h_j^{j+1} is the storey height between two adjacent floors and d_j^U is the allowable interstorey drift; eq. (1c) defines the constraint on overall building drift, in which δ_m is the lateral deflection at the roof level, H is the building height and d_m^U is the allowable drift at the top of the building; eqs. (1d) are sizing constraints on member cross-section areas, in which A_i^L and A_i^U are corresponding specified upper and lower bounds for member i.

Constraints eqs. (1b) and (1c) are but implicit functions of the member sizing variables A_i. It remains to express these constraints as explicit functions of the A_i in order to facilitate computer solution of the design optimization problem. To that end, by the principle of virtual work, the displacement at any point of interest of a three-dimensional building framework having i=1,2,...,n members can be expressed as

$$\delta_j = \sum_{i=1}^{n} \int_0^{L_i} \left(\frac{F_X f_X}{EA} + \frac{F_Y f_Y}{GA_Y} + \frac{F_Z f_Z}{GA_Z} + \frac{M_X m_X}{GI_X} + \frac{M_Y m_Y}{EI_Y} + \frac{M_Z m_Z}{EI_Z} \right)_i dx \tag{2}$$

where: E, G = axial and shear elastic material moduli; A, A_Y, A_Z = axial and shear areas for the cross-section; I_X, I_Y, I_Z = torsional and flexural moments of inertia for the cross-section; F_X, F_X, F_Z = axial and shear forces, f_X, f_Y, f_Z = virtual axial and shear forces; M_X, M_Y, M_Z = torsional and flexural moments; m_X, m_Y, m_Z = virtual torsional and flexural moments.

Now, for commercial standard steel sections, the cross-section properties A_Y, A_Z, I_X, I_Y and I_Z may all be expressed in terms of the axial area A as follows, [3]

$$1 / A_Y = C_{AY} (1 / A) + C'_{AY} \tag{3a}$$

$$1 / A_Z = C_{AZ} (1 / A) + C'_{AZ} \tag{3b}$$

$$1 / I_X = C_{IX} (1 / A) + C'_{IX} \tag{3c}$$

$$1 / I_Y = C_{IY} (1 / A) + C'_{IY} \tag{3d}$$

$$1 / I_Z = C_{IZ} (1 / A) + C'_{IZ} \tag{3e}$$

where the coefficients C and C' are determined by linear regression analysis and have different values depending on the type and size of the section (e.g. see Tables 1 to 4 for W14 and W24 sections from the AISC-LRFD design manual [2]). Having the relations eqs. (3), the displacement δ_j from eq. (2) can be concisely expressed solely in terms of the member cross- section areas A_i as

$$\delta_j = \sum_{i=1}^{n} \left(\frac{C_{ij}}{A_i} + C'_{ij} \right) \tag{4}$$

where, from eqs. (2) and (3), the coefficients C_{ij} and C'_{ij} are given by

$$C_{ij} = \int_0^{L_i} \left(\frac{F_X f_X + M_Y m_Y C_{IY} + M_Z m_Z C_{IZ}}{E} + \frac{F_Y f_Y C_{AY} + F_Z f_Z C_{AZ} + M_X m_X C_{IX}}{G} \right)_i dx \tag{5a}$$

$$C'_{ij} = \int_0^{L_i} \left(\frac{M_Y m_Y C'_{IY} + M_Z m_Z C'_{IZ}}{E} + \frac{F_Y f_Y C'_{AY} + F_Z f_Z C'_{AZ} + M_X m_X C'_{IX}}{G} \right)_i dx \tag{5b}$$

Recognizing, now, that the interstorey drift constraints eqs. (1b) have the same form as the overall building drift constraints eq. (1c), from eqs. (1) and (4) the structural design optimization problem may be concisely stated explicitly in terms of the member cross-section area A_i as

$$\textit{Minimize:} \qquad W(A_i) = \sum_{i=1}^{n} \rho_i L_i A_i \tag{6a}$$

$$\textit{subject to:} \qquad d_j(A_i) = \sum_{i=1}^{n} \left(\frac{c_{ij}}{A_i} + c'_{ij} \right) \le d_j^U \qquad (j=1,2,...,m) \tag{6b}$$

$$A_i^L \le A_i \le A_i^U \qquad (i=1,2,...,n) \tag{6c}$$

where the coefficients

$$c_{ij} = (C_{ij+1} - C_{ij}) / h_j^{j+1} \tag{7a}$$

$$c'_{ij} = (C'_{ij+1} - C'_{ij}) / h_j^{j+1} \tag{7b}$$

for the m-1 interstorey drift constraints eqs. (1b), while the coefficients

$$c_{im} = C_{im} / H \tag{7c}$$

$$c'_{im} = C'_{im} / H \tag{7d}$$

for the overall building drift constraints eq. (1c).

OPTIMALITY CRITERIA ALGORITHM

Temporarily ignore the sizing constraints eqs. (6c). The resulting design optimization problem eqs. (6a) and (6b) involving drift constraints alone can be reformulated as the minimization of the Lagrangian function

$$L(A_i,\lambda_j) = \sum_{i=1}^{n} \rho_i L_i A_i + \sum_{j=1}^{m} \lambda_j \left[\sum_{i=1}^{n} \left(\frac{c_{ij}}{A_i} + c'_{ij} \right) - d_j^U \right] \tag{8}$$

where the Lagrange multipliers are such that $\lambda_j > 0$ if constraint j is active, or $\lambda_j = 0$ if constraint j is inactive. Temporarily assume that all drift constraints are active.

Differentiate eq. (8) with respect to the sizing variables A_i, and rearrange terms to obtain the stationary conditions, [4,5]

$$\sum_{j=1}^{m} \lambda_j \frac{c_{ij}}{\rho_i L_i A_i^2} = 1 \qquad (i=1,2,\ldots,n) \tag{9}$$

Eq. (9) are optimality criteria having the physical meaning that the weighted sum of the virtual strain energy densities for each member i is equal to unity for the optimum structure.

A recursive algorithm is applied to find the sizing variables A_i satisfying eqs. (9) at the optimum. To that end, multiply both sides of eqs. (9) by A_i^{η} and take the ηth root to obtain

$$A_i = A_i \left(\sum_{j=1}^{m} \lambda_j \frac{c_{ij}}{\rho_i L_i A_i^2} \right)^{\frac{1}{\eta}} \qquad (i=1,2,\ldots,n) \tag{10}$$

Eq. (10) can be viewed as the basis for an exponential recursive relation or, upon applying a first-order binomial expansion, the linear recursive relation

$$A_i^{\nu+1} = A_i^{\nu} \left[1 + \frac{1}{\eta} \left(\sum_{j=1}^{m} \frac{\lambda_j c_{ij}}{\rho_i L_i A_i^2} - 1 \right)_{\nu} \right] \qquad (i=1,2,\ldots,n) \tag{11}$$

where $\nu+1$ and ν indicate successive iterations, and η is a step-size parameter that controls convergence of the recursive process, [4,5].

In order to apply eqs. (11) to find the new sizing variable $A_i^{\nu+1}$, the current values of the Lagrange multipliers λ_j^{ν} must first be determined. To that end, consider the change $(d_k^{\nu+1}-d_k^{\nu})$ in the k-th drift constraint due to changes $(A_i^{\nu+1}-A_i^{\nu})$ in the sizing variables, i.e.,

$$\left(d_k^{\nu+1} - d_k^{\nu} \right) = \sum_{i=1}^{n} \left(\frac{\partial d_j}{\partial A_i} \right)_{\nu} \Delta A_i \tag{12}$$

where, from eqs. (6b) for drift constraint k,

$$\left(\frac{\partial d_k}{\partial A_i} \right)_{\nu} = - \left(\frac{c_{ik}}{A_i^2} \right)_{\nu} \tag{13}$$

and, from Eqs. (11),

$$A_i^{\nu+1} - A_i^{\nu} = \frac{A_i^{\nu}}{\eta} \left(\sum_{j=1}^{m} \frac{\lambda_j c_{ij}}{\rho_i L_i A_i^2} - 1 \right)_{\nu} \tag{14}$$

Suppose that drift constraint k becomes active after the design perturbation, i.e., that $d_k^{v+1} = d_k^U$ in eq. (12). Substitute from eqs. (13) and (14) into eq. (12) and rearrange terms to obtain for all k=1,2,...,m drift constraints,

$$\sum_{j=1}^{m} \lambda_j \sum_{i=1}^{n} \left(\frac{c_{ik} c_{ij}}{\rho_i L_i A_i^3}\right)_v = \sum_{i=1}^{n} \frac{c_{ik}}{A_i^v} - \eta\,(d_k^U - d_k^v) \qquad (k=1,2,...,m) \quad (15)$$

Eqs. (15) are a system of m simultaneous linear equations in terms of m Lagrange variables λ_j.

Eqs. (11) and (15) together form the basis for the Optimality Criteria algorithm, as follows:

1. Adopt an initial set of sizing variables A_i^v (i=1,2,...,n).
2. For the current A_i^v, analyze the structure and establish the coefficients c_{ij} (i=1,2,...,n; j=1,2,...,m).
3. For the current A_i^v, use the Gauss-Seidel iterative technique to solve eqs. (15) for the corresponding set of Lagrange multipliers λ_j^v (j=1,2,...,m); any λ_j^v value found to be negative at any stage of the solution process is temporarily set to zero (i.e., the corresponding drift constraint is passive).
4. For the current λ_j^v, apply eqs. (11) to find the new set of sizing variables A_i^{v+1}; any A_i^{v+1} value found to violate one of the bounds A_i^L or A_i^U in eqs. (6c) is temporarily set to that bound value (i.e., the sizing variable is passive).
5. Check convergence of Gauss-Seidel process: if all $A_i^{v+1} = A_i^v$ and $\lambda_j^{v+1} = \lambda_j^v$, go to step 6; otherwise, return to step 3.
6. Check convergence of design process: if structure weight is the same for two successive stages, terminate with the minimum weight structure; otherwise, return to step 2.

DESIGN EXAMPLE

Consider the 12-storey one-bay planar steel frame having the structural geometry, material properties and horizontal loading shown in Figure 1. This deliberately simple structure behaves like a tall building under lateral loads by virtue of the fact that its height-to-width aspect ratio is 7.2.

The allowable interstorey drift for each of the twelve storeys is 1/500, while the allowable overall building drift is 1/400. The column member section sizes are restricted to lie in the range of AISC standard sections W14X90 to W14X730, [2]. Similarly, the beam member section sizes are restricted to lie in the range of AISC standard sections W24X55 to W24X492. As such, the coefficients C and C' in Tables 1 to 4 apply to formulate the drift constraint eqs. (6b) explicitly in terms of cross-section areas A_i. Moreover, the cross-section areas corresponding to the limiting W-shape sections noted above define the bounds A_i^L and A_i^U in the sizing constraint eqs. (6c).

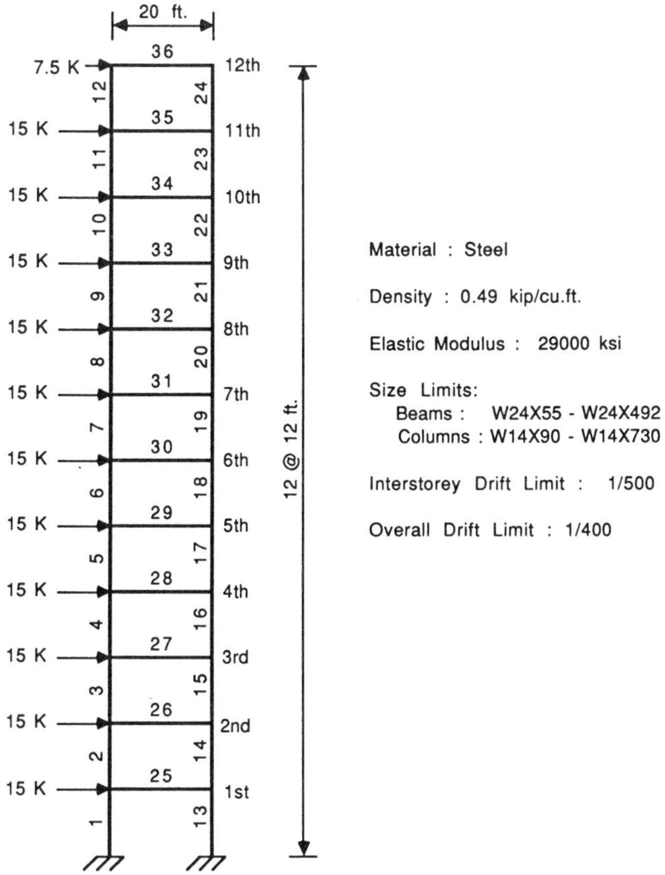

Figure 1. A 12-Storey and One-Bay Frame

The design optimization problem has thirteen drift constraints and thirty-six sizing constraints. Four different sets of initial sizes are selected for the frame members so as to investigate the effect of different initial sizes on the final optimal design. The first scheme A starts with the minimum limiting sizes for both the columns (W14X90) and the beams (W24X55). The second scheme B uses the maximum sizes for the columns (W14X730) and the beams (W24X492). The third scheme C uses the minimum size for the columns (W14X90) and the maximum size for the beams (W24X492). Lastly, the fourth scheme D starts with the maximum column size (W14X730) and the minimum beam size (W24X55).

The optimal design results found using the Optimality Criteria algorithm are illustrated in Figures 2 and 3. It is noted from Figure 2 that the various designs found using the four different schemes for initial member sizes all converge to the same final minimum structural weight of 152.2 kips; Figure 3 illustrates that the initial lateral deflection profiles for the four design schemes were all quite different from the final optimal profile to which they all converged. This indicates, for this example at least, that the final optimal design obtained by the Optimality Criteria method is not affected by the starting member sizes. Convergence was found to occur after the second design cycle for all cases except scheme D, which required one more design cycle to converge. Such quick convergence for schemes A, B and C can be explained by observing that their internal force distributions are quite insensitive to changes in member sizes. However, scheme D starts with very weak beams and strong columns and, thus, the frame essentially behaves as two vertical cantilevers with little frame action being developed; hence, the internal force distribution for the initial design is considerably different from that for the final optimal design. Nevertheless, reasonable sizes are obtained after the first design cycle for scheme D and convergence occurs rapidly after the third design cycle.

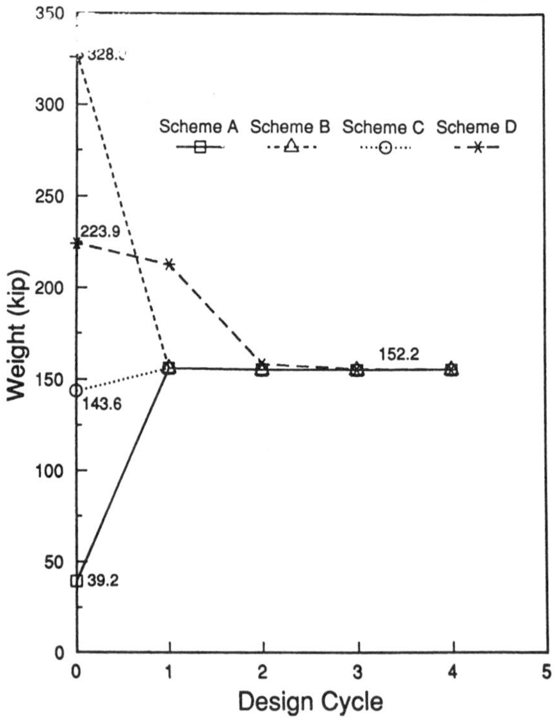

Figure 2. Design History of the 12-Storey One-Bay Planar Frame

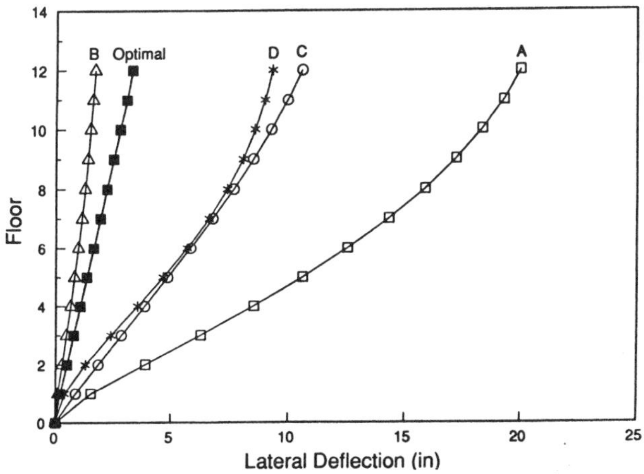

Figure 3. Lateral Deflection Profiles of the 12-Storey One-Bay Planar Frame

CONCLUDING REMARKS

As for the illustrative example, the OC iterative design technique will generally converge quite rapidly for building structures because the member force distributions for such structures are somewhat insensitive to changes in member sizes. The efficiency of the OC technique is mainly influenced by the number of constraints and is weakly dependent on the number of variables. From the foregoing, the OC technique provides an effective optimal design method for tall building structures involving many sizing variables and relatively few drift constraints.

ACKNOWLEDGEMENTS

This work was sponsored by the National Sciences and Engineering Research Council of Canada, and is based upon research conducted by the second author under the supervision of the first author for the degree of Doctor of Philosophy in Civil Engineering at the University of Waterloo.

REFERENCES

1. ETABS, Computer and Structures, Berkeley, California, USA.
2. Manual of Steel Construction - Load and Resistance Factor Design, American Institute of Steel Construction, Inc., Chicago, Illinois, First Edition, 1986.
3. Chun-Man Chan, "Design Optimization of Tall Steel Buildings", PhD Thesis, University of Waterloo, Canada (in progress 1991).
4. Berke, L. and Knot, N.S., "Use of Optimality Criteria Methods for Large Scale Systems", AGARD-LS-70, pp. 1-1 to 1-29, Sept. 1974.
5. Knot, N.S., Berke, L. and Venkayya, V.B., "Comparison of Optimality Criteria Algorithms for Minimum Weight Design of Structures", AIAA/ASME 19th Structures, Struct. Dynamics and Material Conf., Bethesda, Md., pp. 37-46, 1978.

Table 1. Relationships Between Cross-Section Area A and Shear Areas A_Y and A_Z for AISC-LRFD W14 Sections

AISC-LRFD W14 sections - Relationships of $1/A_Y$ vs $1/A$ and $1/A_Z$ vs $1/A$								
Sub-group	Sections	No. of Sect.	$\frac{1}{A_Y} = \frac{C_{AY}}{A} + C'_{AY}$		Max. % err.	$\frac{1}{A_Z} = \frac{C_{AZ}}{A} + C'_{AZ}$		Max. % err.
			C_{AY}	C'_{AY}		C_{AZ}	C'_{AZ}	
1	W14X22 - 26	2	1.4353	0.0953	0.00	2.5618	-0.0962	0.00
2	W14X30 - 38	3	1.6343	0.0851	0.98	2.0845	-0.0440	0.92
3	W14X43 - 53	3	3.0049	0.0011	0.26	1.5729	-0.0069	0.02
4	W14X61 - 82	4	3.8145	-0.0199	0.98	1.3837	0.0001	0.39
5	W14X90 - 132	5	4.7256	-0.1580	0.86	1.2934	-0.0005	0.34
6	W14X145 - 176	3	4.9442	-0.0163	0.01	1.2636	-0.0000	0.12
7	W14X193 - 257	4	4.6414	-0.0096	0.60	1.2761	-0.0003	0.16
8	W14X283 - 426	6	4.4197	-0.0068	0.24	1.2794	-0.0003	0.33
9	W14X455 - 730	6	4.1086	-0.0046	0.24	1.2632	-0.0002	0.14

Table 2. Relationships Between Cross-Section Area A and Moment of Inertias I_Z, I_Y and I_X for AISC-LRFD W14 Sections

AISC-LRFD W14 sections - Relationships of $1/I_Z$ vs $1/A$, $1/I_Y$ vs $1/A$ and $1/I_X$ vs $1/A$												
Sub-group	Sections	No. of Sect.	$\frac{1}{I_Z} = \frac{C_{IZ}}{A} + C'_{IZ}$		Max. % err.	$\frac{1}{I_Y} = \frac{C_{IY}}{A} + C'_{IY}$		Max. % err.	$\frac{1}{I_X} = \frac{C_{IX}}{A} + C'_{IX}$		Max. % err.
			C_{IZ}	C'_{IZ}		C_{IY}	C'_{IY}		C_{IX}	C'_{IX}	
1	W14X22 - 26	2	0.03924	-0.00102	0.00	1.27364	-0.05339	0.00	82.5200	-7.95304	0.00
2	W14X30 - 38	3	0.03533	-0.00057	0.63	0.57069	-0.01380	0.81	56.9611	-3.87809	3.50
3	W14X43 - 53	3	0.03197	-0.00020	0.08	0.31402	-0.00280	0.03	28.3980	-1.31202	1.40
4	W14X61 - 82	4	0.02988	-0.00011	0.31	0.18108	-0.00079	0.50	17.5920	-0.54004	3.65
5	W14X90 - 132	5	0.02903	-0.00010	0.42	0.07805	-0.00019	0.38	13.1454	-0.26187	5.69
6	W14X145 - 176	3	0.02852	-0.00008	0.11	0.06892	-0.00014	0.07	6.7058	-0.09228	1.58
7	W14X193 - 257	4	0.02808	-0.00008	0.10	0.06792	-0.00012	0.14	3.6161	-0.03552	2.68
8	W14X283 - 426	6	0.02714	-0.00007	0.31	0.06738	-0.00012	0.13	1.5664	-0.00966	5.11
9	W14X455 - 730	6	0.02444	-0.00004	0.62	0.06345	-0.00008	0.30	0.6118	-0.00220	7.09

Table 3. Relationships Between Cross-Section Area A and Shear Areas A_Y and A_Z for AISC-LRFD W24 Sections

AISC-LRFD W24 sections - Relationships of $1/A_Y$ vs $1/A$ and $1/A_Z$ vs $1/A$								
Sub-group	Sections	No. of Sect.	$\frac{1}{A_Y} = \frac{C_{AY}}{A} + C'_{AY}$		Max. % err.	$\frac{1}{A_Z} = \frac{C_{AZ}}{A} + C'_{AZ}$		Max. % err.
			C_{AY}	C'_{AY}		C_{AZ}	C'_{AZ}	
1	W24X55 - 62	2	1.3930	0.0214	0.00	3.0906	-0.0494	0.00
2	W24X68 - 84	3	1.4284	0.0308	0.43	2.5173	-0.0302	0.43
3	W24X94 - 103	2	1.8574	0.0128	0.00	2.0490	-0.0109	0.00
4	W24X104 - 131	3	2.3280	0.0072	0.15	1.7555	-0.0051	0.01
5	W24X146 - 192	4	2.4865	0.0045	0.42	1.6577	-0.0031	0.21
6	W24X207 - 306	5	2.8978	-0.0030	0.24	1.5181	-0.0006	0.15
7	W24X335 - 492	5	2.8104	-0.0024	0.58	1.5129	-0.0005	0.49

Table 4. Relationships Between Cross-Section Area A and Moment of Inertias I_Z, I_Y and I_X for AISC-LRFD W24 Sections

AISC-LRFD W24 sections - Relationships of $1/I_Z$ vs $1/A$, $1/I_Y$ vs $1/A$ and $1/I_X$ vs $1/A$											
Sub-group	Sections	No. of Sect.	$\frac{1}{I_Z} = \frac{C_{IZ}}{A} + C'_{IZ}$		Max. % err.	$\frac{1}{I_Y} = \frac{C_{IY}}{A} + C'_{IY}$		Max. % err.	$\frac{1}{I_X} = \frac{C_{IX}}{A} + C'_{IX}$		Max. % err.
			C_{IZ}	C'_{IZ}		C_{IY}	C'_{IY}		C_{IX}	C'_{IX}	
1	W24X55 - 62	2	0.01409	-0.00013	0.00	0.79294	-0.01458	0.00	38.7217	-1.54277	0.00
2	W24X68 - 84	3	0.01343	-0.00012	0.18	0.38935	-0.00521	0.40	28.1628	-0.87574	2.20
3	W24X94 - 103	2	0.01196	-0.00006	0.00	0.24887	0.00019	0.00	15.9686	-0.38637	0.00
4	W24X104 - 131	3	0.01101	-0.00004	0.07	0.13718	-0.00062	0.02	15.6784	-0.30399	1.96
5	W24X146 - 192	4	0.01067	-0.00003	0.16	0.12006	-.000025	0.58	7.5777	-0.10349	3.69
6	W24X207 - 306	5	0.00994	-0.00002	0.18	0.12000	-0.00025	0.15	3.1290	-0.02687	5.37
7	W24X335 - 492	5	0.00978	-0.00002	0.41	0.11472	-0.00020	0.66	1.2685	-0.00674	5.90

SECTION 7: OPTIMUM DESIGN OF COMPOSITE MATERIALS

Buckling Optimization of Fiber-Composite Laminate Shells with Large Deformation

H.-T. Hu

National Center for Composite Materials Research, University of Illinois, Urbana, Illinois 61801, U.S.A.

ABSTRACT

The buckling strength of fiber-composite laminate shells with a given material system is maximized with respect to fiber orientations. While a modified Riks nonlinear incremental algorithm is utilized to calculate the buckling load and to study the postbuckling behavior of the composite shells, a sequential linear programming method together with a simple move limit strategy is used to optimize the buckling strength of the shells. Results of this optimization study for simply supported composite cylindrical shells subjected to external hydrostatic compression and with different laminate layups, $[\pm\theta/90_2/0]_S$ and $[\theta/\phi/90_2/0]_S$, are presented.

INTRODUCTION

Applications of fiber composite materials (Fig. 1) to advanced shell structures such as aircraft fuselages, deep submersibles and surface ships have been increased rapidly in recent years. These composite laminate shells in service are commonly subjected to various kinds of external loading which may induce buckling. In many situations buckling is an undesirable phenomenon. Hence, structural instability becomes a major concern in safe and reliable design of the advanced composite shells. The buckling strength of fiber composite shells heavily depends on ply orientations, e.g. Sun and Hansen [1], Hu and Wang [2]. Therefore, the proper selection of appropriate fiber orientations for a given composite material system to achieve the maximum buckling strength of composite shells becomes a crucial problem.

Researches on the subject of structural optimization have been reported by many investigators, e.g. Schmit [3]. Among various optimization schemes, the sequential linear programming method is one of the most popular approaches in solving structural optimization problems and it has been successfully applied to many large scale structural problems, e.g. Vanderplaats [4], Haftka et al. [5], Zienkiewicz and Champbell [6]. Hence, in this paper the sequential linear programming method has been adopted and used together with a simple move-limit strategy to optimize the buckling strength of fiber-composite laminate cylindrical shells with respect to the fiber orientations.

This study uses a modified Riks nonlinear incremental algorithm implemented in the ABAQUS finite element program [7] to calculate the buckling loads and to study the postbuckling behavior of composite shells. For the purpose of comparison, optimization based on linearized buckling analysis is also carried out. In this paper, first the linearized buckling analysis, the nonlinear buckling analysis, the constitutive matrix formulation for composite shells, and the sequential linear programming method are briefly discussed. Then the results of the buckling optimization for simply supported composite cylindrical shells subjected to external hydrostatic compression and with different laminate layups, $[\pm\theta/90_2/0]_S$ and $[\theta/\phi/90_2/0]_S$, are presented. Finally, conclusions obtained from the study are given.

LINEARIZED BUCKLING ANALYSIS

In a finite-element modeling scheme for nonlinear problems, the load-displacement relationship for a structure can be expressed in an incremental form as follows:

$$[K]_t \, d\{U\} = d\{P\} \tag{1}$$

where $[K]_t$ is a tangent stiffness matrix, $d\{U\}$ an incremental nodal displacement vector and $d\{P\}$ an incremental nodal force vector.

If it is assumed that the linear theory of small deformation before buckling holds, the linearized buckling formulation, e.g. Cook et al. [8], then leads to a tangent stiffness matrix with the following expression:

$$[K]_t = [K]_L + [K]_\sigma \tag{2}$$

where $[K]_L$ is a linear stiffness matrix and $[K]_\sigma$ a geometric stiffness matrix dependent upon stresses.

The bifurcation solution for the linearized buckling problem then may be determined from the following eigenvalue equation:

$$([K]_L + \lambda[K]_{\sigma_0}) \, \{\psi\} = \{0\} \tag{3}$$

where λ is an eigenvalue and $\{\psi\}$ an eigenvector. The critical buckling load P_{cr} can be found from $P_{cr} = \lambda P_0$ where P_0 is the nominal load which corresponds to the stress state σ_0.

NONLINEAR BUCKLING ANALYSIS

In the finite element program ABAQUS, the nonlinear response of structures is modeled through the incremental updated Lagrangian formulation, e.g. Bathe [9]. In order to model the potential decrease in loads and displacements as the solution evolves, a modified Riks nonlinear incremental algorithm [7] in

ABAQUS is used to construct the equilibrium solution path. In this algorithm, the nonlinear procedure is based on a motion of a given distance along the tangent line (defined by the tangent stiffness matrix) to the current solution point. Then search for an equilibrium solution in the plane, that passes through the current solution point and that is orthogonal to the same tangent line, can be carried out using an iterative algorithm.

In order to model bifurcation from the prebuckling path to the postbuckling path, a geometric imperfection of the composite shell can be introduced by superimposing a small fraction of the lowest eigenmode, which is determined by a linearized buckling analysis, to the original nodal coordinates of the shell as follows:

$$\{I\} = \{O\} + \varepsilon t\{\psi\} \tag{4}$$

where $\{I\}$ is the resulting imperfect nodal coordinates of the shell, $\{O\}$ is the original nodal coordinates of the shell, ε is a scaling coefficient, t is the thickness of the shell, and $\{\psi\}$ is the normalized lowest eigenmode. In this study, ε is taken to be 0.001 for all the nonlinear buckling analyses.

CONSTITUTIVE MATRIX FOR FIBER-COMPOSITE LAMINAE

The elements used in the numerical analyses are eight-node isoparametric shell elements with six degrees of freedom per node (three displacements and three rotations). The shell formulation is based on Mindlin-type displacement field assumptions which allows transverse shear deformation [7].

During a finite element analysis, the constitutive matrices of composite materials at element integration points must be calculated before the stiffness matrices are assembled from element level to structural level. For fiber-composite laminate materials, each lamina can be considered as an orthotropic layer in a plane stress condition. The stress-strain relations for an orthotropic lamina in the material coordinates (Fig. 1) at an integration point can be written as

$$\{\sigma'\} = [Q_1'] \{\varepsilon'\} \tag{5}$$

$$\{\tau_t'\} = [Q_2'] \{\gamma_t'\} \tag{6}$$

$$[Q_1'] = \begin{bmatrix} \dfrac{E_{11}}{1-\nu_{12}\nu_{21}} & \dfrac{\nu_{12}E_{22}}{1-\nu_{12}\nu_{21}} & 0 \\ \dfrac{\nu_{21}E_{11}}{1-\nu_{12}\nu_{21}} & \dfrac{E_{22}}{1-\nu_{12}\nu_{21}} & 0 \\ 0 & 0 & G_{12} \end{bmatrix}, \quad [Q_2'] = \begin{bmatrix} \alpha_1 G_{13} & 0 \\ 0 & \alpha_2 G_{23} \end{bmatrix} \tag{7}$$

where $\{\sigma'\} = \{\sigma_1, \sigma_2, \tau_{12}\}^T$, $\{\tau_t'\} = \{\tau_{13}, \tau_{23}\}^T$, $\{\varepsilon'\} = \{\varepsilon_1, \varepsilon_2, \gamma_{12}\}^T$, $\{\gamma_t'\} = \{\gamma_{13}, \gamma_{23}\}^T$. The α_1 and α_2 are shear correction factors and are taken to be 0.83 in this study. The constitutive equations for the lamina in the element coordinates become

$$\{\sigma\} = [Q_1]\{\varepsilon\}, \qquad [Q_1] = [T_1]^T [Q_1'] [T_1] \tag{8}$$

$$\{\tau_t\} = [Q_2]\{\gamma_t\}, \qquad [Q_2] = [T_2]^T [Q_2'] [T_2] \tag{9}$$

$$[T_1] = \begin{bmatrix} \cos^2\phi & \sin^2\phi & \sin\phi\cos\phi \\ \sin^2\phi & \cos^2\phi & -\sin\phi\cos\phi \\ -2\sin\phi\cos\phi & 2\sin\phi\cos\phi & \cos^2\phi-\sin^2\phi \end{bmatrix} \tag{10}$$

$$[T_2] = \begin{bmatrix} \cos\phi & \sin\phi \\ -\sin\phi & \cos\phi \end{bmatrix} \tag{11}$$

where $\{\sigma\} = \{\sigma_x, \sigma_y, \tau_{xy}\}^T$, $\{\tau_t\} = \{\tau_{xz}, \tau_{yz}\}^T$, $\{\varepsilon\} = \{\varepsilon_x, \varepsilon_y, \gamma_{xy}\}^T$, $\{\gamma_t\} = \{\gamma_{xz}, \gamma_{yz}\}^T$, and ϕ is measured counterclockwise from the element local x-axis to the material 1-axis. Assume $\{\varepsilon_o\} = \{\varepsilon_{xo}, \varepsilon_{yo}, \gamma_{xyo}\}^T$ are the in-plane strains at the mid-surface of the section and $\{\kappa\} = \{\kappa_x, \kappa_y, \kappa_{xy}\}^T$ are the curvatures. The in-plane strains at a distance, z, from the mid-surface become:

$$\{\varepsilon\} = \{\varepsilon_o\} + z\{\kappa\} \tag{12}$$

If h is the total thickness of the section, the stress resultants, $\{N\} = \{N_x, N_y, N_{xy}\}^T$, $\{M\} = \{M_x, M_y, M_{xy}\}^T$ and $\{V\} = \{V_x, V_y\}^T$, can be defined as

$$\{N\} = \int_{-h/2}^{h/2} \{\sigma\}dz = \int_{-h/2}^{h/2} [Q_1](\{\varepsilon_o\}+z\{\kappa\})dz \tag{13.a}$$

$$\{M\} = \int_{-h/2}^{h/2} z\{\sigma\}dz = \int_{-h/2}^{h/2} z[Q_1](\{\varepsilon_o\}+z\{\kappa\})dz \tag{13.b}$$

$$\{V\} = \int_{-h/2}^{h/2} \{\tau_t\}dz = \int_{-h/2}^{h/2} [Q_2]\{\gamma_t\}dz \tag{13.c}$$

If there are n layers in the layup, the constitutive matrix for composite materials at an element integration point can be written as a summation of integrals over the n laminae in the following form:

$$\begin{Bmatrix} \{N\} \\ \{M\} \\ \{V\} \end{Bmatrix} = \sum_{j=1}^{n} \begin{bmatrix} (z_{jt}-z_{jb})[Q_1] & \frac{1}{2}(z_{jt}^2-z_{jb}^2)[Q_1] & [0] \\ \frac{1}{2}(z_{jt}^2-z_{jb}^2)[Q_1] & \frac{1}{3}(z_{jt}^3-z_{jb}^3)[Q_1] & [0] \\ [0]^T & [0]^T & (z_{jt}-z_{jb})[Q_2] \end{bmatrix} \begin{Bmatrix} \{\varepsilon_o\} \\ \{\kappa\} \\ \{\gamma_t\} \end{Bmatrix} \tag{14}$$

where z_{jt} and z_{jb} are the distance from the mid-surface of the section to the top and the bottom of the j-th layer respectively. The [O] is a 3 by 2 matrix with all the coefficients equal to zero.

SEQUENTIAL LINEAR PROGRAMMING

A general optimization problem may be defined as the following:

Minimize: $f(\underline{x})$ (15.a)

Subjected to: $g_i(\underline{x}) \leq 0, \quad i = 1, ..., r$ (15.b)

$h_j(\underline{x}) = 0, \quad j = r+1, ..., m$ (15.c)

$p_k \leq x_k \leq q_k, \quad k = 1, ..., n$ (15.d)

where $f(\underline{x})$ is an objective function, $g_i(\underline{x})$ are inequality constraints, $h_j(\underline{x})$ are equality constraints, and $\underline{x} = \{x_1, x_2, ..., x_n\}^T$ is a vector of design variables. If a particular optimization problem requires maximization, we simply minimize $-f(\underline{x})$.

The concept of a sequential linear programming is that given a feasible solution $\underline{x}_o = \{x_{o1}, x_{o2}, ..., x_{on}\}^T$ for an optimization problem, a linear programming problem may be established by expanding the nonlinear functions about $\underline{x}_o$ in a Taylor series, and ignoring terms of order higher than the linear ones. With this approximation, the optimization problem, Eqs. (15.a-15.d), becomes:

Minimize: $f(\underline{x}) \approx f(\underline{x}_o) + \nabla f(\underline{x}_o)^T \, \delta\underline{x}$ (16.a)

Subjected to: $g_i(\underline{x}) \approx g_i(\underline{x}_o) + \nabla g_i(\underline{x}_o)^T \, \delta\underline{x} \leq 0$ (16.b)

$h_j(\underline{x}) \approx h_j(\underline{x}_o) + \nabla h_j(\underline{x}_o)^T \, \delta\underline{x} = 0$ (16.c)

$p_k \leq x_k \leq q_k$ (16.d)

where $\delta\underline{x} = \{x_1 - x_{o1}, x_2 - x_{o2}, ..., x_n - x_{on}\}^T$. A solution for Eqs. (16.a-16.d) may be easily obtained by the simplex method, e.g. Kolman and Beck [10]. After obtaining an initial approximate solution for Eqs. (16.a-16.d), say $\underline{x}_1$, we can linearize the original problem, Eqs. (15.a-15.d), at $\underline{x}_1$ and solve the new linear programming problem. The process is repeated until a convergent solution is obtained.

Although the procedure for a sequential linear programming is simple, difficulties may arise during the iterations. First, the optimum solution for the approximate linear problem may violate the constraint conditions of the original optimization problem. Second, in a nonlinear problem, the true optimum solution may appear between two constraint intersections. A straightforward successive linearization in such a case may lead to an oscillation of the solution between the widely separated values. Difficulties in dealing with such problems may be avoided by imposing a "move limit" on the linear approximation, e.g. Vanderplaats [4], Haftka et al. [5], Zienkiewicz and Champbell [6]. The concept of a move limit is that a set of box-like admissible constraints are placed in the range of $\delta\underline{x}$. In general, the choice of a suitable move limit depends on experience and also on the results of previous steps. Once a proper move limit is chosen at the beginning of the sequential linear programming procedure, this move limit should gradually approach to zero as the iterative process continues, e.g. Vanderplaats [4], Zienkiewicz and Champbell [6], Esping [11].

The algorithm of a sequential linear programming with selected move limits may be summarized as follows: (1) Linearize the nonlinear objective function and associated constraints with respect to an initial guess $\underline{x}_0$. (2) Impose move limits in the form of $-\underline{S} \leq (\underline{x} - \underline{x}_0) \leq \underline{R}$, where $\underline{S}$ and $\underline{R}$ are properly chosen positive constraints. (3) Solve the approximate linear programming problem to obtain an initial optimum solution $\underline{x}_1$. (4) Repeat the process by redefining $\underline{x}_1$ with $\underline{x}_0$ until either the subsequent solutions do not change significantly (i.e., true convergence) or the move limit approaches to zero (i.e., forced convergence).

NUMERICAL ANALYSIS

Buckling Optimization of Composite Shell with One Design Variable

In this section, a simply supported fiber-composite laminate cylindrical shell (Fig. 2) with laminate layup $[\pm\theta/90_2/0]_s$ under external hydrostatic compression is investigated. The objective of this study is to determine the optimal fiber angle θ to maximize the buckling load p_{cr} of the shell and to compare the result of the optimization using nonlinear buckling analysis with that using linearized buckling analysis.

Based on the sequential linear programming method, in each iteration the current, linearized optimization problem becomes:

Maximize: $$p_{cr}(\theta) \approx p_{cr}(\theta_0) + (\theta-\theta_0)\left.\frac{\partial p_{cr}}{\partial \theta}\right|_{\theta=\theta_0} \qquad (17.a)$$

Subjected to: $$0^o \leq \theta \leq 90^o \qquad (17.b)$$

$$-r \times q \times 0.5^s \leq (\theta-\theta_0) \leq r \times q \times 0.5^s \qquad (17.c)$$

where θ_0 is a solution in the current iteration. The r and q are the size and the reduction rate of the move limit. In this study, the values of r and q are selected to be 10° and $0.9^{(N-1)}$, where N is a current iteration number. To control the oscillation of the solution, a parameter 0.5^s is introduced in the move limit, where s is the number of oscillation of the derivative $\partial p_{cr}/\partial\theta$ that has taken place before the current iteration. The value of s increases by 1 if the sign of $\partial p_{cr}/\partial\theta$ changes. Whenever oscillation of the solution occurs, the range of the move limit is reduced to half of its current value, which is similar to a bisection method, e.g. Maror [12]. This expedites the solution convergent rate very rapidly.

The derivative $\partial p_{cr}/\partial\theta$ in Eq. (17.a) may be approximated by using a forward finite-difference method as follows:

$$\frac{\partial p_{cr}}{\partial \theta} \approx \frac{[p_{cr}(\theta_o+\Delta\theta)-p_{cr}(\theta_o)]}{\Delta\theta} \tag{18}$$

In order to determine the value of $\partial p_{cr}/\partial\theta$ in Eq. (18), two buckling analyses are needed to compute $p_{cr}(\theta_o)$ and $p_{cr}(\theta_o+\Delta\theta)$ in each iteration. In this study, the value of $\Delta\theta$ is selected to be 1° in most iterations.

Important numerical results obtained in optimization study are given in Fig. 3, which shows the fiber orientation θ and the associated critical buckling pressure p_{cr} determined in each iteration for the shell. The initial values of θ are selected to be 90° for linearized buckling analysis as well as nonlinear buckling analysis. Both solutions converged within 12 iterations. Though, the optimal critical buckling pressure, 24.7 ksi, computed by using linearized buckling analysis is lower than that, 25.5 ksi, computed by using nonlinear buckling analysis, the optimal value of θ converges to 60.8^o for both analyses.

For the optimization study using the nonlinear buckling analysis, in each iteration two nonlinear buckling analyses to calculate the derivative information of Eq. (18) and two linearized buckling analyses to generate the initial imperfections for the nonlinear analyses are required. Since the trends in finding the optimal fiber orientation are the same using both buckling analyses, it is suggested that the calculation of the derivative using the nonlinear buckling analysis may be substituted by that using the linearized buckling analysis. The result is that in each iteration only one nonlinear buckling analysis to evaluate the critical buckling pressure and two linearized buckling analyses to compute the derivative are needed. The elimination of one massive nonlinear buckling analysis will significantly reduce the computer time for the entire optimization calculation.

Figure 4 shows the load-end displacement curves for the composite shell associated with the first iteration and the final iteration (optimal solution), which are computed by using the nonlinear buckling analysis. It can be seen that under the optimal condition, not only the critical buckling pressure of the shell is increased but also the post buckling strength of the shell is greatly improved.

Buckling Optimization of Composite Shell with Two Design Variables

In this section, the composite laminate shell with the same geometry, end conditions and loading conditions as that in the previous section but with laminate layup $[\theta/\phi/90_2/0]_s$ is investigated. Here, the constraint on the fiber angle $\pm\theta$ set in the previous section has been released. The objective of this study is then to find the optimal fiber orientations θ and ϕ, and to examine how the change of the fiber angle constraint will influence the optimal critical buckling pressure p_{cr}.

Based on the sequential linear programming method, in each iteration the current, linearized optimization problem becomes:

Maximize: $$p_{cr}(\theta,\phi) \approx p_{cr}(\theta_o,\phi_o) + (\theta-\theta_o)\frac{\partial p_{cr}}{\partial \theta}\bigg|_{\theta=\theta_o,\phi=\phi_o} + (\phi-\phi_o)\frac{\partial p_{cr}}{\partial \phi}\bigg|_{\theta=\theta_o,\phi=\phi_o} \qquad (19.a)$$

Subjected to:
$$0^o \le \theta \le 90^o \qquad (19.b)$$
$$-90^o \le \phi \le 0^o \qquad (19.c)$$
$$-r_1 \times q_1 \times 0.5^{s_1} \le (\theta-\theta_o) \le r_1 \times q_1 \times 0.5^{s_1} \qquad (19.d)$$
$$-r_2 \times q_2 \times 0.5^{s_2} \le (\phi-\phi_o) \le r_2 \times q_2 \times 0.5^{s_2} \qquad (19.e)$$

where θ_o and ϕ_o are the solutions in the current iteration. The values of r_1 and r_2 (the sizes of move limits) are selected to be 10°. The values of q_1 and q_2 (the reduction rates of move limits) are selected to be $0.9^{(N-1)}$, where N is a current iteration number. The s_1 and s_2 are the number of oscillation of the derivatives $\partial p_{cr}/\partial\theta$ and $\partial p_{cr}/\partial\phi$. The derivative terms in Eq. (19.a) may be approximated with the following finite-difference forms:

$$\frac{\partial p_{cr}}{\partial \theta} \approx \frac{[p_{cr}(\theta_o+\Delta\theta,\phi_o)-p_{cr}(\theta_o,\phi_o)]}{\Delta\theta} \qquad (20.a)$$

$$\frac{\partial p_{cr}}{\partial \phi} \approx \frac{[p_{cr}(\theta_o,\phi_o+\Delta\phi)-p_{cr}(\theta_o,\phi_o)]}{\Delta\phi} \qquad (20.b)$$

In this optimization study, three linearized buckling analyses are used to calculate the derivative information in Eqs. (20.a) and (20.b), and one nonlinear buckling analysis is used to evaluate p_{cr} in Eq. (19.a). The values of $\Delta\theta$ and $\Delta\phi$ are selected to be 1° in most iterations.

Important numerical results obtained in optimization study are given in Fig. 5, which shows the fiber orientations θ and ϕ, and the associated critical buckling pressure p_{cr} determined in each iteration for the shell. The initial values of θ and ϕ are selected to be 90° and -90°. After twelve iterations, the optimal values of θ and ϕ converge to 63.4^o and -59.0^o respectively and the optimal critical buckling pressure converges to 25.6 ksi. Figure 6 shows the load-end displacement curves for the composite shell associated with the first iteration and the final iteration (optimal solution). Again, it can be seen that under the optimal condition, not only the critical buckling pressure of the shell is increased but also the post buckling strength of the shell is greatly improved.

Comparing the optimal critical buckling pressures p_{cr}, 25.6 ksi, obtained from this optimization analysis with that, 25.5 ksi, obtained from previous optimization analysis, one can observe that the increase in the optimal critical buckling pressure is very insignificant. Figure 7 shows the load-end displacement curves under optimal fiber angle conditions for the composite shell with $[\pm\theta/90_2/0]_s$ laminate layup and $[\theta/\phi/90_2/0]_s$ laminate layup. It can be seen that the postbuckling strengths are also almost the same for the shell with these two different laminate layups. Therefore, it can be concluded that the optimal buckling behavior of the composite shell with $[\pm\theta/90_2/0]_s$ laminate layup is about the same as that of the composite shell with $[\theta/\phi/90_2/0]_s$ laminate layup. Hence, the optimization of the composite shell using two design variables may be undesirable since it costs more computation time.

CONCLUSIONS

From the optimization results obtained in this study, the following conclusions can be drawn:

1. For the optimization of a simply supported $[\pm\theta/90_2/0]_s$ composite shell under external hydrostatic compression using the sequential linear programming formulation, the trends in finding the optimal fiber orientation are the same for using linearized buckling analyses and for using nonlinear buckling analysis. Hence, the calculation of the derivative information using the nonlinear buckling analysis may be substituted by that using the linearized buckling analysis.

2. The optimal buckling behavior of the simply supported $[\pm\theta/90_2/0]_s$ composite shell under external hydrostatic compression is about the same as that of the composite shell with $[\theta/\phi/90_2/0]_s$ laminate layup. Therefore, the optimization of the composite shell using two design variables may be undesirable since it costs more computation time.

ACKNOWLEDGEMENTS

This research work was financially supported by the Office of Naval Research under grant number N00014-86-K-0799.

REFERENCES

1. Sun, G. and Hansen, J. S. Optimal Design of Laminated-Composite Circular-Cylindrical Shells Subjected to Combined Loads, Journal of Applied Mechanics, ASME, Vol. 55, pp. 136-142, 1988.
2. Hu, H.-T. and Wang, S. S. Optimization for Buckling Resistance of Fiber-Composite Laminate Shells with and without Cutouts, pp. 1300-1312, Proceedings of the 31st AIAA/ASME/ASCE/AHS/ASC Structures, Structural Dynamics and Materials Conference, Long Beach, California, 1990.
3. Schmit, L. A. Structural Synthesis - Its Genesis and Development, AIAA Journal, Vol. 19, pp. 1249-1263, 1981.

4. Vanderplaats, G. N. Numerical Optimization Techniques for Engineering Design with Applications, McGraw-Hill, New York, 1984.
5. Haftka, R. T., Gurdal, Z. and Kamat, M. P. Elements of Structural Optimization, Second Revised Edition, Kluwer Academic Publishers, Dordrecht, The Netherlands, 1990.
6. Zienkiewicz, O. C. and Champbell, J. S. Shape Optimization and Sequential Linear Programming, Chapter 6, Optimum Structural Design (Ed. Gallagher, R. H. and Zienkiewicz, O. C.), pp. 109-126, Wiley, New York, 1973.
7. Hibbitt, Karlsson & Sorensen, Inc. ABAQUS Theory Manual and User Manual, Version 4-8, Providence, Rhode Island, 1990.
8. Cook, R. D., Malkus, D. S. and Plesha, M. E. Concepts and Applications of Finite Element Analysis, Third Edition, Chapter 14, John Wiley & Sons, Inc., New York, 1989.
9 Bathe, K.-J. Finite Element Procedures in Engineering Analysis, Chapter 6, Prentice-Hall, New Jersey, 1982.
10. Kolman, B. and Beck, R. E. Elementary Linear Programming with Applications, Chapter 2, Academic Press, Orlando, 1980.
11. Esping, B. J. D. Minimum Weight Design of Membrane Structures, Computers and Structures, Vol. 19, pp. 707-716, 1984.
12. Maror, M. J. Numerical Analysis, Second Edition, Chapter 2, Macmillan Publishing Company, New York, 1987.

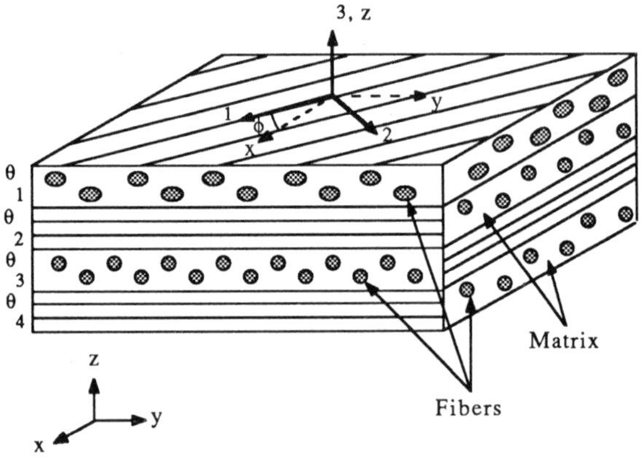

Figure 1 Material and element coordinate systems for fiber composite laminate

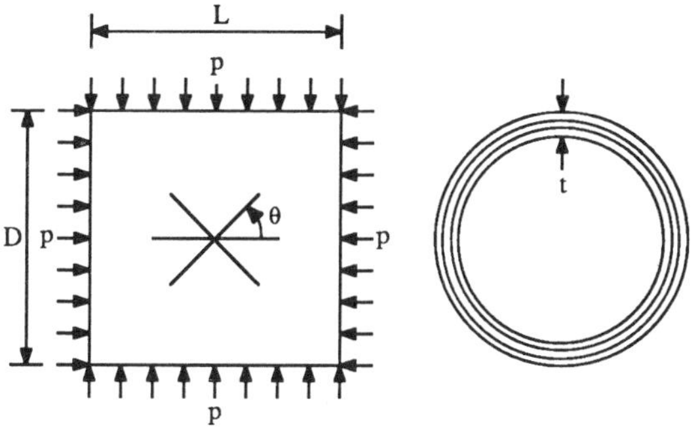

Laminate layup:
$[\pm\theta/90_2/0]_s$

Composite shell geometry:
L = 8 in.
D = 8 in.
t = 0.5 in.

Ply constitutive properties:
E_{11} = 20 Msi
E_{22} = 2.1 Msi
$G_{12} = G_{13}$ = 0.85 Msi
G_{23} = 0.51 Msi
ν_{12} = 0.21

Figure 2 Cylindrical composite laminate shell under external hydrostatic compression

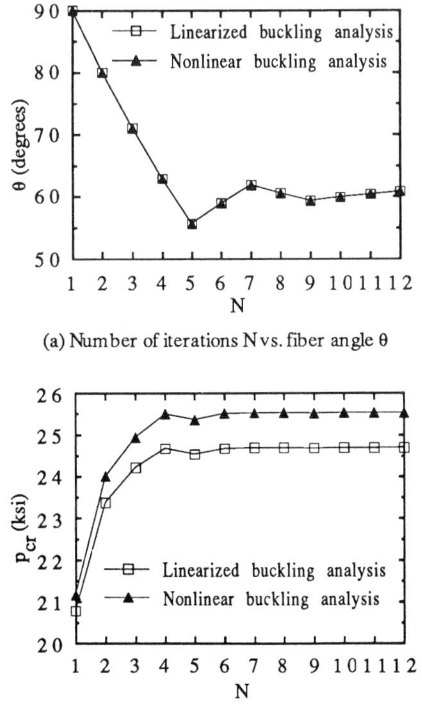

(a) Number of iterations N vs. fiber angle θ

(b) Number of iterations N vs. critical pressure p_{cr}

Figure 3 Buckling optimization of simply supported $[\pm\theta/90_2/0]_s$ composite shell under hydrostatic compression, (a) number of iterations N vs. fiber angle θ, and (b) number of iterations N vs. critical pressure p_{cr}

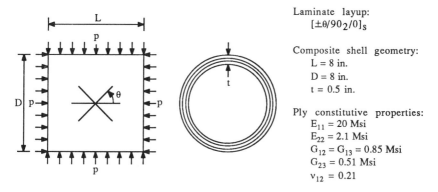

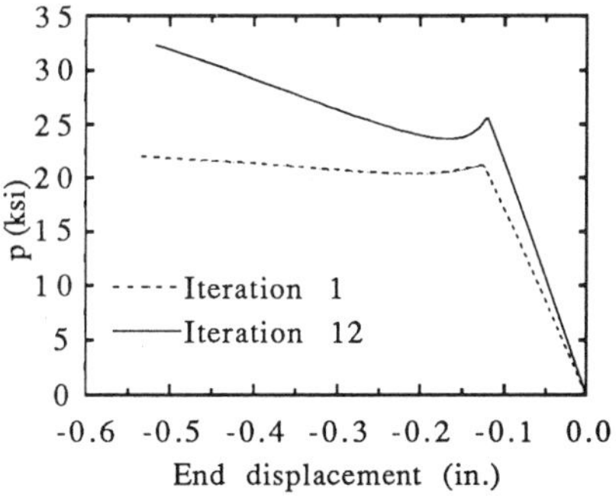

Figure 4 Load-end displacement curves for simply supported $[\pm\theta/90_2/0]_s$ composite shell under hydrostatic compression

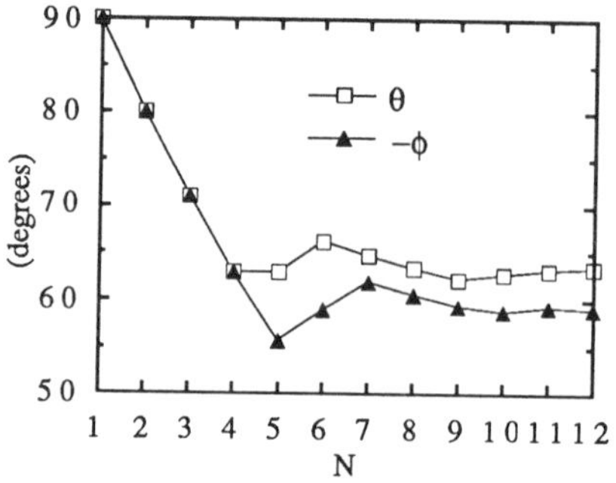

(a) Number of iterations N vs. fiber angles θ and φ

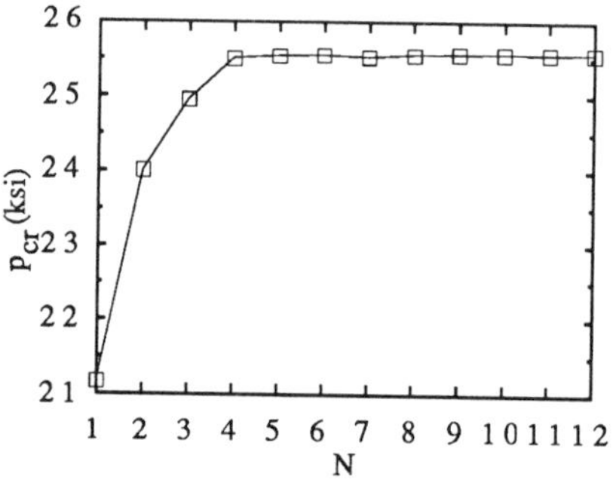

(b) Number of iterations N vs. critical pressure p_{cr}

Figure 5 Buckling optimization of simply supported $[\theta/\phi/90_2/0]_s$ composite shell under hydrostatic compression, (a) number of iterations N vs. fiber angles θ and φ, and (b) number of iterations N vs. critical pressure p_{cr}

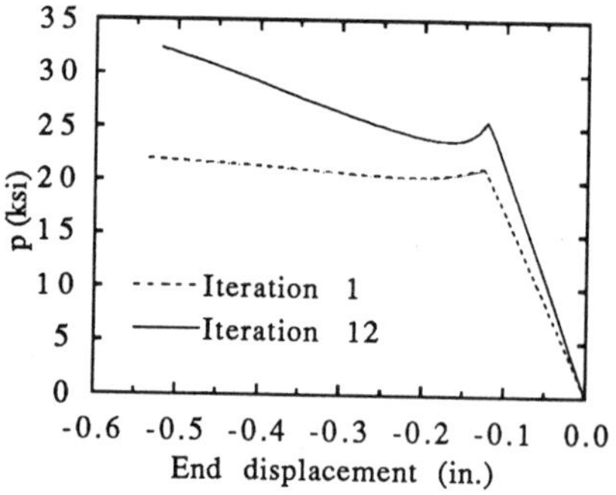

Figure 6 Load-end displacement curves for simply supported $[\theta/\phi/90_2/0]_s$ composite shell under hydrostatic compression

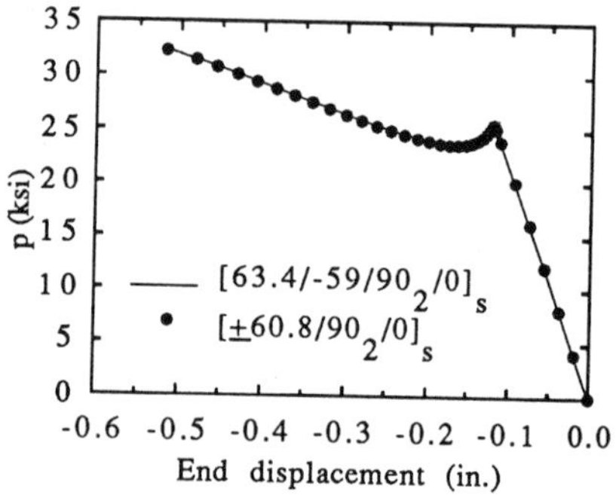

Figure 7 Load-end displacement curves for simply supported composite shell with $[63.4/-59/90_2/0]_s$ and $[\pm 60.8/90_2/0]_s$ laminate layups

Optimum Design of a Linearly Tapered Laminated Cantilever Beam

P.I. Rodriguez (*), M.V. Bower (**)

(*) National Aeronautics and Space Administration, Marshall Space Flight Center, Huntsville, Alabama, U.S.A.

(**) University of Alabama in Huntsville, Huntsville, Alabama, U.S.A.

ABSTRACT

The design of an optimum symmetric laminated orthotropic cantilever beam is presented. The width and thickness are linear functions of the length of the beam. Classical laminated plate theory is used. The objective function is the beam weight which is minimized subject to constraints of stress and deflection. Numerical results are given for various examples. Comparison with an isotropic beam is presented.

INTRODUCTION

In a recent paper by one of the authors [1], a solution to the minimum weight design of an isotropic cantilever beam of variable cross section was presented. As an extension of this work, the procedure is applied to symmetric laminates of orthotropic materials. The solution to the structural problem is obtained using classical laminated plate theory. It is necessary to note that some important assumptions are made in the presented formulation. One assumption is that the transverse shear strains between laminae are negligible. This assumption is adequate if the laminate is thin. It has been shown [2,3], however, that if the laminate is thick (length-to-thickness ratio less than 10), deflections and internal stress distributions become very innacurate. Another assumption is that the laminae orientation is such that the interlaminar normal stresses are minimized at the free edges. These assumptions are included to minimize concerns with free edge delaminations. Both assumptions will be addressed in the optimization solution for the weight minimization problem.

In the present formulation, the coupling stiffness matrix is neglected since only symmetric laminates are analyzed. The bending and extensional matrices, however, are both expressed as functions of the length of the beam (x-coordinate). The resultant expressions for each of the terms of these matrices are polynomials with coefficients that depend on the thickness of the tapered laminae, length of the beam and location of the tapered laminae along the depth of the beam (z-coordinate). In order to calculate the slope and deflections of the designed beams the bending stiffness matrix must be inverted. Since each of the terms in this matrix is a cubic polynomial, the inversion must be done in closed form if one is to obtain an exact expression for the inverted stiffness term needed $\left(D_{11}\right)$. For this paper the inversion is done numerically and then fit with an 8th degree polynomial. Once the curve fitting is performed, the integration of the Euler-Bernoulli equation is done in closed form to obtain the desired slopes and deflections.

The optimization technique used is the Modified Method of Feasible Directions. This method is explained in detail by Vanderplaats [4] and is commercially available in the DOT optimization software. The code is accessed through a CALL statement in a FORTRAN 77 header program containing the structural design problem.

LAMINATED BEAM EQUATIONS

Whitney [5] explains how the bending analysis of a laminated beam can be performed using the classical beam theory by simply replacing the bending stiffness EI with an equivalent bending stiffness $E_x^k I$. Where E_x^k is the effective bending modulus of the beam and k is the kth ply of the laminate. In the present formulation, the assembled extensional and stiffness matrices of the laminate are of the form

$$K_{ij} = \begin{bmatrix} K_{11} & K_{12} & K_{13} & K_{14} & 0 & 0 \\ K_{21} & K_{22} & K_{23} & K_{24} & 0 & 0 \\ K_{31} & K_{32} & K_{33} & K_{34} & 0 & 0 \\ K_{41} & K_{42} & K_{43} & K_{44} & 0 & 0 \\ 0 & 0 & 0 & 0 & K_{55} & 0 \\ 0 & 0 & 0 & 0 & 0 & K_{66} \end{bmatrix} \tag{1}$$

where $K_{ij} = A_{ij}$ for extensional stiffnesses and $K_{ij} = D_{ij}$ for bending stiffnesses.

The individual terms of the extensional and stiffness matrices are expressed as prescribed by the classical laminated plate theory. The difference in the present formulation is that only the innermost laminae are not tapered and thus have extensional and stiffness parameters that do not depend on x. The outer plies are tapered and the stiffness parameters must be expressed as variables of x. In this manner we have,

$$A_{ij} = \sum_{k=k_m}^{k_n} \bar{Q}_{ij}^{(k)} \left(z_k - z_{k-1}\right) + \sum_{k=k_n}^{n} \bar{Q}_{ij}^{(k)} \left\{z_k - \left[z_k - t_k\left(1 - \frac{x}{L}\right)\right]\right\}$$
$$+ \sum_{k=k_m}^{m} \bar{Q}_{ij}^{(k)} \left\{\left[z_{k-1} + t_k\left(1 - \frac{x}{L}\right)\right] - z_{k-1}\right\} \tag{3}$$

and

$$D_{ij} = \sum_{k=k_m}^{k_n} \frac{\bar{Q}_{ij}^{(k)}}{3} \left(z_k^3 - z_{k-1}^3\right) + \sum_{k=k_n}^{n} \frac{\bar{Q}_{ij}^{(k)}}{3} \left\{z_k^3 - \left[z_k - t_k\left(1 - \frac{x}{L}\right)\right]^3\right\}$$
$$+ \sum_{k=k_m}^{m} \frac{\bar{Q}_{ij}^{(k)}}{3} \left\{\left[z_{k-1} + t_k\left(1 - \frac{x}{L}\right)\right]^3 - z_{k-1}^3\right\} \tag{4}$$

Where $\bar{Q}_{ij}^{k}$ are the reduced stiffnesses for the kth lamina. Laminae k_m through m are the tapered laminae at the bottom of the laminate. Laminae k_n through n are the tapered laminae at the top of the laminate. L represents the length of the taper and t_k is the thickness of the individual laminae. Equation (4) has been compared to an equation derived by Ashton [6] for the analysis of plane anisotropic rectangular plates with non-uniform thickness and width dimensions. For the specific case of a plate with constant width and linearly varying thickness, Whitney [5] expresses Ashton's equation in the following form,

$$D_{ij} = \left(D_{ij}\right)_{x=b}\left[1 + 3\alpha\left(\frac{x}{b}\right) + 3\alpha^2\left(\frac{x}{b}\right)^2 + \alpha^3\left(\frac{x}{b}\right)^3\right] \tag{5}$$

where b is the length of the beam, α is the ratio of increase in thickness, $(t_{base}/t_{tip} - 1)$, and $(D_{ij})_{x=0}$ is the bending stiffness at the tip (end) of the beam. In Equation (5) it is assumed that the plate is fabricated of an orthotropic material with the principal material axes at 45 degrees to the plate edges. For the case under consideration a four ply 45/45/s was examined. Material properties are found in Table 1. Figure 1. shows a plot of both equations (4) and (5).

Since the derivation of the current solution is based on symmetrically laminated plates, the extensional stiffness matrix A_{ij} will not play a role in the solution of the beam bending problem. This would not be the case if the solution were to include the effects of transverse shear deformations [7]. This is a topic that will be addressed in a later paper.

Taking into consideration the assumptions just mentioned, the constitutive relationships for the bending of a symmetric laminated beam can be expressed in its inverted form as

$$\begin{Bmatrix} \gamma_x \\ \gamma_y \\ \gamma_z \\ \gamma_{xy} \\ \gamma_{yz} \\ \gamma_{zx} \end{Bmatrix} = \begin{bmatrix} D_{11} & D_{12} & D_{13} & D_{14} & 0 & 0 \\ D_{21} & D_{22} & D_{23} & D_{24} & 0 & 0 \\ D_{31} & D_{32} & D_{33} & D_{34} & 0 & 0 \\ D_{41} & D_{42} & D_{43} & D_{44} & 0 & 0 \\ 0 & 0 & 0 & 0 & D_{55} & 0 \\ 0 & 0 & 0 & 0 & 0 & D_{66} \end{bmatrix}^{-1} \begin{Bmatrix} M_x \\ M_y \\ M_z \\ M_{xy} \\ M_{yz} \\ M_{zx} \end{Bmatrix} \tag{6}$$

For the application of the beam theory, the following assumptions are made,

$$M_y = M_z = M_{xy} = M_{yz} = M_{zx} = 0 \tag{7}$$

From Equations (6) and (7), the typical Euler-Bernoulli equation for beam bending can be obtained

$$\gamma_x = -\frac{\partial^2 w}{\partial x^2} = D_{11}^* M_x \tag{8}$$

where D_{11}^* is the first term of the inverted bending stiffness matrix. Since we will only design beams with high length-to-width ratios, we can express the deflection as

$$w = w(x) \tag{9}$$

and Equation (8) becomes

$$-\frac{d^2 w}{dx^2} = D_{11}^* M_x \tag{10}$$

For the sake of simplicity our studies will be restricted to laminates with four plies. In this manner we have two inner plies of constant thickness throughout the length of the beam and two outer plies of linearly tapered thickness. Equation (4) can be conveniently expressed as

$$D_{ij} = {}_1\beta_{ij}\, x^3 + {}_2\beta_{ij}\, x^2 + {}_3\beta_{ij}\, x + {}_4\beta_{ij} \tag{11}$$

where

$${}_1\beta_{ij} = \frac{\left(\bar{Q}_{ij}^{(4)} a_b - \bar{Q}_{ij}^{(1)} a_t\right)}{3} \tag{12a}$$

$${}_2\beta_{ij} = \frac{\left(\bar{Q}_{ij}^{(4)} b_b - \bar{Q}_{ij}^{(1)} b_t\right)}{3} \tag{12b}$$

$${}_3\beta_{ij} = \frac{\left(\bar{Q}_{ij}^{(4)} c_b - \bar{Q}_{ij}^{(1)} c_t\right)}{3} \tag{12c}$$

$${}_4\beta_{ij} = \frac{\bar{Q}_{ij}^{(2)}}{3}\left(z_2^3 - z_1^3\right) + \frac{\bar{Q}_{ij}^{(3)}}{3}\left(z_3^3 - z_2^3\right) + \frac{\bar{Q}_{ij}^{(1)} z_1^3}{3} - \frac{\bar{Q}_{ij}^{(1)} d_t}{3} + \frac{\bar{Q}_{ij}^{(4)} d_b}{3} - \frac{\bar{Q}_{ij}^{(4)} z_3^3}{3} \tag{12d}$$

and

$$a_t = \left(\frac{t_1}{L}\right)^3 \tag{13a}$$

$$b_t = \frac{3z_1t_1^2}{L^2} - \frac{3t_1^3}{L^2} \tag{13b}$$

$$c_t = \frac{3z_1^2t_1}{L} - \frac{6z_1t^2}{L} + \frac{3t_1^3}{L} \tag{13c}$$

$$d_t = z_1^3 - 3z_1^2t_1 + 3z_1t_1^2 - t_1^3 \tag{13d}$$

$$a_b = -\left(\frac{t_4}{L}\right)^3 \tag{13e}$$

$$b_b = \frac{3z_3t_4^2}{L^2} + \frac{3t_4^3}{L^2} \tag{13f}$$

$$c_b = -\frac{3z_3^2t_4}{L} - \frac{6z_3t_4^2}{L} - \frac{3t_4^3}{L} \tag{13g}$$

$$d_b = z_3^3 + 3z_3^2t_4 + 3z_3t_4^2 + t_4^3 \tag{13h}$$

As can be seen from Equation (11), each of the stiffness terms of Equation (2) can be expressed as a cubic equation. Inverting such a matrix in closed form becomes a bookkeeping nightmare and would defeat the purpose of the task at hand. Inverting Equation (1) numerically and plotting D_{11}^*

as a function of x leads to a smooth curve that can easily be fitted with a higher order polynomial for the desired accuracy. In this manner we have,

$$D_{11}^{*} = \sum_{i=1}^{9} \Gamma_i \, x^{i-1} \tag{14}$$

In Equation (8), M_x is the moment per unit width of the beam. Since the type of beams under analysis have variable width, this term becomes,

$$M_x = \frac{M}{W_1 x + W_2} = \frac{P(L - x)}{W_1 x + W_2} \tag{15}$$

where

P = applied load at free end

W_e = beam width at free end

W_b = beam width at fixed end

$$W_1 = \frac{(W_e - W_b)}{L} \tag{16a}$$

$$W_2 = W_b \tag{16b}$$

L = length of beam

Equation (8) can now be expressed as

$$-\frac{d^2 w}{dx^2} = \sum_{i=1}^{9} \Gamma_i \, x^{i-1} \left[\frac{P(L - x)}{W_1 x + W_2}\right] \tag{17}$$

Performing the necessary multiplications, Equation (17) can be expressed in the same manner as Equation (14). This way the necessary integrations can be performed very easily.

$$-\frac{d^2 w}{dx^2} = \sum_{j=1}^{9} \lambda_j \, x^{j-1} \tag{18}$$

Equation (18) can be integrated twice to obtain the slope and deflection of the beam at any point along the length. Since the origin is located at the base (fixed end) of the beam, the boundary conditions of zero slope and deflection allow the elimination of constants of integration. In this manner we have

$$-\frac{dw}{dx} = \sum_{j=1}^{9} \frac{\lambda_j \, x^{j}}{j} \tag{19}$$

$$-w = \sum_{j=1}^{9} \frac{\lambda_j \, x^{j+1}}{j+1} \tag{20}$$

From Equations (8) and (18) the stresses in each lamina area are determined by,

$$(\sigma_i)^{(k)} = \left[\bar{Q}_{ij}\right]^{(k)} \left\{(\epsilon_j) + z(\gamma_j)\right\} \tag{21}$$

In order to verify the accuracy of the latter approach, a numerical example has been performed to compare the stress and deflection distributions of an isotropic beam tapered in width and length with those obtained using Equations (20) and (21). The material properties selected can be found in Table 1. The calculations for the deflection and stress distributions of the isotropic beam are performed following the procedure in reference [1].

TABLE 1. MATERIAL PROPERTIES FOR VERIFICATION RUN

ISOTROPIC	ORTHOTROPIC
$E = 9.9$ msi	$E_{ii} = 9.9$ msi
$\nu = 0.33$	$\nu_{ij} = 0.33$
$\sigma_y = 35$ ksi	$\sigma_{all} = 35$ ksi
$G = 3.8$ msi	$G_{ij} = 3.8$ msi
--------	$\theta = 0^o$
6061-T6 AL [8]	6061-T6 AL [8]

Table 2. shows the geometry and load parameters used for the verification run.

TABLE 2. DESIGN PARAMETERS FOR VERIFICATION RUN

APPLIED LOAD	200 LBS.
BEAM LENGTH	15 IN.
WIDTH AT ORIGIN (BASE)	4.0 IN.
WIDTH AT END (TIP)	2.5 IN.
THICKNESS AT ORIGIN	0.5 IN.
THICKNESS AT END	0.25 IN.

For the orthotropic solution, four laminae were used. The lamina thickness was 0.125 in. each with four plies at the origin (base) linearly tapering to two plies at the end (tip). Figures 2. and 3. are plots comparing the stress and deflection distributions of the isotropic and orthotropic solutions. As expected, due to the input material properties and equal geometry both solutions agree very well. The maximum stress in both cases was $\sigma_{max} = 19853$ psi and the maximum deflection was $\delta_{max} = 1.029$ in.

INTERLAMINAR SHEARING STRESSES

Because laminated composite materials have relatively weak interlaminar shear strength, it is important to know these stresses if one is to design a structurally sound beam. In a study done by Pipes and Pagano [9], it was noted that the interlaminar shear stress σ_{xz} shows evidence of a possible singularity at the free edges. This singularity causes the shear stresses to approach infinity, presenting delamination problems at these edges. In order to minimize the free edge effects Whitney [5] recommends large ratios of half width to lamina thickness. For the beams of variable thickness this constraint will be expressed as

$$\frac{w_{ave}}{2t_{i_{tip}}} \geq 10 \tag{22}$$

where $t_{i_{tip}}$ is the thickness of a ply at the tip (free end) of the beam. w_{ave} is the average width of the tapered beam. Figure 4. shows the dimensions and geometry of the beams under analysis. Figure 5. shows an element of the beam with the width and taper variations on either side of the element. Equilibrium of the shaded element leads to

$$\sigma_{xz}\left(w + \frac{dw}{2}\right)dx = \int_{-z_1}^{-(t+dt)/2} \left(\sigma_x + d\sigma_x\right)\left(w + dw\right)dz - \int_{-z_1}^{-t/2} \sigma_x w \, dz \tag{23}$$

We should note that in Equations (23), (24), (28) and (29) w is the width dimension of the element of area. Combining terms and neglecting products of small quantities we obtain

$$\sigma_{xz} w \; dx = w \int_{-z_1}^{-(t+dt)/2} \left(\sigma_x + \frac{d\sigma_x}{dx} dx \right) dz + \int_{-z_1}^{-(t+dt)/2} \sigma_x \, dw \, dz - \int_{-z_1}^{-t/2} \sigma_x w \, dz \tag{24}$$

Because we are only dealing with pure bending of beams, the ϵ_j strains can be neglected. In this manner equation (21) becomes

$$\{\sigma_i\}^{(k)} = z \left[\bar{Q}_{ij} \right]^{(k)} \{\gamma_j\} \tag{25}$$

Substituting equation (6) into equation (25) leads to

$$\{\sigma_i\}^{(k)} = z \left[\bar{Q}_{ij} \right]^{(k)} \left[D^*_{ij} \right] \{M_j\} \tag{26}$$

or in expanded form

$$\begin{aligned} \sigma_x^{(k)} &= z\, M_x \left[\bar{Q}_{11}^{(k)} D^*_{11} + \bar{Q}_{12}^{(k)} D^*_{12} + \bar{Q}_{13}^{(k)} D^*_{13} + \bar{Q}_{14}^{(k)} D^*_{14} \right] \\ &= z \left(MQD \right)_x \end{aligned} \tag{27}$$

Taking the derivative of Equation (27) with respect to x yields,

$$\frac{d\sigma_x^{(k)}}{dx} = z\bar{Q}_{ij}^{(k)} \left[\frac{\Gamma_1 P w_1 (x - L_o)}{(w_1 x + w_2)^2} - \frac{\Gamma_1 P}{w_1 x + w_2} \right] + z\bar{Q}_{ij}^{(k)} \sum_{n=2}^{9} \left\{ \frac{\Gamma_n P x^{n-2} \left[(n-1)L_o - nx \right]}{w_1 x + w_2} + \frac{\Gamma_n w_1 P x^{n-1} (x - L_o)}{(w_1 x + w_2)^2} \right\} \tag{28}$$

Substituting Equations (27) and (28) into (24) we obtain, after combining all necessary terms

$$\sigma_{xz}^{k} = \frac{t_x}{4} \left(MQD \right)_x \frac{dt_x}{dx} + \left(\frac{t_x^2}{8} - \frac{z_k^2}{2} \right) \left\{ \frac{d}{dx} \left[\left(MQD \right)_x \right] + \frac{\left(MQD \right)_x}{w_x} \frac{dw}{dx} \right\} \tag{29}$$

where t_x is the thickness of the laminate at location x along the length of the beam. M_x is defined in equation (15) and z is the thickness coordinate with origin at the center of the laminate. It should be noted that M_x and D^*_{ij} in equation (27) are functions of x only. Furthermore the transverse shear stress σ_{xz} changes in magnitude along the length of the beam.

The location of maximum shear stress is at the free end of the cantilever. However, because the shearing stress is not necessarily zero at the top and bottom free surfaces of the tapered beam, they must be taken into consideration when calculating the maximum combined stresses due to to bending. Timoshenko [11] shows that a similar distribution exists for beams of constant width and variable thickness. This interesting fact is of no major consequence when dealing with isotropic (metallic) beams but it can lead to premature delamination of composite tapered beams if not considered appropriately.

WEIGHT OPTIMIZATION

The general problem statement for the minimization of a function of several variables subject to conditions of constraint is

$$\text{MINIMIZE:} \quad F(X) \tag{30}$$

$$\text{SUBJECT TO:} \quad g_j(X) \leq 0 \qquad j = 1,m \tag{31}$$

$$h_k(X) = 0 \qquad k = 1,n \tag{32}$$

Where X is the vector containing the design variables, F is the objective function (function to be minimized), g_j are the inequality constraints and h_k are the equality constraints. In order to limit the region of search for the optimum, side constraints are imposed on the problem. This is done by imposing upper and lower bounds on the search values of the design variables

$$x_i^{lower} \leq x_i \leq x_i^{upper} \tag{33}$$

The objective function for weight minimization is the volume of the beam. For a four ply laminated beam of variable width and depth this function is

$$W(X) = A_1[x(2) + x(3)][x(5) + x(6)] + B_1[x(1) + x(4)]x(6) + C_1[x(1) + x(4)]x(5) \tag{34}$$

where

x(1) = thickness of the top lamina
x(2) = thickness of the second lamina
x(3) = thickness of the third lamina
x(4) = thickness of the fourth lamina
x(5) = beam width at free end
x(6) = beam width at fixed end

$$A_1 = \frac{\rho L}{2} \tag{35}$$

$$B_1 = \frac{\rho L}{3} \tag{36}$$

$$C_1 = \frac{\rho L}{6} \tag{37}$$

The constraints imposed on the beams under investigation are as follows

$$x(1) - x(4) = 0 \tag{38}$$

$$x(2) - x(3) = 0 \tag{39}$$

$$\frac{\delta}{\delta_{all}} - 1 = 0 \tag{40}$$

$$\frac{\sigma}{\sigma_{Tsai-Hill}} - 1 \leq 0 \tag{41}$$

$$\frac{\sigma_{xz}}{\sigma_{xz-all}} - 1 \leq 0 \tag{42}$$

$$1 - \frac{w_{ave}}{20\,x(2)} \leq 0 \tag{43}$$

Constraint equations (38) and (39) assure ply thickness symmetry with respect to the beam midplane. Constraint equation (40) specifies the desired stiffness of the beam. Equations (41) and (42) assure stress levels are below bending and transverse shear allowables respectively. Equation (43) is a geometry constraint that minimizes the free edge delamination effects. Note that w_{ave} = [x(5) + x(6)]/2.

NUMERICAL EXAMPLES

As a first example we shall attempt to optimize the weight of a tapered beam using the laminated plate approach. The individual ply material properties will be set equal to those of isotropic Titanium 6Al-4V. In this manner the optimization of the selected beam can be compared to the results of reference [1]. Table 3 shows the result of the optimization. The design and material data is as follows

Applied Load	2946 lbs. (1336 kg)
Beam length	29.25 in. (74.30 cm)
Maximum desired deflection	7.45 in. (18.92 cm)
Maximum allowable stress	104 ksi (716 MN/sq-m)*
Density	0.16 lb/cu-in (2.56 kg/cu-m)

* - The allowable stress for this example was set at 104 ksi for purposes of comparison with reference [1]. The allowable for the example in Table 4 was 120 ksi per reference [8].

As can be seen, the optimization of an isotropic beam using the orthotropic beam solution method compares very well with the method of reference [1]. Note that $t_e = t_2 + t_3$ and $t_o = t_1 + t_4 + t_e$.

Our second example will compare the optimum design of isotropic titanium 6Al-4V [8] with that of high strength graphite-epoxy (V_f = 0.62) [10]. The allowable working stress for the graphite-epoxy beam is obtained from the Tsai-Hill failure criteria. For both the isotropic and orthotropic cases a factor of safety of 1.4 was used.

Table 3. COMPARISON OF OPTIMIZATION SOLUTIONS

ISOTROPIC BEAM OPTIMIZATION (Reference [1])

Design	w_e	w_o	t_e	t_o	Wt.	Def.	Stress	Iter.
Initial	6.500	6.500	.351	.905	19.905	7.395	102270	--
Final	6.213	6.172	.339	.928	18.360	7.464	104429	46

ORTHOTROPIC BEAM SOLUTION (This Paper)

Design	w_e	w_o	t_1	t_2	Wt.	Def.	Stress	Iter.
Initial	6.500	6.500	.277	.175	19.905	7.395	102270	--
Final	6.150	6.150	.293	.172	18.330	7.459	104431	44

TABLE 4. ISOTROPIC VS. ORTHOTROPIC WEIGHT OPTIMIZATION

ISOTROPIC BEAM OPTIMIZATION

Design	w_e	w_o	t_e	t_o	Wt.	Def.	Stress	Iter.
Initial	6.500	6.500	.351	.905	19.905	7.395	102270	--
Final	4.554	3.295	.313	1.165	13.155	7.457	120575	137

TABLE 4. (Continuation)

ORTHOTROPIC BEAM SOLUTION [5/-5]s

Design	w_e	w_o	t_1	t_2	Wt.	Def.	Stress	Iter.
Initial	6.500	6.500	.277	.175	6.675	5.562	102551	--
Final	4.820	4.820	.310	.170	5.138	7.453	123242#	172
Initial	4.000	6.500	.245	.210	5.885	6.217	107627	--
Final	3.950	5.940	.231	.205	5.315	7.465	123640#	85

- Based on the Tsai-Hill failure criteria, the maximum allowable stress for this material with a [5/-5]s lay-up is 123 ksi.

CONCLUSION

The developed method for designing composite laminated tapered beams compares well with the solution for isotropic beams. The present method is more versatile since both isotropic and orthotropic beams can be designed. The present method also considers the importance of the varying distribution of the interlaminar shear stress along the length of the tapered beams.

As can be seen from Table 4., the weight savings from the optimized orthotropic beam can be as much as 61% when compared to the Titanium beam. Obviously many different combinations of widths and thicknesses can produce acceptable results. This is because the 6-dimensional space of the objective function has many local minima. It should be noticed that for the cases analyzed the [0/0]s lay-up did not yield acceptable results due to the high uniaxial stiffness of the graphite-epoxy $\left(E_{11} = 20.8 \text{ msi}\right)$. The second case of the orthotropic solutions in Table 4. shows the convergence to a different set of width and ply thicknesses. Notice that the final stresses are slightly above the Tsai-Hill allowable. This can be corrected by changing internal tolerance parameters in the DOT program.

REFERENCES

1. Rodriguez, P.I., Minimum Weight Design of a Leaf Spring Tapered in Thickness and Width for the Hubble Space Telescope-Space Support Equipment, NASA TM-4233, 1990.

2. Whitney, J.M., Stress Analysis of Thick Laminated Composite and Sandwich Plates, Journal of Composite Materials, Vol. 6, pp. 426-440, 1972.

3. Jones, R.M., Mechanics of Composite Materials, McGraw Hill Book Co., Washington D.C., 1975.

4. Vanderplaats, G.N., Numerical Optimization Techniques for Engineering Design: with Applications, McGraw Hill Book Co., New York, 1984.

5. Whitney, J.M., Structural Analysis of Laminated Anisotropic Plates, Technomic Publishing Co., Pennsylvania, 1987.

6. Ashton, J.E., Analysis of Anisotropic Plates II, Journal of Composite Materials, Vol. 3, pp. 470-479, 1969.

7. Whitney, J.M., The Effect of Transverse Shear Deformation on the Bending of Laminated Plates, Journal of Composite Materials, Vol. 3, pp. 534-547, 1969.

8. MIL-HDBK-5C, Metallic Materials and Elements for Aerospace Vehicle Structures, Military Standardization Handbook, 1976.

9. Pipes, R.B. and Pagano, N.J., Interlaminar Stresses in Composite Laminates Under Uniform Axial Extension, Journal of Composite Materials, Vol. 4, pp. 538-548, 1970.

10. DOD/NASA Advanced Composites Design Guide, 1983

11. Timoshenko, S., Strength of Materials Part II - Advanced Theory and Problems, D. Van Nostrand Co., Toronto, 1956

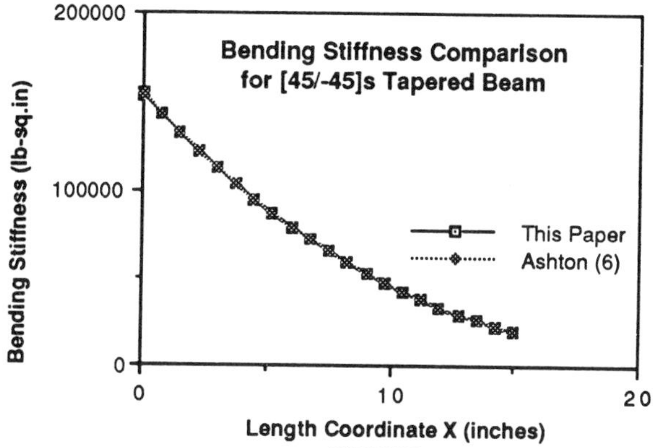

FIGURE 1.

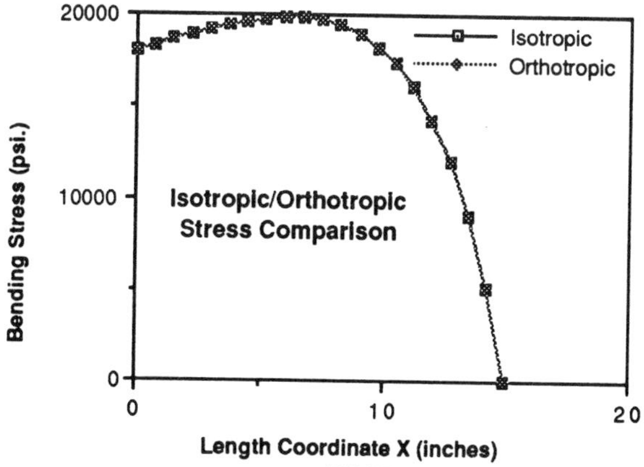

FIGURE 2.

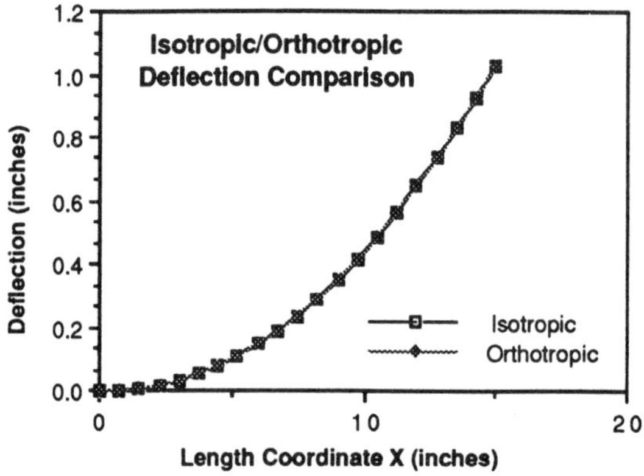

FIGURE 3.

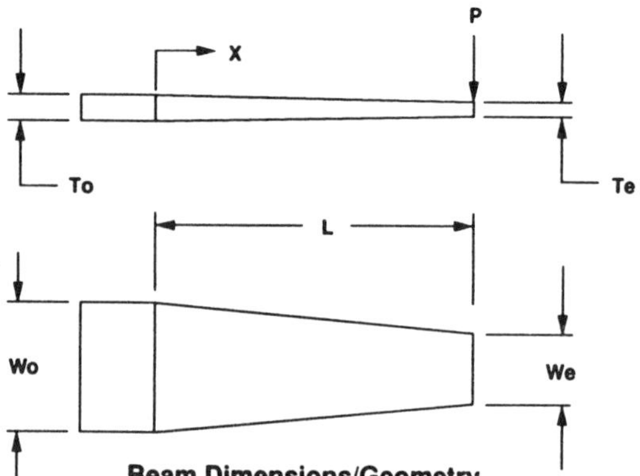

Beam Dimensions/Geometry

FIGURE 4.

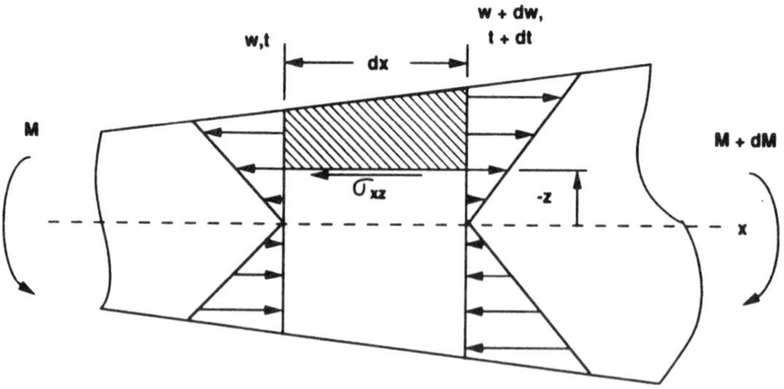

Element of Area for Transverse Shear Stress

FIGURE 5.

Optimum Design of Rectangular Panels with a Hole

C.S. Han (*), B.P. Wang (**)
(*) Department of Mechanical and Materials Engineering, Wright State University, Dayton, Ohio 45435, U.S.A.
(**) Department of Mechanical Engineering, The University of Texas at Arlington, Arlington, Texas 76019, U.S.A.

ABSTRACT

The optimum design of composite panels with a circular or an elliptical hole is considered in this paper. The objective is to find the best hole location, size and orientation so as to minimize the maximum tangential strain along the circumference of the hole. The problem is treated as a shape optimization problem, and the hole boundary parameterization method is developed in order to maintain the required hole shape. The parameterized relation is further utilized to transform the sensitivity data. The optimization problem is solved iteratively using p-method finite element analysis and linear programming. The results of numerical examples indicate that substantial reduction in the maximum tangential strain is achieved by adjusting the size, location, and orientation of a circular and elliptical hole.

INTRODUCTION

The optimum design of a composite panel with a circular or an elliptical hole requires combinations of finite element solution, sensitivity calculation, boundary parameterization, mesh generation, and numerical optimization. This type of design procedures, is called shape optimization process. Zienkiwicz and Campbell in 1973[1], and Ramarkrishnan and Francavilla in 1974[2], initiated the shape optimization without incorporating boundary parameterization and mesh generation techniques. Since then substantial literature has been developed in the field of shape design sensitivity analysis and optimization of structural components.

Still there are several factors that continue to prohibit the development of a generalized shape optimization program. The first reason may be that the structural response data (such as displacement, strain, stress, frequency, etc) of these structures, caused by external loading, are generally implicit functions of the design variables; there are no linear or other explicit relationships between responses and design variables. The second reason may be that the external boundary of the structure is modified at each stage of the optimization process; the finite element mesh must be appropriately updated in order to maintain the design accuracy. The third reason may be that associating the constraints with the finite element meshes or nodes is unattractive, since the finite element mesh will change during the design process. When the boundary shape changes, there will be a possibility that the number of elements and nodes will increase or decrease. This makes it very difficult to uniformly monitor the design process.

An alternative approach to overcome the difficulties arising in shape optimization is to use p-version finite element method together with an explicit boundary parameterization method. P-version finite element method[**3-4**] has the capability of improving the accuracy in the solution data as the polynomial order increases, because it uses hierarchic shape functions composed of Legendre polynomials in addition to nodal shape functions. The advantage of using p-method in hole shape optimization is that relatively coarse meshes can be used for the optimization, which will substantially reduce the effort of incorporating complicated mesh generation procedures. Furthermore shape sensitivity data which will be obtained from the p-method will also be of increasing accuracy as the polynomial order increases.

Explicit boundary parameterization methods are developed for a circle and an ellipse by using properly selected design variables. The selected design variables can be used to describe expansion, contraction, rotation, and translation of the boundary shape. Points along the boundary are indicated by the use of geometric parameters. When design variables are specified, coordinates of boundary points are evaluated by substituting numerical values to the geometric parameters. During the optimization process, geometric parameters remain the same while design variables are changing continuously to improve the design. Uniform design monitoring can be achieved once the geometric parameters are set at uniform intervals in the beginning of the optimization.

The stiffness matrix in p-version finite element method is differentiated analytically to calculate shape sensitivity data. In order to make the optimization problem more general, shape sensitivity data are calculated with respect to the X and Y coordinates of the boundary nodal points, and the sensitivity data then are transformed to the design variable dependent sensitivity data using the parameterized relationship and the chain rule of differentiation.

Linear programming, which is available in the IMSL, is used to solve the shape optimization problem.

Two test problems are used to illustrate the effectiveness of the developed optimization program for the hole shape optimization problem, which requires the prescribed hole shape with different size, orientation, and location. Prescribed shape is maintained during the optimization process, and fast convergence is achieved because the parameterization used excludes feasible designs which does not meet the shape requirement.

BOUNDARY PARAMETERIZATION AND DESIGN VARIABLES

Boundary parameterization plays an important role in shape optimization. Boundary for a continuum structure consists of infinite number of points, and we cannot use all of the points as design variables. The purpose of boundary parameterization in shape optimization is to create a design model with a small number of design variables. Geometrical description of particular shapes are made by several different methods. Oda and Yanazaki[5] use the pattern transformation method for the shape optimization. Iman[6-7] describes a geometrical model by using the design element method. Botkin and Bennett[8-9] use the boundary design element concept by using a polynomial curve, key nodes, straight line segment and circular arc.

The design model is explicitly written by the design variables, and optimization problem only takes care of the mathematical model which is described by the design variables. Analysis model ,which is a finite element mesh, is constantly updated based on the new values of design variables as a result of the optimization.

The parameterized relation is further utilized for the transformation of nodal coordinates dependent on structural sensitivity data to design variable dependent sensitivity data by using the chain rule of differentiation. When a commercially available finite element code produces sensitivity data for the nodes which are indicated by the user as input data, the sensitivity data can be easily transformed into design variable dependent sensitivity by utilizing the parameterized relation.

Typical hole shapes, such as a circular hole or an elliptical hole, are parameterized using design variables and geometrical parameters. The design variables are used to describe the translation, rotation, and size of the hole contour, while geometric parameters are used to indicate the location of the boundary points. Detailed description of the method will be explained in the following two subsections.

Parameterization of a circular hole

A circular hole boundary can be explicitly described by using a radius and its center position coordinates. Change of the radius relates the size of the circular hole, while change of the hole center position relates the movement of the circular hole. Three design variables are used for the explicit representation of any circular hole. Circular hole boundary points are indicated by using a special parameter named geometric parameters. **Figure 1** shows the parameterization of the circular boundary.

Any point on the circular hole boundary can be expressed as

$$X_j = X_c + R\cos\theta_j \tag{1}$$

$$Y_j = Y_c + R\sin\theta_j \tag{2}$$

In this case, the design variables are,

$D_1 = X_c$: Center coordinates of X in circle

$D_2 = Y_c$: Center coordinates of Y in circle

$D_3 = R$: Radius of circle

and geometric parameters are,

θ_j : Locations on the circle in radian

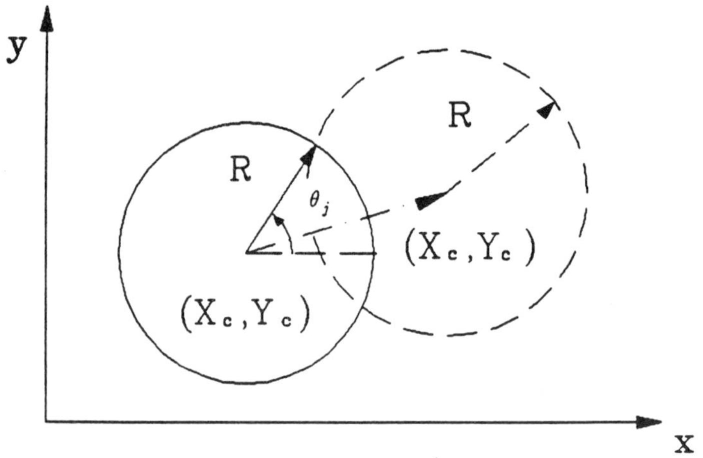

Figure 1. Circular Hole Boundary Parameterization.

A circle can be discretized into infinitely many points using the geometric parameters. They all have the same design variables but different values in geometric parameters.

Parameterization of an elliptic hole
An internal elliptic hole boundary is described by using two radii, center position, and rotation of the closed contour with respect to a fixed reference. The change of major and minor radius is used for describing expansion and contraction of the elliptic hole boundary. Rigid body movement of the ellipse is described by its center point coordinates, while rigid body rotation of the ellipse is described by its rotation angle with respect to the center. Points along the ellipse boundary are indicated by geometric parameters. **Figure 2** shows the parameterization method for an elliptic hole boundary.

Any points on the boundary can be expressed as

$$X_j = X_e + (R(\theta_j)cos\theta_j)cos\psi - (R(\theta_j)sin\theta_j)sin\psi \tag{3}$$

$$Y_j = Y_e + (R(\theta_j)cos\theta_j)sin\psi + (R(\theta_j)sin\theta_j)cos\psi \tag{4}$$

while $$R(\theta_j) = \left(\frac{a^2 b^2}{a^2 sin^2\theta_j + b^2 cos^2\theta_j}\right)^{\frac{1}{2}}$$

In this case the design variables are,

$D_1 = X_e$: Center coordinate of X in ellipse

$D_2 = Y_e$: Center coordinate of Y in ellipse

$D_3 = a$: Length of major radius in ellipse

$D_4 = b$: Length of minor radius in ellipse

$D_5 = \psi$: Rotation angle of ellipse

and the geometric parameters are

θ_j : Locations on the ellipse in radian

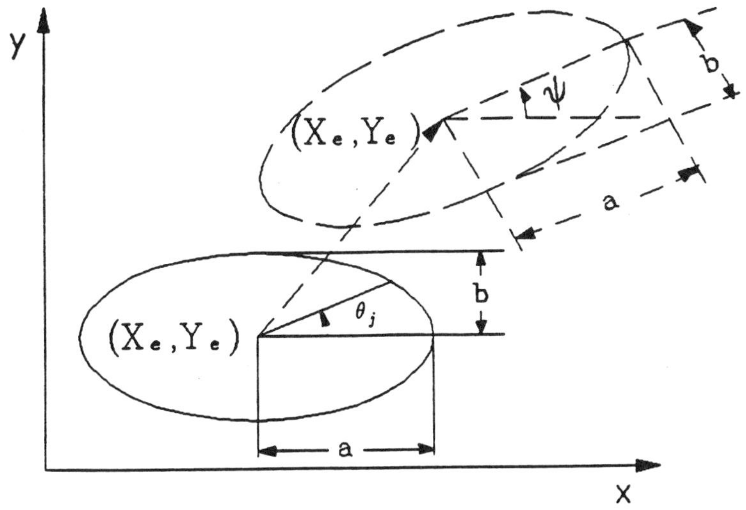

Figure 2. Elliptic Hole Boundary Parameterization.

SENSITIVITY CALCULATION

Design sensitivity analysis is a procedure for determining which directions in design variable changes can produce a better design. It is a step to calculate ratios of performance measures with respect to design variable changes.

There are two basic approaches, namely, direct sensitivity analysis (DSA)[10-12] and variational design sensitivity analysis(VDSA)[13-16]. The DSA approach is carried out via implicit differentiation of the discretized analysis equation with respect to shape design variables. On the other hand, the VDSA approach involves implicit differentiation of the functionals of the variational equation of the continuum analysis problem. The relationship between DSA and VDSA is made by Yang and Botkin[17], and they concluded that both methods may be commutative at least for problems involving shape design changes on traction free boundary. In this paper, DSA is used.

Sensitivity analysis outputs from large scale, general purpose, commercially available programs have been produced by several papers[18-20]. These sensitivity outputs are not yet applicable to shape optimization.

In the governing equation for displacement in finite element analysis,

$$[K](U) = (F) \tag{5}$$

the stiffness matrix $[K]$ is a function of nodal coordinates, but (F) is not a function of nodal coordinates for a fixed applied boundary force. For the purpose of design sensitivity analysis, the differentiation of the governing equation with respect to α_j, which represents the coordinates of a nodal point, X_j or Y_j, is carried out:

$$\frac{\partial[K]}{\partial\alpha_j}(U) + [K]\frac{\partial(U)}{\partial\alpha_j} = \frac{\partial(F)}{\partial\alpha_j} \tag{6}$$

If the loaded boundary is not affected by design changes, (F) is not a function of α_j, and $\partial(F)/\partial\alpha_j$ equals zero. Then, the **Equation 6** is reduced and rearranged to **Equation 7**.

$$\frac{\partial(U)}{\partial\alpha_j} = -[K]^{-1}\frac{\partial[K]}{\partial\alpha_j}(U) \tag{7}$$

Structural response sensitivities for displacement can be computed after evaluating the derivatives of the stiffness matrix. In this paper , the derivative of the stiffness matrix with respect to the nodal coordinate variable, α_j, is computed analytically. This is achieved by assembling all the derivatives of element stiffness matrices as follows.

$$\frac{\partial[K]}{\partial\alpha_j} = \sum\int\int(\frac{\partial B^T}{\partial\alpha_j}CB|J| + B^T C\frac{\partial B}{\partial\alpha_j}|J| + B^T CB\frac{\partial|J|}{\partial\alpha_j})t_z d\xi d\eta \tag{8}$$

where B is a matrix relating element strains to assumed displacement field parameters, and composed of derivative of shape function N_i and inverse of Jacobian matrix, C is a material property matrix which describes the relationship between stresses and strains, and $|J|$ is the determinant of Jacobian matrix J which represent the mapping between global coordinates and local coordinates. The integral in **Equation 8** is integrated numerically by using the Gauss quadrature rule. Detailed procedures are shown in the Han[21].

From the results in **Equation 7**, strain and stress sensitivity data can be evaluated as follows:

$$\frac{\partial(\epsilon)_e}{\partial\alpha_j} = B\frac{\partial(u)_e}{\partial\alpha_j} + \frac{\partial B}{\partial\alpha_j}(u)_e \tag{9}$$

$$\frac{\partial(\sigma)_e}{\partial\alpha_j} = C(B\frac{\partial(u)_e}{\partial\alpha_j} + \frac{\partial B}{\partial\alpha_j}(u)_e) \tag{10}$$

Sensitivity data of tangential strain and tangential strain along the circumference of an elliptic hole are obtained using Mohr's circle[22] , and are shown in the following equations.

$$\epsilon_{Ti} = \frac{1}{2}(\epsilon_{xi} + \epsilon_{yi}) + \frac{1}{2}(\epsilon_{xi} - \epsilon_{yi})cos(2\theta_{Ti}) + \frac{1}{2}\gamma_{xyi}sin(2\theta_{Ti}) \tag{11}$$

$$\frac{\partial\epsilon_{Ti}}{\partial\alpha_j} = \frac{1}{2}(\frac{\partial\epsilon_{xi}}{\partial\alpha_j} + \frac{\epsilon_{yi}}{\partial\alpha_j}) + \frac{1}{2}(\frac{\partial\epsilon_{xi}}{\partial\alpha_j} - \frac{\partial\epsilon_{yi}}{\partial\alpha_j})cos(2\theta_{Ti}) + \frac{1}{2}\frac{\gamma_{xyi}}{\partial\alpha_j}sin(2\theta_{Ti}) \tag{12}$$

where

$\theta_{Ti} = \theta_{ni} + 90$

$\theta_{ni} = tan^{-1}[\frac{(y_i - y_0)/b^2}{(x_i - x_0)/a^2}]$

Note that $b = a$ for a circular hole.

The structural response sensitivity data for tangential strain with respect to the design variable, D_k, for hole boundary points indicated by the geometric parameters, is obtained by the following relation using the chain rule of differentiations.

$$\frac{\partial\epsilon_{Ti}}{\partial D_k} = \sum_{j=1}^{nd}(\frac{\partial\epsilon_{Ti}}{\partial X_j}\frac{\partial X_j}{\partial D_k} + \frac{\partial\epsilon_{Ti}}{\partial Y_j}\frac{\partial Y_j}{\partial D_k}) \tag{13}$$

These data are used as an input data to the optimization routine.

PROBLEM STATEMENT AND SOLUTION OF OPTIMIZATION PROBLEM

The optimization problems considered in this paper are the minimization of maximum tangential strain along the hole boundary. Specifically, these problems can be stated as:

Find the design variable, D_k to minimize the maximum tangential strain Max $|\epsilon_{Ti}|$.

Mathematically this mini-max problem can be converted into the following minimization problem.

Find D_k to minimize D_{n+1}

subject to the following inequality constraints:

$$|\epsilon_{Ti}| \leq D_{n+1}, \qquad i = 1 \; to \; m$$

and, side constraints:

$$D_{k\,min} \leq D_k \leq D_{k\,max}$$

note that $D_{n+1} = \text{Max}\ |\epsilon_{Ti}|$

This problem can be formulated as a linear programming problem after using the following linearization for the tangential strain.

$$\epsilon_{Ti} = \epsilon_{T0} + \sum_{k=1}^{n} \frac{\partial \epsilon_{Ti}}{\partial D_k}(D_k - D_{k0}) \tag{14}$$

In the above formulation, subscript n is the number of design variables and m is the number of design monitoring points.

The above, linear programming(LP) problem can be solved by using the simplex method[23] based routine DDPLS in the IMSL library.

After solving the optimization problem, the analysis model is updated following the design variable changes. Reconnection of the finite element meshes may be necessary to prevent distorted meshes which cause substantial errors in structural response and sensitivity data. The optimization cycles are terminated when a certain number of iteration have been performed or a convergence criterion is met.

NUMERICAL EXAMPLES

Two test problems are presented to demonstrate the effectiveness of using p-version finite element method together with explicit boundary parameterization to hole shape optimization problems.

Example 1 Optimal size and location of a circular hole

A plate made of a composite material is subject to normal and tangential traction forces. The composite plate is considered to be orthotropic material with the following properties, $E_{11} = 11883000$ psi, $E_{22} = 4963000$ psi, $E_{33} = 2554000$ psi and $\nu_{12} = 0.413$. The plate is square with 10 inches each side and 0.1 inches thick. The problem is to obtain an optimal hole shape to minimize the maximum tangential strain around the hole under the loading shown. Three design variables, D_1 , and D_2 for center location, and D_3 for hole radius, are used to explicitly describe the circular hole configuration. **Figure 3** shows the geometry, boundary condition, load condition, and design variables. Vertical short lines along horizontal edge represent the edge being fixed in Y-direction, and short horizontal lines along vertical edge shows the edge being fixed in X-direction. All the tangential traction forces are 2000 lb/in and normal traction forces are 5000 lb/in and 3000 lb/in respectively. The design variables are constrained to change between the following bounds.

$3.5 \leq D_1 \leq 6.5$

$3.5 \leq D_2 \leq 6.5$

$1.5 \leq D_3 \leq 2.5$

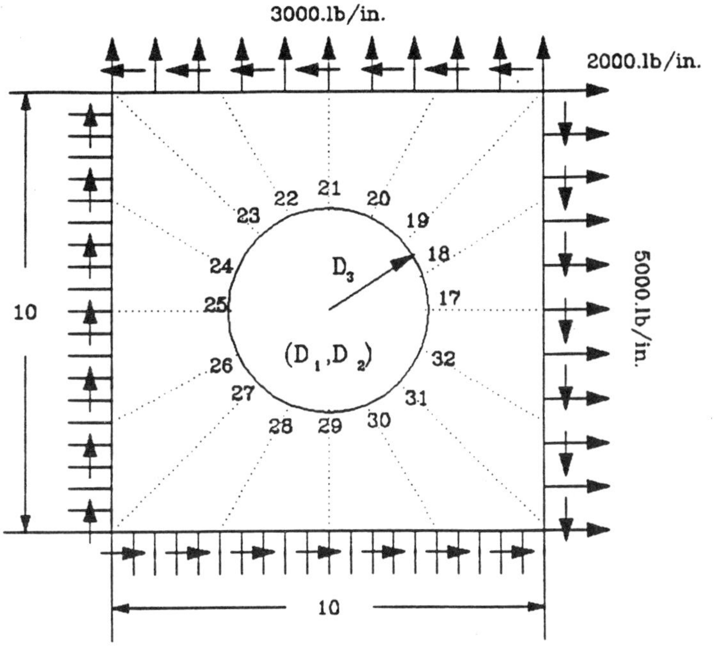

Figure 3. Model for Circular Hole Shape Optimization Prob. for Minimizing Max. Tangential Strain.

In initial design $D_1 = 5.0$ inches and $D_2 = 5.0$ inches from the lower left corner of the plate ,and $D_3 = 2.5$ inches. respectively. Final hole shape, obtained after 5 iterations, is shown in **Figure 4**. The figure shows the size and center location have changed, but it still remains a circular shape. **Table 1** shows the tangential strain distribution for 16 strain monitoring points after each iteration. The ratio of maximum and minimum tangential strain in the initial design is 6, but after 5 iteration, it becomes 2.4, which shows tangential strain distribution becomes rather smooth. **Figure 5** shows the history of maximum tangential strain on the circumference of the hole.

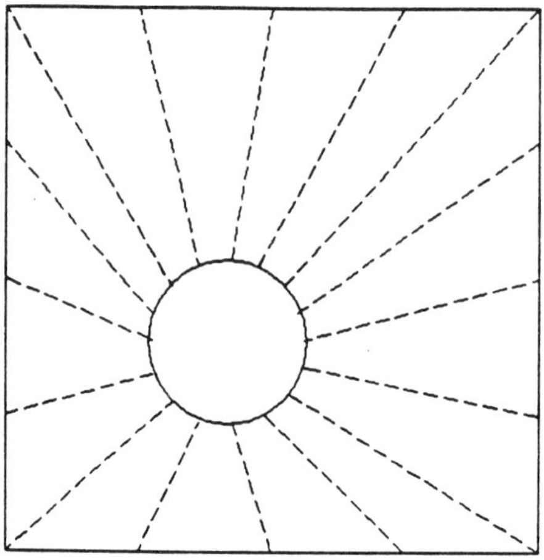

Figure 4. Final Circular Hole Shape for Minimizing Max. Tangential Strain.

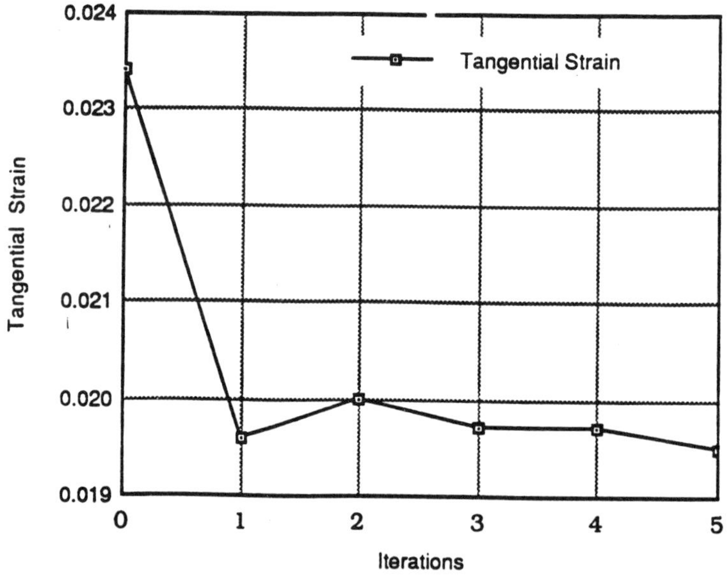

Figure 5. Change in Max. Tangential Strain on Circular Hole.

Table 1. Tangential Strain Distribution around Circular Hole.

Iter / Point	Init	1	2	3	4	5
17	0.0052	0.0070	0.0100	0.0072	0.0093	0.0076
18	0.0174	0.0158	0.0176	0.0159	0.0170	0.0159
19	0.0234	0.0196	0.0200	0.0197	0.0197	0.0195
20	0.0222	0.0189	0.0187	0.0189	0.0185	0.0187
21	0.0135	0.0125	0.0118	0.0124	0.0119	0.0122
22	0.0061	0.0072	0.0062	0.0070	0.0064	0.0068
23	0.0077	0.0091	0.0107	0.0087	0.0102	0.0090
24	0.0111	0.0125	0.0145	0.0122	0.0138	0.0125
25	0.0168	0.0136	0.0176	0.0139	0.0166	0.0144
26	0.0185	0.0161	0.0195	0.0165	0.0186	0.0169
27	0.0174	0.0169	0.0182	0.0170	0.0179	0.0171
28	0.0166	0.0166	0.0183	0.0160	0.0176	0.0162
29	0.0153	0.0146	0.0165	0.0140	0.0155	0.0141
30	0.0172	0.0150	0.0141	0.0152	0.0141	0.0149
31	0.0169	0.0121	0.0110	0.0124	0.0111	0.0120
32	0.0039	0.0067	0.0085	0.0068	0.0083	0.0071

Example 2 Optimal size and location of an elliptic hole

The hole shape considered in this example is an ellipse. It is described explicitly by five design variables as shown in **Figure 6**. Design variables D_1 and D_2 are used for the elliptic hole center location, while design variables D_3 and D_4 are two radii of the ellipse, and D_5 is for the rotation of ellipse. It is also the problem of reducing tangential strain concentration subject to applied boundary traction forces. The plate is made of composite material and overall material property is the same as **Example 1**. Five design variables are used to explicitly represent the elliptic hole boundary and they are constrained to vary between the following ranges.

$$4.5 \leq D_1 \leq 5.5$$

$$4.5 \leq D_2 \leq 5.5$$

$$2.0 \leq D_3 \leq 3.0$$

$$1.0 \leq D_4 \leq 2.0$$

$$0.0 \leq D_5 \leq 3.14$$

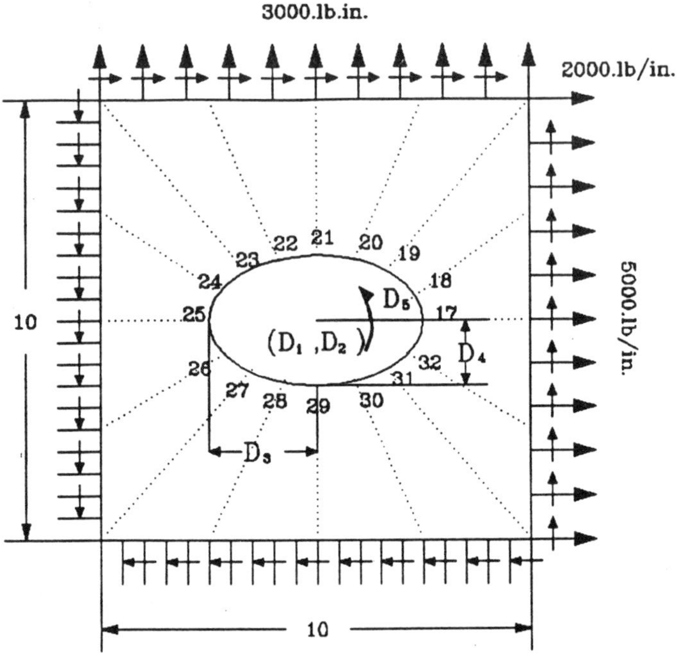

Figure 6. Model for Elliptic Hole Shape Optimization Prob. for Minimizing Max. Tangential Strain.

Initially $D_1 = 5.0$ inches and $D_2 = 5.0$ inches from the lower left corner of the plate, $D_3 = 2.5$ inches and $D_4 = 1.5$ inches, and $D_5 = 0.75$ radian from the line which connect center of ellipse and lower right corner of the plate to the line connecting the center to the point 17. **Figure 7** shows optimum design after iteration 8, and the design variables are $D_1 = 5.18$ inches, $D_2 = 4.5$ inches, $D_3 = 2.0$ inches, $D_4 = 1.0$ inch, and $D_5 = 0.386$ radian respectively. After several iterations, the hole shape changed with different center position, size, and orientation, but it remains as an elliptic shape. **Table 2** shows the tangential strain distribution for 16 design monitoring points after each iterations. The maximum tangential strain has reduced to factor of 2.17 and the ratio of maximum to minimum tangential strain is 6.7 in the initial design and 2.47 after iteration 8. **Figure 8** is the change in maximum tangential strain.

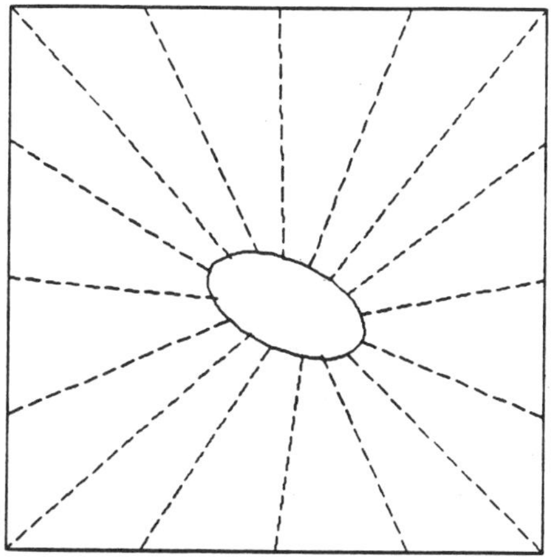

Figure 7. Final Elliptic Hole Shape for Minimizing Max. Tangential Strain.

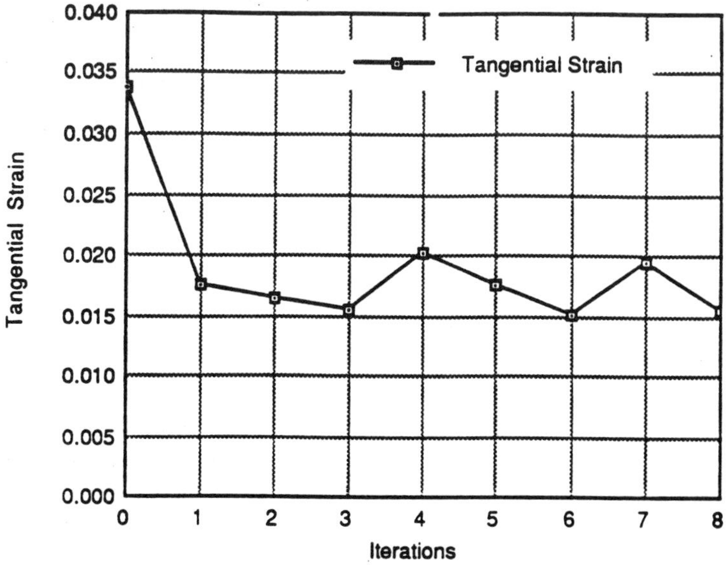

Figure 8. Change in Max. Tangential Strain on Elliptic Hole.

Table 2. Tangential Strain Distribution Around Elliptic Hole.

Iter / Point	0	1	2	3	4	5	6	7	8
17	0.0204	0.0094	0.0137	0.0089	0.0104	0.0095	0.0089	0.0103	0.0091
18	0.0268	0.0197	0.0100	0.0113	0.0192	0.0171	0.0131	0.0195	0.0154
19	0.0154	0.0128	0.0120	0.0125	0.0124	0.0125	0.0125	0.0125	0.0125
20	0.0097	0.0093	0.0087	0.0090	0.0091	0.0091	0.0091	0.0092	0.0091
21	0.0059	0.0078	0.0071	0.0076	0.0077	0.0078	0.0077	0.0077	0.0078
22	0.0040	0.0065	0.0066	0.0067	0.0064	0.0066	0.0068	0.0064	0.0067
23	0.0053	0.0070	0.0078	0.0079	0.0067	0.0072	0.0078	0.0068	0.0075
24	0.0092	0.0099	0.0118	0.0105	0.0088	0.0095	0.0104	0.0092	0.0100
25	0.0338	0.0135	0.0125	0.0111	0.0152	0.0132	0.0116	0.0152	0.0123
26	0.0223	0.0176	0.0016	0.0121	0.0201	0.0176	0.0136	0.0195	0.0156
27	0.0123	0.0100	0.0107	0.0103	0.0103	0.0103	0.0103	0.0103	0.0102
28	0.0094	0.0080	0.0067	0.0076	0.0080	0.0079	0.0077	0.0081	0.0078
29	0.0087	0.0065	0.0055	0.0059	0.0064	0.0063	0.0061	0.0066	0.0063
30	0.0123	0.0069	0.0045	0.0057	0.0063	0.0065	0.0061	0.0068	0.0064
31	0.0120	0.0116	0.0063	0.0114	0.0109	0.0113	0.0115	0.0110	0.0114
32	0.0064	0.0130	0.0163	0.0155	0.0124	0.0135	0.0150	0.0125	0.0143

A little oscillation is detected in the maximum tangential strain history shown in **Figure 8**. This is caused by error of using linear approximation for nonlinear constraints. In order to analyze the reason why tangential strain increases at iteration 4 and iteration 7 during the optimization process , 4 sets of intermediate design variables are used for detailed analysis and it is shown in **Appendix**.

CONCLUSION

A method for hole shape optimization is proposed and tested in this research. The effectiveness of the proposed approach is summarized below.

First, boundary nodes along the circumference of a hole are used for the design monitoring, and the nodes are indicated by the geometric points. Uniform design monitoring can be achieved when geometric parameters are set at uniform intervals in the beginning of the optimization.

Second, nodal point dependent sensitivity data can be easily transformed to the design variable dependent sensitivity data by using the parameterized relation and the chain rule of differentiation. It is predicted that once commercially available finite element code produces nodal point dependent sensitivity data for a continuum structure, the data can be easily transformed to the particular shape optimization problem when appropriate boundary parameterization is developed.

Third, when the parameterization is used with p-version finite element method in optimization, reconnection of the elements is only required when the elements are distorted, because p-version finite element method is proved to have improving accuracy as polynomial order increases[21]. This is in contrast with optimization using h-version finite element method which requires remeshing after each iteration.

Fourth, the proposed boundary parameterization requires a small number of design variables, and it excludes a lot of unacceptable designs, thus optimization converges fast without producing erroneous shapes.

APPENDIX

This Appendix is devoted to a detailed analysis for elliptic hole tangential strain minimization problems at iteration number 4 and number 7, where the maximum value of tangential strain increases, as shown in Figure 10. In order to analyze why this happens, analysis results for 4 sets of intermediate values of design variables are provided to investigate the behavior of tangential strain. Tangential strains at 4 critical design monitoring points are plotted in Figure 9 and Figure 10. Both of the figures show that the tangential strains at design monitoring point 32, at starting of iteration 4 and 7, are maximum. This strain has a negative gradient, while the other points which have lower strains, but have positive gradients. When optimization algorithm is carried out to reduce the maximum strain for design monitoring point 32, the other points tend to increase tangential strains.

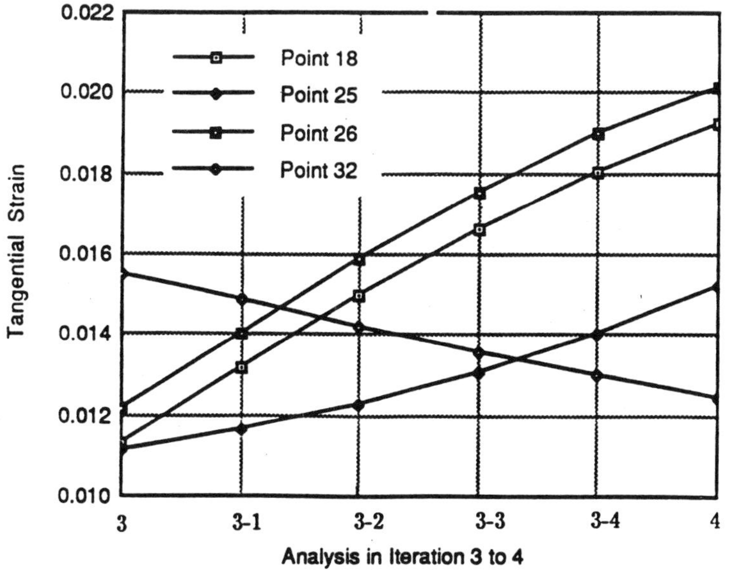

Figure 9. Elliptic Hole Tangential Strain Minimization Prob. Detailed Analysis 1.

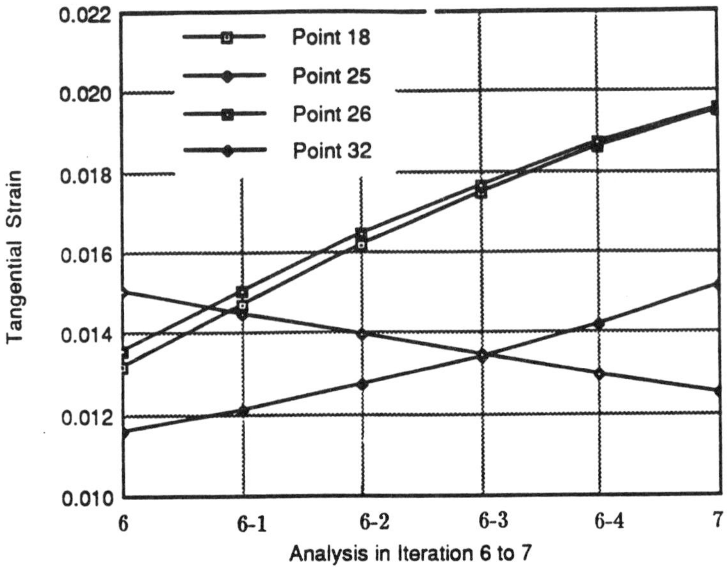

Figure 10. Elliptic Hole Tangential Strain Minimization Prob. Detailed Analysis 2.

One way to solve the problem is to reduce move limits. It should be noted that a small move limit slows down the optimization process and sometimes leads the design variables to settle on the point which is not minimum point.

REFERENCES

1. O.C. Zienkiewicz and J.S. Campbell, "Shape Optimization and Sequential Linear Programming in Optimal Structural Design", Wiley, New York, 1973

2. C.V. Ramakrishnan and A. Francavilla, "Structural Shape Optimization using Penalty Functions", Journal of Structural Mechanics, Vol. 3, No. 4, pp. 403-422, 1974-1975.

3. I. Babuska and B. Szabo, "On The Rate of Convergence of The Finite Element Method", Int. J. Num. Meth. Eng. 18, pp. 323-341, 1982.

4. B. Szabo and I. Babuska, Finite Element Analysis, To be published by John Wiley & Son's,1991.

5. J. Oda and K. Yamazaki, "Pattern Transformation Methods for Shape Optimization and Its Application to Spoked Rotary Disks", Int. Sym. on Optimum Structural Design, Tucson, AZ, pp. 4-29 to 4-35, 1981.

6. M.H. Imam, "Minimum Weight Design of 3-D Solid Components", Proceedings of the Second Int. Computer Eng. Conf., ASME, Vol. 3, San Diego, Calif., pp. 119-126, Aug. 1982.

7. M.H. Imam, "Three-Dimensional Shape Optimization", Int. Journal for Numerical Methods in Engineering, Vol. 18, pp. 661-673, 1982.

8. J.A. Bennett and M.E. Botkin, "Structural Shape Optimization with Geometric Description and Adaptive Mesh Refinement", AIAA Journal, Vol. 23, No. 1, pp. 459-464, 1982.

9. M.E. Botkin and J.A. Bennett, "Shape Optimization of Three-Dimensional Folded-Plate Structures", AIAA Journal, Vol. 23, No. 1, pp. 1804-1810, Nov. 1985.

10. M.E. Botkin, "Shape Optimization of Plate and Shell Structures", AIAA Journal, Vol. 20, No. 1, pp. 268-273, 1981.

11. P. Penderson and C.L. Laursen, "Design for Minimum Stress Concentration by Finite Elements and Linear Programming", Journal of Structural Mechanics", Vol. 10, No. 4, pp. 375-391, 1982-1983.

12. S.Y. Wang, Y. Sun, and R.H. Gallaghar, "Sensitivity Analysis in Shape Optimization of Continuum Structures" Computer & Structures, Vol. 20, No. 5, pp. 855-867, 1985.

13. K. Dems and Z. Mroz, "Variational Approach By Means of Adjoint Systems to Structural Optimization and Sensitivity Analysis-1", Int. J. for Solids Structures, Vol. 19, No. 8, pp. 677-692, 1983.

14. K. Dems and Z. Mroz, "Variational Approach by Means of Adjoint Systems to Structural Optimization and Sensitivity Analysis-2", Int. J. for Solids Structures, Vol. 20, No. 6, pp. 527-552, 1984.

15. E.J. Haug and Rousselet, "Design Sensitivity Analysis in Structural Mechanics-1 Static Response Variations", Journal of Structural Mechanics, Vol. 8, No. 1, pp. 17-41, 1980.

16. K.K. Choi and E.J. Haug, "Shape Design Sensitivity Analysis of Elastic Structures", Journal of Structural Mechanics, Vol. 11, No. 2, pp. 231-269, 1983.

17. R.J. Yang and M.E. Botkin, "Comparison Between the Variational and Implicit Differentiation Approaches to Shape Design Sensitivities", AIAA Journal, Vol. 24, No. 6, pp. 1027-1032, June 1986.

18. G.L. Giles and J.L. Rogers, "Implementation of Structural Response Sensitivity Calculation in a Large-Scale Finite Element Analysis Systems", AIAA/ASME/ASCE/AHS, 23rd Structures, Structural Dynamics, and Materials Conf., pt. 2, AIAA No.82-0714, pp. 348-359, 1982.

19. B. Prasad and J.F. Emerson, "General Capability of Design Sensitivity for Finite Element Systems", AIAA/ASME/ASCE/AHS, 23rd Structures, Structural Dynamics, and Materials Conf., pt. 2, AIAA No. 82-0686, pp. 175-186, 1982.

20. D.V. Wallerstein, "Design Enhancement Tools in MSC/NASTRAN", Recent Experiences in Multidisciplinary Analysis and Optimization, NASA, CP-2327, pp. 505-526, 1984.

21. C.S. Han, "Shape Optimization In 2-D Continuum Structure By Using P-version Finite Element Method", Ph.D. Dissertation, The University Of Texas, at Arlington, 1989.

22. E.P. Popov. "Mechanics of Materials", Textbook, Prentice-Hall, Second Edition, 1976.

23. J. S. Arora, "Introduction to Optimum Design", Textbook, McGraw-Hill, 1989.

On Maximum Stiffness of Orthotropic Shells

M. Rovati (*), A. Taliercio (**), C. Cinquini (***)
(*) Department of Structural Mechanics, University of Trento, I-38050 Mesiano di Povo, Trento, Italy
(**) Department of Structural Engineering, Politecnico di Milano, I-20133 Milano, Italy
(***) Department of Structural Mechanics, University of Pavia, I-27100 Pavia, Italy

ABSTRACT

Assuming the specific elastic strain energy as a meaningful measure of the stiffness/flexibility of a structural element, in this paper orthotropic shell structures of general shape will be considered and extreme values of the stored strain energy density sought. The role of design variable will be played by the local orientation of the principal directions of orthotropy with respect to a given reference frame, defined on the midsurface of the shell. Assuming the strain field as prescribed, the optimization problem will be formulated in variational form as a *min max* problem and the necessary condition for stationarity will be obtained as well. Special features of such optimality condition will be discussed and the analogies with similar problems in plane elasticity pointed out.

INTRODUCTION

The use of composite materials has received, in last years, great attention in research and practical applications; due to the inherent tailoring of mechanical properties of these materials, they appear, regardless of economic factors, as the most appropriate materials to fulfil requirements such as reduced weight, increased specific strength and performance. Because of their superior mechanical properties, design of composite materials represents a promising field of engineering, especially in the design of lightweight structural elements having specific stiffness and strength requirements. An efficient composite structural design can be achieved not only by sizing the cross sectional areas and member thicknesses, but also by local or global tayloring of the material properties through selective use of fiber orientation or number and stacking sequence of laminae in a laminate. In last years all these aspects have attracted the attention of many researchers so that now a considerable amount of literature on the subject is available.

Up to now, attention has been paid in particular to optimization problems concerning orthotropic structural elements in plane elasticity and orthotropic plates in bending.

After the pioneering works of Banichuk [6] and Kartevelishvili, Kobelev [7], mention should also be made of the more recent results obtained by Pedersen [9], [10] and Sacchi Landriani, Rovati [8], concerning optimality criteria for membranes of maximum stiffness and their mechanical interpretation. These results have been also extended to the three-dimensional elastic case by Seregin, Troitskii [11] and Rovati, Taliercio [12], and to the plastic case by Banichuk, Kobelev [13] and Sacchi Landriani, Rovati [14]. Then, several numerical applications have been made, as in Pedersen [15], Thomsen, Olhoff [16] and Bendsøe [17], [18], obtaining interesting results of practical importance.

In this paper orthotropic shells of general shape are considered, and the optimal orientations of principal directions of orthotropy are sought in order to maximize (or minimize) the elastic stiffness of the structure. After some geometrical, kinematical and statical preliminaries, the problem is formulated in variational form assuming as design variable the local orientation of the material symmetry axes with respect to a suitable reference frame defined on the shell midsurface. The elastic strain energy density is assumed as the parameter representative of the structural stiffness of the shell. The necessary condition for extreme values of the specific strain energy is then obtained in general form, and the discussion of those particular cases of major mechanical interest is made.

GEOMETRICAL PRELIMINARIES AND GOVERNING EQUATIONS

Consider a shell of general shape in an orthogonal reference frame x^i (i=1,2,3); let the midsurface be described by means of a set of parametric relations

$$x^i = f^i\,(\zeta^1, \zeta^2) \tag{1}$$

where parameters ζ^k (k=1,2) denote arbitrary curvilinear coordinates defined on the midsurface. The distance of two neighbouring points on the midsurface can be obtained from the well known first fundamental form of a surface

$$ds^2 = A_{hk} d\zeta^h d\zeta^k \qquad h,k = 1,2 \tag{2}$$

where the metric tensor has been defined

$$A_{hk} = f^i|_h f_i|_k \qquad h,k = 1,2. \tag{3}$$

(Here and heieinafter a vertical bar in mathematical expressions denotes covariant o contravariant derivatives). In particular, if ζ^k are orthogonal curvilinear coordinates, now denoted by η^k, the metric tensor becomes diagonal and Equation (2) is simplified in the form

$$ds^2 = A_{11} d\eta^1 d\eta^2 + A_{22} d\eta^1 d\eta^2 \tag{4}$$

and the Lamé parameters (or measure numbers) can be defined as

$$\mathcal{A}_1 = \sqrt{A_{11}} \qquad \mathcal{A}_2 = \sqrt{A_{22}}. \tag{5}$$

Let $\underline{e}_i$ (i=1,2,3) be an orthonormal basis associated to the reference frame x^i. Denoting by $\underline{r}$ the position vector of any point P lying on the midsurface of the shell, i.e.

$$\underline{r} = f^i \underline{e}_i \qquad i = 1,2,3 \tag{6}$$

derivatives of such a vector with respect to the curvilinear coordinates ζ^k allow the definition of unit vectors $\underline{\tau}^k$ (k=1,2) on the plane tangent to the midsurface as

$$\underline{\tau}_k = \frac{\underline{r}|_k}{|\underline{r}|_k|} = \frac{\underline{r}|_k}{\mathcal{A}_k} \qquad k = 1,2. \tag{7}$$

In such a way, a right-handed frame is completely defined by setting

$$\underline{\tau}_3 = \underline{\tau}_1 \wedge \underline{\tau}_2. \tag{8}$$

The state of strain at each point of the shell is then assumed to be given by the deformation of the midsurface. This deformation is completely described by the displacement functions u^i (i=1,2,3). From these scalar functions it is possible (see Niordson [1]) to calculate local measures of the deformation of the midsurface. These are the strain (membrane) components E_{hk}, which in linearized form read

$$E_{hk} = \frac{1}{2}(u_k|_h + u_h|_k) - d_{hk}u^3 \qquad h,k = 1,2 \tag{9}$$

and the linearized bending (flexural curvatures) components

$$K_{hk} = u^3|_{hk} + d_{hm}u^m|_k + d_{km}u^m|_h + u^m d_{mh}|_k - d_{km}d_k^m u^3 \qquad h,k,m = 1,2. \tag{10}$$

In Equations (9) and (10) the symmetric tensor d_{hk} represents the tensor of the geometric curvatures of the midsurface. It is worth noting that the six strain components (9) and (10) are not independent, but they have to satisfy three compatibility equations (see Niordson [1]).

If the shell is subjected to a prescribed external load $\underline{q} = q^i\underline{\tau}_i$ (i=1,2,3), the equilibrium equations read (h,k,m=1,2)

$$\begin{cases} N^{hk}|_h + 2d_m^k M^{mh}|_h + M^{mh}d_m^k|_h + q^k = 0 \\ M^{hk}|_{hk} - d_{hm}d_k^m M^{hk} - d_{hk}N^{hk} - q^3 = 0. \end{cases} \tag{11}$$

Assume that the material forming the shell is linearly elastic, homogeneous and locally orthotropic, in the sense that the orientation of the principal directions of orthotropy is allowed to change smoothly along the midsurface of the shell. In terms of characteristics, the stress–strain relationship takes the form (h,k,r,s=1,2)

$$\begin{cases} N^{hk} = A^{hkrs}E_{rs} + B^{hkrs}K_{rs} \\ M^{hk} = B^{hkrs}E_{rs} + D^{hkrs}K_{rs}. \end{cases} \tag{12}$$

In the case under examination, where the shell is supposed to be homogeneous across the thickness and not to be a laminate, the midsurface of the shell is a symmetry surface for both geometry and elastic properties, so that the constitutive equations can be rewritten in the simpler form

$$\begin{cases} N^{hk} = A^{hkrs}E_{rs} \\ M^{hk} = D^{hkrs}K_{rs}. \end{cases} \tag{13}$$

In such a case the coupling between membrane and flexural effects vanishes.

OPTIMIZATION PROBLEM AND STATIONARITY CONDITION

In order to define the design variable in a suitable way, some reference systems on the shell midsurface must be considered; the most natural choice to identify the principal directions of orthotropy h^k (k=1,2) is to refer them to a fixed frame, representative of the geometry of the shell, namely the principal directions of geometrical curvature γ^k. Conventionally α will denote the angle between the direction of the greatest stiffness (h^1) and the direction of the greatest curvature (γ^1). On the other hand, for computational convenience, it will be preferable to refer the material symmetry axes to either principal directions of membrane strains μ^k or to the principal directions of flexural curvature χ^k. If ψ denotes the angle between the axis of numerically larger flexural curvature χ^1 and the direction h^1, ϕ the angle between the direction of numerically larger membrane strain μ^1 and h^1, and θ the angle between h^1 and μ^1, then $\alpha = \psi + \phi + \theta$. Nevertheless, it turns out that for a prescribed value of α the orientations ψ and ϕ of membrane strains and flexural curvatures appear as a result in the analysis process, thus for optimization purposes the angle θ (or the angle $\theta+\phi$) can be assumed as control variable.

Assuming θ as design variable, the problem can be stated as the search for that (or those) value(s) of θ minimizing (or maximizing) the work performed by the external loads, i.e. maximizing (resp. minimizing) the stiffness of the shell. Thus the problem can be written as follows:

PROBLEM 1. *Find, according to equilibrium, compatibility and constitutive law:*

$$\mathcal{J}_0 \equiv \min_\theta \mathcal{J}(\theta) = \min_\theta \frac{1}{2} \int_\Omega q^i \overline{u}_i \mathcal{A}_1 \mathcal{A}_2 d\eta^1 d\eta^2 \qquad i = 1,2,3 \tag{14}$$

where $\overline{u}_i$ are the actual components of the displacement vector for θ prescribed. The functional $\mathcal{J}(\theta)$ to be minimized (or maximized) is equal, but opposite in sign, to $\mathcal{U}_0 = \min_{u_i} \mathcal{U}$ where $\mathcal{U}(u_i, \theta)$ is the total potential energy and where functions u_i belong to the class of compatible displacement fields. In such a way, the constrained *min* problem (14) can be rewritten (see Sacchi Landriani, Rovati [9]) as an unconstrained *min max* problem as

PROBLEM 2. *Find:*

$$\mathcal{J}_0 = \min_\theta \max_{u_i} \left(\int_\Omega q^i u_i \mathcal{A}_1 \mathcal{A}_2 d\eta^1 d\eta^2 - \int_\Omega \mathcal{E} \mathcal{A}_1 \mathcal{A}_2 d\eta^1 d\eta^2 \right) \tag{15}$$

where

$$\mathcal{E} = \frac{1}{2} \left(N^{hk} E_{hk} + M^{hk} K_{hk} \right) \qquad h, k = 1,2 \tag{16}$$

is the specific strain energy which, making use of the stress–strain relationship (13), in matrix form referred to the principal directions of orthotropy can be written as

$$\mathcal{E} = \frac{1}{2} \begin{Bmatrix} E_{11} \\ E_{22} \\ 2E_{12} \\ K_{11} \\ K_{22} \\ 2K_{12} \end{Bmatrix}^T \begin{pmatrix} A_0^{1111} & A_0^{1122} & 0 & 0 & 0 & 0 \\ A_0^{1122} & A_0^{2222} & 0 & 0 & 0 & 0 \\ 0 & 0 & A_0^{1212} & 0 & 0 & 0 \\ 0 & 0 & 0 & D_0^{1111} & D_0^{1122} & 0 \\ 0 & 0 & 0 & D_0^{1122} & D_0^{2222} & 0 \\ 0 & 0 & 0 & 0 & 0 & D_0^{1212} \end{pmatrix} \begin{Bmatrix} E_{11} \\ E_{22} \\ 2E_{12} \\ K_{11} \\ K_{22} \\ 2K_{12} \end{Bmatrix}. \tag{17}$$

Subscript "0" in the elastic coefficient matrix is adopted to characterize the stiffness parameters referred to the principal directions of orthotropy. Equation (17) explicitly reads

$$\begin{aligned} \mathcal{E} = \frac{1}{2} \big(& A_0^{1111} E_{11}^2 + A_0^{2222} E_{22}^2 + 4A_0^{1212} E_{12}^2 + 2A_0^{1122} E_{11} E_{22} \\ & + D_0^{1111} K_{11}^2 + D_0^{2222} K_{22}^2 + 4D_0^{1212} K_{12}^2 + 2D_0^{1122} K_{11} K_{22} \big) . \end{aligned} \tag{18}$$

It must be noticed that in functional (15) the only term in which the design variable θ appears is the strain energy density $\mathcal{E}$. Then stationarity of $\mathcal{J}$ with respect to θ reads

$$\frac{\partial \mathcal{E}}{\partial \theta} = 0. \tag{19}$$

Now it is convenient to express the specific strain energy (18) in terms of principal strains E_I, E_{II}, K_I, K_{II}. The well known rotation formulae are

$$\begin{cases} E_{11} = \frac{1}{2} \left[(E_I + E_{II}) + (E_I - E_{II}) \cos 2\theta \right] \\ E_{22} = \frac{1}{2} \left[(E_I + E_{II}) - (E_I - E_{II}) \cos 2\theta \right] \\ E_{12} = -\frac{1}{2} (E_I - E_{II}) \sin 2\theta \\ K_{11} = \frac{1}{2} \left[(K_I + K_{II}) + (K_I - K_{II}) \cos 2(\theta + \phi) \right] \\ K_{22} = \frac{1}{2} \left[(K_I + K_{II}) - (K_I - K_{II}) \cos 2(\theta + \phi) \right] \\ K_{12} = -\frac{1}{2} (K_I - K_{II}) \sin 2(\theta + \phi). \end{cases} \tag{20}$$

The derivatives of strain components with respect to design variable θ are

$$\begin{cases} E_{11,\theta} = -(E_I - E_{II})\sin 2\theta = 2E_{12} \\ E_{22,\theta} = (E_I - E_{II})\sin 2\theta = -2E_{12} \\ E_{12,\theta} = -(E_I - E_{II})\cos 2\theta = -(E_{11} - E_{22}) \\ K_{11,\theta} = -(K_I - K_{II})\sin 2(\theta + \phi) = 2K_{12} \\ K_{22,\theta} = (K_I - K_{II})\sin 2(\theta + \phi) = -2K_{12} \\ K_{12,\theta} = -(K_I - K_{II})\cos 2(\theta + \phi) = -(K_{11} - K_{22}). \end{cases} \tag{21}$$

Therefore the stationarity condition is

$$\begin{aligned} \frac{\partial \mathcal{E}}{\partial \theta} = & \frac{1}{2}\left(2A_0^{1111}E_{11}E_{11,\theta} + 2A_0^{2222}E_{22}E_{22,\theta} + 8A_0^{1212}E_{12}E_{12,\theta} + 2A_0^{1122}E_{11,\theta}E_{22}\right. \\ & + 2A_0^{1122}E_{11}E_{22,\theta} + 2D_0^{1111}K_{11}K_{11,\theta} + 2D_0^{2222}K_{22}K_{22,\theta} + 8D_0^{1212}K_{12}K_{12,\theta} \\ & \left. + 2D_0^{1122}K_{11,\theta}K_{22} + 2D_0^{1122}K_{11}K_{22,\theta}\right) = 0, \end{aligned} \tag{22}$$

or, by virtue of Equations (21),

$$\begin{aligned} \frac{\partial \mathcal{E}}{\partial \theta} = & \, 2E_{12}\left[A_0^{1111}E_{11} - A_0^{2222}E_{22} - \left(A_0^{1122} + 2A_0^{1212}\right)(E_{11} - E_{22})\right] \\ & + 2K_{12}\left[D_0^{1111}K_{11} - D_0^{2222}K_{22} - \left(D_0^{1122} + 2D_0^{1212}\right)(K_{11} - K_{22})\right] = 0. \end{aligned} \tag{23}$$

Finally, the necessary stationarity condition can also be written in terms of principal membrane and flexural strains, as

$$\begin{aligned} \frac{\partial \mathcal{E}}{\partial \theta} = & (E_I - E_{II})\sin\theta\cos\theta\left[A_E + B_E\cos 2\theta\right] \\ & + (K_I - K_{II})\sin(\theta + \phi)\cos(\theta + \phi)\left[A_K + B_K\cos 2(\theta + \phi)\right] = 0. \end{aligned} \tag{24}$$

where

$$\begin{cases} A_E = \left(A_0^{1111} - A_0^{2222}\right)(E_I + E_{II}) \\ B_E = \left(A_0^{1111} + A_0^{2222} - 2A_0^{1122} - 4A_0^{1212}\right)(E_I - E_{II}) \\ A_K = \left(D_0^{1111} - D_0^{2222}\right)(K_I + K_{II}) \\ B_K = \left(D_0^{1111} + D_0^{2222} - 2D_0^{1122} - 4D_0^{1212}\right)(K_I - K_{II}). \end{cases} \tag{25}$$

SOME REMARKS ABOUT OPTIMAL ORIENTATIONS

Equation (24) is a fourth-order algebraic equation, with $\mathrm{tg}\theta$ as unknown

$$\begin{aligned} & \mathrm{tg}^4\theta\,(-A_K + B_K\cos 2\phi)\sin 2\phi + 2\mathrm{tg}^3\theta\,\{(E_I - E_{II})(A_E - B_E) \\ & + (K_I - K_{II})\left[\cos 2\phi A_K + B_K\left(\sin^2 2\phi - \cos^2 2\phi\right)\right]\} - 6\mathrm{tg}^2\theta\,(K_I - K_{II})\,B_K\cos 2\phi\sin 2\phi \\ & + 2\mathrm{tg}\theta\{(E_I - E_{II})(A_E - B_E) + (K_I - K_{II})\left[\cos 2\phi A_K - B_K\left(\sin^2 2\phi - \cos^2 2\phi\right)\right. \\ & \left. + \frac{1}{2}A_K\sin 2\phi\right]\} + (K_I - K_{II})(A_K + B_K\cos 2\phi)\sin 2\phi = 0. \end{aligned} \tag{26}$$

Solution of Equation (26) allows to obtain all the orientations θ at which functional (15) is stationary. On the other hand, on the basis of results obtained with reference to plane elastic problems (see Pedersen [12] and Sacchi Landriani, Rovati [10]) for materials characterized by low shear stiffness, only some particular solutions, which can be discussed in a easier way making reference to Equation (24), are of major interest.

Omitting the extremely particular case where $K_I = K_{II}$ and $E_I = E_{II}$ simultaneously, condition (24) can be certainly be satisfied when

- $E_I = E_{II}$, with $K_I \neq K_{II}$ such that Equation (24) reads

$$\sin(\theta+\phi)\cos(\theta+\phi)\left[2B_K\cos^2(\theta+\phi)+A_K-B_K\right]=0 \qquad (27)$$

from which

$$\begin{cases}\sin(\theta+\phi)=0\\ \cos(\theta+\phi)=0\\ \cos^2(\theta+\phi)=-\frac{A_K-B_K}{2B_K} \qquad 0<\frac{A_K-B_K}{2B_K}<1.\end{cases} \qquad (28)$$

The first two solutions indicate collinearity between principal directions of orthotropy and principal directions of flexural curvature. The third solution depends on the given strain field and on the material properties. It is not difficult to show that, in the particular case $D_0^{1111} = D_0^{2222}$, such a solution is $(\theta+\phi) = \pm\frac{\pi}{4}$.

- $K_I + K_{II}$, with $E_I \neq E_{II}$ such that

$$\sin\theta\cos\theta\left(2B_E\cos^2\theta+A_E-B_E\right)=0 \qquad (29)$$

which is satisfied for

$$\begin{cases}\sin\theta=0\\ \cos\theta=0\\ \cos^2\theta=-\frac{A_E-B_E}{2B_E} \qquad 0<-\frac{A_E-B_E}{2B_E}<1.\end{cases} \qquad (30)$$

In this case, apart from the third intermediate solution, stationarity of functional (15) is characterized by collinearity of principal directions of orthotropy and principal directions of membrane strain.

These solutions are of interest for shells where a pure flexural or membrane strain field exists. The several possible orientations of orthotropy axes for the stationarity of functional (15) also possess a mechanical interpretation. In fact, it can be shown that if in condition (24) the two addenda vanish independently (as in the cases just discussed), then it turns out that principal directions of membrane strain are collinear with principal directions of membrane stress, and principal directions of flexural curvature are collinear with principal directions of bending moment. This can be proved by writing, for instance, the condition of collinearity between principal directions of membrane strain/stresses

$$\begin{Bmatrix} N^I\\ N^{II}\\ 0\\ M^{11}\\ M^{22}\\ M^{12}\end{Bmatrix} = \begin{pmatrix} A^{1111} & A^{1122} & A^{1112} & 0 & 0 & 0\\ A^{1122} & A^{2222} & A^{2212} & 0 & 0 & 0\\ A^{1112} & A^{2212} & A^{1212} & 0 & 0 & 0\\ 0 & 0 & 0 & D^{1111} & D^{1122} & D^{1112}\\ 0 & 0 & 0 & D^{1122} & D^{2222} & D^{2212}\\ 0 & 0 & 0 & D^{1112} & D^{2212} & D^{1212}\end{pmatrix} \begin{Bmatrix} E_I\\ E_{II}\\ 0\\ K_{11}\\ K_{12}\\ 2K_{12}\end{Bmatrix}. \qquad (31)$$

Then the condition of collinearity reads

$$A^{1112}E_I + A^{2212}E_{II} = 0 \qquad (32)$$

which, making use of the rotation formulae for the stiffness coefficients, returns the first addendum of Equation (24). An analogous consideration can be obviously made for the second addendum.

CONCLUDING REMARKS

In this paper orthotropic shells of general shape have been considered and the problem consisting of the search for that optimal distributions of material symmetry axes maximizing (or minimizing) the shell stiffness, has been formulated. The variational formulation adopted allows to compute a stationarity condition which provides several values of the design variable at which the functional under consideration attains extreme values. This optimality condition has been rewritten in algebraic form in view of a possible explicit general solution. Then attention has been paid to some particular solutions which lead themselves also to meaningful mechanical interpretations. In particular it must be noticed that solutions (28) and (30) show a strict correspondence with the optimal solutions of the same problem in plane elasticity (see Sacchi Landriani, Rovati [10]), even though in the case of shells the number of solutions increases. Moreover, some optimal orientations of orthotropy axes provide collinearity of material axes with principal directions of membrane strain or flexural curvature. Then, intermediate solutions are also possible, but they show an explicit dependence on the material properties and on the given strain field. Furthermore, by virtue of the hypothesis stating that the strain field is prescribed, the optimal solutions are not related to the intrinsic geometry of the shell, the influence of which appears in the definition of the strain themselves through Equations (9) and (10). Finally, even though the most significant solutions have been given, more general orientations of orthotropy axes are possible. Also possible is a characterization of the nature of the extreme values for functional (15) through suitable checks of the second variations of the functional itself. As in the plane elastic case, a strong dependence on the stiffness coefficients is expected.

ACKNOWLEDGEMENTS

The present work has been made possible by a financial support from M.U.R.S.T. (Italian Ministry for University and Research), which is here gratefully acknowledged.

REFERENCES

1. Niordson F.I., *Shell Theory*, North–Holland, Amsterdam, 1985.
2. Naghdi P.M., *The Theory of Shells and Plates*, in *Mechanics of Solids* (C. Truesdell Ed.), Springer–Verlag, Berlin, 1984.
3. Flügge W., *Stresses in Shells*, Springer–Verlag, Berlin, 1973.
4. Vinson J.R., Sierakowski R.L., *The Behaviour of Structures Composed of Composite Materials*, Martinus Nijhoff Publ., Dordrecht, 1986.
5. Broutman L.J., Krok R.H., *Composite Materials*, Vol.7, *Structural Design and Analysis, Part 1*, (C.C. Chamis Ed.), Academic Press, New York, 1975.
6. Banichuk N.V., *Optimization Problems for Elastic Anisotropic Bodies*, Arch. Mech., Vol. 33, N. 6, 347–363, 1981.
7. Kartevelishvili V.M., Kobelev V.V., *Rational Schemes for Reinforcing Laminar Plates from Composite Materials*, P.M.M., Vol. 48, N. 1, pp. 68–88, 1984.
8. Sacchi Landriani G., Rovati M., *Optimal Design for Two-Dimensional Structures made of Composite Materials*, ASME Journal of Engineering Materials and Technology, Vol. 113, pp. 88–92, 1991.
9. Pedersen P., *On Optimal Orientation of Orthotropic Materials*, Structural Optimization, Vol. 1, pp. 101–106, 1989.
10. Pedersen P., *Bounds on Elastic Energy in Solids of Orthotropic Materials*, Structural Optimization, Vol. 2, pp. 55–63, 1990.
11. Seregin G.A., Troitskii V.A., *On the Best Position of Elastic Symmetry Planes in an Orthotropic Body*, P.M.M., Vol. 45, pp. 139–142, 1981.
12. Rovati M., Taliercio A., *Optimal Orientation of the Symmetry Axes of Orthotropic 3-D Materials*, Proc. *International Conference on Engineering Optimization in Design Processes*, Karlsruhe, FRG, 3–4 September 1990.

13. Banichuk N.V., Kobelev V.V., *On Optimal Plastic Anisotropy*, P.M.M., Vol. 51, N. 3, pp. 489–495, 1987.
14. Sacchi Landriani G., Rovati M., *Optimal Limit Design of Fiber Reinforced Orthotropic Bodies*, Proc. COMETT Course on *Computer Aided Optimal Design of Structures*, University of Pavia, Italy, 4–8 September 1989.
15. Pedersen P., *Combining Material and Element Rotation in One Formula*, Communications in Applied Numerical Methods, Vol. 6, pp. 549–555, 1990.
16. Thomsen J., Olhoff N., *Optimization of Fiber Orientation and Concentration in Composites*, Proc. CEEC Course on *Design and Analysis of Structures made of Composite Materials*, Milano, Italy, 28 May – 1 June 1990.
17. Bendsøe M.Ph., *Optimal Shape Design as a Material Distribution Problem*, Proc. COMETT Course on *Computer Aided Optimal Design of Structures*, University of Pavia, Italy, 4–8 September 1989.
18. Bendsøe M.Ph., *Composites as a Basis for Topology Optimization*, Proc. CEEC Course on *Design and Analysis of Structures made of Composite Materials*, Milano, Italy, 28 May – 1 June 1990.

SECTION 8: EXPERT SYSTEMS AND INTEGRATED PACKAGES IN OPTIMUM DESIGN

The Using of Fuzzy Set Theories in Optimum Structural Design and Modelling

J.L. Chen, E.Y. Lai

Department of Mechanical Engineering, National Cheng Kung University, Tainan, Taiwan 70101, Republic of China

ABSTRACT

This paper describes a study on applying the fuzzy set theories to optimum structural design problems with imprecise information. The fuzzy membership function building methods and fuzzy optimum structural design approaches are evaluated. The use of fuzzy reasoning and expert system techniques for optimum design modelling task is proposed. Examples are solved to illustrate the ideas.

INTRODUCTION

Many real world engineering problems exist a vast amount of imprecise information. However, the designer is often forced to describe their design problems in precise mathematical terms rather than in terms of the imprecise information. The fuzzy set theories can be used to deal with imprecise information quantitatively. The present paper deals with the study of using fuzzy set theories to perform optimum design modeling and optimization.

In the beginning of this paper, several fuzzy membership function building methods are studied. Next, a comparison of existing fuzzy optimum structural design methods are described. This study includes both single- and multi-objective fuzzy optimization methods. Finally, based on the experience of previous studies, an approach that uses expert system technique and fuzzy reasoning to aid the designer in optimum structural design modelling is presented. An expert system for fuzzy optimum structural design modelling is developed. This system can aid the designer in transferring his design problem into fuzzy models and choosing suitable membership functions and fuzzy optimum structural design methods.

The fuzzy optimum structural design model obtained from this approach can describe the optimum design problem more reasonably. Futhermore, this fuzzy optimum design model can be solved by a suitable fuzzy optimum structural design method. Some fuzzy optimum structural design problems are demonstrated to illustrate this idea. The present study offers a new methodology to aid the designer in handling optimum structural design problems with imprecise information.

FUZZY SET THEORIES AND MEMBERSHIP FUNCTIONS

The fuzzy set theories were initiated by Zadeh [1] in 1965. In this theory, the fuzziness is indicated by fuzzy set. In such a set, the membership relation of an element can not be answered simply with "yes" or "no", for it is a gradually changing process. The characteristic function of fuzzy set is called membership function which is indicated by $\mu_A(x)$.

$$\mu_A(x) = \begin{cases} 1 & \text{if } g_m(X) \le b_m^U \\ f(x) & \text{if } b_m^U \le g_m(X) \le b_m^U + d_m^U \\ 0 & \text{if } g_m(X) \ge b_m^U + d_m^U \end{cases}$$

The schematic representation of membership function is shown in Figure 1.

For solving the fuzzy optimum structural design problem, the constructing of membership function, f(x), is the first job for the designer. Although a simple linear membership function is most commonly used, other forms of membership function may be more suitable to represent the actual situations. Several fuzzy membership function building methods are studied in this paper.

1. Existing mathematical functions [2]

(a) Linear function: $f(x) = [(b_m^U + d_m^U) - g_m(X_i)] / d_m^U$

(b) Sine curve I: $f(x) = \frac{1}{2} - \frac{1}{2}\sin\left[\frac{\pi}{d_m^U}\left(g_m(X_i) - b_m^U - \frac{1}{2} d_m^U\right)\right]$

(c) Sine curve II: $f(x) = 1 - \sin\frac{\pi}{2}\left[\frac{1}{d_m^U}\left(g_m(X_i) - \frac{1}{2} d_m^U\right)\right]$

(d) Exponential function: $f(x) = \dfrac{e^{-k\delta} - e^{-k}}{1 - e^{-k}}$

(e) Hyperbolic function [2]
(f) Quadratic function [2]
(g) Logarithmic function [2]

2. Standard functions [3]
   (a) S function [3]
   (b) Z function [3]
   (c) T function [3]
3. Linguistic assessment [4]
4. Using statistical data [5]

FUZZY OPTIMUM STRUCTURAL DESIGN METHODS

In recent years, several fuzzy optimum structural design methods [6-10] have been developed. In this study, five fuzzy optimum structural design methods were chosen for evaluation.

1. Level-cuts method [6]
2. Two-phase method [7]
3. Fuzzy expected value method [8]
4. Fuzzy set theory method [9,10]
5. Fuzzy goal programming method [11]

The fuzzy set theory method can deal with both single- and multi-objective optimum design problems. The fuzzy goal programming method is used for solving multi-objective optimization problems. Other methods can only solve single-objective optimal design problems. Since most methods have been described in the existing publications [6-10], this section will only describe the fuzzy goal programming methods.

For a multi-objective structural optimization problem, the membership function of each objective function is defined as follows:

$$\mu_m = 1. \qquad \text{if } f_m \le f_m^{min}$$

$$= \frac{-f_m + U_m}{U_m - f_m^{min}} \qquad \text{if } f_m^{min} \le f_m \le U_m$$

$$= 0. \qquad \text{if } f_m \ge U_m$$

where f_m^{min} is the goal of the mth objective function. Two models are proposed for fuzzy goal programming.

(a) Fuzzy decision goal programming model [12]

Find $\lambda = \mu_D(X) = \min_{m=1}^{M} \mu_m$

to maximize λ

and subject to $\frac{\sigma_m}{\sigma_m^a} \le 1.$

$$\mu_m \le 1. \qquad X, \mu_m \ge 0.$$

(b) Additive goal programming model [13]

Find $\mu_D(X) = \sum_{m=1}^{M} \mu_m$

to maximize $\mu_D(X)$

and subject to $\frac{\sigma_m}{\sigma_m^a} \leq 1.$

$\mu_m \leq 1.$

$X, \mu_m \geq 0.$

FUZZY OPTIMUM DESIGN MODELLING EXPERT SYSTEM

The fuzzy optimum structural design modelling process can be divided into three steps:

(a) whether the design problem is appropriate to use fuzzy sets theory,
(b) evaluation membership function type, and
(c) evaluation fuzzy structural optimum design method.

During modelling process, the selection of membership function and fuzzy optimum structural design method will be affected by both the designer's subjectivity on observing the facts and the imprecision of knowledge. Therefore, the fuzzy reasoning logic was chosen for inference reasoning during modelling step (b) and (c).

In this study, the fuzzy optimum design modelling expert system is written in Turbo Prolog computer language [14] and operates on an PC-AT personal computer. Due to the fuzziness of linguistic assessment, the possibility of attribute is divided into five members (not, maybe not, unknown, maybe yes, and yes) in this system, as indicated in Figure 2.

A combination of the production rule and the object-attributes-values (O-A-V) knowledge representation scheme was used as knowledge representation approach in this expert system. For each rule in the knowledge base, a fuzzy relation matrix [15] can be obtained. Based on the fuzzy logic reasoning [15], the inference engine can draw the fuzzy conclusions from the facts and the fuzzy relation matrix associated with related rules. For fuzzy decision making from the fuzzy conclusions, the first step is using fuzzy weighted union method [16] for multi-attribute decision making. The second step is ranking fuzzy numbers and it can be done by using gravity center criterion method [17] to obtain possibility index (PI). The designer can evaluate the conclusions based on these PI values.

EXAMPLES

A three bar truss, as shown in Figure 3, was chosen as an example to evaluate the different membership functions. The weight and vertical deflection of the loaded joint are considered as first and second objective functions, respectively. The fuzzy data of this problem is given in Table 1. The fuzzy multi-objective optimization was solved by using the fuzzy set theory method. The numerical results by using different membership functions are shown in Table 2.

For evaluating the different fuzzy optimum structural design methods, the welded beam structure [18], as shown in Figure 4, was adopted. Details of optimum design formulation of this problem can be found in Reference 18. The product cost is selected as the objective function to be minimized. For multi-objective optimal design formulation, the buckling load is chosen as another objective function to be maximized. The allowable shear stress and allowable normal stress are $\tau_d = 13600$ psi and $\sigma_d = 30000$ psi , respectively. The transition zones for shear stress and normal stress are $d_\tau^U = 1400$ psi and $d_\sigma^U = 2000$ psi , respectively. Linear function is used in all membership functions. The numerical results of fuzzy single- and multi-objective optimization of this problem are given in Table 3 and 4, respectively.

The single objective welded beam structure example was chosen to demonstrate the fuzzy optimum design modelling expert system. Due to local change of temperature during the welding process, the value of allowable stresses may be changed. Therefore, this problem can be considered to use fuzzy optimum structural design methods. The results of inference reasoning for membership functions and fuzzy optimum design methods from this expert system are summarized in Table 5 and 6, respectively. The suggestions from the expert system are using linear membership function and fuzzy set theory method or level cuts method.

CONCLUDING REMARKS

This paper demonstrates the application of fuzzy set theories in fuzzy optimum structural design. Based on the results from the comparison of different membership functions, it is found that there is no significant difference in results between the linear membership function and other complicated membership functions. However, defining the transition zones should be done with care and it is an area needing further investigation. Several fuzzy optimum structural design methods are also investigated. Each method has its own characters. Nevertheless, for problems with imprecise information, the results of fuzzy optimum design is always better than the one from traditional optimum design methods. Finally, the developing fuzzy optimum design modelling expert system demonstrated its capabilities to aid the designer choosing suitable membership function and fuzzy optimum design method during modelling process.

ACKNOWLEDGMENTS

This work is supported by a grant from the national science council, Republic of China, Grant No. NSC-79-0401-E006-43.

REFERENCES

1. Zadeh, L.A. Fuzzy Sets, Information and Control, Vol. 8, pp. 338-353, 1965.
2. Dhingra, A.K., Rao, S.S., and Kumar, V. Nonlinear Membership Function in the Fuzzy Optimization of Mechanical and Structural Systems, in Proceedings of the AIAA/ASME/ASCE/AHS/ASC 31st Structures, Structural Dynamics and Materials Conference, Part I, pp. 403-413, Long Beach, CA, USA, April 2-4, 1990.
3. Zedeh, L.A. A Conceptual Approach to Fuzzy Quantifies in Natural Languages, Computer Math. appl., Vol. 9, pp. 149-184, 1983.
4. Hinkle, A.J. and Yao, J.T.P. Linguistic Assessment of Welded Structures with Fatigue Damage, International Journal of Approximate Reasoning, Vol. 2, pp. 47-63, 1988.
5. Civanlar, M.R. and Trussell, H.J. Constructing Membership Functions Using Statistical Data, Fuzzy Sets and Systems, Vol. 18, pp. 1-13, 1986.
6. Wang, G.-Y. and Wang, W.-Q. Fuzzy Optimum Design of Structures, Engineering Optimization, Vol. 8, pp. 291-300, 1985.
7. Xu, C. W. Fuzzy Optimization of Structures by the Two-Phase Method, Computers & Structures, Vol. 31, pp. 575-580, 1989.
8. Wang, G.-Y. Fuzzy Information and Structural Soft Design, in Proceedings of the 12th IMACS World Congress'88 on Scientific Computation, Vol. 3, pp. 433-435, Paris, July 18-22, 1988.
9. Rao, S.S. Optimum Design of Structures in a Fuzzy Environment, AIAA Journal, Vol. 25, pp. 1633-1636, 1987.
10. Rao, S.S. Multi-objective Optimization of Fuzzy Structural Systems, International Journal for Numerical Methods in Engineering, Vol. 24, pp. 1157-1171, 1987.
11. Chen, J.L. and Lai, E.Y. Fuzzy Goal Programming Methods for Optimum Structural Design, 1991, in preparation.
12. Rubin, P.A. and Narasimhan, R. Fuzzy Goal Programming with Nested Priorities, Fuzzy Sets and Systems, Vol. 14, pp. 115-129, 1984.
13. Tiwari, R.N., Dharmar, S., and Rao, J.R. Fuzzy Goal Programming - An Additive Model, Fuzzy Sets and Systems, Vol. 24, pp. 27-34, 1987.
14. Turbo Prolog User's Guide, Version 2.0, Borland International, 1988

15. Zadeh, L.A. Fuzzy Logic, IEEE Transactions on Computer, pp. 83-93, April 1988.
16. Dong, W.M. and Wong, F.S. Fuzzy Weighted Averages and Implementation of the Extension Principle, Fuzzy Sets and Systems, Vol. 21, pp. 183-199, 1987.
17. Bortolan, G. and Degani, R. A Review of Some Methods for Ranking Fuzzy Subsets, Fuzzy Sets and Systems, Vol. 15, pp. 1-15, 1985.
18. Sandgren, E. Gim, G., and Ragsdell, K.M. Optimal Design of a Class of Welded Structures Based on Design for Latitude, ASME Journal of Mechanisms, Transmissions, and Automation in Design, Vol. 107, pp. 482-487, 1985.

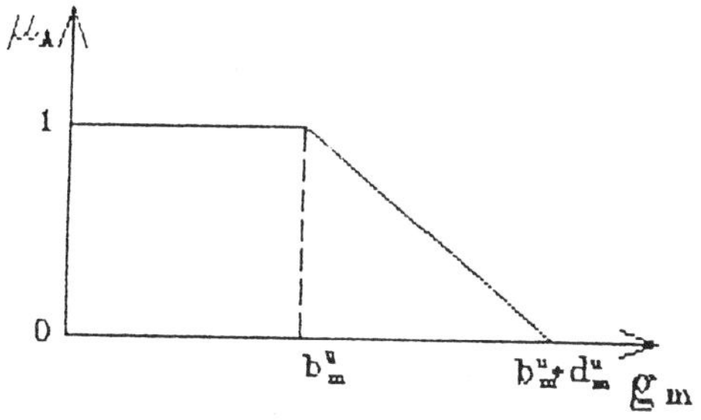

Fig. 1. The schematic representation of membership function.

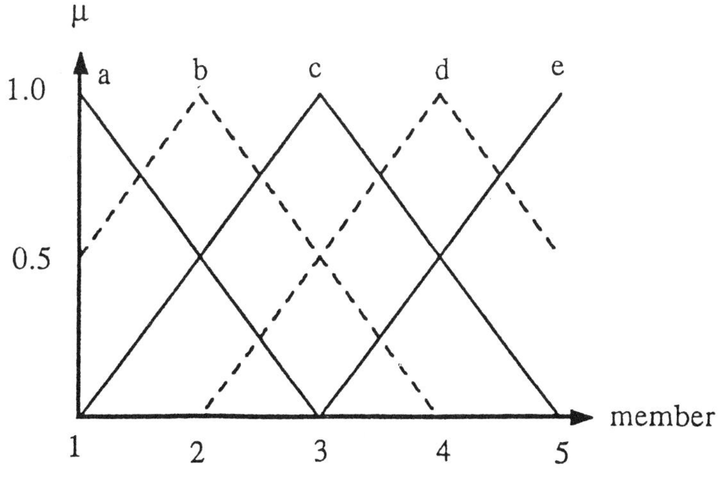

a:not b:maybe not c:unknown d:maybe yes e:yes

Fig. 2. Membership functions of the possibility of attribute.

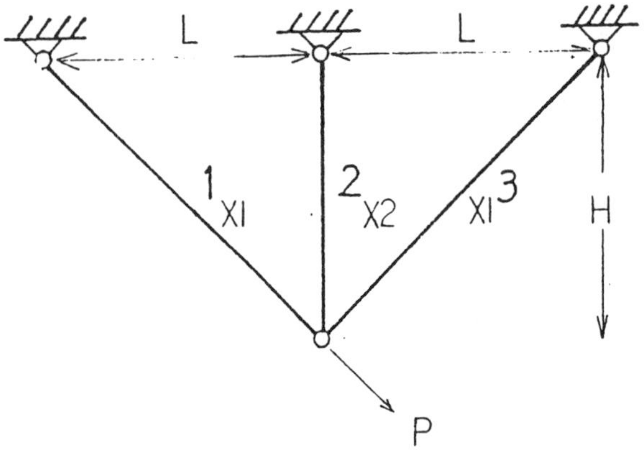

Fig. 3. The three bar truss.

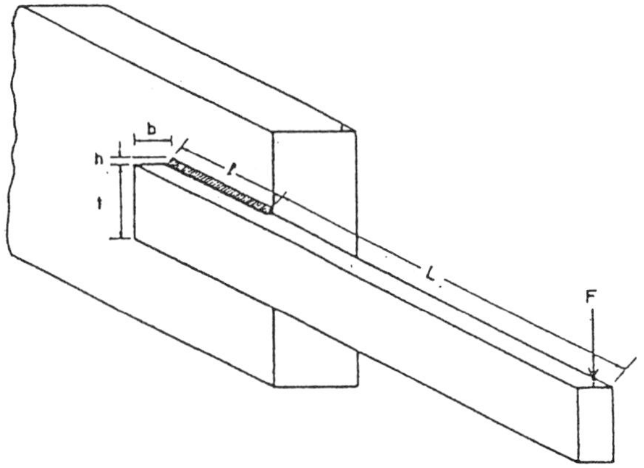

Fig. 4. The welded beam structure.

Table 1. The fuzzy data of three bar truss.

	$\mu = 0$	$\mu = 1$	dm(distance)
f1	19.1421	2.6335	16.5086
f2	14.6719	1.6568	13.0151
σ 1,σ 2	24	20	4
σ 3	-18	-15	3

dm = fi(μ=0) - fi(μ=1) ,i=1,2 or dm = σj(μ=0)-σj(μ=1) ,j=1,3

Table 2. Numerical results of different membership functions.

	λ^*	x 1*	x 2*	σ 1*	σ 2*	σ 3*	f1*	f2*
linear	0.8484	0.5795	3.4871	19.069	3.6291	-15.440	5.1263	3.6290
sinI	0.9267	0.5240	4.0081	20.698	3.2298	-17.468	5.4901	3.2298
sinII	0.7683	0.5517	3.5385	19.923	3.5997	-16.323	5.0992	3.5997
exp	0.7798	0.5627	3.5171	19.576	3.6122	-15.964	5.1088	3.6122
hyper	0.8987	0.5760	3.7947	19.043	3.3656	-15.677	5.4239	3.3655
quad	0.8487	0.5792	3.4913	19.072	3.6253	-15.447	5.1299	3.6253
log	0.9362	0.5943	3.4677	18.557	3.6401	-14.918	5.1524	3.6401
S stand	0.9528	0.5812	3.4871	18.686	3.6367	-15.050	5.1462	3.6367
statis	8.9E-4	0.5574	3.9171	19.565	3.2507	-16.315	5.5328	3.2507
lingui	0.8512	0.5637	3.5188	19.542	3.6100	-15.933	5.1133	3.6100

Table 3. Numerical results of single-objective optimization of the welded beam structure.

methods	α*	x1*=h*	x2*=l*	x3*=t*	x4*=b*	f*=cost
optimum		0.2348	6.3714	8.4681	0.2346	2.333
level cuts	0.1	0.2205	6.1664	8.4430	0.2209	2.148
two phase	0.9417	0.2274	6.5009	8.5807	0.2273	2.295
fuzzy expected	0.2	0.2463	5.6160	8.1598	0.2477	2.260

Table 4. Numerical results of multi-objective optimization of the welded beam structure.

methods	λ^*	x1*=h*	x2*=l*	x3*=t*	x4*=b*	f1*	f2*
fuzzy sets	0.99	0.1290	8.7016	12.898	0.1289	1.9765	1138.1
a	c	0.1130	10.526	14.102	0.1083	1.9499	1138.4
b	d	0.1127	10.578	14.113	0.1081	1.9529	1138.6

a: decision goal b: additive goal
c: uf1*=0.9880,uf2*=0.9791 d:uf1*=0.9875,uf2*=0.9798

Table 5. Results of inference reasoning for membership functions.

membership function type	evaluation	value
linear	maybe positive	0.68182
statistical histogram	negative	0.1875
probability density function	negative	0.1875
domain expert	unknown	0.43085
linguistic assessment	maybe negative	0.34375
exponential or hyperbolic	unknown	0.5266
standard function	maybe negative	0.375
quadratic or logarithmic	negative	0.1875
sin function	negative	0.1875

Table 6. Results of inference reasoning for fuzzy optimum design methods.

fuzzy structural optimum design	evaluation	value
fuzzy sets theory	maybe positive	0.75714
level cuts method	maybe positive	0.72436
fuzzy expected value	unknown	0.41489
two phase method	unknown	0.43055
fuzzy fully stressed design	unknown	0.44444
fuzzy goal programming	maybe negative	0.27272

Knowledge-Based Software Components in Structural Optimization

K. Lehner, D. Hartmann

Computational Engineering, Department of Structural Engineering, Ruhr-Universität-Bochum, D-4600 Bochum, Germany

Key words: Structural optimization, Knowledge Based Systems, Prolog

ABSTRACT

Although structural optimization is usually associated with numerical optimization, it encompasses much more than just solving mathematical problems. Numerical optimization methods have been investigated extensively, and today there is a wide variety of different methods, each with its own advantages and disadvantages. But as the user-friendliness of computer software increases and the entire process of modeling structural optimization tasks is considered, the creation, transformation and evaluation of the underlying concepts, rather than just the brute numerical power, are becoming more and more a dominant factor to finding appropriate solutions.

In this paper, a layer-based model of the total structural optimization process is presented. Each layer contains individual components, most of which are of non-numerical nature. Therefore, each component has to be implemented using appropriate tools (programming languages) and must be tested separately.

Two of these non-numerical components have been implemented in Prolog [1] and are discussed in this paper:

Component 1: Selection of an appropriate optimization concept. Selecting a viable alternative is a typical, knowledge-based task: Given a package of numerical optimization methods, the user must select the appropriate method (or set of methods) while considering various factors such as the available hardware (main memory), the type of constraints (linear, non-linear) or the dimension of the search space. A frame-based expert system reflecting the underlying knowledge domain (hierarchy of numerical optimization methods) has been designed for selecting a set of numerical optimization methods (see also [2], [3]).

Component 2: Interpretation of optimization results. Conventionally, numerical programs such as numerical optimization packages output numbers, key words and other items of text to be read and understood by the user of the program. If, however, a numerical optimization package is to be used not just as a stand-alone program, but rather as one of many components in the structural optimization process, the output of the optimization package must be "understood" and processed by software. Using paradigms from the theory of formal languages, a Prolog-program has been created using the "Definite Clause Grammar" option found in most interpreters.

INTRODUCTION

The purpose of structural optimization is to help the designer of a structural system find an optimum design that meets the ideals of the designer along with all technological requirements. Although trial and error can be used to improve existing solutions, a more formal approach based on structural optimization is more economical.

A formal definition of optimization can be given as follows: Let the vector $\mathbf{x} = (x_1, x_2, \ldots x_n)^T$ be the optimization vector containing the n optimization variables. Minimize the objective function

$$f(x)$$

subject to q equality constraints

$$h(x) = h(h_1(x), h_2(x), \ldots, h_q(x)) = 0$$

and p inequality constraints

$$g(x) = g(g_1(x), g_2(x), \ldots, g_p(x)) \leq 0 .$$

The optimization vector can also be restricted by a lower bound $\mathbf{x}^l$ and an upper bound $\mathbf{x}^u$ (side constraints):

$$x_i^l \leq x_i \leq x_i^u , \quad i = 1(1)n.$$

A Layered Model of Structural Optimization

Due to the complexity of structural optimization problems, a single optimization model is not sufficient. While the model should be as "realistic" as possible (in the sense that all relevant physical and technological aspects should be included), the actual optimization is performed using standard numerical optimization methods which operate in a purely mathematical environment irrespective of any engineering background.

Three distinct models can be defined to reflect the different aspects of structural optimization:

1. The *structural optimization model* defines the fundamentals of the optimization model from an engineering point of view. This includes the necessary technological and structural requirements to be met. That is to say, structural parameters govern the optimization variables, objective function and constraints used.

2. The *structural computational model* is based upon conventional structural analysis techniques (e.g. finite element methods) to compute structural quantities values such as stress, strain, deflection, etc.

3. The *mathematical optimization model* represents a formal abstraction of the structural optimization model and contains all the information required solely for numerical optimization, including information on the starting vector, choice of optimization method, method of calculating gradients, etc.

Based on the above categories of models, the entire modelization process of structural optimization can be split into five steps or stages:

1. Define the structural optimization model.
2. Define the mathematical optimization model.
3. Solve the mathematical optimization model.

4. Evaluate the mathematical optimization model.
5. Evaluate the structural optimization model.

Step 4 differs from step 5 in that within step 4 the results are checked from a mathematical point of view while in step 5 these are interpreted in view of the structural model. For example, we might detect that constraint #7 is slightly violated in step 4. An interpretation of this fact in step 5 may lead to the essential conclusion that the stress in a beam exceeds a maximum value in association with the buckling of the web. The structural optimization steps are illustrated in Fig. 1.

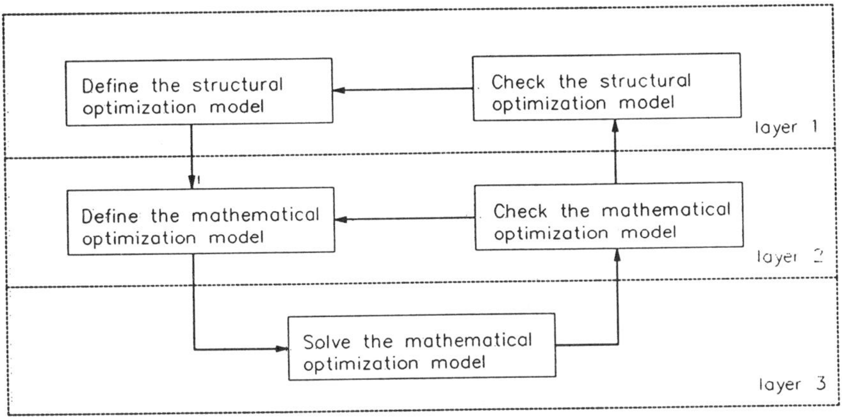

Figure 1. Structural optimization process layers.

The prime advantage of considering separate steps (and substeps) is modularity and flexibility. Particularly a suitable programming language can be used for each individual substep. In this paper, two substeps in layer 2 (mathematical optimizational model) will be described in some detail: First, the SELECT expert system has been implemented to assist the user in the selection of suitable numerical optimization methods from the ADS package of optimization methods [4]. Second, an evaluation program (ADSeval) that automatically checks the output of the ADS-package has been implemented.

THE ADS OPTIMIZATION PACKAGE

The actual numerical solution of the mathematical optimization problem is carried out by the ADS program. ADS is a package containing a plethora of optimization methods from which the user can select a set of methods specific to his optimization problem. The solution of a given optimization problem encompasses three levels:

1. *Strategy.* By means of the strategy a constrained problem is transformed into a sequence of approximate, unconstrained subproblems. Each subproblem may be solved using various techniques for unconstrained problems. A strategy is not required for an unconstrained problem, however, in practice most optimization problems are constrained.

2. *Optimizer.* The optimizer performs the actual function minimization of either the original problem (if unconstrained) or the sequence of approximate problems using ordinary hill-climbing techniques.

3. *One-dimensional Line Search.* The one-dimensional line search routines are called from the optimizer and sometimes from the strategy to find the minimum along a search vector.

To solve a specific optimization problem, the user selects a strategy, optimizer and line search routine suitable for his problem. It must be noted that not all combinations are permitted. For example, a constrained one-dimensional search is not allowed when minimizing unconstrained functions.

For a knowledge-based approach it is efficient to order the set of potential strategies, optimizers and line search routines (collectively called optimization methods) hierarchically. For example, the strategies available in ADS have been ordered as shown in Fig. 2.

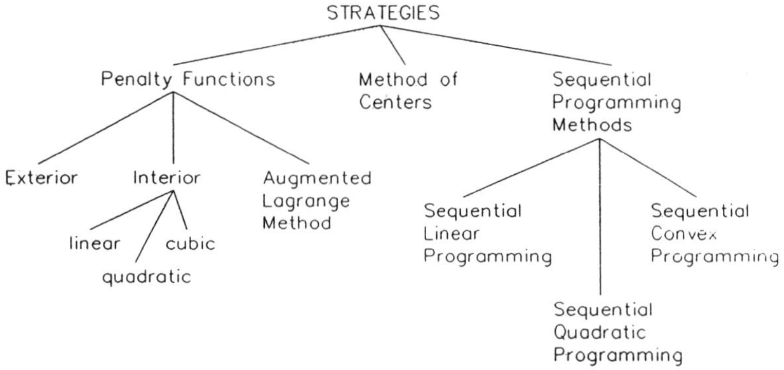

Figure 2. Strategies available in the ADS optimization package.

Since the underlying knowledge domain (i.e. optimization methods) is represented in a hierarchical format, the knowledge representation of the SELECT expert system is hierarchical as well.

THE SELECT EXPERT SYSTEM

The SELECT expert system assists a user in the selection of appropriate methods from the ADS optimization package. In the following, the philosophy of the selection process used in the expert system is presented, followed by the details of the realization. Also, an example is demonstrated.

Formal Description of the Selection Process

The knowledge used in the SELECT expert system is based upon two distinct basic sets:

1. A set of available optimization methods

$$S = \{S_1, S_2, \dots S_n\},$$

for example:
S_1: "no strategy method"
S_2: "exterior penalty function"
S_3: "linear extended interior penalty function"
etc.

2. A set of properties to describe and distinguish optimization methods:

$$P = \{P_1, P_2, \dots P_m\}.$$

For example, the items P_i may take the following form:
P_1: "There are less than 50 optimization variables."
P_2: "Relative minima are known to exist."
P_3: "Function evaluations are not expensive."
P_4: "The objective function and constraints are non-linear."
etc.

Short Term Memory
Properties can be viewed as propositions in the sense of a propositional calculus. In the SELECT expert system, the properties are presented to the user in question form. In conventional propositional calculus, only the Boolean values *true* or *false* are valid. This may be acceptable for a property like P_1, as there are only two possible cases (more than 50 optimization variables or less then 50 variables). On the other hand, properties like P_3 or P_4 cannot simply be labeled true or false: The term "expensive" is too vague and the phrase "non-linear" may encompass distinct degrees of non-linearity (e.g. quadratic, polynominal, exponential). Further difficulties arise when a property possesses no value at all. For example, if the user doesn't know that a relative minima exists, no value can be assigned to P_2.

The above considerations imply that a multi-valued logic is required . Therefore, in the SELECT expert system integers between 0 and 10 were designated to represent "truth values" The number 0 represents *false* and 10 *true*. Within this range, the value 5 means "indeterminate" or "don't know".

Consequently, a given optimization problem can be described adequately if a truth value between 0 and 10 is assigned to individual properties. Formally, this can be described by a partial function g that maps a subset of the properties to truth values:

$$g : P \rightarrow K; \; g(P_i) = A_i.$$

where

$$P_i \in P = \{P_1, P_2, \dots P_m\},$$
$$A_i \in K = [0, 10]$$
$$i \in \{1, 2, \dots m\}.$$

As an example, assume that the following optimization problem is given: $\mathbf{x} = (x_1, x_2, \dots x_{20})^T$ and $f(\mathbf{x})$ is quadratically nonlinear. Moreover, it is not known that relative minima do exist. For this problem, a possible definition of the function g may take the form:

$$g(P_1) = 10, \quad g(P_2) = 5, \quad g(P_3) = 2, \quad g(P_4) = 8.$$

In general, the function g is only partial because not all properties need be relevant and assigned a truth value. Therefore, the definition of g may be considered the short term memory of the SELECT expert system because only the current optimization problem is described.

Long Term Memory
In contrast to the short term memory, the knowledge and experience associated with the applications of optimization methods is encapsulated as long term memory in the expert system knowledge base. In this context, a function h maps each optimization method into the power set of properties:

$$h : S \to 2^P; \quad h(S_i) = \{P_{i1}, P_{i2}, \dots P_{ik}\}$$

where

$$S_i \in S;$$
$$P_{i1}, P_{i2}, \dots P_{ik} \in P;$$
$$i \in \{1, 2, \dots n\}; \; k \in \{1, 2, \dots m_i\}.$$

To give an example, the properties P_1, ... P_4 defined above can be used to describe the Augmented Lagrange Method (S_{alm}):

$$h(S_{alm}) = \{P_1, P_2, P_3, P_4\}.$$

Contrary to function *g*, function *h* is complete because all available methods are defined. To weigh optimization methods, an additional function *f* is introduced that assigns each property of a method a pair of "confidence values" or weighing factors:

$$f : SxP \to K^2; \quad f(S_i, P_j) = (Mb, Md)$$

where

$$S_i \in S;$$
$$E_j \in h(S_j);$$
$$Mb, Md \in K.$$

The pair (Mb, Md) weighs each optimization method with respect to a given property. The value Mb (measure of belief) weighs the optimization method S_i when property P_j is completely satisfied. The value Md (measure of disbelief), on the other hand, weighs the optimization method S_i when the property P_j does not apply.

In the case of the Augmented Lagrange Method, the function *f* may be defined as follows:

$$f(S_{alm}, P_1) = (2, 4), \quad f(S_{alm}, P_2) = (2, 4),$$

$$f(S_{alm}, P_3) = (3, 0), \quad f(S_{alm}, P_4) = (6, 0),$$

The last line can be interpreted as follows: If the objective function and the constraints are non-linear, then the Augmented Lagrange Method is only moderately acceptable (i.e., it is not completely inefficient, but there are more efficient methods). This is expressed by the value Mb = 6. However, if the the objective function and the constraints are linear, then the Augmented Lagrange Method should *not* used, because in this special case other optimization methods (such as the Sequential Linear Programming method) are definitely more adequate. Thus, Md equals 0. This example also demonstrates the importance of a pair of weighing factors: The acceptability of an optimization method with respect to a property can be totally independent to its negation.

Computation of the Total Weighing Factor

To find the "best" set of optimization methods for a given optimization problem, the weighing factors of the user input (i.e. a definition of the function *g*) have to be associated with the knowledge of the optimization methods (i.e. a definition of the functions *h* and *f*). Hence, a computational scheme is used that meets the following requirements: The weight of an optimization method

- should be independent of the number of properties,
- should represent its applicability for a given optimization problem and
- should be independent of the order in which the properties are listed.

A detailed computation scheme which satisfies all the above requirements is defined in [5].

The Knowledge Base

The knowledge base of the SELECT expert system is divided into the long term memory (to describe the knowledge domain) and the short term memory (the context representing the user input to describe a specific optimization problem). The knowledge representation is based on the frame paradigm. The weights of the methods and the property descriptions are stored in frames, called method-frames and property-frames, respectively.

Method Frames. An example for a method frame that describes the Augmented Lagrange Method is the following:

```
istrat5 ::
        notMore50DV             : 2/4 &
        relMinExist             : 2/4 &
        notFuncEvalExpen : 3/0 &
        conObjHighNonLin : 6/0.
```

As mentioned above, the underlying knowledge domain (optimization methods) is hierarchical. Thus the frames used to describe the knowledge domain are also organized in a hierarchical structure, achieved by adding an "a-kind-of" attribute (ako-attribute) which points to the predecessor in the hierarchy. For example, the frame for the linear extended interior penalty function contains an ako-attribute pointing to the interior penalty function frame representing its parent frame.

Property Frames. Apart from storing truth values to specify a particular optimization problem, property frames also contain further attributes for organizational purposes. For example, the two properties

"There are more than 50 optimization variables."

and

"There are more than 200 optimization variables."

classify the size of the search space determined by the maximum number of optimization variables. If during a session the user states that there are less than 50 optimization variables, then obviously the second property should not be presented to the user. Thus, the frames for these two properties should contain attributes to reflect this dependency. This can be done by including an oppo-attribute to define the opposite of a particular frame. Further attributes for properties include:

- *onlyIf-attribute.* An onlyIf-attribute contains the name of another property which is evaluated only if the user-defined rating value is greater than 5, i.e. if this property is "true".

- *notOnlyIf-attribute.* This is similar to the above case except that the user-defined rating value should be below 5.

- *dontKnow-attribute.*If a user-defined value of 5 (don't know) is entered, then the property frame of attribute value should be evaluated.

The aim of the inference component of the SELECT expert system is to combine the values input by the user (short term memory) with the knowledge domain of available optimization method (long term memory). Details are given in [5].

Example: Three Bar Truss

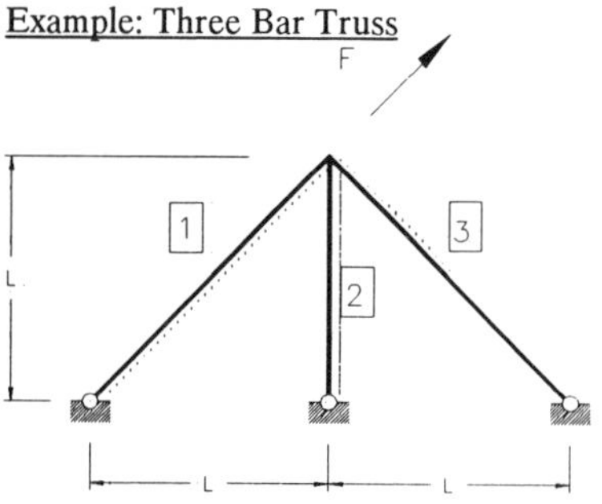

Figure 3. Three bar truss

Figure 3 represents a symmetrical simple truss consisting of 3 bars. The weight (or, equivalently, the volume) of the structural system is to be minimized. The cross sectional areas of bars 1 to 3 are variables, where x_1 is the cross sectional area of bars 1 and 3, and x_2 is the cross sectional area of bar 2. The formal, mathematical definition of the optimization problem is: Minimize the objective function

$$Q = c_1 x_1 + c_2 x_2; \quad c_1 = (2\sqrt{2})L; \quad c_2 = L;$$

subject to the stress constraints

$$g_1(x) = \frac{F\,(x_2+\sqrt{2}\cdot x_1)}{\sqrt{2}\cdot x_1^2 + 2x_1x_2} - \sigma_t, \quad g_2(x) = \frac{F\,\sqrt{2}\cdot x_1}{\sqrt{2}\cdot x_1^2 + 2x_1x_2} - \sigma_t, \quad g_3(x) = \frac{F\,x_2}{\sqrt{2}\cdot x_1^2 + 2x_1x_2} - \sigma_c,$$

where F is the load, σ_t the allowable tensile stress and σ_c allowable compressive stress.

Starting the SELECT expert system, the following menu is presented:

```
?- main.
Choose an option:
  1  List relevant questions  L  List all question info
  p  Print method names       P  Print all method info
  c  Change a question CF     a  Enter all question CF's
  e  Evaluate methods         n  Start a new problem
  r  Reread frame info        s  Save ques. frames
  x  Exit
a
```

Assuming that the user wishes to enter truth values for properties, the abbreviation "a" is chosen (user input is underlined). The system requests the following user input:

```
Are there more than 50 design variables?
  0(no) .. 5(don't know) .. 10(yes) or def  [default: none] 0.
Are relative minima known to exist?
  0(no) .. 5(don't know) .. 10(yes) or def  [default: none] 5.
Are function evaluations expensive?
  0(no) .. 5(don't know) .. 10(yes) or def  [default: none] 3.
Is computer storage very limited?
  0(no) .. 5(don't know) .. 10(yes) or def  [default: none] 1.
Are the constraints and obj. function well-conditioned?
  0(no) .. 5(don't know) .. 10(yes) or def  [default: none] 7.
Do the constraint values increase sharply?
  0(no) .. 5(don't know) .. 10(yes) or def  [default: none] 5.
Is the final design fully constrained?
  0(no) .. 5(don't know) .. 10(yes) or def  [default: none] 5.
Is a feasible starting point required?
  0(no) .. 5(don't know) .. 10(yes) or def  [default: none] 2.
     .
     .                    :
     .
```

After the short term memory has been input, the user may decide to evaluate the knowledge to find a combination of strategy, optimizer and line search routine to use. Thus, option "e" is chosen:

```
Choose an option:
  1  List relevant questions          L  List all question info
```

```
  p  Print method names                 P  Print all method info
  c  Change a question CF               a  Enter all question CF's
  e  Evaluate methods                   n  Start a new problem
  r  Reread frame info                  s  Save ques. frames
  x  Exit
e
```

The progress is displayed on the screen:

```
checking strategy
evol,istrat0,istrat1,istrat2,istrat3,istrat4,istrat5,istrat6,istrat7,istrat8,istrat9,
checking optimizer
iopt1,iopt2,iopt3,iopt4,iopt5,
checking lineSearch
ioned1,ioned2,ioned3,ioned4,ioned5,ioned6,ioned7,ioned8,
checking combinations

12 combinations found.
```

The "best" combination in the current example is the exterior penalty function strategy, the Broydon-Fletcher-Goldfarb-Shanno optimizer and the golden section line search routine.

```
Nr.: 1
    istrat1/0.415  (exterior penalty function method)
    iopt3/-0.00718  (Broydon-Fletcher-Goldfarb-Shanno Algorithm)
    ioned1/0.126  (Golden Section Method only)
```

An "explanation" can be generated to give the grounds for this selection:

```
Next/Quit/Explain/# e.
  Average CF is 0.17794
    CF for istrat1 is 0.415
      Are relative minima known to exist?, CF = 0.35
      Are there less than 50 design variables?, CF = 0.1
    CF for iopt3 is -0.00718
      Are function evaluations inexpensive?, CF = 0.32
      Is computer storage very limited?, CF = -0.17
      Are the constraints or obj. function highly non-linear?, CF = -0.2
    CF for ioned1 is 0.126
      Are function evaluations inexpensive?, CF = 0.32
      Are the constraints or obj. function highly non-linear?, CF = -0.2
```

Further combinations are possible:

```
Next/Quit/Explain/# n.

Nr.: 2
    istrat1/0.415  (exterior penalty function method)
    iopt3/-0.00718  (Broydon-Fletcher-Goldfarb-Shanno Algorithm)
    ioned2/0.102  (Golden Section Method / polynominal interpolation)

Next/Quit/Explain/# n.

Nr.: 3
    istrat2/0.282  (linear extended interior penalty function method)
    iopt3/-0.00718  (Broydon-Fletcher-Goldfarb-Shanno Algorithm)
    ioned1/0.126  (Golden Section Method only)
```

ADSeval PROGRAM

The output of the ADS optimization package consists of an extensive listing of numbers representing intermediate results as well as termination criteria, status reports and other items of information, i.e. it represents a quantitative analysis of the mathematical optimization model. This information serves the experienced user as a source of knowledge for a qualitative analysis of the mathematical optimization model. In order to facilitate the evaluation of the structural analysis model, the following mathematical background information is of help:

- Does the objective function decrease monotonously?
- Is the series of step widths of the **x**-vector monotonous?
- Are there redundant constraints?
- Is the optimum fully constrained?
- Are side constraints active?
- What type of termination criteria occurred?

Therefore, the main purpose of the ADSeval program is to extract "mathematical knowledge" from an ADS output file to assist the user in answering questions associated with the mathematical optimization model. Since the output of the ADS optimization package is designed to be printed and read by the user, it has no formal structure. For example, the text "FINAL OPTIMIZATION RESULTS" heads the results of the line search routine, the optimizer, the strategy and the entire optimization run as well. Thus, to interpret this phrase the context within the the output file has to be considered. A further example: A single value can have many names: The value of the objective function, for example, is variously labeled "OBJ", "OBJECTIVE FUNCTION", "OBJECTIVE FUNCTION VALUE" or "TRUE OBJECTIVE FUNCTION". Also, key words are misspelled (e.g., "tolerance" is sometimes spelled "tolorance"), thus increasing the effort needed to find values.

If we consider an ADS output file to be "written" by the ADS optimization package, then we must detect the implicit rules which are built into the ADS program to generate output files. The theory of formal languages can provide a theoretical basis for such a description. Once the rules which form the output file (grammar) have been defined, we can analyze an ADS output file using standard formal language techniques.

Grammars can be classified according to the type of language they can generate. So-called type 0 grammars can generate any language. However, evaluation of the rules is in general too inefficient to be of practical value. At the other extreme, so-called regular languages have very simple grammatical rules, but languages they can generate are too limited. A type of grammar which lies between these extremes is called a context free grammar. Context free grammars can be implemented directly by the the Definite Clause Grammar (DCG) notation found in most Prolog interpreters. Grammar rules can be coded directly using Prolog elements such as atoms, predicates and lists. Some examples of grammar rules written in the Prolog DCG notation will be given.

As noted above, the state of analysis must be discovered and remembered. A careful inspection of many ADS output files shows that two slightly different phrases introduce strategy and optimizer values:

```
%%%     save various states, mutually exclusive
topLevel -->
        "-- INITIAL DESIGN",
        { setState( strat/initialDesign) }.
topLevel -->
        "INITIAL DESIGN",
        { setState( opt/initialDesign) }.
```

The final optimization results must be evaluated depending on the current state of analysis:

```
topLevel -->
        "FINAL OPTIMIZATION RESULTS",
        { getState( opt/_) },
```

```
        { setState( opt/finalResults) }.
topLevel -->
        "FINAL OPTIMIZATION RESULTS",
        { getState( strat/_) },
        { setState( top/finalResults) }.
```

Duplicate entries are sometimes needed:

```
topLevel -->
        "UNCONSTRAINED MINIMIZATION IS COMPLETE",
        { setState( strat/unconMiniComplete) }.
topLevel -->
        "UNCONSTRAINED MINIMIZATIN IS COMPLETE",  %%% spelling error!
        { setState( strat/unconMiniComplete) }.
```

Store the decision vector by reading in its components (printed row-wise) and saving it. This rule uses other rules and Prolog predicates (not shown here) that transform real numbers from ADS ASCII-notation to binary form.

```
topLevel -->
    "DECISION VARIABLES (X-VECTOR)",
    { getState( [strat/initialDesign, opt/finalResults]) },
    { readVector( Vector), gc( full) },
    { saveVector( xVector, Vector) }.
```

Find a termination criteria and store it:

```
topLevel -->
    "MAXIMUM K-T RESIDUAL",
    { addValue( optTermCrit, 'Kuhn-Tucker conditions') }.
```

Without going into detail too much, the summary output created by the ADSeval program has the following appearance (the interested reader may be referred to [5]:

```
Summary:
Note: The objective function is non-monotonous.
The step length of the x-vector decreases monotonously.
Termination criteria for the optimizer are
  absolute convergence and too many iterations at opt. level.
Termination criteria for the strategy are
  absolute convergence.
There are no active side constraints.
Constraints 5 and 11 are inactive.
    (Constraints 5 and 11 are possibly redundant.)
Constraints 3, 8 and 10 are finally violated.
    (Constraints 8 and 10 are always violated.)
Constraints 1 - 4, 6 - 10 and 12 - 15 are active or violated.
The optimum is approached from the infeasible region.
  (Constraints 1 - 4, 6 - 10 and 12 - 15 are violated.)
The final design is not fully constrained.
  (Constraints 5 and 11 are not active (or violated) at the optimum.)
```

SUMMARY

This paper is intended to demonstrate how a complex process such as the modelization of structural optimization can be managed using knowledge engineering techniques. The amount of detailed information needed to fully describe even small engineering optimization problems is so great that it seems impractical to solve a problem in a single step. Rather, separate models have to be created according to different aspects of the optimization problem. The breakdown of the structural engineering process into subprocesses permits each subprocess to be designed separately, implemented with a programming language of choice and tested in a stand-alone manner.

In particular, the definition of the underlying mathematical model of the structural optimization problem is elucidated. A prototype expert system is discussed that assists an engineer, even an engineer not quite familiar with the details of structural optimization, to select an appropriate optimization method. Also, an evaluation program to analyze ADS output files and report a qualitative summary has been implemented.

Further research is necessary and may address the following objective.

- Object-oriented paradigms for more realistic knowledge representation and inference (i.e. extending frames to objects).

- Coupling "off the shelf", standard engineering software to form an integrated environment by transforming the "languages" created by various programs.

- Use of symbolic computations for computing the gradients used in the optimization.

- New hardware concepts (such as transputers) can parallelize knowledge based and numerically oriented software.

It should be pointed out that the sensible use of hardware and software to solve complex engineering problems such as structural optimization problems requires that we not just look at part of the problem (e.g. numerical aspects), but that we rather view the problem as a whole, including the many non-numerical questions that arise in the course of work.

REFERENCES

[1] Clocksin, W.F., Mellish, C.S., Programming in Prolog, Springer Verlag, Berlin, 1985

[2] Hartmann, D., Selection and Evaluation of Structural Optimization Strategies by Means of Expert Systems" in "Finite boundary elements & expert systems", A. Niku Lari (Ed.), Vol 5., pp. 39-55, Pergamon Press, Oxford, 1986

[3] Hartmann, D., Knowledge Acquisition for Expert Aided Structural Optimization by Means of Transformation Methods and Primal Methods, GAMM Seminar "Structural Optimization", Univ. of Siegen, Siegen, Germany, 1988

[4] Vanderplaats, G. N., Numerical Optimization techniques for Engineering Design, McGraw-Hill, New York, 1984

[5] Lehner, K., "The Use of Knowledge Based Systems in Structural Optimization" (German), Ph.D. Thesis, Ruhr-University-Bochum, Bochum, Germany, 1991

A Computer Expert System for Composite Plate and Strut Design

S.K. Morton (*), J.P.H. Webber (**), C.M.L. Wu (**)
(*) Department of Engineering Mathematics,
(**) Department of Aerospace Engineering,
University of Bristol, Bristol, England

ABSTRACT

In the field of structural component design using advanced fibre-reinforced composite materials (FRMs) optimisation of the amount of material with respect to weight, cost, stiffness, strength and less quantitative aspects such as ease of manufacture, reparability, etc. is crucial. In this paper a computer expert system that will allow a user to design, test and compare candidate FRM solutions to particular design problems is presented. The user has three top-level options: to design a plate using a cyclical approach and redesign heuristics; to design a strut, including end-fittings, by optimising with respect to the buckling load; or to assess the relative merits of previous designs based on various design criteria using support logic. Results are presented for various plate and strut designs and for the relative assessment of comparable designs with different materials and fibre orientation configurations in the case of plates, or cross-sectional shapes in the case of struts.

KEY WORDS : Composite Materials, Design, Expert Systems, Uncertainty, Support Logic.

1. INTRODUCTION

The use of advanced fibre-reinforced composite materials has generally not lived up to the expectations engendered by their many apparent performance advantages due to technical problems in manufacture and to the complexity of the underlying mathematical models applied in the design process. The nature of such materials entails limitations to the kind of designs that can be constructed, so that the design process is more tied up with the manufacturing process than is the case with metals, for example. Apart from quantitative aspects, the assessment of putative designs must therefore consider qualitative aspects such as manufacturability; also reparability and other operational factors, where there is inherent uncertainty. Any computer-based expert system that would be of use to a designer using composite materials will therefore have to be able to do at least three things: readily access and apply analysis programs implementing the mathematical theories as and when required; represent and reason with various forms of uncertainty; and perform a systematic assessment of designs taking into account both quantitative and qualitative desiderata. Beyond that, in order to fulfil a role as adviser to the designer, a user-friendly and flexible man-machine interface is essential.

Various examples of this kind of system have been developed as aids to aerospace designers. For example, the computer package HYPERCOMPOSITES is an extensive and powerful system that has been developed over a number of years and which represents the implementation of the composite design and analysis procedures of Tsai [1]. The research reported in this paper follows from that of Webber [2], who presents some design optimisations for composite laminates using an interactive program written in BASIC, and of Burden and Lipton [3] who developed that work to incorporate different failure criteria and ply degradation due to non-catastrophic ply failure.

The principal components of the CODEX (COmposite Design EXpert) system described herein are the plate design, strut design, and assessment sub-systems. Given a prescribed loading case, the plate and strut design sub-systems are applied to the user-selected materials knowledge, and the resulting designs stored in directories accessible for subsequent assessment. The implementation of the system is menu-based. At the top level the user has the plate design, strut design and assessment options as well as sundry auxiliary options. After selection of the option the relevant sub-system is called and further menus guide the user towards the particular operation he wishes to perform.

2. PLATE DESIGN

The nature of fibre-reinforced composite materials is such that they are extremely strong in the fibre direction but relatively weak in directions transverse to the fibre direction. The usual solution to this problem when designing a plate from such materials is to construct the plate as a laminate comprising layers of material with the fibres oriented at different angles. Thus, while each layer is weak in the transverse fibre direction, the laminate as a whole has all-round strength. Given a configuration of fibre orientations, i.e. the layers and their orientations, of a putative laminate, the design problem becomes that of finding the set of layer thicknesses that gives the laminate the strengths required for a particular set of applied multi-directional in-plane loads $\underline{N}=[N_x,N_y,N_{xy}]$ and moments $\underline{M}=[M_x,M_y,M_{xy}]$.

Due to the complexity of the underlying theory there is no easy way to determine the required thicknesses for a given load case, lay-up and material. Instead an **heuristic** approach is taken; that is, a method which experience has shown leads to a "good" solution in most cases, but which is not theoretically guaranteed to do so in all cases. This consists of an initial netting analysis phase (section 2.1), followed by a design-and-test cycle driven by redesign rules (section 2.2). Finally the resulting laminate is analysed to find its theoretical first ply failure (FPF), ultimate failure, and delamination initiation loads (section 2.3).

The plate design and analysis procedures of CODEX to be discussed are an extension of those described by Webber and Morton [4]. Throughout this section we assume that the only ply fibre orientations considered are 0°, 90° and ±45°, and that the plate is always balanced. The materials available to the system are shown on the right. The assumed values for the mechanical properties of these are given by Webber and Morton [4].

CFRPHS914 - carbon fibres in 914 resin;
CFRPHS913 - carbon fibres in 913 resin;
CFRPEEK - carbon fibres in PEEK resin;
GFRPE913 - E-glass fibres in 913 resin;
KFRP49913 - Kevlar 49 fibres in 913 resin;
SCRAL- silicon carbide fibres in an aluminium matrix;
AAL168 - aluminium alloy.

2.1 Netting Analysis

The netting analysis is an approximate method which enables us to specify an initial design in terms of a set of layer thicknesses, from which it is hoped a satisfactory design will be efficiently derived using the redesign rules of the design cycle. Although the netting analysis for in-plane loads is straightforward, that for applied moments is more complex due to the effect on the bending strains of a ply's distance from the neutral plane. We now present a brief summary of the method outlined by Morton and Webber [5].

2.1.1 In-plane loads A netting analysis is a simplification of laminated plate theory where the individual plies are assumed to withstand the applied loads independently in their respective fibre directions. Thus the plies with fibres at 0° orientation withstand the N_x load, and the plies with fibres at 90° orientation withstand the N_y load. Additionally since a pure shear load resolves into direct loads in the ±45° directions, the ±45° plies are deemed to withstand such a load. The thicknesses obtained from the analysis for each fibre orientation are such that the strain in the fibre direction at the relevant load equals the failure strain. In such an approach we are ignoring the transverse and shear stiffnesses and strengths of the material, and the mechanical dependence between the layers; but this does allow us to find an analytical approximation to the problem.

2.1.2 Moments The situation with applied moments is somewhat more complicated than with applied loads, since the strain in any ply depends on its distance from the neutral plane of the laminate as well as on the magnitude of the moment(s). If we are given a prescribed configuration of ply orientations, e.g. $[0°/45°/-45°/90°]_{sym}$, the moment netting analysis can be performed in two stages. Firstly an initial thickness can be obtained for each of the 0°, 90° and ±45° plies by considering a single ply of that fibre orientation, and finding the thickness required for the relevant moment. Thus, in a manner similar to the load netting analysis, the 0° and 90° layers relate to the bending moments M_x and M_y, and the ±45° layers to the twisting moment M_{xy}. In general the thickness values derived from this initial stage of the moment netting analysis will, when assigned to the plies in the prescribed lay-up configuration, be too great; in other words the plate will have been over-designed. This is because the overall bending stiffness of the resulting lay-up is increased when the plies are further away from the neutral plane. Therefore we need to reduce the thicknesses so that taken together in the given lay-up, each of the applied moments will be resisted by the plies with fibre orientations in the corresponding direction. Since the thickness of one sort of ply determines the distance from the neutral plane of another sort of ply, an iterative procedure must be followed to derive the final moment netting analysis thicknesses.

2.2 Design and Analysis

2.2.1 Design Heuristics There are a number of heuristic design guidelines that apply to the lay-up configuration in general and to the individual ply thicknesses. The former relate to the order of the fibre orientations in the lay-up, and are all satisfied by each of the twelve lay-ups that the system user may choose. The latter relate to the relative and absolute thicknesses of the plies. These heuristics must be observed at each stage of the design-and-test cycle, and especially after the initial thicknesses have been derived from the netting analysis. The more important fibre orientation heuristics are derived from [6].

When analysing an initial or intermediate design we must use the same prescribed failure criterion on each cycle. There are many such described in the literature: for example [7] provides a summary of the most plausible criteria for failure in a single ply of fibre-reinforced composite material. The failure criteria currently available in CODEX are maximum strain, maximum stress, Tsai-Hill and Tsai-Wu. The first two assume that the failure modes are independent, i.e. failure occurs when the failure strains or stresses are exceeded in any direction. The Tsai-Hill and Tsai-Wu criteria (also referred to as quadratic criteria) assume that the strengths in the different modes are inter-dependent, and are single-valued functions.

2.2.2 Redesign Heuristics At each stage of the design-and-test procedure the intermediate plate designs must be improved by a selective thickening of plies. The complexity of the

Table 1 : Redesign heuristics

	ACTION	CONDITIONS		
	THICKEN LAYER	FAILURE		LAYERS EXISTING
Rule #	fibre angle/°	fibre angle/°	mode	fibre angle/°
1A	0	90	transverse	0
1B	90	90	transverse	not 0
2A	90	0	transverse	90
2B	0	0	transverse	not 90
3A	±45	90	shear	±45
3B	90	90	shear	not ±45
4A	±45	0	shear	±45
4B	0	0	shear	not ±45
5	±45	±45	transverse	
6A	0, 90	±45	shear	0, 90
6B	±45	±45	shear	not 0, not 90
7A	0	0	fibre	
7B	90	90	fibre	
7C	±45	±45	fibre	

laminated plate theory does not allow for an analytic approach in the general case. Instead a number of rules derived from the subjective observations of human composite design "experts" are employed to decide on the layers to be thickened given a certain mode of failure. These are referred to as the redesign heuristics, and are represented by an action (or head) and a set of conditions (or tail): the action specifies the fibre orientation of the layers to be thickened, while the tail specifies the failure mode and failing layer fibre orientation, as well as some subsidiary condition(s) to do with the existence of layers of particular fibre orientations. In the CODEX plate design sub-system the heuristics currently used are shown in Table 1. As an example consider rule number 1A: here the failure condition is transverse failure in a 90° layer, and the subsidiary condition is existence of 0° plies; the action is to thicken those same 0° plies. Similarly rule 1B deals with the same failure situation when there are no 0° plies, when we instead thicken the 90° plies. For a fuller description see Morton and Webber [5]. In each case the component of the thickness corresponding to the load netting analysis is increased by one over the load factor, while that corresponding to the moment netting analysis is increased by the square root of one over the load factor to avoid over-design.

2.3 Free Edge Analysis

Although the laminate analysis discussed so far assumes no out-of-plane stresses, it is well known that delamination sometimes occurs at a free edge due to interlaminar normal and shear stresses arising from a mismatch of elastic properties between plies. To predict the values of these stresses and hence the delamination initiation load, a finite element approach such as that of Herakovich [8] will supply detailed information about the stress distributions at and near the free edge. However this is a very time-consuming method, and in order to provide a designer with a simple and efficient way of estimating these stresses, analytic approximations to the stress distributions have been devised.

As part of the plate design sub-system, a free edge analysis module has been implemented based on the model of Kassapoglou and Lagace [9]. This is a general solution which can cope with any number of plies efficiently to give estimates of the interlaminar stresses. The principal assumptions are that the functional dependences of the stresses on the y (transverse co-ordinate) and z (out-of-plane co-ordinate) values can be separated, and that they can be written in a form involving two exponential parameters. Furthermore these functions are chosen so that they satisfy the overall equilibrium equations, the stress-free conditions at the free edge, and also stress continuity between adjacent plies. The parameters of these functions are then determined by minimising the plate complementary energy.

Having established the stress state in the boundary layer near the free edge the next task is to predict the applied load at which delamination initiation occurs. At the free edge itself the analytic approach predicts very high stresses which do not, in practice, occur. Instead average stresses, taken over an "averaging distance", d, measured from the free edge, are compared with certain empirical strength parameters using some failure criterion. The value of d is usually taken to be the value of y at which the interlaminar normal stress is zero. The failure criterion used in the system is the quadratic stress criterion as proposed by Brewer and Lagace [10], for which it is necessary to have values for the interlaminar failure stresses.

The user may also select the ply ordering required based on the predicted delamination initiation load. For a given set of ply fibre orientations the value of this load may vary considerably with the order of the plies; therefore, with the condition that 45° and -45° plies must be together, a free edge analysis is carried out for each possible lay-up configuration, and the corresponding delamination initiation load evaluated. For example if the user had initially chosen the $[0°/45°/-45°/90°]_{sym}$ lay-up, the free edge analysis is carried out for this and for the five other fibre orientation orderings possible. Once the user has selected the lay-up configuration required, this becomes the design lay-up. The FPF, ultimate failure and single load analyses are unaffected by the swapping around of fibre orientations.

3. STRUT DESIGN

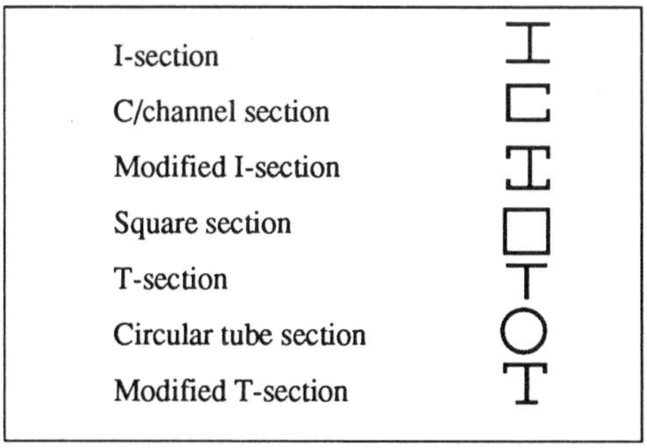
I-section
C/channel section
Modified I-section
Square section
T-section
Circular tube section
Modified T-section

The second principal component of CODEX is a strut design sub-system that optimises with respect to minimum weight. Orthotropic composite struts are designed to include about 10% of 90° fibres. The transverse fibres are sandwiched in the middle of the lay-up by the 0° fibres to ensure that everywhere in the strut is balanced about its own local mid-plane. The compressive load on the strut is purely axial. Three types of end constraints are considered: both ends simply supported; one end simply supported and the other fixed; and both ends fixed. The details of the formulation for the solution of least weight struts as explained in section 3.1 can be found in Wu and Webber [11]. Given the compressive force, the end constraint conditions, the strut shape, and the required material, a solution is provided as the cross-sectional dimensions of the strut. Seven cross-sections of struts are considered and they are shown above. In addition the seven materials considered in the laminated plate design are all available for the strut design.

The design of laminated struts using the method in section 3.1 does not take into account the effect of joint attachments at their ends. Practically, a strut has end fittings and they should be sound enough to take the structural loads. This enhancement to the strut design sub-system is discussed in section 3.2.

3.1 Minimum weight optimisation of struts

Struts of constant cross-section are considered. If the length of a strut is given, the minimum weight of the strut corresponds to its minimum area for a chosen material. The following formulation aims at minimising the strut area so that for a specified compressive load, the strut is within the failure criteria for structural failure, overall buckling, and local buckling. As in the plate design sub-system the user can choose from the maximum stress, maximum strain, Tsai-Hill or Tsai-Wu failure criteria for structural failure.

The minimum area of a strut is obtained when the critical buckling stresses for the strut in overall buckling and for the local components of the strut in local buckling are the same. This leads to a set of simultaneous equations to be solved for each strut. The solution of these equations is given in Wu and Webber [11] for each strut shape. The dimensions of the strut with respect to the minimum area are also found and a check is then carried out to see whether the strut also comes within the chosen structural failure criterion. If not, the minimum area is then based on this criterion.

The solutions for the strut section thicknesses described above are usually not multiples of commercially available prepreg thicknesses. For a practical design it is necessary to obtain a feasible solution for minimum weight so that cross-sectional thicknesses are integer numbers of prepreg thicknesses. In general, when the number of layers in the cross-section is increased, the width of the laminate may be decreased so that the strut is still within the limit for overall and local buckling. To obtain the feasible solution, the analytical solution for the exact dimensions of the strut cross-section is used to provide an initial sizing of the strut by taking the nearest number of layers for each thickness. From there on an iteration is carried out. The starting point of the iteration is the minimum area from the initial sizing, and for each iteration a small increment of area is added. During each iteration various sets of dimensions of the strut are used. For each of these sets of dimensions checks are carried out to see whether the strut comes within the chosen structural failure criterion. For a strut of given dimensions and lay-up geometry, an FPF analysis is carried out. The applied load at FPF can then be used to calculate the critical stress at FPF. If this critical stress is found to be larger than that evaluated from the optimisation on overall and local buckling, a re-sizing procedure is carried out.

3.2 End Fittings

As an extension to the strut design sub-system, procedures to deal with end fittings have been incorporated. As data on bolted joints are limited for composite materials, the current procedures only consider CFRPHS914, CFRPHS913, GFRPE913, SCRAL and AAL168. Also, as a first attempt on design analysis with bolted joints, only the I-section and circular tube strut shapes are considered. The details of the bolted joint formulation can be found in Wu [12].

The bolted joints have the configurations shown in Figure 1. For the joint with an attachment using the web of the I-section strut and that with end caps for the circular tube strut (see Figures 1(a) and 1(d)), an eccentricity is introduced due to the nature of the single-lap shear joint. This eccentricity results in a bending moment on the strut. This moment has to be taken by the strut, and it can be assumed in the I-section case that the flanges of the strut can be designed to take the bending. After this bending moment has been taken into account, the strains in the strut are recalculated. A check is then carried out to make sure that the section is within the structural failure criterion. If not, the section dimensions are adjusted to accommodate the bending moment. In the circular tube case, the tube thickness has to be increased to take the bending moment.

For the I-section struts three attachment methods are available in the system, as shown in Figures 1(a) to 1(c). Referring to Figure 1(a), the first method, J1, attaches the web of the strut to the web of a T-section beam as a single lap shear joint. Figure 1(b) shows that the second method, J2, uses clamping pads to attach the web of the strut to the web of a T-section beam. The third method, as shown in Figure 1(c), joins the flanges of the strut to the flanges of an inverted U-section beam. It is assumed that the web width of the T-section beam and the flange width of the U-section beam are wide enough for the joint installation. It is a common design practice in bolted joint design to have the thickness of the plate at least the same as the substrate thickness at the joint. This means that the plate thickness may need to be padded up locally to the required thickness. The padding material is now taken as the same material as the plate, with ply orientations in pairs of ±45°. Depending on the original thickness of the plate, the composition of ±45° plies in the plate can be varied. For the purposes of bolted joint design, the web and flange of the I-section strut and the extended flange of the cap of the circular tube strut are treated as plates. This allows the data from the literature to be used.

For the circular tube strut only one type of end fitting is available. Figure 1(d) shows that the ends of the strut are bond-fitted with titanium end caps which have extended flanges to be bolted onto the web of a T-section beam. This is effectively the same joining method as in Figure 1(a) and is also called joint J1.

The bearing strength of a bolted joint varies with the width of the plate, the edge distance from the bolt centre to the ends, the percentage of the ±45° layers, the thickness of the joint, the diameter of the bolt, stacking sequence, bolt fit, material properties, joint geometry such as single or double lap shear joint, and pitch between bolts. The current design procedures consider these parameters as far as possible to provide a realistic design.

The data for CFRPHS914 are obtained from Collings [13], those for CFRPHS913 and GFRPE913 from Kretsis and Matthews [14], and the SCRAL data from Webber [15]. It is worth noting that some data for the above materials are not fully available from the literature and so they are estimated.

The joint is designed with a view to obtaining a minimum overall strut weight, which includes padded materials, bolts, and clamping pad or end caps if applicable. To achieve this, the diameter of the bolt and the width of the plate are varied in the search for the most efficient strut. The width and thickness of the plate are increased to avoid local buckling. In the current design situation, the length of the gap in the structure across which the strut is to be attached is given by the user, as opposed to the actual strut length as in the design analysis without the bolted joint of section 3.1. The actual strut length depends on the number of bolts used, the pitch between bolts and the distance between the last row of bolts and the strut end. The diameter of the bolt is varied from the minimum bolt diameter selected by the user to the maximum size of bolt available by the optimisation program. The possible number of columns of bolts is also examined. For each variation of these parameters the number of bolts and the joint length are calculated. The joint length is then added to the length of the gap across which the strut is to be attached giving the total strut length. The least weight strut can then be selected. As the strut length changes with the number of rows of bolts used, an iteration is required for the best configuration.

The above end fitting analysis is carried out for each strut dimension within the iteration process for minimum cross-sectional area described in section 3.1. This ensures that the strut design has the minimum weight after the incorporation of the bolted joint.

4. DESIGN ASSESSMENT

For a particular plate or strut design task the user may have created several candidate solutions using the respective CODEX sub-systems. Which to choose may be evident from the properties of the designs stored in the design files. The user can select the "display designs" option from the top-level CODEX menu and specified designs will be displayed. On the other hand the choice may not be easily made on this basis, especially when there are more than two or three designs to choose from, and proper account must be taken of the designer's relative priorities amongst the designs' properties. The assessment sub-system in CODEX allows the user to determine the relative suitabilities of a set of candidate designs using support logic. The positive and negative aspects of a design are weighted and combined to give an overall assessment of its relative worth, where the weights themselves may be set and modified by the user. In addition explanation and what-if facilities are available. The assessment sub-system employs a number of graphics routines for the display of its output.

4.1 Support Logic

The man-machine interface and the top-level control structures of CODEX are written in the Artificial Intelligence systems programming language FRIL [16]. Essentially it has all the power of PROLOG but can also deal with uncertain information, using the methodology of **support logic**. The representation and semantics of support logic are a generalisation of those of logic programming with the added feature that uncertainties in the form of **support pairs** are attached to facts and rules and combined during inference.

A program written in FRIL consists of facts and rules. A program is "run" by loading it into the FRIL environment, where it becomes a knowledge base, and asking a query of this knowledge base. The query is answered by inference from facts to rules. The uncertainty values are propagated according to the calculus of Baldwin [17] and a support for the query expression is thus derived. The query does not succeed (true) or fail (false) as in PROLOG; rather the support pair(s) returned to the query represent the extent of truth or falsity of the answer(s), ranging from (1 1) (absolutely true), through (0 1) (absolutely unknown), to (0 0) (absolutely false). For more details and examples of the operation of FRIL in the context of material assessment, see Morton and Webber [18].

4.2 Assessment rules

Uncertainty is inherent to the procedure of assessment between competing designs: firstly in the relative importances of different aspects of a plate or strut design type, represented as **conditional support pairs** ascribed to support logic rules; and secondly in the semantic unification between point values of design properties and **fuzzy sets** such as *rel_low* (see [16]).

For example the rules for the assessment of a plate might be as shown on the right. The high positive conditional supports on the rules for weight, FPF and ultimate failure mean that these are the most important aspects for plate design comparison. The moderate conditional supports on the rules for delamination, cost, manufacturability and environmental factors indicate that these are less important aspects. The low conditional supports on the three laminate stiffness rules show that these are of relative indifference to the overall assessment of a plate.

```
((good_plate P D)(good_plate_wt P D)) :((0.9 1)(0 0.1))
((good_plate P D)(good_plate_fpf P D)) :((0.8 1)(0 0.2))
((good_plate P D)(good_plate_ult P D)) :((0.8 1)(0 0.4))
((good_plate P D)(good_plate_delam P D)) : ((0.5 1) (0 0.5))
((good_plate P D)(good_plate_cost P D)) :((0.5 1)(0 0.8))
((good_plate P D)(good_plate_stiffX  P D)) :((0.2 1)(0 1))
((good_plate P D)(good_plate_stiffY  P D)) :((0.2 1)(0 1))
((good_plate P D)(good_plate_stiffXY P D)) :((0.1 1)(0 1))
((good_plate P D)(good_plate_manuf P D)) :((0.4 1)(0 0.2))
((good_plate P D)(good_plate_environ P D)) :((0.4 1)(0 0.7))
```

When a support logic query such as qs((good_plate myplate myplate$1)) is asked, the variables P and D are instantiated, and each of the rules is invoked. They in turn invoke other rules which ultimately invoke the support logic facts representing the features of the design type myplate and of the particular design myplate$1. For example the good_plate_wt rule is

((good_plate_wt P D)(rel_low_wt P D)) :((0.7 1)(0 0.5))

which in turn calls

((rel_low_wt P D) (design_weight D W)
(rel_weight P W WREL)
(match rel_low WREL)) :((1 1)(0 0))

Here the relative weight is found arithmetically and then matched with the fuzzy set *rel_low* as defined in the accompanying sketch.

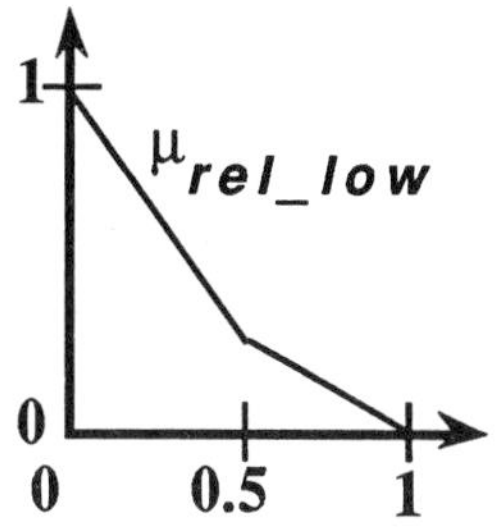

The weight rule is one of the rules relating to a quantitative aspect of a plate. The exceptions are the rules for manufacturability and the environmental factor which embody qualitative aspects. For example the rules for the manufacturability of a strut invoke the four predicates called high_strut_precision, few_holes_notches, many_components_required and little_material_waste. The high_strut_precision rule in turn invokes information about the plate's constituent materials: currently aluminium alloy is deemed to be the only material allowing high machine precision. The few_holes_notches and many_components_required rules pick up the information for the relevant strut type initially input by the user. There are two little_material_waste rules: one for aluminium alloy for which there is little waste if the extrusion method is used (this information must be requested from the user); and one for composites, for which the amount of material waste is governed by the proportion of ±45° plies in the plate or strut laminates. There are similar rules for the environmental aspect of plates and struts relating to moisture absorption, operating temperature, etc..

In CODEX an assessment is carried out on a set of designs of the same design type so that rules involving relative values, e.g. rel_low_wt, are applicable. Therefore the result from the assessment is a set of support pairs representing the support for each design being a "good plate" or "good strut". For clarity these support pairs are not only output as numbers but in bar chart form.

4.3 Explanation

Once an assessment has been performed it may not be immediately obvious to the user why one candidate design has been preferred to others or vice versa. For greater perspicuity of the system and reassurance for the user therefore a simple explanation facility is incorporated into the assessment sub-system. Following an assessment the user may choose one particular design and one particular rule and ask for clarification of how the supports on that rule were obtained. For example after the assessment resulting in the text output above, a request for an explanation leads to the rules being displayed with the relevant arguments instantiated to strutX and strutX$2, i.e.

(good_strut strutX strutX$1) : (0.8542 0.9061)
(good_strut strutX strutX$2) : (0.0126 0.0309)
(good_strut strutX strutX$3) : (0.0451 0.0671)
(good_strut strutX strutX$4) : (0.7464 0.8000)

Rule 1 : ((good_strut strutX strutX$2)

(good_strut_wt strutX strutX$2)): ((0.9 1) (0 0.1))
Rule 2 : ((good_strut strutX strutX$2)
(good_strut_fpf strutX strutX$2)): ((0.8 1) (0 0.2))
Rule 3 : ((good_strut strutX strutX$2)
(good_strut_cost strutX strutX$2)): ((0.5 1) (0 0.8))
Rule 4 : ((good_strut strutX strutX$2)
(good_strut_stiffX strutX strutX$2)): ((0 1) (0 1))
Rule 5 : ((good_strut strutX strutX$2)
(good_strut_manuf strutX strutX$2)): ((0 1) (0 1))
Rule 6 : ((good_strut strutX strutX$2)
(good_strut_environ strutX strutX$2)): ((0 1) (0 1))

The user is now asked to indicate which rule he wishes to be clarified; in this case rule 3 is selected, so that the following information is displayed :-

Rule 3 for good_strut is:
((good_strut strutX strutX$2)
(good_strut_cost strutX strutX$2)) : ((0.5 1) (0 0.8))
Inferences from body of rule 3 for good_strut are:
((good_strut_cost strutX strutX$2)) : (0.7793 0.7793)
Inferences from rule 3 for good_strut are:
(good_strut strutX strutX$2) : (0.3897 0.9559)

The last support pair shown is the result of combining the support from the body (tail) of the rule with the conditional support on the rule.

Clearly for different design types the designer's priorities may be different, e.g. for one plate type he may regard the stiffness as very important and the ultimate failure load factor of marginal relevance, and for another he may hold the opposite view. Therefore there is the facility in the CODEX assessment sub-system for the user to change the conditional supports on the assessment rules himself. In order to see what the effect of changing the conditional supports on the assessment rules will be on a particular design type, there is a "what-if" facility in the CODEX assessment sub-system. This allows the user to choose a particular rule for which the support is to be varied and see the effect on the overall supports derived for a set of designs. This effect is displayed as a set of plots as shown in Figure 2 for the plate type "myplate", where the cost rule has been varied. For example if the original rule is

((good_plate P D)(good_plate_cost P D)) :((0.5 1)(0 0.8))

then assessment is carried out for each design for myplate using the rule

((good_plate P D)(good_plate_cost P D)) :((α 1)(0 0.8))

where α takes values 0.0, 0.1, 0.2, ..., 1.0.

5. RESULTS

Example runs and results in tabular and graphic form of the CODEX plate design, strut design and design assessment sub-systems can be found in other papers [5; 11; 12; 19]. In this section we merely discuss a few design trends that have emerged from application of the system.

5.1 Plate weight v. load for different load/moment combinations

In Figure 3 are graphs showing the total plate thicknesses of designs obtained using CODEX for a wide range of complex loads (reproduced from [5]). In each case the material used is CFRPHS914, the lay-up is $[0°/+45°/-45°/90°]_{sym}$, and failure is predicted using the maximum strain criterion. Figures 3(a) and 3(b) show the effects of loading cases excluding applied moments: Figure 3(a) shows the effect of varying N_y while keeping N_x fixed and Figure 3(b) shows the effect of varying N_{xy} for different values of N_y while keeping N_x fixed. The approximately linear relationship between laminate thickness and applied load is evident from these graphs, as it is to a lesser extent when an M_x moment is added as shown in Figure 3(c) where N_x and N_{xy} are fixed and M_x and N_y are varied. The slight "wobbles" in the graphs are attributable to the adjustment of ideal thicknesses to integral multiples of the available prepreg thicknesses.

With more complex loads the situation is less straightforward as shown in Figure 3(d) where M_y and N_y are varied. The design thickness drops off as M_y is increased from 250 Nmm/mm to 500 Nmm/mm for a given value of N_y. This can be explained by the bending effect cancelling out the stretching effect at the bottom of the plate. In Figure 3(e) the effect of adding a twisting moment M_{xy} as well is shown for different values of N_y while keeping the other loads fixed. The variation in the thickness here has no apparent trend, except a convergence to a value of 5.3-5.4 mm when M_{xy} becomes large with respect to M_x. In Figure 3(f) the in-plane loads are set to zero and the unusual and unpredictable effect of varying the twisting moments for different values of M_y here is even more pronounced.

5.2 Strut weight v. load for different shapes/materials

In Figure 4 are shown the weights of struts of different cross-sectional shape and material derived using CODEX. (The weights of struts obtained using CFRPHS913 and CFRPEEK are not shown since they are within 1% of those obtained using CFRPHS914.) It is clear that the worst material with respect to weight is GFRPE913 followed by AAL168; KFRP49913 and SCRAL are about the same, while CFRPHS914 is the best for every shape. However this does not take into account the relative costs of the materials, nor the other quantitative and

Table 2 : Weights of CODEX-designed struts with and without end fittings

weight (kg)	I				○	
	no joint	J1	J2	J3	no joint	J1
CFRPHS914	0.131	0.158	0.175	0.142	0.050	0.106
GFRPE913	0.229	0.265	0.288	0.250	0.116	0.190
SCRAL	0.150	0.168	0.205	0.161	0.091	0.131
AAL168	0.169	0.217	0.236	0.181	0.075	0.239

qualitative aspects of the design: for a proper comparison the assessment sub-system of section 4 should be used.

Also apparent is that the ordering of the shapes with respect to weight is more or less the same for each material, namely: T and I being the worst; C, modified I, square and modified T being in the middle; and the tube being the best. However this is only half the story since the additional weight of the end fittings required for each shape is not included, and this is particularly relevant to the tube, where a satisfactory attachment is not always easy to obtain. Thus, if we consider the strut design results when the procedures for the joint designs discussed in section 3.2 are included for two cross-sectional shapes, we see from Table 2 that the increase in weight due to the end fittings is proportionally much greater for the tube than for the I-shape, and for AAL168 the necessary end fitting takes the weight above that for any of the joints possible for the I-shape. Of the three joints J3 is the most efficient and J2 the least efficient in terms of weight as might be expected from their construction. (Data relevant to the end fitting procedures is not available for CFRPEEK or KFRP49913.)

6. CONCLUSIONS

The computer expert system CODEX provides a user-friendly and efficient environment for the interactive design of plates and struts using a range of composite materials. A high level assessment sub-system allows the user to measure the relative merits of competing candidate solutions to plate and strut design problems by using support logic.

ACKNOWLEDGEMENT

This research has been carried out as part of a project jointly funded by the Science and Engineering Research Council and British Aerospace, Filton, Bristol, grant number GR/E78791.

REFERENCES

[1] Tsai, S.W., Composites Design (4th edition), Think Composites, Dayton, Paris and Tokyo, 1988.

[2] Webber, J.P.H., "Some laminated plate optimum design studies using an interactive computer programme", Aeronautical Journal of the Royal Aeronautical Society, Vol.92, pp. 107-114, 1988.

[3] Burden, P.I., & Lipton, B.J., "An interactive computer program for the design of composite laminated plates", Department of Aerospace Engineering Report no. 368, University of Bristol, Bristol.

[4] Webber, J.P.H., & Morton, S.K., "An expert system for laminated plate design using composite materials", Computers and Structures, Vol.37, pp. 1051-1067, 1990.

[5] Morton, S.K., & Webber, J.P.H., "Heuristic methods in the design of composite laminated plates", Dept. of Aerospace Engineering internal report no. 418, University of Bristol, Bristol, 1990.

[6] British Aerospace (Warton), "E.F.A. joint structures team working document: CFC design rules", internal document ref. JST-D-01, 1988 (private communication).

[7] Engineering Sciences Data Unit, "Failure criteria for an individual layer of a fibre reinforced composite laminate under in-plane loading", ESDU Item no. 83014, 1986 (private communication).

[8] Herakovich, C.T., "Free edge effects in laminated composites", Chapter 4, Handbook of Composites (Eds. Herakovich, C.T., & Tarnopol'skii, Y.M.), Vol. 2, Elsevier, Amsterdam, 1989.

[9] Kassapoglou, C., & Lagace, P.A., "An efficient method for the calculation of interlaminar stresses in composite materials", Journal of Applied Mechanics, Vol.53, pp. 744-750, 1986.

[10] Brewer, J.C., & Lagace, P.A., "Quadratic stress criterion for initiation of delamination", Journal of Composite Materials, Vol.22, pp. 1141-1155, 1988.

[11] Wu, C.M.L., & Webber, J.P.H., "Minimum weight optimization of composite laminated struts", to appear in Engineering Optimization, 1991.

[12] Wu, C.M.L., "Bolted joints in a laminated composite strut design expert system", Dept. of Aerospace Engineering internal report no. 415, University of Bristol, Bristol, 1990.

[13] Collings, T.A., "The strength of bolted joints in multi-directional CFRP laminates", Composites, Vol.8, pp. 43-55, 1977.

[14] Kretsis, G. & Matthews, F. L., "The strength of bolted joints in glass fibre/epoxy laminates", Composites, Vol.16, pp. 92-102, 1985.

[15] Webber, J.P.H., "Single pin bearing tests on metal matrix (SiC fibre/Al 6061) laminated plates", Department of Aerospace Engineering, University of Bristol, Bristol, 1987 (private communication).

[16] Baldwin, J.F., Martin, T.P., & Pilsworth, B.W., FRIL manual, FRIL Systems Ltd., Bristol, 1988.

[17] Baldwin, J.F., "Evidential Support Logic Programming", Fuzzy Sets and Systems, Vol.24, pp. 1-26, 1987.

[18] Morton, S.K., & Webber, J.P.H., "Uncertainty reasoning applied to the assessment of composite materials for structural design", Engineering Optimization, Vol.16, pp. 43-77, 1990.

[19] Wu, C.M.L., Webber, J.P.H., & Morton, S.K., "A knowledge based expert system for laminated composite strut design", to appear in The Aeronautical Journal, 1991.

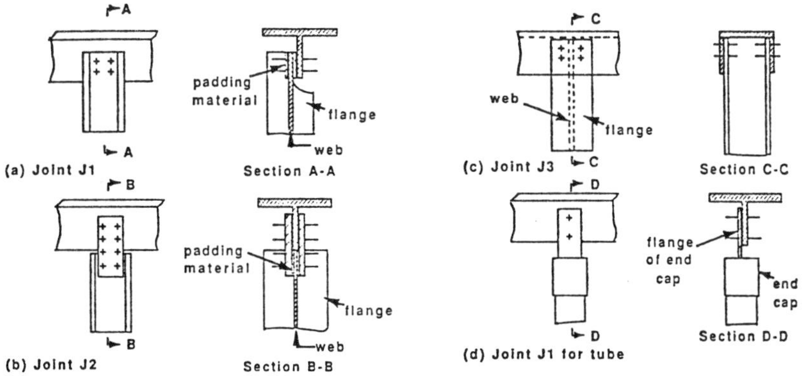

Figure 1: Bolted joint configurations for struts

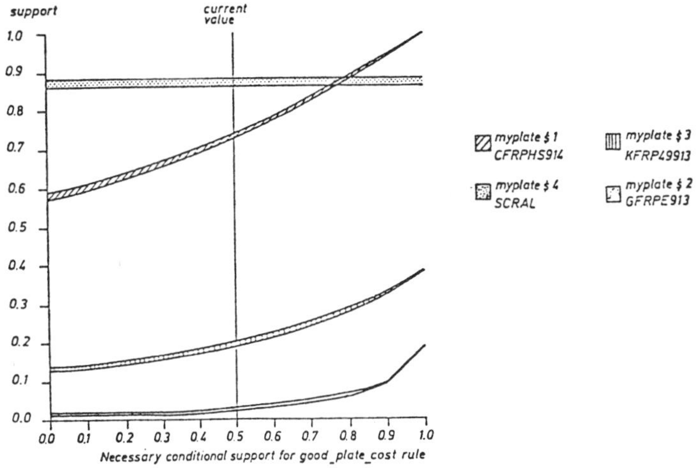

Figure 2: Effect of varying conditional support on good_plate_cost rule on overall support for designs of plate type "myplate"

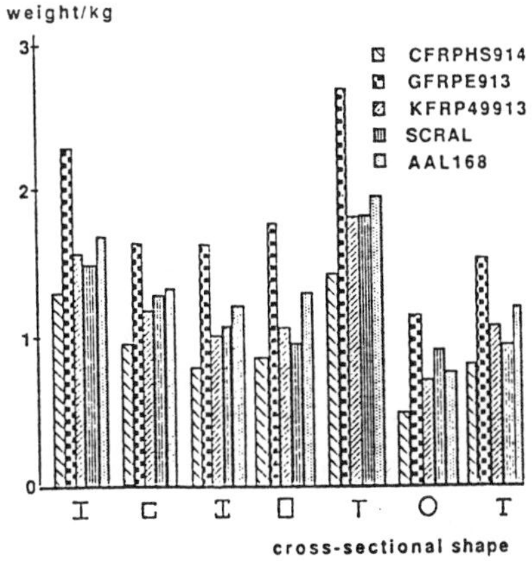

Figure 4: weights of struts derived using CODEX with Nx=10000 N, L=1m, both ends fixed, and using the maximum stress failure criterion

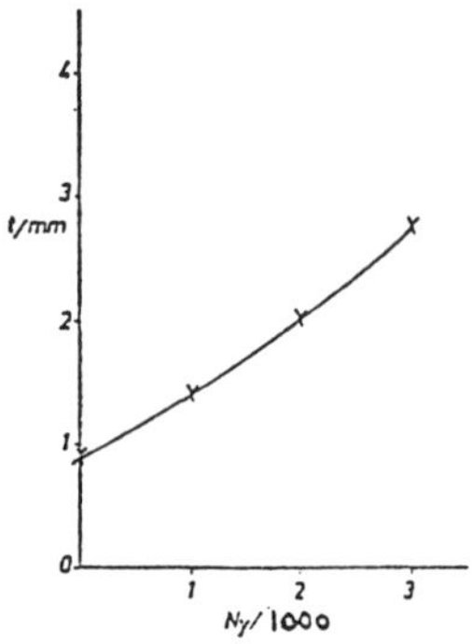

(a) $\underline{N}$ = [1000,Ny,0] N/mm ; $\underline{M}$ = $\underline{0}$ Nmm/mm

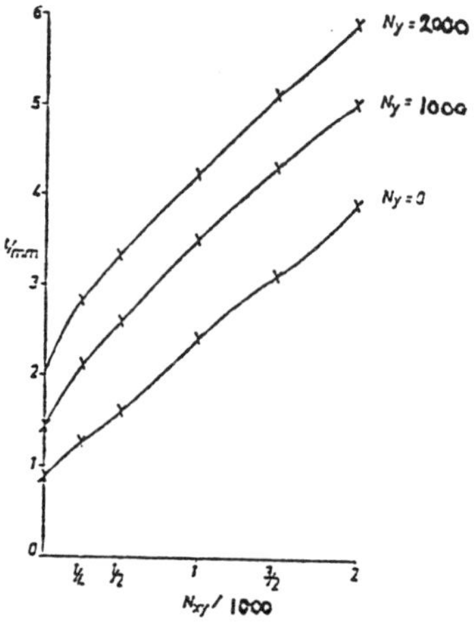

(b) $\underline{N}$ = [1000,Ny,Nxy] N/mm ; $\underline{M}$ = $\underline{0}$ Nmm/mm

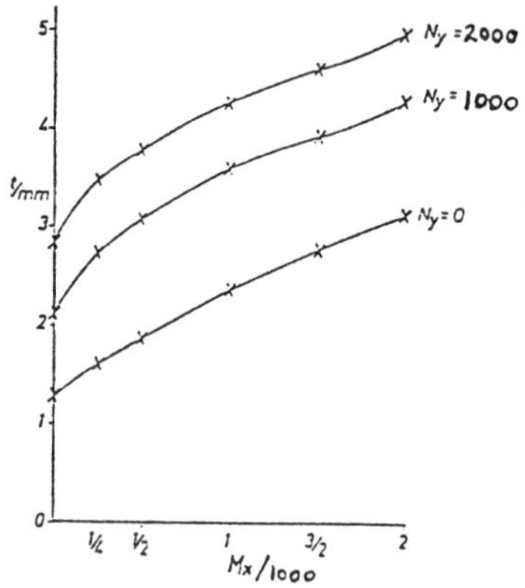

(c) $\underline{N}$ = [1000,Ny,250] N/mm ; $\underline{M}$ = [Mx,0,0] Nmm/mm

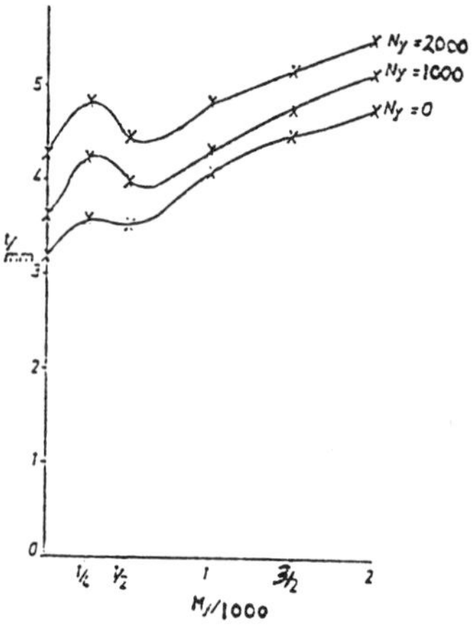

(d) $\underline{N}$ = [1000,Ny,250] N/mm ; $\underline{M}$ = [1000,My,0] Nmm/mm

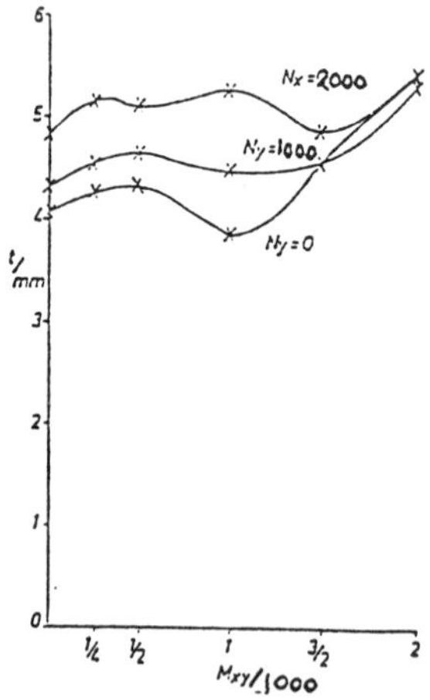

(e) $\underline{N}$ = [1000,Ny,250] N/mm ; $\underline{M}$ = [1000,1000,Mxy] Nmm/mm

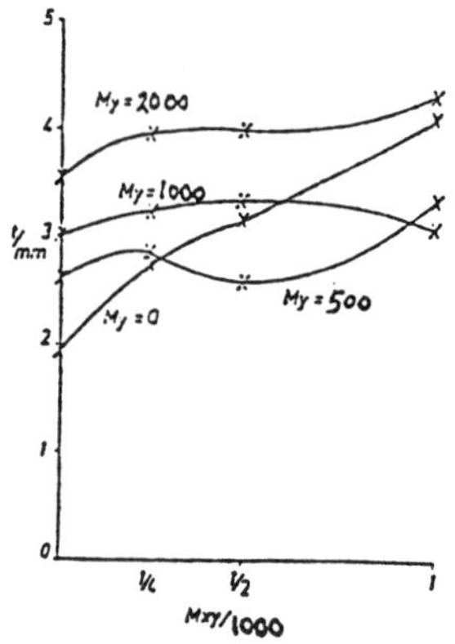

(f) $\underline{N}$ = $\underline{0}$ N/mm ; $\underline{M}$ = [1000,My,Mxy] Nmm/mm

Figure 3: CODEX plate designs for various loads

Development of a Distributed Computing Environment for Optimization-Based CAD of Structures

M.A. Austin, B.K. Voon

Systems Research Center and Department of Civil Engineering, University of Maryland, College Park, MD 20742

ABSTRACT

Networking technology, an ensemble of loosely coupled workstation processors, and new algorithms are being used to compute engineering simulations concurrently. A prototype environment for Distributed Numerical Computing (DNC) has been developed. Details are given on a new algorithm for integration of smooth structural dynamics equations. We assess the algorithm's performance in the DNC environment. The paper concludes with a description of future work.

INTRODUCTION

Now that engineering workstations with network connectivity are readily available in the marketplace, opportunities exist for the formulation of new algorithms and software tools that exploit concurrency as a means of increasing computational speed. The strong need for this research dates back to the early 1980's when considerable work was done to better represent real world design problems, and to capitalize on the emergence of engineering workstations. At U.C. Berkeley, for example, Nye et al. [14] proposed an optimization algorithm called the Phase I-II-III Method of Feasible Directions. This algorithm has been successfully applied to a wide variety of engineering problems including the design of chemical polymers [3], integrated circuits [15] and earthquake resistant buildings [1,4]. For problems of a realistic size, however, the quality of user interaction has often been very poor, with time consuming engineering (structural) simulations - and hence iterations of optimization - severely restricting the size of the problems studied. Together these limitations have not only limited the appeal of our work to others, but also restrained the scope of problems that could be practically investigated.

Researchers at the Systems Research Center, University of Maryland, are attempting to mitigate this problem by focusing their work in two areas. First, a new class of FSQP optimization algorithms has been formulated [16]. These algorithms have superlinear convergence properties, and therefore require fewer iterations to converge than the Phase I-II-III Method of Feasible Directions [14]. Still, this leaves the problem of having to compute engineering simulations. Our experience indicates that for optimization problems of a realistic size, more than 90% of the computational effort is dedicated to the calculation of engineering system behavior. Indeed, upwards of 98-99% of the total computational effort is consumed by structural simulations during the optimal design of earthquake resistant structures [1]. Consequently, any efforts to speed up the engineering simulations will also improve user interaction and reduce design turn-around time.

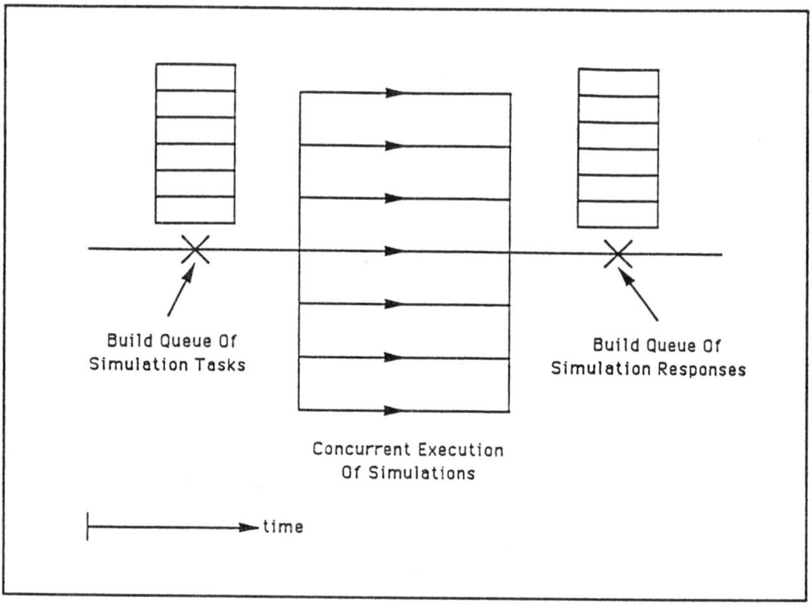

Figure [1] : Algorithm Structure

Objectives and Scope

Rather than simply rely on faster hardware to increase computational speed, we are attempting to exploit networking technology and an ensemble of loosely coupled workstation processors to compute engineering simulations concurrently. Applications of interest include the asynchronous solution of large sets of numerical equations [5], and parallel solutions to the integration of structural dynamics equations [3]. The long-term objective is to study problems in optimal design/control of large flexible space structures, and earthquake resistant structures.

The preliminary stages of work have focussed on the formulation and writing of software components to setup, execute, and monitor multiple simulations running concurrently on separate workstations. Our test application is a distributed version of a Newmark-Extrapolation algorithm for integrating smooth structural dynamics equations; see [3] for details. The purposes of this paper are to: (a) motivate the ideas leading to our selected network topology, (b) report on the current state of software development, (c) briefly outline the formulation of the new Newmark-Extrapolation algorithm, and (d) present numerical results of the algorithm in the concurrent computing environment.

Overview of Algorithm Structure

Our research direction stems from the simple observation that optimization and numerical algorithms frequently require many simulations that differ only slightly in their input data. Indeed, Figure [1] shows that the same algorithms often contain critical points that cannot be passed until a complete block of simulations is finished. Since the

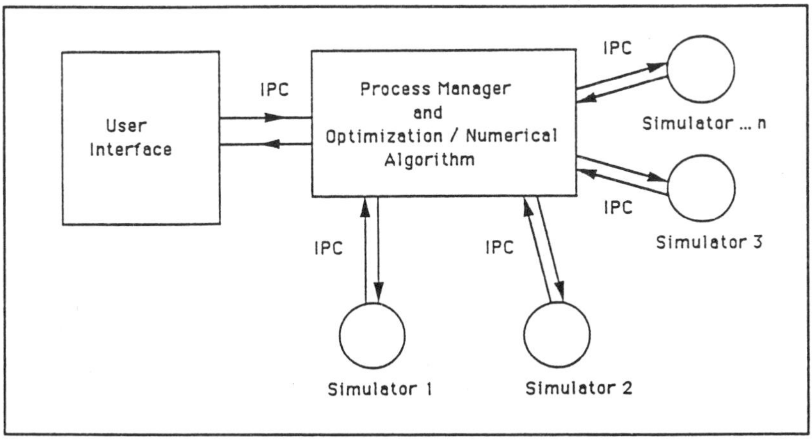

Figure [2] : Network Topology of DNC

quantity of input information that distinguishes simulations is minimal, a practical way of increasing computational speed is to setup multiple simulator packages distributed over a workstation network and concurrently compute individual components of the simulation block. This is coarse grained parallelism.

We start each simulation block by building a FIFO (first-in first-out) queue of initial condition data for the simulation components. Packets of simulation requests are then distributed to remote simulators via the network. When a remote simulation finishes, the essential features of the system response/behavior are sent back to the algorithm and temporarily stored in a response queue. The block of simulations is complete when the length of the simulation response queue equals the initial length of the simulation task queue.

DNC ENVIRONMENT

The prototype implementation of our computing environment is called DNC, an acronym for Distributed Numerical Computing. The vehicle for this work is the workstation model consisting of a high resolution bit mapped screen, a multiwindow user interface model with mouse and keyboard input, multitasking, and network connectivity. The network topology of the DNC environment is shown in Figure [2]. Its main components are a user interface, the process manager/optimization algorithm module, and several numerical simulator packages. All components are being written in C running under UNIX.

User Interface

The adopted philosophy of this work is that a designer and computer should be complementary as they work together to complete a design. Indeed, there is a growing body of evidence to suggest that ease of use is at least as important as functionality [9] in determining the likely success of an application program; this is particularly the case where new styles of design are being proposed. Consequently, a user interface that

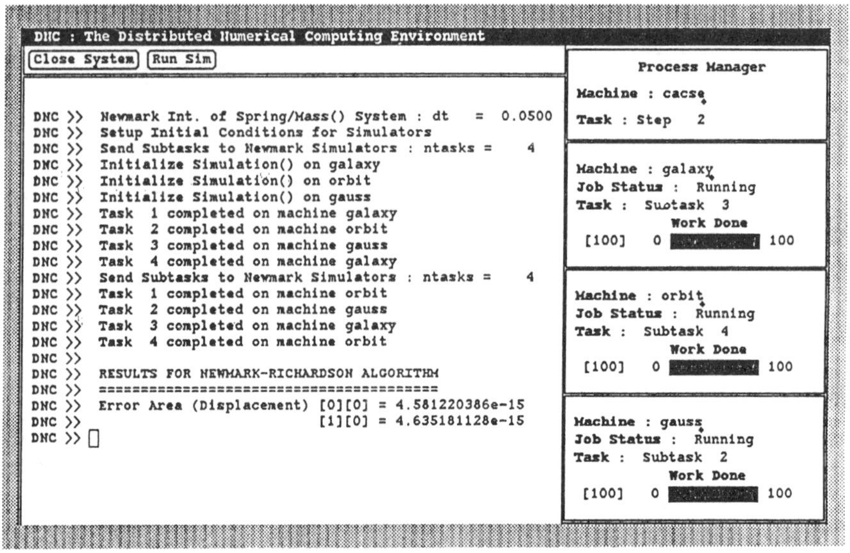

Figure [3] : DNC User Interface

supports multiple styles of interaction with graphics is considered indispensable for this work.

Figure [3] is a screendump of the DNC User Interface developed under the SunView Window systems [18]. As already mentioned in the opening sections, an important purpose of the interface is to report on computational activities at the process manager and remote numerical simulators. Subwindows dedicated to this task appear down the right hand side of the interface. Each window contains the machine name, job status, plus a slider for percentage of work done. The second function of the interface is to accept sequences of user commands. The current version of DNC accepts mouse button events, and very simple commands via the TTY window and keyboard input. Instructions of the latter type are passed using the UNIX tool YACC [11]. When a post command action includes the mailing of a message to the process manager - possibly requesting a numerical simulation - the user interface should remain unblocked for the processing of further keyboard/mouse events. Unfortunately, standard event-based systems do not behave in this way. A callback function embedded within a standard base window event handler will wait for the arrival of numerical results on an incoming socket before releasing the manager to other tasks. SUN's interposition mechanism [18] overcomes this problem by allowing client programs, such as our user interface, to register event sensitive interposer functions with the window notifier. The subsequent arrival of incoming data on the process manager/user interface socket triggers an interception of window manager control, and a callback to the interposer function. After the interposer function finishes reading the socket, control of events is returned to the base window event handler. Since the interception of window manager control occurs only when data arrives on the socket, users are given the impression that the interface is completely decoupled from the process manager and simulator nodes.

Process Manager

The purpose of the process manager is to coordinate and monitor computational activities for an ensemble of loosely coupled processors over which InterProcess Communication (IPC) [12] is used to parallelize simulation components. Instead of coding the process manager as a single UNIX process, our approach has been adopted to construct the process manager from lightweight processes. Support for the concurrent execution of these processes is provided by a Threads package [7]. The benefits are evident. Not only does the programming interface provided by Threads insulate the user from such details as the number of processors being used, but it contains the tools to build programming constructs for the synchronization of individual threads (using semaphores and monitors), to perform I/O, and to respond intelligently to interrupts and runtime exceptions. It is noted, however, that a Threads approach to programming is appropriate for the solution of problems that demand low level tightly coupled processes and shared memory; the internal details of the process manager match these requirements. The distributed simulation components on the other hand, consist of asynchronous processes and no shared memory.

Figure [4] shows a schematic of Thread components within the Process Manager. The caretaker is connected to the process manager/user interface IPC socket. It serves the purpose of receiving task requests from the user interface, and invoking optimization/numerical threads as directed. There is a natural tendency to think of the optimization/numerical algorithm code as a C function; it is, in fact, created by the caretaker as a separate lightweight process. A series of dispatcher threads act as an interface between the tightly coupled thread processes within the manager and loosely coupled simulation processes over the network via IPC. IPC socket connections to each of the remote simulators are established as part of the DNC startup procedure, and are subsequently available to handle all data and messages to/from the remote simulators. For example, messages are automatically forwarded to the user interface, and incoming simulation response data appended to the response queue. Runtime support for the synchronization of events among the dispatchers and optimization/numerical algorithms is handled by a single monitor. Finally, we note that only three segments of code need to be written for the process manager, since each of the dispatchers is distinguished by its initial data.

Remote Simulators

The remote simulators are connected via IPC to the dispatcher threads in the process manager. Each simulator listens for requests on the IPC socket and carries out numerical calculations as required. Messages indicating the status of intermediate calculations are sent to the user interface via the process manager. When a simulation component is finished, a packet of response information is sent to the dispatcher, and subsequently stored on the simulation response queue.

DISTRIBUTED NUMERICAL ALGORITHM

Prototype versions of this environment are being tested with a new algorithm for integrating the equations of motion for smooth dynamical systems. This algorithm is a combination of Newmark Integration [13] and Richardson Extrapolation [17]; its development was motivated as the result of observations reported in Austin et al. [2]. It is well suited to our development because it may be naturally partitioned for concurrent simulations as already described above. Below is a brief review of the Newmark method,

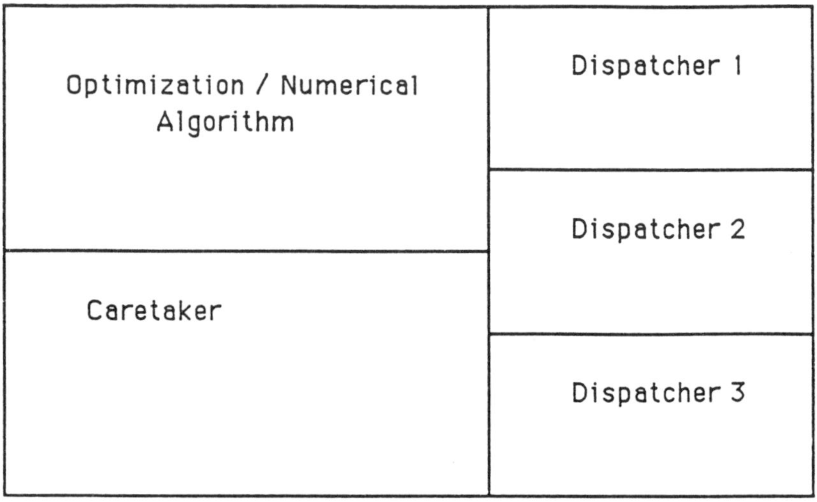

Figure [4] : Components of Process Manager

followed by a step-by-step guide to the new algorithm. The interested reader is referred to [3] for a detailed discussion.

Standard Newmark Method

Newmark integration methods [10,13] approximate the time dependent response of linear and nonlinear 2nd-order equations by insisting that equilibrium be satisfied only at a discrete number of points (or timesteps). If t_q and t_{q+1} are successive timesteps in the integration procedure for a linear structural system, then the two equations of equilibrium that must be satisfied are:

$$[m]\,[\ddot{x}_q] + [c]\,[\dot{x}_q] + [k]\,[x_q] = P_q \tag{1.1}$$

and

$$[m]\,[\ddot{x}_{q+1}] + [c]\,[\dot{x}_{q+1}] + [k]\,[x_{q+1}] = P_{q+1} \tag{1.2}$$

where $[m]$, $[c]$ and $[k]$ are $(n \times n)$ mass, damping and stiffness matrices, respectively, and P_q is a $(n \times 1)$ vector of external forces applied at the nodal degrees of freedom. Let's also assume that (1.1) has been solved and we are interested in trying to solve (1.2). At each timestep there are 3n unknowns corresponding to the displacement, velocity, and acceleration of each component of x. Since we only have n equations, the natural relationship existing between the acceleration and velocity, and velocity and displacement must be enforced to reduce the number of unknowns to n. That is:

$$\dot{x}(t) = \dot{x}(t_q) + \int_{t_q}^{t} \ddot{x}(\tau)d\tau \tag{2.1}$$

$$x(t) = x(t_q) + \int_{t_q}^{t} \dot{x}(\tau) d\tau \tag{2.2}$$

where $\dot{x}(t_q)$ and $x(t_q)$ are the velocity and displacement at timestep t_q, and $\dot{x}(\tau)$ and $\ddot{x}(\tau)$ are unknown functions. The heart of the Newmark method is to assume the acceleration behaves in a prescribed manner - constant or perhaps a linear variation - within the timestep. For example, if the acceleration is prescribed to be constant within the timestep t_q to t_{q+1}, then the function:

$$\ddot{x}(t) = \left[\frac{\ddot{x}_q + \ddot{x}_{q+1}}{2} \right] = \ddot{x}_q + \left[\frac{\ddot{x}_{q+1} - \ddot{x}_q}{2} \right] = \ddot{x}_q + \left[\frac{\Delta \ddot{x}_q}{2} \right] \tag{3}$$

is simply the average of the endpoint accelerations in equations (2). It follows that approximations to the velocity and displacement will be linear and parabolic, respectively. Substituting equations (3) into (2) and rearranging terms gives:

$$[M] [\Delta \ddot{x}_q] = [\Delta P] \tag{4}$$

where $\Delta t = t_{q+1} - t_q$

$$[M] = [m] + \frac{\Delta t}{2} [c] + \frac{\Delta t^2}{4} [k]$$

$$[\Delta P] = [P_{q+1}] - [m] \ddot{x}_q - [c] [\dot{x}_q + \Delta t \ddot{x}_q] - [k] \left[x_q + \Delta t \dot{x}_q + \frac{\Delta t^2}{2} \ddot{x}_q \right]$$

Equations (4) are for linear systems; see [10] for well known extensions of these equations for nonlinear stiffness and damping.

Proposed Newmark-Extrapolation Integration Algorithm

As noted in the previous section, $\ddot{x}(\tau)$ is unknown, and the key source of approximation for the Newmark method. The following algorithm is motivated by the observation that improved numerical estimates of response will result if (2.1) and (2.2) were evaluated with increased precision across the time interval $t \in [t_q, t_{q+1}]$. Instead of simply refining the steplength, our idea is to obtain increased precision via extrapolation. The step by step algorithm for P levels of extrapolation using the Rhomberg sequence of refinement is:

[1a] Initialization : Identify all components of velocity and displacement. For each component, dynamically allocate memory for a $(P \times P)$ extrapolation table.

[1b] Select a timestep Δt. In some cases this may be the maximum steplength for algorithm stability, while in other instances it may be a suitably small Δt needed to draw a smooth graph of response.

[2a] Outer Loop : For Newmark Integration, q = 1 to nsteps

[2b] Set $t_q = [q-1] \cdot \Delta t$

[3a] Inner Loop 1 : For i = 1 to P

[3b] Incremental timestep $\Delta t_i = \left[\frac{\Delta t}{2^{(i-1)}}\right]$

[3c] Form $[M]$ for Δt_i

[4a] Inner Loop 2 : For k = 1 to $2^{(i-1)}$

[4b] Calculate ΔP at time $t = t_q + \mathrm{k} \cdot \Delta t_i$

[4c] Solve $[M] \cdot [\Delta \ddot{x}_q] = [\Delta P]$

[4d] Update : $\ddot{x}_{q+1} = \ddot{x}_q + \Delta \ddot{x}_q$

$$\dot{x}_{q+1} = \dot{x}_q + \ddot{x}_q \cdot \Delta t_i + \left[\frac{\Delta t_i}{2}\right] \Delta \ddot{x}_q$$

$$x_{q+1} = x_q + \dot{x}_q \cdot \Delta t_i + \left[\frac{\Delta t_i^2}{2}\right] \ddot{x}_q + \left[\frac{\Delta t_i^2}{4}\right] \Delta \ddot{x}_{q+1}$$

[4e] End Inner Loop 2

[3d] Put displacement x_{q+1} and velocity $\dot{x}_{q+1}$ components in u_j^1 of Extrapolation Table [1]. Here j is the number of intervals taken to integrate across the Δt time-step, and the order of refinement is 1.

No Steps	Stepsize	$0[\Delta t^2]$	$0[\Delta t^4]$	$0[\Delta t^6]$	$0[\Delta t^8]$
1	Δt	u_1^1			
2	$\frac{\Delta t}{2}$	u_2^1	u_2^2		
4	$\frac{\Delta t}{4}$	u_4^1	u_4^2	u_4^3	
8	$\frac{\Delta t}{8}$	u_8^1	u_8^2	u_8^3	u_8^4

Table [1] : Extrapolation Tableau

[2c] End Inner Loop 1 : Note : Each component of this loop may be computed in parallel.

[1c] For the Rhomberg sequence of internal steplengths it is relatively straight forward to show that successive applications of the formula:

$$\mathrm{u}_{(r)}^{(s)} = \left[\frac{4^{(s-1)}\mathrm{u}_{(r)}^{(s-1)} - \mathrm{u}_{(r-1)}^{(s-1)}}{4^{(s-1)} - 1}\right] \tag{5}$$

will systematically eliminate higher order coefficients in the error polynomial; the interested reader is referred to [3,8] and [17] for detailed explanations of the error polynomial. Use equation (5) to calculate columns 2-P of each extrapolation table.

The lower most right entry of Table [1] - element [4][4] - is taken as the starting displacement and velocity components of the next time-step.

[1d] Backsubstitute displacements/velocities from step [1c] into equation (1.2) and solve for acceleration vector $\ddot{x}_{q+1}$.

[1e] Set $q = q + 1$.

[1f] End Outer Loop

The success of this algorithm pivots on the existence of an asymptotic expansion for the global error of the underlying numerical method. Unfortunately, restrictions apply for its existence. First, the underlying equations of motion must be continuously differentiable. This restriction allows smooth geometric nonlinearities, but automatically excludes this algorithm for systems having sharp material nonlinearities. Second, the basic numerical method must be stable. Although explicit numerical integrators, such as Euler, are much less computational work than implicit schemes, they are notorious for being only conditionally stable. On the other hand, the implicit trapezoidal scheme used in Newmark is A-stable [10].

NUMERICAL EXPERIMENT

The mechanics of the new algorithm and DNC environment have been tested by working through a very simple test application, damped SDOF subject to external loads:

$$\ddot{x}(t) + 4\dot{x}(t) + 13x(t) = \left[\frac{1}{3}\right] e^{-2t} \cdot \sin[3t] \tag{6}$$

If x(0) = 1 and $\dot{x}(0) = -2$, then Laplace transforms gives an analytic solution:

$$x(t) = e^{-2t} \cdot \cos[3t] + \left[\frac{e^{-2t}}{54}\right] [\sin(3t) - 3t\cos(3t)] \tag{7}$$

The scope of this study is restricted to simulations with four levels of stepwise refinement matching the Rhomberg sequence, i.e. Δt, $\Delta t/2$, $\Delta t/4$, and $\Delta t/8$, i.e.:

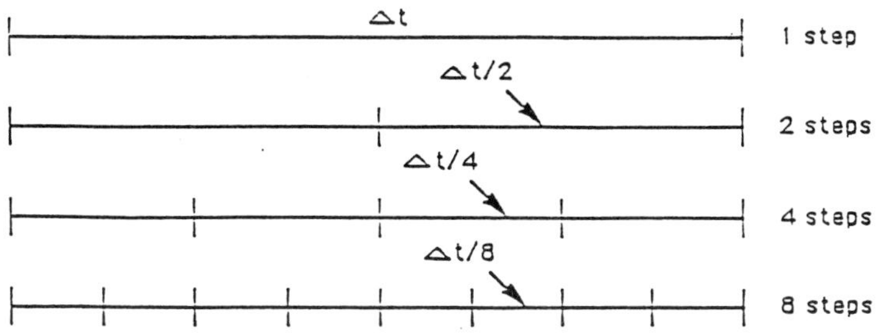

Behavior of DNC Environment

When a user manually requests a design simulation, a message is first sent from the user interface to the caretaker thread. The caretaker thread creates a detached thread for the optimization/numerical algorithm. To illustrate the asynchronous behavior of the Newmark-Extrapolation algorithm in the DNC environment, consider the verbose tracing of activities taken from typical iteration i.e.:

```
Setup Initial Conditions on Remote Simulators;
Initialize Simulation() on galaxy;
Initialize Simulation() on orbit;
Initialize Simulation() on gauss;
Start Iteration 1;
   Add Item 1 to Task Queue : t = 0.03000 seconds : nsteps = 1;
   Add Item 2 to Task Queue : t = 0.01500 seconds : nsteps = 2;
   Add Item 3 to Task Queue : t = 0.00750 seconds : nsteps = 4;
   Add Item 4 to Task Queue : t = 0.00375 seconds : nsteps = 8;
      Send Task 1 to "galaxy";
      Send Task 2 to "orbit";
      Send Task 3 to "gauss";
      Task 1 finished on "galaxy";
      Send Task 4 to "galaxy";
      Task 2 finished on "orbit";
      Task 3 finished on "gauss";
      Task 4 finished on "galaxy";
   Rhomberg Calculation;
End Iteration 1;
```

Before the iterations of numerical integration can proceed, packets of application-specific data must be posted to each remote simulator. In our application, this includes the mass and stiffness matrices.

Each iteration of numerical integration begins with the building of a task queue containing initial conditions, step lengths, and number of integration steps. When the task queue is completely assembled, control of execution is passed from the Numerical Algorithm thread to the dispatchers. First, three simulation tasks are sent to remote simulators galaxy, orbit, and gauss, respectively. We do not assume a proiri that each workstation will be running at the same speed. Instead, the dispatchers test for the arrival of message information and simulation response data on the socket. Messages are automatically forwarded to the DNC User Interface. Simulation responses are read and stored on the response queue; the next item is then fetched from the task queue and sent to the simulator. When the length of the simulation response queue equals the initial length of the task queue, control of execution passes back to the numerical algorithm thread. Equation (5) is used to fill in columns 2-4 of Table [1] for each component of velocity and displacement.

In our example script, tasks 1, 2, and 3 are sent to the remote simulators at approximately the same time. Task 1 requires approximately half the computational effort as 2, and hence finished first. This assumes approximately equal computational speed for each remote simulator. In some of our other simulations - not shown here - the variation in workstation speed has been significant and task 2 has finished before task 1.

Performance of Numerical Algorithm

In order to benchmark the accuracy of the proposed algorithm against traditional approaches (for equivalent computational work per unit increment of time), the timestep for the standard integration procedures is divided by 15 - i.e. stepsize $\Delta t/15$.

The responses of both the standard Newmark and new extrapolation methods are compared against analytic solutions. To capture the accumulation of differences between analytic and numerical response quantities, we define the area of component error as:

$$\text{Error Area} = \text{stepsize} \cdot \sum_{i=1}^{i=N} |\text{Analytic}(t) - \text{Numerical}(t)|_i \tag{8}$$

where N equals the total number of integration timesteps over $[T_o, T_1]$, and problem specific error components include acceleration, velocity, displacements, angular momentum and energy. All calculations were done in double precision arithmetic on a SUN workstation.

Table [2] shows the error areas for 200 timesteps of the proposed algorithm at Δt = 0.03 seconds, and 3000 timesteps of the standard Newmark algorithm using Δt = 0.002 seconds.

Error Component	Standard Newmark	Proposed Algo
Displacement	2.32×10^{-6}	8.067×10^{-16}
Velocity	9.30×10^{-6}	3.223×10^{-16}
Acceleration	3.36×10^{-6}	1.235×10^{-15}

Table [2] : Error Area Accumulation after 200 Time Steps

CONCLUSIONS AND FUTURE WORK

The DNC environment increases computational speed by exploiting networking technology and an ensemble of workstation processors to partition and distribute computations over a local area network. So far emphasis has been placed on the development and testing of DNC components, including mechanisms for mailing messages throughout the system. A new algorithm that combines ideas from Newmark's Method and Richardson Extrapolation has been formulated. DNC has been used to solve the SDOF test problem to a level of computational accuracy approaching machine precision.

Work is currently underway to replace the numerical algorithm with the FSQP optimization algorithm [16]. A new finite element package written exclusively in C will be placed at the simulation nodes. Future work will include extensions of the user interface to handle the graphical description of structures. Software tools will also be developed to graphically present the results and performance of optimization algorithms.

REFERENCES

[1] Austin M.A., Pister K.S., Mahin S.A., "Probabilistic Limit States Design of Moment-Resistant Frames under Seismic Loading," Journal of the Structural Division , ASCE, August 1987.

[2] Austin M.A., Krishnaprasad P.S, and Wang L.S., "On Symplectic and Almost Poisson Integration of Rigid Body Systems," To be submitted to Journal of Computational Physics, 1991.

[3] Austin M.A., "High Order Integration of Smooth Dynamical Systems: Theory and Numerical Experiments," Journal of the Engineering Mechanics Division , ASCE; To be Submitted, March 1991.

[4] Balling R.J, Ciampi V., Pister K.S., and Polak E., "Optimal Design of Seismic-Resistant Planar Steel Frames," Report No EERC 81-20, Earthquake Engineering Research Center, University of California, Berkeley, December, 1981.

[5] Bertsekas D.P., Tsitsiklis J.N., Parallel and Distributed Computation : Numerical Methods, Prentice-Hall Inc, 1989.

[6] Butala D., Choi K.Y., Fan M.K.H., "MultiObjective Dynamic Optimization of Semibatch Free Radical Copolymerization Process with interactive CAD tools," Systems Research Center, University of Maryland, Technical Report TR 87-166, College Park, MD 20742.

[7] Doeppner T.W. Jr., "A Threads Tutorial," Computer Science Technical Report CS-87-06, Brown University, March 1987.

[8] Deuflhard P., "Recent Progress In Extrapolation Methods for Ordinary Differential Equations," SIAM Review, Vol. 27, No. 4. pp 505-535, December 1985.

[9] Gould J.D., and Lewis C., "Designing for Usability," Communications of the ACM, March 1987, pp. 229-233.

[10] Hughes T.R.J., The Finite Element Method : Linear Static and Dynamic Finite Element Analysis, Prentice-Hall Inc, 1987.

[11] Johnson S.C., "YACC - Yet Another Compiler Compiler," Computer Science Technical Report 32, AT&T Bell Laboratories, Murray Hill, New Jersey, 1975.

[12] Joy. W., Fabry R., Leffler S., A 4.3BSD Interprocess Communication Primer, Computer Systems Research Group, University of California, Berkeley, CA 94720.

[13] Newmark N.M., "A Method for Computation of Structural Dynamics," Proceedings of American Society of Civil Engineers, 85 EM3 (1959) pp 67-94.

[14] Nye W.T., Tits A.L., "An Application-Oriented, Optimization-Based Methodology for Interactive Design of Engineering Systems," International Journal of Control , Vol. 43, No. 6, pp 1693-1721, 1986.

[15] Nye W.T., Riley D.C., Sangiovanni-Vincentelli A.L., Tits A.L., "DELIGHT.SPICE: An Optimization-Based System for the Design of Integrated Circuits," IEEE Trans. CAD Integrated Circuits and Systems, CAD-7, pp 501-520.

[16] Panier E. and Tits A., 'On Feasibility,Descent and Superlinear Convergence In Inequality Constrained Optimization', Systems Research Center, University of Maryland, Technical Report SRC-TR-89-27, College Park, MD 20742, 1989.

[17] Richardson, L.F., "The Deferred Approach to the Limit," Phil. Trans., A, Vol. 226, 1927, pp 299-349.

[18] Sun Microsystems, SunView 1 Programmer's Guide, Copyright 1982,1988, Sun Microsystems, Inc.

An Advanced System for Electric Design

A. Longo (*), J. Unzueta (*), A. Alvarez (*),
T. Gutierrez (*), J.A. Tárrago (**)
*(*) Analysis & Design Department, LABEIN,
C. de Olabeaga, 16, 48013 Bilbao - Spain
(**) ETSII e IT, U.P.V., Alda. Urquijo s/n
48013 Bilbao - Spain*

ABSTRACT

An advanced integrated design system called SIDOMAT is presented. The system follows the research tradition in shape optimization (Haftka [1], Ding [2]). Some important enhancements are included which make the system suitable for industrial application. Geometric Representation is based on a Variational Geometry approach, Analysis is performed by means of Boundary Element Method, a continuous Material Derivative approach is used to perform Sensitivity Analysis, and Generalized Approximation Concepts are included in the optimization phase. An application example is given.

INTRODUCTION

Shape optimal design problems involve the determination of the shape of a component (or system) in order to minimize an objective function subject to some constraints. Main steps related to conventional C.A.E. application to the Design Process are also involved in Shape Optimization problems (i.e. Geometric Definition, Mesh Generation, Analysis and Displaying of results). Also additional elements play a decisive role in optimization, such as Sensitivity Analysis and Mathematical Programming methods. A main difficulty of Shape Optimization is the integration of such elements, in particular, the combination of a suitable geometric representation, capable of efficiently representing changing geometries, and the numerical method used to perform the analysis (e.g. Boundary Elements, Finite Elements).

As it was pointed out by Schmidt [3], research advances in optimum shape design have lead to the conclusion that, in order to develop a system useful in an industrial environment, it is crucial to distinguish between Design Model and Analysis Model.

Design Model should be placed above Analysis Model in the design hierarchy. Its shape must be defined by a small number of natural design variables, closely related to designer and manufacturer philosophy. Then, the Analysis Model should be automatically created from Design Model, and shape changes during optimization process should directly affect the Design Model. In this context, Design Model is independent of analysis techniques used, which is a major difference between early shape optimization applications (Zienkiewicz [4]) and modern shape optimization techniques.

In this paper, an optimum design system which follows this philosophy is presented. The system, called SIDOMAT, is strongly based in a powerful tool for describing changing geometries (i.e. Variational Geometry).

Also other components, like Sensitivity and Nonlinear Programming modules, have been carefully studied in order to take advantage of most recent developments. In this direction, a Continuous Material Derivative approach has been developed and implemented which allows sensitivity evaluation with a small computational effort. Also a robust generalization of Approximation Concepts, well founded from a theoretical point of view, has been incorporated for the solution of the Nonlinear Optimization problem.

As a result, the system is very friendly and easy to use, and optimization results are meaningful from a practical point of view.

DESIGN MODEL

The Design Model implemented in SIDOMAT has four components:

* Geometry–Topology: Includes geometric entities (e.g. points) and topological entities (e.g. regions).

* Dimensions: Constraints between geometric entities are defined by means of dimensions (e.g. distances, lengths, radii angles).

* Function: Information about materials and boundary conditions, assigned to topological entities (e.g. Boundary Dirichlet condition attached to an edge of the model), and also some data for mesh generation (i.e. element size).

* Specifications: Information about Design Variables (related to dimensions), Objective Function and Constraints (related to Design Variables and topology).

Geometry-Topology
The definition of 2D (or axisimetric) models can be done by means of the following geometric entities: points, straight lines, circular arcs, B-splines. Topological entities are vertex, edges and loops. A region is composed by a Father loop and a number of Child loops (holes).

The interactive graphic module of SIDOMAT, called GV2, handles a 2D winged-edge structure that allows, by means of Euler Operators, the automatic generation of topological information from a geometric data introduced by means of a mouse. That means that regions are correctly defined independently of the order in which geometric data are defined, even if a model with various materials is defined. Also any geometric change updates topology automatically (e.g. an exterior boundary can be defined or modified after a "hole" was drawn).

Dimensions
In order to control the shape of a model, SIDOMAT uses constraints between geometric entities (i.e. points), defined by means of Dimensions. There are two kinds of Dimensions: Global and Local.

Global Dimensions. Are used to establish constraints between points which correspond to vertex and arc centres of the model (Key Points or Model Variables). These constraints have the form

$$\varphi_k \ (L_k, \ x_i) = 0 \qquad k = 1, \ 2n; \quad i = 1, \ n \qquad (1)$$

where
φ_k: constraint k
L_k: dimension k
x_i: key point i

For a given value of dimensions, the equations (1) can be solved by the Newton-Raphson Method, in order to find the coordinates of key points, and so the correct shape of the model. Advantage of Jacobian Sparsity is taken, as it was suggested by Light [6].

Also, differentiating equations (1), derivatives of key points coordinates with respect to dimensions (i.e. velocities) can be evaluated.

Assuming that (1) have the form

$$\varphi_k(L_k, \ x_i) \equiv f_k(L_k) + g_k(x) = 0 \qquad (2)$$

it holds that

$$\delta\varphi_k \equiv (\partial f_k/\partial L_k)\ \delta L_k + \nabla^t{}_x g_k\ \delta x = 0 \qquad (3)$$

which can be expressed as

$$[-\ 1/(\partial f_k/\partial L_k)]\ \nabla^t{}_x g_k\ \delta x = \delta L_k \qquad (4)$$

and in matrix form

$$K\delta x = \delta L \qquad (5)$$

where K is the Rigidity matrix of the dimensional scheme. The inverse of matrix K gives the velocities of key points with respect to dimensions.

Currently, five types of constraints have been implemented: 1) Distance between points, 2) Distance between points in horizontal or vertical direction, 3) Distance from point to a segment defined by two points, 4) Angle between two pairs of points, 5) Equality of distances between two pairs of points.

These five types of constraints can be related to standard dimensions used in engineering drawings, such as distances between points, lengths of straight edges, radii of arcs, angles between edges or distances from point to edge. Also, technological constraints such as parallelism, orthogonality or tangencies can be handled, by means of constraints related to "dimensions" of a fixed value.

In SIDOMAT, the designer can define explicit dimensions like in any standard drawing package. The system automatically interprets the dimension type from data picked with a mouse. For example, if the user picks two edges and a third point on the screen, the dimension is interpreted as an angle between edges, that dimension is drawn and the related constraint is created (figure 1).

Before any dimension is modified, the system performs a verification process based on a sequential Gram-Smidt factorization of Rigidity Matrix. During this process, technological constraints are internally inferred from the geometric sketch, redundant dimensions are identified and displayed and internal constraints are added in underconstrained geometries. As an example, if the user sketches a fillet, tangencies are automatically identified during verification, and the fillet remains for any dimensional (shape) modification.

Local Dimensions. Are used to control the local shape of boundaries defined by B-Splines. The first and end points of a B-spline are assumed to be vertex of the model. So, the curve is internally related to a local coordinate system defined by its first and last control points, which are controlled by global dimensions. Directions of movement defined with respect to the local coordinate system can be assigned to other control points. Local dimensions are movements of control points in that directions:

$$p_i = p_o + \theta \; (l\xi_i + s_i d_i) \qquad (6)$$

where:

p_i: global coordinates of control point i
p_o: global coordinates of initial control point
θ : rotation matrix of local coordinate system
l : distance between first and last control points
ξ_i: local coordinates of control point i for initial configuration
s_i: local dimension associated to control point i
d_i: local components of movement direction

The velocity of p_i with respect to local dimension s_i is

$$\partial p_i/\partial s_i = \theta d_i \qquad (7)$$

and with respect to the global dimension L_k

$$\partial p_i/\partial L_k = \partial p_o/\partial L_k + (\partial\theta/\partial L_k)(l\xi_i+s_i d_i) + \theta \; (\partial l/\partial L_k) \; \xi_i \qquad (8)$$

When dimensioning, if the user picks a control point and an additional point, movement direction is created and referred to local system. It is displayed as an arrow, and can be used to move the control point along it.

Function

Since regions are automatically created when sketching, materials can be assigned by picking any interior point of a desired region. Boundary conditions (Dirichlet, Newman or Float) and data about size of elements are assigned to edges.

Specifications

Design Variables are selected by picking desired dimensions (global and/or local). There is a library of functions that can be defined either as Objective Function or Constraints.

- Electric field modulus in a point over the boundary.
- Tangential component of electric field in a point over the boundary.
- Mean error of electric field modulus with respect to a selected value along a portion of the boundary.
- Mean error of tangential electric field with respect to a selected value along a portion of the boundary.
- Area (volume if axisimetric) of one or more regions.
- A dimension.

Also bound and linear constraints can be specified for Design Variables.

ANALYSIS MODEL

Once a Design Model has been created, it is automatically converted to an Analysis Model. This task is carried out by an Automatic Mesh Generator which performs the following tasks:

- Create discretized geometric model of points, elements and regions.
- Boundary conditions and specifications are reassigned to discretized model.
- Velocities previously calculated by expressions (5), (7) and (8) are used in order to assign velocities to points of the mesh. This is performed using parametric equations of edges.

Electrostatic analysis is carried out by the Boundary Element Method. As a result, electric field and values of Objective Function and Constraints are obtained.

Sensitivity Analysis is performed using the same Analysis Model (with different Boundary conditions) by solving the associated problem:

$$\nabla^2 \emptyset' = 0 \qquad \text{in } \Omega \qquad (9)$$

$$\emptyset' = -\nabla\emptyset V \qquad \text{in } \Gamma_\emptyset \qquad (10)$$

$$\nabla\emptyset' n = (\partial Vn/\partial s)\nabla\emptyset s - nD^2\emptyset V - V_s\nabla\emptyset(\partial n/\partial s) \quad \text{in } \Gamma_q \qquad (11)$$

$$\emptyset^{i\prime} = \emptyset^{i\prime} + \nabla\emptyset^j V - \nabla\emptyset^i V \qquad \text{in } \Gamma_I^{i/j} \qquad (12)$$

$$\nabla\emptyset^{i\prime} n^i = -\nabla\emptyset^{j\prime} n^j - V \sum_{i,j} [(D^2\emptyset)n] - \sum_{i,j} [\nabla\emptyset n] \ \text{in } \Gamma_I^{i/j} \qquad (13)$$

where apostrophe means partial derivative with respect to a design variable, a point means total derivative, and

$\emptyset$: electric potential
$\nabla\emptyset$: electric field
V : velocity field with respect to a design variable
n : normal to the boundary
D^2: hessian matrix
s : tangential coordinate of boundary
Ω : domain
$\Gamma_\emptyset$: Dirichlet boundary
Γ_q: Newman boundary
Γ_I: Interface boundary (i and j indicate regions that share the interface)

The solution of equations (9) to (13) gives the partial derivative of state variables with respect to a design variable. With this information, it is straight forward to obtain sensitivities of Objective Function and Constraints.

These sensitivities correspond to derivatives with respect to Dimensions. That means that Design Variables handled by the Optimization Module will be no more displacements of points but meaningful quantities from a designer point of view. Also, since velocities were evaluated by equations (5), (7) and (8), sensitivity results contain information about technological constraints (e.g. tangencies), and no additional constraints nor linking are needed.

MATHEMATICAL PROGRAMMING

The Nonlinear Programming problem involved in the Optimization stage is solved by means of the formulation of a Conservative Approximate Problem.

$$\min_{x,s} f(x) + cs \tag{14}$$

$$\text{subject to } g_i(x) - s \leq 0 \tag{15}$$

$$L(x) \leq 0 \tag{16}$$

$$x_l \leq x \leq x_u \tag{17}$$

$$s \geq 0 \tag{18}$$

where f and g_i are conservative approximations of Objective Function and constraints, built from their values and derivatives, L are linear constraints, x_l and x_u are bound constraints, c is a relaxation parameter and s is a relaxation variable.

A line search is performed along the direction given by (14) to (18), using the Non Differentiable Penalty Function.

$$P(x,c) = F(x) + cV(x) \tag{19}$$

where c is the parameter used in (14), F(x) is the Objective Function and V(x) is the maximum constraint violation.

An Active Set Strategy and specific algorithms for updating c parameter are also included.

This algorithm is, in fact, a generalization of Approximation Concepts which was formulated taking as a basis the well-known SQP Pschenichnyi's algorithm. It has been implemented in a C module called ACPM, and more details can be found in Longo [7] and Longo [8].

Results of Optimization phase feed equation (1), and the new shape corresponding to new dimension values is determined.

Figure 1 shows a data flow diagram of the whole SIDOMAT system. The user has access to the Design Model by means of an interactive graphic module, which allows the creation of an approximate geometric sketch, assign Dimensions, Materials and Boundary conditions, and also the selection of design variables and the definition of the objective function and constraints. Also this interactive graphic module displays design results, such as Objective Function, Constraint values, Optimization history and so on. Of course, there is not a problem in using a standard postprocessor to visualize the Analysis Model and results, as in fact it is being done in SIDOMAT (Figure 3).

EXAMPLE

Figures 2 to 5 show an application example. Figure 2 shows the geometry modeled by means of straight lines and arcs. It is an axisimetric electric insulator surrounded by a gas atmosphere. Over the insulator, there is an electrode conected to ground, and the opposite side, there is another electrode connected to a given potential.

Dimensions selected as Design Variables are showed. The optimization problem consists of determining V1, V2 and V3, in order to minimize V3. There are two behaviour constraints. Constraint 1 corresponds to a limit of 0.8 in the electric modulus and constraint 2 to a limit of 0.08 in the electric modulus.

Figure 3 shows the Analysis Model for initial and final shape (with potential isolines). The capability of the system in order to preserve technological constraints (e.g. tangencies) must be pointed out, while optimization problem deals with a small number of variables, meaningful from a design point of view.

Initial shape corresponds to values of 70º, 15 and 100 for V1, V2 and V3. Final shape corresponds to values of 45º, 11 and 83.41. Variables V1 and V2 reach their lower limits. The solution obtained is, in fact, not very realistic, but a more practical design can be obtained by increasing the lower limit of angle V1. Figures 4 and 5 show Objective Function and Maximum Constraint Violation history. At the solution, only constraint 1 is active.

CONCLUSIONS

General philosophy of a Shape Optimization System has been presented. A novel feature is that it uses Variational Geometry as a method for representing changing shapes. This lead to a well posed optimization problem, from a designer point of view, and also simplifies the data definition. Boundary Elements combined with a Continuous Material Derivative method have been used for Analysis & Sensitivity and Approximation Concepts for Optimization. As a result, a system suitable for a practical use has been obtained. Currently, the system is being tested against real life design problems, and we can conclude that the global methodology is satisfactory. With respect to Variational Geometry, it should be pointed out that it performs very well in the optimization process, and it is also a powerful tool for interactive shape definition.

REFERENCES

1. Haftka R.T. and Grandhi R.V. Structural Shape Optimization - A survey, Computer Methods in Applied Mechanics and Engineering. 57, pp 91-106, 1986.

2. Ding Y. Shape Optimization of Structures: A Literature Survey. Computers & Structures. Vol 24, No 6, pp 985-1004, 1986.

3. Schmit L.A. Concluding Remarks in The Optimum Shape (Ed. Bennett J.A., Botkin M.E.). Plenum Press, 1986.

4. Zienkiewicz O.C. and Campbell J.S. Shape Optimization and Sequential Linear Programming in Optimum Structural Design (Ed. Gallagher R.H., Zienkiewicz O.C.) pp 109-126. Wiley, New York, 1973.

5. Rajan S.D., Belegundu A.D. and Budiman J. An Integrated System for Shape Optimal Design. Computers & Structures. Vol 30, No 1/2, pp 337-346, 1988.

6. Light R.A. and Gossard D.C. Variational Geometry: A New Method for modifying part geometry for Finite Element Analysis. Computers and Structures, Vol 17, No 5/6, 1983.

7. Longo A., Nó M. and Aizpitarte M. A New Method for Optimal Design in Engineering Based on Approximation Concepts and a Set Active Strategy. In NUMETA 90. Vol. one, pp 492-499. Proceedings of the International Conference on Numerical Methods in Engineering: Theory and Applications/Swansea/7-11 January 1990. (Ed. Pande G.N. and Middleton J.) Elsevier Applied Science, 1990.

8. Longo A., Nó M., Aizpitarte M. and Unzueta J. Conservative Approximations in Nonlinear Optimization. Theory and Examples. Computers & Structures. To appear.

Acknowledgement. This research was carried out under project OCIDE 132135.

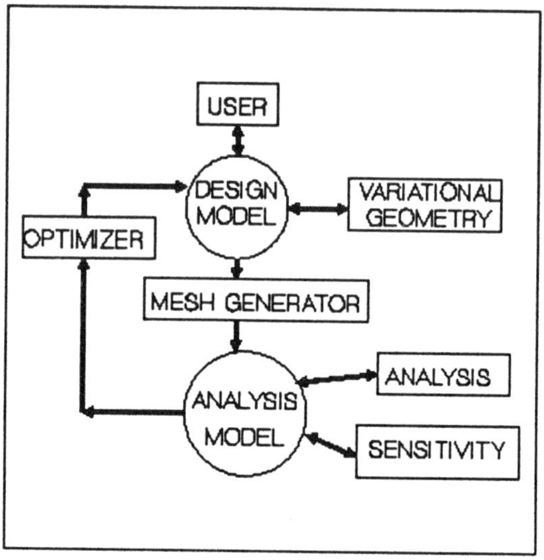

figure 1. SIDOMAT data flow

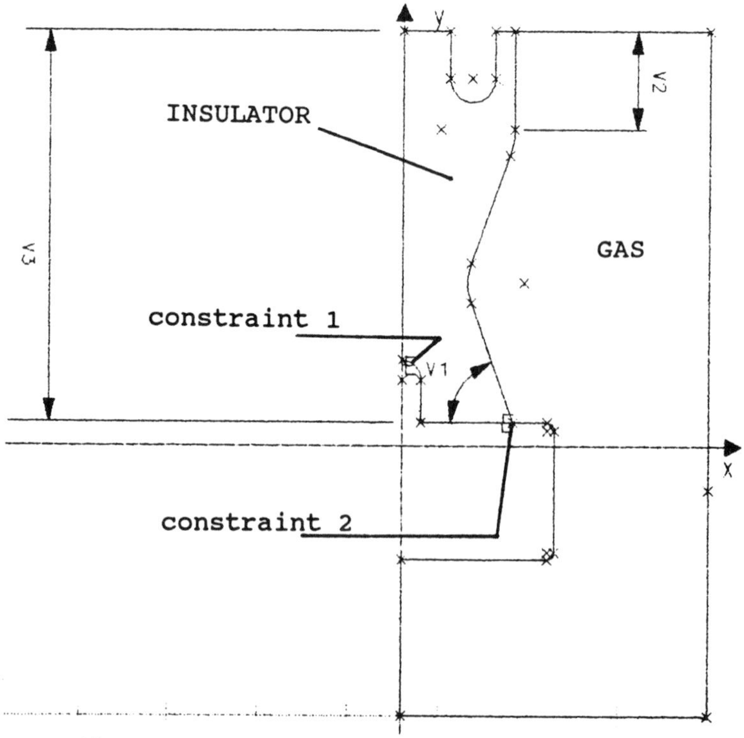

figure 2. Insulator : Design Model

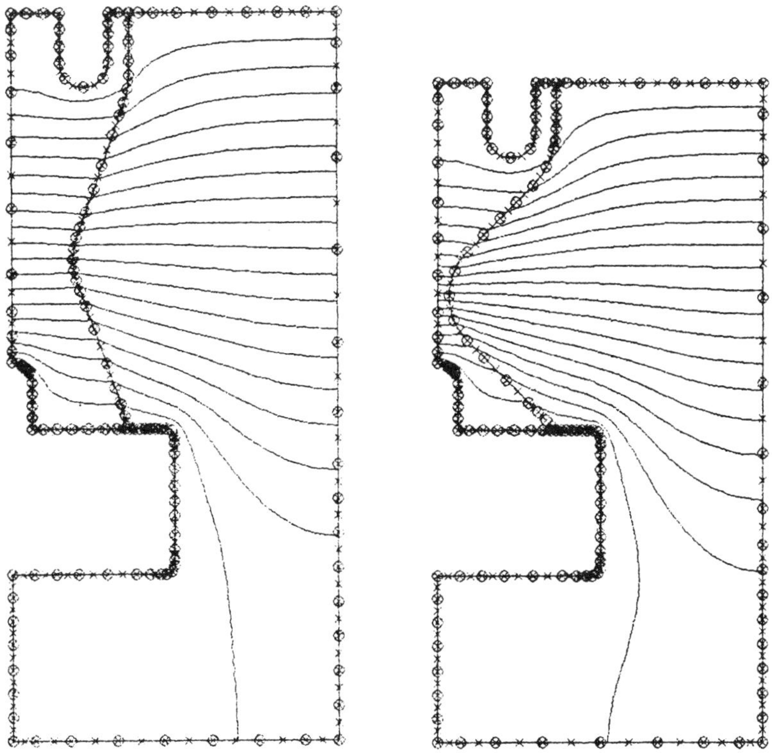

a) b)

figure 3. Insulator : Analisis Model.
a) Initial Shape b) Final Shape

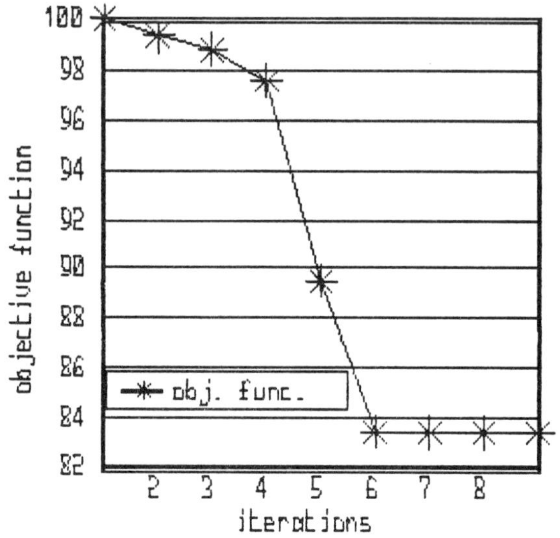

figure 4. Insulator: Objective Function versus optimization iterations.

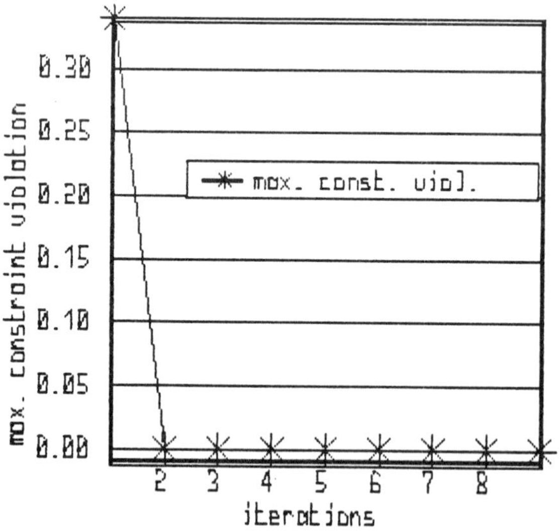

figure 5. Insulator : Maximum Constraint Violation versus optimization iterations.

MCADS and Shape Optimization of Thermo-Stressed Continuum Structures

Y. Gu, G. Cheng

Research Institute of Engineering Mechanics, Dalian University of Technology, Dalian 116024, China

ABSTRACT

Efficient numerical techniques to bridge optimization and analysis developed in package MCADS have been presented. They include an integrated idea of structural modeling for shape optimization and a geometry based approach, the semi-analytic sensitivity analysis with considering influence of load and boundary condition variations caused by shape change, improved algorithm of sequential approximation and a strategy with variable decomposition, and application oriented facilities of software. Shape optimization applications of thermo-stressed continuum structures using MCADS are illustrated.

1. INTRODUCTION

It is being recognized that the structural optimization really valuable to industry should be based on commercially available analysis packages to make use of their versatile capabilities dealing with complicated structures. This facility of optimization should be further integrated with the environment of computer aided design (CAD) [1,2,7]. From this point of view, researches on efficient numerical techniques linking up optimization method, finite element analysis and CAD functions and developments of applicable softwares are important in computational mechanics. The present paper describes this effort and some approaches of structural modeling, sensitivity analysis, optimization algorithm, and programming facility developed in the integrated system MCADS.

In the field of shape optimization of continuum components and structures, the structural modeling and the sensitivity analysis are two critical processes. Since they depend on the description and the control of structural shape, a basic issue

*The project supported by National Science Foundation of China

of optimization, analysis and design. In this paper, structural modeling is presented from an integrated point of view and classified into procedures of geometry, design, analysis and perturbation modelings. A geometry based modeling approach, making use of shape parameterization and control with "natural variable" and automatic mesh generation with "design elements", is proposed. The semi-analytic method of sensitivity analysis, using a FEM program as "black box", has been further studied with respect to variations of load and boundary conditions caused by shape change. It is shown by numerical experiment, that this special kind of shape sensitivity is usually not able tobe ignored, e.g. in the case of shape optimization of thermo-stressed structures.

The algorithm of sequential linear/quadratic programming has been improved in MCADS by the approximate line search and the adaptive move limit, and combined with the goal programming to treat multi-objective design. Moreover, a decomposition strategy carried out in variable space has been proposed. User oriented facilities has been employed in programming. The optimization solver is transparent to engineers who are possibly not experienced in optimization, The automatic iteration is also assisted with user's interactive design via checking and modifying design model. Furthermore, a programming interface is provided for user to deal with such complicated problems as particular application dependent and not able to be expressed with definite formulations. This way, the optimization is robust and its design model can be extended easily.

Resulting from techniques mentioned above, MCADS has been developed for computer aided optimum design of structures. It possesses a rich element library (about twenty types of element include membrane, plate, shell, brick, axisymmetric element, beam, bar, etc.), various facilities of load and condition modeling, pre- and post-processing functions for data of analysis and optimization, and two kinds of shape and sizing design variables. Therefore, the system is applicable to general structures, particularly to the optimum shape design of continuum structures. Its microcomputer version can also solve engineering problems with quite large scale. Two application examples of MCADS for optimum shape design of thermo-stressed continuum structures, turbine engine disk and train wheel, are illustrated.

2. STRUCTURAL MODELING OF SHAPE OPTIMIZATION AND A GEOMETRY BASED APPROACH

Within the shape optimization, the structural modeling involves considerations of geometric design, finite element analysis, sensitivity analysis and optimization. Different from the conventional size optimization, a parameterization ideahas been received recently in shape optimization. That is to build design model with "natural variable", i.e. some geometric parameters which control the shape of structural boundary. Then generate analysis model from the design model, and modify mesh when shape changes during optimization. Some implementations of this idea have been presented. However, there is not a generalized approach suitable to various cases of shape optimum design. This is one of principle issues making optimization package non versatile to general structures as finite element analysis. The entire structural modeling of shape optimization includes such respects as (a) Shape description of structural boundary. (b) Determination of design variables and their relationship with shape. (c) discretization of finite element model and its modification with shape change for reanalysis. (d) Perturbation of local part of the analysis model for the sensitivity analysis. From an integrated point of view, the structural modeling with parameterization approach should be presented as the following procedures.

1. First of all, to describe structural boundary shape with those curves and surfaces usually used in design, e.g. line, arc, spline, quadratic surface, etc. The requirements of deign and manufacture, as well as of analysis modeling, should be satisfied in this geometric modeling procedure.

2. Then to select design parameters, such as vertex of line, center and radius of arc, key nodes of interpolation curve or surface, as variables. These parameters are so-called "natural variables" which control structural shape but don't required relating to analysis model directly.

3. To discrete structure from above geometric model into analysis model and to build a relationship between two models enables a later update with former. Bedsides mesh, the load and boundary condition related to boundary shape need also be changed.

4. To determine a perturbation model for the sensitivity analysis with semi-analytic method. That allows a local part of analysis model (mesh, load and boundary condition) modified with respect to a small perturbation of each design variable.

Above four procedures of geometry, design, analysis and perturbation modeling compose the entire process of structural modeling of shape optimization. In an integrate system, they are linked up with each other through data interfaces, but independent in implementation. Thus each modeling procedure can be implemented by means of several methods so long as its function is provided. For instance, the structural shape may be described by manual prepared data or geometric modeling software, the mesh generation may use different methods. The perturbation model is necessary because of widespread usage of the semi-analytic method in shape sensitivity analysis.

It is important to differentiate two kinds of control relationship of variable to analysis model and perturbation model. In general, the analysis model A(m,l) is indirectly controlled by design model D(x) through geometric model G(g). This relationship can be symbolically expressed as follows

$$D(x) \longrightarrow G(g) \longrightarrow A(m,l) \qquad (1)$$

x, g and m represent design variables, geometric parameters, and mesh respectively, l represents the influence of load and boundary conditions to analysis model. The change of A(m,l) with respect to a new D(x) may be quite large, especially within start several iterations of optimization. Whereas, the change of perturbation model (part of analysis model A(m,l)) is usually small in sensitivity analysis. Additionally, analysis and perturbation model are modified at different stage with different compute consuming. Therefore, these two models can be controlled by design model via different methods. For example, the mesh perturbation for sensitivity analysis could move nodes near to changed boundary only. But the mesh regeneration for reanalysis should move internal nodes, and can even change numbers of node and element (e.g. free meshing approach). The modeling of G(g) and A(m,l), and their relationship are able be unified to make optimization versatile to general structures.

A geometry based approach of structural modeling is proposed in MCADS for shape optimization. Its processes are (1) Geometric model is described with commonly used curves and surfaces, such as lines, arcs, splines, quadratic surfaces and cylinders. Structural shape is parameterized with geometric parameters such as vertex positions of lines, radii and centers of arcs, key node coordinates of interpolation curves and surfaces. (2) The "natural variables" of design model can be selected directly from those changeable geometric parameters mentioned above, or any other parameters the designer assigned to control the above geometric parameters. The structural domain is further divided into some basic geometry entities, regular shaped sub-regions. Those sub-regions near to changeable boundaries are defined as "design elements", and their edges and shape are controlled by the natural variables. (3) Analysis model is produced from above geometric model. Automatic mesh generation is finished by means of mapping method taking regular shaped sub-regions as mapped elements. The analysis model, includes mesh, load and boundary conditions, is controlled by design variables through geometric model. (4) The perturbation model of sensitivity analysis is defined to make a local part of mesh, load and boundary condition within some design elements updated with respect to one design variable.

The essential parameters of this modeling approach are geometric parameters which are most interesting to designer and link up design model and analysis model. The analysis modeling in MCADS is carried out by a commercial program MESHG. It produces data of mesh, load, boundary support as well as control relationship between geometry and analysis models. An interface between MESHG and a commercial software of solid modeling, GEMS, is being developed. The design modeling is application related and defined by user prepared data or programming interface. For 2D structures whose boundary can be fitted with lines and arcs, MCADS provides a simple and unified method to describe geometry and design models, and build their relationship automatically by program itself.

The cross-section shape optimization of a train's wheel is an example. Its geometric model is shown in Figure 1. The changeable boundary of middle part is fitted with splines, the fix boundary of hub and rim are lines. The z-coordinates of key nodes of splines are chosen as design variables. Whole domain is divided into three sub-regions, i.e. mapped element of mesh generation, and the No.1 is design element. Generated mesh of axisymmetric brick elements is shown in Figure 3. Same updating of mesh is assigned within this design element for analysis and perturbation models. Another example is shape optimum design of turbine engine disk. The actual design of cross-section shape is described with lines and arcs shown in Figure 4. The vertexes, centers and radii denoted as L_i, H_i, A_i, B_i, R_i are basic geometry parameters, and 18 parameters of them are taken as design variables. Mapped element division is shown in Figure 2. The design variables control corners and edge middle points of mapped elements, and then, control mesh regeneration.

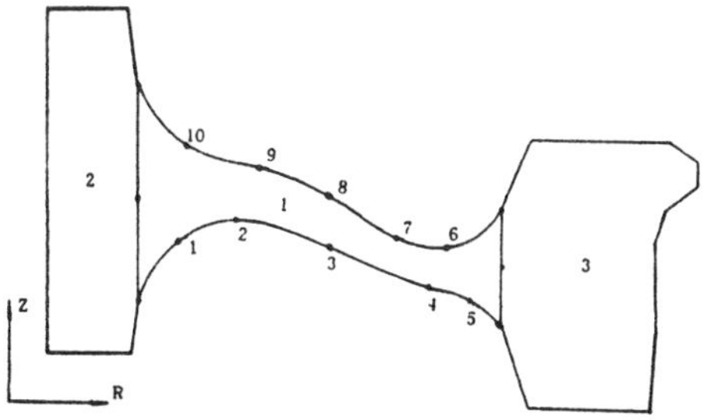

Figure 1.

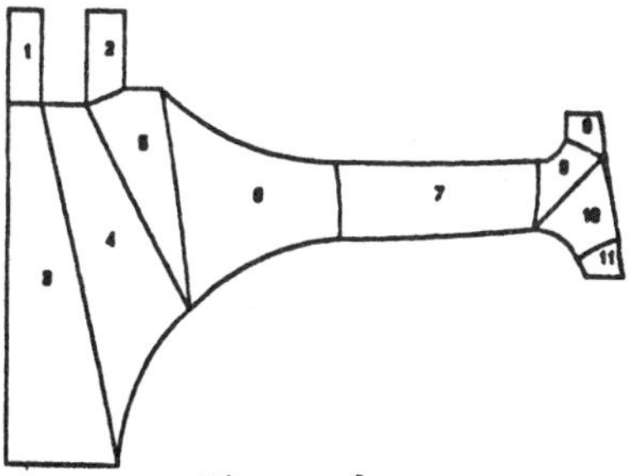

Figure 2.

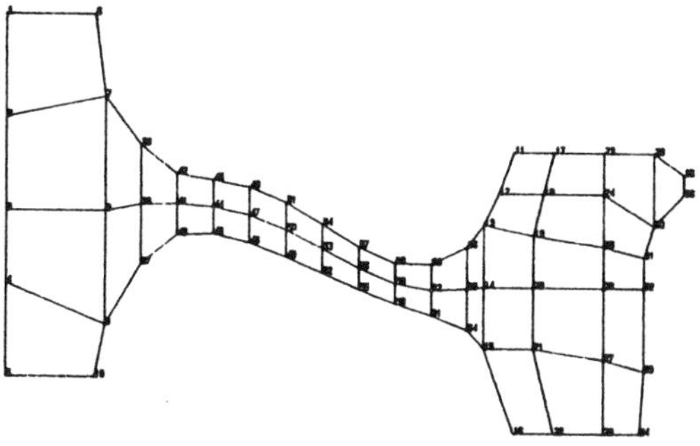

Figure 3. Axisymmetric element Mesh of wheel.

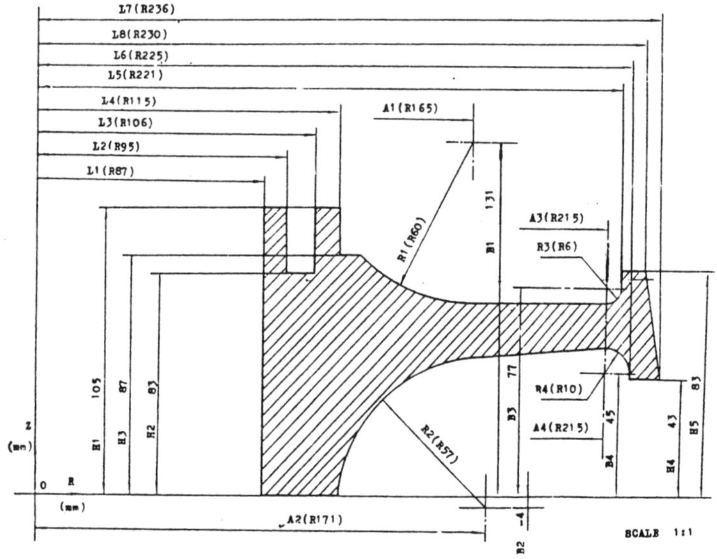

Figure 4. Design model of turbine engine disk.

3. SEMI-ANALYTIC SENSITIVITY ANALYSIS AND LOAD VARIATION

The semi-analytic method of sensitivity analysis has been employed by most of large scale packages of optimization. Since this method avoids difficult formulating and programming of complicated relationship between various elements and shape variables, commercial FEM packages can be used as analyzer of optimum design. So long as the perturbation model is built and updated with design variable, sensitivities can be computed easily by analyzer. The scheme of semi-analytic sensitivity analysis of MCADS is using a commercial FEM package, DDJ-W, as a half-open "black box". That means the data bank of DDJ-W is changed with perturbation model, and its subroutinesare called to compute responses. The accuracy of semi-analytic method for shape sensitivity analysis is being investigated. An alternative forward/backward difference [1,3] and a second order correction [4] have been proposed to reduce the error with respect to mesh refinement. According to studies and applications of MCADS, the semi-analytic method is accurate and applicable for shape optimization of structures usually encountered in engineering and composed of membrane, brick and axisymmetric elements. Further accuracy improvement on semi-analytic method for general cases, particularly for plate, shell and beam elements, is necessary.

The sensitivity analysis is related to the structural modeling. Replacing analysis model A(m,l) of equation (1) with perturbation model P(m,l), the sensitivity of response R is expressed as follow

$$\frac{\partial R}{\partial x} = \frac{\partial R}{\partial P} * \frac{\partial P}{\partial G} * \frac{\partial G}{\partial x} = (\frac{\partial R}{\partial m}\frac{\partial m}{\partial G} + \frac{\partial R}{\partial l}\frac{\partial l}{\partial G}) * \frac{\partial G}{\partial x} \quad (2)$$

Within semi-analytic method based on structural modeling idea mentioned in above section, $\partial G/\partial x$ means shape variation with respect to design variable. The $\partial m/\partial G$ means mesh perturbation within local region and may be different from regeneration of whole mesh. The second item within bracket indicates the sensitivity of R with respect to variation of load and some boundary conditions caused by shape change. For instance, shape related volume load (gravity and centrifugal force), pressure distributed normal to boundary, thermo-load produced by uneven temperature field, and boundary support give this special kind of sensitivity. This part of sensitivities is usually ignored in analytic method since computing isdifficult. An advantage over analytic method, is the semi-analytic method is capable of evaluating this special kind sensitivity effectively by means of including load and boundary condition variations into perturbation model.

Our research and applications of MCADS have shown that the variations of load and boundary condition with respect to shape updating may posses great influence to design sensitivity. Ignoring this influence may produce noticeable error to sensitivity and make iteration not converge. Here, the numerical experiment on load variation of thermo-stressed structure is illustrated for this issue. The train wheel, whose design model and element mesh have been shown in Figures 1 and 3, subjects to thermo-load caused by brake. This thermo-load is shape dependent and updated in perturbation model, recomputing its equivalent node forces. Maximum thermal stresses arising at nodes 43,49,40,63,60 and maximum radial displacement at node 34 are considered. The increments of thermal stresses and displacement with respect to shape change by giving a perturbation $\Delta x=1.0$ to design variables in turn are given in Table 1. The increment values on lines denoted by D_T and D_{NT} are computed by structural reanalysis. The increment values on lines denoted by S_T and S_{NT} are linear approximations with first order sensitivities calculated via semi-analytic method. Subscripts N and NT means reanalysis or sensitivity analysis is carried out with and without thermo-load variation respectively. The numerical result shows (a) The influence of thermo-load variation is significant for sensitivities of structural stresses and displacements with respect even to small shape change. (b) The semi-analytic method is accurate for shape sensitivity analysis of thermo-stressed structures composed of axisymmetric brick elements. The shape optimization of turbine engine disk, presented in section 5, has also shown that the variations of thermo-load and centrifugal force are not able ignored in sensitivity analysis. Same issue exists in the case where boundary condition is shape dependent.

Table 1. Increments of thermal stresses and displacement.

		σ_{43}	σ_{46}	σ_{40}	σ_{63}	σ_{60}	u_{34}
Initial value		91.798	77.396	57.937	63.958	56.989	1.6937
$\Delta x_1=1.0$	D_T	-0.2632	-2.3835	3.1606	-0.1363	-0.1154	0.00023
	D_{NT}	-0.0656	-3.4452	4.3411	0.0614	-0.2565	0.00059
	S_T	-0.2585	-2.3837	3.2855	-0.1316	-0.1111	0.00011
	S_{NT}	-0.0557	-3.4781	4.4698	0.0734	-0.2497	0.00060
$\Delta x_2=1.0$	D_T	1.9073	0.9979	-2.1929	-0.7598	-0.0329	0.00614
	D_{NT}	4.1755	3.2377	-1.8452	-1.5590	0.1503	0.00729
	S_T	1.9502	1.2803	-2.1294	-0.7631	-0.0006	0.00608
	S_{NT}	4.3168	3.5599	-1.7286	-1.5428	0.2053	0.00730
$\Delta x_3=1.0$	D_T	-2.2026	0.6814	-1.6457	0.1865	0.7666	0.00556
	D_{NT}	-2.5066	1.7575	-2.0589	2.4451	2.7149	0.02764
	S_T	-2.2272	0.7277	-1.6497	0.2352	0.7005	0.00520
	S_{NT}	-2.4057	1.9791	-1.9676	2.5263	2.5871	0.02785
$\Delta x_4=1.0$	D_T	0.6396	-0.2021	0.5147	0.7164	0.6633	-0.00036
	D_{NT}	-0.3343	-1.0171	-0.5444	3.3942	-1.8204	0.04743
	S_T	0.6077	-0.2174	0.4914	0.9285	0.5664	-0.00003
	S_{NT}	-0.2497	-0.9394	-0.0922	4.1579	-2.2200	0.04750
$\Delta x_5=1.0$	D_T	-0.1799	0.2709	-0.1843	0.8399	-0.3611	-0.00001
	D_{NT}	-0.1962	0.3844	-0.2079	-3.6641	-2.3054	0.04787
	S_T	-0.2426	0.2409	-0.2265	0.4937	-0.8310	-0.00003
	S_{NT}	-0.2518	0.3763	-0.2476	-3.9478	-2.8664	0.04750
$\Delta x_6=1.0$	D_T	0.5201	0.2407	0.3560	-3.1865	1.4938	0.00265
	D_{NT}	0.6530	0.4493	0.4189	-9.6704	-0.2139	0.05068
	S_T	0.5502	0.2620	0.3776	-3.0853	1.6495	0.00035
	S_{NT}	0.7578	0.5093	0.4967	-12.527	0.2757	0.04866
$\Delta x_7=1.0$	D_T	-0.1855	-0.0393	-0.0777	1.4369	-3.1492	0.00356
	D_{NT}	-0.6439	-1.0526	-0.2356	0.0336	-10.663	-0.00837
	S_T	-0.1886	-0.0839	-0.0669	1.4164	-3.2214	0.00143
	S_{NT}	-0.6290	-1.1059	-0.1929	0.0504	-10.847	-0.01031
$\Delta x_8=1.0$	D_T	0.6504	-0.3003	0.4226	-0.0737	-0.2707	-0.00103
	D_{NT}	-1.2793	-1.4844	-0.7724	1.0069	-0.4058	-0.00348
	S_T	0.4964	-0.2892	0.3659	-0.0588	-0.3985	-0.00135
	S_{NT}	-1.4497	-1.5121	-0.8095	1.0879	-0.4915	-0.00349
$\Delta x_9=1.0$	D_T	-0.9210	-1.1121	0.5750	-0.2496	-0.6696	-0.00309
	D_{NT}	-0.9659	-0.6966	0.1555	-0.1486	-0.2070	-0.00210
	S_T	-0.9609	-1.0589	0.5683	-0.2480	-0.6806	-0.00338
	S_{NT}	-1.0569	-0.8364	0.1878	-0.1616	-0.2353	-0.00235
$\Delta x_{10}=1.0$	D_T	-0.9543	0.7400	-0.7797	0.0422	-0.0731	-0.00081
	D_{NT}	-0.5138	0.5015	-0.3662	0.1038	-0.0492	-0.00065
	S_T	-0.9271	0.7588	-0.7794	0.0386	-0.0775	-0.00087
	S_{NT}	-0.6257	0.5156	-0.3481	0.0948	-0.0477	-0.00046

4. OPTIMIZATION ALGORITHM AND USER ORIENTED FACILITIES

The basic optimization algorithm of MCADS is sequential linear programming or quadratic programming (if the second order sensitivity of objective function can be computed, e.g. weight minimization). This algorithm is improved by approaches of approximate line search and adaptive move limit [5]. That is to evaluate new design Xn ($Xn=Xo+\beta d$) obtained by solving linear or quadratic programming from old design Xo with the following Goldstein criterion

$$\beta(1-\alpha)\nabla\Phi(Xo)d \le \Phi(Xn)-\Phi(Xo) \le \beta\alpha\nabla\Phi(Xo)d \quad (3)$$

d=Xn-Xo is search direction, β is step length of search with 1.0 as initial value, parameter α is chosen from (0, 0.5). The evaluation function Φ is objective function, and $\nabla\Phi$ is its gradient. The design Xn is accepted when formulation (3) is satisfied. If $\Phi(Xn)-\Phi(Xo) < \beta(1-\alpha)\nabla\Phi(Xo)d$, then design Xn is also accepted and the move limit of design variables is enlarged for next iteration. If $\Phi(Xn)-\Phi(Xo) > \beta\alpha\nabla\Phi(Xo)d$, then cut step length β by half and search in the direction d until Goldstein criterion is satisfied, and the move limit of next iteration is reduced. The inner-iterations of line search is limited to 3-5 times. This algorithm has made good use of the descent direction solved by linear/quadratic programming, and adjusted move limit adaptively. To deal with problems of unfeasible design and multi-objective optimization, the goal programming method is employed by means of adding constraint functions, which deviate constraint condition seriously, into evaluation function Φ with weight factors. This way, an optimization problem can be defined as multi-objective design so long as to give lower bounds for some constraint functions. The solver of this algorithm is designed as an independent module of optimization algorithm library to enable other algorithms be introduced too.

A variable decomposition strategy is proposed in MCADS. Different from the transformation of design space [6], the decomposition is carried out in the variable space. The design variables are divided into several groups, so that the original optimization problem is reduced to some sub-problems with fewer variables. Each sub-problem with only one group of variables is solved within each iteration. The different influence of variables to the design model as well as convenience of design and manufacture can be considered in group division of variables. This variable decomposition algorithm is not only useful to solve large scale problems of optimization effectively, but also suitable to practical design process dealing with design parameters respectively.

An user and application oriented idea has been followed in programming of MCADS. The optimization solver is transparent to user. Its detailed procedures such as approximate line search, move limit adjustment, unfeasible design treatment etc., are implemented by solver itself without user's assistant. Therefore, the program is robust for those engineers being short of knowledge and experience of optimization. In other hand, the automatic iteration of sequential approximation algorithm is combined with interactive design of user. At any stage of iteration, user is permitted to check and change design interactively. The design model can be modified by updating, deleting, or adding variables, objective and constraint functions. This interaction facility enables user to employ his experience of engineering and make creative design.

Particularly, an application oriented programming interface has been developed. The treatments of application related special problems are localized within some subroutines of user open module, i.e. programming interface. For example, description of control relationship between design and geometry models for complicated shape, updating of some kinds of changeable load and boundary conditions, computations of stiffness and stress of beam elements with complicated cross section, and definition of particular constraints, etc. Some of these application dependent problems can not be foreseen at programming stage and expressed with unified scheme. However, user is able to deal with these problems by rewriting the source program of some subroutines of the programming interface. Additionally, a mechanism is prepared to store and access particular data which is needed in these modifiable subroutines and defined by user. By means of this user's programming interface, MCADS program is flexible for user to extend design model himself for various complicated optimization problems of practical engineering. For example, the programming effort of shape optimization for turbine engine disk stated in the paper is modifying only two subroutines of this interface, updating centrifugal force and describing control relationship between design variable and geometric model.

5. SHAPE OPTIMIZATIONS OF THERMO-STRESSED STRUCTURES

Two practical applications of MCADS for the shape optimization of thermo-stressed continuum structures coupling stress field with temperature field are presented here.

Example 1. The shape optimization of train wheel. The thermo-load of high temperature caused by brake is considered since it produces noticeable concentration of thermal stress. The mechanical load is checked separately. The load bearing of each wheel of a new heavy train is designed to increase from 21 tones to 25 tones and the radius of wheel is kept. Then temperature, thermo-load and thermal stress will be increased. The optimization objective is to reduce stress concentration by means of shape modification of wheel cross-section. The structural modeling has been stated in section 2, and as mentioned in section 3, the variation of thermo-load with shape is taken into account both in analysis and perturbation models. The optimum design is obtained after 5 iterations and the maximum thermal stress of wheel has been reduced from 91.8 to 67.9 under 25-tone load bearing. This value is even lower than the maximum thermal stress 77.6 of old design with 21-tone load bearing. The cross-section shape and distribution of thermal stress (Mises stress) of optimal design under 25-tone load bearing is shown in Figure 5. The temperature field is assumed unchanged during optimization and recomputed for optimal design. The results is that the temperature is decreased with optimization too, and the optimal design is safe. It is also shown that the temperature field is less sensitive than thermal stress and displacement to the shape change.

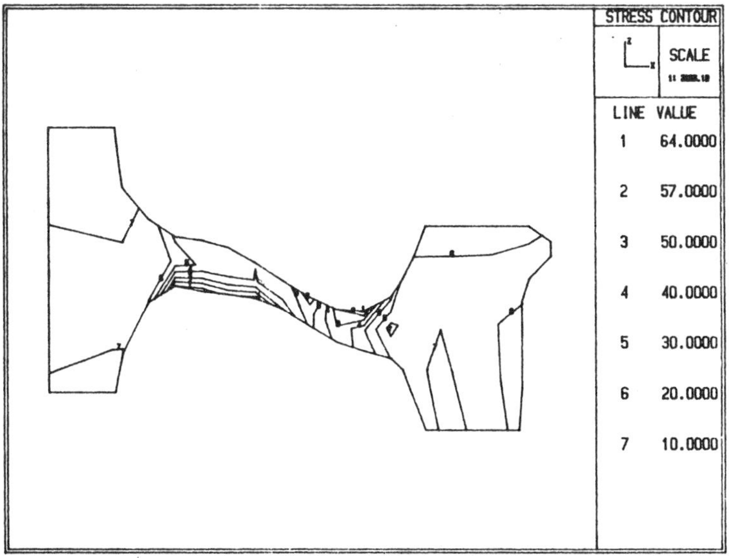

Figure 5. Optimum design of wheel.

Example 2. The shape optimization of turbine engine disk. The disk subjects to thermo-load caused by high temperature and centrifugal forces caused by high-speed rotating. Its geometric model and design model has been described in section 2. The 8-node axisymmetric brick elements are used in analysis model. By means of the programming interface, the control relationship between design variables and geometric model is defined in a subroutine, and the computing of centrifugal forces is defined by another subroutine. Two load cases are all shape dependent and updated both in structural reanalysis and sensitivity calculation so that make sensitivity accurate and iteration stable. The objective of design optimization is to reduce both structural weight and stress level. This bi-objective problem is solved by goal programming approach mentioned in section 4. Various design criteria on structural strength and usage life related to considerations of all facts on fatigue, fracture, creep and manufacture have been studied and expressed in forms of stress constraints. Thus the structural design model is very close to the real state of turbine engine disk, and requirements proposed in ENSIP (Engine Structural Integrity Program) of USA are satisfied. The strategy of variable decomposition stated in above section is used to solve this problem. Eighteen variables are divided into three groups to make boundaries of two ends and the middle part of cross-section are controlled by one group variable respectively. The optimization procedure shows that this variable decomposition is successful. After shape optimization, the structural weight and maximum Mises stress of the turbine engine disk have been reduced 18.52% and 14.91% respectively. The boundary shapes and the distributions of Mises stress of optimal design is shown in Figure 6.

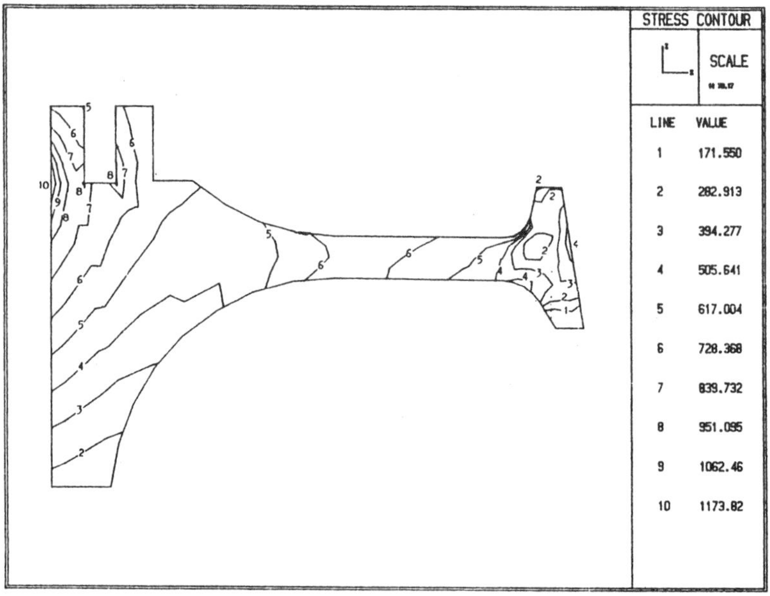

Figure 6. Optimum design of disk.

REFERENCES

1. Gu,Y. and Cheng,G. Structural Shape Optimization Integrated with CAD Environment, Structural Optimization, Vol.2, pp.23-28, 1990.
2. Gu,Y. and Cheng,G. Integration of FEM, Optimization and CAD on Microcomputers, in Numerical Techniques for Engineering Analysis and Design (Ed. Pande,G.N. and Middleton,J.), S-10, Proceedings of Int. Conf. NUMETA'87, Swansea,UK, 1987. Martinus Nijhoff, Dordrecht, 1987.
3. Cheng,G., Gu,Y. and Zhou,Y. Accuracy of Semi-Analytic Sensitivity Analysis, Finite Element Analysis and Design, Vol.6, pp.113-128, 1989.
4. Cheng,G., Gu,Y. and Wang,X. Improvement of Semi-Analytic Sensitivity Analysis and MCADS, Proc. of Conf. on Engineering Optimization of Design Processes, Karlsruhe, FRG, 1990.
5. Cheng,G. and Gu,Y. Applications of SQP to Structural Dynamic Optimization, Vibration and Shock (in Chinese), pp.12-20, 1986.
6. Gu,Y. and Cheng, G. A two-Level Optimization Strategy with Transformation of Design Space, Applied Mechanics (in chinese), Vol.3, pp.7-18, 1986.
7. Bennett,J.A. and Botkin,M.E.(Ed.). The Optimum Shape, Proc. of General Motors Symposium on The Optimum Shape: Automated structural Design, Plenum Press, New York, 1986.

AUTHORS' INDEX